TABLES OF PHYSICAL AND CHEMICAL CONSTANTS

Tables of Physical and Chemical Constants

and some Mathematical Functions

Originally compiled by
G. W. C. KAYE, O.B.E., M.A., D.Sc., F.R.S.
and
T. H. LABY, M.A., Sc.D., F.R.S.

**Now prepared under the direction
of an Editorial Committee**

**Longman
London and New York**

Longman Group Limited
Longman House, Burnt Mill, Harlow
Essex CM20 2JE, England
Associated companies throughout the world

*Published in the United States of America
by Longman Inc., New York*

First published 1911
Fourteenth Edition 1973
Fifteenth Edition 1986

British Library Cataloguing in Publication Data
Kaye, G.W.C.
    Tables of physical and chemical constants and
    some mathematical functions.——15th ed.
    1. Physics——Tables   2. Chemistry——Tables
    3. Chemical elements——Tables
    I. Title   II. Laby, T.H.
    530.8   QC61

ISBN 0-582-46354-8

Printed in Great Britain by Bath Press Avon.

PREFACE TO THE FIFTEENTH EDITION

With this new edition, Kaye and Laby celebrates its seventy-fifth birthday. Since 1911 the volume and range of physical and chemical data required for everyday laboratory purposes have enormously increased, and this has been reflected in the size of recent editions. Successive Editorial Boards have always been at pains to ensure that the intentions of the original authors should be maintained: the primary criterion for the inclusion of material is that it should be of value not only to specialists but more generally to scientists working in a variety of fields.

In this edition all the material has been scrutinised and revised as necessary to take account of new results. Several completely new sections have been added covering, for example, wavelength standards, cosmic rays, atomic radii, calorific values of fuels, nuclear fusion and radioactive series. To accommodate this additional material, one or two topics which are now of little interest have been omitted: the most noteworthy is the set of mathematical tables. It was with regret that the Board decided that the time had come to part with these, but pocket calculators now enable arithmetical computations to be carried out and provide values of the trigonometrical and exponential functions to many more decimal places than it is possible to print.

Dr G. W. C. Kaye, FRS, one of the original authors, was Superintendent of the Physics Department of the UK National Physical Laboratory and after he died in 1941 other physicists at that laboratory contributed to the 9th edition, which was then in preparation. This close association has continued to the present day: all the members of the present Editorial Board and many of the contributors to this edition are, or were formerly, members of UK national laboratories, particularly NPL and the Atomic Energy Research Establishment.

The provision of standards of measurement and of high-accuracy data is an important factor in economic well-being and has long been accepted as a responsibility of government in industrialised countries. We may perhaps express the hope that future governments will continue to support long-term programmes in national laboratories for the generation of scientific data which only those laboratories can provide.

A.E.B.   11.5.85

EXTRACT FROM PREFACE TO FIRST EDITION

The need for a set of up-to-date English physical and chemical tables of convenient size and moderate price has repeatedly impressed us during our teaching and laboratory experience. We have accordingly attempted in this volume to collect the more reliable and recent determinations of some of the important physical and chemical constants.

To increase the utility of the book, we have inserted, in the case of many of the sections, a brief *résumé* containing references to such books and original papers as may profitably be consulted.

Attention has been paid to the setting and accuracy of the mathematical tables; these are included merely to facilitate calculations arising out of the use of the book, and limitations of space have cut out all but a few of the more essential functions.

We began this book while at the Cavendish Laboratory, Cambridge, and Dr G. A. Carse shared in its inception. To Mr G. F. C. Searle, F.R.S., we feel we owe much for his encouragement and suggestions when the scope of the book was under consideration. . . .

G.W.C.K.
T.H.L.

September, 1911

ACKNOWLEDGEMENTS

We are grateful to the following for permission to reproduce copyright material. Each extract is acknowledged by footnote in the appropriate place in the text:

Academic Press Ltd.; The American Chemical Society; American Institute of Physics; British Standards Institution; Mr. Harrison Brown and the American Physical Society; Butterworth & Co. (Publishers) Ltd.; Cambridge University Press; Dr. N. F. Chamberlain and Plenum Publishing Corpn; Chance Pilkington Optical Works; The Clarendon Press; Commissariat de L'Energie Atomique; Department of Trade & Industry – National Physical Laboratory; Professor H. K. Drickamer and Butterworth Inc.; the late Professor Sir Ronald A. Fisher, Dr. Frank Yates, Rothamsted and Oliver & Boyd Ltd.; W. H. Freeman & Co. Publishers; Mr. F. A. Gould; Mr. T. Vickers and the Institute of Physics; ICSU Committee on Data for Science and Technology; International Union of Pure and Applied Chemistry; Inter-science Publishers Ltd.; the editor of the Journal of Chemical Education; Macmillan & Co. Ltd.; McGraw-Hill Book Co.; Modern Plastics Inc.; National Academy of Sciences, Washington DC; National Bureau of Standards, Washington DC; D. Van Nostrand Co. Inc.; Pergamon Press Ltd.; Prentice-Hall Inc.; Reinhold Publishing Co.; Reynolds Metal Co.; Royal Society of Chemistry; Springer-Verlag and Dr. E. Pretsch; Taylor and Francis Ltd; University of Chicago Press; John Wiley & Sons Inc.; Wykeham Publications Ltd.

PUBLISHERS' NOTE

CONTRIBUTORS AND INITIALS

D.A.	D. Ambrose, B.Sc., Ph.D.
J.A.	J. Asher, B.Sc., D.Phil.
A.E.B.	A. E. Bailey, M.A.
D.R.B.	D. R. Barraclough, B.Sc., A.R.C.S.
S.J.B.	S. J. Bennett, M.A., Ph.D.
E.C.B.	The late Sir Edward Bullard, M.A., Ph.D., D.Sc., F.R.S.
J.H.C.	J. H. Calderwood, M.Eng., Ph.D., D.Sc.
F.J.J.C.	F. J. J. Clarke, B.Sc., Ph.D.
F.C.	F. Close, B.Sc., D.Phil.
A.H.C.	A. H. Cook, M.A., Ph.D., F.R.S.
J.M.C.	J. M. Corsan, M.A., D.Phil.
N.E.B.C.	N. E. B. Cowern, M.A., D.Phil.
T.E.C.	T. E. Cranshaw, M.A., Ph.D.
J.G.C.	J. G. Cunninghame, B.Sc.
M.D.	M. Debenham
M.E.D.	The Rev. M. E. Delany, B.Sc., Ph.D.
J.A.D.	J. A. Dennis, B.Sc., Ph.D.
D.D-H.	D. Dew-Hughes, D.Sc., C.Eng.
C.H.D.	C. H. Dix. B.Sc.
L.E.D.	L. E. Drain, M.A., D.Phil.
A.E.D.	A. E. Drake
D.J.S.F.	D. J. S. Finlay, B.Sc., Ph.D.

E.J.G.	E. J. Gillham, M.A.
J.H.S.G.	J. H. S. Green, B.Sc., Ph.D.
E.F.G.H.	E. F. G. Herington, Ph.D., D.Sc.
M.J.H.	M. J. Hickman
A.M.H.	A. M. Hillas, Ph.D.
J.T.H.	J. T. Houghton, M.A., D.Phil., F.R.S.
D.J.	D. Jakeman, B.Sc., Ph.D.
O.N.J.	O. N. Jarvis, B.Sc., Ph.D.
O.C.J.	O. C. Jones, M.A., D.Phil.
R.J.K.	R. J. King, B.Sc., M.Sc., Ph.D.
D.J.E.K.	D. J. E. Knight, M.A., D.Phil.
J.W.L.	J. W. Leake, B.Eng. Ph.D.
S.L.L.	S. L. Lewis
M.L.M.	M. L. McGlashan, Ph.D., D.Sc.
M.F.M.	M. F. Markham, B.Sc.
H.D.M.	H. D. Megaw, M.A., Ph.D., Sc.D.
L.W.N.	L. W. Nickols, B.Sc.
B.E.J.P.	B. E. J. Pagel, M.A., Ph.D.
D.P.	D. Parker, M.A., D.Phil.
B.W.P.	B. W. Petley, Ph.D.
D.M.P.	D. M. Poole, B.Sc., M.Sc., D.Phil.
J.E.P.	The late J. E. Prue, M.A., D.Phil.
T.J.Q.	T. J. Quinn, B.Sc., D.Phil.
J.B.R.	J. B. Rands, B.Sc.
E.D.v.R.	The late E. D. van Rest, M.A.
C.P.R.	C. P. Richards, B.Sc., Ph.D.
J.C.R.	J. C. Riviere, M.Sc., Ph.D.
B.R.	B. Rose, M.A., Sc.D.
W.R.C.R.	W. R. C. Rowley, B.Sc., Ph.D.
M.P.S.	M. P. Seah, B.Sc., Ph.D.
D.I.S.	D. I. Simpson
M.G.S.	M. G. Sowerby, B.Sc., Ph.D.
P.V.	P. Vigoureux, D.Sc.
J.T.R.W.	J. T. R. Watson, B.Sc.
D.W.	D. West, B.A., Sc.D.
G.A.W.	G. A. Wilkins, Ph.D.
I.W.	I. Williams, B.Sc.

TABLE OF CONTENTS

GENERAL PHYSICS

CHEMISTRY

MATHEMATICAL FUNCTIONS

NOTE ON THE USE OF PAGE HEADINGS

The user of this book will find that the Index is the key to what it contains. Headings in heavy type on each page are intended to assist him in finding his way in the tables, within the sections listed briefly above, and without continual reference to the Index. When parts of more than one table are to be found on a single page, the heading refers in general to the first table which *starts* on that page.

NOTE ON LABORATORY SAFETY

To ensure safe working conditions in laboratories of all kinds is a matter of great importance. In many cases they are covered by regulations for industrial premises, such as the Health and Safety at Work Act in Britain or the Occupational Safety and Health Act in the USA. Some of the tables in this book can be of value in determining levels of exposure to various types of hazard and it seemed to the Editorial Board that it might be helpful to some readers to add a note on safety information in this edition.

For obvious reasons we cannot give a comprehensive list of safety requirements, nor can we accept any responsibility for the use made of the information, but it seems worthwhile to mention a few references of general interest known to us.

Standards for Safety in Laboratories (AS 2243) is published by the Standards Association of Australia (Sydney, NSW). Another general publication of interest is *Safety Aspects of Laboratory Instrumentation* (Conference held in London on 25 May 1982) published by Scientific Symposia Ltd, London, 1982.

Electrical hazards are very important. The UK national code of safety for electrical installations of all kinds (not only in laboratories) is set out in the 15th edition of the *IEE Wiring Regulations*, published by the Institution of Electrical Engineers, London, 1981; a Commentary and Guide to the Regulations are also available. Two publications particularly concerned with electrical-testing laboratories are: 'Safety for electrical measurement laboratories', by A. Brandolini, G. Gola and E. Tironi, in *TE Int (Italy)*, **3**, 1979, pp. 28–31; and 'Safety considerations in light electrical testing', by R. W. Nettleton, in *SERT J (GB)*, **6**, 1972, pp. 101–3.

Hazards arising from the use of lasers are discussed in section 1.7.6 of this book. British Standard BS4703:1983 classifies laser hazard levels.

From a wide range of material on chemical hazards, we may mention the following:

Hazards in the Chemical Laboratory, L. Bretherick (ed.) (3rd edn), Royal Society of Chemistry, London, 1981.

Safety in the Chemical Laboratory, N. V. Steere, American Chemical Society, Easton, 1974

CRC Handbook of Laboratory Safety, The Chemical Rubber Co., Cleveland, Ohio.

EH40: Occupational exposure limits, 1984, is published by the UK Health and Safety Executive (HMSO, 1984). The Executive also issues a *Toxic Substances Bulletin* which gives interim statements between the annual revisions of EH40.

A useful survey of health risks in the use of mercury (which is more dangerous than some laboratory users seem to think) is given by D. J. More and A. E. Timbs, *Chem Brit*, **20**, 1984, p. 622.

Guidance on the safe use of ionising radiations and radioactive materials is given in the publications of the International Commission on Radiological Protection (ICRP). Some of their publications of particular relevance are:

Publication 26: *Recommendations of the International Commission on Radiological Protection* (1977) together with the Statement from the 1978 Stockholm meeting of the ICRP (*Annals of the ICRP*, **2**, 1 (1978)).

Publication 30: *Limits for Intake of Radionuclides by Workers* (1979).

Publication 35: *General Principles of Monitoring for Radiation Workers* (1982).

Publication 36: *Protection against Ionising Radiation in the Teaching of Science* (1983).

The various recommendations of the ICRP have been embodied in the Directive of the European Communities of the 15 July 1980, laying down the basic safety standards for the health protection of the general public and workers against the dangers of ionising radiation. This directive is binding on the countries of the European Community and will be embodied in the UK legislation due to be introduced in 1985–6. The legislation will be accompanied by a substantial body of guidance as to its application in different areas, include those of research and teaching.

We cannot impress too strongly on those setting up and operating laboratories the need to be aware of the regulations and codes of practice relevant to their work and to make sure that they are followed.

A.E.B.

11.5.85,

1.1 Units

1.1.1 The international system of units (SI)

History

In the second half of the nineteenth century the centimetre, gram and second were in fairly general use as base units for scientific work even in such countries as the UK and the USA where the foot and the pound were employed for commerce and engineering. As a result, the units required by the rapidly emerging science of electricity were based on the centimetre, gram and second, with which they formed a coherent system known as the CGS electromagnetic system. A system of units is said to be coherent when derived units are formed from the base units without the insertion of factors of proportionality other than unity. There was also the CGS electrostatic system, but the only quantities frequently expressed in electrostatic units were electric charge, electric potential, and capacitance.

The young but fast-growing electrical industry soon found that many CGS electromagnetic units were of an extremely inconvenient size for its needs. Accordingly, in 1881, international agreement was reached to fix the practical unit of potential, to be called the volt, at 10^8 CGS units (which is approximately equal to the e.m.f. of a primary cell), and the unit of resistance, the ohm, at 10^9 CGS units (which is approximately the resistance of a column of mercury 1 m long and 1 mm^2 in cross-section). The unit of electric current, the ampere, was made a tenth of the CGS unit. A coherent system of practical electric units was thus secured which, however, was not coherent with the mechanical units based on the centimetre and gram. The practical electric units suited the needs of telegraphy, which was then the main electrical industry, and they also happen to be convenient for heavy electrical engineering and for electronics.

The magnetic units, however, were left at their CGS values, presumably because the CGS unit of magnetic flux density, subsequently called 'gauss', is of the order of the flux density of the Earth's field, and, as it was suitable for geomagnetism, there seemed no point in changing it for a unit 10^4 times larger. Coherence was thereby lost to electromagnetism as it had already been lost to the system embracing the mechanical units and the practical electric units.

Whereas the electric units, by the agreement of 1881, were chosen to be of suitable magnitude for everyday use, and whereas the centimetre and the second have acceptable sizes, the gram is too small for the practical needs of man, which are better served by a unit nearer the size of the pound or the kilogram. Moreover, the CGS unit of force, the dyne, and the unit of energy, the erg, are much too small. On the other hand, the unit of energy provided by the practical electric units, the volt-ampere-second, called the joule—which equals 10^7 ergs—is of a satisfactory size.

These considerations—the advantages of coherence and the fortuitous circumstance that a mechanical system based on the metre and the kilogram has precisely the same unit of energy as is provided by the practical electric units—led G. Giorgi in 1902 to propose a system based on the **metre**, the **kilogram**, the **second**, and one of the practical electric units. He pointed out that if magnetic field strength were expressed as amperes per metre instead of 4π times amperes per metre, which is the definition corresponding to that of the CGS unit, the number π would disappear from most electric and magnetic formulae involving rectilinear geometry, but would appear, as is to be expected, in those involving cylinders or spheres.

The International Electrotechnical Commission eventually chose the **ampere** as the fourth base unit of the MKSA or 'Giorgi' system, and in 1948 the 9th General Conference of Weights and Measures† recommended it for science and technology, as well as for commerce and industry. This system admirably covers mechanics and electromagnetism, but it does not provide for other branches of science such as heat. In 1960, in the hope of securing world-wide uniformity in the units employed

† The General Conference of Weights and Measures (CGPM) is the authority set up by the Metre Convention of 1875 to promote and improve the metric system, and to secure international uniformity in metric units and standards of measurement. It consists of delegations from the member nations (of which there were 46, including the UK, in 1982), which meet every few years, the 15th, 16th and 17th Conferences having been held in 1975, 1979, and 1983 the International Bureau of Weights and Measures (BIPM), Sèvres (near Paris), is the central office and laboratory of the organization, and is managed, under the authority of the General Conference, by the International Committee of Weights and Measures (CIPM) consisting of 18

in natural science, the 11th CGPM added to the units metre, kilogram, second and ampere, the **kelvin** for thermodynamic temperature, the **candela** for luminous intensity, and the radian and steradian for plane and solid angle. The first two joined the original four in being called 'base' units, and the last two were called 'supplementary' units. Any unit formed from two or more base units is called 'derived'. The radian and steradian are regarded as derived units. The MKSA system thus broadened is called the International System of Units, often abbreviated to SI, and is the most satisfactory system of units we have had so far, in that it caters for the commercial and industrial activities of man as well as for the needs of science. In 1971, the 14th CGPM added the **mole**, the unit of amount of substance used in chemistry, to the list of base units, thus making them seven in all.

Definitions of some SI units

The seven base quantities, each with its unit and unit symbol, are listed below.

SI base quantities and units

Quantity	Name of unit	Unit symbol
Length	metre	m
Mass	kilogram	kg
Time	second	s
Electric current	ampere	A
Thermodynamic temperature	kelvin	K
Amount of substance	mole	mol
Luminous intensity	candela	cd

The SI base units are defined as follows:

The metre is the length of the path travelled by light in vacuum during a time interval of $1/299\,792\,458$ of a second.

The kilogram is the unit of mass; it is equal to the mass of the international prototype of the kilogram.

The second is the duration of $9\,192\,631\,770$ periods of the radiation corresponding to the transition between the two hyperfine levels of the ground state of the caesium-133 atom.

The ampere is that constant current which, if maintained in two straight parallel conductors of infinite length, of negligible circular cross-section, and placed 1 metre apart in vacuum, would produce between these conductors a force equal to 2×10^{-7} newton per metre of length.

The kelvin, unit of thermodynamic temperature, is the fraction $1/273.16$ of the thermodynamic temperature of the triple point of water.

The mole is the amount of substance of a system which contains as many elementary entities as there are atoms in 0.012 kilogram of carbon 12.

When the mole is used, the elementary entities must be specified and may be atoms, molecules, ions, electrons, other particles, or specified groups of such particles.

The candela is the luminous intensity, in a given direction, of a source that emits monochromatic radiation of frequency 540×10^{12} hertz and that has a radiant intensity in that direction of $(1/683)$ watt per steradian.

The SI supplementary units are defined thus:

The radian is the plane angle between two radii of a circle which cut off on the circumference an arc equal in length to the radius.

The steradian is the solid angle which, having its vertex in the centre of a sphere, cuts off an area of the surface of the sphere equal to that of a square with sides of length equal to the radius of the sphere.

members, each from a different nation. The International Committee meets yearly and is responsible for recommending proposals for approval by the General Conference. Eight specialist advisory committees assist the International Committee in planning co-operative programmes of research, and in the preparation of recommendations on units of measurement, on length (definition of the metre), mass, time (definition of the second), temperature, electricity, photometry and radiometry, and ionizing radiations.

Derived units. The table below lists some of the more common SI derived quantities, each with its unit and unit symbol. The composite symbols in the last column are to some extent indicative of the definition of the quantity.

Quantity	Unit	Symbol	
Supplementary			
Plane angle	radian	rad	
Solid angle	steradian	sr	
Derived			
Area	square metre		m^2
Volume	cubic metre		m^3
Frequency	hertz	Hz	s^{-1}
Density	kilogram per cubic metre		$kg\,m^{-3}$
Concentration	mole per cubic metre		$mol\,m^{-3}$
Velocity	metre per second		$m\,s^{-1}$
Angular velocity . . .	radian per second		$rad\,s^{-1}$
Acceleration	metre per second squared		$m\,s^{-2}$
Angular acceleration . . .	radian per second squared		$rad\,s^{-2}$
Force	newton	N	$m\,kg\,s^{-2}$
Pressure, stress	pascal	Pa	$N\,m^{-2}$
Viscosity (dynamic). .	pascal second		$Pa\,s$
Viscosity (kinematic) . . .	metre squared per second		$m^2\,s^{-1}$
Energy, work, quantity of heat .	joule	J	$N\,m$
Power, radiant flux	watt	W	$J\,s^{-1}$
Quantity of electricity . . .	coulomb	C	$A\,s$
Potential difference, electro-motive force	volt	V	$W\,A^{-1}$
Electric field strength . . .	volt per metre		$V\,m^{-1}$
Electric resistance . . .	ohm	Ω	$V\,A^{-1}$
Electric conductance . . .	siemens	S	$Ω^{-1}$
Capacitance	farad	F	$C\,V^{-1}$
Magnetic flux	weber	Wb	$V\,s$
Magnetic flux density . . .	tesla	T	$Wb\,m^{-2}$
Inductance	henry	H	$Ω\,s$
Magnetic field strength . . .	ampere per metre		$A\,m^{-1}$
Magnetomotive force . . .	ampere	A	
Wave number*	1 per metre		m^{-1}
Activity (of a radionuclide). .	becquerel	Bq	s^{-1}
Absorbed dose	gray	Gy	$J\,kg^{-1}$
Dose equivalent	sievert	Sv	$J\,kg^{-1}$
Luminous flux	lumen	lm	$cd\,sr$
Luminance	candela per square metre		$cd\,m^{-2}$
Illuminance	lux	lx	$lm\,m^{-2}$
Heat flux density, irradiance .	watt per square metre		$W\,m^{-2}$
Heat capacity, entropy . .	joule per kelvin		$J\,K^{-1}$
Specific heat capacity, specific entropy	joule per kilogram kelvin		$J\,kg^{-1}\,K^{-1}$
Thermal conductivity . . .	watt per metre kelvin		$W\,m^{-1}\,K^{-1}$
Molar energy	joule per mole		$J\,mol^{-1}$
Molar entropy, molar heat capacity	joule per mole kelvin		$J\,mol^{-1}\,K^{-1}$

* Wave numbers in the infra-red are still often expressed in cm^{-1}.

Multiples. Prefixes may be used, instead of powers of 10, to express certain decimal multiples of the units. Their names and symbols are listed below.

Factor	Name	Symbol	Factor	Name	Symbol
10^{18}	exa	E	10^{-1}	deci	d
10^{15}	peta	P	10^{-2}	centi	c
10^{12}	tera	T	10^{-3}	milli	m
10^{9}	giga	G	10^{-6}	micro	μ
10^{6}	mega	M	10^{-9}	nano	n
10^{3}	kilo	k	10^{-12}	pico	p
10^{2}	hecto	h	10^{-15}	femto	f
10	deca	da	10^{-18}	atto	a

An exponent attached to a symbol containing a prefix indicates that the multiple of the unit is raised to the power expressed by the exponent.

Example: $1 \text{ cm}^3 = 10^{-6} \text{ m}^3$; $1 \text{ cm}^{-1} = 10^2 \text{ m}^{-1}$.

Compound prefixes should not be used, e.g. p not $\mu\mu$. Names of multiples of the unit of mass are formed by attaching prefixes to the word 'gram'.

1.1.2 Realization of SI units

The metre

In 1975 the 16th CGPM, having examined results of recent measurements of the frequencies and wavelengths of several laser lines, recommended the use of the value 299 792 458 m s^{-1} for the speed of light in vacuum. In 1983, further measurements having shown no cause for changing this value, the 17th CGPM confirmed it, and re-defined the metre in terms of it and of the second defined on p. 2. The new definition, also given on p. 2, supersedes the definition based on the wavelength of a spectral line of krypton.

In order to facilitate realization of the metre by laboratory workers according to the new definition, CIPM has recommended the procedures to follow. The methods fall into two classes: (1) a time of flight measurement, using the relation $l = c.t$; (2) an interferometric measurement, using a wavelength derived from a frequency by the relation $\lambda = c/f$. The first method is suitable for long distances, the second for the laboratory and small-scale engineering. To avoid the need for frequency measurements by individual workers using the second method, CIPM has listed the wavelengths: (a) of five stabilized laser radiations, with operating conditions and uncertainties; (b) of the former krypton-86 standard; and (c) of the other discharge tube standards recommended in 1963.

In most cases what is required is not realization of the unit itself, but rather measurement of a particular distance, e.g. the length of a rod or the separation of two optical flats. Interferometers, usually of the Fabry–Perot or of the Michelson type, are employed to determine the number of waves covering the distance. The krypton-86 line gives satisfactory fringes at path differences less than half a metre, but stabilized lasers, for instance those listed by CIPM, which have much narrower lines, have increased the workable distance to over 100 m.

The kilogram

Realization of the kilogram consists in making weights equal in mass to the mass of the prototype of the kilogram. The prototype is kept by BIPM, and the various nations adhering to the Metre Convention have copies which from time to time they send to Sèvres for comparison with the prototype or with the Bureau's copies of the prototype.

Methods of adjusting weights are well known, as also are balances for comparing them. Similarly, the production of masses which are multiples of the unit, and their calibration by means of balances, are standard practice. With reliable weights—for example, of platinum–iridium or of stainless steel—and good balances, and with a great deal of care, masses of the order of 1 kg can be compared to 1 in 10^9.

The second

The second of time was formerly defined as $1/86\ 400$ of the mean solar day. Its present definition in terms of the radiation corresponding to the hyperfine transition of the caesium-133 atom was adopted at the 13th General Conference of Weights and Measures in 1966, the duration of 9 192 631 770 periods of this radiation being chosen in order to secure as close agreement as possible with astronomical definitions.

The transition of the caesium atom is observed in an atomic beam equipment designed to eliminate the major causes tending to broaden or shift the line. The atoms pass through a system of magnets in which they are deflected, and through an electrical resonator supplied with an alternating field at a frequency derived by synthesis from a quartz oscillator. When the applied frequency equals that of the spectral line, transitions between the states are induced and deflection of the atoms is reversed. A resonant line is thus observed and, if the frequency deviates from that of the spectral line, an error signal is produced and applied to the quartz oscillator to pull its frequency to the correct value. The quartz oscillator acts as the working standard, as it did when time was based on astronomical measurements.

Although the unit has been defined in terms of the caesium line, similar lines of other atoms can be measured with great accuracy, and can then be used as standards. The frequency of the hydrogen line, for example, is 1 420 405 751.77 Hz.

It is not necessary for other than highly specialized organizations to own caesium beam or similar standards with which to realize the second, for many countries broadcast frequency and time signals by which any laboratory having a suitable radio receiver can calibrate its wavemeters and electronic counters (see p. 152). These transmissions are, in general, accurate to 1 in 10^{11}, and corrections to 1 in 10^{12} are published later.

The ampere

The definition of the ampere is simple and satisfies legal requirements, but the configuration it specifies precludes it from being applied as it stands to the measurement of electric current. However, as it is consistent with electromagnetic theory, other more practical configurations can be employed. The force per unit current between current-carrying coils of circular or helical shape can be calculated from the dimensions of the coils by fairly simple, exact formulae, and the current becomes known if the force is also measured with, for instance, a balance. These current balances, or dynamometers, are only used at rare intervals of, say, a few years. As their accuracy of a few p.p.m. is far superior to the precision of pointer or even reflecting ammeters, they are fed with a current stabilized by equalizing the volt drop RI across a standard resistor to the e.m.f. E of a Weston cell. The current I in the balance is then E/R, where *a priori* neither E nor R is known in absolute value. As explained below, however, R can be measured independently in terms of the metre and the second, and the two measurements thus yield E and R. In the intervals between measurements with current balances, the ampere is maintained, to a precision of about 1 p.p.m., by means of standard resistors and Weston cells.

The monitoring, though not the realization, of the ampere—i.e. of the resistors and cells which serve to maintain it—may also be effected by an application of nuclear magnetic resonance. The current, nominally equal to E/R as for the current balance, is carried by a coil of shape such that the magnetic field near its centre is uniform, and the frequency of precession of protons in that field is observed. If the dimensions of the coil do not change from year to year, or if the change is measured and allowed for, the frequency of precession provides a check on the stability of E/R. Measurement of a frequency is less laborious than accurate 'weighing' of a current.

The volt and the ohm

The volt, or watt/ampere, could, according to its definition, be realized from the ampere by a measurement of dissipation of energy in the form of heat. However, as energy is not easily measured to an accuracy of 1 p.p.m., it has been the practice to realize the volt, or the ohm, by an independent measurement involving electromagnetism. Whichever is realized, the other is obtained from it and the ampere by Ohm's law.

In the past the ohm has in general been chosen for the independent measurement. If we remember that the quantities ωL, ωM, $1/\omega C$—where ω, L, M and C stand for angular frequency, self-inductance, mutual inductance and capacitance—all have the dimensions of R, we see that a method of determining R is to compare it with one of those quantities. Self- and mutual inductors and capacitors have all been used. They are designed so that their values may be calculated from their linear dimensions. Accuracies of 1 in 10^7 and even better have been claimed for the calculable capacitor method of realizing the ohm. As the metre and the second are defined in terms of atomic constants, the ohm also depends on those constants.

The calculable capacitor, being fairly easy to use once it has been adjusted, is also suitable for monitoring standard resistors. The Weston cells which serve to maintain the volt can be monitored independently by an application of the Josephson effect in superconductivity. The p.d. between two superconductors separated by a narrow gap assumes values which are multiples of $hf/2e$, where f is the frequency of electromagnetic waves irradiating the superconductors, h is Planck's constant and e the electronic charge. For the mere purpose of monitoring the Weston cell from year to year, it is not even necessary to know the value of the constant $h/2e$.

Standard resistors for preserving the ohm are in general of nominal value 1 Ω. The e.m.f. of Weston cells depends slightly on temperature and on the acidity of the electrolyte. The figures below are representative of the variation with temperature.

E.M.F. of saturated (0.05 N) Weston cell

Temperature/°C	Electromotive force/V	Temperature/°C	Electromotive force/V
0	1.018 97	25	1.018 40
5	1.018 98	30	1.018 14
10	1.018 92	35	1.017 84
15	1.018 80	40	1.017 51
20	1.018 62		

International electric units. From the beginning of the century until 1948, the electric and magnetic units were not derived directly from the base units of length, mass and time, but from internationally agreed values for the electrochemical equivalent of silver and for the specific resistance of mercury. These units were called 'international units'. As one would not, for accurate work, use tables of constants published before 1948, the need for conversion factors hardly ever arises. The figures given by the International Bureau of Weights and Measures are

$$1 \text{ 'mean international ohm'} = 1.000\ 49\ \Omega$$
$$1 \text{ 'mean international volt'} = 1.000\ 34\ V$$

The kelvin

The unit of thermodynamic temperature, the kelvin, is defined by assigning the value 273.16 K to the temperature of the triple point of water. The determination of other temperatures in terms of this unit requires measurements, by means of a gas thermometer for example, which can be evaluated in accordance with the thermodynamic definition of temperature. Absolute measurements of this kind are difficult, and their accuracy is usually less than the reproducibility attainable by non-absolute methods based on the fixed points of various substances, and interpolation or extrapolation by instruments such as resistance thermometers, thermocouples, and optical pyrometers. This situation has led to the adoption, by international agreement, of a practical temperature scale, IPTS-68, based on the use of these more reproducible methods, and adjusted to conform to the best available knowledge of the thermodynamic temperatures of fixed points (see section 1.5, p. 44).

The mole

Although the mole is defined in terms of number of entities, it is usually realized by weighing rather than by counting. A mole of atoms of an element X, for example, is obtained by weighing an amount, in grams, of X, equal to its relative atomic mass (atomic weight); similarly a mole of molecules of a substance Y is obtained by weighing an amount, in grams, of Y, equal to its relative molecular mass (molecular weight).

In the case of perfect gases, 1 mole of molecules occupies the same volume, independently of the particular gas, at any given temperature and pressure. This relation provides a method for measuring equal amounts of substance of perfect gases. The method of volume comparison can be extended to non-perfect gases, because the corrections to apply are well known.

Ratios of amounts of substance liberated in electrolytic reactions can be determined by measuring the corresponding quantities of electricity. For example, 1 mole of Ag and 1 mole of Cu(1/2) are deposited on a cathode by the same quantity, approximately 96 485 C, of electricity.

The candela

The definition of the candela on p. 2 was promulgated by the 16th CGPM, in 1979, to replace that based on black-body radiation. The secondary standards of luminous intensity are tungsten-filament lamps powered by a specified direct current. They are calibrated by comparison with the monochromatic radiation prescribed in the definition, account being taken of their 'relative spectral luminous efficiencies', $V(\lambda)$ (see section 1.7.3., p. 86), recommended by CIPM. This experimentally determined function is a relation between the sensitivity of the average eye and the wavelength of the light falling on it. There are two such functions, $V(\lambda)$ for photopic vision, $V'(\lambda)$ for scotopic vision. By their means photometric quantities are defined in purely physical terms as quantities proportional to the sum or integral of a spectral power distribution, weighted according to a specified function of wavelength.

1.1.3 Relations between SI and other units

CGS units

Many books and papers on electromagnetism were written before the general adoption of SI. The following table gives the relations between the units, and the conversion factor by which the number expressing the value of the quantity in CGS units must be multiplied to express its value in SI units.

Electric and magnetic units

Quantity and symbol	SI unit		Conversion factors	
	Name and symbol	Defining equation	CGSm	CGSe
1. Electric current, I	ampere A	$F_z = 10^{-7} I^2 (\mathrm{d}N/\mathrm{d}z)$‡	10	$10/c$†
2. Current density, J	A m^{-2}		10^5	$10^5/c$
3. Electromotive force, potential difference, V	volt V	P§$= IV$	10^{-8}	$10^{-8}c$

Electric and magnetic units

Quantity and symbol	SI unit		Conversion factors	
	Name and symbol	Defining equation	CGSm	CGSe†
4. Electric field strength, E . .		V m^{-1}	$E = V/l$	10^{-6}	$10^{-6}c$
5. Resistance, R	ohm Ω		$R = V/I$	10^{-9}	$10^{-9}c^2$
6. Conductance, G	siemens S		$G = 1/R$	10^9	$10^9/c^2$
7. Volume resistivity, ρ . . .		Ω m		10^{-2}	$10^{-11}c^2$
8. Conductivity, γ		S m^{-1}	$\gamma = 1/\rho = J/E$	10^2	$10^{11}/c^2$
9. Electric charge, Q	coulomb C		$Q = It$	10	$10/c$
10. Capacitance, C	farad F		$C = Q/V$	10^9	$10^9/c^2$
11. Electric flux density, D . .		C m^{-2}	$D = Q/l^2$ ‖	10^5	$10^5/c$
12. Permittivity, ϵ		F m^{-1}	$\epsilon = D/E$		$10^{11}/4\pi c^2$
13. Magnetic field strength, H .		A m^{-1}	$\oint H\,\mathrm{d}l = nI$	$10^3/4\pi$	
14. Magnetic flux, ϕ	weber Wb		$E = \mathrm{d}\phi/\mathrm{d}t$	10^{-8}	
15 Magnetic flux density, B . .	tesla T		$B = \phi/l^2$ ‖	10^{-4}	
16. Inductance, L, M	henry H		$M = \phi/I$	10^{-9}	
17. Permeability, μ		H m^{-1}	$\mu = B/H$	$4\pi \times 10^{-7}$	

† c = velocity of light in free space in cm s^{-1} = 2.997 924 58 × 10^{10}.

‡ N denotes Neumann's integral for two linear circuits each carrying the current I; F_z in the force between the two circuits in any direction z, the circuits being in a vacuum; r is the distance between the vector elements $\mathrm{d}\mathbf{s}_1$, $\mathrm{d}\mathbf{s}_2$.

$$N = \oint_1 \oint_2 \frac{\mathrm{d}\mathbf{s}_1 . \mathrm{d}\mathbf{s}_2}{r}$$

§ P denotes power.
‖ l^2 denotes area.

The proportionality constants in the following equations are determined by use of the defining equations (1) to (17) in the table above.

For the mutual inductance M of two linear circuits, for which Neumann's integral is N, in a medium of permeability μ:

$$M = \frac{\mu}{4\pi} N \qquad \qquad \ldots(18)$$

For the force F_z due to currents I_1 and I_2 in the above two circuits:

$$F_z = I_1 I_2 \frac{\mathrm{d}M}{\mathrm{d}z} = \frac{\mu}{4\pi} I_1 I_2 \frac{\mathrm{d}N}{\mathrm{d}z} \qquad \ldots(19)$$

For the capacitance δC of an element of volume of length δl and cross-sectional area δA in a uniform dielectric medium of permittivity ϵ, the electric field having the direction of l:

$$\delta C = \epsilon \frac{\delta A}{\delta l} \qquad \qquad \ldots(20)$$

In the SI, Maxwell's equations for propagation in a medium are

$$\nabla \times H = \dot{D} + J, \qquad -\nabla \times E = \dot{B} \qquad \ldots(21)$$

where the dot means d/dt, J is current density, and the other quantities are as in the table. The velocity of propagation of an electromagnetic wave in the medium is given by

$$v^2 = 1/\mu\epsilon \qquad \qquad \ldots(22)$$

In free space, where J is zero, and where D and B are linked with E and H by the electric and magnetic constants ϵ_0 and μ_0, the equations may be written

$$c^2 \nabla \times B = \dot{E}, \qquad -\nabla \times E = \dot{B} \qquad \ldots(23)$$

The velocity of propagation c is given by

$$c^2 = 1/\mu_0 \epsilon_0 \qquad \dots(24)$$

The values of μ_0, ϵ_0 and of the characteristic impedance Z_0 of free space are

$$\mu_0 = 4\pi \times 10^{-7}\ \text{H m}^{-1} = 1.256\,637\,\mu\text{H m}^{-1}$$
$$\epsilon_0 = 8.854\,19\,\text{pF m}^{-1}$$
$$Z_0 = \sqrt{(\mu_0/\epsilon_0)} = 376.730\,\Omega$$

Coulomb's law of force f between electric charges Q, Q' a distance r apart, and its analogue for magnetic poles ϕ, ϕ', can be deduced from Maxwell's equations. The expressions are

$$f = \frac{QQ'}{4\pi\epsilon r^2} \qquad \dots(25)$$

$$f = \frac{\phi\phi'}{4\pi\mu r^2} \qquad \dots(26)$$

Other mechanical units

In the gravitational systems of units sometimes still used in engineering and aerodynamics, the units of length and time are base units, but the third base unit is a unit of force, namely the force due to a conventional value of gravity, called standard gravity, acting on a specified mass. The unit of mass is thus a derived unit, and since unit mass is given unit acceleration when acted on by unit force, the unit of mass is equal to the standard value of gravity multiplied by the specified mass.

In the metric gravitational system, standard gravity is $9.806\,65$ m s^{-2}, the unit of force is called kilogram-force, symbol kgf, the specified mass is the kilogram, and the unit of mass is therefore $9.806\,65$ kg. In Germany and some other European countries the same unit of force is called kilopond, symbol kp.

In the foot–pound–second gravitational system the base unit of force is the pound-force, symbol lbf, standard gravity is 32.174 ft s^{-2}, and the derived unit of mass, the slug, is now 32.174 lb, although originally it was 32.2 lb, based on 32.2 ft s^{-2} for gravity.

Conversion factors

The following table gives conversion factors for British and other units which are still used in some fields, but which, for the most part, are likely to disappear in course of time. Exact conversion factors are in bold type.

Unit name and symbol		SI equivalent		Reciprocal
Length				
ångström . . . . Å		**0.1**	nm	**10**
fermi		**1**	fm	
micron . . . . . . μ		**1**	μm	
yard . . . . yd		**0.914 4**	m	1.093 61
foot . . . . ft		**30.48**	cm	0.032 808 4
inch . . . . in		**2.54**	cm	0.393 701
statute mile		**1.609 344**	km	0.621 371
nautical mile (international) .		**1.852**	km	0.539 957
fathom (6 ft)		**1.828 8**	m	0.546 807
X unit		0.100 2	pm	
astronomical unit (AU) . .		0.149 6	Tm	
parsec . . . . pc		30.857	Pm	

Conversion factors *(contd)*

Unit name and symbol	SI equivalent		Reciprocal
Area			
hectare . . . . . . ha	1	hm^2	
barn . . . . b	100	fm^2	
square yard . . . . . yd^2	0.836 127	m^2	1.195 99
square foot . . . . . ft^2	9.290 30	dm^2	0.107 639
square inch . . . . . in^2	6.451 6	cm^2	0.155 000
square statute mile	2.589 99	km^2	0.386 102
acre	0.404 686	hm^2	2.471 05
Volume			
stere . . . . st	1	m^3	
litre l	1	dm^3	
cubic yard . . . . . yd^3	0.764 555	m^3	1.307 95
cubic foot . . . . . ft^3	0.028 316 8	m^3	35.314 7
cubic inch . . . . . in^3	16.387 1	cm^3	0.061 023 7
gallon (UK) . . . . . gal	4.546 09	dm^3	0.219 969
gallon (US) = 231 in^3 . . .	3.785 41	dm^3	0.264 172
pint (UK) . . . . . pt	0.568 261	dm^3	1.759 75
barrel (US) = 42 gal (US) . . bbl	0.159 0	m^3	6.290
Plane angle			
right angle = $\pi/2$ rad . . .	1.570 796	rad	0.636 620
degree = 1/90 right angle . . °	$\pi/180$	rad	57.295 8
minute = (1/60)° . . . . . '	$\pi/10\ 800$	rad	3437.75
second = (1/60)' . . . . "	$\pi/648$	mrad	206.265
grade = 0.01 right angle . .	$\pi/200$	rad	63.662 0
Mass			
tonne . . . . t	1	Mg	
metric carat	0.2	g	5
pound (avdp) . . . . lb	0.453 592 37	kg	2.204 62
grain . . . . . . gr	0.064 798 91	g	15.432 4
ounce (avdp) . . . . oz	28.349 5	g	0.035 274 0
ounce (troy) = 480 gr . . .	31.103 5	g	0.032 150 7
ton = 2240 lb	1.016 05	Mg	0.984 207
slug	14.593 9	kg	0.068 521 8
Specific volume			
cubic foot per pound . . . ft^3 lb^{-1}	0.062 428	m^3 kg^{-1}	16.018 5
Density			
pound per cubic foot . . lb ft^{-3}	16.018 5	kg m^{-3}	0.062 428
pound per cubic inch . . lb in^{-3}	27.679 9	Mg m^{-3}	0.036 127 3
pound per gallon (UK) . . lb gal^{-1}	99.776 3	kg m^{-3}	0.010 022 4
slug per cubic foot . . . slug ft^{-3}	0.515 379	Mg m^{-3}	1.940 32
Consumption, yield			
gallon (UK) per mile . . .	2.824 8	dm^3 km^{-1}	0.354 01
bushel per acre	0.089 87	m^3 ha^{-1}	11.127
Moment of inertia			
lb in^2	2.926 40	kg cm^2	0.341 717
slug ft^2	1.355 82	kg m^2	0.737 562
Moment of section			
in^4	41.623 1	cm^4	0.024 025 1

Conversion factors (*contd*)

Unit name and symbol	SI equivalent		Reciprocal
Time			
minute . . . . . . min	**60**	s	
hour = 60 min . . . . . h	**3600**	s	
day = 24 h . . . . . d	**86 400**	s	
Velocity			
foot per second . . . . ft s^{-1} $\{$	**0.304 8**	m s^{-1}	3.280 84
	1.097 28	km h^{-1}	0.911 344
mile per hour . . . . mile h^{-1}	**1.609 344**	km h^{-1}	0.621 371
knot (international)	**1.852**	km h^{-1}	0.539 957
Acceleration			
gal . . . . . . Gal	**1**	cm s^{-2}	
Force			
dyne . . . . . . dyn	**10**	μN	**0.1**
poundal . . . . . pdl	0.138 255	N	7.233 01
pound-force . . . . lbf	4.448 22	N	0.224 809
ton-force . . . . . tonf	9.964 02	kN	0.100 361
kilogram-force . . . . kgf	**9.806 65**	N	0.101 972
Force per unit length			
lbf in^{-1}	0.175 127	kN m^{-1}	5.710 15
tonf ft^{-1}	32.690 3	kN m^{-1}	0.030 590 1
Torque			
pound-force foot . . . . lbf ft	1.355 82	N m	0.737 562
Pressure, stress			
bar . . . . . . bar	**0.1**	MPa	**10**
lbf in^{-2}	6.894 76	kPa	0.145 038
lbf ft^{-2}	47.880 3	Pa	0.020 885 4
tonf in^{-2}	15.444 3	MPa	0.064 749 0
standard atmosphere . . . atm	**0.101 325**	MPa	9.869 23
torr $\}$			
mmHg $\}$ . . . . . torr	0.133 322	kPa	7.500 62
inHg	3.386 39	kPa	0.295 300
Work, energy, heat			
erg . . . . . . erg	**0.1**	μJ	**10**
electronvolt . . . . eV	0.160 219	aJ	6.241 46
calorie (IT) . . . . . cal	**4.186 8**	J	0.238 846
British thermal unit . . . Btu	1.055 06	kJ	0.947 817
therm = 10^5 Btu	0.105 506	GJ	9.478 17
foot poundal . . . . ft pdl	0.042 140 1	J	23.730 4
foot pound-force . . . . ft lbf	1.355 82	J	0.737 562
Power			
horsepower = 550 ft lbf s^{-1} . hp	0.745 700	kW	1.341 02
cheval vapeur = 75 m kgf s^{-1} . ch, CV	0.735 499	kW	1.359 62
Temperature			
degree Celsius . . . . . °C	**1**	K	
t °C	**273.15** $+ t$	K	
degree Fahrenheit . . . °F	5/9	K	1.8
t °F	$5(t-32)/9$	°C	
degree Rankine	5/9	K	1.8
Heat transmission			
Btu h^{-1}	0.293 071	W	3.412 14

Conversion factors (*contd*)

Unit name and symbol		SI equivalent		Reciprocal
Viscosity (dynamic)				
poise	P	0.1	Pa s	10
lb ft^{-1} s^{-1} = pdl s ft^{-2} . . .		1.488 16	Pa s	0.671 969
slug ft^{-1} s^{-1} = lbf s ft^{-2} . .		47.880 3	Pa s	0.020 885 4
Viscosity (kinematic)				
stokes	St	1	cm^2 s^{-1}	
ft^2 s^{-1}		9.290 30	dm^2 s^{-1}	0.107 639
Electric stress				
kilovolt per inch ⎫				
volt per mil ⎬	kV in^{-1}	0.393 701	kV cm^{-1}	2.54
Magnetism				
gauss	Gs, G	100	μT	0.01
oersted	Oe	(1/4π)	kA m^{-1}	4π
maxwell	Mx	10	nWb	0.1
gamma	γ	1	nT	
Light				
stilb	sb	10	kcd m^{-2}	0.1
phot	ph	10	klx	0.1
Radioactivity				
curie	Ci	37	GBq	0.027 027
röntgen	R	0.258	mC kg^{-1}	3.875 97
rad	rad	10	mGy	0.1
rem	rem	10	mSv	0.1

References

(1) *The International System of Units*, 1982 (translation of a BIPM publication), HMSO.
(2) BS 5555 : 1981 (ISO 1000–1981) SI units.
(3) *Quantities, Units and Symbols*, 1975 (The Royal Society).
(4) BS 350 : 1974 (1983), *Conversion Factors*.   P.V.

1.1.4 International specifications for units

Following the adoption of SI units by the CGPM, their incorporation in international standard specifications has been extensively discussed by the International Organization for Standardization (ISO). In recent years, a series of international standards has been issued, dealing with the definitions of quantities, units of measurement, and symbols. These are published as British Standards by the British Standards Institution. The following are now available:

ISO 1000–1981 (BS 5555 : 1981): *Specification for SI units and recommendations for the use of their multiples and of certain other units.*

ISO 31–1981 (BS 5775 : 1982): *Specification for quantities, units and symbols.*
   Part 0 : *General principles.*
   Part 1 : *Space and time.*
   Part 2 : *Periodic and related phenomena.*
   Part 3 : *Mechanics.*
   Part 4 : *Heat.*
   Part 5 : *Electricity and magnetism.*

Part 6 : *Light and related electromagnetic radiations.*
Part 7 : *Acoustics.*
Part 8 : *Physical chemistry and molecular physics.*
Part 9 : *Atomic and nuclear physics.*
Part 10 : *Nuclear reactions and ionising radiations.*
Part 11: *Mathematical signs and symbols for use in the physical sciences and technology.*
Part 12 : *Dimensionless parameters.*
Part 13 : *Solid state physics.*

A.E.B.

1.2 Fundamental constants

1.2.1 Velocity of electromagnetic waves

Modern determinations of c_0, the speed of electromagnetic radiation in vacuo, obtain c_0 through the relation $c_0 = \nu\lambda$, where ν is the frequency either of the radiation itself or of an impressed high-frequency modulation, and λ is the corresponding wavelength. Several measurements based on this principle were carried out in the period 1946 to 1957, using microwave radiation or modulated light; the most accurate was a measurement by microwave interferometry (Froome, Nature 1958), which gave the value $c_0 = 299\ 792.50 \pm 0.10$ km s^{-1}. More recently the development of methods of stabilizing and measuring the frequency of infra-red lasers has led to a number of measurements of much higher accuracy, and in 1973 the following value was recommended by the International Committee of Weights and Measures: $c_0 = 299\ 792.458 \pm 0.001$ km s^{-1}. This value was based on measurements of the frequency and wavelength of a helium-neon laser radiation, stabilized to an infra-red transition of methane at about 3.39 μm, and was later confirmed by similar work using radiation of about 9.3 μm wavelength from a stabilized carbon dioxide laser (*Metrologia*, 1974, **10**, 9). This latter value has now been fixed by the new definition of the metre adopted by the CGPM in 1984; any changes in the measured value will in future be ascribed to discrepancies between the maintained metre and the metre of the SI definition.

B.W.P.

1.2.2 The constant of gravitation

Newton's law of gravitation states that the force between two masses M_1 and M_2 at a distance d is GM_1M_2/d^2. The constant G is not known to much better than 1 in 10^4; the most recent value is that of Luther and Towler (*P.R. Lett.* **48**, 121, 1982):

$$G = 6.668(2) \times 10^{-11}\ \text{m}^3\ \text{kg}^{-1}\ \text{s}^{-2}$$

A.H.C.

1.2.3 Atomic constants

The values of e, h, N_A etc., given on pages 15, 16, are based on the values of Cohen and Taylor, *J. Phys. Chem. Data*, 1973, **2**, 663–774.

The best values of constants such as the mass and charge of the electron are not found by measuring each separately (e.g. as in Millikan's oil-drop experiment). The values recommended in the above paper were obtained by a weighted 'least-squares' statistical treatment of selected experimental data. The values depend on comparatively few experimental results and the number of 'unknowns' was reduced by treating certain combinations of the constants as being exact (or auxiliary constants) for the purposes of the evaluation. This group included quantities such as the Rydberg constant $\mu_0 c^3 m_e e^4/8h^3$, the ratio μ_p'/μ_B between the proton magnetic moment (in water) μ_p' and the Bohr magneton $\mu_B (= eh/4\pi m_e)$, the Josephson effect value of $2e/h$ in maintained voltage units, and set relationships between e, h, and m_e. The unknowns included the spectroscopic fine-structure constant α $((\mu_0 c^2/4\pi)e^2/h)$, the ratio $\bar{R}$ of the maintained ohm to the SI ohm, and the ratio K of the maintained ampere to the SI ampere. Some of the experimental data can only be interpreted in terms of α, etc. by invoking very sophisticated quantum electrodynamic calculations and so, as time passes, improved values are obtained for α as higher order terms are taken into account.

Since the 1973 evaluation by Cohen and Taylor it has become apparent that the recommended value of K is probably in error by between five and around seven parts per million. In addition there have been further accurate measurements such as the gravitational constant, the gas constant, the Avogadro constant, and the Rydberg constant. The advent of ion traps has led to increased accuracy of measurement of the electron magnetic moment anomaly $g_e - 2$, and of the ratio of the proton to electron mass. Better direct determinations are also in hand for the realizations of the ampere and ohm, while the quantized Hall resistance in MOSFET semi-conductors at low temperatures provides information concerning $\bar{R}$ and α.

As a result of the method of evaluation there are correlations between the output values, so that the full variance and co-variance matrix should be used for computing values that are not given in the table. Although the accuracy is normally well ahead of user requirements, it is always important to specify which evaluation one has used and also to ensure that they come from a consistent set of constants. Results from different evaluations may differ and consequently should not be mixed, otherwise the constants will not form a consistent set.

Values accepted formerly. Other conventional values of e, h, etc. may be encountered in books, and some of these earlier values are given below. The first is essentially Millikan's oil-drop value of e, 4.77×10^{-10} esu, and the second the 'X-ray grating' value 4.802. The numbers in brackets are standard errors. It will be seen that the accuracies claimed have sometimes appeared over-optimistic in retrospect, and that besides the supposed random errors of observation some unsuspected systematic error has been present. (In Millikan's value of e, accepted from 1917 to 1935, it was an inaccuracy in the assumed viscosity of air.) Consequently, it may be thought prudent to regard the standard errors given here with caution. A further evaluation by Cohen and Taylor is expected shortly.

Author of discussion	$e \times 10^{19}/C$	$h \times 10^{34}/(J\ s)$	Kaye and Laby
Birge, 1929	1.591 1 (0.002 4)	6.547 (0.012)	7th ed.
Birge, 1942	1.602 03 (0.000 50)	6.624 2 (0.003 5)	10th ed.
DuMond and Cohen, 1963 . . .	1.602 10 (0.000 02)	6.625 6 (0.000 16)	13th ed.
Taylor et al., 1969	1.602 192 (0.000 007)	6.626 20 (0.000 05)	14th ed.
Cohen and Taylor, 1973 . . .	1.602 189 2 (0.000 004 6)	6.626 176 (0.000 036)	14th ed. (1978 reprint)
Cohen and Taylor, 1985 . . .	1.602 178 (0.000 002)	6.626090 (0.000 004)	approx. values

Table of fundamental constants

(At the time of going to press these are the recommended values, which are still of Cohen and Taylor, 1973.)

			SI unit	Standard error (parts in 10^6)
Principal constants				
Speed of light in vacuum	c	$2.997\ 924\ 580 \times 10^8$	m s^{-1}	—
Planck constant	h	$6.626\ 176 \times 10^{-34}$	J s	5.4
Planck constant $(h/2\pi)$	$\hbar$	$1.054\ 588\ 7 \times 10^{-34}$	J s	5.4
Elementary charge	e	$1.602\ 189\ 2 \times 10^{-19}$	C	2.9
Mass of electron	m_e	$9.109\ 534 \times 10^{-31}$	kg	5.1
Mass of electron in atomic mass units . .		$5.485\ 802\ 6 \times 10^{-4}$	u	0.38
Avogadro constant	N_A, L	$6.022\ 045 \times 10^{23}$	mol^{-1}	5.1
Atomic mass unit, 10^{-3} kg mol^{-1} $N_A{}^{-1}$.	u	$1.660\ 565\ 4 \times 10^{-27}$	kg u^{-1}	5.1
Faraday constant	$F\ (=N_A e)$	$9.648\ 456 \times 10^4$	C mol^{-1}	2.8
Constant of gravitation	G	$6.672\ 0 \times 10^{-11}$	N m^2 kg^{-2}	615
Spectroscopy and atoms				
Planck constant	h	$4.135\ 701 \times 10^{-15}$	eV s	2.6
Planck constant $(h/2\pi)$	$\hbar$	$6.582\ 173 \times 10^{-16}$	eV s	2.6
Charge/mass ratio of electron	e/m_e	$1.758\ 804\ 7 \times 10^{11}$	C kg^{-1}	2.8

Table of fundamental constants (*contd*)

			SI unit	Standard error (parts in 10^6)
Fine structure constant	α	7.297 350 6 $\times 10^{-3}$		0.82
Fine structure constant, reciprocal . . .	α^{-1}	137.036 04		0.82
Scale of the fine structure	$\alpha^2 R_\infty$	5.843 658 $\times 10^2$	m^{-1}	1.64
Rydberg constant (fixed nucleus) . .	R_∞	1.097 373 177 $\times 10^7$	m^{-1}	0.075
Rydberg constant (hydrogen H) . .	R_H	1.096 775 854 $\times 10^7$	m^{-1}	0.075
Bohr radius $(4\pi/\mu_0 c^2)\hbar^2/m_e e^2$	a_0	5.291 770 6 $\times 10^{-11}$	m	0.82
Compton wavelength of electron . . .	λ_C	2.426 308 9 $\times 10^{-12}$	m	1.6
Compton wavelength of electron $\div 2\pi$.	$\hbar/m_e c$	3.861 590 5 $\times 10^{-13}$	m	1.6
Classical 'radius' of electron $(\mu_0 c^2/4\pi)e^2/mc^2$	r_e	2.817 938 $\times 10^{-15}$	m	2.5
Thomson cross-section $8\pi r_e^2/3$	σ_T	6.652 244 8 $\times 10^{-29}$	m	4.9
Zeeman effect, μ_B/hc		46.686 04	$m^{-1} T^{-1}$	2.4
Bohr magneton $e\hbar/2m_e$	μ_B	9.274 078 $\times 10^{-24}$	$J T^{-1}$	3.9
Nuclear magneton $e\hbar/2m_p$	μ_N	5.050 824 $\times 10^{-27}$	$J T^{-1}$	3.9
Ratio of masses proton/electron . .	m_p/m_e	1.836 151 52 $\times 10^3$		0.38
Gyromagnetic ratio of proton . .	$\gamma_p = \mu_p/\frac{1}{2}\hbar$	2.675 198 7 $\times 10^8$	$s^{-1} T^{-1}$	2.8
in H_2O	γ_p'	2.675 130 1 $\times 10^8$	$s^{-1} T^{-1}$	2.8
in H_2O (cycles)	$\gamma_p'/2\pi$	4.257 602 $\times 10^7$	$Hz T^{-1}$	2.8

Conversion factors for Mass, Energy and Wavelength

Energy and mass

Electron volt	1.602 189 2 $\times 10^{-19}$	J	2.9
Atomic mass unit	931.501 6	MeV	2.8
1 MeV	1.073 535 5 $\times 10^{-3}$	u	2.8
Rest-mass of electron	0.511 003 4	MeV	2.8
1 eV per molecule	9.648 456 $\times 10^7$	$J kmol^{-1}$	2.8

Frequency, wavelength and energy

Quantum energy $\div$ wave number	1.986 477 6 $\times 10^{-25}$	J m	5.4
Energy $\times$ wavelength	1.239 852 0 $\times 10^{-6}$	eV m	2.6
Wave number $\div$ energy	8.065 478 8 $\times 10^5$	$eV^{-1} m^{-1}$	2.6
Quantum energy $\div$ frequency . .	4.135 701 $\times 10^{-15}$	$eV Hz^{-1}$	2.6
Frequency $\div$ energy	2.417 969 6 $\times 10^{14}$	$Hz eV^{-1}$	2.6

Thermal constants

Molar gas constant	R	8.314 41	$J K^{-1} mol^{-1}$	32
Loschmidt constant (number of molecules in 1 m^3 of ideal gas at stp). . . .		2.686 754 $\times 10^{25}$	m^{-3}	32
Boltzmann constant R/N_A	k	1.380 662 $\times 10^{-23}$	$J K^{-1}$	32
Boltzmann constant		8.617 35 $\times 10^{-5}$	$eV K^{-1}$	32
Energy kT for $T = 273.15$ K		0.023 538 3	eV	32
Stefan constant (σ)†	$2\pi^5 k^4/15 c^2 h^3$	5.670 32 $\times 10^{-8}$	$W m^{-2} K^{-4}$	125
Constant in Planck formula (c_1)† . .	$2\pi h c^2$	3.741 832 $\times 10^{-16}$	$W m^2$	5.4
Constant in Planck formula (c_2)† . .	hc/k	1.438 786 $\times 10^{-2}$	m K	31

† See also p. 84.

Note: Magnetic moments are defined so that mechanical energy $= -\mu.B$; the unit is $1 J T^{-1} = 1 J Wb^{-1} m^2 = 1 A m^2$ (1 tesla = 1 weber $m^{-2} = 10^4$ gauss).

B.W.P.

1.3 Measurement of mass, pressure and other mechanical quantities

1.3.1 Mass, volume and density

Correction of weighings for buoyancy

If a substance is weighed in air and found to be balanced by mass standards ('weights') of value M, the true mass of the substance is

$$M + Md_a \left[\left(\frac{1}{d} \right) - \left(\frac{1}{d_s} \right) \right]$$

where d is the density of the substance, d_s is the density of the mass standards and d_a is the density of the air. In practice, the members of an ordinary set of mass standards, including the fractions, are standardized by being assigned values which are equal to the masses of reference standards of density 8000 kg m^{-3} which they would balance in air. The value 8000 kg m^{-3} is a conventional value close to the density of the larger weights in a set. The buoyancy of any group of weights from such a set should be calculated by taking this value for d_s. The following table has been prepared on this basis, d_a being taken to be 1.2 kg m^{-3}.

Density of substance weighed	Buoyancy correction (mg per gram of substance)	Density of substance weighed	Buoyancy correction (mg per gram of substance)
$d/\text{kg m}^{-3}$	$d_s = 8000 \text{ kg m}^{-3}$ $d_a = 1.2 \text{ kg m}^{-3}$	$d/\text{kg m}^{-3}$	$d_s = 8000 \text{ kg m}^{-3}$ $d_a = 1.2 \text{ kg m}^{-3}$
500	+2.250	6 000	+0.050
1000	+1.050	8 000	0
1500	+0.650	10 000	−0.030
2000	+0.450	12 500	−0.054
2500	+0.330	15 000	−0.070
3000	+0.250	20 000	−0.090
4000	+0.150	22 000	−0.096

Density of ambient air (unit = 1 kg m^{-3}) 50% rel. humidity, 0.04% CO_2 by volume

Air pressure, P/kPa (1 kPa = 10 mb)	Air temperature, t/°C												
	6	8	10	12	14	16	18	20	22	24	26	28	30
80	0.997	0.989	0.982	0.974	0.967	0.960	0.953	0.946	0.939	0.931	0.924	0.917	0.910
81	1.009	1.002	0.994	0.987	0.979	0.972	0.965	0.958	0.950	0.943	0.936	0.929	0.922
82	1.022	1.014	1.006	0.999	0.991	0.984	0.977	0.969	0.962	0.955	0.948	0.941	0.933
83	1.034	1.026	1.019	1.011	1.004	0.996	0.989	0.981	0.974	0.967	0.959	0.952	0.945
84	1.046	1.039	1.031	1.023	1.016	1.008	1.001	0.993	0.986	0.978	0.971	0.964	0.956
85	1.059	1.051	1.043	1.036	1.028	1.020	1.013	1.005	0.998	0.990	0.983	0.975	0.968
86	1.071	1.064	1.056	1.048	1.040	1.032	1.025	1.017	1.009	1.002	0.994	0.987	0.979
87	1.084	1.076	1.068	1.060	1.052	1.044	1.037	1.029	1.021	1.014	1.006	0.998	0.991
88	1.096	1.088	1.080	1.072	1.064	1.056	1.049	1.041	1.033	1.025	1.018	1.010	1.002
89	1.109	1.101	1.093	1.084	1.076	1.068	1.061	1.053	1.045	1.037	1.029	1.022	1.014
90	1.121	1.113	1.105	1.097	1.089	1.081	1.073	1.065	1.057	1.049	1.041	1.033	1.025
91	1.134	1.126	1.117	1.109	1.101	1.093	1.085	1.076	1.068	1.061	1.053	1.045	1.037
92	1.146	1.138	1.129	1.121	1.113	1.105	1.096	1.088	1.080	1.072	1.064	1.056	1.048
93	1.159	1.150	1.142	1.133	1.125	1.117	1.108	1.100	1.092	1.084	1.076	1.068	1.060
94	1.171	1.163	1.154	1.146	1.137	1.129	1.120	1.112	1.104	1.096	1.088	1.079	1.071
95	1.184	1.175	1.166	1.158	1.149	1.141	1.132	1.124	1.116	1.107	1.099	1.091	1.083
96	1.196	1.188	1.179	1.170	1.161	1.153	1.144	1.136	1.128	1.119	1.111	1.103	1.094
97	1.209	1.200	1.191	1.182	1.174	1.165	1.156	1.143	1.139	1.131	1.122	1.114	1.106
98	1.221	1.212	1.203	1.195	1.186	1.177	1.168	1.160	1.151	1.143	1.134	1.126	1.117
99	1.234	1.225	1.216	1.207	1.198	1.189	1.180	1.172	1.163	1.154	1.146	1.137	1.129
100	1.246	1.237	1.228	1.219	1.210	1.201	1.192	1.184	1.175	1.166	1.157	1.149	1.140
101	1.259	1.250	1.240	1.231	1.222	1.213	1.204	1.195	1.187	1.178	1.169	1.160	1.152
102	1.271	1.262	1.253	1.243	1.234	1.225	1.216	1.207	1.198	1.190	1.181	1.172	1.163
103	1.284	1.274	1.265	1.256	1.246	1.237	1.228	1.219	1.210	1.201	1.192	1.184	1.175
104	1.296	1.287	1.277	1.268	1.259	1.249	1.240	1.231	1.222	1.213	1.204	1.195	1.186
105	1.309	1.299	1.290	1.280	1.271	1.261	1.252	1.243	1.234	1.225	1.216	1.207	1.198
106	1.321	1.312	1.302	1.292	1.283	1.273	1.264	1.255	1.246	1.236	1.227	1.218	1.209

Corrections for humidity (unit = 1 kg m^{-3})

Relative humidity, R%	Air temperature, t/°C		
	10	20	30
20	+0.002	+0.003	+0.006
25	+0.001	+0.002	+0.005
30	+0.001	+0.002	+0.004
35	+0.001	+0.001	+0.003
40	0	+0.001	+0.002
45	0	0	+0.001
50	0	0	0
55	0	−0.001	−0.001
60	−0.001	−0.001	−0.002
65	−0.001	−0.002	−0.003
70	−0.001	−0.002	−0.003
75	−0.002	−0.003	−0.004
80	−0.002	−0.003	−0.005

The figures in the main table relate to air of 50% relative humidity and containing 0.04% carbon dioxide by volume. For other states of humidity a correction should be applied from the subsidiary table. The tables are based on the following expression recommended by an international working group formed by the International Bureau of Weights and Measures (BIPM).(P. Giacomo, *Metrologia*, 1982):

$$d = 3.483\ 53 p(1 - 0.378 x_v)/-ZT$$

where d = density of moist air (kg m^{-3})
  p = atmospheric pressure (kPa)
  x_v = the mole fraction of water vapour present in the air
  Z = compressibility factor
  T = temperature (K)

and where x_v and Z are given by the following expressions:

$$x_v = hf \frac{p_{sv}}{p} \times 10^{-5}$$

where h = relative humidity (%)
  f = 'enhancement factor'—a correction for departure from a perfect gas

  $= 1.000\ 62 + 3.14 \times 10^{-5} p + 5.6 \times 10^{-7} t^2$

  p_{sv} = saturation vapour pressure of water

  $= \exp\ (1.281\ 180\ 5 \times 10^{-5} T^2 - 1.950\ 987\ 4 \times 10^{-2} T$

  $+ 34.049\ 260\ 34 - 6.353\ 631\ 1 \times 10^3\ T^{-1})$

and $Z = 1 - \dfrac{1000 p}{T}(1.624\ 19 \times 10^{-6} - 2.8969 \times 10^{-8} t + 1.0088 \times 10^{-10} t^2)$

  $+ (5.757 \times 10^{-6} - 2.589 \times 10^{-8} t) x_v$

  $+ (1.9297 \times 10^{-4} - 2.285 \times 10^{-6} t) x_v^2$

  $+ \dfrac{p^2}{T^2} \times 10^6 (1.73 \times 10^{-11} - 1.034 \times 10^{-8} x_v^2)$

where t = temperature (°C) $(= T - 273.15)$

Density of dry CO_2-free air

The density in kg m^{-3} of dry air free from carbon dioxide may be derived from the BIPM expression referred to above, taking the mole fractions of CO_2 and water vapour as zero. The expression then becomes

$$d = 3.482\ 95\ p/ZT$$

where $Z = 1 - 100 p\ (1.624\ 19 \times 10^{-6} - 2.8969 \times 10^{-8} t + 1.0088 \times 10^{-10} t^2)$

  $+ \dfrac{p^2}{T^2} \times 1.73 \times 10^{-5}$

Reduction of gaseous volumes to s.t.p.

The volume at s.t.p. (0 °C and 1 atm) $= vp/[101.325\,(1+\gamma t)]$, where v and t °C are the observed volume and temperature, p is the observed pressure in kPa (1 kPa = 10 mb) and γ is the coefficient of cubical expansion at constant pressure of the gas concerned. For 'permanent' gases $\gamma = 0.003\ 67 \pm 0.000\ 01$ approximately. For coefficients of other gases, see section 2.5, p. 215.

If it is desired to find the volume of dry gas at s.t.p., v and p being measured when the gas contains water vapour whose pressure is e kPa, p in the above expression must be replaced by $(p-e)$.

Volumetric calibration of vessels with water or mercury

The volume content (V_t) of a vessel in cm^3 at the temperature of determination (t °C) is given by

$$V_t = Wf,$$

where W is the apparent mass in grams of contained water or mercury at t°C when weighed in air at t°C and 1 atm (101.325 kPa) against mass standards adjusted on the basis of density 8000 kg m^{-3}, and f is a factor which includes allowances for the densities of the liquids and the appropriate buoyancy corrections. Density tables for water and mercury are given in section 1.4.1, p. 29.

Temperature of liquid, t/°C	10	11	12	13	14	15
f Water	1.001 39	1.001 48	1.001 58	1.001 70	1.001 83	1.001 97
Mercury	0.073 685	0.073 698	0.073 712	0.073 725	0.073 738	0.073 752

Temperature of liquid, t/°C	16	17	18	19	20
f Water	1.002 13	1.002 29	1.002 47	1.002 65	1.002 85
Mercury	0.073 765	0.073 778	0.073 792	0.073 805	0.073 819

Temperature of liquid, t/°C	21	22	23	24	25
f Water	1.003 06	1.003 28	1.003 51	1.003 75	1.004 00
Mercury	0.073 832	0.073 845	0.073 859	0.073 872	0.073 886

An additional buoyancy correction is required when weighings with water are made at air pressures other than 101.325 kPa. The values of f (for water) must be increased ($+$), or decreased ($-$), by the following amounts (unit 0.000 01):

Pressure kPa (1 kPa = 10 mb)	97	98	99	100	101	102	103	104	105
Correction to f (water)	-5	-4	-3	-2	0	$+1$	$+2$	$+3$	$+4$

No additional buoyancy correction is required for mercury. The above tables give the volume content of the vessel at the temperature of determination (t). At any other temperature (T) the volume V_T is given by

$$V_T = V_t\,(1 + \gamma(T-t))$$

where γ is the coefficient of cubical expansion of the material of the vessel.

Example. Let the apparent mass of water contained in a vessel at 10 °C, when weighed in air at 10 °C and pressure 102 kPa, be 100.00 g.

Then volume of vessel at 10 °C is

$$100.00\,(1.001\,39 + 0.000\,01) = 100.14\,\mathrm{cm}^3$$

The same vessel, if made of glass ($\gamma = 0.000\,027\,\mathrm{K}^{-1}$, assumed), would contain at 20 °C

$$100.14(1 + 0.000\,027(20 - 10)) = 100.17\,\mathrm{cm}^3$$

Comprehensive correction tables are given in BS 1797:1968.

Measurement of density

The density of a solid or liquid specimen is normally found by determining in air the apparent mass of a particular volume of the specimen and the apparent mass of an equal volume of water. The latter is obtained in the case of a solid specimen by weighing it when immersed in water and thus measuring its apparent decrease in mass due to the water displaced, and in the case of a liquid by using the same container (bottle or pyknometer) for the water as for the liquid. The ratio, r, of the two apparent masses thus determined is approximately equal to the density of the specimen in $\mathrm{g\,cm}^{-3}$ ($1\,\mathrm{g\,cm}^{-3} = 1\,000\,\mathrm{kg\,m}^{-3}$); the true density is equal to $r(d_{\mathrm{w}} - d_{\mathrm{a}}) + d_{\mathrm{a}}$, where d_{w} is the density of the water at its observed temperature $t°\mathrm{C}$ (see p. 29) and d_{a} is the density of the air (see p. 18); the result will be in the same units as those used for d_{w} and d_{a}. The correction to be applied to r at 20 °C and 1 atm is of the order of 1 part in 350 of the density. For temperatures ranging from 12 °C to 25 °C and pressures from 97 kPa (970 mb) to 104 kPa, the correction ranges over about 1 part in 400 of the density.

r is usually termed the relative density in air with reference to water, and is denoted by $d\,t/t'$ *in air*, where t and t' are the temperatures of the material and water respectively (usually $t = t'$). The true relative density, viz. $d\,t/t'$ *in vacuo*, is equal to $r(1 - d_{\mathrm{a}}) + d_{\mathrm{a}}$ when d_{a} is expressed in $\mathrm{g\,cm}^{-3}$.

Hydrometers

• *Density* hydrometers usually indicate density, $\mathrm{kg\,m}^{-3}$ (or $\mathrm{g\,cm}^{-3}$) at 20 °C or 15 °C when readings are taken at 20 °C or 15 °C respectively. In hot countries 27 °C may be used.
• *Relative density* (or specific gravity) hydrometers usually indicate density at 60 °F relative to water at 60 °F ($d\,60/60\,°\mathrm{F}$) when readings are taken at 60 °F.
• *Twaddle* hydrometers indicate ($d\,60/60\,°\mathrm{F} - 1)200$ when readings are taken at 60 °F.
• *Baumé* scales are related to relative density by various arbitrary formulae.

Temperature correction. To obtain from the reading R (density or relative density) of a soda-glass hydrometer, standard at $t°\mathrm{C}$, the density or relative density $d\,\theta°\mathrm{C}/t°\mathrm{C}$, of the liquid at the temperature $\theta°\mathrm{C}$ of observation, subtract $R \times 0.000\,025\,(\theta - t)$.

Surface tension. The surface tensions of aqueous solutions are often reduced considerably ($20\,\mathrm{mN\,m}^{-1}$ or more) by contaminating films. Change of reading due to $-1\,\mathrm{mN\,m}^{-1}$ change in surface tension $= +4000/nld\,\mathrm{kg\,m}^{-3}$, where n = density ($\mathrm{kg\,m}^{-3}$) or relative density $\times 1000$ (i.e. 'degrees'), l = length of scale (mm) corresponding to 10 $\mathrm{kg\,m}^{-3}$ or 10 'degrees' relative density, and d = stem diameter (mm).

J.B.R.

1.3.2 Barometry

Barometric units

The pascal (Pa) is the name given to the SI unit of pressure, the newton per square metre. Following the 8th Congress of the World Meteorological Organization the hectopascal (hPa) will, from 1 January 1986, be the preferred unit for the measurement of pressure for meteorological purposes.

However, barometers in current use usually bear millibar (mbar), conventional millimetre of mercury (mmHg) or conventional inch of mercury (inHg) scales.

$$1\ \text{hPa} = 1\ \text{mbar} = 100\ \text{Pa}$$
$$1\ \text{mmHg} = 133.322\ 4\ \text{Pa}$$
$$1\ \text{inHg} = 3\ 386.39\ \text{Pa}$$

Barometers bearing the above scales should be graduated to measure pressures directly in their respective units when the barometer is subject to standard conditions, viz. 0 °C and gravity 9.806 65 m/s^2. However, corrections taking into account the actual temperature of the instrument and the acceleration due to gravity at the place of observation have normally to be applied. (See BS 2520 : 1983, *Barometer conventions and tables, their application and use.*)

Note: the reference condition known as a standard atmosphere (atm) is defined as 101 325 Pa, which is equal to 760 mmHg to within 1 part in 7 million. It should not now be used as a unit of pressure, but only to define a standard reference environment.

Calculation of pressure from Fortin barometer readings

$$\text{Pressure} = \frac{g}{9.806\ 65}\left[R + c - R\left(\frac{(\beta - \alpha)t}{(1 + \beta t)}\right)\right]$$

where R is the barometer reading
    c is its calibration correction (see note 1 below)
    g is the acceleration due to gravity at the point of observation in m s^{-2} (see note 2 below)
    β is the coefficient of expansion of mercury (taken to be 0.000 181 8/°C)
    α is the coefficient of linear expansion of the scale (taken to be 0.000 018 4/°C for brass)
    t is the temperature of the barometer in °C

At normal room temperatures the correction for instruments with brass scales approximates closely to the simpler expression:

$$\text{Pressure} = \frac{g}{9.806\ 65}\left[R + c - 0.000\ 163Rt\right]$$

Calculation of pressure from Kew pattern barometer readings

$$\text{Pressure} = \frac{g}{9.806\ 65}\left[R + c - R\left(\frac{(\beta - \alpha)t}{(1 + \beta t)}\right) - \frac{V}{A}f(\beta - 0.000\ 030)t\right]$$

where the symbols are as above and

$\dfrac{V}{A}$ is a geometrical factor in millimetre units which should be inscribed on the barometer;

f is a unit conversion factor: 1.333 for a hPa or mbar scale,
                                 1.000 for a mmHg scale,
                                 0.039 37 for an inHg scale.

The value of 0.000 030/°C in the equation represents an expansion coefficient which allows for a steel cistern and the glass tube. At normal room temperatures the correction for instruments with brass scales approximates closely to the simpler expression:

$$\text{Pressure} = \frac{g}{9.806\ 65}\left[R + c - 0.000\ 163Rt - 0.000\ 152\frac{V}{A}ft\right]$$

Note 1: Although both Fortin and Kew pattern barometers are fundamental in operation, calibration is necessary to evaluate any corrections arising from capillarity effects, scale errors, inadequate reference vacuum, etc.

Note 2: The value of *g* may be calculated in terms of geographical latitude and height above sea level using the formula in section 1.9.4.

Calculation of pressure at different heights

A mercury barometer measures atmospheric pressure at the level of its lower mercury surface. Should a pressure value at a different level be required, an allowance has to be made for the hydrostatic pressure exerted by the intervening vertical air column. To correct for small height differences, such as from floor to floor in a laboratory, the pressure at height *H* metres *above* the barometer's lower mercury surface is given by

$$P_{\mathrm{H}} = P - \rho_{\mathrm{a}} g H / U$$

where   *P* is the pressure at the barometer's lower surface

ρ_{a} is the density of the intervening vertical air column in $\mathrm{kg/m^3}$ (see section 1.3.1)

U is a factor which converts the height correction term from pascals to the pressure units used.

Note: This expression is only correct for small height differences and provided there are no other causes of pressure differential such as wind or air conditioning fans.

Capillary depression

Capillary forces tend to depress the surface of mercury columns by an amount which depends on the column diameter, the surface tension and the angle of contact between the mercury surface and the boundary wall. With the exception of primary instruments, barometer manufacturers usually make an allowance for this effect by off-setting the scale (under standard conditions, the mercury surface in an 8-mm diameter tube can be depressed by up to about 1 mm). However, long-term changes in capillary depression do occur; this is usually because the angle of contact and the surface tension change as the mercury surface becomes contaminated. Changes in capillary depression are most easily detected by calibrating the barometer against a more accurate instrument.

Barometers used as primary instruments should either have their readings corrected for capillary depression (Gould and Vickers, 1952) or be of sufficient diameter to render the correction negligible (with a 40-mm diameter tube, the capillary depression is about 0.0001 mm).

For mercury U-tubes, it should not be assumed that the capillary depression is the same in each limb even though the limbs may be of the same diameter.

Reference

F. A. Gould and T. Vickers, 'Capillary depression in mercury barometers and manometers', *J. Sci. Instrum.*, 1952, **29**, 85–7.

D.I.S.

1.3.3 The measurement of high pressures

The Practical Pressure Scale at high pressures

Beyond the range conveniently attainable by primary standards, there are several practical techniques for the measurement of pressure which are now recognized internationally: fixed points (phase transitions), equations of state and phenomena such as the ruby fluorescence shift. Two international conferences have issued recommendations on a practical pressure scale. At a symposium at the US National Bureau of Standards[1], held in 1968, agreement was reached on the pressure values associated with certain phase transitions up to 10 GPa. A Task Group, appointed under the auspices of the International Association for the Advancement of High Pressure Science and Technology at the 6th AIRAPT Conference (Boulder, Colorado, 1977), is charged with making recommendations for an International Practical Pressure Scale; the first such recommendations were made at the 8th AIRAPT Conference (Uppsala, Sweden, 1981)[2].

Basic group of reference pressures associated with phase transitions of pure substances

(Means of identification of transition: electrical resistance or volume change)

Substance	Type of transition	Pressure/GPa	Temperature/°C
Mercury.	Solid-liquid equilibrium	0.7569±0.0002*	0
Bismuth	Polymorphic (Bi I–II)	2.550±0.006*	25
Thallium	Polymorphic (Tl II–III)	3.67±0.05†	25
Barium	Polymorphic (Ba I–II)	5.53±0.12†	25
Bismuth	Polymorphic (Bi III–V)	7.67±0.18†	25

* Values recommended at the NBS symposium, from reference 1.

† Derived from Decker et al.[3]; these values differ very slightly from those recommended in reference 1, because some of the contributory results were based on the equation of state of NaCl and have subsequently been adjusted to take into account the revision of the equation of state of NaCl by Decker in 1971.

The values given in the table represent the best estimate of the true transition pressure; the uncertainty associated with each value represents an estimate of how close the true transition pressure lies to the value given. It must be stressed that the degree of reproducibility of a transition pressure will depend on the experimental conditions and the method of realization of the transition. For detailed discussions of these topics see the review by Decker et al.[3] and Chapter 3, by Bean, of reference 4.

Mercury solid-liquid equilibrium at other temperatures

It has been recommended [2] that the best representation of the mercury melting curve up to 1.2 GPa is a third-order polynomial based on the work of Molinar et al.[5]:

$$p = 1.932\,835 \times 10^{-2}d + 1.706\,8 \times 10^{-6}d^2 + 6.0867 \times 10^{-8}d^3 \, \text{GPa}$$

where $d = T - 234.309$.

Extrapolation of the polynomial above 1.2 GPa is *not* recommended.

Additional reference pressures associated with phase transitions

Substance	Type of transition and method of detection	Pressure/GPa	Temperature/°C
Tin	I–II; elec. resist. change	9.4†	25
Barium	II–III; elec. resist. change	11.8–12.2‡	25
Lead	I–II; elec. resist. change	13.4†	25
Zinc sulphide . .	Elec. resist. change (rising pressure)	15.4†	25
Gallium phosphide	Elec. resist. change (rising pressure)	22.0†	25
Sodium chloride .	Volume change	29.6†	25

† Bean (Ch. 3) in ref. 4.

‡ From Hall et al. and Drickamer, see ref. 1.

Pressure scale as defined by the equation of state of sodium chloride

The AIRAPT Task Group report of 1981[2] recommended that the Decker equation-of-state data for sodium chloride[6] be used as the practical reference standard in the pressure range below its phase transition (i.e. 29.6 GPa).

Calculated pressures versus compression for NaCl at 25 °C

Compression		Pressure	Compression		Pressure
Linear $-\Delta a/a_0$	Volume $-\Delta V/V_0$	GPa	Linear $-\Delta a/a_0$	Volume $-\Delta V/V_0$	GPa
0.001	0.0030	0.071	0.068	0.1904	8.379
0.002	0.0059	0.144	0.070	0.1956	8.772
0.004	0.0119	0.293	0.072	0.2008	9.175
0.006	0.0178	0.446	0.074	0.2059	9.590
0.008	0.0238	0.604	0.076	0.2111	10.017
0.010	0.0297	0.767	0.078	0.2162	10.457
0.012	0.0355	0.936	0.080	0.2213	10.908
0.014	0.0414	1.109	0.082	0.2263	11.372
0.016	0.0472	1.288	0.084	0.2314	11.850
0.018	0.0530	1.472	0.086	0.2364	12.340
0.020	0.0588	1.662	0.088	0.2414	12.845
0.022	0.0645	1.858	0.090	0.2464	13.364
0.024	0.0702	2.060	0.092	0.2513	13.897
0.026	0.0759	2.268	0.094	0.2563	14.445
0.028	0.0816	2.482	0.096	0.2612	15.085
0.030	0.0873	2.703	0.098	0.2661	15.508
0.032	0.0929	2.930	0.100	0.2710	16.183
0.034	0.0985	3.164	0.102	0.2758	16.795
0.036	0.1041	3.405	0.104	0.2806	17.424
0.038	0.1097	3.653	0.106	0.2854	18.070
0.040	0.1152	3.909	0.108	0.2902	18.735
0.042	0.1207	4.172	0.110	0.2950	19.417
0.044	0.1262	4.443	0.112	0.2997	20.119
0.046	0.1317	4.722	0.114	0.3044	20.840
0.048	0.1371	5.009	0.116	0.3091	21.581
0.050	0.1426	5.304	0.118	0.3138	22.342
0.052	0.1480	5.608	0.120	0.3185	23.125
0.054	0.1534	5.921	0.122	0.3231	23.929
0.056	0.1587	6.243	0.124	0.3277	24.755
0.058	0.1641	6.574	0.126	0.3323	25.603
0.060	0.1694	6.915	0.128	0.3369	26.476
0.062	0.1747	7.266	0.130	0.3414	27.372
0.064	0.1799	7.626	0.132	0.3460	28.292
0.066	0.1852	7.997	0.134	0.3505	29.238

The above table is based on that given by Decker in reference 3. These data are derived from more extensive input data than those previously published in reference 6 and cover a greater pressure range; data from the two sources differ by $\sim 0.2\%$, with an associated uncertainty for the scale of $\pm 1.1\%$ to $\pm 2.4\%$ for the range up to 20 GPa.

Data for other temperatures, namely 0°, 100°, 200°, 300°, 500° and 800 °C, may be found in reference 6.

Pressure measurement using the ruby fluorescence shift

The shift in the wavelength of the R_1 fluorescence line of ruby is now used extensively as a pressure indicator. The following equations are taken from reference 2, where $\Delta\lambda$ is the wavelength shift in nanometres:

(a) for pressures up to 29 GPa, based on the work of Piermarini et al.[7,8] who used the equation of state of NaCl as the pressure standard:

$$p = (2.740 \pm 0.016)\Delta\lambda;$$

(b) for pressures up to 100 GPa, based on the work of Mao *et al.*[9] who used as their pressure standard isothermal equations of state derived from shock wave experiments:

$$p = 380.8[(\Delta\lambda/\lambda_0 + 1)^5 - 1]$$

where λ_0 is the wavelength at 0.1 MPa.

For the uncertainties associated with these data reference must be made to the original source.

Pressure dependence of melting temperatures of some pure metals in the range 0–20 GPa

(Derived from Akella and Kennedy[10])

Pressure	Melting temperature/°C		
GPa	Gold	Silver	Copper
0	1062	959	1083
1	1120	1019	1121
2	1178	1077	1159
3	1235	1132	1196
4	1290	1184	1232
5	1340	1234	1266
6	1385	1283	1297
7	1441	1328	1332
8	1488	1372	1364
9	1536	1412	1396
10	1584	1450	1428
11	1629	1486	1458
12	1674	1521	1486
13	1718	1558	1512
14	1762	1592	1540
15	1804	1628	1564
16	1845	1660	1588
17	1884	1692	1610
18	1922	1722	1632
19	1956	1752	1654
20	1988	1778	1676

References

(1) E. C. Lloyd (ed.), 'Accurate Characterisation of the High-Pressure Environment', *Proceedings of Symposium at the National Bureau of Standards, Gaithersburg,* 1968, NBS Special Publication 326, 1971.

(2) C. M. Backman, T. Johannisson and L. Tegner (eds.), Task Group Report, 'Toward an International Practical Pressure Scale', *Proceedings of the 8th AIRAPT Conference,* Arkitektkopia, Uppsala, Sweden, 1982, Vol. I, 144–51.

(3) D. L. Decker, W. A. Basset, L. Merrill, H. T. Hall and J. D. Barnett, *J. Phys. Chem. Ref. Data,* 1972, **1**, 773–836 (NBS).

(4) G. N. Peggs (ed.), *High Pressure Measurement Techniques,* 1983 (Applied Science Publishers, London).

(5) G. F. Molinar, V. Bean, J. Houck and B. Welch, *Metrologia,* 1980, **16**, 21–9.

(6) D. L. Decker, *J. Appl. Phys,* **42**, 3239–44, 1971.

(7) G. J. Piermarini, S. Block, J. D. Barnett and R. A. Forman, 1975, *J. Appl. Phys.,* **46**, 2774–80.

(8) G. J. Piermarini and S. Block, *Rev. Sci. Instrum.,* 1975, **46**, 973–9.

(9) H. K. Mao, P. M. Bell, J. W. Shaner and D. J. Steinberg, 1978, *J. Appl. Phys.,* **49**, 3276–83.

(10) J. Akella and G. C. Kennedy, *J. Geophys. Res.,* 1971, **76**, 4969–77.

S.L.L.

1.3.4 Hygrometry

Relative humidity

The relative humidity is the ratio (expressed as a percentage) of the pressure of the water vapour actually present to the saturation pressure at the same air temperature. For most practical purposes, this is equal to the ratio of the mass of water vapour actually present in unit volume of moist air (absolute humidity) to the mass in unit volume of saturated air at the same temperature. (For definitions see BS 1339.) For a table of saturation pressures see section 2.4.1, p. 203.

Chemical hygrometer

The values below are grams of water contained in a cubic metre (m^3) of saturated air at a total pressure of 101 325 Pa (1013 mb).

Temperature/°C	0	1	2	3	4	5	6	7	8	9
0	4.85	5.20	5.55	5.95	6.35	6.80	7.25	7.75	8.25	8.80
10	9.40	10.00	10.65	11.35	12.05	12.80	13.60	14.45	15.35	16.30
20	17.30	18.35	19.40	20.55	21.75	23.05	24.35	25.75	27.20	28.75
30	30.35	32.05	33.80	35.60	37.55	39.55	41.65	43.90	46.20	48.60

Wet- and dry-bulb hygrometer

For this instrument the water-vapour pressure p is given in terms of the actual or dry-bulb temperature t and the wet-bulb temperature t_w by the relation

$$p = p_w - AH(t - t_w)$$

where p_w is the saturation water-vapour pressure at temperature t_w, A is a constant and H is the total atmospheric pressure. The relative humidity expressed as a percentage is $100p/p_s$, where p_s is the saturation water-vapour pressure at temperature t. The values of A depend on the speed of the air passing the wet bulb and are

A = 0.000 666/°C for moving air as in the Assmann ventilated psychrometer.

A = 0.000 80/°C in a Stevenson screen as used by the Meteorological Office.

(See BS 1339: 1964 for other values of A.)

Wet- and dry-bulb humidity tables

The tables below are for use with forced-ventilated instruments.

For more complete tables see *Hygrometric Tables for Aspirated Psychrometer Readings*, 1961 (M.O. 265c, London, HMSO): for range −35 °C to 60 °C see *Aspirations-Psychrometer-Tafeln*, 1963 (Braunschweig, F. Vieweg & Sohn); for range −30 °C to 60 °C see 'Institution of Heating and

Ventilating Engineers Guide 1970—Book C' (London), properties of humid air from −10 °C to 60 °C. See also *Measurement of Humidity*, Notes on Applied Science, No. 4, 1970 (HMSO).

Dry-bulb temperature/°C	Wet-bulb depression/°C										
	0.5	1.0	1.5	2.0	2.5	3.0	3.5	4.0	4.5	5.0	5.5
	Relative humidity (%)										
10	94	88	82	76	71	65	60	54	49	44	39
12	94	89	83	78	73	68	63	57	53	48	43
14	95	90	84	79	74	70	65	60	56	51	47
16	95	90	85	81	76	71	67	62	58	54	50
18	95	91	86	82	77	73	69	65	60	56	52
20	96	91	86	83	78	74	70	66	62	59	55
22	96	92	87	83	79	76	72	68	64	61	57
24	96	92	88	84	80	77	73	69	66	62	59
	6.0	6.5	7.0	7.5	8.0	8.5	9.0	9.5	10.0	10.5	11.0
10	34	29	24	19	14	9	5	—	—	—	—
12	38	34	29	24	20	16	11	7	3	—	—
14	42	38	33	29	25	24	17	13	9	5	2
16	46	41	37	34	30	26	22	18	15	11	8
18	49	45	41	37	34	30	27	23	20	16	13
20	51	48	44	41	37	34	30	27	24	21	18
22	54	50	47	44	40	37	34	31	28	25	22
24	56	52	49	46	43	40	37	34	31	28	26

Humidities over saturated salt solutions

Saturated solutions of various salts in water are used to obtain known relative humidities of air circulated in sealed enclosures which are maintained at a uniform and constant temperature.

See BS 3718: 1964, *Laboratory Humidity Ovens (Non-injection Type)*.

Saturated salt solution	Temperature/°C										
	0	5	10	15	20	25	30	35	40	50	60
	Relative humidity (%)										
Potassium sulphate	99	98	98	97	97	97	96	96	96	96	96
Potassium nitrate	97	96	95	94	93	92	91	89	88	85	82
Potassium chloride	89	88	88	87	86	85	84	83	82	81	80
Ammonium sulphate . . .	83	82	82	81	81	80	80	80	79	79	78
Sodium chloride	76	76	76	76	76	75	75	75	75	75	75
Sodium nitrite	—	—	—	—	66	65	63	62	62	59	59
Ammonium nitrate . . .	77	74	72	69	65	62	59	55	53	47	42
Sodium dichromate . . .	60	59	58	56	55	54	52	51	50	47	—
Magnesium nitrate	60	58	57	56	55	53	52	50	49	46	43
Potassium carbonate . . .	—	—	47	44	44	43	43	43	42	—	—
Magnesium chloride . . .	35	34	34	34	33	33	33	32	32	31	30
Potassium acetate	25	24	24	23	23	22	22	21	20	—	—
Lithium chloride	15	14	13	13	12	12	12	12	11	11	11
Potassium hydroxide . . .	—	14	13	10	9	8	7	6	6	6	5

M.J.H.

1.4 Mechanical properties of materials

1.4.1 Densities

Density of water (unit $1\ \mathrm{kg\,m^{-3}}$)

Pure air-free water under a pressure of 1 atmosphere.

Temp./°C	0	2	4	6	8	10	12	14	16	18
0	999.84	999.94	999.97	999.94	999.85	999.70	999.50	999.24	998.94	998.59
20	998.20	997.77	997.30	996.78	996.23	995.65	995.03	994.37	993.69	992.97
40	992.22	991.44	990.63	989.79	988.93	988.04	987.12	986.18	985.21	984.22
60	983.20	982.16	981.10	980.01	978.90	977.77	976.62	975.44	974.25	973.03
80	971.80	970.54	969.27	967.97	966.65	965.32	963.97	962.59	961.20	959.79
100	958.36									

Values up to 40 °C, Chappuis (1907) and Thiesen, Scheel and Diesselhorst (1900), recalculated by P. H. Bigg, *Br. J. appl. Phys.*, 1967.

Above 40 °C derived from the following expression by G. S. Kell, *J. Chem. Engng. Data*, 1975.

$$\text{density at } t\,°\mathrm{C}$$
$$= (999.839\,52 + 16.945\,176t - 7.987\,040\,1 \times 10^{-3}t^2 - 46.170\,461 \times 10^{-6}t^3$$
$$+ 105.563\,02 \times 10^{-9}t^4 - 280.542\,53 \times 10^{-12}t^5)/(1 + 16.879\,850 \times 10^{-3}t)$$

This is based on measurements of density and thermal expansion by Kell and Whalley (1965), Gildseth *et al.* (1972) and Steckel and Szapiro (1963). The values given by Kell's expression in the range 0 °C to 40 °C do not differ by more than 0.002 kg m^{-3} from Bigg's values.

The temperature ($t_m/°\mathrm{C}$) of maximum density at different pressures (p) measured in atmospheres is given by

$$t_m = 3.98 - 0.0225\,(p-1)$$

Density of heavy water

Calculated by Kell, *J. Chem. Engng. Data*, 1967, from measurements by Chang and Tung (1949), Shrader and Wirtz (1951) and Steckel and Szapiro (1963).

$t/°\mathrm{C}$ $\rho/\mathrm{kg\,m^{-3}}$	5 1105.62	10 1105.99	15 1105.87	20 1105.34	25 1104.45	30 1103.23	35 1101.73	40 1099.96	45 1097.94	50 1095.70
$t/°\mathrm{C}$ $\rho/\mathrm{kg\,m^{-3}}$	55 1093.25	60 1090.60	65 1087.77	70 1084.75	75 1081.58	80 1078.24	85 1074.75	90 1071.12	95 1067.36	100 1063.46

Density of mercury (unit $1\ \mathrm{kg\,m^{-3}}$)

Under a pressure of 1 atmosphere.

Temp./°C	0	2	4	6	8	10	12	14	16	18
−20	13 644.59	13 639.62	13 634.66	13 629.70	13 624.75	13 619.80	13 614.85	13 609.90	13 604.96	13 600.02
0	13 595.08	13 590.15	13 585.21	13 580.29	13 575.36	13 570.44	13 565.52	13 560.60	13 555.69	13 550.78
20	13 545.87	13 540.96	13 536.06	13 531.16	13 526.26	13 521.36	13 516.47	13 511.58	13 506.69	13 501.81
40	13 496.92	13 492.04	13 487.17	13 482.29	13 477.42	13 472.55	13 467.68	13 462.81	13 457.95	13 453.09
60	13 448.23	13 443.37	13 438.52	13 433.66	13 428.81	13 423.96	13 419.12	13 414.27	13 409.43	13 404.59
80	13 399.75	13 394.92	13 390.08	13 385.25	13 380.42	13 375.59	13 370.77	13 365.94	13 361.12	13 356.30

°C	0	20	40	60	80	100	120	140	160	180
100	13 351.48	13 303.4	13 255.4	13 207.6	13 159.8	13 112.1	13 064.5	13 016.9	12 969.2	12 921.5
300	12 873.7	12 825.8	12 777.8							

Based on:

$$\text{Density at } 0\,°C = 13\,595.08\ \text{kg m}^{-3}$$

and the expression

$$d_t = d_0/(1 + \alpha t)$$

where d_t is the density at $t\,°C$ and

$$10^8\alpha = 18\,152.53 + 0.5898\,t + 31.507 \times 10^{-4}\,t^2 + 2.405 \times 10^{-6}\,t^3$$

The formula for α relates to temperatures expressed in terms of the IPTS-68 temperature scale and, together with the density at $0\,°C$, has been derived from the density at $20\,°C$ determined by Cook and the expansion formula of Beattie, both of which were related to the IPTS-48 scale. (A. H. Cook, *Phil. Trans. Roy. Soc. A*, 1961, J. A. Beattie *et al.*, *Proc. Amer. Acad. Arts Sci.*, 1941; and A. H. Cook, *Br. J. appl. Phys.*, 1956.)

Approximate densities of commonly used materials*

Substance	Density g cm^{-3}	Substance	Density g cm^{-3}	Substance	Density g cm^{-3}
Acetone	0.8	Cronite	8.1	Mica	2.8
Agate	2.6	Diamond	3.5	Mild steel	7.9
Alcohol	0.8	Duralumin	2.8	Milk	1.03
Alni	6.9	Ebonite	1.2	Monel	8.8
Alnico	7.1	Ebony	1.2	Mumetal	8.8
Aluminium-bronze		Elinvar	8.1	Mycalex	2.4
(8% Al)	7.7	Emery	4.0	Naphtha	0.8
Amber	1.1	Gas carbon	1.9	Nickel-chromium	8.4
Asbestos	2.4	Gelatine	1.3	Nickel-silver	8.8
Ash (timber)	0.75	German silver	8.4	Nimonic	8.2
Asphalt	1.4	Glass (soda)	2.5	Oak	0.7
Balsa wood	0.2	,, (Pyrex)	2.23	Olive oil	0.9
Bamboo	0.4	,, (lead)	3–4	Paraffin oil	0.8
Bearing metal (80% Sn)	7.3	Glycerine	1.3	Paraffin wax	0.9
Beech	0.75	Gold (22 carat)	17.5	Permalloy C	8.6
Beeswax	0.95	,, (9 carat)	11.3	Petroleum	0.8
Beryllium-copper	8.2	Granite	2.7	Phosphor-bronze	8.9
Bone	1.9	Graphite	1.6–2.3	Pine (white)	0.5
Borax	1.7	Gunmetal	8.2	Pitch	1.1
Boxwood	1.0	'Heavy alloy'†	16.8–18.0	Plaster of Paris	1.8
Brass (60/40)	8.4	Ice	0.92	Plastics	see p. 287
,, (70/30)	8.5	Inconel	8.5	Platinum-iridium	
Brightray	8.4	Invar	8.0	(90/10)	21.5
Butter	0.9	Ivory	1.8	Porcelain	2.3
Carbon steel (<1% C)	7.8	Keramot	1.6	Quartz (crystal)	2.6
Cast iron	7.0–7.4	Lard	0.9	Resin	1.1
Castor oil	0.95	Lignum vitae	1.3	Sand (dry)	1.6
Cedarwood	0.55	Linseed oil	0.95	Sealing wax	1.8
Charcoal	0.4	Lo Ex	2.7	Sea-water	1.03
China clay (kaolin)	2.6	Magnalium	2.6	Silica (fused)	
Coal (anthracite)	1.6	Mahogany	0.8	translucent	2.1
,, (bituminous)	1.4	Manganin	8.5	transparent	2.2
Constantan	8.9	Marble	2.7	Silicon iron	6.9
Cork	0.25	Mazak (No. 2)	6.7	Silver sand	2.6
Corundum	4.0	Methylated spirit	0.8	Slate	2.8

Approximate densities of commonly used materials (*contd*)

Substance	Density g cm^{-3}	Substance	Density g cm^{-3}	Substance	Density g cm^{-3}
Soft solder (70% Sn 30% Pb)	8.3	Thiokol	1.4	White spirit . . .	0.85
Stainless iron (12% Cr)	7.7	Tungsten carbide (6% Co) . . .	15.0	Wrought iron . . .	7.8
Stainless steel . . .	7.8	Tungsten carbide		Xylol	0.85
Supermalloy . . .	8.9	(12% Co) . . .	14.2	Y-alloy	2.8
Tar	1.0	Turpentine	0.85		
Teak	0.85	Wax (soft red) . . .	1.0		

* For densities of the elements and chemical compounds, see under the Chemistry Section of this book.   J.B.R.
† Tungsten with metallic additives.

1.4.2 Elasticities

The following is a list of constants in common use, any two of which are sufficient to define the elastic properties of a homogeneous isotropic solid.

The two fundamental constants are those which relate change of volume and change of shape to external stress. They are, respectively, the bulk modulus K and the shear modulus G.

For many practical purposes, the following constants are commonly used:

Young's modulus, or longitudinal elasticity, E.
Poisson's ratio, $\sigma =$ lateral contraction per unit breadth divided by the longitudinal extension per unit length.
Compressibility, $\kappa = 1/K$.
Longitudinal modulus, M, which is the longitudinal modulus for zero lateral strain and determines the velocity of ultrasonic stress pulses in solids.

For a homogeneous isotropic solid, the following relations between these constants exist:

$$G = \frac{E}{2(1+\sigma)} \quad \cdots \quad (a) \qquad\qquad K = \frac{E}{3(1-2\sigma)} \quad \cdots \quad (b)$$

$$K = \frac{1}{3}\frac{EG}{(3G-E)} \quad \cdots \quad (c) \qquad\qquad M = K + \frac{4}{3}G \quad \cdots \quad (d)$$

The value of Poisson's ratio is usually positive and lies between 0 and $+\frac{1}{2}$, but in some cases it may be negative.

Elasticities of metals and alloys

Material 20 °C	E GPa	G GPa	σ	K GPa
Aluminium	70.3	26.1	0.345	75.5
Bismuth	31.9	12.0	0.330	31.3
Cadmium	49.9	19.2	0.300	41.6
Chromium	279.1	115.4	0.210	160.1
Copper	129.8	48.3	0.343	137.8
Gold	78.0	27.0	0.44	217.0
Iron (soft)	211.4	81.6	0.293	169.8
Iron (cast)†	152.3	60.0	0.27	109.5
Lead†	16.1	5.59	0.44	45.8
Magnesium	44.7	17.3	0.291	35.6

Elasticities of metal and alloys (contd)

Material 20 °C	E GPa	G GPa	σ	K GPa
Nickel (unmag., soft)† . .	199.5	76.0	0.312	177.3
„ („ hard)† . .	219.2	83.9	0.306	187.6
Niobium	104.9	37.5	0.397	170.3
Platinum	168.0	61.0	0.377	228.0
Silver	82.7	30.3	0.367	103.6
Tantalum	185.7	69.2	0.342	196.3
Tin	49.9	18.4	0.357	58.2
Titanium	115.7	43.8	0.321	107.7
Tungsten	411.0	160.6	0.280	311.0
Vanadium	127.6	46.7	0.365	158.0
Zinc	108.4	43.4	0.249	72.0
Brass (70 Zn, 30 Cu) . . .	100.6	37.3	0.350	111.8
Constantan	162.4	61.2	0.327	156.4
Hidurax Special†‡ . . .	144.5	54.4	0.333	144.1
Invar (36 Ni, 63.8 Fe, 0.2 C) .	144.0	57.2	0.259	99.4
Nickel Silver§	132.5	49.7	0.333	132.0
Steel (Mild)	211.9	82.2	0.291	169.2
„ ($\frac{3}{4}$% C)	210.0	81.1	0.293	168.7
„ ($\frac{3}{4}$% C hardened) . .	201.4	77.8	0.296	165.0
„ Tool‖	211.6	82.2	0.287	165.3
„ Tool (hardened)‖ . .	203.2	78.5	0.295	165.2
„ Stainless†† . . .	215.3	83.9	0.283	166.0
Tungsten Carbide† . . .	534.4	219.0	0.22	319.0

† Approx. value or values for materials of variable composition.
‡ Cu–Ni alloy with Al, Fe and Mn additions.
§ Approx. % composition: Cu 55, Ni 18, Zn 27.
‖ Oil hardening non-deforming tool steel of approx. % composition:
     C 0.98, Mn 1.03, Cr 0.65, W 1.01, V 0.1.
†† Approx. % composition: C 0.2, Si 0.5, Mn 0.7, Ni 2, Cr 18.

Elasticities of miscellaneous materials

Material 20 °C	E GPa	G GPa	σ	K GPa
Glass (Heavy Flint) . . .	80.1	31.5	0.27	57.6
Glass (Crown)	71.3	29.2	0.22	41.2
Quartz (fused)	73.1	31.2	0.170	36.9

Note. Many of the values in the above tables are from measurements made at the NPL and are reproduced from G. Bradfield, Notes on Applied Science No. 30, *Use in Industry of Elasticity Measurements with the Help of Mechanical Vibrations*, 1964 (H.M.S.O., London).

Elasticities for woods

All woods are elastically anisotropic and in general there are nine independent elastic constants. The values in the table below are for some common woods and give the three principal values of Young's modulus measured along the grain E_L, in a radial direction E_R and tangential direction E_T. The values have been supplied by R. F. S. Hearmon (Forest Products Research Laboratory). Further information is given in a report by R. F. S. Hearmon, *Elasticity of Wood and Plywood*, Forest Products Research Special Report No. 7, 1948 (H.M.S.O., London).

Wood	Relative density	E_L GPa	E_R GPa	E_T GPa
Ash	0.7	16	1.6	0.9
Balsa	0.2	6	0.3	0.1
Beech	0.7	14	2.2	1.1
Birch	0.6	16	1.1	0.6
Mahogany	0.5	12	1.1	0.6
Oak	0.7	11	—	—
Walnut	0.6	11	1.2	0.6
Teak	0.6	13	—	—
Douglas Fir	0.5	16	1.1	0.8
Scots Pine	0.5	16	1.1	0.6
Spruce	0.4–0.5	10–16	0.4–0.9	0.4–0.6

Elasticities for plastics

All plastics are visco-elastic and consequently the elasticity varies considerably with temperature and strain rate. The table below gives approximate values at 20 °C for very slow rates of strain.

Plastic	E GPa	Plastic	E GPa
Nylon 6 (cast) . . .	2.4–3.1	Polyethylene (high density)	0.4–1.3
Nylon 6 (moulded) . .	0.8–3.1	Polyimide	~ 3.1
Nylon 66	1.2–2.9	Polymethylmethacrylate .	2.4–3.4
Polybenzoxazole . . .	~ 3.5	Polypropylene . . .	1.1–1.6
Polycarbonate . . .	2.4	Polystyrene	2.7–4.2

Temperature coefficients of elastic constants for some materials

At 15 °C	Temperature coefficient α in $E_t = E_{15}\{1 - \alpha(t - 15)\}$ $G_t = G_{15}\{1 - \alpha'(t - 15)\}$	
	α for E	α' for G
Aluminium	4.8×10^{-4}	5.2×10^{-4}
Brass	3.7×10^{-4}	4.6×10^{-4}
Copper	3.0×10^{-4}	3.1×10^{-4}
German silver		6.5×10^{-4}
Gold	4.8×10^{-4}	3.3×10^{-4}
Iron	2.3×10^{-4}	2.8×10^{-4}
Phosphor-bronze		3.0×10^{-4}
Platinum	0.98×10^{-4}	1.0×10^{-4}
Quartz fibre	-1.5×10^{-4}	-1.1×10^{-4}
Silver	7.5×10^{-4}	4.5×10^{-4}
Steel	2.4×10^{-4}	2.6×10^{-4}
Tin		5.9×10^{-4}

Tensile strengths of materials

The elastic limit is always exceeded before the breaking stress is reached. The process of drawing the material into a wire increases the strength, and the finer the wire the greater the breaking stress. (See Poynting and Thomson's *Properties of Matter*.) Cold working generally tends to increase the breaking stress of the material at the expense of its ductility.

For crushing and shearing strengths see Ewing's *Strength of Materials* or one of the Engineering *Pocket Books*.

For engineering properties (i.e. proof stress, breaking stress, fatigue and creep properties, etc.) of materials at normal and elevated temperatures see S. L. Hoyt's *Metals and Alloys Data Book*.

Substance	Tensile strength MPa	Substance	Tensile strength MPa
Metals		**Miscellaneous**	
Aluminium (cast)	90–100	Catgut	420
,, (rolled)	90–150	Glass	30–90
Brass ⌠ 66% Cu ⌡ (cast)	150–190	Hemp rope	60–100
,, ⌠ 34% Zn ⌡ (rolled)	230–270	Leather belt	30–50
Calcium	42–60	Silk fibre	260
Cobalt	260–750	Spider thread	180
Copper (cast)	120–170	Woods† :	
,, (rolled)	200–400	Ash, beech, oak, teak, mahogany	60–110
Gun metal (90% Cu, 10% Sn)	190–260	Fir, pitch-pine	40–80
Iron (cast)	100–230	Red or white deal	30–70
,, (wrought)	290–450	White or yellow pine	20–50
Lead (cast)	12–17	Quartz fibre (fused)	~1000
Magnesium (cast)	60–80		
,, (extruded)	170–190	**Wires**	
Phosphor-bronze (cast)	180–280	Aluminium	200–450
Steel (castings)	400–600	Brass	350–550
,, (mild) (0.2% C)	430–490	Copper (hard-drawn)	400–460
High-carbon spring steel:		,, (annealed)	280–310
(annealed)	700–770	Duralumin	400–550
(tempered)	930–1080	German silver	460
(nickel) (5% Ni)	800–1000	Gold	200–250
(nickel-chromium)	1000–1500	Iron (charcoal, hard-drawn)	540–620
Soft solder	55–75	,, (annealed)	460
Tin (cast)	20–35	Molybdenum	1100–3000
Zinc (rolled)	110–150	Nickel	500–900
		Palladium	350–450
Plastics		Phosphor-bronze (hard-drawn)	690–1080
Nylon 6	76–97	Platinum	330–370
Nylon 66	62–83	Pt + 10% Rh	630
Polyacetal	~69	Silver	290
Polybenzoxazole	82–117	Steel (ordinary)	~1100
Polycarbonate	55–65	,, (tempered)	1550
Polyethylene	21–35	,, (pianoforte, hard-drawn)	1860–2330
Polyimide	69–104	Tantalum	800–1100
Polymethylmethacrylate	50–76	Tungsten	1500–3500
Polypropylene	30–40	Zirconium (annealed)	260–390
Polystyrene	34–52	,, (hard-drawn)	1000

† Along the grain.

Bulk moduli of elements

Element	K GPa	Element	K GPa	Element	K GPa	Element	K GPa
Aluminium† .	75.5	Chlorine		Molybdenum	231.0	Selenium . .	8.3
Antimony . .	42.0	(liq) . . .	1.1	Nickel†		Silicon . . .	100.0
Arsenic . . .	22.0	Chromium† .	160.1	(soft) . .	177.3	Silver . . .	103.6
Bismuth . . .	31.3	Copper . .	137.8	(hard) . .	187.6	Sodium . .	6.3
Bromine . . .	1.9	Gold . . .	217.0	Palladium . .	182.0	Sulphur . .	7.7
Cadmium . .	41.6	Iodine . . .	7.7	Phosphorus		Thallium . .	43.0
Caesium . . .	1.6	Iron † . .	169.8	(red) . .	10.9	Tin . . .	58.2
Calcium . . .	17.2	Lead† . . .	45.8	Phosphorus		Zinc† . . .	72.0
Carbon‡		Lithium . .	11.1	(white) . .	4.9		
(diamond) .	542.0	Magnesium .	44.7	Platinum . .	228.0		
Carbon		Manganese .	118.0	Potassium . .	3.1		
(graphite) .	33.0	Mercury . .	25.0	Rubidium . .	2.5		

† G. Bradfield, *Notes on Applied Science*, No. 30, 1964 (HMSO).
‡ Measurements made by M. F. Markham, NPL.

Bulk moduli of liquids

As the pressure increases, K increases. In general a rise in temperature decreases the bulk modulus of a liquid; water, however, shows a maximum value of K at about 50 °C. (See Poynting and Thomson's *Properties of Matter* and Bridgman's *The Physics of High Pressure*).

Liquid	Temp. °C	K GPa	Liquid	Temp. °C	K GPa
Acetic acid, 1–16 atm . . .	20	1.45	Mercury:		
Amyl alcohol, 8 atm	17.7	1.12	8–37 atm	20	26.2
Benzene, 8 atm	17.9	1.10	100–200 atm	15	30.0
Butyl alcohol, 8 atm	17.4	1.13	Methyl acetate, 8–37 atm . .	14.3	1.04
Butyl alcohol, *iso-*, 8 atm . . .	17.9	1.03	Methyl alcohol, 37 atm . .	14.7	0.97
Carbon bisulphide, 8–37 atm .	15.6	1.16	Olive oil	20.5	1.60
Carbon tetrachloride	20	1.12	Paraffin oil	14.8	1.62
Chloroform, 100–200 atm . .	20	1.1	Pentane	20	0.318
Ether:			Petroleum	16.5	1.46
1–50 atm	0	0.689	Propyl alcohol, 8 atm . . .	17.7	1.04
900–1000 atm	0	1.56	Propyl alcohol, *iso-*, 8 atm .	17.8	0.983
900–1000 atm	198	0.703	Turpentine	19.7	1.280
Ethyl acetate, 8–37 atm . . .	13.3	0.974	Water:		
Ethyl alcohol:			1–25 atm	15	2.05
1–500 atm	0	1.32	900–1000 atm	15	2.75
150–200 atm	310	0.024	900–1000 atm	198	1.81
Ethyl bromide, 8–37 atm . .	99.3	0.343	2500–3000 atm	14.2	3.88
Ethyl chloride, 8–37 atm . .	15.2	0.662	Water (sea)	—	2.32
Glycerine	20.5	4.03			

M.F.M.

1.4.3 Viscosities

Viscosities of liquids

The **dynamic viscosity**, η, of a (Newtonian) fluid is given by $\eta = \tau \div \mathrm{d}v/\mathrm{d}r$; $\tau =$ shearing stress between two planes parallel with the direction of flow, $\mathrm{d}v/\mathrm{d}r =$ velocity gradient at right angles to the direction of flow. The dimensions of dynamic viscosity are $ML^{-1}T^{-1}$, and the SI unit is Pa s.

Kinematic viscosity, v, is the ratio of the dynamic viscosity to the density, ρ. The dimensions of kinematic viscosity are $L^2 T^{-1}$ and the SI unit is $m^2\ s^{-1}$.

Fluidity, ϕ, is the reciprocal of the dynamic viscosity. η of liquids decreases with the temperature approximately according to the equation $\log \eta = A + B/T$, where A and B are characteristic constants and T is the absolute temperature. Values of A and B for a large number of liquids are given by Barrer, *Trans. Farad.*, 1943, **39**, 48.

(i) Viscosity of water

Data from Kestin, Sokolov and Wakeham, *J. Phys. Chem. Ref. Data*, 1978, **7**(3), 941.

Temp °C	η mPa s	Temp. °C	η mPa s	Temp. °C	η mPa s	Temp. °C	η mPa s
0	1.792	20	1.0020	50	0.5471	90	0.3150
5	1.519	25	0.8902	60	0.4670	100	0.2821
10	1.307	30	0.7973	70	0.4046	125	0.2217
15	1.138	40	0.6526	80	0.3551	150	0.1818

(ii) Viscosities of various liquids, η in m Pa s

Liquid	−100 °C	−50 °C	0 °C	25 °C	30 °C	50 °C	75 °C	100 °C
Acetic acid	—	—	—	1.116	1.037	0.792	0.591	0.457
Acetone	—	—	0.402	0.310	0.295	0.247	0.200	0.165
Aniline	—	—	9.450	3.822	3.298	1.982	1.201	0.808
Benzene	—	—	—	0.603	0.562	0.436	0.332	0.263
Bromobenzene	—	—	1.592	1.065	0.995	0.780	0.605	0.488
n-Butane	0.747	0.339	0.197	0.157	0.150	—	—	—
Carbon disulphide	2.132	0.796	0.445	0.357	0.343	—	—	—
Carbon dioxide	—	0.227	0.098	0.057	—	—	—	—
Carbon tetrachloride	—	—	1.341	0.912	0.851	0.662	0.503	0.395
Chloroform	—	1.514	0.709	0.536	0.510	0.422	0.341	0.281
Di-ethyl ether	1.545	0.544	0.288	0.224	0.214	0.179	0.146	0.119
Ethanol	98.96	8.318	1.873	1.084	0.983	0.684	0.459	0.323
Ethyl acetate	—	1.284	0.575	0.428	0.406	0.332	0.264	0.215
Ethyl formate	—	1.060	0.504	0.381	0.362	0.299	0.240	0.196
n-Hexane	—	0.782	0.387	0.296	0.282	0.235	0.190	0.155
n-Hexadecane	—	—	—	3.044	2.729	1.852	1.245	0.896
Mercury	—	—	1.616	1.528	1.497	1.401	1.322	1.255
Methane	0.0357	—	—	—	—	—	—	—
Methanol	—	2.258	0.797	0.543	0.507	0.392	0.294	0.227
Nitrobenzene	—	—	—	1.842	1.688	1.244	0.915	0.710
n-Octane	—	1.837	0.719	0.516	0.486	0.390	0.306	0.247
Oil castor	—	—	—	700	451	125	42.0	16.9

(ii) **Viscosities of various liquids, η in mPas** *(contd)*

Liquid	$-100\,°C$	$-50\,°C$	$0\,°C$	$25\,°C$	$30\,°C$	$50\,°C$	$75\,°C$	$100\,°C$
Oil olive	—	—	—	67.0	54.0	25.8	9.4	7.0
n-Pentane.	1.300	0.498	0.273	0.214	0.205	0.173	0.140	0.115
n-Propane.	0.421	0.215	0.127	0.099	0.094	—	—	—
Sulphuric acid . . .	—	—	—	23.8	20.1	11.7	6.6	4.1
Toluene	—	2.124	0.768	0.551	0.520	0.420	0.334	0.272

For more data see: Engineering Science Data (ESDU) Physical Data, Chemical Engineering, Vol. 3, *Viscosity*, 1966–83 (London) Landolt-Börnstein, 1969, Vol. II, *Properties of Matter in its Aggregated States*, Part 5a Viscosity and Diffusion, 6th edn (Springer Verlag, Berlin).

(iii) **Viscosity of aqueous glycerol solutions**

Data from Segur and Oberstar, *Ind. Eng. Chem.*, 1951, **43**, 2117, corrected to value for water at 20 °C of 1.002 mPas.

Density 20° kg/l	% weight glycerol	η/Pas		
		20°	30°	40°
1.2611	100	1.408	0.610	0.283
1.2588	99	1.146	0.498	0.234
1.2562	98	0.936	0.408	0.195
1.2534	97	0.763	0.339	0.165
1.2508	96	0.622	0.280	0.142
1.2482	95	0.521	0.236	0.121
1.2085	80	0.0599	0.0338	0.0207
1.1254	50	0.00598	0.00420	0.00309
1.0459	20	0.00175	0.00135	0.00107
1.0215	10	0.00131	0.00103	0.000823

(iv) **Viscosity of aqueous sucrose solutions**

Data from Bingham and Jackson, *National Bureau of Standards, Bulletin*, 1918, **14**, 59, corrected to recent value for the viscosity of water (p. 36).

Relative density 20°/4 °C	% weight sucrose	η/Pas		
		15 °C	20 °C	25 °C
1.3790	75	4.039	2.328	1.405
1.3472	70	0.7469	0.4816	0.3216
1.3163	65	0.2113	0.1472	0.1054
1.2865	60	0.07949	0.05849	0.04003
1.2296	50	0.01953	0.01543	0.01240
1.1764	40	0.007463	0.006167	0.005164
1.1270	30	0.003757	0.003187	0.002735

(v) Relative viscosities of some aqueous solutions

For a complete list see International Critical Tables (1928) and R. H. Stokes and R. Mills, *Viscosity of Electrolytes and Related Properties*, 1965, (Pergamon Press, London). (The latter covers 1929–63.)

Substance	Temp./°C	Relative viscosity	Substance	Temp./°C	Relative viscosity
Ammonia . . .	25	1.02	Potassium chloride . .	17.6	0.98
Ammonium chloride .	17.6	0.98	Potassium iodide . .	17.6	0.91
Calcium chloride . .	20	1.31	Sodium hydroxide . .	25	1.24
Hydrochloric acid . .	15	1.07	Sulphuric acid . . .	25	1.09

(vi) Viscosity of liquid metals and molten salts

η in mPa s

Liquid	100 °C	400 °C	600 °C	700 °C	800 °C	1100 °C	1200 °C
Aluminium	—	—	—	2.96	2.66	—	—
Gold	—	—	—	—	—	5.13	4.64
Lead	—	2.32	1.55	1.37	1.24	—	—
Potassium	0.458	0.224	0.172	0.155	0.141	—	—
Sodium	0.680	0.286	0.215	0.192	0.174	—	—
Tin	—	1.33	1.04	0.95	0.89	0.78	0.77
Potassium chloride	—	—	—	—	1.096	—	—
Potassium nitrate	—	2.09	1.07	—	—	—	—
Sodium chloride	—	—	—	—	1.500†	—	—
Sodium nitrate	—	1.91	—	—	—	—	—

† 816 °C.

For more data on molten salts see G. J. Janz, *NSRDS-NBS 15, report*, Natl. Bur. Standards, Washington, 1968. For liquid metals see Smithell's *Metal Reference Book*, 6th edn, 1983, p. 14.2, (Butterworths, London).

Viscosities of glasses and minerals

$\log_{10}(\eta/\text{Pa s})$

Material	900 °C	1000 °C	1100 °C	1200 °C	1300 °C	1400 °C	1600 °C	1800 °C	2000 °C
Plate glass	4.00	3.03	2.41	1.87	1.46	1.07	—	—	—
Medium flint glass . .	3.9	2.8	1.9	1.4	0.9	0.7	—	—	—
Silica	—	—	14.6	12.7	11.8	9.7	8.2	4.7	3.4
Olivine	—	—	—	2.5	1.5	1.2	—	—	—
Diorite	—	—	—	3.1	2.3	1.8	—	—	—
Diopside.	—	—	—	—	0.52	0.43	—	—	—

Viscosities of liquids at high pressures

(i) Relative viscosity of water

Pressure/MPa	2.2 °C	10 °C	20 °C	30 °C	50 °C	75 °C	100 °C
0.1	1.000	1.000	1.000	1.000	1.000	1.000	1.000
49	0.946	0.969	0.990	0.998	1.021	1.029	1.043
98	0.926	0.957	0.990	1.008	1.046	1.063	1.085
196	0.940	0.982	1.023	1.053	1.104	1.137	1.170
294	0.993	1.037	1.081	1.116	1.174	1.217	1.256
392	1.072	1.185	1.163	1.195	1.253	1.302	1.349
588	1.296	1.330	1.367	1.386	1.439	1.492	1.546
784	—	—	1.629	1.642	1.664	1.708	1.757
981	—	—	—	1.950	1.936	1.948	1.986

(ii) Relative viscosities of various liquids

Ratio $= \eta_p/\eta_0$ at same temperature.

Liquid	Temp./°C	Pressures/MPa					
		98	100	392	400	784	1177
Acetone	30	1.68	—	4.03	—	9.70	—
	75	1.65	—	3.55	—	7.36	13.7
Benzene	30	2.22	—	—	—	—	—
	50	—	2.06	—	—	—	—
	75	2.07	2.00	—	—	—	—
	100	—	1.98	—	7.40	—	—
Carbon disulphide	30	1.45	—	3.23	—	6.92	15.5
	75	1.50	—	3.14	—	6.25	11.8
Di-ethyl ether	30	2.11	—	6.20	—	18.2	46.8
	75	1.87	—	5.28	—	12.8	27.1
Ethanol	30	1.59	—	4.14	—	10.5	24.5
	75	1.64	—	4.28	—	9.48	18.3
n-Hexane	25	—	2.10	—	—	—	—
	30	2.15	—	8.20	—	32.7	—
	75	2.33	2.21	7.91	7.41	24.8	69.7
	100	—	2.26	—	7.22	—	—
n-Hexadecane	50	—	2.89	—	—	—	—
	75	—	2.69	—	—	—	—
	100	—	2.61	—	15.0	—	—
Methanol	30	1.47	—	2.96	—	5.62	9.95
	75	1.46	2.74	—	—	4.77	7.69
n-Octane	25	—	2.31	—	12.9	—	—
	30	2.12	—	12.3	—	—	—
	75	2.20	2.24	8.97	9.19	35.7	—
	100	—	2.25	—	8.56	—	—
n-Pentane	30	2.07	—	7.03	—	22.9	70.2
	75	2.25	—	7.33	—	20.3	48.1
Toluene	30	1.87	7.89	—	—	50.0	—
	75	1.86	—	6.33	—	24.6	109

Data from P. W. Bridgman, *Proc. Am. Acad. Arts Sci.*, 1926, **61**, 57–99, except at 100 and 400 MPa which comes from K. J. Young, PhD Thesis, University of Glasgow, Nov. 1980, and Dymond, J. H., Robertson, J., and Isdale, J. D., *Int. J. Thermophysics*, 1981, **2**(2), 133–54; ibid., 1981, **2**(3), 223–36.

Viscosities of gases and vapours

(i) Viscosities at normal pressure

Units: μPa s

Gas	0 °C	20 °C	50 °C	100 °C	200 °C	300 °C	400 °C	500 °C	600 °C
Air	17.3	18.2	19.6	22.0	26.1	29.8	33.2	36.4	39.4
Ammonia	9.2	9.9	11.0	13.0	16.8	20.6	24.4	28.2	31.9
Argon	21.0	22.3	24.2	27.3	32.8	37.7	42.2	46.4	50.4
Benzene	7.0	7.5	8.1	9.4	12.0	—	—	—	—
Carbon dioxide . .	13.7	14.7	16.1	18.5	23.0	27.1	30.8	34.2	37.4
Carbon monoxide .	16.6	17.4	18.8	21.0	25.2	29.0	32.5	35.6	38.6
Chlorine. . . .	12.3	13.2	14.5	16.9	21.0	25.0	—	—	—
Chloroform. . . .	9.4	10.1	11.1	12.8	16.2	19.5	—	—	—
Ethylene	9.7	10.3	11.2	12.8	15.4	17.9	—	—	—
Helium	18.7	19.6	21.0	23.2	27.3	31.2	34.8	38.4	41.8
Hydrogen	8.4	8.8	9.4	10.4	12.1	13.7	15.3	16.9	18.4
Krypton.	23.4	25.0	27.4	31.2	38.0	44.2	49.9	55.2	60.2
Methane	10.3	11.0	11.9	13.5	16.3	18.8	21.1	23.3	25.3
Neon	29.8	31.3	33.6	37.0	43.2	48.9	54.3	59.4	64.4
Nitrogen	16.6	17.6	18.9	21.2	25.1	28.6	31.9	34.9	37.8
Nitrous oxide . . .	13.7	14.7	16.1	18.4	22.9	27.0	30.7	34.0	37.0
Oxygen	19.5	20.4	21.8	24.4	29.3	33.7	37.6	41.3	44.7
Steam	9.2	9.7	10.6	12.4	16.2	20.3	24.5	28.6	32.6
Sulphur dioxide . .	11.6	12.6	14.0	16.4	20.9	25.1	29.0	32.6	36.1
Xenon	21.2	22.8	25.1	28.8	35.7	42.0	47.9	53.4	58.6

For viscosity of gases μPa s is a convenient size of unit. From the kinetic theory the viscosity is expected to be independent of pressure and to vary as the square root of the absolute temperature. The first is true except at very low and at high pressures; the second requires correction. Dividing the kinetic theory expression by a correction factor which is a linear function of the reciprocal of the absolute temperature leads to Sutherland's formula $\eta = K T^{3/2}/(T+C)$, where K and C are constants characteristic of the gas. For higher accuracy a polynomial can be used instead of a linear factor.

(ii) Viscosities of some gases at high pressure

η/μPa s (all at 300 K)

Gas	Pressure/MPa				
	2	5	10	20	30
Air	18.7	19.3	20.5	23.7	27.5
Argon	23.3	24.0	25.7	30.5	36.4
Helium	19.9	19.9	20.0	20.1	20.3
Hydrogen	8.98	9.01	9.09	9.31	9.59
Methane.	11.6	12.3	14.0	19.2	24.7
Nitrogen.	18.3	18.9	20.1	23.2	26.8
Oxygen	20.9	21.5	22.9	27.1	32.2

(iii) Viscosity of nitrogen at high pressure

$\eta/\mu Pa\,s$

Temperature/K	Pressure/MPa				
	5	10	20	30	50
200	14.6	17.6	26.4	34.9	48.9
250	16.7	18.5	23.1	28.4	38.8
300	18.9	20.1	23.2	26.8	34.4
350	20.9	21.9	24.2	26.9	32.7
400	22.9	23.7	25.5	27.6	32.3
500	26.5	27.1	28.4	29.9	33.2

J.T.R.W.

1.4.4 Mean velocity, free path and size of molecules

From the Maxwell–Boltzmann distribution law for molecular velocities, the mean square and mean velocities are

$$\overline{v^2} = 3kT/m; \qquad \bar{v} = (8\overline{v^2}/3\pi)^{1/2}$$

From the kinetic theory of gases, assuming that molecules interact like hard spheres, it follows that:

$$\eta = (5/16\,\sigma^2)(mkT/\pi)^{1/2} \qquad \tau = l/\bar{v} = 4\eta/5p$$

$$l = m/(\pi\rho\sigma^2\sqrt{2})$$

where k = Boltzmann's constant σ = molecular diameter
   T = absolute temperature m = mass of molecule
   p = pressure l = mean free path
   ρ = density τ = mean time between collisions.
   η = viscosity

A more exact theory uses the Lennard–Jones intermolecular potential. (See Hirschfelder, Curtiss and Bird's *Molecular Theory of Gases and Liquids*, 1954 (Wiley, New York).)

Equations are obtained for viscosity and second virial coefficient, B (i.e. first order departure from the perfect gas law) from which the molecular diameter, σ, can be estimated more accurately than from the simple equation for viscosity given above.

Gas	$\bar{v}/(\text{m s}^{-1})$	l/nm	τ/ps	σ/pm	
	at 0 °C and atmospheric pressure			η	B
Argon	380	62.6	165	342	340
Benzene	272	148.2	545	527	—
Carbon dioxide . .	362	39.0	108	390	407
Carbon monoxide .	454	58.6	129	371	376
Chlorine	285	27.4	96	440	—
Chloroform . . .	220	161.0	732	543	—
Ethylene	454	34.3	75	423	452
Helium	1202	173.6	144	258	256
Hydrogen	1694	110.6	65	297	293
Methane	600	48.1	80	380	382
Neon	535	124.0	232	279	275
Nitrogen	454	58.8	130	375	370
Nitrous oxide . . .	362	38.7	107	388	459
Oxygen	425	63.3	149	354	358
Sulphur dioxide . .	300	27.4	91	429	—

J.T.R.W.

1.4.5 Surface tensions

Surface tension is caused by the inward attraction of molecules at a boundary. This attraction produces curvature of free liquid surfaces, and causes a pressure difference to exist at the curved boundary: $\Delta p = \gamma(1/R_1 + 1/R_2)$, where Δp = pressure difference, R_1 and R_2 = principal radii of curvature. γ, the *surface tension* at a liquid-gas boundary, is usually measured in mN/m. At liquid–liquid boundaries there is *interfacial tension*; at liquid–solid boundaries *adhesion tension*.

Temperature variation. γ decreases with increase of temperature and vanishes at the critical temperature. For many liquids $-\mathrm{d}[\gamma(M/\rho)^{2/3}]/\mathrm{d}t = 2.12$, where M = molecular weight; ρ = density (g cm^{-3}); t = temperature (°C) (Eötvös, 1886). (See Adam, *The Physics and Chemistry of Surfaces*.)

Surface tensions of various liquids

$\gamma = a - bT$, (mN m^{-1})

Inorganic

Substance	Temp. °C	γ mN m^{-1}	Equation constants		
			a	b	Range °C
Aluminium	700	900.0	1145	0.35	660–800
Gold	1100	1120.0	1692	0.52	1065–1200
Lead	350	444.5	467.7	0.066	344–650
Mercury.	25	485.5	490.6	0.2049	5–200
Potassium	65	110.9	115.0	0.0625	65–500
Sodium	100	209.9	220.7	0.1080	100–500
Lead chloride . . .	520	135.3	199.8	0.124	510–580
Potassium chloride . .	780	100.3	160.4	0.0770	770–970
Potassium nitrate . . .	340	111.0	136.5	0.0750	340–500
Sodium chloride . . .	810	113.3	171.5	0.0719	805–970
Sodium fluoride . . .	1000	185.2	267.2	0.082	1000–1080
Sodium nitrate. . . .	320	119.2	138.8	0.0613	320–600
Sodium sulphate . . .	900	194.5	239.6	0.0501	900–1080
Oxygen	−184	13.40	−33.72	0.2561	−202 to −184
Nitrogen.	−183	5.99	−35.48	0.2266	−195 to −183

More data on molten salts in G. J. Janz, *NSRDS–NBS 28*, report, Natl. Bur. Standards, Washington, 1969. For liquid metals see Smithell's *Metal Reference Book*, 6th edn, 1983, p. 14.2 (Butterworths, London).

Organic

Substance	Temp. °C	γ mN m^{-1}	Equation constants		
			a	b	Range °C
Acetic acid	20	27.59	29.58	0.0994	20–90
Acetone	20	23.46	26.26	0.112	20–50
Aniline	20	42.67	44.83	0.1085	15–90
Benzene	20	28.88	31.50	0.1287	10–80
n-Butanol	20	25.39	27.18	0.08983	10–100
Carbon disulphide . .	20	32.32	35.29	0.1484	10–50
Carbon tetrachloride .	20	27.04	29.49	0.1224	15–105

Organic (*contd*)

Substance	Temp. °C	γ mN m^{-1}	Equation constants		
			a	b	Range °C
Chloroform	20	27.32	29.91	0.1295	15–75
Di-ethyl ether	20	17.10	18.92	0.0908	15–30
Ethanol	20	22.39	24.05	0.0832	10–70
Ethyl acetate	20	23.97	26.29	0.1161	10–100
Glycerol	20	63.4	65.17	0.088 45	20–150
n-Hexane	20	18.40	20.44	0.1022	10–60
n-Octane	20	21.62	23.52	0.095 09	10–120
Methanol	20	22.50	24.00	0.0773	10–60
Methyl acetate . . .	20	25.37	27.95	0.1289	10–60
Phenol	40	39.27	43.54	0.1068	40–140
n-Propanol	20	23.71	25.26	0.0777	10–90
Toluene	20	28.52	30.90	0.1189	10–100

For more data see J. J. Jasper, *J. Phys. Chem. Ref. Data*, 1972, **1**(4), p. 841–1010.

Surface tension of water against air

Temp./°C	0	10	15	20	25	30	40	50	60	70	80	100
γ/(mN m^{-1})	75.7	74.2	73.5	72.75	72.0	71.2	69.6	67.9	66.2	64.4	62.6	58.8

Surface tensions of aqueous salt solutions

Usually greater than that of water. Approximately $\gamma = \gamma H_2O + M . \Delta\gamma$, where M is concentration. Values below are for $M = 1$ mole dm^{-3} at 20 °C.

Salt	$\Delta\gamma$/(mN m^{-1})
KCl	1.4
NaCl	1.64
Na$_2$CO$_3$	2.7
NaNO$_3$	1.2
Na$_2$SO$_4$	2.7

Interfacial tensions of liquids at 20 °C

Liquids	γ/(mN m^{-1})	Liquids	γ/(mN m^{-1})
Water against:		Mercury against:	
Benzene	35	Acetone	390
Carbon tetrachloride . . .	45	Benzene	357
Chloroform	28	Chloroform	357
Di-ethyl ether	10	Di-ethyl ether	379
Heptylic acid	7	n-Heptane	379
n-Heptane	51	Oleic acid	322
n-Octane	51		
Olive oil	20		
Paraffin oil	48		

J.T.R.W.

1.5 Temperature and heat

1.5.1 The International Practical Temperature Scale of 1968 (IPTS-68)

The history of the IPTS-68

The International Temperature Scale was adopted in 1927 to overcome the practical difficulties of the direct realization of thermodynamic temperatures by gas thermometry and to unify existing temperature scales. It was introduced by the Seventh General Conference on Weights and Measures with the intention of producing a practical scale of temperature which was easily and accurately reproducible and which gave as nearly as possible thermodynamic temperatures. The Scale was revised in 1948, amended in 1960 (the numerical values of temperature remaining the same as in 1948) and revised again in 1968. (See *The International Practical Temperature Scale of 1968*, 1969 (HMSO).) The 1968 revision reduced the lower limit of the Scale from 90.18 K to 13.81 K and the values assigned to the defining fixed points were modified where necessary to conform as nearly as possible to thermodynamic temperatures. The approximate differences between the values of temperature given by IPTS-68 and its predecessor, IPTS-48 are given below.

The second edition of the IPTS-68 was adopted by the Fifteenth General Conference on Weights and Measures in 1975 (see *The International Practical Temperature Scale of 1968*, amended edition of 1975, HMSO 1976). It constitutes only an amendment of the first edition of the Scale, not a replacement. Any measured temperature T_{68} remains unchanged by this amendment of IPTS-68.

IPTS-68 extends down to 13.81 K, the triple point of hydrogen. To meet a need for an agreed practical temperature scale at lower temperatures, there now exists the 1976 Provisional 0.5 K to 30 K Temperature Scale known as EPT-76. Temperatures on this Scale have the symbol T_{76} (see *Metrologia*, **15**, 65–68, 1979).

The unit of temperature

The unit of the fundamental physical quantity known as thermodynamic temperature, symbol T, is the kelvin, symbol K, defined as the fraction $1/273.16$ of the thermodynamic temperature of the triple point of water.

For historical reasons, connected with the way temperature scales were originally defined, it is common practice to express a temperature in terms of its difference from that of a thermal state 0.01 kelvins lower than the triple point of water. A thermodynamic temperature, T, expressed in this way is known as a Celsius temperature, symbol t, defined by

$$t = T - 273.15 \, \text{K}$$

The unit of Celsius temperature is the degree Celsius, symbol °C, which is, by definition, equal in magnitude to the kelvin. A difference of temperature may be expressed in kelvins or degrees Celsius.

The International Practical Temperature Scale of 1968 (IPTS-68) has been constructed in such a way that any temperature measured on it is a close approximation to the thermodynamic temperature. Moreover, such measurement are more easily made and more reproducible than direct measurements of thermodynamic temperatures. The IPTS-68 uses both International Practical Kelvin Temperatures, symbol T_{68}, and International Practical Celsius Temperatures, symbol t_{68}. The relation between T_{68} and t_{68} is the same as that between T and t, i.e.,

$$t_{68} = T_{68} - 273.15 \text{K}$$

The units of T_{68} and t_{68} are the kelvin, symbol K, and the degree Celsius, symbol °C; that is, the names of the units are the same as those used for the thermodynamic temperatures T and t.

Approximate differences $(t_{68} - t_{48})$, **in kelvins, between the values of temperature given by the IPTS-68 and the IPTS-48**

t_{68}/°C	0	−10	−20	−30	−40	−50	−60	−70	−80	−90
−100	0.022	0.013	0.003	−0.006	−0.013	−0.013	−0.005	0.007	0.012	
−0	0.000	0.006	0.012	0.018	0.024	0.029	0.032	0.034	0.033	0.029

t_{68}/°C	0	10	20	30	40	50	60	70	80	90
0	0.000	−0.004	−0.007	−0.009	−0.010	−0.010	−0.010	−0.008	−0.006	−0.003
100	0.000	0.004	0.007	0.012	0.016	0.020	0.025	0.029	0.034	0.038
200	0.043	0.047	0.051	0.054	0.058	0.061	0.064	0.067	0.069	0.071
300	0.073	0.074	0.075	0.076	0.077	0.077	0.077	0.077	0.077	0.076
400	0.076	0.075	0.075	0.075	0.074	0.074	0.074	0.075	0.076	0.077
500	0.079	0.082	0.085	0.089	0.094	0.100	0.108	0.116	0.126	0.137
600	0.150	0.165	0.182	0.200	0.23	0.25	0.28	0.31	0.34	0.36
700	0.39	0.42	0.45	0.47	0.50	0.53	0.56	0.58	0.61	0.64
800	0.67	0.70	0.72	0.75	0.78	0.81	0.84	0.87	0.89	0.92
900	0.95	0.98	1.01	1.04	1.07	1.10	1.12	1.15	1.18	1.21
1000	1.24	1.27	1.30	1.33	1.36	1.39	1.42	1.44		

t_{68}/°C	0	100	200	300	400	500	600	700	800	900
1000		1.5	1.7	1.8	2.0	2.2	2.4	2.6	2.8	3.0
2000	3.2	3.5	3.7	4.0	4.2	4.5	4.8	5.0	5.3	5.6
3000	5.9	6.2	6.5	6.9	7.2	7.5	7.9	8.2	8.6	9.0
4000	9.3									

Principle of the IPTS-68

The IPTS-68 is based on the assigned values of the temperatures of a number of reproducible equilibrium states (**defining fixed points**) and on standard instruments calibrated at those temperatures. Interpolation between the fixed point temperatures is provided by formulae used to establish the relation between indications of the standard instruments and values of International Practical Temperature.

The defining fixed points are established by realizing specified equilibrium states between phases of pure substances. These equilibrium states and the values of the International Practical Temperature assigned to them are given below, together with the defining fixed points of the EPT-76 and a selection of secondary reference points between about 63 K and 3700 K.

In the IPTS-68 the standard instrument used from 13.81 K to 630.74 °C is the platinum resistance thermometer. The thermometer resistor must be strain-free, annealed pure platinum. The resistance ratio $W(T_{68})$, defined by

$$W(T_{68}) = R(T_{68})/R(273.15 \text{ K})$$

where R is the resistance, must not be less than 1.392 50 at $T_{68} = 373.15$ K. Below 0 °C the resistance–temperature relation of the thermometer is found from a reference function† and specified deviation equations. From 0 °C to 630.74 °C two polynomial equations provide the resistance–temperature relation.

The standard instrument used from 630.74 °C to 1064.43 °C is the platinum-10% rhodium/platinum thermocouple, the e.m.f.–temperature relation of which is represented by a quadratic equation.

Above 1337.58 K (1064.43 °C) the International Practical Temperature of 1968 is defined by the Planck law of radiation with 1337.58 K as the reference temperature and a value of 0.014 388 metre kelvin for c_2.

† The reference function and full details of the deivation equations are given in *The International Practical Temperature Scale of 1968*, amended edition of 1975 (HMSO, 1976).

Defining fixed points of IPTS-68(75) and of EPT-76, and some Secondary reference points

Equilibrium state	T_{68}(K)	T_{76}(K)
†Cd superconducting transition point . . .		0.519
†Zn superconducting transition point . . .		0.851
†Al superconducting transition point. . . .		1.1796
†In superconducting transition point. . . .		3.4145
†^{4}He boiling point.		4.2221
†Pb superconducting transition point . . .		7.1999
*†Triple point of equilibrium hydrogen	13.81	13.8044
*†Boiling point of equilibrium hydrogen at a pressure of 33 330.6 pascal (25/76 standard atmosphere)	17.042	17.0373
*†Boiling point of equilibrium hydrogen . . .	20.28	20.2734
†Ne triple point	24.5622	24.5591
*†Ne boiling point	27.102	27.102

Equilibrium state	T_{68}(K)	t_{68}(°C)
*O_2 triple point	54.361	−218.789
N_2 triple point	63.146	−210.004
N_2 boiling point	77.344	−195.806
*Ar triple point	83.798	−189.352
*O_2 condensation point	90.188	−182.962
Kr triple point	115.764	−157.386
CO_2 sublimation point	194.673	−78.477
Hg triple point	234.308	−38.842
H_2O freezing point	273.15	0
*H_2O triple point	273.16	0.01
Ga triple point	302.924	29.774
*H_2O boiling point	373.15	100
In freezing point	429.784	156.634
*Sn freezing point	505.1181	231.9681
Bi freezing point	544.592	271.442
Cd freezing point	594.258	321.108
Pb freezing point	600.652	327.502
*Zn freezing point	692.73	419.58
S boiling point	717.824	444.674
Cu-Al eutectic melting point	821.41	548.26
Sb freezing point	903.905	630.755
Al freezing point	933.61	660.46
Ag-Cu eutectic melting point	1052.73	779.58
Ag freezing point	1235.08	961.93
*Au freezing point	1337.58	1064.43
Cu freezing point	1358.03	1084.88
Ni freezing point	1728	1455
Co freezing point	1768	1495
Pd freezing point	1827	1554
Pt freezing point	2042	1769
Rh freezing point	2236	1963
Ir freezing point	2720	2447
W freezing point	3695	3422

All except the triple points and the hydrogen boiling point at 33 330.6 Pa are at a pressure of 101 325 Pa (1 standard atmosphere). Most of the unattributed numbers are taken from the more recent reports of Working Group 2 of the CCT.

* Defining point of IPTS-68(75).

† Defining point of EPT-76.

T.J.Q.

1.5.2 Thermoelectric thermometry

The data given in these tables are representative of the behaviour of modern thermocouple materials and are identical to those given in BS 4937 (1973) and IEC 584–1 (1977), which also include the polynomial expressions from which the tables are derived. The calibration of a particular thermocouple may be obtained by measurement at a few points over the temperature range and by plotting a curve of difference from these reference tables.

Thermocouple tables

Note that the temperature for a given entry is obtained by adding the corresponding temperature in the top row to that in the left or right-hand column, regardless of whether the latter is positive or negative. In every case the reference junction is at 0 °C.

Temp. °C	0	10	20	30	40	50	60	70	80	90	Temp. °C
					e.m.f. in microvolts						
Platinum 10% rhodium/platinum, Type S											
−100						−236	−194	−150	−103	−53	−100
0	0	55	113	173	235	299	365	432	502	573	0
100	645	719	795	872	950	1 029	1 109	1 190	1 273	1 356	100
200	1 440	1 525	1 611	1 698	1 785	1 873	1 962	2 051	2 141	2 232	200
300	2 323	2 414	2 506	2 599	2 692	2 786	2 880	2 974	3 069	3 164	300
400	3 260	3 356	3 452	3 549	3 645	3 743	3 840	3 938	4 036	4 135	400
500	4 234	4 333	4 432	4 532	4 632	4 732	4 832	4 933	5 034	5 136	500
600	5 237	5 339	5 442	5 544	5 648	5 751	5 855	5 960	6 064	6 169	600
700	6 274	6 380	6 486	6 592	6 699	6 805	6 913	7 020	7 128	7 236	700
800	7 345	7 454	7 563	7 672	7 782	7 892	8 003	8 114	8 225	8 336	800
900	8 448	8 560	8 673	8 786	8 899	9 012	9 126	9 240	9 355	9 470	900
1000	9 585	9 700	9 816	9 932	10 048	10 165	10 282	10 400	10 517	10 635	1000
1100	10 754	10 872	10 991	11 110	11 229	11 348	11 467	11 587	11 707	11 827	1100
1200	11 947	12 067	12 188	12 308	12 429	12 550	12 671	12 792	12 913	13 034	1200
1300	13 155	13 276	13 397	13 519	13 640	13 761	13 883	14 004	14 125	14 247	1300
1400	14 368	14 489	14 610	14 731	14 852	14 973	15 094	15 215	15 336	15 456	1400
1500	15 576	15 697	15 817	15 937	16 057	16 176	16 296	16 415	16 534	16 653	1500
1600	16 771	16 890	17 008	17 125	17 243	17 360	17 477	17 594	17 711	17 826	1600
1700	17 942	18 056	18 170	18 282	18 394	18 504	18 612				1700
Platinum 13% rhodium/platinum, Type R											
−100						−226	−188	−145	−100	−51	−100
0	0	54	111	171	232	296	363	431	501	573	0
100	647	723	800	879	959	1 041	1 124	1 208	1 294	1 380	100
200	1 468	1 557	1 647	1 738	1 830	1 923	2 017	2 111	2 207	2 303	200
300	2 400	2 498	2 596	2 695	2 795	2 896	2 997	3 099	3 201	3 304	300
400	3 407	3 511	3 616	3 721	3 826	3 933	4 039	4 146	4 254	4 362	400

Thermocouple tables (contd)

Temp. °C	0	10	20	30	40	50	60	70	80	90	Temp. °C
					e.m.f. in microvolts						

Platinum 13% rhodium/platinum, Type R (contd)

Temp. °C	0	10	20	30	40	50	60	70	80	90	Temp. °C
500	4 471	4 580	4 689	4 799	4 910	5 021	5 132	5 244	5 356	5 469	500
600	5 582	5 696	5 810	5 925	6 040	6 155	6 272	6 388	6 505	6 623	600
700	6 741	6 860	6 979	7 098	7 218	7 339	7 460	7 582	7 703	7 826	700
800	7 949	8 072	8 196	8 320	8 445	8 570	8 696	8 822	8 949	9 076	800
900	9 203	9 331	9 460	9 589	9 718	9 848	9 978	10 109	10 240	10 371	900
1000	10 503	10 636	10 768	10 902	11 035	11 170	11 304	11 439	11 574	11 710	1000
1100	11 846	11 983	12 119	12 257	12 394	12 532	12 669	12 808	12 946	13 085	1100
1200	13 224	13 363	13 502	13 642	13 782	13 922	14 062	14 202	14 343	14 483	1200
1300	14 624	14 765	14 906	15 047	15 188	15 329	15 470	15 611	15 752	15 893	1300
1400	16 035	16 176	16 317	16 458	16 599	16 741	16 882	17 022	17 163	17 304	1400
1500	17 445	17 585	17 726	17 866	18 006	18 146	18 286	18 425	18 564	18 703	1500
1600	18 842	18 981	19 119	19 257	19 395	19 533	19 670	19 807	19 944	20 080	1600
1700	20 215	20 350	20 483	20 616	20 748	20 878	21 006				1700

Platinum 30% rhodium/platinum 6% rhodium, Type B

Temp. °C	0	10	20	30	40	50	60	70	80	90	Temp. °C
0	0	−2	−3	−2	−0	2	6	11	17	25	0
100	33	43	53	65	78	92	107	123	140	159	100
200	178	199	220	243	266	291	317	344	372	401	200
300	431	462	494	527	561	596	632	669	707	746	300
400	786	827	870	913	957	1 002	1 048	1 095	1 143	1 192	400
500	1 241	1 292	1 344	1 397	1 450	1 505	1 560	1 617	1 674	1 732	500
600	1 791	1 851	1 912	1 974	2 036	2 100	2 164	2 230	2 296	2 363	600
700	2 430	2 499	2 569	2 639	2 710	2 782	2 855	2 928	3 003	3 078	700
800	3 154	3 231	3 308	3 387	3 466	3 546	3 626	3 708	3 790	3 873	800
900	3 957	4 041	4 126	4 212	4 298	4 386	4 474	4 562	4 652	4 742	900
1000	4 833	4 924	5 016	5 109	5 202	5 297	5 391	5 487	5 583	5 680	1000
1100	5 777	5 875	5 973	6 073	6 172	6 273	6 374	6 475	6 577	6 680	1100
1200	6 783	6 887	6 991	7 096	7 202	7 308	7 414	7 521	7 628	7 736	1200
1300	7 845	7 953	8 063	8 172	8 283	8 393	8 504	8 616	8 727	8 839	1300
1400	8 952	9 065	9 178	9 291	9 405	9 519	9 634	9 748	9 863	9 979	1400
1500	10 094	10 210	10 325	10 441	10 558	10 674	10 790	10 907	11 024	11 141	1500
1600	11 257	11 374	11 491	11 608	11 725	11 842	11 959	12 076	12 193	12 310	1600
1700	12 426	12 543	12 659	12 776	12 892	13 008	13 124	13 239	13 354	13 470	1700
1800	13 585	13 699	13 814								1800

Nickel–chromium/copper-nickel, Type E

Temp. °C	0	10	20	30	40	50	60	70	80	90	Temp. °C
−300				−9 835	−9 797	−9 719	−9 604	−9 455	−9 274	−9 063	−300
−200	−8 824	−8 561	−8 273	−7 963	−7 631	−7 279	−6 907	−6 516	−6 107	−5 680	−200
−100	−5 237	−4 777	−4 301	−3 811	−3 306	−2 787	−2 254	−1 709	−1 151	−581	−100
0	0	591	1 192	1 801	2 419	3 047	3 683	4 329	4 983	5 646	0
100	6 317	6 996	7 683	8 377	9 078	9 787	10 501	11 222	11 949	12 681	100

Thermocouple tables (contd)

Temp. °C	0	10	20	30	40	50	60	70	80	90	Temp. °C
					e.m.f. in microvolts						

Nickel–chromium/copper-nickel, Type E (contd)

Temp. °C	0	10	20	30	40	50	60	70	80	90	Temp. °C
200	13 419	14 161	14 909	15 661	16 417	17 178	17 942	18 710	19 481	20 256	200
300	21 033	21 814	22 597	23 383	24 171	24 961	25 754	26 549	27 345	28 143	300
400	28 943	29 744	30 546	31 350	32 155	32 960	33 767	34 574	35 382	36 190	400
500	36 999	37 808	38 617	39 426	40 236	41 045	41 853	42 662	43 470	44 278	500
600	45 085	45 891	46 697	47 502	48 306	49 109	49 911	50 713	51 513	52 312	600
700	53 110	53 907	54 703	55 498	56 291	57 083	57 873	58 663	59 451	60 237	700
800	61 022	61 806	62 588	63 368	64 147	64 924	65 700	66 473	67 245	68 015	800
900	68 783	69 549	70 313	71 075	71 835	72 593	73 350	74 104	74 857	75 608	900
1000	76 358										1000

Iron/copper–nickel, Type J

Temp. °C	0	10	20	30	40	50	60	70	80	90	Temp. °C
−300									−8 096		−300
−200	−7 890	−7 659	−7 402	−7 122	−6 821	−6 499	−6 159	−5 801	−5 426	−5 036	−200
−100	−4 632	−4 215	−3 785	−3 344	−2 892	−2 431	−1 960	−1 481	−995	−501	−100
0	0	507	1 019	1 536	2 058	2 585	3 115	3 649	4 186	4 725	0
100	5 268	5 812	6 359	6 907	7 457	8 008	8 560	9 113	9 667	10 222	100
200	10 777	11 332	11 887	12 442	12 998	13 553	14 108	14 663	15 217	15 771	200
300	16 325	16 879	17 432	17 984	18 537	19 089	19 640	20 192	20 743	21 295	300
400	21 846	22 397	22 949	23 501	24 054	24 607	25 161	25 716	26 272	26 829	400
500	27 388	27 949	28 511	29 075	29 642	30 210	30 782	31 356	31 933	32 513	500
600	33 096	33 683	34 273	34 867	35 464	36 066	36 671	37 280	37 893	38 510	600
700	39 130	39 754	40 382	41 013	41 647	42 283	42 922	43 563	44 207	44 852	700
800	45 498	46 144	46 790	47 434	48 076	48 716	49 354	49 989	50 621	51 249	800
900	51 875	52 496	53 115	53 729	54 341	54 948	55 553	56 155	56 753	57 349	900
1000	57 942	58 533	59 121	59 708	60 293	60 876	61 459	62 039	62 619	63 199	1000
1100	63 777	64 355	64 933	65 510	66 087	66 664	67 240	67 815	68 390	68 964	1100
1200	69 536										1200

Copper/copper–nickel, Type T

Temp. °C	0	10	20	30	40	50	60	70	80	90	Temp. °C
−300				−6 258	−6 232	−6 181	−6 105	−6 007	−5 889	−5 753	−300
−200	−5 603	−5 439	−5 261	−5 069	−4 865	−4 648	−4 419	−4 177	−3 923	−3 656	−200
−100	−3 378	−3 089	−2 788	−2 475	−2 152	−1 819	−1 475	−1 121	−757	−383	−100
0	0	391	789	1 196	1 611	2 035	2 467	2 908	3 357	3 813	0
100	4 277	4 749	5 227	5 712	6 204	6 702	7 207	7 718	8 235	8 757	100
200	9 286	9 820	10 360	10 905	11 456	12 011	12 572	13 137	13 707	14 281	200
300	14 860	15 443	16 030	16 621	17 217	17 816	18 420	19 027	19 638	20 252	300
400	20 869										400

Nickel–chromium/nickel–aluminium, Type K

Temp. °C	0	10	20	30	40	50	60	70	80	90	Temp. °C
−300				−6 458	−6 441	−6 404	−6 344	−6 262	−6 158	−6 035	−300
−200	−5 891	−5 730	−5 550	−5 354	−5 141	−4 912	−4 669	−4 410	−4 138	−3 852	−200
−100	−3 553	−3 242	−2 920	−2 586	−2 243	−1 889	−1 527	−1 156	−777	−392	−100

Thermocouple tables (contd)

Temp. °C	0	10	20	30	40	50	60	70	80	90	Temp. °C
					e.m.f. in microvolts						

Nickel–chromium/nickel–aluminium, Type K (contd)

Temp. °C	0	10	20	30	40	50	60	70	80	90	Temp. °C
0	0	397	798	1 203	1 611	2 022	2 436	2 850	3 266	3 681	0
100	4 095	4 508	4 919	5 327	5 733	6 137	6 539	6 939	7 338	7 737	100
200	8 137	8 537	8 938	9 341	9 745	10 151	10 560	10 969	11 381	11 793	200
300	12 207	12 623	13 039	13 456	13 874	14 292	14 712	15 132	15 552	15 974	300
400	16 395	16 818	17 241	17 664	18 088	18 513	18 938	19 363	19 788	20 214	400
500	20 640	21 066	21 493	21 919	22 346	22 772	23 198	23 624	24 050	24 476	500
600	24 902	25 327	25 751	26 176	26 599	27 022	27 445	27 867	28 288	28 709	600
700	29 128	29 547	29 965	30 383	30 799	31 214	31 629	32 042	32 455	32 866	700
800	33 277	33 686	34 095	34 502	34 909	35 314	35 718	36 121	36 524	36 925	800
900	37 325	37 724	38 122	38 519	38 915	39 310	39 703	40 096	40 488	40 879	900
1000	41 269	41 657	42 045	42 432	42 817	43 202	43 585	43 968	44 349	44 729	1000
1100	45 108	45 486	45 863	46 238	46 612	46 985	47 356	47 726	48 095	48 462	1100
1200	48 828	49 192	49 555	49 916	50 276	50 633	50 990	51 344	51 697	52 049	1200
1300	52 398	52 747	53 093	53 439	53 782	54 125	54 466	54 807			1300

Thermal e.m.f. of chemical elements relative to platinum

E.M.F. in microvolts; cold junctions at 0 °C
(At the hot end, the current flows from the platinum to the element when the sign is positive)

Element	− 100 °C	+ 100 °C	Element	− 100 °C	+ 100 °C	Element	− 100 °C	+ 100 °C
Lithium	− 1000	+ 1820	Aluminium	− 60	+ 420	Iron	− 1940	+ 1980
Sodium	+ 290	—	Carbon	—	+ 700	Cobalt	—	− 1330
Potassium	+ 780	—	Silicon	+ 37 170	− 41 560	Nickel	+ 1220	− 1480
Rubidium	+ 460	—	Germanium	− 26 620	+ 33 900	Iridium	− 350	+ 650
Calcium	− 130	− 510	Tin	− 120	+ 420	Rhodium	− 340	+ 700
Cerium	—	+ 1140	Lead	− 130	+ 440	Palladium	+ 480	− 570
Magnesium	− 90	+ 440	Antimony	—	+ 4 890	Molybdenum	—	+ 1450
Zinc	− 330	+ 760	Bismuth	+ 7 540	− 7 340	Tungsten	− 150	+ 1120
Cadmium	− 310	+ 900	Copper	− 370	+ 760	Tantalum	− 100	+ 330
Mercury	—	+ 45	Silver	− 390	+ 740	Thorium	—	− 130
Indium	—	+ 690	Gold	− 390	+ 780			
Thallium	—	+ 580						

Thermal e.m.f. of alloys relative to platinum

E.M.F. in microvolts; cold junctions at 0 °C and hot junctions at 100 °C

Alloy	e.m.f.
Chromel	+ 2810
Alumel	− 1290
Constantan	− 3510
Manganin (84 Cu, 4 Ni, 12 Mn)	+ 610
Phosphor-bronze (96 Cu, 3.5 Sn, 0.3 P)	+ 550
Solder (50 Sn–50 Pb)	+ 460
Stainless steel (18–8)	+ 440
Nickel–chromium (80 Ni, 20 Cr)	+ 1140
,, ,, (60 Ni, 24 Fe, 16 Cr)	+ 850
Yellow brass (70 Cu, 30 Zn)	+ 600
Beryllium copper ($97\frac{1}{2}$ Cu, $2\frac{1}{2}$ Be)	+ 670

Absolute thermoelectric power of platinum

Temperature/K	300	400	500	600	700	800	900	1000	1100	1200
Thermoelectric power/$(\mu V \ K^{-1})$	-5.05	-7.66	-9.69	-11.33	-12.87	-14.38	-15.97	-17.58	-19.03	-20.56

T.J.Q.

1.5.3 Optical pyrometry

Optical pyrometers (whether visual or photoelectric) are normally calibrated in terms of blackbody radiation and, when sighted on an unenclosed or freely-radiating surface, measure an apparent or spectral radiance temperature, i.e. the temperature of a blackbody having the same spectral radiance. The spectral radiance temperature T_r for a wavelength λ is related to the true temperature T of the radiating body by the equation $(1/T) - (1/T_r) = (\lambda \log \varepsilon(\lambda))/c_2$ where $\varepsilon(\lambda)$ is the spectral emittance of the body and c_2 is the second radiation constant in the Planck equation, equal to $0.014\,388$ m K.

Emissivity corrections (°C) for spectral pyrometer having effective wavelength of 0.65 μm
$c_2 = 0.014\,388$ m K

Observed temp./°C	Spectral emissivity								
	0.1	0.2	0.3	0.4	0.5	0.6	0.7	0.8	0.9
600	87	59	44	33	25	18	12	8	4
800	135	91	67	50	37	27	19	12	6
1000	194	130	95	71	53	39	27	17	8
1200	267	177	128	96	71	52	36	22	10
1400	353	232	168	125	93	67	46	29	13
1600	453	295	212	157	117	85	58	36	17
1800	570	368	263	195	144	104	72	44	21
2000	704	450	321	236	174	126	86	53	25
2500	1124	700	493	360	264	190	130	80	37
3000	1690	1022	709	513	374	267	182	112	52

Normal spectral emissivities

The emissivity of a material is a function of its surface shape and texture, its temperature and the wavelength. The figures in the table below refer to a smooth, polished surface and are given only as a guide to the values that might be encountered in practice. Much more detailed information on metals, alloys and non-metallic solids is to be found in Y. S. Touloukian and D. De Witt (eds), *Thermophysical Properties of Matter*, 1972, vols 7, 8 and 9, (Plenum Press).

Material	Emissivity			T/K
	Wavelength			
	0.65 μm	1.0 μm	5.0 μm	
Aluminium	(0.1 at 2 μm)		0.05	500
Aluminium oxide	0.1	0.06	0.39	1200
(recrystallized alumina) . . .	0.15	0.07	0.43	1400
	0.25	0.1	0.46	1600
	0.4	0.2	0.6	1800

Normal spectral emissivities (*contd*)

Material	Emissivity			T/K
	Wavelength			
	0.65 µm	1.0 µm	5.0 µm	
Beryllia†	0.5	0.35	0.8	1100
Chromium	0.35	—	—	1550
Cobalt	0.35	0.25	—	1300
Gold	0.15	0.05	0.03	1000
Hafnium	0.45	—	—	1200
Iridium	0.3	0.23	0.1	1500
Iron	0.35	0.3	0.15	1400
Stainless steel	0.33	0.3	0.2	1200
,, ,, (oxidized)	0.8	0.8	0.7	1200
Kanthal A (oxidized)	0.85	0.85	0.75	1300
Molybdenum	0.4	0.3	0.15	2000
Magnesia†	0.25	0.2	0.37	1400
Nickel	0.45	0.35	0.15	1100
Nickel (oxidized)	0.88	0.84	0.75	1300
Nickel/20% chromium	0.4	—	—	1200
,, ,, (oxidized)	0.9	(0.85)	(0.8)	1200
Niobium	0.4	0.32	0.2	2000
Osmium	0.43	—	—	1500
Palladium	0.35	—	—	1500
Platinum	0.35	0.25	0.08	1200
Platinum 13% rhodium	0.28	—	—	1100
Rhenium	0.4	0.35	0.2	1400
Rhodium	0.2	—	—	1400
Ruthenium	0.34	—	—	1500
Silicon	0.4	0.25	0.25	1500
Silver	0.05	0.05	0.05	1000
Tantalum	0.41	0.3	0.18	2200
Thoria†	0.4	0.35	0.37	1400
Titanium	0.6	0.3	0.18	1000
Tungsten	0.43	0.38	0.12	2400
Uranium	0.3	—	—	1100
Vanadium	—	0.65	0.28	800
Zirconium	0.5	0.45	0.3	1600
Zirconia†	0.4	0.2	0.45	1400

† These materials have an emissivity which increases with temperature in a way similar to that of aluminium oxide.

T.J.Q.

1.5.4 Thermal expansion

Coefficients of cubical expansion of gases

The volume coefficient, α, at constant pressure is defined by $v_t = v_0(1 + \alpha t)$; the pressure coefficient, β, at constant volume is defined by $p_t = p_0(1 + \beta t)$, where v_t and p_t are the volume and pressure respectively corresponding to t °C, the initial volume and pressure (v_0, p_0) being measured at 0 °C. Both α and β depend on the initial pressure of the gas, and for a perfect gas $\alpha = \beta$.

(a) **Gases used in thermometry.** $p_0 = 0.1333$ MN/m². Temperature interval 0–100 °C.

Coefficient	Gas				
	He	H_2	N_2	Air	Ne
$\alpha/(10^{-3}\ \mathrm{K}^{-1})$	3.6580	3.6588	3.6735	3.6728	3.6600
$\beta/(10^{-3}\ \mathrm{K}^{-1})$	3.6605	3.6620	3.6744	3.6744	3.6617

(b) **Other gases** and various initial pressures. Temperature interval 0–100 °C except where otherwise stated.

Gas	$p_0/(\mathrm{MN\ m}^{-2})$	$\alpha/(10^{-3}\ \mathrm{K}^{-1})$	$\beta/(10^{-3}\ \mathrm{K}^{-1})$
Air	0.031 (0–1067 °C)	—	3.6643
	2.67		3.887
O_2	0.024–0.031 (0–1067 °C)	—	3.6652
	0.088	—	3.6738
	10.1	4.86	—
CO	0.031 (0–1067 °C)	—	3.6648
	0.1013	3.669	3.667
CO_2	0.032 (0–1067 °C)	—	3.6756
	0.069	3.7073	3.6981
	0.133	3.7410	3.7262
N_2O	0.1013	3.719	3.676
SO_2	0.1013	3.903	3.845
NH_3	0.1013 (0–50 °C)	3.854	—
A	0.069	—	3.6680
N_2	20.3	4.34	—
	101	2.18	—

T.J.Q.

Coefficients of cubical expansion of liquids

The following table gives values of $\alpha = (1/V)(dV/dt)$ at $t = 293$ K ($20\,°$C). The expansion coefficient generally increases with rising temperature.

Liquid	$\dfrac{\alpha}{10^{-5}\,\mathrm{K}^{-1}}$	Liquid	$\dfrac{\alpha}{10^{-5}\,\mathrm{K}^{-1}}$
Acetic acid	107	Ethyl bromide	141
Acetone	143	Ethylene glycol	57
Alcohol, methyl	118	Glycerol (glycerine) . . .	49
Alcohol, ethyl	109	Mercury†	18.2
Aniline	85	Methyl iodide	120
Benzene	121	n-Pentane	158
Bromine	112	Sulphuric acid (100%) . .	56
Carbon bisulphide	119	Toluene	107
Carbon tetrachloride . . .	122	Turpentine	96
Chloroform	127	m-Xylene	99
Ether, ethyl	163	Water‡	21

† See also section 1.4.1, page 29 (Density of mercury)
‡ See also section 1.4.1, page 29 (Density of water)

Coefficients of linear expansion of solids

The coefficient of linear thermal expansion (expansivity) of a solid is defined by the equation $\alpha = (1/L)(dL/dt)$. The expansivities of most solids increase with temperature and can be represented by equations of the form $\alpha = a + bt + ct^2$ over limited temperature ranges. Values of α are given below for a number of elements, alloys and compounds at four temperatures. In addition, approximate values are given for the expansion coefficient at $20\,°$C of several miscellaneous materials of general or special interest.

Further thermal expansion data for a wide range of materials are listed in Y. S. Touloukian and C. Y. Ho (eds), *Thermophysical Properties of Matter*, vols 12, 13, IFI/Plenum.

Elements	$\dfrac{\alpha}{10^{-6}\,\mathrm{K}^{-1}}$			
	100 K	293 K (20 °C)	500 k	800 K
Aluminium	12.2	23.1	26.4	34.0
†Antimony	9.1	11.0	11.7	11.7
Beryllium	1.3	11.3	15.1	19.1
†Bismuth	12.3	13.4	12.7	—
Boron	—	4.7	5.4	6.2
†Cadmium	26.9	30.8	36.0	—
Carbon (glassy)	—	3.1	3.3	3.6
Carbon (diamond)	0.05	1.0	2.3	3.7
Chromium	2.3	4.9	8.8	11.8
†Cobalt	6.8	13.0	15.0	15.2
Copper	10.3	16.5	18.3	20.3
Germanium	2.4	5.7	6.5	7.2
Gold	11.8	14.2	15.4	17.0
Indium	25.4	32.1	—	—

† Where anisotropic crystalline forms exist, the value given applies to the polycrystalline form.

Coefficients of linear expansion of solids (*contd*)

Elements	$\dfrac{\alpha}{10^{-6}\,\mathrm{K}^{-1}}$			
	100 K	293 K (20 °C)	500 k	800 K
Iridium	4.4	6.4	7.2	8.1
Iron	5.6	11.8	14.4	16.2
Lead	25.6	28.9	33.3	—
†Magnesium	14.6	24.8	29.1	35.4
Nickel	6.6	13.4	15.3	16.8
Niobium	5.2	7.3	7.8	8.2
Palladium	8.0	11.8	13.2	14.5
Platinum	6.6	8.8	9.6	10.3
Rhodium	5.0	8.2	9.3	10.8
Silicon	−0.4	2.6	3.5	4.1
Silver	14.2	18.9	20.6	23.7
Tantalum	4.8	6.3	6.8	7.2
†Thallium	25.2	29.9	34.7	—
†Tin	16.5	22.0	27.2	—
†Titanium	4.5	8.6	9.9	11.1
Tungsten	2.6	4.5	4.6	5.0
†Uranium	10.0	13.9	16.9	24.3
Vanadium	5.1	8.4	9.9	10.9
†Zinc	24.5	30.2	32.8	—

† Where anisotropic crystalline forms exist, the value given applies to the polycrystalline form.

Alloys and compounds	$\dfrac{\alpha}{10^{-6}\,\mathrm{K}^{-1}}$			
	100 K	293 K (20 °C)	500 K	800 K
Aluminium bronze (90Cu + 5Al + 4.5Ni)	12–14	15.9	18.1	20.3
Brass (67Cu + 33Zn)	—	17.5	20.0	22.5
Bronze (85Cu + 15Sn)	—	17.3	19.3	21.9
Cast iron (Fe + 3C + 2Si)	—	11.9	13.1	14.5
Constantan (65Cu + 35Ni)	11.2	15.0	17.4	19.2
Cupro-nickel (65Ni + 30Cu + 1.5Fe + 1Mn)	9.8	12.7	15.4	18.2
Duralumin (94Al + 4to5Cu)	13.1	21.6	27.5	30.1
Gallium arsenside	1.9	5.7	6.5	7.1
Gallium phosphide	—	4.7	5.5	6.0
Inconel	8.7	11.6	14.4	17.6
Indium antimonide	2.8	5.0	6.1	—
Magnesium fluoride, // axis	3.9	14.5	17.0	19.2
Magnesium fluoride, ⊥ axis	1.4	9.5	11.5	15.8
Magnesium fluoride, poly	2.2	11.1	13.3	16.8
Nickel steels (64Fe + 36Ni – Invar)	1.4	0.13	5.1	17.1
(50Fe + 50Ni)	—	9.4	9.6	12.5

Coefficients of linear expansion of solids (contd)

Alloys and compounds	α / $10^{-6} K^{-1}$			
	100 K	293 K (20 °C)	500 K	800 K
Phosphor bronze (90Cu + 10Sn + P) . . .	—	17.0	20.0	—
Quartz, // axis	4.0	6.8	11.4	31.4
Quartz, ⊥ axis	9.1	12.2	19.5	37.6
Quartz, polycrystalline	7.3	10.3	16.8	32.2
Silicon carbide	0.14	3.3	4.2	4.9
Stainless steel				
(18Cr + 8Ni)	11.4	14.7	17.5	20.2
(13–17Cr)	6.0	9.5	12.1	13.8
Steel, carbon (0.7–1.4C)	6.7	10.7	13.7	16.2
Stellite (65Co + 25Cr + 10W)	6.9	11.2	14.6	17.2
Tungsten carbide (poly)	—	3.7	4.3	4.8

Miscellaneous materials	α / $10^{-6} K^{-1}$
Building materials	
Brick	3–10
Cement and concrete	7–14
Granite, limestone	4–10
Marble	3–15
Portland stone	approx. 3
Sandstone, slate	5–12
Corundum (Alundum)	5.5
Glass	
borosilicate crown	7–8
dense flint	8–9
Pyrex	2.8
Plastics	see p. 287
Polytetrafluoroethylene (PTFE)	525
Porcelain	2–6
Silica, fused	0.4–0.55
Woods, along grain	3–6
Woods, across grain	35–60
Zerodur	< 0.1

see p. 287

S.J.B.

1.5.5 Specific heat capacities

For molar heat capacities and other thermodynamic properties of elements and chemical compounds, see section 2.10.

Ratio of the principal specific heat capacities for gases and vapours

The ratio γ of the specific heat capacity at constant pressure to that at constant volume is usually determined directly by some method involving an adiabatic expansion, such as the determination of the velocity of sound in the gas. From a knowledge of either (1) the pressure or (2) the temperature immediately following an adiabatic expansion (Clément and Desormes', Lummer and Pringsheim's methods respectively), γ can be deduced from $pv^{\gamma} = \text{const.}$, or $\theta v^{\gamma-1} = \text{const.}$ (See J. K. Roberts, *Heat and Thermodynamics*, 5th edn, 1960, (Blackie, London).)

Gas	Temp./°C	γ	Gas	Temp./°C	γ
Monatomic gases			Nitrogen $\{N_2O_4$ . . .	20	1.172
Helium	0	1.63	peroxide $\{NO_2$	150	1.31
Argon.	0	1.667	H_2S	—	1.340
Neon	19	1.642	CS_2	—	1.239
Krypton	19	1.689	Sulphur dioxide . . . $\{$	16–34	1.26
Xenon	19	1.666		500	1.2
Mercury vapour	310	1.666			
			Polyatomic gases		
Diatomic gases			Methane, CH_4	—	1.313
Air (dry)	−79.3	1.405	Ethane, C_2H_6	—	1.22
Air (dry)	0–17	1.401/2	Propane, C_3H_8	—	1.130
Air (dry)	500	1.357	Acetylene, C_2H_2. . . .	—	1.26
Air (dry)	900	1.32	Ethylene, C_2H_4	—	1.264
Air (dry) (200 atm) . . $\{$	0	1.828	Benzene C_6H_6	20	1.40
	−79.3	2.333	Benzene C_6H_6	99.7	1.105
Hydrogen	4–17	1.407/8	Chloroform $CHCl_3$ . . $\{$	24–42	1.110
Nitrogen	20	1.401		99.8	1.150
Oxygen	5–14	1.400	CCl_4	—	1.130
Carbon monoxide . . .	1800	1.297	Me. alcohol	99.7	1.256
Nitric oxide	—	1.394	Me. bromide	—	1.274
			Me. chloride	19–30	1.279
Triatomic gases			Me. iodide	—	1.286
Ozone	—	1.29†	Et. alcohol	53	1.133
Water vapour	100	1.334	Et. alcohol	99.8	1.134
Carbon dioxide	4–11	1.300	Et. bromide	—	1.188
Carbon dioxide	300	1.22	Et. chloride	22.7	1.187
Carbon dioxide	500	1.20	Et. ether	12–20	1.024
Ammonia, NH_3	—	1.336	Et. ether	99.7	1.112
Nitrous oxide, N_2O . . .	—	1.324	Acetic acid	136.5	1.147

† Extrapolated.

Specific heat capacity of water

The International Committee of Weights and Measures, Paris, 1950, accepted W. J. de Haas's recommended value of $4.1855 \, \text{J g}^{-1} \, ^{\circ}\text{C}^{-1}$ for the specific heat capacity of water at 15°C (c_p (15°C)). Values at other temperatures are then deduced from the equation:

$$\frac{c_p(t \, ^{\circ}\text{C})}{c_p(15 \, ^{\circ}\text{C})} = 0.996 \, 185 + 0.000 \, 287 \, 4 \left(\frac{t + 100}{100}\right)^{5.26} + 0.011 \, 160 \times 10^{-0.036 t}$$

due to Osborne, Stimson and Ginnings.

Water $c_p(t)/(\mathrm{Jg^{-1}°C^{-1}})$

°C	0	1	2	3	4	5	6	7	8	9
0	4.2174	4.2138	4.2104	4.2074	4.2045	4.2019	4.1996	4.1974	4.1954	4.1936
10	4.1919	4.1904	4.1890	4.1877	4.1866	4.1855	4.1846	4.1837	4.1829	4.1822
20	4.1816	4.1810	4.1805	4.1801	4.1797	4.1793	4.1790	4.1787	4.1785	4.1783
30	4.1782	4.1781	4.1780	4.1780	4.1779	4.1779	4.1780	4.1780	4.1781	4.1782
40	4.1783	4.1784	4.1786	4.1788	4.1789	4.1792	4.1794	4.1796	4.1799	4.1801
50	4.1804	4.1807	4.1811	4.1814	4.1817	4.1821	4.1825	4.1829	4.1833	4.1837
60	4.1841	4.1846	4.1850	4.1855	4.1860	4.1865	4.1871	4.1876	4.1882	4.1887
70	4.1893	4.1899	4.1905	4.1912	4.1918	4.1925	4.1932	4.1939	4.1946	4.1954
80	4.1961	4.1969	4.1977	4.1985	4.1994	4.2002	4.2011	4.2020	4.2029	4.2039
90	4.2048	4.2058	4.2068	4.2078	4.2089	4.2100	4.2111	4.2122	4.2133	4.2145

Specific heat capacities of metals, alloys and miscellaneous substances

Metal	$c_p/\mathrm{Jg^{-1}K^{-1}}$					
	Temperature/K					
	77	173	273	373	573	773
Aluminium	0.336	0.743	0.880	0.937	1.00	1.13
Antimony	0.150	—	~0.21	—	—	~0.28
Beryllium	0.080	0.852	1.75	2.01	2.44	~2.8
Bismuth	0.103	0.119	0.123	—	—	—
Cadmium	0.179	0.218	0.229	0.238	0.255	—
Cerium	—	—	0.190	0.199	0.219	—
Chromium	0.117	0.351	0.438	0.481	0.527	0.55
Cobalt	—	—	0.411	0.443	0.490	0.54
Copper	0.195	0.341	0.379	0.397	0.419	~0.43
Erbium	0.192	0.140	0.166	—	—	—
Gadolinium	0.166	—	0.300	0.35	0.38	0.39
Gold	0.097	0.121	0.128	0.131	0.135	0.140
Hafnium	0.101	—	0.142	0.148	0.157	0.164
Holmium	—	—	0.164	—	—	—
Indium	0.191	0.222	0.231	—	—	—
Iron (α)	0.144	0.336	0.435	0.48	0.56	0.67
Lanthanum	0.156	—	—	0.194	0.210	—
Lead	0.114	0.124	0.129	—	—	—
Lithium	1.32	—	3.48	—	—	—
Lutetium	—	0.602	0.64	—	—	—
Magnesium	0.488	0.887	1.00	1.06	1.14	1.27
Manganese (α)	0.201	0.394	0.467	0.502	0.565	0.63
Mercury	0.115	0.132	—	—	—	—
Molybdenum	0.096	0.209	0.242	0.256	0.269	0.280
Nickel	0.163	0.357	0.429	0.465	0.569	0.527
Niobium	0.167	0.247	0.265	0.272	0.28	0.29
Osmium	—	—	0.130	0.132	0.136	—
Palladium	0.124	0.218	0.240	0.249	0.259	—
Platinum	0.085	0.123	0.132	0.135	0.141	0.146
Plutonium	—	—	0.146	~0.17	~0.20	~0.18
Potassium	—	—	0.732	0.812	0.770	—
Praseodymium	0.193	—	0.196	0.200	0.217	~0.24

Specific heat capacities of metals, alloys and miscellaneous substances (*contd*)

Metal	$c_p/\mathrm{Jg^{-1}K^{-1}}$ Temperature/K					
	77	173	273	373	573	773
Rhenium	0.078	0.125	0.138	0.142	0.145	0.146
Rhodium	0.088	0.209	0.238	0.252	0.269	—
Rubidium	—	—	0.346	—	—	—
Ruthenium	0.094	—	0.231	0.237	0.251	—
Samarium	—	—	0.174	0.218	0.263	0.296
Scandium	—	—	0.557	—	—	—
Silver	0.162	0.219	0.235	0.239	0.249	0.257
Sodium	0.90	1.11	1.21	—	—	—
Tantalum	0.095	0.130	0.139	0.143	0.150	0.152
Terbium	0.178	0.255	0.186	0.180	—	—
Thallium	0.115	0.126	0.131	0.138	—	—
Thorium	0.091	—	0.113	0.120	0.130	0.14
Thulium	0.146	0.156	0.159	0.161	—	—
Tin	0.170	0.211	0.221	0.243	—	—
Titanium	0.218	0.438	0.511	0.536	~0.58	~0.62
Tungsten	0.068	0.120	0.133	0.135	0.140	0.143
Uranium	0.085	0.106	0.114	0.119	0.144	0.167
Vanadium	0.188	0.409	0.489	0.510	0.531	0.573
Yttrium	0.200	0.273	0.294	0.304	0.320	—
Zinc	0.251	0.357	0.385	0.402	0.437	—
Zirconium	0.170	0.257	0.273	0.287	0.296	~0.31

Alloy	$c_p/\mathrm{Jg^{-1}K^{-1}}$ Temperature/K					
	77	173	273	373	573	773
Alumel 72Ni, 25Mn, 2Al . .	—	—	0.481	0.496	—	
Aluminium alloys	0.48	0.69	0.82	0.91	1.03	—
94 Al, 4.5 Cu, 15 Mg						
90 Al, 5.5 Zn, 25 Mg						
Brass	0.20	0.34	0.387	0.390	0.448	—
Chromel 90Ni, 10 Cr. . . .	—	—	—	0.473	0.521	—
Constantan 60 Cu, 40 Ni . .	0.175	—	~0.40	0.42	0.45	—
Inconel 75 Ni, 15 Cr + Fe . .	—	—	0.440	0.465	0.506	—
Monel 67 Ni, 29 Cu + Fe . .	—	0.35	0.406	0.448	0.489	0.514
Nichrome 5, 77 Ni, 19.5 Cr. .	—	—	0.432	0.464	0.509	0.548
Nickel steels						
3.5% Ni	—	—	—	0.496	0.569	0.694
5.0% Ni	0.155	0.365	0.445	—	—	—
9% Ni	0.154	0.364	0.454	—	0.557	—
20% Ni	—	—	—	—	0.536	0.654
30% Ni	—	—	—	—	0.546	0.57
50% Ni	—	—	—	—	0.561	0.582
Solder 50 Pb, 50 Sn	0.142	0.167	0.177	—	—	—

Specific heat capacities of metals, alloys and miscellaneous substances (*contd*)

Alloy	$c_p/\mathrm{Jg^{-1}K^{-1}}$ Temperature/K					
	77	173	273	373	573	773
Steels						
Mild steel	—	—	—	0.49	0.56	0.68
Tool steel 1.2% C	—	—	—	0.50	0.57	0.65
Low alloy steels	—	—	—	0.48–0.51	0.55–0.59	0.67–0.70
→ 5% additions	—	—	—			
High alloy steels						
12% Cr	—	—	—	0.48	0.57	0.70
18 Cr/8 Ni stainless steel . .	—	—	—	0.519	0.555	0.611
18 Cr/12 Ni stainless steel . .	0.197	0.401	0.47	—	—	—
24 Cr/20 Ni stainless steel . .	0.195	0.393	0.463	—	—	—
Titanium alloy						
Ti 6 Al 4 V	0.21	0.45	0.527	0.565	0.612	—

Miscellaneous	$c_p/\mathrm{Jg^{-1}K^{-1}}$ Temperature/K					
	77	173	273	373	573	773
Air (dry) c_p	—	1.008	—	1.011	—	1.092
Alumina Al_2O_3	0.061	0.40	0.73	0.91	—	—
Asbestos	—	—	(~0.84)†		—	—
Basalt	—	—	(~0.84–1.0)†		—	—
Boron	0.064	0.48	~0.96	1.37	1.84	2.22
Boron nitride	0.15	0.43	0.72	0.99	1.38	1.60
Calcium silicate	—	—	0.68	0.84	0.97	1.03
(Ca Si O_3)						
Carbon (diamond)	0.008	0.14	0.42	0.77	1.30	1.59
Carbon (graphite)	0.009	0.34	0.64	0.94	1.38	1.60
Cermets						
99% Be, 1% BeO . . .	—	—	—	2.42	2.56	2.70
10% Be, 90% BeO . . .	—	—	—	1.30	1.65	1.83
BN + C	—	—	—	1.04	1.41	1.62
77% C + 23 Si C . . .	—	—	—	0.96	1.27	1.46
20% C + 78% Si C . . .	—	—	—	0.87	1.10	1.22
95% WC + 5% Co . . .	—	—	—	0.24	0.27	0.29
Ebonite	—	—	(1.38)†		—	—
Fluorspar CaF_2	—	—	0.83	0.93	0.99	1.07
Glass crown	—	—	(~0.67)‡		—	—
flint	—	—	(~0.5)‡		—	—
Pyrex	—	—	0.70	0.85	1.1	—
Ice	0.69	1.38	2.10	—	—	—
India rubber	—	—	(1.1–2.0)†		—	—
Magnesia, MgO	0.10	0.57	0.88	1.03	1.12	1.22
Marble, white	—	—	~0.9	—	—	—
Paraffin wax	—	—	~2.9	—	—	—

Specific heat capacities of metals, alloys and miscellaneous substances (*contd*)

Miscellaneous	$c_p/\mathrm{Jg^{-1}K^{-1}}$					
	Temperature/K					
	77	173	273	373	573	773
Porcelain	—	—	—	$(\sim 0.75)_{\mathrm{A}}{}^\S$	—	$(\sim 1.07)_{\mathrm{B}}{}^\S$
Polyethylene	0.54	1.04	~ 1.8	—	—	—
Polymers.	—	—	—	—	—	—
Potassium chloride	0.44	0.63	0.68	—	—	—
Pyroceram	—	—	0.71	0.90	1.03	1.12
Quartz Si O_2	0.19	0.47	0.73	0.86	1.06	1.2
Quartz glass	—	—	0.70	0.83	1.02	1.11
Rubber (natural from						
latex)	0.52	1.0	1.8	—	—	—
Sand	—	—	~ 0.8	—	—	—
Silica (fused)	—	—	—	~ 0.84	—	—
Silicon	0.18	0.49	0.68	0.77	0.85	0.88
Silicon Carbide	0.05	—	—	0.82	1.01	1.12
Sodium Chloride	0.48	0.77	0.84	—	—	—
Teflon	0.30	0.65	1.02	—	—	—
Tungsten carbide	—	—	—	—	~ 0.24	—

† Mean values 293 K to 373 K.   J.M.C.
‡ Mean values 283K to 323 K.
§ Means values A 288 K to 473 K, B 288 K to 1273 k.

1.5.6 Thermal conductivities

Definition and units

The thermal conductivity, λ, of a substance may be defined as the quantity of heat transmitted, due to unit temperature gradient, in unit time under steady conditions in a direction normal to a surface of unit area, when the heat transfer is dependent only on the temperature gradient.

$$\lambda = - q / \frac{\partial T}{\partial n}$$

In this section thermal conductivity values are assembled for metallic, semi-conducting and insulating elements, representative groups of alloys, refractories and miscellaneous constructional and insulating materials, some liquids and some gases. The values are expressed in the SI unit $\mathrm{W\,m^{-1}\,K^{-1}}$ throughout. Factors for converting to other units are as follows:

$$
\begin{aligned}
1\,\mathrm{W\,m^{-1}\,K^{-1}} &= 0.01\,\mathrm{J\,cm\,cm^{-2}\,s^{-1}\,K^{-1}} \\
&= 0.002\,388\,\mathrm{cal\,cm\,cm^{-2}\,s^{-1}\,K^{-1}} \\
&= 0.859\,8\,\mathrm{kcal\,m\,m^{-2}\,h^{-1}\,K^{-1}} \\
&= 0.001\,926\,\mathrm{Btu\,in\,ft^{-2}\,s^{-1}\,{}^\circ F^{-1}} \\
&= 6.933\,\mathrm{Btu\,in\,ft^{-2}\,h^{-1}\,{}^\circ F^{-1}} \\
&= 0.577\,8\,\mathrm{Btu\,ft\,ft^{-2}\,h^{-1}\,{}^\circ F^{-1}} \\
&= 0.000\,160\,5\,\mathrm{Btu\,ft\,ft^{-2}\,s^{-1}\,F^{-1}}
\end{aligned}
$$

More extensive collections of thermal conductivity data will be found in three volumes of the series *Thermophysical Properties of Matter*, 1970 (IFI/Plenum Data Corp., New York, Washington: Vol. 1,

Thermal Conductivity: Metallic Elements and Alloys, by Y. S. Touloukian, R. W. Powell, C. Y. Ho and P. G. Klemens; Vol. 2, *Thermal Conductivity: Nonmetallic Solids*, by Y. S. Touloukian, R. W. Powell, C. Y. Ho and P. G. Klemens; and Vol. 3, *Thermal Conductivity: Nonmetallic Liquids and Gases*, by Y. S. Touloukian, P. E. Liley and S. C. Saxena.

Thermal conductivities of metallic elements

The thermal conductivity values in the table below are for metallic elements in the purest polycrystalline condition for which reliable measurements have been reported. Entries in italics relate to the liquid phase.

$$\lambda/(\mathrm{W\,m^{-1}\,K^{-1}})$$

Metal	Temperature/K					Metal	Temperature/K				
	173.2	273.2	373.2	573.2	973.2		173.2	273.2	373.2	573.2	973.2
Aluminium	241	236	240	233	*92*	Niobium	53	53	55	58	64
Antimony	33	25.5	22	19	27	Osmium	93	88	87	87	—
Beryllium	367	218	168	129	93	Palladium	72	72	73	79	93
Bismuth	11	8.2	7.2	*13*	*17*	Platinum	73	72	72	73	78
Cadmium	100	97	95	89	*45*	Plutonium	4	6	8	—	—
Caesium	37	36	*20*	*20.6*	*17.7*	Potassium	105	104	*53*	*45*	*32*
Cerium	8	11	13	16	—	Praseodymium	9.9	12	*13.4*	—	—
Chromium	120	96.5	92	82	66	Rhenium	52	49	47	44	45
Cobalt	130	105	89	69	53	Rhodium	156	151	147	137	—
Copper	420	403	395	381	354	Rubidium	59	58	*32*	*29*	*22*
Dysprosium	9	10.5	—	—	—	Ruthenium	123	117	115	108	98
Erbium	14	15	—	—	—	Samarium	10	13	13	14	—
Gadolinium	12	10	—	—	—	Scandium	15	16	—	—	—
Gallium†	43	41	*33*	*45*	—	Silver	432	428	422	407	377
Gold	324	319	313	299	272	Sodium	141	142	*88*	*78*	*60*
Hafnium	25	23	22	21	21	Tantalum	58	57	58	58.5	60
Holmium	14	16	17	—	—	Technetium	—	51	50	50	—
Indium	92	84	76	*42*	—	Terbium	11	10.5	—	—	—
Iridium	156	147	145	139	—	Thallium	51	47	44	—	—
Iron	99	83.5	72	56	34	Thorium	55	54	54	56	58
Lanthanum	12	13	14.5	—	—	Thulium	16	17	—	—	—
Lead	37	36	34	32	*21*	Tin	76	68	63	*32*	*40*
Lithium	94	86	82	*47*	*59*	Titanium	26	22	21	19	21
Lutetium	18	17	—	—	—	Tungsten	188	177	163	139	119
Magnesium	160	157	154	150	—	Uranium	24	27	29	33	43
Manganese	7	8	—	—	—	Vanadium	32	31	31	33	38
Mercury	29.5	*7.8*	*9.4*	*11.7*	—	Yttrium	16.5	17	—	—	—
Molybdenum	145	139	135	127	113	Zinc	117	117	112	104	*66*
Nickel	113	94	83	67	71	Zirconium	26	23	22	21	23

† Values for the *a*-axis which approximate to the polycrystal; those for the *b*-axis are 91 and 88 and for the *c*-axis are 16.5 and 16 at 173.2 and 273.2 K.

The thermal conductivities of less pure samples of these elements will be lower than the values given above. Thermal conductivity invariably decreases with decreasing purity; such dependence being weak at ambient and higher temperatures but very strong at cryogenic temperatures. At low temperatures the thermal conductivity of a given metal tends to increase in proportion to the reciprocal of its residual resistivity ρ_0. Many metals, especially good electrical conductors, have thermal conductivities that follow the simple relation

$$\lambda = L_0 \alpha T$$

at very low temperatures and also at temperatures higher than their Debye temperature; L_0 being the Lorentz coefficient 2.45×10^{-8}, σ the electrical conductivity in $\mathrm{S m^{-1}}$ and T the absolute temperature. This behaviour enables the thermal conductivities of metallic samples to be estimated fairly reliably from simple electrical resistivity measurements (see alloys).

Thermal conductivities of single crystals of some non-cubic metals at normal temperature

	$\lambda/(\mathrm{W m^{-1} K^{-1}})$		
Metal	*Thermal conductivity in direction of*		
	c-axis	*a-axis*	*b-axis*
Bismuth	5.4	9.3	
Cadmium	83.05	104	
Dysprosium	11.65	10.25	
Erbium	18.4	12.6	
Gadolinium	10.7	10.3	
Gallium	16.0	40.8	88.3
Holmium	22.1	13.6	
Lutetium	23.3	13.8	
Mercury (at 227.7 K)	33.0	25.9	
Terbium	14.5	9.45	
Thulium	24.2	14.1	
Tin	51.8	74.0	

Thermal conductivities of alloys

As the thermal conductivities of alloys depend strongly on their mechanical and thermal history (heat treatment) as well as on their chemical composition, the values tabulated below should be regarded as typical for the compositions listed. For many groups of alloys the thermal conductivity of a particular sample, near room temperature and above, can be estimated within about 6% from its more easily measured electrical conductivity using the relation $\lambda = L\sigma T + C$. The optimum values for L and C for different alloy types are as follows:

Main constituent metal	L	C
	$(10^{-8}\,\mathrm{W S^{-1} K^{-2}})$	$(\mathrm{W m^{-1} K^{-1}})$
Aluminium	2.22	10.5
Copper	2.39	7.5
Alpha-iron	2.43	9.2
Gamma-iron	2.39	4.2
Magnesium	2.21	9.6
Nickel	2.13	8.4
Nickel–chromium (nimonic type)	2.20	6.0
Titanium	2.30	2.9
Zirconium	2.50	2.2

For more detailed information on the relationship between the thermal and electrical conductivities of alloys see:

(1) R. W. Powell, 'Correlation of metallic thermal and electrical conductivities for both solid and liquid phases', *Int. J. Heat and Mass Transfer*, 1965, vol. 8, 1033–1045.

(2) J. G. Hust and A. F. Clark, 'The Lorentz ratio as a tool for predicting the thermal conductivity of metals and alloys', *Mat. Research and Standards*, 1971, vol. 11, no. 8, p. 22–24.

Thermal conductivities of alloys at ambient and elevated temperatures

Composition (weight percent) $\lambda/(\mathrm{W\,m^{-1}\,K^{-1}})$, Temperature K

Alloy	Al	C	Cr	Cu	Fe	Mg	Mn	Ni	Si	W	Zn	Other	273	373	573	773	973	1273
Aluminium alloys																		
Aluminium	100	—	—	—	—	—	—	—	—	—	—	—	236	240	233	—	—	—
Alpax gamma	87	—	—	—	0.3	0.3	0.3	—	12	—	—	—	188	188	184	—	—	—
Lo-Ex.	85	—	—	1	0.5	0.9	—	1	11.8	—	—	—	172	175	173	—	—	—
Y-alloy	92	—	—	3.8	0.4	1.3	—	1.8	0.4	—	—	—	180	188	194	—	—	—
RR 59	93	—	—	2.3	1.2	1.5	—	1.2	0.9	—	—	—	168	176	186	—	—	—
RR 57	89	—	—	2.2	0.3	2.5	0.5	—	0.3	—	5	—	161	171	178	—	—	—
Copper alloys																		
Copper	—	—	—	100	—	—	—	—	—	—	—	—	403	395	381	—	354	—
Brass	—	—	—	70	—	—	—	—	—	—	30	—	106	128	146	—	—	—
Bronze	—	—	—	90	—	—	—	—	—	—	—	10 Sn	53	60	80	—	—	—
German Silver	—	—	—	62	—	—	—	15	—	—	22	—	23	29	45	—	—	—
Constantan	—	—	—	60	—	—	—	40	—	—	—	—	22	24	27	—	—	—
Manganin	—	—	—	84	—	—	12	4	—	—	—	—	21	26	—	—	—	—
Nickel alloys																		
Nickel	—	—	—	—	—	—	—	100	—	—	—	—	94	83	67	—	71	—
Alumel	2	0.2	—	—	—	0.1	2	95	1	—	—	—	30	32	35	—	—	—
Monel	—	—	—	29.2	1.7	—	1.0	67.1	—	—	—	—	21	24	30	—	43	—
Chromel P	—	—	10	—	—	—	—	90	—	—	—	—	—	19	23	—	—	—
Nichrome	—	0.1	21	—	0.6	0.1	0.65	77.3	0.4	—	—	—	13	14	17	—	21	—
Inconel 600	—	0.1	16	0.3	8	—	0.5	74	0.4	—	—	—	14.6	15.8	19.1	22.1	25.7	27.9
Inconel X-750	0.8	0.04	15	0.05	6.8	—	0.7	73	0.3	—	—	2.5 Ti, 0.8 Nb	11.3	13.0	16.5	20.1	23.6	31
Incoloy 800	0.4	0.05	21	0.5	—	—	1	33	0.7	—	—	0.4 Ti	11.3	12.8	16.4	19.4	22.8	31
Incoloy 802	—	0.3	21	0.5	—	—	1	33	0.4	—	—	—	11.3	13.1	16.2	19.2	22.1	26
Hastelloy R-235	2	0.16	15.5	—	10	—	1	62	1	—	—	5.5 Mo, 2.5 Co	13.9	11.7	14.8	17.6	20	25.5
Nimonic 75	0.3	0.1	20	0.5	5	—	1	bal	1.0	—	—	2 Co, 0.3 Mo, 0.2 Ti	—	17.5	21.0	24.3	—	—
Nimonic 80	1.5	0.07	20	0.2	1	—	1	bal	1.0	—	—	2 Co, 0.3 Mo, 2 Ti	—	12.1	15.5	18.4	23.5	—
Nimonic 90	1.5	0.07	20	0.2	1	—	1	bal	1.5	—	—	17 Co, 0.3 Mo, 2 Ti	—	13.0	16.5	20.0	23.7	—
Nimonic 105	4.5	0.14	14.5	0.2	1	—	1	bal	1.0	—	—	20 Co, 5 Mo, 2 Ti	—	11.6	14.7	17.4	21.2	—
Carbon steels																		
0.08 C, 0.3 Mn	—	0.08	0.045	—	bal	—	0.31	0.07	0.08	—	—	—	59	58	49	40	32	28
0.23 C, 0.6 Mn	—	0.23	—	0.13	"	—	0.64	0.07	0.11	—	—	—	52	51	46	39	32	27
0.42 C, 0.6 Mn	—	0.42	—	0.12	"	—	0.64	0.06	0.11	—	—	—	52	51	46	38	30	27
0.8 C, 0.2 Mn	—	0.84	—	0.02	"	—	0.24	—	0.13	—	—	—	51	49	42	36	31	27

Thermal conductivities of alloys at ambient and elevated temperatures (*contd*)

$\lambda/(\mathrm{W\,m^{-1}\,K^{-1}})$

Alloy	Composition (weight percent)												Temperature K					
	Al	C	Cr	Cu	Fe	Mg	Mn	Ni	Si	W	Zn	Other	273	373	573	773	973	1273
1.2 C, 0.35 Mn	—	1.22	0.11	0.08	bal	—	0.35	0.13	0.16	—	—	—	45	45	40	35	28	26
0.2 C, 1.5 Mn	—	0.23	0.06	0.10	,,	—	1.51	0.04	0.12	—	—	—	46	46	43	37	31	27
Low alloy steels																		
0.3 C, 1 Cr	—	0.32	1.09	0.07	bal	—	0.7	0.07	0.20	—	—	—	49	46	42	36	29	28
0.4 C, 1 Cr, 0.3 Ni	—	0.35	0.88	0.12	,,	—	0.6	0.26	0.21	—	—	—	43	43	41	36	31	28
0.2 C, 0.6 Ni, 0.5 Mo	—	0.20	—	—	,,	—	1.35	0.6	0.25	—	—	0.5 Mo	37	38	37	34	29	—
0.3 C, 0.2 Cr, 3.5 Ni	—	0.33	0.17	0.08	,,	—	0.55	3.47	0.18	—	—	—	36	38	38	34	28	28
0.3 C, 1 Cr, 3.4 Ni	—	0.33	0.80	0.05	,,	—	0.53	3.38	0.17	—	—	—	34	36	37	35	29	28
0.4 C, 1 Cr, 3.6 Ni	—	0.4	0.8	—	,,	—	0.66	3.6	0.2	—	—	—	33	36	37.5	35	28	—
0.5 C, 1 Cr, 3.5 Ni	—	0.34	0.78	0.05	,,	—	0.55	3.53	0.27	—	—	—	33	34	36	34	28	28
0.5 C, 1 Mn, 2 Si	—	0.49	0.04	0.09	,,	—	0.9	0.16	2.0	—	—	—	25	28	31	31	28	26
High alloy steels																		
1.2 C, 13 Mn	—	1.22	0.03	0.07	bal	—	13.0	0.07	0.22	—	—	—	13	15	18	21	23	26
0.3 C, 28 Ni	—	0.28	—	0.03	,,	—	0.9	28.4	0.15	—	—	1 V	13	15	18	21	23	28
4 Cr, 18 W, 1 V	—	0.72	4.26	0.06	,,	—	0.25	0.07	0.30	18.5	—	17 Co	24	26	28	28	27	28
Kovar	—	0.02	—		54	—	0.47	29				—	14.1	14.7	15.6	17.5	19.3	—
Stainless steels																		
304, 321, 347	—	0.05	17.5	—	bal	—	<2	9	<1	—	—	—	14.5	16.5	20	22.5	25.5	29.5
316	—	0.05	17	—	bal	—	<2	12	<1	—	—	2.5 Mo	13.5	15	18.5	21.5	24	28.5
310	—	0.1	24	—	bal	—	<2	20	<1	—	—	—	12	13.5	17.5	20.5	23	—
16 Cr, 20 Ni	—	0.01	16	—	bal	—	1.2	20	0.3	—	—	—	13.6	15.7	18.9	21.5	23.8	26.8
Era–ATV	—	0.5	15	—	bal	—	1.2	27	1.3	2.8	—	—	11	12.5	15.5	—	21.5	—
403, 405, 409	—	0.1	12	—	bal	—	<2	—	<1	—	—	—	—	25	26	27	—	—
430, 434	—	0.05	17	—	bal	—	<2	0.5	<1	—	—	1 Mo	—	22.2	22.9	23.7	24.4	—
410, 420	—	0.3	13	0.1	bal	—	0.5	0.5	0.4	—	—	—	—	—	23.6	24.6	26.3	28
Miscellaneous alloys																		
Platinum 90%, iridium 10%													31	—	—	—	—	—
Platinum 90%, rhodium 10%													38	—	—	—	—	—
Platinum 60%, rhodium 40%													46	51	58	69	—	—
Titanium 92.5%, aluminium 5%, tin 2.5%													7	8.3	10.5	—	—	—
Titanium 96%, aluminium 2%, manganese 2%													9.3	10.5	10.7	—	—	—
Zirconium 93.1%, tin 6.7%, carbon 0.1%													—	8.7	12	—	—	—
Zirconium 97.5%, tin 2.3%, carbon 0.1%													—	11.3	13	—	—	—

Thermal conductivities of alloys used in low temperature applications

$\lambda/(\mathrm{W\,m^{-1}\,K^{-1}})$

Alloy	Composition (Weight percent)												Temperature K					
	Al	C	Cr	Cu	Fe	Mg	Mn	Ni	Si	Sn	Zn	Other	273.2	173	100	50	20	4
Aluminium alloys†																		
Aluminium	99.99	—	—	—	—	—	—	—	—	—	—	—	228	—	295	870	4100	1075
(1C) 1100–0	bal	—	—	1	—	0.05	0.05	—	1.0	—	0.05	—	205	—	228	315	225	45
(N3) 3003 F	bal	—	—	0.12	—	—	1.2	—	—	—	—	—	170	158	143	117	58	11
2219 T81	bal	—	—	6.3	—	0.02	0.3	—	0.20	—	0.1	—	118	—	68	46	26	—
(N8) 5083 0	bal	—	0.15	0.1	—	4.5	0.7	—	0.4	—	0.25	—	110	92	66	39	17	3.3
7039 T61	bal	—	0.2	0.1	—	2.8	0.25	—	0.2	—	4	—	150	—	96	64	30	14.7
Copper alloys																		
Copper	—	—	—	99.96	—	—	—	—	—	—	—	—	400	—	480	1230	3700	1450
ETP Cu	—	—	—	99.95	—	—	—	—	—	—	—	—	395	—	445	880	1320	325
OFHC Cu	—	—	—	99.95	—	—	—	—	—	—	—	—	400	—	460	750	900	200
Brass	—	—	—	70	—	—	—	—	—	—	30	—	106	92	70	45	21	4
Brass	—	—	—	65	—	—	—	—	—	—	35	—	113	—	59	31	14	1
Cupro Nickel	—	—	—	90	—	—	—	10	—	—	—	—	—	—	—	—	—	—
German Silver	—	—	—	62	—	—	—	15	—	—	22	—	23	20	—	24	10	2
Copper Beryllium	—	—	—	98	—	—	—	—	—	—	—	2 Be	—	—	—	—	—	—
Nickel alloys																		
Nickel	—	—	—	—	—	—	—	99.99	—	—	—	—	94	115	154	320	865	138
Inconel X	0.9	0.04	15	—	7	—	0.7	73	0.3	—	—	2.5 Ti	11.4	10	8.7	—	—	—
K Monel	3	0.15	—	30	1	—	0.6	65	0.15	—	—	—	17	14	12	—	—	—
Hastelloy X	—	0.15	22	—	24	—	—	45	—	—	—	9 Mo	9.9	8	6.5	5	3	0.5
Inconel 718	0.4	0.04	18.6	—	18.5	—	—	—	—	—	—	1 Ti, 5 Nb, 3 Mo	10	8.7	7	4.8	2.6	0.4
Steels																		
Armco iron	—	0.02	—	—	99.8	—	0.03	—	—	—	—	—	76	83.6	95	109	65	13
2.5% Ni	—	0.1	—	—	bal	—	0.8	2.5	0.2	—	—	—	38	33	21	—	—	—
3.5% Ni	—	0.1	—	—	bal	—	0.8	3.5	0.2	—	—	—	34	29	19	—	—	—
5% Ni	—	0.1	—	—	bal	—	0.8	5	0.2	—	—	—	31	26	16	—	—	—
9% Ni	—	0.1	—	—	bal	—	0.8	9	0.2	—	—	—	28	23	19	—	—	—
Invar 36% Ni	—	0.07	—	—	bal	—	0.4	36	0.2	—	—	—	13.5	11	7.7	4.3	1.6	1
Stainless steels																		
304/316/321/347	—	0.05	18	—	bal	—	2	8–12	1	—	—	—	14.5	11.5	9	5.5	2	0.3
310	—	0.2	25	—	bal	—	2	20	1.5	—	—	—	11	8.3	6.9	4.3	1.7	—
16 Cr, 20 Ni	—	0.01	16.2	—	bal	—	1.2	20.2	0.28	—	—	1.3 Mo, 0.3 V	13.6	11.5	9.3	6.0	2.4	0.4
15 Cr, 26 Ni	—	0.05	15	—	bal	—	1.4	26	0.4	—	—	—	11.2	—	7.6	5.0	2.2	1.0
Titanium alloys																		
Titanium	—	—	—	—	—	—	—	—	—	—	—	99.9 Ti	22	26	31	40	28	14
Ti 5 Al 2.5 Sn	5	0.1	—	—	—	—	—	—	—	2.5	—	—	7.8	6.1	4.8	3.6	2.0	—
Ti 6 Al 4 V	6	0.05	—	—	0.4	—	0.2	—	—	—	—	4 V	7.0	5.4	4.0	2.5	1.3	—
Ti 13 V 11 Cr 3 Al	3	0.08	11	—	0.3	—	—	—	—	—	—	13 V	7.4	5.0	3.4	1.9	0.9	—

† American Aluminium Association Designation (BS Equivalent in brackets).

Thermal conductivities of elements which are semi-conductors or insulators

$\lambda/(\mathrm{W\,m^{-1}\,K^{-1}})$

Substance	Temperature/K				
	173.2	273.2	373.2	573.2	973.2
Boron	72	32	19	11	10
Carbon:					
Amorphous	1.1	1.5	1.8	2.2	2.5
Diamond.	1700–4900	1000–2600	700–1700	—	—
Graphite	70–220	80–230	75–195	50–130	35–70
Pyrolytic graphite:					
Parallel to planes	3870	2130	1510	936	549
Normal to planes	10.8	6.4	4.4	2.8	1.6
Germanium	113	67	46.5	29	17.5
Iodine	—	0.5	0.4	0.09	—
Phosphorus:					
Black	20	13	—	—	—
White (or yellow)	—	0.25	0.18	0.16	—
Selenium:					
Amorphous	0.23	0.43	—	—	—
Crystalline:					
Parallel to c-axis	6.8	4.8	4.8	—	—
Normal to c-axis	2.0	1.4	1.4	—	—
Silicon	330	168	108	65	32
Sulphur:					
Amorphous	0.18	0.20	—	0.17	—
Polycrystalline	0.39	0.29	0.15	0.17	—
Tellurium:					
Parallel to c-axis	5.1	3.6	2.9	2.4	6.3
Normal to c-axis	2.9	2.1	1.7	1.5	6.3

Thermal conductivities of refractory materials: *Dense, polycrystalline, single-phase compounds*

$\lambda/(\mathrm{W\,m^{-1}\,K^{-1}})$

Material	Chem. formula	Temperature/K				
		(25 °C)	(100 °C)	(500 °C)	(1000 °C)	(1500 °C)
		298	373	773	1273	1773
Oxides:						
Alumina	Al_2O_3	38	35	11	7	6
Aluminosilicate	$Al_6Si_2O_{13}$	—	6	4.5	4	—
Beryllia	BeO	300	220	70	18	14
Calcia	CaO	—	15	8.7	7.8	—
Magnesia	MgO	40	35	16	7	6.5
Spinel	$MgAl_2O_4$	16	15	9	6	—
Silica	SiO_2 (vitreous)†	1.6	1.7	2.1	5.0	—
Thoria	ThO_2	14	12	6	2	2
Titania	TiO_2	—	9.2	4.5	3.3	—
Urania	U_2O	12	8	4.5	3.2	—
Zirconia	ZrO_2 (stabilized)	1.8	1.8	2.0	2.2	2.4
Zircon	$ZrSiO_4$	8	5.8	4.8	4.2	—

Thermal conductivities of refractory materials: *Dense, polycrystalline, single-phase compounds (contd)*

Material	Chem. formula	$\lambda/(\mathrm{W\,m^{-1}\,K^{-1}})$				
		Temperature/K				
		(25 °C)	(100 °C)	(500 °C)	(1000 °C)	(1500 °C)
		298	373	773	1273	1773
Quartz	SiO_2 (single crystal)					
	along c-axis	11	8.3	5	—	—
	normal to c-axis	6.5	5	3.6	—	—
Carbides:						
Boron carbide	B_4C	30	25	21	17	15
Silicon carbide	SiC	110	90	65	45	40
Titanium carbide	TiC†	30	32	36	40	45
Tungsten carbde	WC†	40	—	—	45	50
Zirconium carbide	ZrC	—	—	31	35	—
Nitrides:						
Aluminium nitride	AlN	36	33	23	—	—
Silicon nitride	Si_3N_4 (1% MgO)	30	28	21	14.5	13
Titanium nitride	TiN	—	25	27		
Borides:						
Titanium diboride	TiB_2	—	70	64	—	—
Zirconium diboride	ZrB_2	—	73	67	—	—
Silicon (AXM–5Q)	Si	150	110	45	26	—
Graphite (POCO)	C ($1.77\,\mathrm{Mg\,m^{-3}}$)	108	107	76	—	—

† Values at high temperatures influenced by direct transmission of radiation.

Thermal conductivities of oxide and silicate ceramics: *Commercial products.*

Composition and density may vary, values should be taken as typical of type.

Material type	IEC classification	$\lambda/(\mathrm{W\,m^{-1}\,K^{-1}})$				
		Temperature/K				
		298	373	773	1273	1773
Porcelains and clay-based materials:						
Siliceous	C-110, C-111	1.7–2.1	1.7–2.0	1.8–2.0	1.9–2.0	—
Steatite (normal)	C-220	5.5–6.0	—	2.8–3.7		—
Cordierite (dense)	C-410	1.5–2.5	1.5–2.5	—	—	—
Zircon (dense)	—	7	6	4	3.5	—
Refractory	C-512	2.4	2.3	2.1	2.0	—
Mullite	C-610	2–6	2–6	—	—	—
Oxides:						
Alumina ($>99.5\%$)	C-799	33	29	12	9	7
Alumina 95%	C-795	23	13	9	6	5
Alumina 90%	C-786	17	12	7	5	4
Alumina 85%	C-780	15	12	7	4	3.5
Beryllia $>99.5\%$	C-810	300	220	70	18	14
Magnesia (30% porous)	C-820	10–14	5–8	—	—	—
Thoria (sintered)	—	8–10	6–8	3–5	2–3	—
Titania (sintered)	C-310	2.5–4	—	—	—	—
Urania (sintered)	—	8–10	6.8	4–5	2–3	2
Zirconia (stabilized)	C-830	1.7–2.0	1.7–2.0	1.7–2.0	1.7–2.2	1.8–3.3

For further information on refractory materials consult R. Morrell, *Handbook of Properties of Technical and Engineering Ceramics. Part 1,* 1985 HMSO, London.

For further information on the thermal conductivities of solid materials generally at high temperatures see, for example, R. W. Powell, 'Thermal conductivities of solid materials at high temperatures', *Research,* 1954, vol. 7, 492–501.

Thermal conductivities of miscellaneous solids

The values below are for normal temperature, except where stated (K) and should be regarded as average values for the type of material specified. Values for the commoner polymers will be found on page 287.

Substance	λ $\overline{\mathrm{W\,m^{-1}\,K^{-1}}}$	Substance	λ $\overline{\mathrm{W\,m^{-1}\,K^{-1}}}$
Asbestos cloth	0.125	Ice (268)	2.3
,, insulating board	0.11	,, (173)	3.9
,, paper	0.104	Kapok	0.035
,, wool	0.055	Mica	0.6–0.7
Beeswax	0.25	Mineral wool	0.04
Bitumen	0.17	Paper	0.06
Brick	0.8–1.2	Paraffin wax	0.25
Cardboard	0.21	Plasterboard (gypsum)	0.16
Charcoal	0.2	Plasticine	0.65–0.8
Coal (1400)	0.2	Plastics, solid (see p. 287)	
Concrete: (conductivity increases with density)		Plastics, cellular: (varies with density)	
cellular	0.1–0.2	phenolic foam board	0.031–0.037
lightweight aggregate	0.2–0.6	polystyrene, expanded board	0.031–0.038
dense	0.6–1.8	polystyrene, expanded beads	0.035–0.055
Cotton wool	0.03	polyurethane, gas-filled board (fresh)	0.017–0.020
Cork, baked slab	0.038–0.046	Polyurethane, gas-filled board (aged)	0.027
,, granular	0.04	polyvinyl chloride, rigid foam board	0.035–0.041
Diatomaceous powder	0.07	urea formaldehyde foam	0.030–0.032
Ebonite, solid	0.17	Plywood	0.125
,, cellular	0.03	Porcelain	1.5
Felt	0.04	Pyrophyllite, normal to plane	2.0
Fibreboard, insulating	0.055	Rubber, cellular	0.045
,, hardboard	0.125	,, natural	0.15
Glass, borosilicate crown	1.1	,, silicone	0.25–0.4
,, double extra dense flint	0.55	Sand, silver	0.3–0.4
Glass, light flint	0.85	Silica aerogel powder	0.024
,, Pyrex	1.1	Soil, clay	1.1
Glass fibre: (conductivity varies with density)		Timber, ordinary	0.14–0.17
		,, balsa	0.055
wool blanket	0.035–0.07	Vermiculite granules	0.065
rigid board	0.030–0.036	Wool	0.05

Thermal conductivities of liquids

Values below are for the thermodynamic temperature (K) shown in brackets. Linear relationships mostly hold for range covered. Liquid metal values are given on p. 62.

Liquid	$\lambda/(\mathrm{W\,m^{-1}\,K^{-1}})$	Liquid	$\lambda/(\mathrm{W\,m^{-1}\,K^{-1}})$
Acetone	0.198 (193), 0.146 (333)	Glycerine . . .	0.286 (273), 0.292 (333)
Aniline	0.172 (293)	Medicinal paraffin	0.127 (273), 0.125 (423)
Argon	0.1260 (84.2), 0.1216 (87.3)	Methane . . .	0.2153 (93.2), 0.1964 (108.2)
Benzene . . .	0.147 (293), 0.137 (323)	Methyl alcohol .	0.223 (233), 0.186 (333)
n-Butyl alcohol . .	0.167 (213), 0.106 (353)	Nitrogen . . .	0.1511 (69.1), 0.1480 (71.4)
Carbon monoxide .	0.1589 (72.0), 0.1421 (80.8)	Oil, cylinder . .	0.152 (293), 0.142 (473)
Carbon tetrachloride	0.115 (253), 0.102 (333)	, transformer .	0.136 (273), 0.127 (373)
Dichlorodifluoro-		n-Propyl alcohol .	0.168 (233), 0.148 (353)
methane† . . .	0.09 (253), 0.073 (293)	Toluene . . .	0.159 (193), 0.119 (353)
Ethyl alcohol. . .	0.189 (233), 0.150 (353)	Water	0.561 (273), 0.673 (353)
Ethyl benzene . .	0.152 (193), 0.117 (353)		0.686 (378–433), 0.598 (542)
Ethyl glycol . . .	0.252 (273), 0.264 (373)	Xenon	0.07 (173), 0.05 (223)

Thermal conductivities of some liquids and their vapours ($\lambda/\mathrm{W\,m^{-1}\,K^{-1}}$)

In the table below the thermal conductivities of liquids at their equilibrium saturation pressure are compared with the values for their dilute vapours at the same temperatures.

	He	H_2	A	C_6H_6	H_2O	KNO_3
Temperature (K)	4	20	90	298	373	683
Vapour	1.25×10^{-4}	0.0145	0.0057	0.0070	0.0217	—
Liquid	0.0275	0.1178	0.1198	0.1463	0.6819	0.425

Thermal conductivities of gases

The thermal conductivity of a gas is independent of pressure at normal pressures. It increases at high pressures and decreases at low pressures, e.g. for air below about 1 mmHg. Values are given for a pressure of 1 atm.

$$\lambda/(10^{-2}\,\mathrm{W\,m^{-1}\,K^{-1}})$$

Gas	Temperature/K					Gas	Temperature/K			
	73.2	173.2	273.2	373.2	1273.2		173.2	273.2	373.2	1273.2
Argon . .	—	1.09	1.63	2.12	5.0	Air	1.58	2.41	3.17	7.6
Bromine . .	—	—	0.4	0.6	—	Ammonia . .	—	2.18	3.38	—
Chlorine .	—	—	0.79	1.15	—	Carbon dioxide	—	1.45	2.23	7.9
Fluorine . .	—	1.56	2.54	3.47	—	Carbon monoxide	1.51	2.32	3.04	—
Helium . .	5.95	10.45	14.22	17.77	41.9	Ethane . . .	—	1.80	—	—
Hydrogen .	5.09	11.24	16.82	21.18	—	Ethylene . .	—	1.64	—	—
Krypton. .	—	0.57	0.87	1.15	2.9	R12 (CF_2Cl_2) .	—	0.85	1.35	—
Neon . .	1.74	3.37	4.65	5.66	12.8	Hydrogen sulphide	—	1.2	—	—
Nitrogen .	—	1.59	2.40	3.09	7.4	Methane . .	1.88	3.02	—	—
Oxygen . .	—	1.59	2.45	3.23	8.6	Nitric oxide .	1.54	2.38	—	—
Radon . .	—	—	0.33	0.45	—	Nitrous oxide .	—	1.51	—	—
Xenon . .	—	0.34	0.52	0.70	1.9	Sulphur dioxide .	—	0.77	—	—
						Water vapour .	—	1.58	2.35	—

I.W.

1.6 Acoustics

1.6.1 The velocity and attenuation of sound

Gases and vapours

The attenuation of plane sound waves in neper/unit length is $\alpha = (1/2d)\log_e (I_0/I_d)$, where the initial intensity I_0 has decreased to I_d after traversing distance d. When expressed in decibels per unit length, convenient for practical application, the value is 8.686α. In normal monatomic gases α varies classically as f^2, f being the frequency, but in polyatomic gases relaxation phenomena may dominate over classical absorption. (See Herzfeld and Litovitz, *Absorption and Dispersion of Ultrasonic Waves*, 1959, Academic Press). For gases at moderate pressure the velocity of sound $c = \sqrt{(\gamma p/\rho)}$ or alternatively $c = \sqrt{(\gamma RT/M)}$, p being the ambient gas pressure, γ the ratio of the specific heat at constant pressure to that at constant volume, R the gas constant, T the absolute temperature and M the molecular weight. Thus, to first order, c is proportional to $\sqrt{T}$ and is independent of gas pressure; at higher pressures, however, the value is changed due to contributions from the second and higher virial coefficients. The velocity of sound waves bounded by walls or tubes is less than the free-space value. Velocity dispersion is observed for sound propagation in many polyatomic gases due to molecular relaxation: above the relaxation frequency the rotational or vibrational degrees of freedom are not excited, the specific heats are modified and the velocity of sound is increased. The table below gives experimentally determined values of sound velocity in gases and vapours, selected from the literature. Where known, both the low-frequency and high-frequency limiting values of velocity are given, the latter being in brackets.

Velocity of sound in gases and vapours

Gas	$t/°C$	$c/(\text{m s}^{-1})$	Gas	$t/°C$	$c/(\text{m s}^{-1})$
Air	(see p. 72)		Freon 22 (CHCl F$_2$) . .	17	179 (193)
Acetaldehyde	0	278	Freon 113 (CCl$_2$FCCl F$_2$) .	53	124 (139)
Acetylene	0	329	Helium	0	972.5
Ammonia	30	440	Hydrogen bromide . . .	0	200
Argon	0	307.85	Hydrogen chloride . . .	0	296
Benzene	90	200	Hydrogen iodide . . .	0	157
Bromine	58	149	Hydrogen sulphide . . .	24	309
Carbon dioxide . . .	51	280 (293)	Krypton	30	224
Carbon disulphide . .	35	206	Methane	41	466
Carbon tetrachloride . .	22	133 (146)	Neon	30	461
Chloroform	22	154 (166)	Nitric oxide	16	334
Cyclohexane	30	181 (200)	Nitrogen	29	354.4
Deuterium	0	888 (969)	Nitrous oxide	25	268 (281)
Diethyl ether	40	187	Oxygen	30	332.2
Ethane	31	316 (335)	Sulphur hexafluoride . .	11	133 (147)
Ethylene	20	327	Water	100	477.5
Fluorine	102	332 (339)	Water (6 MPa)	350	571
Freon 11 (CCl$_3$F) . . .	18	143 (154)	Water (heavy)	100	450
Freon 12 (CCl$_2$F$_2$) . .	17	140 (152)			

Attenuation of sound in air

The attenuation of sound in air at 20 °C due to viscous, thermal and rotational loss mechanisms is $1.6 \times 10^{-10}f^2$ dB m^{-1} (see Sivian, *J. Acoust. Soc. Am.*, 1947, **19**, 914–916). However, losses due to vibrational relaxation of oxygen molecules are generally much greater than those due to the classical processes, and the attenuation of sound varies significantly with temperature, water-vapour content and frequency. The table gives values of attenuation in dB km^{-1} for a temperature of 20 °C and a pressure of 1 atmosphere (see Bazley, NPL Report, Ac.74, 1976). For the variation of attenuation with ambient pressure see Harris, *J. Acoust. Soc. Am.*, 1967, **43**, 530–532.

Attenuation of sound in air (dB km^{-1})

f/kHz	Relative humidity (%)								
	10	20	30	40	50	60	70	80	90
1	12	6	5	4	4	5	5	5	6
1.25	18	9	7	6	5	6	6	6	6
1.6	28	14	9	8	7	7	7	7	7
2	41	21	14	11	9	9	9	9	9
2.5	59	31	20	16	13	12	11	11	10
3.15	82	47	31	23	19	17	15	14	14
4	110	72	48	36	29	25	22	20	19
5	139	106	72	54	43	36	32	28	26
6.3	169	153	110	83	66	55	48	42	39
8	197	216	167	128	103	86	74	65	59
10	220	286	241	192	156	130	112	99	89
12.5	243	362	337	281	233	197	170	150	134
16	269	448	469	419	360	310	270	239	214
20	298	526	606	580	520	459	405	362	326
25	338	605	752	775	731	666	601	543	493
31.5	400	697	913	1005	1002	951	883	813	749
40	499	818	1094	1268	1336	1327	1276	1206	1131
50	643	977	1297	1544	1693	1751	1743	1694	1625
63	878	1221	1575	1887	2121	2266	2332	2336	2298
80	1266	1616	1994	2356	2668	2908	3070	3160	3189
100	1841	2194	2587	2983	3352	3672	3927	4112	4231

Velocity of sound in air

The velocity of free progressive sound waves at a frequency of 1 kHz in dry air containing 0.03% CO_2 by volume is 331.45 ± 0.05 m s^{-1} at a temperature of 0 °C and a pressure of 1 atmosphere (see Smith, *J. Acoust. Soc. Am.*, 1953, **25**, 81–86). The effect of water-vapour content on sound velocity is complex. As moisture is added to dry air the velocity at first decreases, owing to dispersion caused by molecular relaxation, but then begins to increase linearly with increasing humidity as the molecular weight of the air/water-vapour mixture and, to a lesser extent, the ratio of specific heats decreases (see Delany, *Acustica*, 1977, **38**, 201–223, also Morfey and Howell, *J. Acoust. Soc. Am.*, 1980, **68**, 1525–1527).

Velocity and attenuation of sound in liquids

The velocity of dilatational waves in unbounded fluids is $c = (\beta_a \rho)^{-1/2}$, β_a being the adiabatic compressibility and ρ the density. In normal fluids the classical attenuation is proportional to f^2, f being the frequency, but relaxation phenomena may dominate; the attenuation generally decreases with increasing temperature. Values for various liquids are given below, selected from numerous sources; f is in Hz and α is in neper m^{-1}. Where known, the frequency at which the attenuation was measured is given in brackets, in MHz.

Velocity and attenuation of sound in liquids

Liquid	t °C	c m s^{-1}	dc/dt m s^{-1} K^{-1}	(α/f^2) 10^{-15} neper m^{-1} Hz^{-2}
Acetic acid	19.6	1173	—	—
Acetone	25	1170	−4.5	∼35
Aniline	25	1640	−3.6	∼50
Argon	−243	840	−6.5	—
Benzene	25	1300	−4.7	870 (<70)
Bismuth	280	1651	−0.13	∼8 (20)
n-Butanol	25	1242	−3.4	85 (25)
Caesium	40	980	−0.31	112 (30)
Carbon disulphide	25	1141	−3.2	∼5600 (3)
Carbon tetrachloride	25	921	−3.0	535
Chloroform	25	984	−3.5	370
Chlorobenzene	25	1270	−3.9	140 (<200)
Cyclohexane	20	1280	−5.4	∼180
Cyclohexanol	25	1465	−3.7	∼500 (<45)
Ethyl alcohol	25	1145	−3.3	51
Freon ($C_5$1–12)	25	524	−3.4	—
Glycerine	25	1920	−1.9	∼3000 (4−12)
n-Hexanol	25	1303	−3.4	—
Hydrogen	−255	1246	−26	—
Indium	160	2313	−0.29	—
Lead	340	1766	−0.28	∼9 (20)
Mercury	25	1449	−0.46	5.6 (100)
Methane	−170	1420	−9.7	—
Methyl alcohol	25	1103	−3.3	32
Naphthalene	100	1248	−2.5	—
Neon	−243	540	−16.9	—
Nitrogen	−202	912	−9.8	—
Oil (castor)	25	1490	−3.1	∼5300 (3)
Oil (lubricating)	25	1461	−3.4	(at 2 MHz)
Oil (sperm)	32	1411	—	—
Oxygen	−202	1056	−7.8	—
n-Pentanol	25	1277	−3.3	—
Potassium	80	1869	−0.49	34 (30)
n-Propanol	25	1207	−3.3	67
Pyridene	25	1417	−4.2	—
Rubidium	50	1247	−0.38	78 (30)
Sodium	110	2520	−0.52	12 (30)
Tin	240	2471	−0.25	∼6 (20)
Toluene	25	1306	−4.3	84 (15−200)
Water (distilled)		see below		
Water (heavy)	20	1383.6	+3.2	32 (50)
Water (sea)		see p. 74		
Zinc	450	2780	−4.3	∼4 (20)

Velocity and attenuation in distilled water

Air-free distilled water is non-dispersive but the temperature dependence of sound velocity is anomalous, mainly because of the temperature dependence of the adiabatic compressibility of the water molecule itself; the maximum velocity occurs at a temperature of approximately 74.16 °C. The pressure coefficient of velocity is approximately +0.156 m s^{-1} atm^{-1} at 20 °C (see Del Grosso and

Mader, *J. Acoust. Soc. Am.*, 1972, **52**, 1442, also Chen and Millero, *J. Acoust. Soc. Am.*, 1976, **60**, 270). The estimated uncertainty in the tabulated values is less than 0.02 m s^{-1}. The attenuation of sound in distilled water is proportional to f^2, at least for the range 3–70MHz, but is considerably in excess of that expected from the shear viscosity alone. The excess is attributed to structural relaxation between an open type of molecular structure, where each water molecule has four nearest neighbours tetrahedrally arranged about it, and a closer type of packing in which there are twelve nearest neighbours (see Litovitz and Carnevale, *J. Appl. Phys.*, 1955, **26**, 816–820). Data in the table are based largely on Pinkerton, *Proc. Phys. Soc.*, 1949, **62**, 129–141, and on Fox and Rock, *Phys. Rev.*, 1946, **70**, 68–73.

Velocity and attenuation of sound in distilled water

$t/°C$	0	10	20	30	40	50	60	70	80	90	100
$c/(\text{m s}^{-1})$	1402.39	1447.27	1482.34	1509.13	1528.86	1542.55	1550.99	1554.80	1554.49	1550.48	1543.11
(α/f^2) $(10^{-15} \text{ neper m}^{-1} \text{ Hz}^{-2})$	54	37	26	19	15	12	9.9	8.6	7.7	7.2	7.1

Velocity and attenuation in sea water

The velocity of sound in sea water is a function of the temperature, the excess pressure, and the salinity, their relative importance being in that order (See Lovett, *J. Acoust. Soc. Am.*, 1978, **63**, 1713–1718); values in the table relate to surface depth and a salinity of 3.5%. The attenuation of sound in sea water is much higher than in distilled water due to complex relaxation phenomena involving $MgSO_4$, $MgCO_3$ and $B(OH)_3$ (see Mellen, Browning and simmonds, *J. Acoust. Soc. Am.*, 1980, **68**, 248–257); values in the table, relating to surface depth, a pH of 8.0 and a salinity of 3.5%, are based on Schulkin and Marsh, *J. Acoust. Soc. Am.*, 1978, **63**, 43 for low frequencies, and Schulkin and Marsh, *J. Acoust. Soc. Am.*, 1962, **34**, 864–865, with correction in *J. Acoust. Soc. Am.*, 1963, **35**, 739 for high frequencies.

Velocity of sound in sea water

$t/°C$	0	5	10	15	20	25	30
$c/(\text{m s}^{-1})$	1449.0	1470.6	1489.8	1506.7	1521.5	1534.4	1545.5

Attenuation of sound in sea water (dB km^{-1})

$t/°C$	*Frequency/kHz*					
	0.5	1	2	5	10	20
0	0.03	0.07	0.14	0.41	1.3	4.6
10	0.02	0.07	0.14	0.33	0.92	3.2
20	0.02	0.06	0.13	0.30	.70	2.2
30	0.01	0.05	0.13	0.29	0.58	1.6

Velocity and attenuation in solids

In isotropic solids, both shear (transverse) and longitudinal waves can be propagated. The velocity of shear waves in an extensive medium is $c_S = \sqrt{(G\rho)} = \sqrt{\{E/2\rho(1+\sigma)\}}$, E being Young's modulus, G the rigidity modulus and σ Poisson's ratio, and this is also the velocity of torsional waves in thin cylindrical bars. The velocity of longitudinal or irrotational waves in an extensive medium is $c_L = \sqrt{\{(K + \frac{4}{3}G)/\rho\}} = \sqrt{\{E(1-\sigma)/\rho(1-2\sigma)(1+\sigma)\}}$, K being the bulk modulus. In straight uniform bars and in tubes thin compared with a wavelength, the velocity of longitudinal waves is $c_E = \sqrt{(E/\rho)}$. Surface waves propagating along the surface of an extensive solid are generally known as Rayleigh waves and propagate with velocity $c_{SR} = ac_S$, where a is the least positive root of the equation

$$\frac{a^6}{8(1-a^2)} + a^2 = \frac{1}{1-\sigma}$$

See Bradfield, *NPL Notes on Applied Science No. 30,* 1964 (HMSO).

Variation of a with Poisson's ratio

σ	0.20	0.25	0.30	0.35	0.40
a	0.9110	0.9194	0.9274	0.9350	0.9422

In anisotropic solids, which may have as many as 21 independent elastic constants, there may exist, for a given direction of the wave normal, three distinct displacement vectors each associated with a distinct plane wave velocity. Of the three waves, one is analogous to the longitudinal and the others to transverse waves in the isotropic case. The directions of the respective displacement vectors are mutually orthogonal and are in general oblique to the wave normal. A generalized Rayleigh-type wave may be propagated, in a limited number of directions, along the surface of an extensive anisotropic medium. See Kolsky, *Stress Waves in Solids*; Musgrave, *Proc. Roy. Soc., A,* 1954, **226**, 339; Synge, *Proc. Roy. Irish Acad., A,* 1956, **58**, 13; Stoneley, *Proc. Roy. Soc., A,* 1955, **232**, 447.

Velocities and attenuation constants for various solids in the region of 20 °C are given below. In the case of metals, factors such as texture, cold work, stress, hardening, tempering and aging can cause significant departures from the values given in the table. Properties of plastics vary considerably with molecular weight, with additives and with temperature. In view of this, many values are approximate and relate to materials of variable composition; the velocities given should therefore only be regarded as typical. The bracketed figures following the attenuation are frequencies in MHz. Composition of materials is given as percentage by weight.

Velocity and attenuation of waves in solids

Material	Velocity/(m s^{-1})				α longit. waves neper m^{-1}
	c_L longitudinal bulk waves	c_E irrot. rod waves	c_S shear waves	c_{SR} Rayleigh waves	
Aluminium	6374	5102	3111	2906	0.40 (10)
ADP crystal, X-cut	6250	—	—	—	10.8 (10)
,, Y-cut	6250				
,, Z-cut	4300	3500	—	—	9.69 (10)
Barium titanate ceramic . . .	4000	—			—
Beryllium.	12890	—	8880	—	—
Bone, human tibia	4000	—	1970	—	460 (2.9)
Brass	4372	3451	2100	1964	—
Brick	—	3650	—	—	—
Butyl rubber/carbon (100/40) . .	1600	—	—	—	133 (0.35)
Cadmium	2780	2400	—	—	—
Cellulose acetate butyrate . . .	2080	—	—	—	103 (2.5)

Velocity and attenuation of waves in solids (*contd*)

Material	Velocity/(m s^{-1})				α longit. waves neper m^{-1}
	c_L longitudinal bulk waves	c_E irrot. rod waves	c_S shear waves	c_{SR} Rayleigh waves	
Chromium	6608	6229	4005	3655	—
Concrete	4250–5250	—	—	—	—
Constantan	5177	4276	2625	2445	—
Copper	4759	3813	2325	2171	—
Cork	—	500	—	—	—
Duralumin	6398	5120	3122	2917	1.23 (10)
Ebonite	2500	—	—	—	—
Glass (crown)	5660	5342	3420	3127	2 (10)
,, (heavy flint)	5260	4717	2960	2731	—
,, (pyrex)	5640	5170	3280	—	—
Gold (hard-drawn)	3240	2030	1200	—	—
Granite	5400	—	—	—	60 (0.6)
Ice (maximum density, polar ice sheets, firn temperature − 20 °C) .	3840	—	—	—	—
Invar (36 Ni, 63.8 Fe, 0.2 C) . .	4657	4216	2658	2447	—
Iron (soft)	5957	5189	3224	2986	—
,, (cast)	4994	4477	2809	2590	—
Lead	2160	1188	700	—	—
Magnesium	5823	5082	3163	2930	—
Manganese	4600	3830	—	—	—
Marble	—	3810	—	—	—
Molybdenum	6475	5636	3505	3248	—
Monel metal	5350	4400	2720	—	—
Neoprene	1510	—	—	—	230 (2.5)
Neoprene/carbon (100/60) . .	1690	—	—	—	—
Nickel (unmag. soft) . . .	5608	4787	2929	2722	—
,, (unmag. hard) . . .	5814	4974	3078	2857	—
Niobium	5068	3497	2092	1970	—
Ni-Span-C	—	4831	2799	—	—
Nylon	2680	—	—	—	11.5 (1)
Perspex	2700	2177	1330	1242	57 (2.5)
Platinum	3260	2800	1730	—	—
Polythene (polyethylene) . .	2000	—	—	—	54 (1)
Polystyrene	2350	1840	1120	1047	23 (2.5)
Polyvinyl chloride . . .	2300	—	—	—	3.5 (0.35)
Polyvinyl chloride acetate . .	2250	—	—	—	1270 (10)
Polyvinyl formal	2680	—	—	—	115 (2.5)
Polyvinylidene chloride . . .	2400	—	—	—	207 (2.5)
Quartz (crystal) X-cut . . .	5720	5440	—	—	0.0127 (10)
,, (fused)	5970	5759	3765	3410	—
Rubber (natural) . . .	1600	—	—	—	15 (0.35)
Rubber/carbon (100/40) . .	1680	—	—	—	36.6 (0.35)
Rubber (RTV silicone) . . .	900–1050	—	—	—	—
Sandstone	2920	2820	1840	—	—
Silica (fused)	5968	5760	3764	—	—
Silver	3704	2806	1698	1592	—
Slate	—	4510	—	—	—
Steel (mild)	5960	5196	3235	2996	—
,, (tool) hardened . . .	5874	5116	3179	2945	4.94 (10)
,, (stainless)	5980	5282	3297	3049	—

Velocity and attenuation of waves in solids *(contd)*

Material	*Velocity*/$(m\ s^{-1})$				α *longit. waves* neper m^{-1}
	c_L *longitudinal bulk waves*	c_E *irrot. rod waves*	c_S *shear waves*	c_{SR} *Rayleigh waves*	
Tantalum	4159	3337	2036	1902	—
Teflon	1400	—	—	—	—
Tin	3380	2626	1594	1491	—
Titanium	6130	5164	3182	2958	—
Tourmaline (crystal) Z-cut . . .	7250	7170	—	—	—
Tungsten (annealed)	5221	4619	2887	2668	—
,, carbide	6655	6223	3984	3643	—
,, (drawn)	5410	4320	2640	—	—
Uranium	3370	—	1940	—	—
Vanadium	6023	4584	2774	2600	—
Wood (Ash) with grain	—	4670	—	—	—
,, across grain	—	1390	—	—	—
,, (Oak) with grain	—	4100	—	—	—
,, (Pine) with grain	—	3600	—	—	—
Zinc (rolled)	4187	3826	2421	2225	—
Zirconium	4650	—	2250	—	—

M.E.D.

1.6.2 Physiological and subjective acoustics

The decibel (dB). Two sounds having intensities I_1 and I_2 differ in intensity level by n decibels where $n = 10 \log_{10} (I_1/I_2)$. Measurement in terms of intensity is seldom convenient and it is usual to express differences in terms of sound pressure, in which case $n = 20 \log_{10} (p_1/p_2)$, p_1 and p_2 being the r.m.s. sound pressures of the two sounds. The **sound pressure level** (s.p.l.) of a sound or noise is $20 \log_{10} (p_1/p_0)$ where p_1 is the r.m.s. sound pressure and p_0 is the reference sound pressure 2×10^{-5} Pa.

Threshold of hearing, for a sound of given character, is the minimum value of the r.m.s. sound pressure which excites the sensation of hearing. Normal threshold of hearing is the modal value of the thresholds of hearing of a group of otologically normal subjects. Normal threshold for binaural listening in a free progressive wave incident from directly ahead of the listener is called **minimum audible field** (MAF) and defined in terms of the sound pressure in the undisturbed wave. Values of MAF in the following table are taken from ISO Recommendation R 226–1961 (see Robinson and Dadson, *Br. J. appl. Phys.*, 1956, **7**, 166). Normal threshold for monaural listening to a sinusoidal sound pressure produced by an earphone is called **minimum audible pressure** (MAP), the sound pressure being measured at the entrance to the ear canal (see Dadson and King, *J. Laryng. Otol.*, 1952, **66**, 366).

Hearing level is measured in decibels by the difference between a listener's threshold of hearing and the audiometric zero, for monaural earphone listening to pure tones. The audiometric zero is specified in terms of the s.p.l. set up in an artificial ear or coupler (see BS 2497: Part 2: 1969 and BS 2497: part 3: (1972).

    The **threshold of hearing** for frequencies above approximately 1 kHz increases with age at a rate which increases with frequency (see Robinson and Sutton, *Audiology*, 1979, **18**, 320–334).

Minimum audible field

Frequency/Hz	MAF dB	Frequency/Hz	MAF dB
20	74.3	500	6.0
30	58.1	700	4.7
40	48.4	1 000	4.2
60	36.8	1 500	—
80	29.8	2 000	1.0
100	25.1	3 000	−2.9
125	—	4 000	−3.9
140	18.9	6 000	4.6
200	13.8	8 000	15.3
250	11.2	10 000	16.4
300	9.4	12 000	12.0
400	7.2	15 000	24.1

The subjective quality known as the **loudness** of a noise or sound is measured by reference to the free field s.p.l. of a pure tone of frequency 1000 Hz which is judged by an observer to be equally loud. The **loudness level** is the value of the s.p.l. so found by a group of normal listeners, and is expressed in **phon**, the value being numerically equal to the free field s.p.l. of the reference tone, a 1000 Hz tone presented as a free progressive plane wave reaching the listener from directly in front. A loudness level of 4 phon corresponds to the threshold of hearing, and above 130 phon pain is felt. For typical loudness levels of common noises see Parkin, *Acustica*, 1957, **7**, 57.

Equal-loudness contours. Loudness levels are frequently shown in the form of parametric curves of constant loudness known as equal-loudness contours. Values in the table below refer to otologically normal listeners between the ages of 18 and 25 (see Robinson and Dadson, *Br. J. appl. Phys.*, 1956, **7**, 166, and ISO Recommendation R 226–1961).

Sone scale. The sone scale is arithmetic and provides a numerical designation of loudness of sounds or noises that is proportional to the subjective magnitude, as estimated by normal observers. For all practical purposes it has been found experimentally that over the range 20 to 120 phons the loudness is given to sufficient accuracy by $S = 2^{(P-40)/10}$, S being the loudness in sones and P the loudness level in phons. The loudness of a sound of 1 sone corresponds to a loudness level of 40 phons, and a twofold change in loudness corresponds to an interval of 10 phons.

Loudness of complex sounds. For experimental data on the loudness of frontally incident and diffuse sound fields for bands of noise, see Robinson and Whittle, *Acustica*, 1964, **14**, 24, and ISO Standard 454–1975. For complex sounds the phon is not a convenient unit since a team of normal observers is required to assess each noise. When direct measurement of loudness level is impracticable, the value may be calculated from the sound pressure levels in a number of contiguous bands of the frequency spectrum. Several methods are available, of varying degrees of complexity (see Zwicker, *Frequenz*, 1959, **13**, 234; Stevens, *J. Acoust. Soc. Am.*, 1961, **33**, 1577 and ISO Standard 532–1975).

Loudness level of pure tones in phons

Sound pressure level/dB	Frequency/Hz												
	25	50	100	200	500	1000	2000	3000	4000	6000	8000	10 000	15 000
0	—	—	—	—	—		3	7	8	—	—	—	—
10	—	—	—	—	9	10	13	16	18	10	—	—	—
20	—	—	—	12	20	20	22	26	27	20	10	9	—
30	—	—	11	25	32	30	32	36	37	30	21	21	14
40	—	1	25	38	43	40	42	46	48	40	32	33	30
50	—	18	38	49	53	50	52	57	58	50	42	44	45
60	—	34	51	61	64	60	62	67	69	61	53	55	58
70	15	49	63	72	74	70	73	78	79	71	63	65	69
80	34	63	75	82	84	80	84	90	90	81	74	75	79
90	53	76	86	92	93	90	95	101	102	92	84	85	88
100	69	88	97	102	102	100	106	113	113	102	94	94	94
110	84	100	107	110	111	110	117	124	125	112	103	103	100
120	98	111	116	119	120	120	129	137	136	123	113	112	103
130	109	121	125	127	128	130	—	—	—	—	—	—	—

M.E.D.

1.6.3 Preferred frequencies for acoustical measurements

For many acoustical measurements it is convenient to adopt constant-percentage increments of frequency. To simplify intercomparison of results it is recommended that when data are to be presented at discrete frequencies, appropriate frequencies from the following series be adopted; the order of preference is indicated by the type of print, in the order, bold, italic and roman. When concerned with frequency bands, the frequencies listed in the table should be the geometric centre frequencies of those bands. Extension of the table in either direction is effected by successive multiplication or division by 1000 (see ISO Standard 266–1975).

Preferred frequencies/Hz	1/1 oct.	1/2 oct.	1/3 oct.	Preferred frequencies/Hz	1/1 oct.	1/2 oct.	1/3 oct.	Preferred frequencies/Hz	1/1 oct.	1/2 oct.	1/3 oct.
16	×	×	×	*180*		×		**2 000**	×	×	×
20			×	*200*			×	*2 500*			×
22.4		×		**250**	×	×	×	2 800		×	
25			×	*315*			×	*3 150*			×
31.5	×	×	×	355		×		**4 000**	×	×	×
40			×	*400*			×	*5 000*			×
45		×		**500**	×	×	×	5 600		×	
50			×	*630*			×	*6 300*			×
63	×	×	×	710		×		**8 000**	×	×	×
80			×	*800*			×	*10 000*			×
90		×		**1000**	×	×	×	11 200		×	
100			×	*1250*			×	*12 500*			×
125	×	×	×	1400		×		**16 000**	×	×	×
160			×	*1600*			×				

M.E.D.

1.6.4 Architectural acoustics

The sound absorption coefficient of a material is $\alpha = (1 - r)$, where r, the **sound energy reflection coefficient**, is the ratio of sound energy reflected from the surface of the material to that incident upon it. Values for a specific material depend upon frequency and upon the angle of incidence of the sound. When the sound field is approximately diffuse the corresponding quantity is called the **reverberation absorption coefficient**; this may be determined in accordance with BS 3638:1963 when the cofficient is designated α(ISO). Since values obtained in accordance with the new procedure are not yet available, those in the following table have been taken from *Sound Absorbing Materials*, by Evans and Bazley, 1960, reprinted 1978 (NPL). Absorption depends on mounting and other details of construction and the following values should be regarded only as typical.

Reverberation absorption coefficients

Material	Thickness (mm)	Frequency/Hz					
		125	250	500	1000	2000	4000
Acoustic plaster	13	0.15	0.20	0.35	0.60	0.60	0.50
Acoustic tiles (perforated fibreboard)	18	0.10	0.35	0.70	0.75	0.65	0.50
Asbestos (sprayed)	25	0.10	0.30	0.65	0.85	0.85	0.80
Brickwork	—	0.02	0.02	0.03	0.04	0.05	0.07
Carpet (Axminster)	8	—	0.05	0.15	0.30	0.45	0.55
Carpet on underlay	14	—	0.05	0.20	0.40	0.60	0.65
Curtain (velour, draped)	—	0.14	0.35	0.55	0.72	0.70	0.65
Glass fibre (resin-bonded)	25	0.10	0.25	0.55	0.70	0.80	0.85
Glass wool (uncompressed)	25	0.10	0.25	0.45	0.60	0.70	0.70
Mineral wool	25	0.10	0.25	0.50	0.70	0.85	0.85
Polystyrene, expanded (rigid backing)	13	0.05	0.05	0.10	0.15	0.15	0.20
Polystyrene, expanded (on 50 mm battens)	13	0.05	0.15	0.40	0.35	0.20	0.20
Polyurethane foam (flexible)	50	0.25	0.50	0.85	0.95	0.90	0.90
Snow	25	0.15	0.40	0.65	0.75	0.80	0.85
Wood panelling (oak, on 25 mm battens)	13	0.20	0.10	0.05	0.05	0.05	0.05

Insulation against airborne noise. The **sound reduction index** (SRI) is the ratio of sound energy incident on a partition to that which is transmitted through the partition, expressed in decibels. The values vary with frequency and angle of incidence; for comparative purposes the mean value over a frequency range 100–3150 Hz is often used (see BS 2750:1980). Simple dependence of SRI on mass and frequency above the fundamental resonance is not generally observed due to 'coincidence' effects associated with flexural waves in the partition (see Cremer, *Akustische Zeit*, 1942, **7**, 81). For single homogeneous partitions the mean sound reduction index (100–3150 Hz) is given roughly by $10 + 15 \log_{10} m$, m being the superficial density in kg m^{-2}.

Typical values given in the table refer to reverberant sound and are taken from Bazley, *The Airborne Sound Insulation of Partitions*, 1966. Reprinted 1978 (NPL).

Sound reduction indices (dB)

Type of partition (figures in brackets are thicknesses in mm)	m $\overline{\text{kg m}^{-2}}$	Frequency/Hz						
		100	200	400	800	1600	3150	Mean
Brick, unplastered (115) . .	220	34	37	40	41	51	57	43
Brick (225) plastered both faces	480	48	40	45	50	61	67	52
Cavity wall, brick (50 × 115) airspace (50) plastered both faces, butterfly ties . . .	480	30	36	45	58	65	82	53
Clinker concrete wall (75) unplastered	100	16	15	20	23	29	38	23
Clinker concrete wall (75) plastered both faces . . .	150	22	33	36	47	55	56	42
Concrete floor (100) reinforced	230	36	40	38	47	55	65	47
Wood joist floor, boarding (22) nailed to joists (225 × 50) plasterboard ceiling (9.5) skim coat plaster	80	16	26	34	35	42	41	32
Fibreboard (12)	4	11	15	18	22	27	30	21
Plasterboard (9.5)	10	13	19	22	28	34	35	25
Plasterboard (9.5) each side of (100 × 50) studs (airspace 100)	25	12	23	32	36	44	46	32
Plywood (6)	4	8	10	15	20	25	28	18
Steel, 16G	12	14	20	25	31	37	43	28
Plate glass (6)	17	18	23	24	30	30	35	27
Window glass (3)	7	18	12	20	26	30	28	22
Window glass, double, airspace (50)	14	19	19	27	36	47	52	33
Window glass, double, airspace (180)	14	28	25	36	43	50	53	39

M.E.D.

1.6.5 Musical acoustics

Standard musical pitch is based on a frequency of 440 Hz for A in the treble stave (a′) (see ISO Standard 16–1975 and BS 880:1950). Wind instruments should be constructed to conform to this frequency at 20 °C room temperature after being blown long enough to reach thermal equilibrium. Tuning C for keyboard instruments (c″) is a minor third of equal temperament above 440 Hz, namely 523.25 Hz.

Diatonic major scale. The frequency ratios of musical intervals in the diatonic major scale are shown below; the upper row shows the ratio between adjacent steps, the lower row the ratio to the keynote.

Diatonic major scale

	Major tone 9/8		Minor tone 10/9		Diatonic semitone 16/15		Major tone 9/8		Minor tone 10/9		Major tone 9/8		Diatonic semitone 16/15	
1 Keynote		1.1250 Major 2nd		1.2500 Major 3rd		1.3333 Perfect 4th		1.5000 Perfect 5th		1.6667 Major 6th		1.8750 Major 7th		2 Octave

The interval between a major and a minor tone, 81/80, is called a *comma*.

A *chromatic semitone* = 135/128 or 25/24 according to whether it divides a major or minor tone.

A *minor interval* is a major interval diminished by a chromatic semitone. An *augmented interval* is a major interval increased by a chromatic semitone. A *diminished interval* is a perfect or minor interval diminished by a chromatic semitone.

Equally tempered major scale. A semitone of equal temperament = $2^{1/12}$ = 1.059 46. A tone of equal temperament = 2 semitones of ET = $2^{1/6}$ = 1.122 46. The table below gives the frequencies (Hz) of musical notes in equal temperament for the octave starting at c′.

Frequencies of notes in equal temperament (Hz)

c′	c′♯	d′	d′♯	e′	f′	f′♯	g′	g′♯	a′	a′♯	b′
261.63	277.18	293.66	311.13	329.63	349.23	369.99	392.00	415.30	440.00	466.16	493.88

Any pitch interval may be expressed in *cents*, a *cent* being defined as the ratio $2^{1/1200}$:1. Thus there are 100 cents to the ET semitone (see Wood, *The Physics of Music*, 1962).

M.E.D.

1.7 Radiation and optics

1.7.1 The electromagnetic spectrum

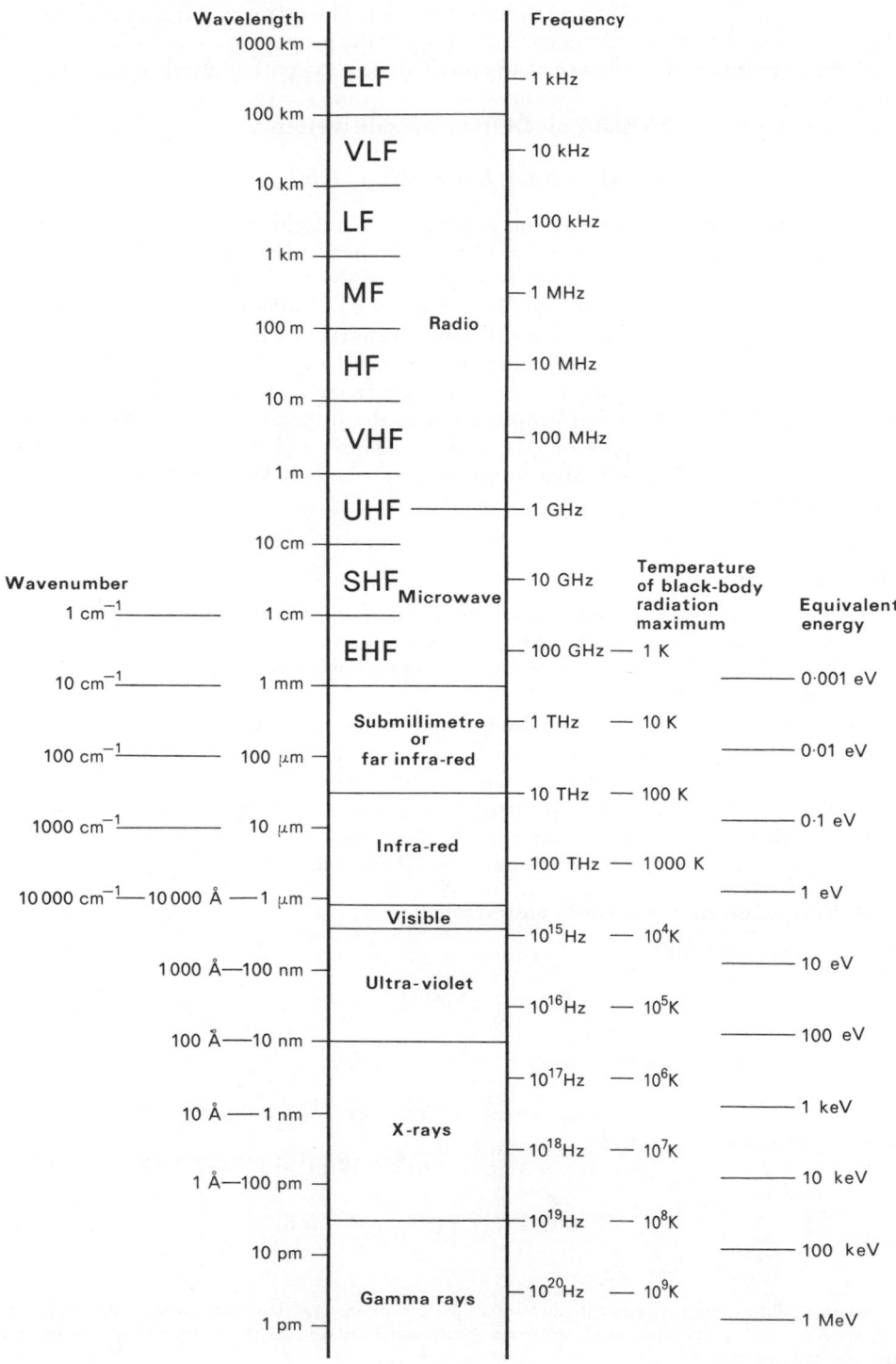

A.E.B.

1.7.2 Thermal radiation

In a field of thermal radiation, the radiant power $\delta^2\varphi$ falling on an infinitesimal area δA, in the range of directions contained by an infinitesimal cone of solid angle $\delta\Omega$, is given by $\delta^2\varphi = L\,\delta A\,\delta\Omega\,\cos\theta$, where θ is the angle between the cone and the normal to δA. The quantity L is known as the **radiance** of the field at δA in the direction of the cone. Radiance has the property that, in a transparent medium of constant refractive index, the radiance at any point along a ray path in the direction of the ray path is constant.†

The total radiant power $\delta\phi$ falling on δA from one side is given by

$$\delta\phi = \delta A \int L \cos\theta\,\mathrm{d}\Omega = E\,\delta A \qquad \dots(1)$$

where the integration is carried out over the appropriate hemisphere on δA. The quantity E which, unlike L, usually depends on the orientation of δA, is known as **irradiance**. If δA is an element of a radiating surface, the equivalent quantity in terms of the radiation passing out through δA is known as **exitance**. The radiation emitted by the whole of a radiation source in a given direction is described by the quantity I known as **radiant intensity**, defined by the equation $\delta\phi = I\,\delta\Omega$, where $\delta\phi$ is the power radiated in a cone of directions of solid angle $\delta\Omega$ about the specified direction.

If the radiation has a continuous spectrum, the power in the frequency interval $\delta\nu$ about ν or the wavelength interval $\delta\lambda$ about λ is proportional to the interval, and is therefore denoted by a spectral power density $\phi(\nu) = \partial\phi/\partial\nu$ or $\phi(\lambda) = \partial\phi/\partial\lambda$. These and derived quantities such as $L(\nu)$ are commonly distinguished from ordinary functions of ν or λ by writing the ν or λ as a subscript. If vacuum wavelengths are used, then $\nu\lambda = c$ and $\phi_\lambda = (c/\lambda^2)\phi_\nu$, etc.

Black-body radiation

The thermal radiation inside a closed cavity, with opaque walls at a uniform temperature T, is known as full or black-body radiation. It has a continuous spectrum and is uniform and isotropic, i.e. the spectral radiance L_ν within the cavity is independent of position and direction. Moreover, L_ν is determined solely by T.‡

This radiation may be observed by making a hole in the cavity wall small enough not to reduce appreciably the radiance of the emerging radiation. A means of testing whether this condition is satisfied is provided by the extended form of Kirchhoff's law, which says that the factor (known as emissivity) by which the radiance in a particular direction is reduced, equals the absorption factor for radiation incident on the hole in the opposite direction. A radiation source of this kind with emissivity sensibly equal to unity is known as a black-body radiator.

Spectral distribution of black-body radiation

According to Planck's formula

$$L_\nu = \frac{2h}{c^2}\,\nu^3(\mathrm{e}^{h\nu/kT} - 1)^{-1} \qquad \dots(2)$$

The spectral radiance is usually expressed in terms of L_λ:

$$L_\lambda = c_1'\lambda^{-5}(\mathrm{e}^{c_2/\lambda T} - 1)^{-1} \qquad \dots(3)$$

where $\qquad c_1' = 2hc^2 = 1.1911 \times 10^{-16}\ \mathrm{W\ m^2\ sr^{-1}}$

and $\qquad c_2 = hc/k = 1.4388 \times 10^{-2}\ \mathrm{m\ K}$

---

† More generally, if the radiation passes without loss through a medium of varying refractive index n, then L/n^2 is constant.

‡ Strictly speaking, L_ν is proportional to n^2, where n is the refractive index inside the cavity, and the subsequent formulae refer to an evacuated cavity.

For E_λ we have, since $E_\lambda = \pi L_\lambda$ for isotropic radiation,

$$E_\lambda = c_1 \lambda^{-5} (e^{c_2/\lambda T} - 1)^{-1}$$

where
$$c_1 = \pi c_1' = 3.7418 \times 10^{-16} \text{ W m}^2 \qquad \cdots (4)$$

The constants c_1 and c_2 are known as the 1st and 2nd radiation constants; more accurate values are given on p. 16.

The total radiance over the spectrum, $L = \int_0^\infty L_\lambda \, d\lambda$, is given by $L = \sigma' T^4$, where

$$\sigma' = \frac{2\pi^4 k^4}{15 c^2 h^3} = 1.8047 \times 10^{-8} \text{ W m}^{-2} \text{ sr}^{-1} \text{ K}^{-4} \qquad \cdots (5)$$

Likewise the total exitance is given by $E = \sigma T^4$, where

$$\sigma = \pi\sigma' = 5.670 \times 10^{-8} \text{ W m}^{-2} \text{ K}^{-4} \qquad \cdots (6)$$

σ is known as the Stefan–Boltzmann radiation constant.

The functions L_ν for different T are of the form $f(T)g(\nu/T)$, so that if normalized to a given maximum value and plotted against ν/T, they reduce to a single curve. The maximum of this curve occurs at

$$\frac{\nu_m}{T} = 5.8787 \times 10^{10} \text{ Hz K}^{-1}$$

This result is more conveniently expressed in terms of the wave number σ used by spectroscopists,[†] $\sigma = 1/\lambda$:

$$\sigma_m/T = 1.9609 \text{ K}^{-1} \text{ cm}^{-1}$$

Likewise the distribution functions in terms of λ reduce to a single function of λT with a maximum at

$$\lambda_m T = 2.8979 \times 10^{-3} \text{ m K}$$

The following table gives the fraction f of the total radiation for temperature T which lies between zero wavelength and that corresponding to various values of λT. It may be used in conjunction with formulae (5) and (6) above to obtain the radiance or exitance in extended spectral bands.

Example: for a black-body at 1000 K the exitance in the band 0.6–0.7 μm ($\lambda T = 600$–700 μm K) is

$$(1.84 \times 10^{-6} - 9.29 \times 10^{-8}) \times 5.67 \times 10^{-8} \times 1000^4 \text{ W m}^{-2}$$

Fraction of total radiation from 0 to λT

$\lambda T/(\mu\text{m K})$	f	$\lambda T/(\mu\text{m K})$	f	$\lambda T/(\mu\text{m K})$	f
600	9.29×10^{-8}	2 500	0.1613	12 000	0.945 05
700	1.84×10^{-6}	3 000	0.2732	14 000	0.962 85
800	1.64×10^{-5}	3 500	0.3829	16 000	0.973 77
900	8.70×10^{-5}	4 000	0.4808	18 000	0.980 81
1 000	3.21×10^{-4}	4 500	0.5643	20 000	0.985 56
1 200	2.13×10^{-3}	5 000	0.6337	30 000	0.995 29
1 400	7.79×10^{-3}	6 000	0.7378	40 000	0.997 92
1 600	1.97×10^{-2}	7 000	0.8081	50 000	0.998 91
1 800	3.93×10^{-2}	8 000	0.8562	75 000	0.999 67
2 000	6.67×10^{-2}	10 000	0.9142	100 000	0.999 86

E.J.G.

Colour-temperature

The colour of an illuminant often resembles that of a black-body radiator at a particular temperature and can conveniently be specified in terms of this temperature. If an exact colour match can be obtained, the temperature is known as the **colour-temperature**. If only an approximate match can be obtained, then the term **correlated colour-temperature** is used. This can be given a precise

† Not to be confused with σ denoting the Stefan–Boltzmann constant.

meaning by representing the colours of the illuminant and of the black-body radiation as points on the CIE (U–V) Uniform Chromaticity Space diagram. The correlated colour-temperature is then defined as the black-body temperature for which the two points are closest together. For correlated colour-temperatures of various illuminants, see p. 87.

With some illuminants, such as the tungsten-filament lamp, both the colour and the relative spectral power distribution resemble that of a black-body radiator at a suitable temperature. In such cases, the term '**distribution temperature**' is proposed for the temperature which gives the best spectral fit. As yet, however, there is no agreement as to what constitutes a best fit, and where this term is used, the criterion of best fit should be stated.

Colour and distribution temperatures are usually expressed in kelvins (K). The reciprocal of the temperature in megakelvins may also be used, the unit being known as the mired: $1 \text{ mired} = 10^{-6} \text{ K}^{-1}$.

O.C.J.

1.7.3 Photometry

The results of photometric matching experiments show that radiation may be evaluated in terms of visual effect by means of a quantity known as luminous flux, ϕ_v, which is related to $\phi_{e\lambda}$, the spectral power distribution of the radiation, by an equation of the form

$$\phi_v = K_m \int_0^\infty \phi_{e\lambda} V(\lambda) \, d\lambda \qquad \qquad \ldots (1)$$

where K_m is a constant, and $V(\lambda)$ a function representing the wavelength-dependent sensitivity of the eye. The exact shape of $V(\lambda)$ depends on the observer, but in order to provide a precise quantitative basis for photometry, an agreed set of values was recommended for adoption by the Commission International de l'Éclairage (CIE) in 1971, and accepted by the Comité International des Poids et Mesures (CIPM) (see p. 7).

The following table gives the values of this function at 10 nm intervals. For the complete tabulation at 1 nm interval refer to CIE 18 (E-1.2) 1970.

Photopic relative luminous efficiency function $\dot{V}(\lambda)$

λ/nm	$V(\lambda)$	λ/nm	$V(\lambda)$	λ/nm	$V(\lambda)$
360	0.000 004	530	0.862 000	680	0.017 000
370	0.000 012	540	0.954 000	690	0.008 210
380	0.000 039				
390	0.000 120	550	0.994 950	700	0.004 102
		555	**1.000 000**	710	0.002 091
400	0.000 396	560	0.995 000	720	0.001 047
410	0.001 210	570	0.952 000	730	0.000 520
420	0.004 000	580	0.870 000	740	0.000 249
430	0.011 600	590	0.757 000		
440	0.023 000			750	0.000 120
		600	0.631 000	760	0.000 060
450	0.038 000	610	0.503 000	770	0.000 030
460	0.060 000	620	0.381 000	780	0.000 015
470	0.090 980	630	0.265 000	790	0.000 007
480	0.139 020	640	0.175 000		
490	0.208 020			800	0.000 004
		650	0.107 000	810	0.000 002
500	0.323 000	660	0.061 000	820	0.000 001
510	0.503 000	670	0.032 000	830	0.000 000
520	0.710 000				

The magnitude of the unit of luminous flux, the lumen, is determined by assigning a value to K_m of equation (1). However, for historical reasons the SI base unit for photometry is the unit of luminous intensity, the candela or lumen per steradian; the formal definition (16th CGPM, 1979) appears on page 2. The rounded frequency 540×10^{12} Hz corresponds to a wavelength of 555.016 nm in standard air, whereas K_m occurs at 555 nm exactly. Thus K_m is precisely 683.002 lm W^{-1}, although the value 683 lm W^{-1} is sufficient for all practical purposes.

The spatial properties of light are described in photometry by quantities analogous to those used in radiometry (see p. 84), with luminous flux replacing radiant power. The same symbols are used for corresponding quantities, distinguished where necessary by the suffixes e (radiation) and v (photometric). These quantities are listed in the following table, together with their radiometric equivalents to indicate the relation between the two systems of measurement.

Photometric quantities

Quantity	Radiometric equivalent	Symbol	Definition	Unit name and symbol
Luminous flux	Radiant power	ϕ		lumen lm
Luminous intensity	Radiant intensity	I	$\partial\phi/\partial\Omega$	candela cd $\equiv$ lm sr^{-1}
Luminance	Radiance	L	$\partial^2\phi/\partial A\,\partial\Omega$	(nit) lm sr^{-1} m^{-2}
Illuminance	Irradiance	$E\big\}$	$\partial\phi/\partial A$	lux lm m^{-2}
Luminous exitance	Radiant exitance	$M\big\}$		lm m^{-2}

Luminance and colour temperature of selected light sources

Light source	Luminance 10^4 lm sr^{-1} m^{-2}	Correlated colour temp. K
Candle	0.5	1930
Paraffin flame (flat wick)	1.25	2055
Acetylene (Kodak burner)	10.8	2360
Tungsten strip lamp (vacuum)	124	2400
,, ,, ,, (gas-filled)	540	2800
,, ,, ,, (,,)	990	3000
,, coil ,, (quartz-halogen)	$\sim$2800	3300
Mercury vapour (low pressure, in glass)	2.3	—
,, ,, (high pressure, compact, 250 W)	2×10^4	—
,, ,, (,, ,, ,, 1000 W)	4×10^4	—
,, ,, (tubular fluorescent, colour matching 40 W)	1.2	6050
,, ,, (tubular fluorescent, warm white 40 W)	1.9	3000
Arc crater (solid carbon)	1.72×10^4	3780
,, ,, (high intensity, 150 A)	8.0×10^4	5000–5500
Xenon arc (compact) 250 W	2.5×10^4	6000
,, ,, ,, 2 kW	1.2×10^5	6000
Clear blue sky	$\sim$0.4	12 000–24 000
Starlit sky	$\sim5 \times 10^{-8}$	—
Zenith Sun (through atmosphere)	$\sim1.6 \times 10^5$	5400
Moon (through atmosphere)	$\sim$0.4	—
Perfect diffuser in sunlight	$\sim$4	
,, ,, ,, moonlight	$\sim10^{-5}$	

Scotopic system of photometry

At low levels of luminance the relative luminous effects of radiations of different wavelength are in general no longer represented even approximately by the values of $V(\lambda)$ given in the above table. This is because a different visual mechanism known as the scotopic or dark-adapted mechanism, as opposed to the photopic or light-adapted mechanism, is dominant. To evaluate electromagnetic radiation with respect to the effect it produces on the scotopic mechanism, an alternative set of weighting factors—the **scotopic relative luminous efficiency function** $V'(\lambda)$—must be used, the values of which were agreed by the CIE in 1951 and accepted by the CIPM in 1976.

Scotopic relative luminous efficiency function $V'(\lambda)$

λ/nm	0	10	20	30	40
300	—	—	—	—	—
400	0.009 29	0.034 84	0.096 6	0.1998	0.3281
500	0.982	0.997	0.935	0.811	0.650
600	0.033 15	0.015 93	0.007 37	3.335×10^{-3}	1.497×10^{-3}
700	1.780×10^{-5}	9.14×10^{-6}	4.78×10^{-6}	2.546×10^{-6}	1.379×10^{-6}

λ/nm	50	60	70	80	90
300	—	—	—	5.89×10^{-4}	2.209×10^{-3}
400	0.455	0.567	0.676	0.793	0.904
500	0.481	0.3288	0.2076	0.1212	0.0655
600	6.77×10^{-4}	3.129×10^{-4}	1.480×10^{-4}	7.15×10^{-5}	3.533×10^{-5}
700	7.60×10^{-7}	4.25×10^{-7}	2.413×10^{-7}	1.390×10^{-7}	—

In the alternative system of photometry obtained by taking $V'(\lambda)$ as basis instead of $V(\lambda)$, all the formal definitions of the previous section are retained, with terms modified by the adjective 'scotopic' and symbols by the addition of a prime. The calculated value for the scotopic maximum luminous efficacy K'_m is $1700.06\,\text{lm}\,\text{W}^{-1}$, which may safely be rounded to $1700\,\text{lm}\,\text{W}^{-1}$.

In the so-called mesopic region between the photopic and the scotopic, the concept of luminous flux as an additive quantity does not apply and there is at present no agreed method of spectral weighting for mesopic vision. The CIE (1971) has suggested the following procedure. For any given luminous measurement in the mesopic region, both the photopic and the scotopic values are determined. The appropriate mesopic quantity, e.g. $E_{mesopic}$, is then obtained by means of the formula

$$E_{\text{mesopic}} = \frac{E' + 17E^2}{1 + 17E}$$

O.C.J.

1.7.4 Colorimetry

Evaluation of colour

The trichromatic system of colorimetry recommended by the CIE in 1931 is used internationally for specifying the colour of light or of illuminated objects. The colour of any radiation is specified by three **tristimulus values** X, Y, Z which represent the amounts of three standard reference stimuli (X), (Y), (Z) which if mixed together by optical superposition would produce the same sensation of light and colour. The tristimulus values for radiation of spectral power distribution $P(\lambda)$ are given by the formulae:

$$X = K \int P(\lambda) . \bar{x}(\lambda) \, d\lambda; \qquad Y = K \int P(\lambda) . \bar{y}(\lambda) \, d\lambda; \qquad Z = K \int P(\lambda) . \bar{z}(\lambda) \, d\lambda$$

where K is a scaling constant and $\bar{x}(\lambda)$, $\bar{y}(\lambda)$ and $\bar{z}(\lambda)$ are known as **spectral tristimulus values**, i.e. the tristimulus values of unit powers of monochromatic radiation as given in the table below. The 'colour quality' of the radiation, as distinct from its amount, is expressed in terms of the **chromaticity co-ordinates** x, y, z, given by:

$$x = X/(X+Y+Z); \qquad y = Y/(X+Y+Z); \qquad z = Z/(X+Y+Z).$$

Of the three tristimulus values, Y represents the luminosity of a colour, and if K_m is used for the constant K above, then Y becomes the photometric flux, since the function $\bar{y}(\lambda)$ is identical to $V(\lambda)$. (See p. 86.)

The colour of a reflecting surface is specified in terms of the tristimulus values X, Y, Z of the radiation reflected when the surface is illuminated with radiation of specified spectral power distribution $P(\lambda)$. Here the absolute values of X, Y, Z are usually just as important as the chromaticity co-ordinates x, y, z, and they are defined as

$$X = \frac{\int P(\lambda) . \bar{x}(\lambda) . \rho(\lambda) \, \mathrm{d}\lambda}{\int P(\lambda) . \bar{y}(\lambda) \, \mathrm{d}\lambda}$$

$$Y = \frac{\int P(\lambda) . \bar{y}(\lambda) . \rho(\lambda) \, \mathrm{d}\lambda}{\int P(\lambda) . \bar{y}(\lambda) \, \mathrm{d}\lambda}$$

$$Z = \frac{\int P(\lambda) . \bar{z}(\lambda) . \rho(\lambda) \, \mathrm{d}\lambda}{\int P(\lambda) . \bar{y}(\lambda) \, \mathrm{d}\lambda}$$

where $\rho(\lambda)$ is the reflectance expressed as a percentage. These definitions give Y the value 100 for a perfect reflector, or more generally Y equals the luminous reflectance (the photometric value); in addition the chromaticity co-ordinates x, y, z for a spectrally non-selective reflector will be those of the illuminant. Similar expressions hold for a transmitting object, with the spectral transmittance $\tau(\lambda)$ replacing $\rho(\lambda)$ above.

The table of spectral tristimulus values allows the chromaticity co-ordinates $x(\lambda)$, $y(\lambda)$, $z(\lambda)$ of the spectral radiations to be calculated immediately since

$$x(\lambda) = \bar{x}(\lambda)/(\bar{x}(\lambda) + \bar{y}(\lambda) + \bar{z}(\lambda)), \text{ etc.}$$

The table is normalized so that the chromaticity co-ordinates of the equi-energy white radiation are each equal to 0.3333. In many cases good accuracy can be obtained by computation at 10 nm intervals from 380 nm provided that the tristimulus values X, Y, Z so obtained are multiplied by 1.0002, 1.0000 and 1.0008 respectively; this has the effect of renormalizing the shortened table for 10 nm intervals to give the same chromaticity co-ordinates for equi-energy white, and hence to minimize spectral sampling errors on average.

The (X, Y, Z) colour space is subjectively rather non-uniform as far as perceived colour differences are concerned. In 1976 the CIE recommended for use two possible non-linear transformations of (X, Y, Z), namely L^*, U^*, V^*) and (L^*, a^*, b^*), which are more nearly uniform for colour differences. Of these (L^*, a^*, b^*) has proved more suitable for the colour manufacturing industries (textiles, plastics, paints, ceramics, paper, etc.) while (L^*, U^*, V^*) has been used in television and photography. However, under the aegis of the Society of Dyers and Colourists, a system based on the (L^*, a^*, b^*) colour space with local modulations of colour tolerance, specified by supplementary equations, is being increasingly used by the colour manufacturing industries. This achieves a better correlation with visual judgement.

1931 CIE colorimetric Standard Observer for subtenses 0.5° to 4°

λ/nm	Spectral tristimulus values			λ/nm	Spectral tristimulus values		
	$\bar{x}(\lambda)$	$\bar{y}(\lambda)$	$\bar{z}(\lambda)$		$\bar{x}(\lambda)$	$\bar{y}(\lambda)$	$\bar{z}(\lambda)$
380	0.0014	0.0000	0.0065	580	0.9163	0.8700	0.0017
385	0.0022	0.0001	0.0105	585	0.9786	0.8163	0.0014
390	0.0042	0.0001	0.0201	590	1.0263	0.7570	0.0011
395	0.0076	0.0002	0.0362	595	1.0567	0.6949	0.0010
400	0.0143	0.0004	0.0679	600	1.0622	0.6310	0.0008
405	0.0232	0.0006	0.1102	605	1.0456	0.5668	0.0006
410	0.0435	0.0012	0.2074	610	1.0026	0.5030	0.0003
415	0.0776	0.0022	0.3713	615	0.9384	0.4412	0.0002
420	0.1344	0.0040	0.6456	620	0.8544	0.3810	0.0002
425	0.2148	0.0073	1.0391	625	0.7514	0.3210	0.0001
430	0.2839	0.0116	1.3856	630	0.6424	0.2650	0.0000
435	0.3285	0.0168	1.6230	635	0.5419	0.2170	0.0000
440	0.3483	0.0230	1.7471	640	0.4479	0.1750	0.0000
445	0.3481	0.0298	1.7826	645	0.3608	0.1382	0.0000
450	0.3362	0.0380	1.7721	650	0.2835	0.1070	0.0000
455	0.3187	0.0480	1.7441	655	0.2187	0.0816	0.0000
460	0.2908	0.0600	1.6692	660	0.1649	0.0610	0.0000
465	0.2511	0.0739	1.5281	665	0.1212	0.0446	0.0000
470	0.1954	0.0910	1.2876	670	0.0874	0.0320	0.0000
475	0.1421	0.1126	1.0419	675	0.0636	0.0232	0.0000
480	0.0956	0.1390	0.8130	680	0.0468	0.0170	0.0000
485	0.0580	0.1693	0.6162	685	0.0329	0.0119	0.0000
490	0.0320	0.2080	0.4652	690	0.0227	0.0082	0.0000
495	0.0147	0.2586	0.3533	695	0.0158	0.0057	0.0000
500	0.0049	0.3230	0.2720	700	0.0114	0.0041	0.0000
505	0.0024	0.4073	0.2123	705	0.0081	0.0029	0.0000
510	0.0093	0.5030	0.1582	710	0.0058	0.0021	0.0000
515	0.0291	0.6082	0.1117	715	0.0041	0.0015	0.0000
520	0.0633	0.7100	0.0782	720	0.0029	0.0010	0.0000
525	0.1096	0.7932	0.0573	725	0.0020	0.0007	0.0000
530	0.1655	0.8620	0.0422	730	0.0014	0.0005	0.0000
535	0.2257	0.9149	0.0298	735	0.0010	0.0004	0.0000
540	0.2904	0.9540	0.0203	740	0.0007	0.0002	0.0000
545	0.3597	0.9803	0.0134	745	0.0005	0.0002	0.0000
550	0.4334	0.9950	0.0087	750	0.0003	0.0001	0.0000
555	0.5121	1.0000	0.0057	755	0.0002	0.0001	0.0000
560	0.5945	0.9950	0.0039	760	0.0002	0.0001	0.0000
565	0.6784	0.9786	0.0027	765	0.0001	0.0000	0.0000
570	0.7621	0.9520	0.0021	770	0.0001	0.0000	0.0000
575	0.8425	0.9154	0.0018	775	0.0001	0.0000	0.0000
580	0.9163	0.8700	0.0017	780	0.0000	0.0000	0.0000

The CIE (X, Y, Z) system of 1931 is valid for observations with uniform areas subtending up to $4°$; for uniform areas subtending larger angles there is a supplementary (X_{10}, Y_{10}, Z_{10}) system recommended by CIE in 1964, based on observations with a $10°$ field.

Standard illuminants for colorimetry

In 1931 the CIE recommended three standard illuminants for the colorimetry of materials. Illuminant A was intended to represent the illumination from incandescent lamps used in general lighting and in projectors. Illuminant B was intended to represent direct sunlight with no sky component, while Illuminant C was intended to represent overcast skylight. Of these, Illuminant B has fallen largely into disuse and is therefore omitted from the specifications below. However in 1964 the CIE introduced Illuminant D_{65} based on the measured spectral power distribution of average overcast skylight or of sunlight mixed with total skylight.

Illuminant A. Black-body radiation corresponding to a value of c_2/T (see p. 000) of $(1.4350 \times 10^{-2}/2848)$ m, or on IPTS-68, to a temperature of approximately 2856 K. The chromaticity co-ordinates of Illuminant A are $(x = 0.4476, y = 0.4074)$.

Illuminant A is realized in the laboratory by Source A, the radiation from a gas-filled tungsten-filament lamp of the same correlated colour temperature.

Illuminant C. The radiation from Illuminant A after selective attenuation in accordance with the published CIE data on the transmittance of the filter described below. The chromaticity co-ordinates of Illuminant C are $(x = 0.3101, y = 0.3162)$ and its correlated colour temperature is 6774 K (IPTS-68).

Illuminant C is realized in the laboratory by Source A combined with a colour filter consisting of a layer 10 mm thick of each of two solutions C_1 and C_2, contained in a double cell made of colourless optical glass.

Solution C_1

Copper sulphate $(CuSO_4 . 5H_2O)$	3.412 g
Mannite $(C_6H_8(OH)_6)$	3.412 g
Pyridine (C_5H_5N)	30.0 ml
Distilled water, to make 1 litre.	

Solution C_2

Cobalt ammonium sulphate $(CoSO_4 . (NH_4)_2SO_4 . 6H_2O)$	30.58 g
Copper sulphate $(CuSO_4 . 5H_2O)$	22.52 g
Sulphuric acid (density 1.835 g ml^{-1})	10.0 ml
Distilled water, to make 1 litre.	

Illuminant D_{65}. A relative spectral power distribution defined and recommended by CIE as representing a phase of daylight with a correlated colour temperature of approximately 6504 K (IPTS-68). This is a much better representation of an average overcast sky than Illuminant C. The chromaticity co-ordinates of Illuminant D_{65} are $(x = 0.3127, y = 0.3290)$. At present Illuminant D_{65} cannot be realized with enough accuracy for many applications, and there is no recommended source specification as yet.

Methods of specifying standard illuminants D_T of different correlated colour-temperatures have been published by CIE: these are spectral power distributions which represent different phases of daylight.

General reference: *Colorimetry* (CIE Publication No. 15, 1971).

Relative spectral power distributions of CIE standard illuminants

λ/nm	(A) $P(\lambda)$	(C) $P(\lambda)$	(D$_{65}$) $P(\lambda)$	λ/nm	(A) $P(\lambda)$	(C) $P(\lambda)$	(D$_{65}$) $P(\lambda)$
375	8.77	27.50	51.0	575	110.80	100.15	96.1
380	9.80	33.00	50.0	580	114.44	97.80	95.8
385	10.90	39.92	52.3	585	118.08	95.43	92.2
390	12.09	47.40	54.6	590	121.73	93.20	88.7
395	13.35	55.17	68.7	595	125.39	91.22	89.3
400	14.71	63.30	82.8	600	129.04	89.70	90.0
405	16.15	71.81	87.1	605	132.70	88.83	89.8
410	17.68	80.60	91.5	610	136.35	88.40	89.6
415	19.29	89.53	92.5	615	139.99	88.19	88.6
420	20.99	98.10	93.4	620	143.62	88.10	87.7
425	22.79	105.80	90.1	625	147.24	88.06	85.5
430	24.67	112.40	86.7	630	150.84	88.00	83.3
435	26.64	117.75	95.8	635	154.42	87.86	83.5
440	28.70	121.50	104.9	640	157.98	87.80	83.7
445	30.85	123.45	110.9	645	161.52	87.99	81.9
450	33.09	124.00	117.0	650	165.03	88.20	80.0
455	35.41	123.60	117.4	655	168.51	88.20	80.1
460	37.81	123.10	117.8	660	171.96	87.90	80.2
465	40.30	123.30	116.3	665	175.38	87.22	81.2
470	42.87	123.80	114.9	670	178.77	86.30	82.3
475	45.52	124.09	115.4	675	182.12	85.30	80.3
480	48.24	123.90	115.9	680	185.43	84.00	78.3
485	51.04	122.92	112.4	685	188.70	82.21	74.0
490	53.91	120.70	108.8	690	191.93	80.20	69.7
495	56.85	116.90	109.1	695	195.12	78.24	70.7
500	59.86	112.10	109.4	700	198.26	76.30	71.6
505	62.93	106.98	108.6	705	201.36	74.36	73.0
510	66.06	102.30	107.8	710	204.41	72.40	74.3
515	69.25	98.81	106.3	715	207.40	70.40	68.0
520	72.50	96.90	104.8	720	210.36	68.30	61.6
525	75.79	96.78	106.2	725	213.27	66.30	65.7
530	79.13	98.00	107.7	730	216.12	64.40	69.9
535	82.52	99.94	106.0	735	218.92	62.80	72.5
540	85.95	102.10	104.4	740	221.67	61.50	75.1
545	89.41	103.95	104.2	745	224.36	60.20	69.3
550	92.91	105.20	104.0	750	227.00	59.20	63.6
555	96.44	105.67	102.0	755	229.59	58.50	55.0
560	100.00	105.30	100.0	760	232.12	58.10	46.4
565	103.58	104.11	98.2	765	234.59	58.00	56.6
570	107.18	102.30	96.3	770	237.01	58.20	66.8

F.J.J.C.

1.7.5   Wavelength standards

Laser standards

Wavelength values of optical radiations may be expressed either as the values in standard air (see below) or in vacuum.

One of the principal means of realization of the metre, according to the 1983 definition, is through the wavelength values of stabilized laser radiations. These vacuum wavelength values are determined by the relation $\lambda = c/f$, from the fixed value for the speed of light and the measured frequencies of the radiations. Five such laser standards are recommended. They form the most precise and stable references for interferometric measurement and spectroscopic investigations. They are emitted by single-mode lasers stabilized to transitions of absorbing molecules contained in cells within or external to the lasers. Their frequency and wavelength values, and their relative uncertainties (99% confidence level), are given in the table. These values apply only when the associated conditions and accepted good practice are followed.

Frequency (MHz)	Vacuum wavelength (nm)	Uncertainty parts in 10^9 ($\pm$)	Absorber molecule	Transition	Component
88 376 181.608	3 392.231 397 0	0.13	CH_4	v_3, P(7)	$F_2^{(2)}$
473 612 214.8	632.991 398 1	1.0	$^{127}I_2$	11–5, R(127)	i
489 880 355.1	611.970 769 8	1.1	$^{127}I_2$	9–2, R(47)	o
520 206 808.51	576.294 760 27	0.6	$^{127}I_2$	17–1, P(62)	o
582 490 603.6	514.673 466 2	1.3	$^{127}I_2$	43–0, P(13)	a_3

Source conditions

3 392 nm: methane pressure $\leqslant 3$ Pa,
           mean one-way axial surface power density $\leqslant 10^4$ Wm^{-2},
           radius of wavefront curvature $\geqslant 1$ m,
           inequality of power in counter-propagating beams $\leqslant 5\%$.
 633 nm: cold-finger temperature of cell 15 °C $\pm$ 1 °C,
           cell-wall temperature between 16 °C and 50 °C,
           one-way intracavity beam power 15mW $\pm$ 10mW,
           modulation amplitude, peak to peak, 6 MHz $\pm$ 1 MHz.
           (Single-frequency He–Ne lasers containing neon-20, when stabilized to the centre of the neon transition, generally emit with a wavelength within ± 1.5 parts in 10^7 of this value.)
 612 nm: cold-finger temperature of cell -5 °C $\pm$ 2 °C.
 576 nm: cold-finger temperature of cell $+6$ °C $\pm$ 2 °C.
 515 nm: cold-finger temperature of cell -5 °C $\pm$ 2 °C.

Discharge tube sources

The former primary wavelength standard and the twelve former secondary standards have been retained as recommended standards for interferometric measurement, with unchanged wavelength values and uncertainties. These values were specified by the International Committee of Weights and Measures in 1963, and are also recognized by the International Astronomical Union for use in spectroscopy (*Trans. IAU*, vol. XII A, 1965). The table shows their vacuum wavelength values in nanometres.

Krypton-86	Mercury-198	Cadmium-114
605.780 210		
645.807 20	579.226 83	644.024 80
642.280 06	577.119 83	508.723 79
565.112 86	546.227 05	480.125 21
450.361 62	435.956 24	467.945 81

Source conditions

Krypton 86. The wavelength of the 605 nm radiation, when emitted by a lamp conforming to the specification below, has an estimated uncertainty (99% confidence) of ± 4 parts in 10^9. The other radiations, under similar conditions, have uncertainties of ± 2 parts in 10^8.

The source is a hot-cathode discharge lamp containing krypton-86 (purity $\not< 99\%$) in sufficient quantity to assure the presence of solid krypton at 64 K, the lamp having a capillary portion with the dimensions: internal diameter 2–4 mm, wall thickness 1 mm. The conditions of operation are:
  (i) the capillary is observed end-on from the anode side of the lamp;
  (ii) the lower part of the lamp, including the capillary, is immersed in a refrigerant bath maintained within 1 K of the triple point of nitrogen;
  (iii) the current density in the capillary is 3 ± 1 mA mm^{-2}.

Mercury-198. The uncertainties of the wavelengths are ± 5 parts in 10^8 when emitted by a high-frequency electrodeless discharge lamp, operated at moderate power with the radiation observed through the side of the capillary. The lamp should be maintained at a temperature below 10 °C and contain mercury-198 (purity $\not< 98\%$) with argon as carrier gas at a pressure between 65 and 133 N m^{-2}. The internal diameter of the capillary should be about 5 mm, with the volume of the lamp preferably > 20 cm^3.

Cadmium-114. The standard wavelengths have an estimated uncertainty of ± 7 parts in 10^8 when emitted by an electrodeless discharge lamp source, maintained at a temperature such that the green line is not reversed and containing cadmium-114 (purity $\not< 95\%$) with argon as carrier gas (pressure about 150 N m^{-2} at ambient temperatures). The radiations should be observed through the side of the capillary part of the tube, having an internal diameter of about 5 mm.

W.R.C.R.

Practical wavelength standards for calibration of spectrophotometers

There are several extensive tables of emission lines but these are not very suitable for practical users of general purpose spectrophotometers. Many of the lines cited cannot be detected or resolved because the dispersion and light grasp of these instruments are not high enough, or because the pressure or electron temperatures in commercial discharge lamps may not allow a line to be separated from neighbours or the continuum background. Some elements, like neon, iron or iodine, have too many lines for easy identification.

For this reason a practical table of useful atomic emission lines is given below: these lines can usually be found and recognized using atomic emission lamps of the kind available commercially for wavelength calibration. Certain doublet or triplet emissions, such as the 365 nm mercury emission group, are included to aid recognition of other lines: these doublets or triplets should only be used for calibration when they are fully resolved by the instrument. Often included in lists of lines recommended by textbooks are strong resonance lines such as the 253.65 nm mercury line. In most mercury lamps this particular line is broadened to the point of being nearly inverted, so that it seems like two broadened lines at perhaps 251 and 256 nm, for example, and can be mistaken for the 253.65 nm and much weaker 257.63 nm lines. However, if the mercury lamp is a genuinely low-pressure one for UV use, this line at 253.65 nm is satisfactory.

Useful emission lines for wavelength calibration (air values)

Wavelength (nm)	Cadmium	Caesium	Helium	Mercury	Potassium	Zinc	Rubidium
	—	—	—	194.23	—	—	—
200							
	214.44	—	—	—	—	202.55	—
	226.50	—	—	—	—	206.19	—
	228.80	—	—	230.21	—	213.86	—
	231.28	—	—	234.56	—	—	—
	232.93	—	238.54	235.25	—	—	—
	—	—	—	239.95	—	—	—
250							
	—	—	251.12	253.65	—	250.20	—
	257.31	—	—	257.63	—	255.80	—
	267.76	—	—	—	—	258.25	—
	274.86	—	273.32	—	—	260.86	—
	283.69	—	—	—	—	—	—
	288.08	—	—	—	—	277.09	—
	—	—	—	—	—	280.00	—
	298.06	—	294.51	296.73	—	—	—
300							
	308.08	—	—	—	—	303.58	—
	—	—	318.77	312.56	—	307.21	—
	325.25	—	320.32	313.16	—	307.59	—
	326.11	—	—	313.18	—	328.23	322.80
	—	—	—	—	321.72	330.26	322.91
	—	—	—	334.15	—	330.29	334.87
	340.37	—	—	339.01	—	334.50	335.09
	346.62	—	—	—	344.64	334.56	—
	346.77	—	—	—	344.74	334.59	—
350							
	—	—	—	—	—	—	358.71
	361.05	—	361.36	365.02	—	—	359.16
	361.29	—	370.50	365.48	—	—	—
	361.44	—	—	366.33	—	—	—
	—	—	388.87	—	—	—	—
	398.20	—	396.47	—	—	—	—
400							
	—	—	402.62	404.66	404.42	—	—
	—	—	412.08	407.78	404.72	—	—
	—	—	414.38	—	—	—	420.18
	—	—	—	—	—	—	421.56
	—	—	438.79	435.84	—	—	—
	441.30	—	443.75	—	—	—	—
	—	—	447.15	—	—	—	—
450							
	—	455.55	—	—	—	—	—
	—	459.32	—	—	—	462.98	—
	467.82	460.38	468.58	—	—	468.01	—
	—	—	471.31	—	—	472.22	—
	479.99	—	—	—	—	481.05	—
	—	—	—	—	—	491.16	—
	—	—	492.19	—	—	492.40	—

Useful emission lines for wavelength calibration (air values)

Wavelength (nm)	Cadmium	Caesium	Helium	Mercury	Potassium	Zinc	Rubidium
500							
	508.58	—	501.57	—	—	—	—
	—	—	504.77	—	—	—	—
	533.80	—	—	—	535.97	530.86	—
	537.90	—	—	—	—	—	—
	—	—	—	546.07	—	—	—
550							
	—	—	—	—	578.26	—	—
	—	—	—	—	580.19	—	572.45
	—	—	—	576.96	581.24	577.22	—
	—	—	587.56	579.07	583.20	589.44	—
600							
	—	621.30	—	623.44	—	—	620.65
	—	—	—	—	—	—	629.86
	—	—	—	—	—	636.23	—
	643.85	—	—	—	—	—	—
650							
	—	658.65	—	—	—	—	—
	—	672.33	667.81	—	—	—	—
	—	697.33	—	—	691.13	—	—
	—	698.34	—	690.75	693.90	692.84	—
700							
	—	—	706.52	—	—	—	—
	—	—	—	—	—	—	728.02
	—	—	728.13	—	—	—	—
	—	—	—	—	—	—	740.84
750							
	—	760.90	—	—	766.49	—	—
	—	—	—	—	769.90	—	—
	—	—	—	—	—	—	780.03
	—	794.40	—	—	—	—	794.76
800							
	—	807.91	—	—	—	—	—
	—	807.98	—	—	—	—	—

F.J.J.C.

1.7.6 Laser radiation

The table which follows lists a selection of laser emissions from the many thousand that have been discovered (see, for example, *Handbook of Laser Science and Technology*, 2 vols, 1982, CRC Press, Boca Raton, Florida, USA). The selection is restricted to those that are commercially available (*Laser Focus Buyers Guide*, 20th edn, 1985, PennWell Publ. Co., Littleton, MA, USA) or are important for other reasons, such as their place in the spectrum or their high power. Whilst most lasers, especially powerful ones, are of fixed frequency, tunable coverage of much of the tabulated spectrum is available through the use of nonlinear devices to add or subtract laser frequencies, or to generate their harmonics. The table gives the vacuum wavelength, the method of excitation, the tunability, and an added note to cover power, pulse characteristics and other matters of interest. The vacuum wavelength λ_{vac} is related to the frequency f by λ_{vac} (μm) $= 299\,792\,458/f$(MHz), from the 1983 definition of the metre (see p. 2).

Gaps in the spectral coverage may be related to strong absorption by atmospheric gases such as CO_2 (at 1.4, 1.9 and 2.6–2.9 μm), or H_2O (at 5.5–7.2 μm and in the far infrared). In parts of the infrared, the air wavelength varies strongly with atmospheric properties such as humidity.

Hazard classes (BS4803 1983, ANSIZ-136.1 (1980), IEC TC-76 (1984–5)) have been established for laser systems based on internationally-agreed limits for exposure of the body to laser radiations (WHO reports ICP/CEP 803 Copenhagen 1977, EHC No. 2 1983). Of these, Class 1 are safe in normal use, and Class 2 emit visible radiation of < 1 mW for which eye-aversion reflexes provide protection. The higher classes 3a, 3b and 4 require special care. In the USA the classes are named respectively I, II, IIIa, IIIb and IV. Wavelengths 0.4–1.4 μm approximately can be focused by the eye on the retina and are therefore particularly dangerous. Blue to ultraviolet wavelengths also require care because of the high photon energy. Electric shock risks exist with most lasers, especially gas lasers, because high voltages and high electrical powers can be involved.

The cost of lasers varies from about £100 to more than £100,000. The cheapest lasers are small dc-excited discharge devices such as the ubiquitous He-Ne gas laser, and possibly semiconductor diode lasers for room temperature use. Such features as cryogenic operation and optical pumping, especially by lasers, add considerably to the cost, as do features such as mode-locking, single-mode output, line-tunability and frequency stabilization.

Laser radiation

λ_{vac}/μm†	Name	Type/excit^n	Tuning	Notes‡
0.157	F_2	gas, dc	ML	p 10mJ/10 ns, 3 lines{ uv lasers are important as pump sources.
0.1934	ArF	excimer, TEA	ML	p 1 mJ–0.5 J/5–20 ns, e 1%
0.249	KrF	excimer, TEA	ML 50 ppm	p 1 mJ–7 J/4–20 ns, e 2%, 2 lines
0.3–1.2	dye	liquid, o–p	T 10%	cw 0.01–1 W, p 10 W – 10 MW/10 fs–1 μs, e 0.3–1%, versatile
0.3251	He–Cd	metal vapour, dc	few ppm	cw 1–20 mW
0.337	N_2	gas, dc	62 lines 0.2%	p 10 μJ–30 mJ/1–10 ns, dye-laser pump
0.351 212	Ar ion	gas, dc	ML few ppm	cw 0.025–0.8 W
0.413 250	Kr ion	gas, dc	ML few ppm	cw 0.05–1.8 W, violet
0.44169	He–Cd	metal vapour, dc	few ppm	cw 3–80 mW, violet
0.45806	Ar ion	gas, dc	ML few ppm	cw 0.1–1 W, blue
0.488 122	Ar ion	gas, dc	ML few ppm	cw 5 mW–6 W, 2–8 μJ/< 10–15 ns, m–l, blue
0.510 696	Cu	metal vapour, dc	few ppm	p 0.2–10 mJ/20–50 ns, e 1%, 1–40 W mean, green
0.514 673	Ar ion	gas, dc	ML few ppm	cw 0.3–8 W, p < 8μJ/< 15 ns, FS (I_2), m–l, green
0.531 014	Kr ion	gas, dc	ML few ppm	cw 0.2–1.5 W, FS (I_2), green
0.532 0	Nd/YAGx2	f-doubled YAG	as 1.064μm	p 0.1–100 mJ/7–20 ns, green
0.543 516	He–Ne	gas, dc	few ppm	cw 0.1–1mW, green
0.568 348	Kr ion	gas, dc	ML few ppm	cw 0.1–1 W, FS (I_2), yellow
0.578 373	Cu	metal vapour, dc	few ppm	p 0.1–5 mJ/20–50 ns, e 0.3%, 1–20 W mean, yellow
0.611 971	He–Ne	gas, dc	few ppm	cw 0.1–2m W, FS (I_2), orange
0.632 991	He–Ne	gas, dc	few ppm	cw 0.1–60 mW, FS (I_2), most common laser, red
0.640 283	He–Ne	gas, dc	few ppm	cw 0.01–0.1 mW, FS (I_2), red
0.64727	Kr ion	gas, dc	ML few ppm	cw 0.5–3.5 W, red
0.6945	Cr/ruby	s/s, o–p (lamp)	0.1% cooled	p 1 mJ–30 J/15 ns–1 ms, the first laser, red
0.73–0.80	alexandrite	s/s, o–p	T 8%	cw 0.6 W, e < 40%, p 0.1–0.7J/100 ns, red–IR

† For approximate air wavelengths in the visible subtract 2.8 parts in 10^4.

Laser radiation

$\lambda_{vac}/\mu m$	Name	Type/excit^n	Tuning	Notes‡
0.75–0.90	GaAlAs	diode, dc (lv)	1% ext cav	cw, 0.1–10mW, p < 10W/10ns–1 ms, e 10% +, red–IR
0.752 75	Kr ion	gas, dc	ML few ppm	cw 0.1–1.2 W
0.82–3.3	F-centre	s/s, o–p	T 20%	cw 0.1–100 mW, temp 77–300 K
0.89–0.90	GaAs	diode, dc (lv)	1% cooled	cw 0.1–10mW, p < 50W/1 ns–1 μs, first diode laser
1.064	Nd/YAG	s/s, o–p (lamp)	0.03%	cw < 50 W, p 10mJ–50J/3–200 ns & > 100μs, e 1%, m–l
1.092 64	Ar ion	gas, dc	ML few ppm	cw 50–200 mW
1.1–1.6	InGaAsP	diode, dc (lv)	0.5%, temp	cw 0.17–mW, p 50 mW/0.1–50 ns, optical fibre use
1.152 590	He–Ne	gas, dc	few ppm	cw 0.1–50 mW, first gas laser, FS (I_2) at 2f
1.315 244	I	gas, o–p (lamp)	6 lines /0.01%	1–10J/10 ns–6μs, m–l, e 0.5%, 0.1 s lifetime
1.319	Nd/YAG	s/s, o–p (lamp)	as 1.064 μm	cw 0.1–6 W, p, m–l
1.523 488	He–Ne	gas, dc	few ppm	cw 0.1–20 mW
1.73	Er/YLF	s/s, o–p (lamp	as ruby	p 5 mJ/0.2 μs
2.026 777	Xe	gas, dc	few ppm	cw 1–10 mW, long laser, non-commercial
2.06	Ho/YLF	s/s, o–p (lamp)	—	cw 5 W multimode
2.395 795	He–Ne	gas, dc	few ppm	cw 0.1–10 mW, non-commercial gas mixture, 3 lines
2.8–3, 3.4–4.1	HF,DF	gas, chem., TEA	ST few ppm	cw 2–25 W, p 0.3–1 J/0.1–1 μs, Lamb dip
3–30	lead salt	diode dc (lv)	T 10%, temp	cw 0.1–0.5 mW, temp 15–90 K, various materials
3.392 231	He–Ne	gas, dc/rf	few ppm	cw 0.1–50 mW, important FS (CH_4)
3.507 986	Xe	gas, dc	few ppm	cw 1–50 mW, FS (H_2CO)
5–6.4	CO	gas, dc/chem.	ST few ppm	cw 2–30 W p 3–10 mJ/1–1000 μs, e < 10%, atm. absorption
9.0–11.0	CO_2	gas, dc	ST few ppm	cw 1–50 W, 1 kW, FS (CO_2), e ~ 5% ~ 1000 lines
9.0–11.0	CO_2 wg	gas, dc/rf	ST 10 ppm	cw 0.1–10W, FS (CO_2, OsO_4), compact
9.0–11.0	CO_2 atm	gas, TEA/o–p, rf	T 1%	p 10mJ–3J/100 ns + 3 μs tail, pressure 1–10 atm
10.3–11.0	N_2O	gas, dc	ST few ppm	cw 0.1 10W, p
10.7–13.3	NH_3	gas, o–p	MLcw 1 ppm	cw 0.2–10 W, e < 28%, p 10 mJ–1.4J/200 ns, Raman process
15.3–16.4	CF_4	gas, o–p	ST few ppm	cw 2mW at 150 K, p 0.1J/3μs, e 1%
27.970 75	H_2O	gas, dc	ML few ppm	cw 1–100 mW, FS (Lamb dip), & 78, 119 μm lines
42.159 1	CH_3OH	gas, o–p	ML few ppm	cw 0.1–10 mW
70.511 6	CH_3OH	gas, o–p	ML few ppm	cw 1–50 mW
81.496 9	NH_3	gas o–p	ML few ppm	cw 1–50 mW
118.834 1	CH_3OH	gas, o–p	ML few ppm	cw 1–200 mW
194.702 7	DCN	gas, dc	ML few ppm	cw 1–200 mW, 2 lines, & 2 at 190 μm

There are about 2 000 o–p lines between ~30 μm and 2 mm but few give > 10 mW. Pumping is by 9–11 μm lines.

continued

Laser radiation

$\lambda_{vac}/\mu m$	Name	Type/excit^n	Tuning	Notes‡
214.579 1	CH_2F_2	gas, o–p	ML few ppm	cw 1–200 mW, & 184 μm strong line
336.557 8	HCN	gas, dc/rf	few ppm	cw 1–200 mW, early sub-mm laser, & 311 μm
385	D_2O	gas, o–p	0.1%	p 10–140 mJ/0.2–1 μs, Raman process
570.568 7	CH_3OH	gas, o–p	ML few ppm	cw 0.1–5 mW ⎫ Microwave devices
963.487 3	C_2H_3Br	gas, o–p	ML few ppm	cw 0.2–5 mW ⎬ reach to wavelengths ⎭ < 500 μm

‡ Typical approximate values are given.

Symbols and abbreviations:
atm: atmosphere (pressure), atmospheric (absorption)
chem: chemical (the laser is pumped by a reaction when gases mix)
cw: continuous-wave
diode: semiconductor-junction diode (carrier injection laser)
e: efficiency (output power/excitation input power)
ext cav: in an external cavity resonator
FS(A): frequency standard when locked to a transition in system A, e.g. in I_2
lv: low voltage, a few volts
λ_{vac}: vacuum wavelength
m–l: can be mode locked (to produce a train of sub-ns pulses)
ML: multiline (the laser can be tuned to different lines)
o–p: optically-pumped
p: pulsed
ppm: parts per million
s/s: solid state (lasing impurity in a uniform host crystal)
ST: step tunable (the laser can be tuned over a regularly-spaced set of lines)
T: continuously tunable (by a frequency-selective element in the cavity)
TEA: transversely-excited atmospheric
temp: temperature
wg: waveguide
YAG: yttrium aluminium garnet
YLF: yttrium lithium fluoride

D.J.E.K.

1.7.7 Refractive index of gases

Refractive index of air

The wavelength λ_{air} of a radiation in air is related to its vacuum value λ_{vac} by $\lambda_{vac} = n\lambda_{air}$, where n is the refractive index. For standard air (dry air at 15 °C and 101 325 Pa, containing 0.03% by volume of carbon dioxide) the refractive index n_s is given by the dispersion equation (Edlén, *Metrologia*, 1966, **2**, 71)

$$(n_s - 1) \times 10^8 = 8\,342.13 + 2\,406\,030(130 - \sigma^2)^{-1} + 15\,997(38.9 - \sigma^2)^{-1}$$

where $\sigma = 1/\lambda_{vac}$ and λ_{vac} is expressed in μm. This equation is based upon observations within the range 200 nm to 2 μm, and is in better agreement with recent measurements than the equation (Edlén, *J. Opt. Soc. Am.*, 1953, **43**, 339) which was recommended by the Joint Commission for Spectroscopy in 1952, although the difference is at most 1.4×10^{-8}.

In the visible region (405–705 nm) the following approximate expression is more convenient and gives a maximum discrepancy of only 1.2×10^{-8},

$$n_s - 1 = 0.047\ 2300(173.3 - \sigma^2)^{-1}$$

For air at a temperature $t\ °C$ and a pressure p Pa, the refractivity is given by the equation

$$n_{tp} - 1 = (n_s - 1) \times \frac{p[1 + p(61.3 - t) \times 10^{-10}]}{96\ 095.4(1 + 0.003\ 661\ t)}$$

The refractivity of water vapour is less than that of air, so that if the air is moist its refractive index will be smaller than the value calculated for dry air. This water vapour term is dependent upon wavelength. In the visible region (405–644 nm) the relationship is

$$n_{tpf} - n_{tp} = -f(4.2918 - 0.0342\ \sigma^2) \times 10^{-10},$$

where n_{tpf} is the refractive index of air containing water vapour at a partial pressure of f Pa, the total pressure still being p. This equation is valid only for conditions not deviating very much from normal laboratory conditions ($t = 20\ °C$, $p = 100\ 000$ Pa, $f = 1500$ Pa).

W.R.C.R.

Refractive indices of gases

Refractive index for the wavelength 589.3 nm (mean of sodium D lines) at a pressure of 101 325 Pa and temperature of 0 °C, relative to a vacuum.

Gas	Refractive index	Gas	Refractive index
Acetone	1.001 090	Helium	1.000 036
Air	1.000 292	Hydrochloric acid	1.000 447
Ammonia	1.000 376	Hydrogen	1.000 132
Argon	1.000 281	Hydrogen sulphide	1.000 634
Benzene	1.001 762	Methane	1.000 444
Bromine	1.001 132	Methyl alcohol	1.000 586
Carbon dioxide	1.000 451	Methyl ether	1.000 891
Carbon disulphide	1.001 481	Nitric oxide	1.000 297
Carbon monoxide	1.000 338†	Nitrogen	1.000 297
Chlorine	1.000 773	Nitrous oxide	1.000 516
Chloroform	1.001 450	Oxygen	1.000 272
Ethyl alcohol	1.000 878	Pentane	1.001 711
Ethyl ether	1.001 533	Sulphur dioxide	1.000 686
		Water vapour	1.000 254

† Value for white light.

M.D.

Refractive indices of gases at radio frequencies

Values below are for dry gases at 0 °C, 101.325 kN m^{-2}. The quoted uncertainty limits are about 3σ (σ = standard error of the mean).

Gas	$(n-1)/10^{-6}$
Air (CO$_2$ free)	288.15 ± 0.1
Deuterium	134.8 ± 0.3
Helium	35.0 ± 0.2
Carbon dioxide	494 ± 1.0
Hydrogen	136.0 ± 0.2
Nitrogen	294.1 ± 0.1
Oxygen	266.4 ± 0.2
Water vapour†	60.7 ± 0.1

† At 20°C, 1.333 kPa (10 mmHg).

Refractive index of moist air at radio frequencies

The following formula has been derived from measured values and the gas laws, and holds over a wide range of conditions:

$$(n-1) \times 10^6 = \frac{103.49}{T} p_1 + \frac{177.4}{T} p_2 + \frac{86.26}{T} \left(1 + \frac{5748}{T}\right) p_3$$

where p_1 = partial pressure of dry air in mmHg
p_2 = partial pressure of carbon dioxide in mmHg
p_3 = partial pressure of water vapour in mmHg
T = thermodynamic temperature in K.

W.R.C.R.

1.7.8 Refractive index of optical materials

Spectral lines for refractometry

Refracting optical materials are specified by their refractive indices for a series of wavelengths. Those listed below were recommended by the International Commission for Optics in 1962 (*Optica Acta*, 1963, **10**, 217), and are quoted to the nearest 0.1 nm for standard air. The source element and the designating letter used for some wavelengths (mainly in the visible spectral region) are shown. A number of other spectral lines at which refractive indices have been frequently quoted is also given, including some commonly used laser wavelengths.

Gas discharge lamps are available as sources over a wide range of wavelengths, and in some instances the spectra of two or three elements may be combined in one lamp. Interference filters give the most complete suppression of unwanted lines.

Source	Wavelength nm	Designating letter	Source	Wavelength nm	Designating letter
ICO 1962			Other commonly used lines		
Recommended standard lines					
Zn	213.9		Hg	365.0	i
Hg	237.8		Hg	365.5	
Hg	269.9		H	434.0	G′
Hg	289.4		H	486.1	F
Hg	296.7		Hg	577.0	
Hg	334.2		Hg	579.0	
Hg	366.3		Na	589.0	} D
Hg	404.7	h	Na	589.6	
Hg	435.8	g	H	656.3	C
Cd	467.8		K	766.5	
Cd	508.6		K	769.9	
Hg	546.1	e	Hg	1357.0	
He	587.6	d	Hg	1367.3	

Source	Wavelength nm	Designating letter	Source	Wavelength nm	Designating letter
ICO 1962			Common laser wavelengths		
Recommended standard lines					
Cd	643.8	C'			
He	667.8				
He	706.5	r	Ar	488.0	
Rb	780.0		Ar	514.5	
Cs	852.1	s	Kr	568.2	
Cs	894.4		He–Ne	632.8	
Hg	1014.0	t	Kr	647.0	
Hg	1128.7		Ruby	694.3	
Hg	1395.1				
Hg	1529.5				
Hg	1813.1			μm	
Hg	1970.1				
He	2058.1		Nd	1.06	
Hg	2325.4		Nd–YAG	1.065	
Recommended secondary lines			He–Ne	3.39	
			CO₂	10.6	
Hg	239.9				
Hg	275.3				
Hg	292.5				
Hg	390.6				
Cd	480.0	F'			
Rb	794.7				

Refractive indices of optical glasses

Optical glasses are available in a wide range of refractive indices and dispersions. Data are usually given in the following way:

Refractive index for the helium d line (587.6 nm), n_d
and for the mercury e line (546.1 nm), n_e

Mean dispersion between the hydrogen F line (486.1 nm) and C line (656.3 nm), $n_F - n_C$
and between the cadmium F' line (480.0 nm) and C' line (643.8 nm), $n_{F'} - n_{C'}$

Reciprocal dispersive power (or constringence), $V_d = (n_d - 1)/(n_F - n_C)$
and $V_e = (n_e - 1)/(n_{F'} - n_{C'})$

Refractive indices are generally quoted for a number of wavelengths. Partial dispersions, for example $(n_d - n_C)$ or $(n_e - n_d)$, and relative partial dispersions, for example $(n_e - n_d)/(n_F - n_C)$ or $(n_e - n_d)/(n_{F'} - n_{C'})$, may also be stated.

A six-figure reference number for each glass is obtained from the first three significant figures of the term $(n_d - 1)$ and the first three significant figures of V_d. For example, BSC 510644 is a borosilicate crown glass with $n_d = 1.509\,70$ and $V_d = 64.44$. Glasses with n_d less than 1.6 and V_d greater than 55, and with n_d greater than 1.6 and V_d greater than 50, are described as crowns and the remainder as flints.

Some data from a selection of optical glasses manufactured by Chance Pilkington Limited are given in the table for borosilicate crown, hard crown, zinc crown, medium barium crown, dense barium crown, telescope flint, barium flint, borate flint, extra light flint, light flint, dense flint, extra dense flint, double extra dense flint, lanthana crown and lanthana flint glasses. The refractive indices are quoted at a temperature of 20 °C and at standard atmospheric pressure (101 325 Pa).

Optical glasses

Some data for Chance Pilkington optical glasses
(For complete range of optical glasses see Chance Pilkington catalogue)

Refractive index	n_d	$n_F - n_C$	V_d	n_e	$n_{F'} - n_{C'}$	V_e	n_i	n_h	n_g	$n_{F'}$	n_C	n_r	n_t
Wavelength nm	587·56	486·13–656·28		546·07	479·99–643·85		365·02	404·66	435·84	479·99	656·28	706·52	1014·00
BSC 510644	1·509 70	0·007 910	64·44	1·511 59	0·007 964	64·24	1·528 82	1·522 90	1·519 41	1·515 62	1·507 27	1·505 86	1·500 41
HC 519604	1·518 99	0·008 590	60·42	1·521 04	0·008 661	60·16	1·540 06	1·533 47	1·529 61	1·525 45	1·516 37	1·514 87	1·509 24
ZC 508612	1·507 59	0·008 300	61·16	1·509 57	0·008 361	60·95	1·527 76	1·521 49	1·517 80	1·513 81	1·505 04	1·503 58	1·497 90
MBC 569561	1·568 83	0·010 135	56·13	1·571 25	0·010 227	55·86	1·594 03	1·586 07	1·581 44	1·576 47	1·565 76	1·564 02	1·557 61
DBC 610573	1·610 29	0·010 660	57·25	1·612 83	0·010 756	56·97	1·636 71	1·628 42	1·623 56	1·618 33	1·607 06	1·605 23	1·598 48
TF 530512	1·530 33	0·010 360	51·19	1·532 80	0·010 464	50·92	1·556 61	1·548 18	1·543 32	1·538 17	1·527 21	1·525 44	1·519 00
BF 606439	1·605 62	0·013 787	43·93	1·608 89	0·013 953	43·64	1·641 81	1·629 90	1·623 17	1·616 12	1·601 52	1·599 24	1·591 29
Bor F 614439	1·614 00	0·013 990	43·89	1·617 33	0·014 139	43·66	1·650 13	1·638 33	1·631 65	1·624 61	1·609 80	1·607 44	1·598 74
ELF 548456	1·547 69	0·012 010	45·60	1·550 54	0·012 146	45·33	1·578 85	1·568 67	1·562 89	1·556 82	1·544 10	1·542 10	1·534 90
LF 579411	1·578 60	0·014 070	41·12	1·581 94	0·014 249	40·84	1·615 95	1·603 55	1·596 60	1·589 34	1·574 44	1·572 14	1·564 19
DF 620364	1·620 04	0·017 050	36·37	1·624 08	0·017 283	36·11	1·666 28	1·650 64	1·642 02	1·633 10	1·615 03	1·612 27	1·602 83
EDF 706300	1·705 85	0·023 528	30·00	1·711 40	0·023 900	29·77	1·773 38	1·749 38	1·736 73	1·724 00	1·699 03	1·695 30	1·682 73
DEDF 805254	1·805 18	0·031 660	25·43	1·812 65	0·032 197	25·24	1·896 45	1·864 26	1·847 03	1·829 69	1·796 08	1·791 16	1·775 12
LAC 651585	1·651 60	0·011 134	58·52	1·654 25	0·011 227	58·27	1·678 98	1·670 42	1·665 39	1·659 98	1·648 21	1·646 28	1·639 04
LAF 717479	1·717 00	0·014 970	47·90	1·720 56	0·015 122	47·65	1·755 05	1·742 81	1·735 78	1·728 33	1·712 49	1·709 95	1·700 63

Optical plastics

Refractive index, n, and reciprocal dispersive power, V, are given for the four most commonly used optical plastics. The data have been gathered from a variety of sources, including the manufacturer, and are subject to change as more information is generated.

Refractive index Wavelength/nm	n_F 486.1	n_D 589.3	n_C 656.3	V
Methyl methacrylate (acrylic)	1.498	1.492	1.489	57.2
Polystyrene (styrene)	1.604	1.590	1.585	30.8
Polycarbonate	1.598	1.584	1.578	30.1
Methyl methacrylate styrene copolymer (NAS)	1.574	1.563	1.558	33.5

Refractive indices of optical cements

Refractive index, n_D^{20}, for the wavelength 589.3 nm (mean of sodium D lines) at a temperature of 20 °C is given for a number of well-tried polymerized cements (*Optica Acta*, 1967, **14**, 401). With the increasing availability of new resins for optical purposes, inspection of manufacturer's data is recommended to determine the most appropriate cement for a specific application.

Cement	n_D^{20}	Cement	n_D^{20}
Solid Canada balsam	1.54	Beetle 8128	1.55
Soft Canada balsam	1.53	BS No. 8	1.59†
Cellulose caprate	1.47	Epikote 817	c.1.57†
Cellulose caprate (plasticized)	1.49	Crystic 191 LV	1.56†
H.T.	1.45	Loctite 357 (UV curing)	1.47†

† Manufacturer's data.

Refractive indices of calcite (Iceland Spar)

(From 200 nm to 706 nm at 18 °C, and from 801 nm to 3324 nm at 20 °C)

Wavelength nm	Refractive index		Wavelength nm	Refractive index	
	O-ray	E-ray		O-ray	E-ray
200	1.902 84	1.576 49	1422	1.635 90	
303	1.719 59	1.513 65	1497	1.634 57	1.477 44
410	1.680 14	1.496 40	1609	1.632 61	
508	1.665 27	1.489 56	1682	1.631 27	
643	1.655 04	1.484 90	1749		1.476 38
706	1.652 07	1.483 53	1761	1.629 74	
801	1.648 69	1.482 16	1849	1.628 00	
905	1.645 78	1.480 98	1909		1.475 73
1042	1.642 76	1.479 85	1946	1.626 02	
1159	1.640 51	1.479 10	2053	1.623 72	
1229	1.639 26	1.478 70	2100		1.474 92
1307	1.637 89	1.478 31	2172	1.620 99	
1396	1.636 37	1.477 89	3324		1.473 92

Refractive indices of crystal quartz at 20 °C

Wavelength nm	Refractive index O-ray	Refractive index E-ray	Wavelength nm	Refractive index O-ray	Refractive index E-ray
185	1.675 78	1.689 88	1541.4	1.527 81	1.536 30
198	1.650 87	1.663 94	1681.5	1.525 83	1.534 22
231	1.613 95	1.625 55	1761.4	1.524 68	1.533 01
340	1.567 47	1.577 37	1945.7	1.521 84	1.530 04
394	1.558 46	1.568 05	2053.1	1.520 05	1.528 23
434	1.553 96	1.563 39	2300	1.515 61	
508	1.548 22	1.557 46	2600	1.509 86	
589.3	1.544 24	1.553 35	3000	1.499 53	
768	1.539 03	1.547 94	3500	1.484 51	
832.5	1.537 73	1.546 61	4000	1.466 17	
991.4	1.535 14	1.543 92	4200	1.456 9	
1159.2	1.532 83	1.541 52	5000	1.417	
1307.0	1.530 90	1.539 51	6450	1.274	
1395.8	1.529 77	1.538 32	7000	1.167	
1479.2	1.528 65	1.537 16			

Refractive indices of synthetic materials at 20 °C

Wavelength nm	Synthetic fused silica	Calcium fluoride	Lithium fluoride	Wavelength nm	Synthetic fused silica	Calcium fluoride	Lithium fluoride
213.86	1.534 30	1.484 96	1.432 29	780.02	1.453 67	1.430 79	1.389 24
237.83	1.514 72	1.472 15	1.422 56	794.76	1.453 41	1.430 63	1.389 07
239.94	1.513 36	1.471 25	1.421 86	852.11	1.452 47	1.430 07	1.388 49
269.89	1.498 04	1.460 98	1.413 93	894.35	1.451 84	1.429 71	1.388 09
275.28	1.495 91	1.459 54	1.412 80	1014.0	1.450 24	1.428 84	1.387 07
289.36	1.490 99	1.456 20	1.410 18	1060.0	1.449 68		1.386 70
292.54	1.489 99	1.455 52	1.409 65	1128.7	1.448 87	1.428 15	1.386 17
296.73	1.488 73	1.454 67	1.408 97	1395.1	1.445 83	1.426 80	1.384 10
334.15	1.479 77	1.448 52	1.404 10	1529.5	1.444 27	1.426 17	1.383 02
366.33	1.474 35	1.444 79	1.401 11	1813.1	1.440 71	1.424 83	1.380 56
404.66	1.469 62	1.441 52	1.398 47	1970.1	1.438 53	1.424 06	1.379 09
435.84	1.466 69	1.439 50	1.396 83	2058.1	1.437 23	1.423 61	1.378 22
467.82	1.464 29	1.437 85	1.395 47	2325.4	1.432 93	1.422 16	1.375 30
479.99	1.463 50	1.437 31	1.395 02	2800.0		1.419 23	1.369 37
486.13	1.463 12	1.437 05	1.394 80	3400.0		1.414 87	1.360 32
508.58	1.461 86	1.436 18	1.394 08	4000.0		1.409 63	1.349 37
546.07	1.460 08	1.434 97	1.393 05	4600.0		1.403 57	1.336 40
587.56	1.458 46	1.433 89	1.392 11	5000.0		1.399 08	1.326 56
632.80	1.457 02		1.391 26	5600.0			1.309 88
643.85	1.456 70	1.432 72	1.391 07	5893.2		1.387 12	
656.28	1.456 37	1.432 50	1.390 87	6000.0			1.297 40
667.82	1.456 07	1.432 31	1.390 69	7071.8		1.368 05	
706.52	1.455 15	1.431 71	1.390 14	8250.5		1.344 40	
769.90	1.453 86	1.430 91	1.389 35	9429.1		1.316 05	

Refractive indices and useful transmission ranges of other optical materials

Values at or near a temperature of 20 °C.

Infrared materials	Wavelength μm	Refractive index	Transmission range/μm
Arsenic trisulphide.	1.00	2.4777	0.6 to 13
	10.00	2.3816	
Barium fluoride.	0.546	1.4759	~0.25 to ~15
	10.346	1.3964	
Cadmium telluride	1.00	2.838	0.9 to ~16
(Irtran 6)	10.00	2.672	
Caesium bromide	1.0	1.6779	~0.3 to ~55
	39.0	1.5624	
Caesium iodide	1.00	1.7572	0.25 to ~80
	50.0	1.6366	
Diamond	0.546	2.4235	~0.25 to >80
Gallium arsenide	10.0	3.135	1 to ~15
Germanium	10.00	4.0032	1.8 to 23
Lead fluoride	0.55	1.7722	0.25 to ~16
	10.00	1.6367	
Magnesium oxide	1.00	1.7227	0.25 to 8.5
(Irtran 5)	8.00	1.4824	
Potassium bromide	0.546	1.5639	~0.25 to 40
	21.18	1.4866	
Potassium chloride.	0.546	1.4932	0.21 to ~30
	20.4	1.389	
Potassium iodide	0.546	1.6731	~0.25 to ~45
	20	1.5964	
Silicon.	10.00	3.4170	1.2 to >15
Silver chloride	1.0	2.0224	0.4 to ~28
	20.0	1.9069	
Sodium chloride	0.50	1.5518	0.21 to >26
	20.0	1.3831	
Sodium fluoride	0.546	1.3264	<0.19 to 15
	10.3	1.233	
Strontium titanate.	0.56	2.4254	0.39 to 6.8
	5.00	2.1221	
Thallium bromo-iodide	0.54	2.6806	0.6 to >40
(KRS 5)	30.0	2.2887	
Zinc selenide	1.00	2.485	0.45 to ~21.5
(Irtran 4)	15.00	2.370	
Zinc sulphide	1.00	2.2907	1.0 to 14.5
(Irtran 2)	12.00	2.1688	
Zinc sulphide	0.546	2.3884	0.4 to 14.5
(Cleartran)	12.00	2.1710	

Birefringent materials	Wavelength μm	Refractive index		Transmission range/μm
		O-ray	E-ray	
Ammonium dihydrogen phosphate	0.546	1.5266	1.4808	0.13 to ~ 1.7
(ADP)	1.014	1.5084	1.4690	
Potassium dihydrogen phosphate	0.546	1.5115	1.4698	0.25 to ~ 1.7
(KDP)	1.014	1.4954	1.4604	
Magnesium fluoride	0.546	1.3786	1.3904	0.11 to 7.5
	1.083	1.3731	1.3846	
Rutile	0.546	2.652	2.958	~0.43 to 6.2
(Titanium dioxide)	1.014	2.484	2.747	
Sapphire	0.546	1.7708	—	0.14 to 6.5
(Aluminium oxide)	1.014	1.7555	1.7479	

Refractive indices of liquids

Suitable media for refractometry.

Refractive index for the wavelength 589.3 nm (mean of sodium D lines) and temperature of 20 °C or close to 20 °C.

Liquid	Refractive index
Water	1.333
Paraldehyde	1.405
Carbon tetrachloride	1.46
Glycerol	1.47
Liquid paraffin	1.48
Toluene	1.497
Benzene	1.501
Ethyl salicylate	1.523
Chlorobenzene	1.525
Methyl salicylate	1.538
Ethyl cinnamate	1.559
Benzyl benzoate	1.568
Aniline	1.586
Quinoline	1.627
α-Monobromonaphthalene	1.660
Mercury potassium iodide	1.717
Methylene iodide	1.737
Methylene iodide and sulphur (saturated)	1.78
Barium mercuric iodide aq.	1.793
Potassium iodide and mercuric iodide aq.	1.82†
Solution 35% by weight CH_2I_2 Solution 31% by weight SnI_4 Solution 16% by weight AsI_3 Solution 8% by weight SbI_3 Solution 10% by weight S	1.868

Refractive indices of liquids (*contd*)

Liquid	Refractive index
Hydrogen disulphide	1.885
Phosphorus in carbon disulphide	1.95†
Yellow phosphorus, 8 parts by weight ⎫	
Yellow sulphur 1 part by weight ⎬	2.06
Methylene iodide 1 part by weight ⎭	
Mercuric iodide in aniline or quinoline	2.2†
Oil, paraffin	1.44
Oil, olive	1.46
Oil, turpentine	1.47
Oil, cedar	1.516
Oil, cloves	1.532
Oil, cinnamon	1.601

† Maximum value obtainable.

Calibration of liquid refractometers

Liquids of known refractive index, n, may be used to calibrate critical angle and Vee-block refractometers throughout the visible spectral region, and over a small range of ambient temperatures using the temperature coefficient of refractive index, dn/dt.

Wavelength nm	Toluene 'Ultrar' grade†		Distilled water	
	$n^{20\,°C}$	dn/dt per °C at 20 °C	$n^{20\,°C}$	dn/dt per °C (20 °C–25 °C)
404.66	1.526 120	− 0.000 597	1.342 742	− 0.000 101
435.84	1.517 830	− 0.000 588	1.340 210	− 0.000 100
479.99	1.509 285	− 0.000 577		
486.13	1.508 315	− 0.000 576	1.337 123	− 0.000 099
546.07	1.500 715	− 0.000 565	1.334 466	− 0.000 098
587.56	1.496 920	− 0.000 560	1.333 041	− 0.000 097
589.00	1.496 800	− 0.000 560	1.332 988	− 0.000 097
589.59	1.496 755	− 0.000 560		
632.80	1.493 680	− 0.000 556	1.331 745	− 0.000 096
643.85	1.493 005	− 0.000 555		
656.28	1.492 285	− 0.000 554	1.331 151	− 0.000 096
706.52	1.489 795	− 0.000 551	1.330 020	− 0.000 095

† Toluene 'Ultrar' grade is marketed by Hopkin and Williams Ltd.

Light transmission, Chance Pilkington optical glass

Internal transmittance, T_{25}, is defined as the ratio of radiant flux reaching the exit surface of a plane parallel plate of glass 25 mm thick to the flux which leaves the entry surface.

Glass type	BSC 510644	HC 519604	ZC 508612	MBC 569561	DBC 610573	LF 579411	DF 620364	EDF 706300	DEDF 805254
Wavelength nm					T_{25}				
360	0.915	0.897	0.747	0.875	0.799	0.816	0.749	—	—
380	0.941	0.930	0.872	0.945	0.930	0.908	0.889	0.342	0.203
400	0.980	0.978	0.954	0.981	0.971	0.969	0.958	0.768	0.609
420	0.982	0.978	0.968	0.987	0.983	0.974	0.972	0.886	0.837
440	0.982	0.976	0.974	0.989	0.985	0.973	0.977	0.922	0.930
460	0.987	0.975	0.982	0.990	0.989	0.978	0.986	0.939	0.968
500	0.989	0.987	0.990	0.995	0.995	0.984	0.992	0.966	0.991

In the wavelength range 500 nm to 1000 nm, $T_{25} \nless 0.970$ for the glasses shown in the Table.
Data taken from Chance Pilkington optical glass catalogue.

References

(1) W. G. Driscoll (ed.), *Handbook of the Optical Society of America*, 1978 (McGraw-Hill Book Co.), Chapter 7.
(2) D. E. Gray (ed.), *Handbook of the American Institute of Physics*, 1972 (McGraw-Hill Book Co.), Section 6.

M.D.

1.7.9 Light reflection

The reflectance R at normal incidence at the interface between two non-absorbing media of refractive index n_0 and n_1 is given by

$$R = [(n_0 - n_1)/(n_0 + n_1)]^2$$

$R(\%)$ for various values of n_1 and $n_0 = 1$ (air) are tabulated below

n_1	1.46	1.52	1.60	1.80	2.0	3.0	4.0
$R(\%)$	3.5	4.3	5.3	8.2	11.1	25.0	36.0

For light at an angle of incidence θ_0 and angle of refraction θ_1 in the two media, the reflectance R depends on the state of polarization of the light and is given by

$$R = [(\mathcal{N}_0 - \mathcal{N}_1)/(\mathcal{N}_0 + \mathcal{N}_1)]^2$$

where for light polarized with the electric vector parallel to the plane of incidence (p vibration, TM wave)

$$\mathcal{N}_0 = n_0/\cos\theta_0 \qquad \mathcal{N}_1 = n_1/\cos\theta_1$$

while for light with the electric vector perpendicular to the plane of incidence (s vibration, TE wave)

$$\mathcal{N}_0 = n_0\cos\theta_0 \qquad \mathcal{N}_1 = n_1\cos\theta_1$$

For incident unpolarized light, the reflectance is the mean of R_p and R_s. Values tabulated below are for an air/glass interface ($n_0 = 1$, $n_1 = 1.52$)

Angle of incidence	R_p (%)	R_s (%)	Angle of incidence	R_p (%)	R_s (%)
0°	4.3	4.3	56.7°	0.0	15.7
15°	3.9	4.7	60°	0.2	18.3
30°	2.7	6.1	75°	10.6	40.8
45°	0.9	9.7	85°	49.2	73.8

The angle at which $R_p = 0$ is given by $\tan\theta_0 = n_1/n_0$ and is known as the Brewster angle.

At grazing incidence ($\theta_0 = 90$), $R_p = R_s = 100\%$.

The reflection from a dielectric surface may readily be modified by the application of thin film coatings. For a single dielectric film located between the two bulk media of refractive indices n_0 and n_1, the reflectance varies with the film thickness d_f and refractive index n_f. When the film optical thickness ($n_f d_f$) equals one quarter of the wavelength of the incident light, the reflectance is a minimum if n_f is between n_0 and n_1, or a maximum if it lies outside this range. The reflectance R at normal incidence is then given by

$$ R = \left(\frac{n_f^2 - n_0 n_1}{n_f^2 + n_0 n_1} \right)^2 $$

Typical refractive indices of common thin film materials (wavelength 550 nm) and the reflectance of quarter-wave films on glass ($n_0 = 1$, $n_1 = 1.52$) are shown below. The reflectances of quarter-wave multi-layer stacks of alternately high and low refractive index films on a glass substrate are also tabulated.

Material	Refractive Index (n_f)	Reflectance (%)
Cryolite (Na_3AlF_6)	1.35	0.8
Magnesium fluoride	1.38	1.3
Silicon dioxide	1.46	2.8
Aluminium oxide	1.63	7.4
Lead fluoride	1.75	11.3
Zirconium dioxide	2.05	22.0
Tantalum pentoxide	2.15	25.5
Zinc sulphide	2.35	32.5
Titanium dioxide	2.2–2.7	27.3–42.9
Multilayer stack (HL . . . LH)		
(ZnS/MgF_2)		
3 layer		68.3
5 layer		87.7
7 layer		95.6
9 layer		98.5
11 layer		99.5

For complete elimination of the reflected beam the refractive index of a single film should equal $\sqrt{n_0 n_1}$, when the reflectance will be zero for the wavelength at which the film is quarter-wave in optical thickness. With the range of thin film materials available, complete suppression of the reflected beam cannot be achieved using a single film on common optical glasses. However, with a multilayer design, high-efficiency antireflection coatings on glass can be produced with a reflectance of less than 0.5% through the visible spectrum.

High refractive index films can usefully be employed as beam splitters. At oblique angles of incidence, appropriate design of a multilayer system allows the intensity and state of polarization of the transmitted and reflected components to be varied over wide limits.

For metals and other opaque media the normal incidence reflectance R is given by

$$R = \frac{(n_1 - n_0)^2 + k_1{}^2}{(n_1 + n_0)^2 + k_1{}^2}.$$

where n_0 is the refractive index of the entrance medium and n_1 and k_1 are the real and imaginary parts of the complex refractive index $n_1 = n_1 - ik_1$ of the absorbing medium. n_1 and k_1 are usually known as the refractive index and absorption index respectively.

Typical optical constants and reflectance (normal incidence) of metals and semi-conductors in air

Material	Wavelength μm	n	k	$R(\%)$
Aluminium	0.22	0.14	2.35	91.8
	0.546	0.82	5.99	91.6
	10.0	25.4	67.3	98.1
Silver	0.55	0.055	3.32	98.2
	10.0	10.7	69.0	99.1
Gold	0.55	0.33	2.32	81.5
	10.0	7.4	53.4	99.0
Rhodium	0.546	1.62	4.63	77.1
Chromium	0.546	2.51	2.66	48.2
Nickel	0.54	1.85	3.27	60.7
Copper	0.55	0.76	2.46	66.8
Steel	0.55	2.4	3.4	58.5
Silicon	0.546	4.05	0.03	36.5
	5.0	3.422	0	30.0
Germanium	0.545	5.15	2.15	51.5
	10.0	4.003	0	36.0

Values of the optical constants exhibit significant variations depending on whether the material is in bulk or thin film form and its method of preparation. The reflectance of a metal surface is altered by the build up of oxide or other surface layers, while very dramatic changes can be produced by the deposition of thicker dielectric films.

Opaque metal films, e.g. aluminium, silver and gold (in the infrared), are commonly used as high reflectance mirrors, while very thin semi-transparent films (i.e. a few nm thick) can be employed as beam splitters, although there is significant light loss due to absorption.

R.J.K.

1.7.10 Optical rotation

Natural optical activity

A number of crystals, liquids, solutions and vapours rotate the electric vector of linearly polarized light passing through them, this property being known as optical activity. The rotation is proportional to the thickness of the medium traversed and its sense is related to the direction of the light beam. Rotations that are clockwise to an observer looking against the direction of the light beam are said to be right handed or dextrorotatory and are considered positive; anticlockwise rotations are said to be left handed or laevorotatory.

In many chemical compounds, both organic and inorganic, the optical activity is due to asymmetry of the structure of the molecule and is retained in solution form. However, some optically active crystals lose this property on fusion or in solution, the optical activity being due not to the individual molecules but to their orientation within the crystal (e.g. crystalline quartz). At wavelengths well away from an absorption band the optical rotation varies approximately with wavelength λ as $1/\lambda^2$.

The specific rotation of a solid is defined as the optical rotation in degrees produced by a 1 mm thickness of the solid. For a compound in solution the specific rotation $[\alpha]$ is given by

$$[\alpha] = 10^4 \alpha/dc$$

where α is the measured rotation in degrees, d the path length of the solution in mm and c the concentration in g/100 cm^3.

Specific rotation (degrees) at 20 °C

Wavelength (nm)	Quartz	Sucrose†
404.7	48.112	—
435.8	41.546	—
480.0	33.674	—
546.1	25.535	78.39
589.3	21.726	66.57
632.8	18.690	57.20

† Sucrose concentration: 26 g/100 cm^3 in water.

Specific rotation values from *Proceedings of the 18th Session of ICUMSA, 1982* (The International Commission for Uniform Methods of Sugar Analysis).

Magnetic rotation of polarized light

Polarized light passing through a medium in which there is a magnetic field H parallel to the direction of propagation suffers a rotation of the plane of vibration $\alpha = rHl$, where l is the length of path and r is a constant for the material known as Verdet's constant. r may be conveniently expressed in terms of rotation per ampere turn. It is considered positive if, as is usually the case, the rotation produced by passage through a solenoid with a core of the material is in the same sense as the circulating current. As a rough guide r may be taken to vary with wavelength λ as $1/\lambda^2$.

Verdet constants in the literature are still frequently expressed in terms of min/oersted/cm; these must be multiplied by the factor 1.2566 to give values in min/A, i.e. as given in the tables below.

Verdet constants

Gases†		Solids		
Gas	$r/(10^{-6}\ \text{min A}^{-1})$	Solid	λ/nm	$r/(10^{-2}\ \text{min A}^{-1})$
Hydrogen	8.867	Fused silica at 25 °C . .	546.1	2.175
Helium	0.667		435.8	3.565
Nitrogen	8.861		334.2	6.48
Oxygen	7.598		302.2	8.29
Carbon dioxide	13.22		284.8	9.59
Methane	24.15			
Propane	50.05	Crystalline quartz at 20 °C	589.3	2.091
iso-Butane	68.12	(along axis)	546.1	2.453
			435.8	3.997
			257.3	13.56

Chance–Pilkington Glasses at 20 °C and 589.3 nm		Rock salt	670.8	3.08
Glass	$r/(10^{-2}\ \text{min A}^{-1})$		404.7	9.74
			259.9	34.03
Hard Crown ($n_D = 1.519$)	2.4	Fluorite (calcium fluoride)	589.3	1.127
Medium Barium Crown ($n_D = 1.572$)	2.5		435.8	2.158
			253.7	7.526
Dense Barium Crown ($n_D = 1.612$)	2.4	Hoya Faraday . . .FR4	632.8	−13.07
Light Flint ($n_D = 1.579$)	3.9	rotator glass FR5	632.8	−31.54
Dense Flint ($n_D = 1.623$)	4.85			
Extra Dense Flint ($n_D = 1.700$)	6.55			
SF 57* ($n_D = 1.846$)	10.3			
Plate Glass ($n_D = 1.52$)	2.3			

Water at 20 °C		Liquids†		
λ/nm	$r/(10^{-2}\ \text{min A}^{-1})$	Liquid	Temp./°C	$r/(10^{-2}\ \text{min A}^{-1})$
		Acetone	20	1.42
589.3	1.645	Methyl alcohol	20	1.17
546.1	1.935	Ethyl alcohol	20	1.41
435.8	3.168	Carbon disulphide	18	5.28
334.2	5.931	Carbon tetrachloride	15	4.03
302.2	7.677	Chloroform	20	2.06
280.4	9.418	Pentane	15	1.48
248.2	13.567	Hexane	15	1.57
		Toluene	28	3.38

† For $\lambda = 589.3$ nm.
* Schott low stress optical coefficient glass.

R.J.K.

1.7.11 Electro-optic materials

The electro-optic effect

In general, an electric field E applied to a transparent material modifies the refractive index n_0 for a particular mode of propagation in accordance with the equation

$$\frac{1}{n^2} = \frac{1}{n_0^2} + rE + RE^2 + \cdots \text{higher terms}$$

In amorphous materials or centro-symmetric crystals the coefficients of odd powers of E are zero, and the first non-zero term RE^2 represents the quadratic **Kerr** effect, usually characterized for a particular material by the Kerr constant:

$$B = \frac{\Delta n}{\lambda_0 E^2} = \frac{n_0^3}{2\lambda_0} R$$

where λ_0 is the free-space wavelength.

For the case of a birefringent crystal, the field E modifies the index ellipsoid $\sum_{i=1}^{3} (x_i^2/n_i^2) = 1$, in the principal axis co-ordinate system, to:

$$\sum_{i=1}^{3} \sum_{j=1}^{3} \sum_{k=1}^{3} \sum_{l=1}^{3} \left(\frac{1}{n_{ij}^2} + r_{ijk}E_k + R_{ijkl}E_kE_l + \cdots \right) x_i x_j = 1$$

where $n_{ij} = n_i$ for $i = j$, $n_{ij} = 0$ for $i \neq j$. From a knowledge of the change in orientation and dimensions of the index ellipsoid, as given by this equation, the field-induced birefringence for a ray of given direction and polarization may be calculated.

In practice the electro-optic effect is either predominantly linear or quadratic in E, and is therefore characterized by either r_{ijk} or R_{ijkl}, depending on the material. Since r_{ijk} is symmetrical in i, j, and R_{ijkl}, symmetrical in i, j and in k, l, pairs of values i, j and k, l are denoted by indices m and n respectively, which run from 1 to 6 according to the scheme:

$$1 \leftrightarrow 1, 1 \quad\quad 2 \leftrightarrow 2, 2 \quad\quad 3 \leftrightarrow 3, 3 \quad\quad 4 \leftrightarrow 2, 3 \quad\quad 5 \leftrightarrow 1, 3 \quad\quad 6 \leftrightarrow 1, 2$$

When measured at low frequencies ($< 10^4$ Hz), the coefficients may include contributions from the elasto-optic effects of piezo-electric and/or electrostrictive strains. The linear electro-optic effect observed at frequencies sufficiently high for these contributions to be negligible is called the **Pockels** effect. The strain effects may contribute up to 50% of the low-frequency effect.

The quadratic effect is exhibited most markedly by materials in which the permittivity is high and varies rapidly with temperature. Here, since the R_{mn} vary accordingly, it is more convenient to replace $R_{mn}E_kE_l$ in the equation of the index ellipsoid by $g_{mn}P_kP_l$ (where $\boldsymbol{P} = \boldsymbol{D} - \epsilon_0\boldsymbol{E}$), and to express the properties of the material in terms of g_{mn}.

Typical parameters for selected electro-optic materials at low frequencies are given in the table on p. 116. Co-ordinate convention: the indicatrix and electro-optic coefficients are referred to the usual crystallographic co-ordinate system as follows. $Ox_3 \equiv Oz$ and is the fourfold axis for cubic and tetragonal symmetries, or the threefold axis for trigonal symmetry; $Ox_1 \equiv Ox$; $Ox_2 \equiv Oy$, except for the trigonal case in which Ox_1 is perpendicular to the mirror plane. For uniaxial crystals $n_1 = n_2 = n_0$, $n_3 = n_e$.

It should be noted that the half-wave voltage V_π is used conventionally to characterize the sensitivity of an electro-optic material. It is the voltage required to obtain one half-wavelength of optical path difference between the two vibration components of a wavefront, using a cube of the material of side 1 cm with specified directions of light and applied field. In the table below, half-wave voltages are quoted for $\lambda = 633$ nm and correspond to the following conditions:

Material symmetry	Light direction	Field direction	Induced birefringence Δn
∞	any	transverse	$B\lambda_0 E^2$
$\bar{4}2$	Oz	Oz	$n_0^3 r_{63} E_3$
$\bar{4}3m$	Ox	Oz	$n_0^3 r_{41} E_3$
$m3m$	Ox	Oz	$n_0^3(g_{11} - g_{12})P_3^2/2$
$3m$	Ox	Oz	$n_3^3 \left\{ r_{33} - \left(\dfrac{n_1}{n_3}\right)^3 r_{13} \right\} E_3/2$
$4mm$	$45°$ to Oz	$45°$ to Oz longitudinal	$(n_3^3/4\sqrt{2})(r_{13} - r_{33})E$

Further reading

For a clear description of tensor representation see J. F. Nye, *Physical Properties of Crystals*, 1964 (Oxford University Press). For discussion of applications of electro-optic materials see A. F. Harvey, *Coherent Light*, 1970 (John Wiley & Sons).

O.C.J.

Properties of selected electro-optic materials

Material	Point-group symmetry	V_π/kV	Coefficient	Value		Refractive index	Relative* permittivity	Transparency range nm
Nitrobenzene	Isotropic ∞	34	B	4400	fm V^{-2}	1.55	35.7	460–1500
CS$_2$	∞	370	B	36	"	1.63	2.62	400–3000
H$_2$O	∞	310	B	52	"	1.331	80	300–1300
BaTiO$_3$	Cubic m3m	0.66	$g_{11}-g_{12}$	+0.13	m^4 C^{-2}	2.4	4500 (hf)	430–7000
KTaO$_3$	m3m	0.95 at 4.2 K	$g_{11}-g_{12}$	0.16	"	2.24	~800 (hf)	
			g_{44}	0.12	"			
KTa$_{0.65}$Nb$_{0.35}$O$_3$	m3m	~0.6	g_{11}	0.136	"	2.29	~10^4	400–5500
			g_{12}	−0.038	"			
			g_{44}	0.147	"			
SrTiO$_3$	Cubic m3m	~10	$g_{11}-g_{12}$	0.14	"	2.38	300 (hf)	400–5500
CuCl	$\bar{4}3m$	6.2	r_{41}	6.1 pm V^{-1}		1.93	8.3 (hf)	400–20 000
Hexamine	$\bar{4}3m$	91	r_{41}	0.8	"	1.59	3.2	300–2000
ZnS	$\bar{4}3m$	10.2	r_{41}	2.1	"	2.36	16	400–1300
Ba$_{0.25}$Sr$_{0.75}$Nb$_2$O$_6$	Tetragonal 4mm	~0.1	r_{13}	64.5	"	$n_o=2.31$	~2×10^4	400–6000
			r_{33}	1340	"	$n_e=2.30$		
(NH$_4$)H$_2$PO$_4$ (ADP)	Tetragonal $\bar{4}2m$	10.5	r_{41}	+24.5	"	$n_o=1.522$	$\epsilon_1=57$	250–1300
			r_{63}	− 8.5	"	$n_e=1.477$	$\epsilon_3=16$	
KH$_2$PO$_4$ (KDP)	$\bar{4}2m$	8.7	r_{41}	− 8.6	"	$n_o=1.507$	$\epsilon_1=42$	250–1600
			r_{63}	−10.5	"	$n_e=1.467$	$\epsilon_3=21$	
KD$_2$PO$_4$ (KDDP)	$\bar{4}2m$	3.5	r_{41}	8.8	"	$n_o=1.51$	$\epsilon_3=50$	250–1800
			r_{63}	−26.4	"	$n_e=1.47$		
LiNbO$_3$	Trigonal 3m	2.8	r_{13}, r_{22}	8.6, 3.4	(hf)	$n_o=2.30$	$\epsilon_1=78$	400–4500
			r_{33}, r_{51}	30.8, 28	(hf)	$n_e=2.21$	$\epsilon_3=32$	
LiTaO$_3$	3m	2.4	r_{13}, r_{22}	7.9, 1	(hf)	$n_o=2.176$	$\epsilon_3=43$ (hf)	
			r_{33}, r_{51}	35.8, 20	(hf)	$n_e=2.180$		

* (hf) denotes value for high-frequency operation.

O.C.J.

1.8 Electricity and magnetism

1.8.1 Electrical resistivities

The electrical resistivity or specific resistance ρ is the resistance between the opposite faces of a metre cube of a material. The values given below are in ohm metre units (Ω m). The reciprocal of ρ is the electrical conductivity.

The electrical resistivity is readily influenced and usually increased by factors such as impurity content, porosity, cold work, irradiation, etc. The values given in the main table relate to samples for which such effects are usually minimal. For similar reasons values are not given below 78.2 K, since the effects of impurities, etc., become increasingly important the lower the temperature and finally determine the so-called residual resistance eventually attained for many metals. Certain metals, however, become superconducting below a certain temperature (see p. 124), and for such metals the transition temperature is entered after the name of the metal.

In the main table, values given in italics are for the liquid phase. All other values are for a poly-crystalline solid sample. For some non-cubic metals, a separate table indicates the degree of electrical resistivity anisotropy reported for single crystals at normal temperature. A similar degree of anisotropy can be expected throughout the temperature range of the particular crystal form.

Semi-conducting elements are also listed separately, and again only for normal temperature. Since the electrical resistivities of these poor conductors are so strongly dependent on the small amounts of impurity that may be added intentionally or be present inadvertently, the published values differ considerably, as may the value for any specific sample. The table indicates an approximate value or a likely range of values.

Other elements which are solid at normal temperature and do not conduct electricity are dealt with under electrical insulators (see p. 122). These include boron, carbon in the form of diamond, iodine, phosphorus and sulphur. Elements which are gases at normal temperature and pressure are also omitted.

Values are also given for a few typical alloys. These are mainly the alloys for which thermal conductivity data were given and further compositional details will be found on page 64 *et seq.*

Resistivities of metallic elements

Metal	$\rho/(10^{-8}\ \Omega\ \text{m})$					
	Temperature/K					
	78.2	273.2	373.2	573.2	973.2	1473.2 (or other)
Aluminium (1.17 K)	0.21	2.50	3.55	5.9	24.7	*32.1*
Antimony	8	39	59	—	*114*	*123.5* (1273 K)
Arsenic	5.5	26	—	—	—	—
Barium	7	36	—	—	—	—
Beryllium	—	2.8	5.3	11.1	26	—
Bismuth	35	107	156	*129*	*155*	*172* (1273 K)
Cadmium (0.54 K)	1.6	6.8	9.8	—	—	*36.3* (873.2 K)
Calcium	0.7	3.2	4.75	7.8	20	—
Cerium	—	73	80	92	110	123 (1053.2 K)
Caesium	4.5	18.8	*43.5*	*67*	*134*	*295*
Chromium	0.5	12.7	16.1	25.2	47.2	80
Cobalt	0.9	5.6	9.5	19.7	48	88.5
Copper	0.2	1.55	2.23	3.6	6.7	*21.3* (1356 K)
Dysprosium	26	89	103	124	156.5	184
Erbium	41	81	103	135	183	216
Europium	60	89	—	—	—	—
Gadolinium	20	126	—	—	—	—

Resistivities of metallic elements (*contd*)

Metal	$\rho/(10^{-8}\ \Omega\ \text{m})$					
	Temperature/K					
	78.2	273.2	373.2	573.2	973.2	1473.2 (or other)
Gallium (1.1 K)	2.75	13.6	27.2	31	—	—
Gold	0.5	2.05	2.88	4.63	8.6	31 (1336 K)
Hafnium (0.35 K)	—	29.6	42.6	—	—	—
Holmium	34	90	105	134	175	203
Indium (3.35 K)	1.8	8.0	12.1	36.7	47	55 (1273 K)
Iridium (0.14 K)	0.9	4.7	6.8	10.8	22	33.5
Iron	0.7	8.9	14.7	31.5	85.5	122; 139 (1823 K)
Lanthanum (4.71 K) . . .	—	54	66	83	105	126 (1173 K)
Lead (7.2 K)	4.7	19.2	27	50	108	126 (1273 K)
Lithium	1.04	8.55	12.4	30	40.5	53
Lutetium	16	54	—	—	—	—
Magnesium	0.62	3.94	5.6	10.0	27.7	28.7 (1173 K)
Manganese	—	138	—	—	—	—
Mercury (4.12 K)	5.8	94.1	103.5	128	214	630
Molybdenum (0.92 K) . . .	0.7	5.0	7.6	12.7	23.3	37.2
Neodymium	—	61	74	93	120	138 (1183 K)
Neptunium	—	—	119.3	—	—	121.3 (811.2 K)
Nickel	0.55	6.2	10.3	22.5	40	109 (1773 K)
Niobium (9.1 K)	3.0	15.2	19.2	27.1	43	59
Osmium (0.65 K)	—	8.1	11.4	17.8	30.4	46
Palladium	1.73	10.0	13.8	21	33	42
Platinum	1.96	9.81	13.6	21.0	34.3	48.3
Plutonium	~150	146	142	109	—	—
Polonium	—	~40	—	—	—	—
Potassium	1.38	6.1	17.5	28.2	66.4	160
Praseodymium	—	65	78	96	118	134 (1123 k)
Promethium†	—	50	64	89	126	—
Protactinium (1.4 K) . . .	6.1	17.7	—	—	—	—
Rhenium (1.7 K)	2.62	17.2	24.9	39.7	63.5	84.4
Rhodium	0.46	4.3	6.2	10.2	20	33
Rubidium	2.2	11.0	27.5	48	99	260
Ruthenium (0.49 K) . . .	1.34	7.1	10.0	15.6	27.8	44.4
Samarium	66	91.4	—	—	—	—
Scandium	—	50.5	75	115	167	198
Silver	0.3	1.47	2.08	3.34	6.1	19.4
Sodium	0.8	4.2	9.7	16.8	39.2	89
Strontium	5	20	30	52.5	94.5	
Tantalum (4.48 K)	2.5	12.3	16.7	25.5	43.0	61.5
Technetium (11.2 K) . . .	—	—	22 6	33.3	51	65
Terbium	27	113	—	—	—	—
Thallium (2.37 K)	3.7	15	22.8	38	85	88 (1073 K)
Thorium (1.37 K)	3.9	14.7	20.8	32.5	53.6	68
Thulium	31	67	—	—	—	—
Tin (3.69 K)	2.1	11.5	15.8	50	60	72
Titanium (0.39 K)	4.6	39	58	90	142	—
Tungsten (0.01 K)	0.6	4.9	7.3	12.4	24	39
Uranium (0.68 K)	11	28	35	47	—	—
Vanadium (5.03 K)	2.6	18.2	25.3	38	—	54 (873.2 K)
Ytterbium	13	27.7	—	—	—	—
Yttrium	15.5	55	—	—	—	—
Zinc (0.85 K)	1.1	5.5	7.8	13.0	—	37 (773.2 K)
Zirconium (0.55 K)	7.3	40	58	88	125	110 (1273 K)

† Estimated values based on R. K. Williams and D. L. McElroy, USAEC, ORNL-TM-1424, 1966.

Resistivities of semi-conducting elements at normal temperature

Carbon	$\rho/(\Omega\ m)$	Other elements	$\rho/(\Omega\ m)$
Amorphous	$\sim 6 \times 10^{-5}$	Germanium	$(1-500)\,10^{-3}$
Graphite	$(3-60) \times 10^{-6}$	Selenium	~ 0.1
Pyrolytic graphite, along planes	$\sim 5 \times 10^{-6}$	Silicon	$(1-600) \times 10^{-1}$
„ „ normal to planes	$\sim 5 \times 10^{-3}$	Tellurium	$\sim 3 \times 10^{-3}$

Resistivities of single crystals of some non-cubic metals at normal temperature

Metal	$\rho/(10^{-8}\ \Omega\ m)$ in direction of		
	c-axis	a-axis	b-axis
Antimony	35.6	42.6	
Beryllium	3.58	3.13	
Bismuth	138	109	
Cadmium	7.79	6.54	
Dysprosium	77.4	98.2	
Erbium	47.0	87.6	
Gadolinium	122	135	
Gallium	55.5	17.3	7.85
Holmium	59.9	101.2	
Lutetium	34.0	75.6	
Magnesium	3.78	4.53	
Mercury (at 227.7 K) . .	17.7	23.5	
Tellurium	56 000	15 400	
Terbium	101	122	
Thulium	46	87	
Tin	13.1	10.0	
Zinc	6.05	5.83	

Resistivities of some typical alloys

Alloy†	$\rho/(10^{-8}\ \Omega\ m)$				
	Temperature/K				
	273.2	373.2	573.2	973.2	1473.2
Alpax gamma‡	3.5	5.0	7.85	—	—
Alumel	28.1	34.8	43.8	53.2	65.1
Brass	6.3	—	—	—	—
Bronze	13.6	—	—	—	—
Chromel P	70.0	72.8	79.3	89.3	100.1
Constantan	49	—	—	—	—
German silver	40	—	—	—	—
Lo-Ex‡	3.95	5.45	8.55		
Manganin	41.5	—	—		

† See thermal conductivity table, p. 64 *et seq.* for further compositional details.
‡ After heat treatment at 573.2 K, followed by air cooling.

Resistivities of some typical alloys (*contd.*)

Alloy†	$\rho/(10^{-8}\ \Omega\ m)$				
	Temperature/K				
	273.2	373.2	573.2	973.2	1473.2
Monel	42.9	50.1	52.5	63.3	—
Nichrome	107.3	108.3	110.0	110.3	—
RR 59‡	3.5	4.9	7.6	—	—
RR 77‡	3.95	5.2	7.8	—	—
Steel, Carbon	17.0	23.2	39.8	93.5	123.1
„　　18/8	66.3	74.3	89.1	109.4	124
„　　Era ATV	98.0	102.7	111.0	122.0	—
„　　Ni–Cr	27.7	33.7	48.7	99.4	122.2
„　　Silicon	41.9	47.0	60.1	105.7	127.1
„　　Stainless	55.0	63.4	80.2	114.1	—
Pt 90%, Ir 10%	24.8	28.0	—	—	—
Pt 90%, Rh 10%	18.7	21.8	—	—	—
Ti 92.5%, Al 5%, Sn 2.5%	155.6	165.6	179.4	—	—
Ti 96.0%, Al 2.0%, Mn 2.0%	110.0	123.1	149.5	—	—
Zr 93.5%, Sn 6.7%, C 0.1%	132.8	139.5	148.3	—	—
Zr 97.5%, Sn 2.3%, C 0.1%	91.5	105.2	123.9	—	—

† See thermal conductivity table, p. 64 *et seq.*, for further compositional details.　　　　C.H.D.

‡ After heat treatment at 573.2 K, followed by air cooling.

1.8.2　Resistance alloys and wire resistances

Copper–manganese alloys ($\sim$84% Cu, 12% Mn with nickel, aluminium or germanium as the remaining constituent). These alloys are sold under various proprietary names, and manganin, the pioneer alloy of this group, was for many years the traditional material for high-grade standard resistors. The resistivity is about $40 \times 10^{-8}\,\Omega$ m and varies approximately parabolically with temperature over the range 0 to 50 °C, with a maximum close to 20 °C. The temperature coefficient can be as low as $3 \times 10^{-6}\,°\text{C}^{-1}$ over the range 15 °C to 20 °C. Its secular stability is very good and, if wires are supported in strain-free conditions, can be less than 1 in 10^7 per year. The thermo-e.m.f. of the alloys against copper is close to zero and may be positive or negative according to composition and heat treatment. Joints between the copper manganese alloys and copper are made most effectively by welding in an atmosphere of argon, and by hard soldering if welding is impracticable.

Copper–nickel alloys ($\sim$55% Cu, 45% Ni). These alloys are manufactured commercially under a wide range of proprietary names, and are used in the construction of standard resistors. The resistivity is about $50 \times 10^{-8}\,\Omega$ m with a temperature coefficient which may lie between $\pm 0.000\ 04\ °\text{C}^{-1}$.

The alloys can be soft-soldered with ease, but their high thermo-e.m.f. against copper ($\sim$40 μV °C^{-1}) is a disadvantage in d.c. resistors, although the effect is usually negligible in a.c.

resistors dropping 1 volt or more. These alloys are also used for current controlling resistors when constancy is more important than low cost.

The table gives resistance of wires in ohms per metre and approximate currents in amperes required to maintain stated temperature rises in straight horizontal wires of nickel–chromium free to radiate in air.

Properties of resistance wires

S.W.G.	Diameter		Copper	Copper–manganese alloys	Copper–nickel alloys	Quaternary alloys	Nickel–chromium		
	mm	in	$\Omega\,m^{-1}$ at 20 °C	$\Omega\,m^{-1}$	$\Omega\,m^{-1}$	$\Omega\,m^{-1}$	$\Omega\,m^{-1}$	Current (A) to maintain temp. rise	
								500 °C	1000 °C
12	2.642	0.104	0.003 12	0.076	0.090	0.243	0.197	38	78
14	2.032	0.080	0.005 32	0.128	0.151	0.410	0.333	26	53
16	1.626	0.064	0.008 31	0.200	0.235	0.640	0.520	19	40
18	1.219	0.048	0.014 8	0.355	0.420	1.14	0.92	13	27
20	0.914	0.036	0.026 3	0.630	0.745	2.03	1.65	8.5	18
22	0.711	0.028	0.043 4	1.05	1.23	3.35	2.72	6.3	13
24	0.559	0.022	0.070 3	1.69	2.00	5.40	4.40	4.5	9.5
26	0.457	0.018	0.105	2.53	3.00	8.1	6.60	3.5	7.0
28	0.376	0.0148	0.155	3.75	4.40	12.0	9.7	2.7	5.5
30	0.315	0.0124	0.221	5.30	6.30	17.1	13.9	2.2	4.5
32	0.274	0.0108	0.292	7.00	8.30	22.5	18.3	1.9	3.5
34	0.234	0.0092	0.402	9.7	11.4	31.0	25.2	1.6	3.0
36	0.193	0.0076	0.589	14.2	16.7	45.5	37.0	1.3	2.3
38	0.152	0.0060	0.945	22.7	27.0	73.0	59.0	1.0	1.7
40	0.122	0.0048	1.48	35.5	42.0	114	92	0.8	1.4
42	0.102	0.0040	2.13	51.0	60.5	164	133	0.65	1.1
44	0.0813	0.0032	3.32	80	94	255	208	—	—
46	0.0610	0.0024	5.91	142	168	455	370	—	—
48	0.0406	0.0016	13.3	320	380	1030	835	—	—
50	0.0254	0.0010	34.0	820	970	2620	2130	—	—

Nickel–chromium alloys ($\sim 80\%$ Ni, 20% Cr). These alloys are also available under a variety of trade names and are used for heater elements as well as for resistors where high accuracy is not required. Their resistivity is about $110 \times 10^{-8}\,\Omega$ m with a temperature coefficient (20–500 °C) of 0.000 06 °C^{-1}. They will operate satisfactorily at temperatures of up to 1100 °C. A ternary alloy (65% Ni, 15% Cr, 20% Fe) is less expensive than nickel–chromium and is satisfactory at temperatures up to 900 °C.

A particular development of 80/20 nickel-chromium for reference or standard resistors is the use of thick-film deposits on glass substrates. These resistors are essentially thermally-compensated strain gauges in which the change of resistance with temperature counteracts the change due to the strain imparted by the substrate. When suitably heat-treated after deposition, such resistors can achieve temperature coefficients of not more than a few p.p.m. and secular stability comparable with that of manganin or quaternary alloys. However, because of their very small size, they can be used at much higher frequencies than is possible with wire-wound types.

Quaternary alloys ($\sim 73\%$ Ni, 21% Cr, 2% Al, with copper, iron, cobalt, manganese or molybdenum as the fourth main constituent). These alloys are again processed and marketed under many proprietary names, each of which is characterized by the fourth constituent; they are increasingly being used for standard resistors, especially those of high value. Their resistivity is about $130 \times 10^{-8}\ \Omega$ m, with a temperature coefficient controllable by heat treatment and which can be made as small as $\pm 0.000\ 002\ °C^{-1}$. The thermo-e.m.f. of the alloys, like that of the copper–manganese alloys, is close to zero and may be slightly positive or negative. The tensile strength of the alloys is high, making possible the drawing of very fine wire. Welding in argon is by far the best method of joining with copper, with hard soldering as an alternative only if welding is not possible.

C.H.D.

1.8.3 Electrical insulating materials

The resistivity of an insulator varies greatly with the purity and surface condition of the material, time of application of the voltage, and in some cases with the magnitude of the applied stress. It usually decreases rapidly as the temperature is increased (in many cases changing by orders of magnitude) and may be reduced temporarily or permanently by irradiation.

Conventionally, resistivity measurements are made one minute after application of the measurement voltage. For almost all insulating materials the current at this time is not a true conduction current, but an absorption current involving storage of charge in the insulator. The equilibrium conduction current may be several orders of magnitude smaller.

Some typical one-minute values for both resistivity and surface resistivity are given below for normal temperatures and 50–60% relative humidity. Surface resistivity is defined as the resistance between opposite edges of a square (the size is immaterial) and the unit is the ohm per square.

Resistivity of insulators

Material	*Volume resistivity* ohm metre	*Surface resistivity* ohm per square
Beeswax (fresh surface)	10^{12}–10^{13}	$\sim 10^{14}$
Boron	10^{10}	
Ceramics		
Alumina	10^{9}–10^{12}	
Porcelain	10^{10}–10^{12}	10^{11} glazed
		10^{9} unglazed
Pyrophyllite	10^{10}–10^{13}	
Special H.F.	10^{14}	
Steatite	10^{11}–10^{13}	
Titanates	10^{6}–10^{13}	
Diamond	10^{10}–10^{11}	

Resistivity of insulators (*contd*)

Material	Volume resistivity ohm metre	Surface resistivity ohm per square
Glass		
Soda-lime	10^9-10^{11}	$10^{10}-10^{12}$
Borosilicate (Pyrex)	10^{12}	
Plate	2×10^{11}	5×10^{10}
Guttapercha	10^7	
Hard rubber (Ebonite, etc.)	$10^{13}-10^{15}$	$10^{10}-10^{18}$
Iodine	10^{13}	
Ivory	10^6	10^9
Marble	10^7-10^9	10^9
Mica, sheet	$10^{11}-10^{15}$	$10^{10}-10^{13}$
,, moulded	10^{13}	5×10^{13}
Mineral oil	$>10^{10}$	
Mineral insulating oil	$10^{11}-10^{15}$	
Paper (dry)	$\sim 10^{10}$	10^9-10^{10}
Paraffin (kerosine, coloured)	$10^{11}-10^{12}$	
Paraffin wax	$10^{13}-10^{17}$	10^{15}
Plastics		
Acrylic (Perspex, etc.)	$>10^{13}$	$>10^{14}$
Alkydes or polyester (no filler) . . .	$10^{12}-10^{13}$	$10^{13}-10^{14}$
Aminos, melamines (cellulose) . . .	$>10^9$	$10^{12}-10^{14}$
,, ,, (mineral) . . .	10^9	$10^{12}-10^{14}$
Casein	10^7-10^8	$10^{10}-10^{11}$
Cellulose acetate	10^8-10^{11}	$10^{11}-10^{12}$
Epoxy cast resin (no filler) . . .	$10^{12}-10^{13}$	$3 \times 10^7 -> 10^{14}$
Phenolics	10^6-10^{12}	10^8-10^{14}
Polyamides (nylon)	10^8-10^{13}	$10^{11}-10^{15}$
P.C.T.F.E.†	10^{16}	$10^{12}-10^{13}$
P.E.T.‡	$10^{15}-10^{17}$	
Polyformaldehyde	$\sim 6 \times 10^{12}$	$>2 \times 10^{13}$
Polypropylene	$10^{13}-10^{15}$	$>10^{15}$
Polystyrene (general purpose) . . .	$10^{15}-10^{19}$	$>10^{14}$
,, (toughened)	$10^{10}-10^{15}$	$>10^{14}$
P.T.F.E.§	$10^{15}-10^{19}$	$4 \times 10^{12}-10^{17}$
Polythene (high density)	$10^{14}-10^{15}$	$10^{12}-10^{17}$
,, (low density)	$10^{14}-10^{18}$	$>10^{14}$
Polyurethanes	10^9-10^{12}	$10^{13}-10^{15}$
P.V.C.‖ (rigid)	$5 \times 10^{12}-10^{13}$	$10^{12}-10^{15}$
,, (flexible)	$5 \times 10^6-5 \times 10^{12}$	$>10^{14}$
Silicone (glass)	10^8-10^{12}	$>10^{11}$
Quartz,‖ optic axis	10^{12}	
,, ⊥ optic axis	10^{14}	
,, fused	$>10^{16}$	3×10^{12}
Shellac	10^7	5×10^{13}
Silicone, oils	10^{12}	
,, rubber	10^9	
Slate	10^5-10^6	10^7
Soil	10^2-10^4	
Sulphur	$10^{14}-10^{15}$	7×10^{15}
Wood (paraffined)	10^8-10^{11}	10^{12}
Water (distilled)	10^2-10^5	

† Polychlorotrifluoroethylene.   § Polytetrafluoroethylene.
‡ Polyethyleneterephthalate.   ‖ Polyvinylchloride.

B.W.P.

1.8.4 Superconducting properties of materials

Superconductivity is the complete loss of electrical resistance at some finite, but low, temperature. The phenomenon is a fairly common one in that a quarter of the elements and over one thousand alloys and compounds are known to be superconductors. The superconducting state is characterized by three parameters: the critical temperature, T_c, the critical magnetic field H_c and the critical current, I_c, or critical current density, J_c.

Critical temperatures vary from 0 K to 23.6 K. Above T_c superconductors behave exactly as normal metals, their resistivity increasing with rising temperature. At T_c, in the absence of any external magnetic field and with no current flowing, the resistivity drops sharply to zero and remains at zero for all temperatures below T_c. In addition to the loss of resistivity, magnetic flux is excluded from the interior of the superconductor; this is known as the Meissner effect. The application of an external magnetic field above a critical value H_c will cause reversion to the normal state. The critical field is a function of the temperature, falling from its value at 0 K, H_c (0) to zero as the temperature rises to T_c. This behaviour is shown for a number of elemental superconductors in the graph. Critical fields extend to 32.7 MAm^{-1} (41 T)†.

† It is common practice, where no confusion can arise, to express magnetic field strength in terms of corresponding flux density for free space, and to use Tesla rather than A m^{-1} as the unit.

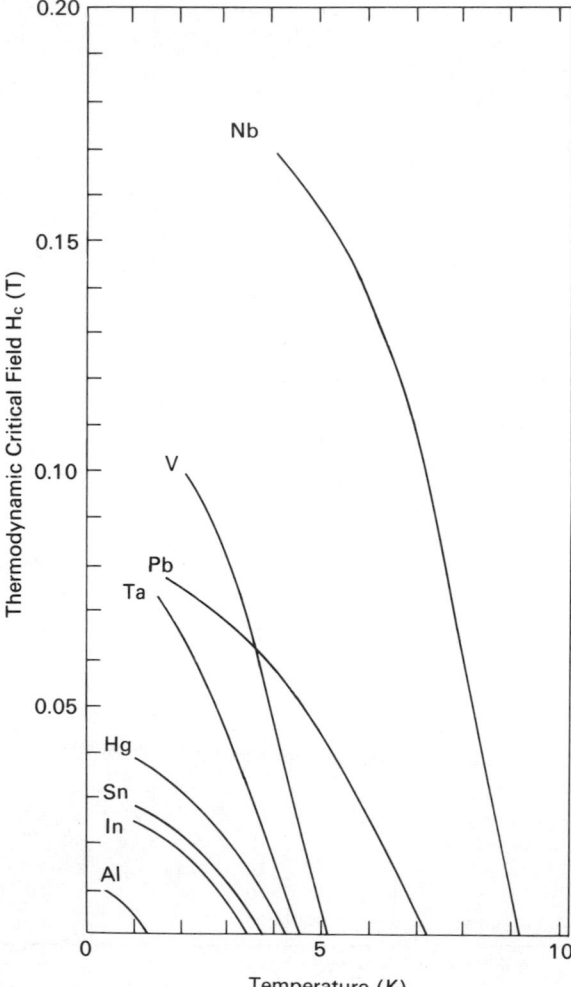

Superconductors fall into two classes: Type I and Type II. The principal difference between them is that a Type II superconductor has two critical fields, a lower critical field H_{c1} and an upper critical field H_{c2}. In a field below H_{c1} a Type II superconductor behaves exactly as a Type I superconductor below its H_c, showing a Meissner effect.

At field strengths between H_{c1} and H_{c2} the flux exclusion is only partial, the superconductor being threaded by flux vortices; associated with each vortex is a single quantum of flux ($h/2e = 2 \times 10^{-15}$ Wb). This is referred to as the mixed state. Above H_{c2}, which is usually many times greater than H_c for a typical Type I superconductor, the Type II superconductor is fully normal. Values of lower and upper critical fields vary with temperature, as H_c, and are also extremely sensitive to composition. The temperature dependence of H_{c2} for some Type II superconductors is shown in the graph. Values of T_c and H_{c2} for various high T_c, high H_{c2} alloys and compounds are listed in the second table.

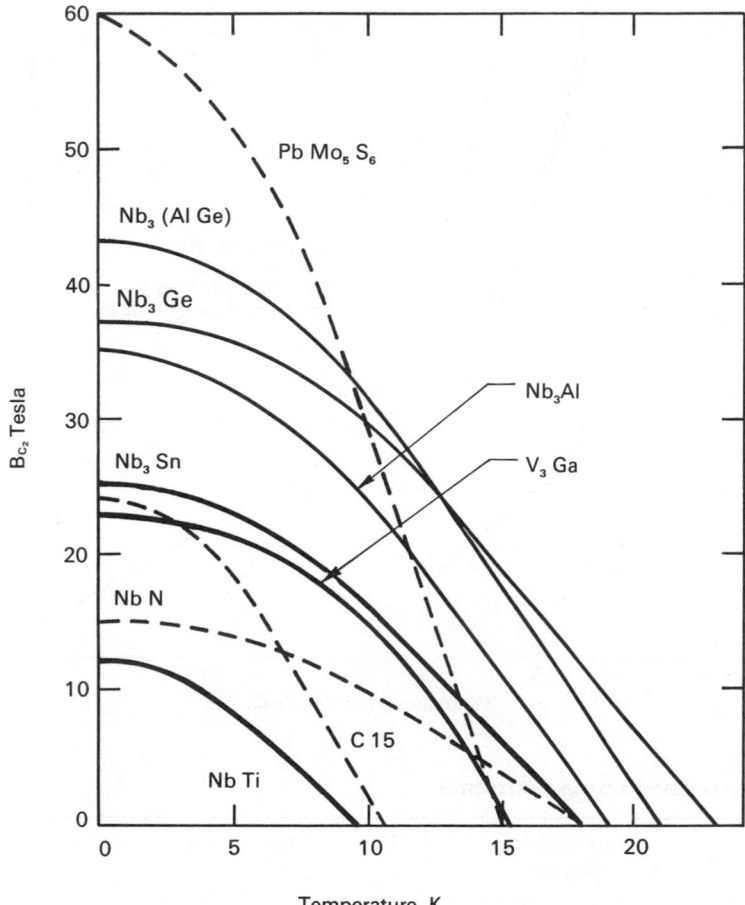

The critical current I_c of a Type I superconductor is just that current which will produce the critical field at the surface of the conductor. For a cylindrical conductor of radius R, $I_c = \frac{1}{2}RH_c$, and the critical current density $\tilde{J}_c = I_c/\pi R^2 = H_c/2\pi R$. The passage of a current down a Type II superconductor in the mixed state exerts a Lorentz force on the flux vortices. If the vortices are allowed to move a resistance is induced. Interactions between the vortices and microstructural features in the superconductor, such as grain-boundaries or precipitates, act so as to resist motion of the vortices. The critical current density $\tilde{J}_c$ in a Type II superconductor is that level of current density at which the

Lorentz force on the vortices overcomes the microstructural resistance to vortex motion, and is much less than $H_{c2}/2\pi R$. $\mathcal{J}_c$ is thus extremely sensitive to the microstructure of the superconductor, in addition to being a function of temperature and magnetic field. Current densities in excess of $10^{10} \mathrm{A\,m^{-2}}$ can be carried by some superconductors. Values of current density at 4.2 K versus magnetic field are shown for some commercially available superconductors in the graph.

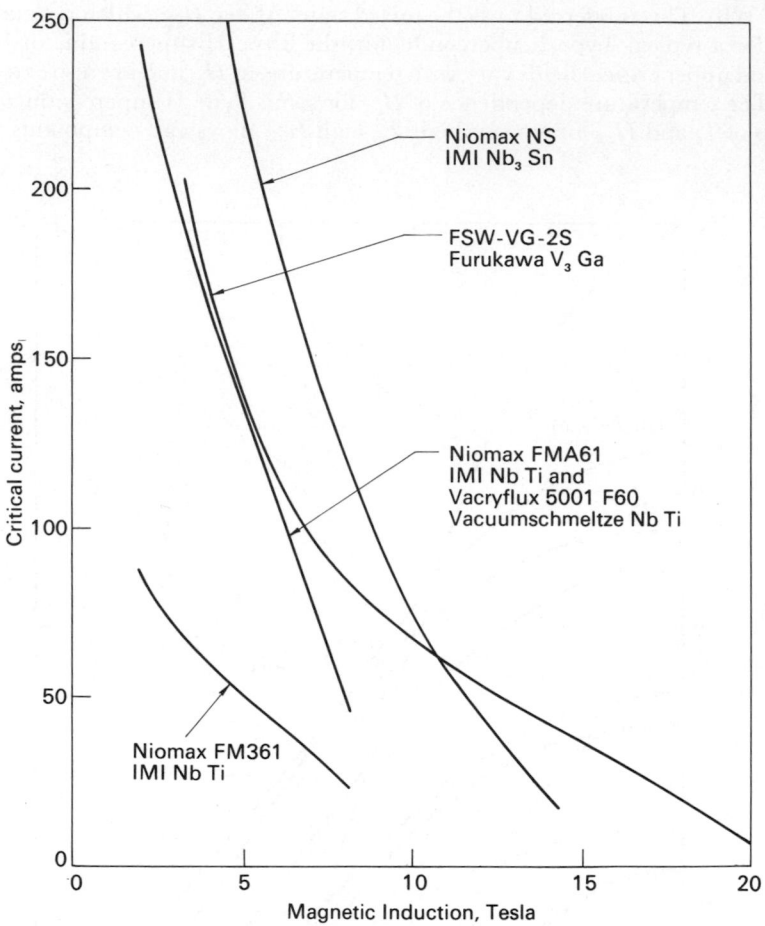

Properties of superconducting elements

Element	Periodic table group no.		T_c, K	Type	$H_c(0)$, mT	$H_{c1}(0)$, mT	$H_{c2}(0)$, mT
Berylium	II A		0.026	—	—	—	—
Lanthanum.	III A	(α)	4.88	I	80	—	—
		(β)	6.00	I	160	—	—
Titanium	IV A		0.4	I	5.6	—	—
Zirconium	—	—	0.61	I	4.7	—	—
Hafnium	—		0.128	I	1.27	—	—
Thorium	—	—	1.38	I	16.0	—	—
Vanadium	V A		5.4	II	—	26	268
Niobium.	—		9.25	II	—	173	405

Properties of superconducting elements (*contd.*)

Element	Periodic table group no.		T_c, K	Type	$H_c(o)$, mT	$H_{c1}(o)$, mT	$H_{c2}(o)$, mT
Tantalum	—		4.47	II	—	45	200
Protactinium	—		1.4	—	—	—	—
Molybdenum	VI A		0.92	—	9.6	—	—
Tungsten	—		0.015	I	0.115	—	—
Technetium	VII A		7.8	II	—	116	312
Rhenium	—		1.7	I	20	—	→
Ruthenium	VIII A		0.49	I	6.9	—	—
Osmium	—		0.66	I	7.0	—	—
Iridium	—		0.11	I	1.6	—	—
Americium	—	(α)	0.6	—	—	—	—
		(β)	1.0				
Zinc	II B		0.85	I	5.4	—	—
Cadmium	—		0.517	I	2.8	—	—
Mercury	—	(α)	4.154	I	41.1	—	—
		(β)	3.949	I	33.9	—	—
Aluminium	III B		1.75	I	10.5	—	—
Gallium	—	(α)	1.083	I	5.8	—	—
		(β)	5.9, 6.2	I	56	—	—
		(γ)	7.62	II	—	—	> 300
		(δ)	7.85	II	—	—	
Indium	—		3.41	I	28.2	—	—
Thallium	—		2.38	I	17.8	—	—
Tin	IV B		3.72	I	30.5	—	—
Lead	—		7.2	I	80.3	—	—
Lutetium	VII B		0.1	I	35.0	—	—

Many other elements become superconducting when deposited as thin films, when in the amorphous state, or under high pressure.

Properties of some high T_c superconducting alloys and compounds

Structure	Material	T_c, K	$H_{c2}(4.2) T$
A2 (bcc)	$Mo_{0.66}Re_{0.34}$	11.8	1.2
	$Nb_{0.75}Ti_{0.25}$	9.8	10.0
	$Nb_{0.44}Ti_{0.56}$	9.0	14.1
	$Nb_{0.75}Zr_{0.25}$	10.9	8.3
	$Nb_{0.67}Zr_{0.33}$	11.0	>8.3
B1 (NaCl)	Mo C	14.3	5.2
	Nb C	11.1	1.7
	Nb N	17.3	15.0
	Nb $C_{0.3}N_{0.7}$	17.8	12.5
	(Nb, Ti) N	18.0	13.5
	Ta C	10.35	0.5
	Ta N	14.3	—
	W C	10.0	—
Hexagonal	Mo N	14.8	—

Properties of some high T_c superconducting alloys and compounds (*contd.*)

Structure	Material	T_c, K	$H_{c2}(4.2)T$
A15 (βW)	$Nb_3 Al$	19.1	29.5
	$Nb_3(Al_{0.7}Ge_{0.3})$	20.7	41
	$Nb_3 Ga$	20.7	33
	$Nb_3 Ge$	23.6	37
	$Nb_3 Si$	18.6	14.5
	$Nb_3 Sn$	18.3	26
	$V_3 Ga$	15.3	25
	$V_3 Si$	17.1	23
C15 (Laves phase) . . .	$V_2 H_f$	9.2	20
	$V_2(Hf_{0.5}Zr_{0.5})$	10.1	24
	$V_{1.83}Nb_{0.13}Hf_{0.75}$	10.4	25
	$V_{1.81}Ta_{0.19}Hf_{0.71}$	10.0	26
Chevrel Phase	$PbMo_{5.1}S_6$	14.4	51
Hexagonal	$Li_{0.3}Ti_{1.1}S_2$	13	—
D5c(bcc)	$Y_2 C_3$	11.5	—
	$La_2 C_3$	11.0	—
	$(Y,Th)_2 C_3$	17.0	—
	$(La,Th)_2 C_3$	14.3	—
Spinel (fcc)	$Li Ti_2 O_4$	13.7	20
Hydride	$(Pd_{0.55}Cu_{0.45})H_{0.7}$	16.6	—

Values of T_c and H_{c2} are dependent upon composition, in many cases degree of long-range order, and method of preparation. The above represent optimum values for the quoted composition or for the stoichiometric compound.

D.D-H.

1.8.5 Dielectric properties of materials

The absolute complex permittivity of a material is represented by the symbol ε, where $\varepsilon = \varepsilon' - j\varepsilon''$. This is related to the dimensionless relative complex permittivity ε_r, where $\varepsilon_r = \varepsilon_r' - j\varepsilon_r''$, by the expression $\varepsilon = \varepsilon_0\varepsilon_r$, ε_0 being the permittivity of free space, a fixed constant given by $\varepsilon_0 = 8.85 \times 10^{-12}\,Fm^{-1}$. In general, ε depends on temperature and, to a lesser extent, pressure. It is also frequency dependent, although ε' and ε'' cannot vary independently with frequency, since their frequency variations are connected through the Kramers–Krönig relationship: a drop in ε' with increasing frequency is necessarily associated with a peak in ε''. Except for exceedingly high applied fields, ε is independent of the magnitude of the applied electric field for all dielectric materials used in practice, excluding ferro-electrics.

A capacitor filled with a dielectric material has a real capacitance ε_r' times greater than would have a capacitor with the same electrodes in vacuum. The dielectric-filled capacitor would also have a power dissipation W per unit volume at each point when, resulting from an applied voltage, a sinusoidal electric field of frequency f and root-mean-square value E exists at that point. This power dissipation is given by $W = 2\pi f E^2 \varepsilon''$. Thus ε'' is a measure of the energy dissipation per period, and for this reason it is known as the loss-factor.

The complex permittivity ε is often represented in the Argand plane with ε' as abscissa and ε'' as ordinate, giving a curve with frequency as parameter. The join of any point on this curve to the origin therefore represents the complex conjugate ε^* of the complex permittivity ε where $\varepsilon^* = \varepsilon' + j\varepsilon''$. Unfortunately, the use of the symbol ε^* to represent complex permittivity is widespread and has become established in the literature, and care is needed if confusion over signs is to be avoided. The join to the origin makes an angle δ with the abscissa, such that $\tan \delta = \varepsilon''/\varepsilon'$. Thus W may be rewritten as $W = 2\pi f E^2 \varepsilon' \tan \delta$. Hence δ is known as the loss angle, and $\tan \delta$ is known as the loss tangent.

The application of a sinusoidal voltage of root-mean-square value V to the dielectric-filled capacitor results in a current flow in the external circuit which leads the voltage by a phase angle or power-factor angle φ, where φ is the complement of δ. Thus, the power dissipation in the capacitor, given by $IV \cos \varphi$, may also be expressed as $IV \sin \delta$. Since in most cases in engineering practice δ is small, $\sin \delta \simeq \tan \delta$ and the power dissipation is given to a good approximation by $IV \tan \delta$. It should be noted that no such approximation is involved in the expression for W in the previous paragraph.

When the wavelength of electro-magnetic radiation is in the optical region, the velocity v of propagation through a loss-free transmitting medium of refractive index n is given by $v = c/n$, where c is the velocity in free space. The velocity is also given by $v = c/(\mu_r \varepsilon_r')^{\frac{1}{2}}$, where μ_r is the relative permeability. Thus for loss-free non-magnetic materials, for which $\mu_r = 1$, $\varepsilon_r' = n^2$. However, in general losses do occur, and the material is characterized by a complex refractive index $\hat{n}$ given by $\hat{n} = n - jk$, where k is the absorption coefficient. Then $\varepsilon_r = \hat{n}^2$, or $\varepsilon_r' - j\varepsilon_r'' = (n - jk)^2$, from which it follows that $\varepsilon_r' = n^2 - k^2$ and $\varepsilon_r'' = 2nk$. Nevertheless, when the loss is small, so that $k < < n$, then $\varepsilon_r' \simeq n^2$. The use of these relationships allows values of ε_r at high frequencies to be derived from optical measurements. As the frequency is reduced, specially designed interferometers (infra-red), free radiation methods (sub-millimetric wavelengths), wave-guides, coaxial lines and resonant cavities (centimetric wavelengths), and Q meters and bridges (radio frequencies to d.c.) have all been used. Time-domain spectroscopy, involving an analysis of the response of the medium to a step-function field, is capable in principle, and has had some success in practice, in giving a rapid measurement of ε over a very wide frequency spectrum.

The relative permittivity is directly related to the electronic, atomic and orientational polarization of the material. The first two of these are induced by the applied field, and are caused by displacement of the electrons within the atom, and atoms within the molecule, respectively. The third only exists in polar materials, i.e. those with molecules having a permanent dipole moment. Electronic and atomic polarization are temperature independent, but orientational polarization, depending on the extent to which the applied field can order the permanent dipoles against the disordering effect of the thermal energy of their environment, varies inversely with absolute temperature. All of these polarization mechanisms can only operate up to a limiting frequency, after which a further frequency increase will result in their disappearance. Because of the spring-like nature of the forces involved, this is accompanied by an absorption of the resonance type for electronic and atomic polarization, but for orientational polarization the disappearance, accompanied by a broader peak in the loss factor, is more gradual, because the mechanism involved is of the relaxation type, and may involve a broad distribution of relaxation times. Indeed, the decline in ε' may be so gradual that ε'' may appear almost constant, and be correspondingly small, over a wide frequency range. This applies particularly to some polymers commonly used in engineering practice, many of which are polar. Those which are non-polar, usually with $\varepsilon_r' < 2.5$, show nearly constant values of ε' and ε'' over the entire electrical frequency spectrum.

The frequency at which these mechanisms drop out is related to the inertia of the moving entities involved. Typically, electronic polarization persists until a frequency of about 10^{16} Hz, atomic polarization until about 10^{13} Hz, while the dispersion for orientational polarization may lie anywhere within a wide frequency range, say 10^2–10^{10} Hz, depending on the material and its temperature. In addition to these polarization mechanisms, the existence of interfacial effects such as macroscopic discontinuities in the material, or blocking at the electrodes, causes the trapping of charge carriers, and such phenomena, as well as the inclusion in the dielectric of impurities giving rise to conducting regions, result in behaviour classified under the general heading of Maxwell–Wagner effects. These give rise to an effective polarization and associated loss, the frequency behaviour of which is similar to that of orientational polarization, with a dispersion region which may lie in the region of 1 Hz or lower.

When orientational polarization is operative, it is usually the dominant polarization mechanism present. The classical theory of this mechanism is due to Debye. For a single relaxation time τ, the variation of ε_r with angular frequency ω is given by the Debye equation, $(\varepsilon_r - \varepsilon_\infty)/(\varepsilon_S - \varepsilon_\infty) = (1 - j\omega\tau)/(1 + \omega^2\tau^2)$, where ε_S and ε_∞ are the relative permittivities at frequencies much lower and much higher (but not high enough to involve any reduction in atomic or electronic polarizations) respectively than the anomalous dispersion region. Equating real and imaginary parts gives $(\varepsilon_r' - \varepsilon_\infty)/(\varepsilon_S - \varepsilon_\infty) = 1/(1 + \omega^2\tau^2)$ and $\varepsilon''/(\varepsilon_S - \varepsilon_\infty) = \omega\tau/(1 + \omega^2\tau^2)$.

If ε'' is plotted against ε', the Cole–Cole plot results. This is a semicircle if the Debye equation is obeyed. Frequently experimental results yield a circular arc, rather than a semicircle, with its centre below the abscissa. Such behaviour can be predicted if a suitable distribution of relaxation times is postulated, though no satisfactory physical reason for doing so has yet been established. There is a variety of other shapes obtained in practice, such as the skewed arc in which the high frequency end of the arc approximates to a straight line.

The permittivity of many substances changes not only with frequency and temperature, but also with specimen age and history. Two specimens of nominally the same material may have significantly different permittivities because of different manufacturing processes, different amounts of oxidation, and different inclusions, some of which might have been deliberately introduced, e.g. anti-oxidants. For such reasons, tables of values should be used as an indication of the magnitudes to be expected, and not as a source of precise data which can be repeated by accurate measurements on particular test specimens, except in cases in which the physical and chemical state of both the reference material and the test specimen are very closely specified. The properties of ferroelectric materials depend on so many factors that it is inappropriate to include them in tables of data. Generally, they have permittivities of the order of a thousand, strongly dependent on applied voltage and temperature, and exhibit considerable power loss.

References

(1) A. R. von Hippel, *Dielectrics and Waves*, 1954 (Chapman and Hall, London).
(2) H. Fröhlich, *Theory of Dielectrics*, 2nd edn, 1958 (Clarendon Press, Oxford).
(3) V. V. Daniel, *Dielectric Relaxation*, 1967 (Academic Press, London).
(4) Nora E. Hill, Worth E. Vaughan, A. H. Price, Mansel Davies, *Dielectric Properties and Molecular Behaviour*, 1969 (van Nostrand Reinhold Company Ltd., London).
(5) C. J. F. Bötcher, Vol. 1, *Dielectrics and Static Fields*, 2nd edn, 1973 (Elsevier Scientific Publishing Company, Amsterdam).
(6) C. J. F. Bötcher and P. Bordewijk, Vol. 2, *Dielectrics in Time Dependent Fields*, 2nd edn, 1978 (Elsevier Scientific Publishing Company, Amsterdam).

Tables of relative permittivity and loss tangent

Temperature (t) is in °C, and frequency (f) in Hz, k$(\times 10^3)$, M$(\times 10^6)$, G$(\times 10^9)$. Temperature coefficient of ε_r' is denoted by a $= 10^5 \, d\varepsilon_r'/\varepsilon_r' dt$ and density in g cm^{-3} by d. For non-cubic crystals, the symbols $\perp$, $\parallel$, indicate measurements with field respectively perpendicular to and parallel to the c-axis. Ranges of quantities are indicated by the numerical limits of the range, separated by a solidus. For commercial materials, the values should be regarded as examples only, since some vary greatly with composition and purity. This applies also to the loss angle of some pure materials, which may depend on traces of impurity. The ranges of ε_r' and tan δ, however, are intended to indicate not these variations, but only the variation within the stated ranges of temperature and/or frequency. However, because data relating to different temperatures and frequencies often have to be taken from more than one source, even for what is nominally the same material, it is commonly impossible to be certain of the cause of the variations.

Solids

Material	Remarks	$t/°C$	f/Hz	ε_r'	$10^4 \times \tan \delta$
Cellulose (see also paper)					
Cellophane	unplasticized	20	50/1 M	7.6/6.7	100/650
		−30/70	50	7.2/8.0	100/150
Paper fibres	calculated	20	50	6.5	50
Ceramics					
Alumina	pure	20/100	50/1 M	8.5	20/5
	pure, porosity 1 %	20	1 M	10.8	
Calcium titanate	$a = -200$	20	1 M	150	3
Lead zirconate	$a = +140$	20	1 M	110	30
Magnesium titanate		20/150	50/1 M	14	1/4
Porcelain	h.v. electrical	20/100	50/1 M	5.5	300/80
Rutile	$a = -80$	20	1 M/1 G	80	3/8
	$a = -40$	20	1 M/1 G	40	15/30
	$a = -2$	20	1 M/100 M	12	30
	$a = +6$	20	1 M/100 M	15	1
Steatite	$a = +13$	20	1 M/1 G	6	20
(low loss)	$a = +13$	20	1 M/1 G	6	2
Strontium titanate	$a = -300$	20	1 M	200	5
Strontium zirconate	$a = +12$	20	1 M	38	3
Crystals (single, inorganic)					
Alkali halides					
LiF		20/25	1 k/10 G	8.9/9.1	2
LiCl		20	1 k/1 M	11.8/11.0	
LiBr		20	1 k/1 M	13.2/12.1	
LiI		20	1 k/1 M	16.8/11.0	
NaF		20	1 k/1 M	5.1/6.0	
NaCl		20/25	1 k/10 G	6.1/5.9	5/1
NaBr		20	1 k/1 M	6.5/6.0	
NaI		20	1 k/1 M	7.3/6.6	
KF		20	1 k/1 M	5.3/6.0	
KCl		20	1 k/10 G	4.9/4.8	
KBr		20/25	1 k/10 G	5.0/4.9	2/7
KI		20	1 k/1 M	5.1/5.0	
RbF		20	1 k	6.5	
RbCl		20	1 k	4.9	
RbBr		20	1 k	4.9	
RbI		20	1 k	4.9	
Calcite	$CaCO_3 \perp$	20	1 k/10 k	8.5	
	‖	20	1 k/10 k	8.0	
Diamond	C	20	500/100 M	5.7/5.5	
Fluorite	CaF_2	20	10 k/2 M	7.4/6.8	
Gallium Arsenide		20	1 k	12	
Germanium		20	1 k	16.3	
Iodine		17/22	100 M	4.0	
Mica, muscovite (best) . . .		20/100	50/100 M	7.0	10/2
Periclase	MgO	25	100/100 M	9.7	3
Quartz	$SiO_2 \perp$	20	1 k/30 M	4.5/4.3	2
	‖	20	1 k/30 M	4.5/4.3	2
Ruby	Al_2O_3	17/22	10 k	13.3	
		17/22	10 k	11.3	

Solids (*contd*)

Material	Remarks	$t/°C$	f/Hz	ε_r'	$10^4 \times \tan \delta$
Crystals (*contd*)					
Rutile	$TiO_2 \perp$	20	50/100 M	86	100/2
	‖	17/22	100 M	170†	
Sapphire	$Al_2O_3 \perp$	20	50 /1 G	9.4	2
	‖	20	50/1 G	11.6	2
Selenium		17/22	100 M	6.6	
Silicon . .		20	1 k	11.7	
Sulphur	rhombic (100)	25	1 k	3.8	5
	(010)	25	1 k	4.0	5
	(001)	25	1 k	4.4	5
Urea	$CO(NH_2)_2$	17/22	400 M	3.5	
Zircon	$ZrSi O_4 \perp, \parallel$	17/22	100 M	12	
Glasses					
Borosilicate	normal	20	1 k/1 M	5.3	50/40
	low alkali	20	1 M	5	30
	very low alkali	20	50/100 M	4	15/5
Fused quartz		20/150	50/100 M	3.8	10/1
Lead		20	1 k/1 M	6.9	17/13
Soda	average	20	1 M/100 M	7.5	100/80
Minerals					
Amber		20	1 M/3 G	2.8/2.6	2/90
Asbestos (chrysotile) . . .	purified, 50% R.H.	25	50/1 M	5.8/3.1	1800/250
	board	20	1 M	3	2200
Bitumen	Gilsonite	25	50/100 M	2.7/2.55	60/10
		20	1 k	3.5	300
Granite		20	1 M	8	
Gypsum		20	10 k	5.7	
Marble	pure, dry	20	1 M	8	400
Sand	dry	20	1 M	2.5	
	15% water	20	1 M	9	
Sandstone		20	1 M	10	
Soil	dry	20	1 M	3	
	moist	20	1 M	10	
Sulphur	cast	20	3 G/10 G	3.4	7/14
Paper and Pressboard					
(see also cellulose)					
Unimpregnated, dry					
Kraft (cable)	$d = 0.9$	20/90	50	2.1	15/20
Kraft (tissue)	$d = 0.8$	20/90	1 k	1.8	10/15
	$d = 1.2$	20/90	1 k	3.0	25/35
Kraft		20	1 k	2.9	45
Rag (cotton)	$d = 0.6$	20/90	50/50 k	1.7	8/65
Impregnated, mineral oil ($\varepsilon_r' = 2.2$)					
Kraft (cable)	$d = 0.8$	20/90	50	3.2	18/20
	$d = 1.2$	20/90	50	4.7	28/33
Kraft (tissue)	$d = 0.9$	20	50	3.6	22
	$d = 1.1$	20	50	4.3	27
Rag (cotton)	$d = 0.9$	20	50	3.5	13
	$d = 1.1$	20	50	4.2	18

† Varies critically with stoichiometric ratio.

Solids (*contd*)

Material	Remarks	$t/°C$	f/Hz	ε_r'	$10^4 \times \tan \delta$
Paper and Pressboard (*contd*)					
Impregnated (Pentachlordiphenyl)					
Kraft (tissue)	$d = 0.9$	20	50	5.7	33
	$d = 1.1$	20	50	6.0	39
Fibre		20	1 M	4.5	500
Pressboard	dry $d = 0.8$	20	50	3.2	80
Plastics (non-polar, synthetic)					
Poly-					
ethylene		20	50/1 G	2.3	2/3
isobutylene		20	50/3 G	2.2	2/5
4-methylpentene (TPX) . .		20	100/10 k	2.1	2/1
(dimethyl)phenyloxide (PPO)		25	100/1 M	2.6	4/7
propylene		20	50/1 M	2.2	5
styrene		20	50/1 G	2.6	2/5
tetrafluoroethylene (PTFE) .	teflon	20	50/3 G	2.1	2
Plastics (polar, synthetic)					
Poly-					
amides	typical Nylon	20	50/100 M	4/3	200
carbonates	typical	20	50/1 M	3.2/3.0	10/100
ethyleneterephthalate . .		20	50/100 M	3.2/2.9	20/150
imides	typical	20	1 M	3.4	
methylmethacrylate . .		20	50/100 M	3.4/2.6	600/60
vinylcarbazole		20	50/100 M	2.8	5/10
vinylchloride	unplasticized	20	50/100 M	3.2/2.8	200/100
Plastics (miscellaneous)					
Aniline resin	unfilled	20	3 G	3.5	500
	paper filled	20	1 M/1 G	5/4	600/300
		100	1 M	6	800
Cellulose acetate		20	1 M/1 G	3.5	300/400
Cellulose triacetate . . .		20	50/100 M	3.8/3.2	100/300
Ebonite	unfilled	20	1 k/1 G	3/2.7	90/30
	filled (Mg CO_3)	20	50/1 G	4.1/3.8	100/180
Epoxy resin		25	1 k/100 M	3.6/3.5	200
Melamine resin		20	3 G	4.7	400
Phenolic resin	fabric filled	20	1 M	5.5	500
	paper filled	20	1 M/1 G	5	300/800
		140	1 M/10 M	6	800/400
	wood filled	20	1 M	5	400
Urea resin	paper filled	20	1 M	6	300
Vinyl acetate (poly-) . . .	plasticized	20	1 M/10 M	4	500
Vinyl chloride (poly-) . .	plasticized	20	1 M/10 M	4	600
(PVC)					
Rubbers					
Natural	crepe	20/80	1 M/10 M	2.4	15/100
	vulcan, soft	20	1 M/10 M	3.2	280/200
Butadiene/styrene	unfilled	20/80	50/100 M	2.5	5/70
(GR-S)	compounded	20/80	50/100 M	2.5	10/200

Solids (*contd*)

Material	Remarks	$t/°C$	f/Hz	ε'_r	$10^4 \times \tan \delta$
Rubbers (*contd*)					
Butyl	unfilled	20	50/100 M	2.4	35/10
Chloroprene	Neoprene	20	1 k/1 M	6.5/5.7	300/900
Silicone	filled 67% TiO_2	20	50/100 M	8.6/8.5	50/10
Silicone	unfilled	25	1 k/100 M	3.2/3.1	
Waxes, etc.					
Chlornaphthalene					
(tri and tetrachlor-) . . .		20	50/100 M	5.4/4.2	7/2700
Ozokerite		20	50/100 M	2.3	5/10
Paraffin wax		20	1 M/1 G	2.2	2
Petroleum jelly		20/60	50	2.1/1.9	1/5
Rosin	colophony	20	3 G	2.4	6
Wood (% **water**)					
Balsa 0%		20	50/3 G	1.4/1.2	40/140
Beech 16%	$d = 0.62$	20	1 M/100 M	9.4/8.5	580/830
Birch 10%	$d = 0.63$	20	1 M/100 M	3.1	400/800
Douglas fir 11%	$d = 0.45$	15	1 M/10 M	3.2	520/810
compressed	$d = 0.64$	15	1 M/10 M	4.3	570/950
Mahogany		20	3 G	2	340
Scots pine 15%	$d = 0.61$	20	1 M/100 M	8.2/7.3	590/940
Walnut 0%		20	10 M	2	350
Walnut 17%		20	10 M	5	1400
Whitewood 10%	American	20	1 M/100 M	3	400/750

References

(1) A. R. von Hippel, *Dielectric Materials and Applications*, 1954 (Chapman and Hall).
(2) *Handbook of the American Institute of Physics*, 2nd edn, 1963 (McGraw-Hill Book Co.).
(3) *The International Critical Tables*.
(4) *CRC Handbook of Chemistry and Physics*, 1983 (CRC Press Inc., Florida).

Liquids

The permittivities in this table, except when a frequency is stated, are 'static' values, relating to frequencies high enough to exclude ionic conductivity, but below any region of dispersion. For non-polar liquids ($\varepsilon'_r \simeq 2$) the lower limit depends only upon ionic impurities, while the higher is usually above 10 GHz. Within these limits, the permittivity is constant, and the loss unlikely to exceed a few times 10^{-4}. For polar liquids, denoted by 'P', the lower frequency limit depends both upon purity and the intrinsic dissociation of the liquid, while the upper limit varies sharply with temperature. The two limits may overlap, and the permittivity is then nowhere constant, nor the loss tangent small. For these reasons, frequency and loss tangent are quoted only for a few liquids of controlled purity which are used for electrical purposes. More extensive data, and on many more liquids, are given in the references below. The permittivity of liquids is easier to determine accurately than that of solids, and probable accuracy is indicated by the number of places quoted in the Table. Temperature coefficients ($a = 10^5 d\varepsilon'_r / \varepsilon'_r dt$) are given, but their accuracy is sometimes low.

Material	Remarks	$t/°C$	f/Hz	ε_r'	$10^4 \times \tan \delta$
Castor oil		20	1 k	4.5	
Chlordiphenyl (tri-)		$-10/100$	50/20 k	7/5	2000/2
(penta-)		0/100	50	5.2/4.3	700/3
Paraffin oil	medicinal	20	1 k	2.2	1
Silicone fluid	0.65 cS	20	50/3 G	2.2	2/19
	1000 cS	20	50/3 G	2.78/2.74	1/100
Transformer oil	BS 138	20	50/100 M	2.2	1/42
		20	100 M/10 G	2.2	42/8

Material	$t/°C$	ε_r'	a	Material	$t/°C$	ε_r'	a
Alcohols (primary)				$CHCl_2F$	28	5.34 P	
Methanol	25	32.6 P	-599	$CHClF_2$	24	6.11 P	
Ethanol	25	24.3 P	-622	$(—CCl_2F)_2$	25	2.52	
Propanol	25	20.1 P	-676	$(—CClF_2)_2$	25	2.26	
Butanol	20	17.1 P	-741	$(—CH_2Cl)_2$†	20	10.66 P	-550
Pentanol	25	13.9 P	-530	$(=CCl_2)_2$	25	2.30	-85
Hexanol	25	13.3 P	-806	$CCl_2=CHCl$	20	3.4 P	
				F-pentane	20	4.24 P	
Hydrocarbons				F-benzene	25	5.42 P	
n-Pentane	20	1.84_4	-87	Cl-benzene†	20	5.70_8 P	-299
n-Hexane	20	1.89_0	-82				
n-Heptane	20	1.92_4	-73	**Miscellaneous**			
n-Octane	20	1.94_8	-67	Aniline	20	6.89	-341
n-Nonane	20	1.97_2	-68	Acetone	25	20.7 P	-472
n-Decane	20	1.99_1	-65	Diethylketone	20	17.0 P	-520
n-Undecane	20	2.00_5	-62	Diethylether	20	4.34 P	-500
n-Dodecane	20	2.01_4	-60	Cyclohexanone	20	18.3 P	
Benzene†	20	2.283_6	-88	Nitrobenzene	25	34.8 P	-518
Cyclopentane	20	1.96		CS_2	20	2.64	-101
Cyclohexane†	20	2.025_0	-79				
Toluene	20	2.39	-102	**Liquid gases**	T/K		
				Argon	82	1.53	-220
(Chloro/Fluoro)-				Helium	4.19	1.04_8	
hydrocarbons					2.06	1.05_5	
CCl_4	20	2.24	-89	Hydrogen†	20.4	1.22_8	-280
CCl_3F	29	2.28		Nitrogen	70	1.45	-200
CCl_2F_2	29	2.13		Oxygen†	80	1.50_7	-160
$CClF_3$	-30	2.3					
$CHCl_3$	20	4.80 P	-368				

† These liquids have been recommended as standards for calibrations. The values given apply to especially pure and anhydrous samples. Cyclohexane is preferred at normal temperatures.

References

(1) *Dielectric Constants of Pure Liquids*, National Bureau of Standards Circular No. 514, 1951.
(2) *Dielectric Dispersion Data for Pure Liquids and Dilute Solutions*, National Bureau of Standards Circular No. 589, 1958.

Water

Water is strongly polar, with a region of dispersion, at 20 °C, centred around 17 GHz. It is also intrinsically dissociated, so that even de-ionized water cannot be treated as a dielectric at frequencies much below 1 MHz. Measurements at the high frequencies of the dispersion range contain, before about 1953, many errors shown by values of ε' and ε'' mutually inconsistent with the simple Debye equations. It is established that water obeys these with some accuracy, the value of the dispersion coefficient α in the Cole–Cole equation not exceeding 0.05.

At frequencies higher than about 1 MHz (where the loss tangent passes through a minimum of about 5×10^{-3}) values of ε' and ε'' can be calculated from the Debye equations given in the introduction to this section, with accuracy better than is obtainable by interpolation in a table. The necessary values of ε_s, ε_∞ and τ follow, chosen to give the best fit with internally consistent data.

The static relative permittivity as a function of temperature, within the range 0–60 °C, is given with an accuracy of ± 0.1 unit by

$$\varepsilon_s = 88.15 - 41.4\theta + 13.1\theta^2 - 4.6\theta^3$$

where $\theta =$ Celsius temperature$/100$ °C.

The relaxation time as a function of temperature is as follows, with an accuracy of about $\pm 2\%$:

$t/°C$	0	10	20	30	40	50	60
τ/ps	17.7	12.6	9.2	7.1	5.7	4.8	3.9

The 'infinite frequency' dielectric constant, ϵ_∞, occurs in the infra-red, and cannot be directly measured electrically. A value of 5.0 is appropriate for use with the foregoing data. ε_r decreases continuously from this value throughout the infra-red to a value of 1.8 in the optical region. The Debye equations therefore become increasingly inaccurate for $\omega\tau \gg 1$.

References

(1) *Dielectric Dispersion Data for Pure Liquids*, National Bureau of Standards Circular No. 589, 1958, Table 3.
(2) E. H. Grant and R. Shack, *Br. J. appl. Phys.*, 1967, **18**, 1807.
(3) J. B. Hasted, *Aqueous Dielectrics*, 1975 (Chapman and Hall, London).

Gases and vapours

The values relate, excepting the final entry, to a pressure of one standard atmosphere, and hold for all frequencies below the start of the infra-red spectrum. Other values may be calculated over a limited range of temperature and pressure, for non-polar permanent gases, by assuming that $(\varepsilon_r - 1)$ is proportional to density. This does not hold for polar gases, but if the polarity is strong (e.g. water vapour) a close approximation is

$$(\varepsilon_r - 1) \propto \text{pressure}/(\text{absolute temperature})^2$$

provided that the vapour is not near its condensation point, under the conditions either of the data used, or of the desired result. This relation can safely be used, for example, to obtain values for damp air, the densities and pressures involved being then the partial values, and the contributions from the two components additive. The relation should not be applied to mixtures of two polar vapours.

Values of relative permittivity may also be obtained from the data on refractive indices at radio frequencies by using the relation $\mu_r\varepsilon_r = n^2$ which applies to non-absorbing gases. $\mu_r = 1$ for all gases except O_2 where $\mu_r = 1 + 1.9 \times 10^{-6}$.

Relative permittivity of gases and vapours (*contd*)

Material	$t/°C$	$10^4 \, (\varepsilon_r - 1)$	Material	$t/°C$	$10^4 \, (\varepsilon_r - 1)$
Air (dry)	20	5.36_1	Nitrous oxide . . .	25	10.3
Nitrogen	20	5.47_4	Ethylene	25	13.2
Oxygen	20	4.94_3	Carbon disulphide . .	29	29.0
Argon	20	5.16_7	Benzene	100	32.7
Hydrogen	0	2.72	Methanol	100	57
Deuterium	0	2.69_6	Ethanol	100	78
Helium	0	0.7	Ammonia	1	71
Neon	0	1.3	Sulphur dioxide . . .	22	82
Carbon dioxide . . .	20	9.21_6	Water	100	60
Carbon monoxide . .	25	6.4	Water (10 mmHg) . .	20	1.21_4

J.H.C.

1.8.6 Magnetic properties of materials

Many magnetic properties of materials are expressed in terms of the magnetic field strength H, magnetic flux density and the magnetic polarization J (sometimes referred to as the intrinsic magnetic induction B_i). The SI units of H and B are, respectively, ampere per metre $(A\,m^{-1})$ and tesla (T).

The relation between the quantities expressed in SI units is:

$$B = \mu_0 H + J$$

in which μ_0 is $4\pi \times 10^{-7}$ H m^{-1}, the permeability of free space. The absolute permeability, $\mu \, (= B/H)$ and the volume susceptibility $\kappa \, (= J/\mu_0 H)$, are thus related by the equation:

$$\mu = \mu_0 \, (1 + \kappa)$$

The mass susceptibility χ is equal to κ/ρ, where ρ is the density. The relative permeability $\mu_r = \mu/\mu_0$ is the permeability of the material relative to that of a vacuum and is the value given in the tables.

In ferromagnetic materials as H is increased steadily from zero the permeability changes and is at first relatively small, its value being defined as the initial permeability, then reaches a maximum value, and finally decreases towards μ_0 as the polarization tends towards a limiting value $(B - \mu_0 H)$. The flux density remaining when H is reduced to zero is the remanent flux density and the negative H needed to reduce B to zero is the coercive force. The remanent flux density and coercive force for a cycle which proceeds to saturation are called the remanence, B_r, and the coercivity, $_BH_c$. In an open magnetic circuit the variation of J with H is usually measured and the coercivity is then denoted by $_JH_c$.

When a ferromagnetic material is taken through a cycle of magnetization there is a loss of energy as heat due to the combined effects of hysteresis and induced eddy currents. The hysteresis loss per unit volume, $Q_h = \oint H \, dB$, has been shown empirically to vary as $B_{max}^{1.6}$ over a limited range of peak flux density of up to about 1 T for high saturation materials, and 0.5 T for low saturation materials. This relationship, known as the Steinmetz law, is nevertheless only approximate. Some indication of the second loss, namely the eddy current power loss, may be calculated from standard formulae once certain relevant physical parameters are known. In their present forms, however, these formulae are only approximate. The total power losses that will be dissipated in laminar material when an alternating flux is developed in it has a direct bearing on the efficiency that can be realized in equipment such as transformers and electric motors and should therefore be known accurately. Accordingly the power losses of representative forms of typical materials are measured and some of these are given in Table (1) in terms of power loss per unit mass.

Many magnetic properties of ferromagnetic materials depend greatly on previous history, state of strain, temperature, and size, perfection and orientation of crystals, and the effect of small traces of impurity may be enormous.

When heated, ferromagnetic materials become paramagnetic at a temperature known as the (ferromagnetic) Curie point.

text continued on page 147

Symbols used in tables:

B = magnetic flux density

B_r = remanence

H = magnetic field strength

$_BH_c$ = induction coercive force, coercivity

$_JH_c$ = magnetization coercive force, coercivity

J = magnetic polarization

$J_s = (B - \mu_0 H)_s$ = saturation polarization

Q_h = hysteresis loss per unit volume per cycle

t_c = Curie temperature

ρ = resistivity

μ_r = relative magnetic permeability

μ_i = initial relative magnetic permeability

(1) Soft (low coercivity) materials

Material (approx. % composition, balance iron)	Flux density B/T for $H/(\text{A m}^{-1}) =$							
	1.0	10	20	50	100	1000	5000	50 000
Ferromagnetic elements								
Iron, high purity (single crystals in preferred direction)	1.50	1.99	2.00	2.01	—	—	—	—
Armco iron	—	0.005	0.010	0.060	0.45	1.55	1.72	2.20
Electrolytic iron	—	0.020	0.48	0.90	1.25	1.68	1.74	2.20
Cast iron (annealed)	—	—	—	—	0.02	0.60	0.86	1.45
Swedish iron (annealed)	—	0.010	0.020	0.15	0.70	1.52	1.72	2.20
Nickel	—	—	—	—	0.10	0.45	0.55	0.68
Nickel (wrought)	—	—	—	—	—	—	—	—
Cobalt	—	—	—	—	0.010	0.21	0.70	1.22
Steels (solid)								
Carbon steel (annealed) 1% C	—	—	—	—	0.02	0.75	1.54	2.00
Cast steel	—	0.004	0.010	0.050	0.22	1.38	1.68	2.12
Constructional steels: 0.3% C, 1% Ni	—	—	—	—	0.13	1.32	1.68	2.12
0.4% C, 3% Ni, 1.5% Cr	—	—	—	—	0.03	0.75	1.67	2.08
Mild steel, 0.1% C	—	0.004	0.010	0.055	0.35	1.46	1.74	2.15
Steels (sheet)†								
Grain oriented silicon steels with preferred magnetic properties in direction of rolling of the parent strip:								
Unisil-H, 30M2H 2.9% Si	—	1.21	1.55	1.72	1.79	1.93	2.00	2.04
Unisil, 27M4 ⎫	—	0.95	1.32	1.56	1.67	1.86	1.96	—
30M5 ⎬ 3.1% Si	—	0.75	1.24	1.56	1.67	1.86	1.96	—
35M6 ⎭	—	0.70	1.22	1.53	1.63	1.84	1.94	—
Non-oriented silicon steels:								
Transil 300–35 2.9% Si	—	0.10	0.13	0.54	0.86	1.46	1.65	—
Losil 400–50 2.4% Si	—	0.06	0.08	0.40	0.82	1.48	1.69	—
Losil 800–65 1.6% Si	—	—	0.05	0.40	0.65	1.53	1.73	—
Non-oriented, non-silicon steel:								
Newcor 1000–65	—	—	0.03	0.38	1.20	1.59	1.75	—
Amorphous iron–boron alloys (metallic glass)								
Metglas‡ 2605 S–3	—	—	—	—	—	—	—	—
Metglas‡ 2605 SC	—	—	—	—	—	—	—	—

Relative permeability μ_r			$\dfrac{J_s}{T}$	$\dfrac{H_c}{A\,m^{-1}}$	$\dfrac{B_r}{T}$	$\dfrac{t_c}{°C}$	$\dfrac{\rho}{10^{-8}\,\Omega m}$	Specific total loss for $J=1.5\,T$, $f=50\,Hz$ W/kg	Specific apparent power for $J=1.5\,T$, $f=50\,Hz$ VA/kg
Initial		Maximum							
$\dfrac{\mu_r}{1000}$	$\dfrac{H}{A\,m^{-1}}$	$\dfrac{\mu_r}{1000}$							
—	—	1500	2.16	12	—	770	10	—	—
0.25	—	7	2.16	80	1.3	770	11	—	—
—	—	—	2.16	30	—	770	10	—	—
—	—	—	1.70	400	—	—	—	—	—
—	—	—	2.16	70	—	770	—	—	—
—	—	—	0.615	400	—	358	9	—	—
0.25	—	2	0.6	120	0.3	358	7	—	—
—	—	—	1.76	950	—	1115	9	—	—
—	—	—	2.00	600	—	—	—	—	—
—	—	—	2.10	250	—	—	—	—	—
—	—	—	2.10	250	—	—	—	—	—
—	—	—	2.05	500	—	—	—	—	—
—	—	—	2.15	150	—	—	10	—	—
—	—	93	2.00	6	—	745	45	1.12 ($J=1.7\,T$)	1.55 ($J=1.7\,T$)
—	—	75	2.00	7	—	745	48	0.84	1.16
—	—	59	2.00	7	—	745	48	0.89	1.32
—	—	58	2.00	7	—	745	48	1.00	1.39
—	—	8	2.00	40	—	745	48	2.95	28
—	—	7	2.03	40	—	748	44	3.60	19
—	—	5	2.08	70	—	758	34	6.50	14
—	—	8	2.15	50	—	770	12	8.00	11
—	—	—	1.58	8	0.7	405	125	0.15 ($J=1.7\,T$)	0.20 ($J=1.0\,T$)
—	—	—	1.61	5	1.1	370	125		

(1) Soft (low coercivity) materials (*contd.*)

Material (approx. % composition, balance iron)		Flux density B/T for $H/(\mathrm{A\,m^{-1}}) =$							
		1.0	10	20	50	100	1000	5000	50 000
Nickel iron alloys									
Supermumetal§ / Nilomag 771¶	70–80% Ni with small	0.45	0.72	0.75	0.76	0.78	—	—	—
Mumetal Plus§	amounts of other	0.30	0.70	0.72	0.75	0.77	—	—	—
Mumetal§ / EPC 20¶	elements	0.18	0.60	0.65	0.72	0.75	—	—	—
Nilomag 641¶	65% Ni + small amount of other elements, oriented	—	—	—	—	—	—	—	—
Nilomag 471¶ / Super Radiometal§	50% Ni + small	0.06	1.02	1.12	1.25	1.4	1.62	—	—
Radiometal 4550§	amounts of other elements	0.01	0.48	0.75	1.05	1.18	1.62	—	—
Satmumetal§		0.20	1.15	1.25	1.3	1.35	—	—	—
HCR alloy§	+ oriented structure	—	0.3	1.0	1.46	1.50	1.55	—	—
Radiometal 36§	35% Ni	—	0.15	0.35	0.72	0.90	1.2	—	—
Hyperm 36	36% Ni	constant permeability alloy							
R2799§	30% Ni, temperature compensating alloy	—	—	—	—	—	0.1	0.2	—
Cobalt–iron alloys§									
Permendur 24	24% Co	—	0.002	0.01	0.02	0.05	1.45	1.85	2.3
Permendur 49	49% Co	—	0.01	0.04	0.13	0.33	1.85	2.3	2.4
Supermendur	49% Co, 2% V	—	—	—	2.05	2.1	2.3	2.35	2.4
Other alloys									
Heusler alloy	61% Cu, 26% Mn, 13% Al . .	—	—	—	—	0.01	0.25	0.38	0.45
Isoperm	30% Ni, 11% Cu	constant permeability alloy							
Perminvar	40% Ni, 25% Co	constant permeability alloy							
Nickel copper	70% Ni, 30% Cu	—	—	—	—	—	0.07	0.10	—

Relative permeability μ_r			$\dfrac{J_s}{T}$	$\dfrac{H_c}{A\,m^{-1}}$	$\dfrac{B_r}{T}$	$\dfrac{t_c}{°C}$	$\dfrac{\rho}{10^{-8}\,\Omega m}$	Specific total loss (W/kg) for $J = 0.5\ T$ at frequencies (Hz) of:		
Initial		Minimum								
$\dfrac{\mu_r}{1000}$	$\dfrac{H}{A\,m^{-1}}$	$\dfrac{\mu_r}{1000}$						50	400	2400
140	0.40	350	0.77	0.55	0.5	350	55	—	0.11	2.2
								0.026	1.1	—
80	0.40	300	0.77	0.8	0.45	350	55			
60	0.40	240	0.77	1.0	0.45	350	55	0.010	0.155	3.3
								0.010	0.40	8.8
0.4–1.0	—	200–400	1.4	4	1.35	590	48			
2–11	0.40	50–120	1.6	4–12	0.4–1.2	530	43	—	0.55	—
3–6	0.40	20–50	1.6	12–24	0.4–1.0	530	50	0.12	3.3	—
65	0.40	240	1.5	2.0	0.7	550	60	—	—	—
0.5–1.0	0.40	50–100	1.6	10	1.5	525	40	—	—	—
2	—	15	1.3	12	0.35	180–270	80	0.12	3.3	—
1.75	0.8	6	—	—	—	—	—	—	—	—
—	—	—	0.45	—	—	70	85	—	—	—
0.25	—	2.0	2.35	950	1.65	980	20	0.33	7.1	66
1.0	—	7	2.35	140	1.5	980	47	—	—	—
—	—	70	2.35	20	2.1	980	40	—	—	—
—	—	—	0.48	550	—	330	—	—	—	—
0.06	—	0.065	—	—	—	—	—	—	—	—
0.30	—	1.5	1.55	100	—	715	19	—	—	—
—	—	—	—	—	—	10–100	—	—	—	—

Manufacturers: † British Steel Corporation, Newport, Gwent. ‡ Allied Chemical Corpn., Morris Township, NJ, USA. § Telcon Metals, Crawley. ¶ Henry Wiggin & Co., Hereford.

(1) Soft (low coercivity) materials (*contd.*)

Material	Initial relative permeability μ_i	Frequency range MHz	Loss factor at maximum frequency $\frac{}{10^{-6}}$
Carbonyl iron powder cores‖			
type 100	30	0.1–2	700
type 500	12	1–10	250
type 900	10	1–50	600
type 901	5	10–100	1500
Magnetic iron oxide powder cores‖			
type 910	4	20–300	500 (at 100 MHz)
Iron flake cores‖			
used for interference suppression,	90 at 1 kHz	—	—
relative initial permeability falls rapidly with frequency }	65 at 150 kHz	—	—
Ferrite cores‖			
(a) for radio, TV and low power uses:			
nickel zinc, type F13	650	0.05–1	130
type F14	220	0.1–2	50
type F16	125	1–10	100
type F22	19	5–40	500
manganese zinc, type F10	5000	0.01–0.1	12
type F8 .	1500	0.05–0.5	80
type F11	600	0.1–1	50
(b) perminvar, high frequency low power uses			
type F25	50	5–40	300
type F29	12	10–200	1000
(c) manganese zinc for high power uses			
type F6	1500	—	—
type F5	2000	—	—
(d) manganese zinc, high stability, low loss, telecommunications uses			
type P10	2000	—	—
type P11	2200	—	—
type P12	2200	—	—

Loss factor at 100 kHz / 10^{-6}	Temperature factor / 10^{-6}/°C	Flux density B/T for $H/(\text{A m}^{-1})$ = 800	Power loss density for $B = 0.2\ T$, $f = 16\ kHz$ / mW/cm³	IEC hysteresis coefficient ηB	Disaccommodation factor / 10^{-6}	t_c / °C	ρ / Ωm
—	20	—	—	—	—	—	—
—	12	—	—	—	—	—	—
—	12	—	—	—	—	—	—
—	12	—	—	—	—	—	—
—	40	—	—	—	—	—	—
—	—	—	—	—	—	—	—
—	—	—	—	—	—	—	—
—	—	—	—	—	—	180	300
—	—	—	—	—	—	270	1000
—	—	—	—	—	—	270	1000
—	—	—	—	—	—	500	1000
—	—	—	—	—	—	180	0.5
—	—	—	—	—	—	180	1
—	—	—	—	—	—	220	5
—	—	—	—	—	—	450	1000
—	—	—	—	—	—	500	1000
—	—	0.45	150	—	—	180	1
—	—	0.48	75	—	—	200	1
12	0–2	—	—	2.5	8	150	1
5	0.5–1.5	—	—	0.8	5	150	1
3	0.4–1.0	—	—	0.4	3	150	1

Manufacturer: ‖ Neosid Ltd, Welwyn Garden City, Herts.

Note: For more complete details of soft ferrite materials see E. C. Snelling, *Soft Ferrites, Properties and Applications* (Iliffe Books Ltd., London).

(2) Magnetically hard (permanent magnet) materials

Material	Approx. % composition (balance iron)	Remanence B_r/T
Typical alloys in the Al–Ni–Co series (cast material):		
Alni, Magloy 6 } isotropic	Ni 25, Al 13, Cu 4	0.56
Alnico, Magloy 5 } isotropic	Ni 19, Al 10, Co 12, Cu 6	0.73
Alcomax III, Ticonal 600, Magloy I } anisotropic	Ni 13.5, Al 8, Co 24, Cu 3	1.30
Hycomax II anisotropic	Ni 14.5, Al 7, Co 29, Cu 4.5, Ti 5	0.85
Hycomax III, Ticonal 550 } anisotropic	Ni 14, Al 7.3, Co 34, Cu 3, Ti 5.25	0.90
Columax, Magloy 100X } columnar	Ni 13.5, Al 8, Co 24, Cu 3	1.35
Alcomax III semi-columnar	Ni 13.5, Al 8, Co 24, Cu 3	1.32

Note: The isotropic and anisotropic alloys can also be produced by sintering, in which case the magnetic properties can be up to 20% less than those for cast material.

Ferrites:		
Feroba 1, Ferroxdure 100, Neoperm D1 } sintered isotropic		} 0.22
Feroba 2, Ferroxdure 300, Neoperm E2 } sintered	BaO 5.9 (balance Fe_2O_3)	} 0.39
Feroba 3, Ferroxdure 380, Neoperm E3 } anisotropic	SrO 5.9 (balance Fe_2O_3)	0.37
Neoperm C1, Bonded feroba, Ferroxdur SP50 } bonded isotropic		0.15
Plastiform, Ferroxdur SP170, Feroba } bonded anisotropic		0.25

Rare-earth cobalt alloys:		
Supermagloy } sintered, Vacomax } anisotropic	} Co 66, Sm 43	} 0.80–0.93
Supermagloy, polymer bonded		0.30–0.59

$(BH)_{max}$		Coercivity		$t_c/°C$	Max. operating temperature °C	$\dfrac{\rho}{\mu\Omega m}$
kJ m^{-3}	at H (kAm^{-1})	$_BH_c/$(kAm^{-1})	$_JH_c/$(kAm^{-1})			
10.0	28	46	49	760	550	0.63
13.5	29	45	48	800	550	0.65
43	42	52	53	850	550	0.55
32	60	95	97	850	550	0.50
44	85	127	129	850	550	0.50
60	51	59	60	860	550	0.55
49	45	56	57	860	550	0.55
8	72	135	220	450	180	10
28.5	136	150	160	450	180	10
26	138	240	300	450	180	10
3.2	50	85	175	*	120*	100
11.2	96	175	240	*	120*	100
120–175	—	610–700	1200–1400	725	200	0.70
16–64	—	220–420	480–800	*	60	10 000

* Limited by properties of bonding material.

(3) Feebly magnetic steels and cast irons

Material	Approx. % composition (Balance iron)	Condition	μ_r for $H = 5$ kA/m	$\dfrac{\rho}{\mu\Omega\text{m}}$
Cast irons†				
Nomag	Ni 11, Mn 6	} as cast	1.03–1.05	—
Ni-Resist, type 1	Ni 14, Cu 6, Cr 2		1.03	—
Ductile Ni resist, types D–2, D–2B, D–2C and D–4	Si 0/3, Ni 18/32, Cr 0/5 Mn 0/2.4		1.02–1.1	—
Austenitic stainless steels‡				
AISI type:				
301	Ni 7.8, Cr 17.6	Austenized	1.003	0.68
		19.5% cold reduction	1.15	
		55% cold reduction	14.8	
302	Ni 9.0, Cr 18.4	Austenized	1.003	0.70
		20% cold reduction	1.008	
		44% cold reduction	1.050	
		68% cold reduction	1.59	
304	Ni 10.7, Cr 19.0	Austenized	1.004	0.72
		13.8% cold reduction	1.005	
		32% cold reduction	1.04	
		65% cold reduction	1.55	
305	Ni 11.7, Cr 17.9	Austenized	1.003	—
		18.5% cold reduction	1.004	
		52.5% cold reduction	1.05	
310	Ni 20.7, Cr 24.3	Austenized	1.002	0.94
		64.2% cold reduction	1.002	
316	Ni 13.4, Cr 17.5	Austenized	1.003	0.74
		81% cold reduction	1.007	
321	Ni 10.3, Cr 18.3, Ti 0.68	Austenized	1.003	0.72
		16.5% cold reduction	1.018	
		41.5% cold reduction	1.40	
347	Ni 10.7, Cr 18.4, Co 0.95	Austenized	1.004	0.73
		13.5% cold reduction	1.007	
		40% cold reduction	1.06	
		60% cold reduction	1.25	

† Data taken from *Ni-Resists and ductile Ni-Resists, Engineering properties*, published by INCO Europe Ltd.
‡ C. B. Post and W. S. Eberley, 'Stability of austenite in stainless steels', *Trans. Am. Soc. Metals*, 1947, **39**, p. 868.

(4) Magnetic susceptibilities of paramagnetic and diamagnetic materials

Values are mass susceptibility per kilogram, χ, at 20 °C.

	$\times 10^{-8}$		$\times 10^{-8}$		$\times 10^{-8}$
Common elements					
Hydrogen	-2.49	Aluminium	$+0.82$	Uranium	$+2.19$
Oxygen	$+133.6$	Copper	-0.107	Germanium	-0.15
Helium	-0.59	Silver	-0.25	Silicon	-0.16
Neon	-0.41	Gold	-0.19	Arsenic	-0.39
Argon	-0.60	Platinum	$+1.22$	Indium	-0.14
Krypton	-0.41	Mercury	-0.21	Antimony	-1.09
Xenon	-0.40	Bismuth	-1.70	Tellurium	-0.39
Nitrogen	-0.54	Sulphur	-0.62	Gallium	-0.30
Sodium	$+0.75$	Lead	-0.15	Phosphorus	-1.13
Potassium	$+0.65$				
Common compounds				*Common materials*	
H_2O	-0.90	$NiSO_4 7H_2O$	$+20.1$	Araldite	-0.63
NO	$+59.3$	$NiSO_4 K_2SO_4 7H_2O$	$+13.9$	P.V.C.	-0.75
CO_2	-0.59	$CuSO_4 5H_2O$	$+7.7$	Perspex	-0.5
NH_3	-1.38	$MnSO_4 4H_2O$	$+81.2$	Polyethylene	$+0.2$
HCl	-0.75	$FeSO_4(NH_4)_2$			
H_2SO_4	-0.50	$SO_4 6H_2O$	$+40.6$		
NaCl	-0.64				
$NiCl_2$					
(anhydrous)	$+78.5$				
(in sol)	$+43.0$				

Notes:
(1) To obtain values in CGS units per gramme, the SI values given should be multiplied by $10^3/4\pi$.
(2) For a more complete list of elements, see L. F. Bates, *Modern Magnetism* (CUP).

    Ferrimagnetic materials (ferrites) have all of the above characteristics of ferromagnetic materials. However, due to their high resistivity, soft (low coercivity) ferrites are widely used in high frequency applications, in which case the following parameters are also of interest:

(a) Power loss density—this is another name for specific total power loss, but for ferrite materials the loss is usually expressed per unit volume.

(b) Loss factor—the performance of ferrites at low field strengths is often indicated by the expression $\tan \delta$ where δ is the loss angle, i.e. the phase angle between B and H. However, information regarding power losses is usually given in the form of loss factors normalized to unit permeability, μ, since this facilitates the calculation of loss coefficients of gapped ferrite cores. Hence the loss factor is:

$$\frac{\tan \delta}{\mu} = \frac{\tan \delta_h}{\mu} + \frac{\tan \delta_e}{\mu} + \frac{\tan \delta_r}{\mu}$$

where $\tan \delta_h$, $\tan \delta_e$ and $\tan \delta_r$ are the loss angles for the hysteresis, eddy current and residual losses respectively, all of which are present to a greater or lesser extent and combine to give the total loss, $\tan \delta$.

(c) IEC hysteresis coefficient η_B—in considering recommendations for standard forms of loss expression, the International Electrotechnical Commission agreed the following relationship for the hysteresis coefficient, η_B,

$$\eta_B = \frac{\tan \delta_h}{\mu B_{max}}$$

(d) Temperature factor—the permeability of a magnetic material may change for a variety of reasons, the most obvious being the change of temperature. Over a limited temperature range the relationship between the reversible change in magnetic permeability, $\Delta\mu$, and the corresponding change in temperature, $\Delta\theta$, is given by the temperature coefficient, TC:

$$TC = \frac{\Delta\mu}{\mu\Delta\theta}$$

As with the loss factor, it is usual to normalise the values to unit permeability which gives the loss factor:

$$\text{loss factor} = \frac{\Delta\mu}{\mu^2\Delta\theta}.$$

(e) Disaccommodation factor—the permeability of a magnetic material can also change with time after magnetization. This phenomenon is often called disaccommodation. If the permeabilities μ_1 and μ_2 correspond to times t_1 and t_2 then the disaccommodation is given by:

$$\frac{\mu_1 - \mu_2}{\mu_1} \times 100\%$$

As with the loss and temperature factors, the disaccommodation factor is normalized to unit permeability and is given by:

$$\text{disaccommodation factor} = \frac{\mu_1 - \mu_2}{\mu_1{}^2} \times 100\%.$$

Since the properties may vary considerably from specimen to specimen due to chemical composition and state of heat treatment, the values given are only to be regarded as typical of the materials mentioned. A range of values is indicated by a dash.

A.E.D.

1.9 Astronomy and geophysics

1.9.1. Astronomical and atomic time systems

Time systems

The fundamental reference timescale is now **international atomic time** (TAI); this is intended to provide both an invariable unit of time (the SI second) and an unambiguous, precise method of identifying any instant of time, but it is not convenient for general use. There are still valid, practical reasons for continuing to relate the timescales in general use to the rotation of the Earth by the adoption of a system of standard times each of which differs from that for the prime meridian (zero longitude) by an exact number of hours. (The timescale in use at any given place may not be that of the nearest standard meridian and may differ between summer and winter.) The standard time for the prime meridian is formally known as **coordinated universal time** (UTC), but the name **Greenwich mean time** (GMT) is in widespread use throughout the world. This timescale differs from TAI by an exact number of seconds in such a way that it corresponds closely (to within 1 s) to **universal time** (UT) and so can be used directly in astronavigation as a measure of the rotation of the Earth with respect to the celestial sphere.

The rotation of the Earth on its axis is no longer of value for the measurement of precise time since the length of the day is subject to unpredictable variations. The systems of sidereal and solar time, including universal time, that represent this rotation are, however, still of value for scientific purposes and for the determination of position on the Earth. These timescales may be regarded as angles expressed in time-measure at the rate of 1 day of 24 hours for each complete rotation (or $1^h = 15°$).

Astronomers introduced, in 1955, the system of **ephemeris time** (ET) for use in the development and application of the theories of the motions of the Moon and planets, but ET has now been superseded for such purposes by a system of scales of **dynamical time**. The new system is defined so that the scales are directly related to TAI and may be used with relativistic theories in which the timelike variables depend on the choice of coordinate system and vary with the gravitational potential.

International atomic time

Independent atomic timescales were maintained by several national organizations from 1955 onwards, but the standard timescales continued to be based on universal time (UT) until 1972 and so required the use of offsets in the carrier frequencies of the broadcast transmissions and occasional small step adjustments in phase. In 1967 the General Conference on Weights and Measures (CGPM) adopted the atomic definition of the second as the unit of time interval; the adopted frequency of the chosen caesium transition gave continuity with the previous use of the ephemeris second. In 1971 the CGPM gave formal recognition to the atomic timescale maintained by the Bureau International de l'Heure (BIH) as international atomic time (TAI); the arbitrary epoch of the scale was chosen so that TAI and UT were coincident at 1958 January 1.0.

In forming TAI the BIH combines the results of comparisons of the primary time-standards of three countries and of about 130 secondary time-standards from 20 countries in an attempt to maintain the uniformity of the scale. The Loran-C radio-navigation signals are the main intermediary in the comparisons but it is expected that the use of satellites will lead to further improvements. The current estimate (1984) of the uniformity of the scale is 1.0×10^{-13} in frequency so that the accumulated error in epoch is less than 1 ns/yr.

Coordinated universal time

Coordinated universal time (UTC) was redefined from the beginning of 1972 as an approximation to UT in which the second markers coincide with those of TAI but the numerical values assigned to them are of an adjacent second of UT. At $0^h 00^m 00^s$ on 1 January 1972 the difference $\Delta AT = TAI - UTC$ was made exactly 10 seconds. 'Leap seconds' are added to, or subtracted from, UTC as UT drifts with respect to TAI so that the difference $\Delta UT = UT - UTC$ does not normally exceed 0.7 s. The one-second steps are made on the last second of a month, preferably June or December; the date is announced several months in advance. The following table gives the dates of the beginning of the periods during which the differences ΔAT took the values indicated.

1972 Jan. 1	10 s	1977 Jan. 1	16 s	1983 July 1	22 s
1972 July 1	11 s	1978 Jan. 1	17 s	1985 July 1	23 s
1973 Jan. 1	12 s	1979 Jan. 1	18 s		
1974 Jan. 1	13 s	1980 Jan. 1	19 s		
1975 Jan. 1	14 s	1981 July 1	20 s		
1976 Jan. 1	15 s	1982 July 1	21 s		

Estimates of the differences ΔUT are often given by coding in the broadcast time signals that are used for the dissemination of standard time and frequency.

Sidereal time

Sidereal time represents the orientation of the celestial system of right ascension with respect to the terrestrial system of longitude. Local sidereal time is usually defined as the local hour angle of the equinox. An equivalent definition is that it is the right ascension of the local meridian. The direction of the equinox is the line of intersection of the mean plane of the Earth's orbit around the Sun with the plane of the Earth's equator; it is used as the zero of right ascension. Sidereal time may therefore be determined by observing stars, whose right ascensions are assumed to be known, as they cross the observer's meridian. The local sidereal time (LST) at a place differs from Greenwich sidereal time (GST) by the longitude of the place measured from the Greenwich meridian according to the expression

$$LST = GST + \text{east longitude}$$

where longitude is measured in time at the rate of $15° = 1^h$.

The precession of the Earth's axis of rotation around the pole of the ecliptic causes a slow drift of the equinox with respect to the stars, and so the mean length of the sidereal day is actually slightly shorter than the period of rotation of the Earth with respect to an inertial frame. The luni-solar forces acting on the Earth also cause an oscillatory nutation that is superimposed on the precessional drift. It is therefore necessary to distinguish between apparent (or true) sidereal time and mean sidereal time, where the latter is free from the quasi-periodic irregularities due to nutation; the effects of nutation are computed from adopted numerical series. Observations of local sidereal time are also affected by the motion of the axis of rotation with respect to the axis of figure of the Earth; this 'polar motion' is unpredictable and must be determined from observations at two or more sites. Sidereal time (or UT) and polar motion are now determined from observations of extra-galactic radio sources and of satellites of the Earth, as well as from observations of stars. The results from the various techniques are combined and published by the Bureau International de l'Heure which also decides when leap seconds should be inserted in UTC.

Universal time

Universal time (UT) is formally defined in terms of Greenwich mean sidereal time (GMST) by a conventional expression (see p. 155) that ensures continuity with past practice and yet frees UT from irregularities that would arise from certain effects in orbital motion of the Earth around the Sun. (In effect UT is equivalent to 12^h plus the Greenwich hour angle of a fictitious mean sun.) Even so UT does not provide a uniform timescale since the rate of rotation of the Earth, and hence the length of day in SI units, is subject to secular, irregular and quasi-periodic changes due to many different causes. During the course of a year the length of day varies by about 1 ms (i.e. 1 part in 10^8) about its mean value; during the course of a decade the mean value may change by as much as 4 ms, and during the course of a millenium the mean value increases by about 20 ms.

The notations UT0, UT1, UT2 are now obsolete, but their meanings are as follows:

UT0: 'observed' values to which no corrections for polar motion have been applied.

UT1 = UT: as derived from fully corrected values of GMST.

UT2: 'smoothed' values obtained by applying a conventional expression for the seasonal variation, which has an amplitude of about 30 ms.

The notation UTR has recently been introduced to indicate that the values have been adjusted by removing the effects of certain short-period tidal terms.

Local time and the equation of time

Local mean (solar) time (LMT) differs from UT by the time equivalent of the longitude of the place concerned according to the expression

$$LMT = UT + \text{east longitude.}$$

The time indicated by a simple sundial is known as local apparent (solar) time, and it differs from local mean (solar) time by the so-called **equation of time**. (Local mean time differs from clock time by the time equivalent of the difference in longitude between the place and the appropriate standard meridian; this difference will be 1 hour greater when summer (daylight-saving) time is in force.) The average amounts by which apparent solar time is in advance of mean solar time are indicated in the following table; the actual amounts for a given day may vary by up to $0^m.3$ from these average values.

		m			m			m			m
Jan.	1	− 3.3	Apr.	1	− 3.8	July	1	− 3.8	Oct.	1	+ 10.4
	15	− 9.2		15	0.0		15	− 5.9		15	+ 14.3
Feb.	1	− 13.6	May	1	+ 3.0	Aug.	1	− 6.2	Nov.	1	+ 16.4
	14	− 14.3		15	+ 3.7		15	− 4.4		15	+ 15.3
Mar.	1	− 12.4	June	1	+ 2.0	Sep.	1	+ 0.1	Dec.	1	+ 10.9
	15	− 8.9		15	− 0.4		15	+ 4.9		15	+ 4.7

$$\text{LMT of transit of Sun} = 12^h - \text{equation of time}$$

The principal contributions to the equation of time arise from the eccentricity of the Earth's orbit and from the inclination of the plane of the Earth's orbit to the plane of the Earth's equator. The equation of time is given to a precision of about 1^m by the expression

$$+ 7^m.6 \sin (176°.24 + 0°.9856 \text{ d}) + 9^m.8 \sin (198°.82 + 1°.9712 \text{ d})$$

where d is the interval in days from January 0.

The orientation of the constellations during a given night is specified by the local sidereal time, which is equal to the right ascension of the stars on the meridian. The following table of LST at 0^h LMT (midnight) shows the right ascensions of the stars that will be due south at local midnight on the dates concerned; an approximate allowance must be made for the current difference between clock time and local mean time at the place concerned. LST gains on LMT by about 4 minutes each day.

		h			h			h			h
Jan.	1	6.7	Apr.	1	12.6	July	1	18.6	Oct.	1	0.6
Feb.	1	8.7	May	1	14.6	Aug.	1	20.6	Nov.	1	2.7
Mar.	1	10.6	June	1	16.6	Sep.	1	22.7	Dec.	1	4.7

Ephemeris time

In general terms, ephemeris time is the independent variable of the differential equations for the motions of the Sun, Moon and planets under the influence of their mutual gravitational attractions. Formally the fundamental epoch and the unit of time interval were defined by the first two coefficients in an adopted expression for the geometric mean longitude of the Sun, although in practice they were determined indirectly from observations of the Moon. The ephemeris second is equal to the SI second to within the limits of uncertainty of its determination, and the scale is such that the relation

$$ET = TAI + 32^s_.184$$

can be used for most purposes when interpolating ephemerides whose argument is ephemeris time.

Ephemeris time has been superseded by dynamical time for current purposes in which it is necessary to take into account relativistic effects, but for historical purposes the concept of ephemeris time is still valid. Current estimates of the mean values of $\Delta T = ET - UT$ for the years indicated are given in the following table; the values for the earlier epochs depend critically on the value adopted (in this case $-26''/\text{cy}^2$) for the tidal term in the expression for the Moon's mean longitude.

	h		s		s		s		s
−500	+4.9	1600	+128	1850	+7	1900	−3	1950	+29
0	2.9	1650	50	1860	8	1910	+10	1960	33
+500	1.5	1200	9	1870	+2	1920	20	1970	41
1000	0.6	1750	13	1880	−6	1930	23	1980	51
1500	0.1	1800	25	1890	−7	1940	24	1990	57

$$ET = UT + \Delta T$$

Dynamical time

The International Astronomical Union adopted recommendations in 1976 and 1979 concerning timescales for use in dynamical theories and ephemerides. In particular it specified 'terrestrial dynamical time' (TDT) as the timescale for apparent geocentric ephemerides by giving the relationship to TAI at one instant and by specifying the unit to be the day of 86 400 SI seconds at mean sea level. It also specified 'barycentric dynamical time' (TDB) as the timescale for equations of motion referred to the barycentre of the solar system by stating that there should be only periodic variations between it and TDT. In effect:

$$TDT = TAI + 32^s_.184$$
$$TDB = TDT + 0^s_.001\,658\sin g + \text{smaller terms}$$

where g is the mean anomaly of the Earth in its orbit around the Sun.

G.A.W.

International atomic time

Ephemeris time provides a uniform time scale which is not subject to irregularities due to changes in the Earth's motion. In practice it suffers from the disadvantage that it is accurately known only retrospectively (with a delay of years) because of the time needed to take and process astronomical observations. The development of caesium atomic clocks has made possible the immediate realization of a uniform time scale with high accuracy. The Bureau International de l'Heure (BIH), in Paris, and a number of national laboratories have been maintaining atomic timescales for about twenty-five years. In 1967 the General Conference of Weights and Measures (CGPM) adopted the atomic definition of the second as the unit of time duration. In 1971 the CGPM gave formal recognition to the scale maintained by the BIH as the International Atomic Timescale (TAI).

TAI and Universal Time were coincident on 1 January 1958. UT now lags by about 22 seconds. For practical use Coordinated Universal Time (UTC) has been redefined from 1 January 1972 as an approximation to UT in which the second markers coincide with those of TAI but the numerical values assigned to them are those of the nearest seconds in UT. At $0^h\,0^m\,0^s$ on 1 January 1972 the difference TAI − UTC was made exactly 10 seconds. As Universal Time drifts with respect to Atomic Time, 'leap seconds' are added to or subtracted from UTC so that the difference between UT and UTC does not normally exceed 0.7 second. The one-second steps, if necessary, are made on the last second of a month, preferably June or December. UTC is now used as the basis of broadcast time signals. The seconds markers provide a uniform timescale. The numerical 'labels' attached to them, with known corrections if needed, provide Universal Time for those who require it.

Time and frequency signals

Time and its reciprocal, frequency, are unique among physical quantities in that the magnitude of the units can be disseminated by radio broadcasts, without the need for direct intercomparison of material standards. A number of countries operate standard time and frequency broadcast services: these are available for use by anyone within range of the transmitters who has a suitable radio receiver. The majority of the transmissions are in the MF and HF bands (p. 83) and have an effective range of some thousands of kilometres. A number of stations, notably GBR, Rugby, and WWVL, Boulder, Colorado, operate in the VLF band and can be received virtually anywhere in the world. However, because variable delays occur when signals are propagated via the ionosphere, the full accuracy of the signals can be used only by a receiver within ground-wave range of a transmitter.

In Britain, the service is provided by the National Physical Laboratory, using Post Office transmitters at Rugby and a BBC transmitter at Droitwich. The carriers of these transmitters are all derived from atomic frequency standards and are maintained within ± 2 in 10^{12} of the nominal frequency. The signals are received at Teddington and are compared with the group of atomic clocks which forms the national frequency standard. Corrections to the radiated frequencies are subsequently published in *The Radio and Electronic Engineer* to ± 1 in 10^{13}.

The Droitwich transmission is on 200 kHz; it operates for about 20 hours per day and carries BBC Radio 2 programmes. The Rugby transmissions are effectively continuous and carry time signals. The frequencies used are: GBR, 16 kHz; MSF, 60 kHz, 2.5, 5 and 10 MHz. In practice the transmissions are interrupted for occasional maintenance (four hours each week for GBR and four hours each month for MSF), and the MSF HF service operates for alternate five-minute periods only, to allow time-sharing of the frequencies with other European transmitters.

The time signals on the Rugby transmissions are derived from the same atomic clocks that provide the carriers and have the same accuracy. They indicate Coordinated Universal Time (UTC—see previous section) and the difference $UT - UTC$. Details of the signals and the results of daily calibrations are given in a monthly bulletin issued by the National Physical Laboratory.

A.E.B.

1.9.2 Astronomical units and constants

IAU system of astronomical constants and units

A revised system of values of the principal constants of the solar system was adopted by the International Astronomical Union in 1976, and was introduced (apart from a few small changes in some cases) in the principal national and international almanacs for 1984 onwards. The system differs in one fundamental respect from the 1964 system in that the astronomical unit of time is defined as one day (D) of 86 400 SI seconds, rather than as an ephemeris day. The definitions of the other astronomical units were not changed: the astronomical unit of mass is the mass of the Sun (S), while the astronomical unit of length is that length (A) for which the Gaussian gravitational constant (k) takes the value 0.017 202 098 95 in these units. (The dimensions of k^2 are those of the constant of gravitation, G). The changes in the system of constants were accompanied by the adoption of a new basis for the fundamental reference frame and of the new dynamical timescales. Some of the consequential changes were adopted in 1979/80. The values in the IAU (1976) system of astronomical constants are as follows:

Primary constants
Speed of light $c = 2.997\,924\,58 \times 10^8\,\mathrm{m\,s^{-1}}$
Light–time for unit distance (1 au) $\tau_A = 499.004\,782\,\mathrm{s}$

Equatorial radius for the Earth (note 1) $a_e = 6.378\,140 \times 10^6$ m
Dynamical form-factor for Earth (note 2) $J_2 = 0.001\,082\,63$
Geocentric gravitational constant (note 3) $GE = 3.986\,005 \times 10^{14}$ m^3 s^{-2}
Constant of gravitation $G = 6.672 \times 10^{-11}$ m^3 kg^{-1} s^{-2}
Ratio of mass of Moon to that of Earth $\mu = 0.012\,300\,02$

Derived constants
Unit distance (length of 1 au) $c\tau_A = A = 1.495\,978\,70 \times 10^{11}$ m
Solar parallax $\arcsin(a_e/A) = \pi_\odot = 8''.794\,148$
Constant of aberration (for 2000) $\kappa = 20''.495\,52$
Flattening factor for the Earth (note 2) $f = 0.003\,352\,81$
                                                           $= 1/298.257$
Heliocentric gravitational constant $A^3 k^2/D^2 = GS = 1.327\,124\,38 \times 10^{20}$ m^3 s^{-2}
Ratio of mass of Sun to that of Earth $S/E = 332\,946.0$
Mass of the Sun $S = 1.989\,1 \times 10^{30}$ kg

Notes:
(1) The International Association of Geodesy adopted the value $6.378\,138 \times 10^6$ for its reference ellipsoid in 1979.
(2) See p. 157 for definitions of J_2 and f.
(3) The mass of the Earth is denoted by E.

Other units and the standard epoch

The unit of time in the fundamental formulae for precession (and in similar expressions) is the Julian century of 36 525 days (the tropical century is not to be used). The new standard epoch is designated J2000.0 and is the calendar date 2000 January 1^d.5, which is the Julian date JD 2 45 1545.0

An alternative unit for use in Newtonian dynamics of the solar system and binary stars is the Gaussian year, which is the sidereal period of a particle of negligible mass moving around the Sun in an orbit with a mean distance of 1 au; it is equal to $2\pi/k\ (= 365.256\,898)$ days. Kepler's law for the relative motion of two isolated particles of mass M and m is then, simply,

$$a^3 = P^2(M + m)$$

where a is the semi-major axis of the orbit in au, P is the period in Gaussian years, and the unit of mass is the mass of the Sun.

A convenient unit for the measurement of the distances of nearby stars is the parsec (pc); this is the distance at which 1 au subtends an angle of 1 second of arc. Hence

$$1\ \text{parsec} = 1/(\sin 1'')\ \text{au} = 2.063 \times 10^5\ \text{au} = 3.086 \times 10^{16}\ \text{m}$$

The multiple units kiloparsec (kpc) and megaparsec (Mpc) are more appropriate for most galactic and extragalactic objects respectively. The 'light-year' is normally only used in popular astronomical texts:

$$1\ \text{light-year} = 0.307\ \text{pc} = 6.32 \times 10^4\ \text{au} = 9.46 \times 10^{15}\ \text{m}$$

Constants relating to time

The following relationships hold in all time systems:

1 day = 24 hours = 1440 minutes = 86 400 seconds
1 Julian year = 365.25 days = 8766 hours = 525 960 minutes = 31 557 600 seconds
1 century = 100 years

The following values of the lengths of the year, month and day are expressed in units of 1 day of 86 400 SI seconds; T is measured in Julian centuries from 2000.0.

1 tropical year (equinox to equinox)	$= 365\overset{d}{.}242\ 193 - 0\overset{d}{.}000\ 006\ 1\ T$
	$= 365^d\ 05^h\ 48^m\ 45\overset{s}{.}5 - 0\overset{s}{.}53\ T$
1 sidereal year (fixed star to fixed star)	$= 365\overset{d}{.}256\ 360 + 0\overset{d}{.}000\ 000\ 1\ T$
	$= 365^d\ 06^h\ 09^m\ 09\overset{s}{.}5 + 0\overset{s}{.}01\ T$
1 mean synodic month (new moon)	$= 29\overset{d}{.}530\ 859 = 29^d\ 12^h\ 44^m\ 02\overset{s}{.}9$
1 mean tropical month (equinox)	$= 27\overset{d}{.}321\ 582 = 27^d\ 07^h\ 43^m\ 04\overset{s}{.}7$
1 mean sidereal month (fixed star)	$= 27\overset{d}{.}321\ 661 = 27^d\ 07^h\ 43^m\ 11\overset{s}{.}5$
1 mean solar day (~ 1984)	$= 1\overset{d}{.}000\ 000\ 04 = 86\ 400\overset{s}{.}003$
1 mean sidereal day (~ 1984)	$= 0\overset{d}{.}997\ 269\ 60 = 86\ 164\overset{s}{.}094$
1 mean period of rotation of Earth (~ 1984)	$= 0\overset{d}{.}997\ 269\ 70 = 86\ 164\overset{s}{.}102$

The rate of rotation of the Earth (in 1984) is $7.292\ 115\ 15 \times 10^{-5}\ \mathrm{rad\,s^{-1}}$.

Although the Earth's period of rotation is variable when measured in SI units the following relationships are almost independent of the variability.

1 mean solar day
     = interval between successive transits of the fictitious mean sun
     $= 1\overset{d}{.}002\ 737\ 909 = 24^h\ 03^m\ 56\overset{s}{.}555\ 3 = 86\ 636\overset{s}{.}555\ 3$ of mean sidereal time

1 mean sidereal day
     = interval between successive transits of the mean equinox
     $= 0\overset{d}{.}997\ 269\ 566 = 23^h\ 56^m\ 04\overset{s}{.}090\ 5 = 86\ 164\overset{s}{.}090\ 5$ of mean solar time

1 period of rotation of the Earth with respect to an inertial reference frame
     $= 1\overset{d}{.}000\ 000\ 097 = 24^h\ 00^m\ 00\overset{s}{.}008\ 4 = 86\ 400\overset{s}{.}008\ 4$ of mean sidereal time
     $= 0\overset{d}{.}997\ 269\ 663 = 23^h\ 56^m\ 04\overset{s}{.}098\ 9 = 86\ 164\overset{s}{.}098\ 9$ of mean solar time

Universal time is related to Greenwich mean sidereal time through the expression:

$$\text{GMST at } 0^h\ \text{UT} = 6^h 41^m 50\overset{s}{.}5484 + 8640\ 184\overset{s}{.}8129\ T_u + 0\overset{s}{.}0931\ T_u^2$$

where T_u is measured in centuries of 36 525 mean solar days from 2000 January 1 at 12^h UT (JD 245 1545.0 UT).

Constants for precession and nutation

The precessional motion of the celestial equator (plane normal to the axis of rotation of the Earth) around the ecliptic (mean plane of the Earth's orbit around the Sun) has a period of about 25 725 years, and gives rise to the luni-solar precession in celestial longitude of $50\rlap{.}{''}37$ per year. There is also a slow precessional motion of the ecliptic due to planetary perturbations; this gives both a change in the obliquity (inclination of the ecliptic to the equator) and a motion of the equinox (direction of the line of intersection of equator and ecliptic) along the equator of $0\rlap{.}{''}12$ per year. The combined effect is known as the general precession.

     Obliquity of ecliptic $= 23° 26' 21\rlap{.}{''}45 - 46\rlap{.}{''}81\ T = 23\overset{°}{.}439\ 291 - 0\overset{°}{.}013\ 00\ T$

     Annual general precession in longitude $= 50\rlap{.}{''}2910 + 0\rlap{.}{''}0222\ T$

where T is measured in Julian centuries from 2000.0.

For a star with right ascension α and declination δ the annual precessions in right ascension and declination are $m + n \sin \alpha \tan \delta$ and $n \cos \alpha$, respectively, where

$$m = 3\overset{s}{.}074\ 96 + 0\overset{s}{.}001\ 86\ T, \qquad n = 1\overset{s}{.}336\ 21 - 0\overset{s}{.}000\ 57\ T = 20\rlap{.}{''}0431 - 0\rlap{.}{''}0085\ T$$

The nutation of the axis of rotation of the Earth is normally specified by its effects in celestial longitude and obliquity. The principal terms in the trigonometric series for the nutation are:

In longitude	In obliquity	Period (days)
$-17''.200-0''.017\,T$	$+9''.202+0''.001\,T$	6798
$+0.206$	-0.090	3399
-1.319	$+0.574$	183
-0.227	$+0.098$	13.7

The solar system: Dimensions, etc., of Sun, Moon and planets

Body	Equatorial radius	Mass	Surface gravity	Mean density	Flattening	Period of rotation	Inclination of equator to orbit	Number of known satellites
	(on scale: Earth = 1)†			g cm^{-3}				
Sun . . .	109.12	332 946	27.96	1.41	0.0	$25^{\rm d}\,09^{\rm h}$	$7°.2$	—
Mercury . .	0.382	0.0553	0.38	5.43	0.0	$58^{\rm d}\,16^{\rm h}$	$0°.0$	0
Venus. . .	0.949	0.8150	0.90	5.24	0.0	$243^{\rm d}\,00^{\rm h}$	$177°.3$	0
Earth . . .	1.000	1.0000	1.00	5.52	0.0034	$23^{\rm h}\,56^{\rm m}$	$23°.4$	1
Moon . . .	0.272	0.0123	0.17	3.34	See note 4	$27^{\rm d}\,08^{\rm h}$	$1°.5$	—
Mars . . .	0.532	0.1074	0.38	3.94	0.0052	$24^{\rm h}\,37^{\rm m}$	$25°.2$	2
Jupiter . .	11.19	317.89	2.54	1.33	0.0648	$9^{\rm h}\,50^{\rm m}$	$3°.1$	16
Saturn . .	9.41	95.18	1.07	0.70	0.1076	$10^{\rm h}\,14^{\rm m}$	$26°.7$	17
Uranus . .	3.98	14.50	0.9	1.30	0.030	$15^{\rm h}\,34^{\rm m}$	$97°.9$	5
Neptune . .	3.81	17.24	1.2	1.76	0.026	$18^{\rm h}\,26^{\rm m}$	$29°.6$	2
Pluto . . .	0.23	0.0025	0.05	1.1	0.0	$6^{\rm d}\,09^{\rm h}$	$118°?$	1

† For Earth values see pp. 157, 159.

Notes:

(1) The Sun is a star of spectral type G2 V and absolute magnitude +4.79. The effective temperature of its photosphere is about 5800 K. The radiation emitted per unit area is 63.3 MWm^{-2}, giving a total emission rate of 3.85×10^{26} W. The radiation received at the Earth under standard conditions (i.e. the solar constant) is 1.37 kW m^{-2}.

(2) The periods of rotation of the Sun, Jupiter and Saturn refer to their equatorial visual regions; the periods increase with latitude.

(3) The inclinations of the equators of the Sun and Moon are referred to the ecliptic; the inclination of the equator of the Moon to the plane of the Moon's orbit around the Earth is about 6°.7 and the axis of rotation precesses at the same rate as the line of nodes of the orbit (period 18.6 years).

(4) The mean radius of the Moon may be taken as 1738.0 km, but the irregularities in the surface are of the order of 1/500 of the radius; the principal moments of inertia are all different.

Mean elements of the orbits of the planets

Planet	Mean distance from Sun		Sidereal period	Synodic period	Inclination to ecliptic	Eccentricity	Longitude of node	Longitude of peri-helion
	au	10^6 km	year	day	degree		degree	degree
Mercury . .	0.387	57.9	0.241	116	7.005	0.2056	48.2	77.2
Venus. . .	0.723	108.2	0.615	584	3.394	0.0068	76.5	131.5
Earth + Moon	1.000	149.6	1.000	—	0.0	0.0167	—	102.7
Mars . . .	1.524	227.9	1.881	780	1.850	0.0933	49.4	335.7
Jupiter . .	5.202	778.2	11.862	399	1.305	0.0481	100.3	15.5
Saturn . .	9.56	1431	29.57	378	2.49	0.051	113.5	93
Uranus . .	19.29	2886	84.75	370	0.773	0.047	74.0	177
Neptune . .	30.27	4529	167	367	1.770	0.007	131.6	357
Pluto . . .	39.68	5936	250	367	17.13	0.253	110.2	224

Notes:

(1) The planetary orbits are subject to both secular and periodic perturbations, so that the osculating elements at any instant may differ in the end figures from the elements for 1985 that are given above.

(2) The values given for Earth + Moon refer to motion of their centre of mass, which lies about 4700 km from the centre of the Earth. The mean distance between the centres of the Earth and Moon is 384 400 km; the mean eccentricity of the relative orbit is 0.055 and its mean inclination to the ecliptic is 5°2. The orbit is, however, subject to considerable perturbations. The line of nodes moves round westwards once in 18.6 years, and so the inclination of the orbit to the Earth's equator (and hence the extremes of declination in any month) varies between 28°6 and 18°3.

References

Annual tabulations of astronomical data and ephemerides will be found in: *The Astronomical Almanac* (HMSO, London), *The Handbook of the British Astronomical Association, The Observer's Handbook* (Royal Astronomical Society of Canada), and *Whitaker's Almanac*. The booklet *Astronomical Phenomena* contains extracts from *The Astronomical Almanac*, such as the dates and times of planetary and lunar phenomena and other astronomical data of general interest, and is published by HMSO several years in advance. The volume *Planetary and Lunar Coordinates, 1984–2000* (HMSO) contains heliocentric, geocentric, spherical and rectangular coordinates of the Sun, Moon and planets, eclipse data, and auxiliary data, such as orbital elements and precessional constants, for use in advance of the almanacs and for other purposes.

Extended lists of astronomical data and other information are given in: C. W. Allen, *Astrophysical Quantities*, 1977 (The Athlone Press, London); and K. R. Lang, *Astrophysical Formulae*, 1974 (Springer-Verlag, Berlin).

Full details, with references, about the introduction of the new astronomical constants, timescales and reference frame are given in the *Supplement* to the *Astronomical Almanac* for 1984.

G.A.W.

1.9.3 Physical properties of the Earth

A set of astronomical constants, based mainly on data from artificial satellites and radar observations of the Moon and the planets, was adopted by the International Astronomical Union in 1964 (IAU Reports for 1964). While adopted as conventional values, they also represent the best observational data very closely. In the following table, values given under 'Whole Earth' are IAU primary constants (p) or are derived from them; 'Core' values are obtained from Bullen, *Introduction to the Theory of Seismology*, 1963 (CUP) and Jeffreys, *The Earth*, 5th edn, 1970 (CUP).

The Earth: mechanical properties

Property	Whole Earth	Core
p Equatorial radius, a	6378.160 km	3488 km
Polar radius, c	6356.775 km	3479 km
Polar flattening, $f = (a-c)/a$	1/298.25	1/390
Mean radius $(a^2c)^{1/3}$	6371.02 km	3485 km
Mass M	5.976×10^{24} kg	1.88×10^{24} kg
Mean density	5518 kg m^{-3}	10 720 kg m^{-3}
Moments of inertia in terms of Ma^2		
Polar C/Ma^2	0.3306	0.380
Equatorial A/Ma^2	0.3295	
p GM	$398\ 603 \times 10^9$ m^3 s^{-2}	
p $J_2 = (C-A)/Ma^2$	$1.082\ 65 \times 10^{-3}$	
Dynamical ellipticity $(C-A)/C$	3.275×10^{-3}	
Angular velocity	$7.292\ 115\ 2 \times 10^{-5}$ rad/s	
Surface area	5.101×10^{14} m^2	1.52×10^{14} m^2
Volume	1.083×10^{21} m^3	0.176×10^{21} m^3
Quadrant of meridian	10 002.002 km	5640 km

The Earth: variation of properties with depth

It is not possible to construct a unique model of the variation of mechanical properties within the Earth. The following table is based on models constructed by Haddon and Bullen, *Phys. Earth, Planet. Interiors*, 1969, **2**, and Derr, *J. Geophys. Res.*, 1969, **74**.

Zone	Depth km	Density kg m^{-3}	Elastic wave velocity		Bulk modulus 10^{11} N m^{-2}	Rigidity modulus 10^{11} N m^{-2}	Pressure 10^{11} N m^{-2}	g m s^{-2}
			P km s^{-1}	S km s^{-1}				
	0	2 600	6.00	3.50	0.56	0.32	0	9.82
	60	3 390	7.96	4.62	1.19	0.72	0.018	9.85
	150	3 400	8.15	4.43	1.36	0.66	0.048	9.88
Mantle	350	3 580	8.80	4.58	1.76	0.76	0.116	9.95
	500	3 830	9.65	5.24	2.16	1.05	0.174	9.98
	1000	4 560	11.44	6.36	3.50	1.86	0.388	9.96
	1500	4 825	12.22	6.70	4.26	2.16	0.622	9.92
	2000	5 080	12.78	6.89	5.09	2.42	0.868	10.01
	2886 {	5 520	13.66	7.23	6.44	2.88	} 1.353	10.74
	{	9 900	8.11	0	6.36	0		
	3500	10 680	8.88	0	8.30	0	1.936	9.32
Core	4000	11 410	9.50	0	10.31	0	2.467	7.96
	4500	11 850	9.97	0	11.65	0	2.882	6.64
	5000	12 200	10.30	0	12.96	0	3.231	4.80
	5150 {	12 280	9.86	0	13.05	0	} 3.311	4.62
Inner core	{	13 000	11.18	1	17.03	0.68		
	6371 {	13 500	11.80	1	17.81	0.70	3.691	0

P—Compressional wave;　*S*—Shear wave

The Earth: other physical constants

Land:	area	1.49×10^{14} m^2 (29.2 per cent of Earth's surface)
	mean height	840 m
	greatest height	8840 m
Oceans:	area	3.61×10^{14} m^2 (70.8 per cent of Earth's surface)
	volume	1.37×10^{18} m^3
	mass	1.42×10^{21} kg
	mean depth	3800 m
	greatest depth	10 550 m
Atmosphere:	mass	5.27×10^{18} kg (10^{-6} of Earth's mass)

Heat flow from the interior of the Earth

Heat flows out from the interior of the Earth at an average rate of 0.06 W m^{-2}. Most measurements have been made in the deep oceans. There are large variations in heat flow, from 0.01 to 0.2 W m^{-2} but no significant differences between major continents and oceans.

Area	Equal area average heat flow/(W m^{-2})	
	Average	Standard deviation
World	0.061	0.031
Continents	0.061	0.020
Oceans	0.061	0.033
Atlantic Ocean	0.056	0.025
Indian Ocean	0.055	0.022
Pacific Ocean	0.063	0.036

See W. H. K. Lee, *Phys. Earth Planet. Interiors* 1970, **2**.

A.H.C.

1.9.4 Gravity

The gravity field of the Earth

The potential of the external gravity field of the Earth is usually expressed as a series of spherical harmonic terms:

$$V = -\frac{GM}{r}\left[1 - \sum_{n=2}^{\infty} J_n\left(\frac{a}{r}\right)^n P_n(\cos\theta) + \text{terms depending on longitude}\right]$$

where r = distance from the centre of the Earth

      a = Earth's equatorial radius,† θ = geocentric co-latitude,

     M = mass of the Earth† (see below)

$P_n(\cos\theta)$ = Legendre function of degree n

    $10^6 J_2 = 1082.65$†

    $10^6 J_3 = -2.5$

    $10^6 J_4 = 1.6$

The values of J_n are determined from the behaviour of artificial satellites about the Earth (A.H. Cook, *Phil. Trans. Roy. Soc.*, 1967, **262**).

The corresponding expression for the variation of the acceleration due to gravity over the surface of the (spinning) Earth is

$$g = g_e(1 + \beta_1 \sin^2\phi - \beta_2 \sin^2 2\phi) - 3.086 \times 10^{-6} H \text{ m s}^{-2}$$

where ϕ is the geographical latitude, H is the height above sea level (in metres) and g_e is the value of gravity at the equator.

The values recommended by the International Association of Geodesy are

$$g_e = 9.780\ 318\ 4 \text{ m s}^{-2}$$
$$\beta_1 = 0.005\ 302\ 4$$
$$\beta_2 = 0.000\ 005\ 9$$

(*Sp. Pub. Bull. géod.*, 1970).

The above formula gives the best simple method of calculating g at a place where it has not been measured. It will almost always give results within 10^{-3} and usually within 5×10^{-4} m s^{-2}. The agreement with observation is usually made worse by the application of a correction for the attraction of the land above sea level.

For the definition of standard gravity (standard acceleration) see p. 9.

† See section 1.9.3 above.

Absolute value of the acceleration due to gravity, g

The difference between the accelerations due to gravity at two nearby sites may be compared more accurately (by spring-balance gravity meters) than either can be measured separately. The network of world-wide gravity measurements is therefore based on differential measurements which tie sites to a single site at which the absolute value (m s^{-2}) is believed to be well established. The site hitherto adopted is the Potsdam Geodetic Institute where g was measured by Kühnen and Furtwängler (*Veröff. Preuss. Geodät. Inst.*, 1966) using reversible pendulums of Kater's type. Modern determinations use the direct observation of a body freely moving under the action of gravity. The value of Kühnen and Furtwängler (g at Potsdam $= 9.812\ 74$ m s^{-2}) is in error by 14 p.p.m., whereas the following recent values agree to within 0.1 p.p.m.:

Cook:	Teddington, 1965	9.811 818 1 m s^{-2}
Faller:	Teddington, 1969	9.811 818 6
	Washington, 1969	9.801 024 0
	Sèvres, 1969	9.809 259 6
Sakuma:	Sèvres, 1967–70	9.809 259 6

A.H.C.

1.9.5 Geomagnetism

The Earth's magnetic field corresponds approximately to that of a dipole situated at the centre of the Earth with its axis inclined at an angle of about 11° to the axis of rotation. There are, however, appreciable temporal and spatial departures from this simple model. According to presently accepted theories, the smooth geomagnetic or 'normal' field and its slow secular change are ascribed to fluid motions in the Earth's electrically conducting core. The influence of magnetic constituents of crustal rocks superposes on the normal field anomalies whose magnitude can, in extreme cases, be comparable to that of the normal field.

In addition, changes on the Sun and of its position relative to the Earth cause erratic and often rapid fluctuations in the magnetic field (magnetic storms) occasionally exceeding one-tenth of the normal field value, as well as smaller and more regular diurnal and seasonal variations.

The geomagnetic field at any point is usually defined by three of seven elements: five intensity components, H (parallel to the Earth's surface), Z (vertically downwards), F or T (total, scalar), X (geographic north) and Y (geographic east); and two angles, D (declination or variation, $D = \arctan Y/X$) and I (inclination or dip, $I = \arctan Z/H$).

The main geomagnetic field, due to internal sources, at points on or above the Earth's surface may be represented by the following series:

$$X = \sum_{n=1}^{k} \sum_{m=0}^{n} (g_n^m \cos m\lambda + h_n^m \sin m\lambda)\ (a/r)^{n+2} \frac{dP_n^m(\theta)}{d\theta},$$

$$Y = \sum_{n=1}^{k} \sum_{m=1}^{n} (g_n^m \sin m\lambda - h_n^m \cos m\lambda)\ (a/r)^{n+2}\ m \ \text{cosec}\ \theta\ P_n^m(\theta),$$

$$Z = -\sum_{n=1}^{k} \sum_{m=0}^{n} (g_n^m \cos m\lambda + h_n^m \sin m\lambda)\ (a/r)^{n+2}\ (n+1) P_n^m(\theta),$$

where a is the Earth's mean radius, λ is east longitude, r is radial distance from the Earth's centre and $P_n^m(\theta)$ is an associated Legendre function of the colatitude or north polar distance, θ, of degree n and order m. The set of numerical coefficients (g_n^m, h_n^m), usually expressed in units of nT, constitute a spherical harmonic model of the geomagnetic field.

The associated Legendre functions used in geomagnetism are of the Schmidt quasi-normalized form. They are such that the mean square value of $P_n^m \cos m\lambda$ or $P_n^m \sin m\lambda$ taken over a sphere is $(2n+1)^{-1}$ and may be derived from the relation:

$$P_n^m(c) = \left(\frac{\delta_m (n-m)! \, (1-c^2)^m}{(n+m)!} \right)^{1/2} \frac{d^m}{dc^m} P_n(c)$$

where $c = \cos\theta$

         $\delta_m = 1$ for $m = 0$; 2 for $m \geqslant 1$

and $P_n(c)$ is the Legendre polynomial of degree n.

The International Geomagnetic Reference Field (IGRF) is a set of five spherical harmonic models: four describing the main geomagnetic field at epochs 1965, 1970, 1975 and 1980 and one for the (predicted) annual rate of secular variation for the interval 1980 to 1985. The main-field models extend to $m = n = 10$ (120 coefficients each) and the secular variation model is truncated at $m = n = 8$ (80 coefficients). Field component values for dates differing from the epochs of the main-field models are derived, for dates before 1980, by linear interpolation and, for dates after 1980, by using the secular variation model to up-date the 1980 main-field model. (For further details see D. R. Barraclough, *Nature*, 1981, **294**, 14–15.)

Figures 1 and 2 show contours of the declination and of the secular variation of the declination, respectively, at 1980 derived from the IGRF. A Fortran subroutine for synthesizing field component values from a spherical harmonic model is described by S. R. C. Malin and D. R. Barraclough, *Computers and Geosciences*, 1981, **7**, 401–405.

More detailed world magnetic charts are published by the Hydrographic Department of the Ministry of Defence and are obtainable from Admiralty Chart Agents. The latest charts for all elements are for epoch 1985.

Magnetic elements for London at different epochs

Values from 1850 onwards are for Greenwich. For 1580 the D observation was by Borough and the I value is Norman's observation made about 1576.

Epoch	Declination		Inclination		$H/\mu T$	$Z/\mu T$
	deg	min	deg	min		
1580	11	19 E	71	50	—	—
1665	0	0			—	—
1673		—	73	47	—	—
1719	11	30 W			—	—
1720		—	75	14*	—	—
1816	24	28 W*		—	—	—
1818		—	70	35	—	—
1850	22	24 W	68	47	—	—
1875	19	21 W	67	42	17.97	43.83
1907	16	0 W	66	56	18.55*	43.57
1913	15	15 W	66	50†	18.53	43.32
1929	12	23 W	66	54	18.38	43.06†
1937	11	1 W	66	59	18.34†	43.17
1946	9	37 W	67	2*	18.39	43.37
1975	6	39 W	66	27	19.16	43.98

* Maximum.   † Minimum.

Geomagnetic data 1975

The following table contains the mean values of D, H and Z observed during the year 1975 and their annual rates of change for a selection of permanent magnetic observatories.

Observatory	Lat.	Long. east	D (east)	Annual change	H	Annual change	Z	Annual change
	deg	deg	deg	min yr^{-1}	nT	nT yr^{-1}	nT	nT yr^{-1}
Resolute Bay	+74.7	265.1	−65.9	+118	742	+11	58 578	+31
Bjørnøya	+74.5	19.2	+2.5	+4	9 296	+5	52 780	+44
Point Barrow	+71.3	203.3	+26.0	0	9 830	+4	56 410	+10
Tromsø	+69.7	18.9	0.0	+3	11 400	+10	51 450	+30
College	+64.9	212.2	+28.4	−2	13 022	+10	55 346	+8
Lerwick (UK)	+60.1	358.8	−8.6	+8	14 890	+23	47 753	+34
Magadan	+60.1	151.0	−13.0	0	18 103	+4	52 638	−11
Krasnaya Pakhra . . .	+55.5	37.3	+7.6	0	17 298	+12	48 525	+23
Eskdalemuir (UK) . . .	+55.3	356.8	−9.2	+8	17 200	+29	45 719	+23
Patrony	+52.2	104.4	−2.3	−1	19 602	+5	56 937	−17
Hartland (UK)	+51.0	355.5	−8.5	+8	19 212	+32	43 733	+14
Memambetsu	+43.9	144.2	−8.2	0	26 501	+3	41 325	−6
Fredericksburg	+38.2	282.6	−7.8	−6	20 269	+60	52 031	−136
San Miguel	+37.8	334.4	−13.3	+9	25 542	+26	38 028	−52
Kakioka	+36.2	140.2	−6.5	−1	30 142	+11	34 630	−3
Tucson	+32.2	249.2	+12.5	−5	25 758	−20	43 237	−53
Quetta	+30.2	67.0	+1.2	0	32 919	−23	33 468	+5
Honolulu	+21.3	202.0	+11.6	0	27 810	−17	22 334	−6
Alibag	+18.6	72.9	−0.8	0	38 517	−23	17 434	+18
San Juan	+18.1	293.8	−8.8	−8	27 450	−27	31 752	−157
Muntinlupa	+14.4	121.0	−0.4	−3	39 056	+24	9 695	−1
Nairobi	−1.3	36.8	−1.4	+6	30 569	−26	−15 602	+25
Luanda	−8.9	13.2	−10.1	+8	23 475	−65	−22 539	−57
Apia	−13.8	188.2	+12.3	+2	34 453	−7	−20 134	+7
Tsumeb	−19.2	17.6	−14.8	+10	16 318	−94	−27 072	−2
Vassouras	−22.4	316.3	−18.5	−8	21 489	−63	−10 676	−92
Maputo	−25.9	32.6	−16.1	+2	14 189	−48	−27 550	+81
Gnangara	−31.8	116.0	−3.2	−2	23 608	−37	−53 496	−20
Hermanus	−34.4	19.2	−23.9	+4	11 688	−71	−26 210	+92
Toolangi	−37.5	145.5	+10.8	+3	22 209	−34	−56 308	+7
Kerguelen	−49.4	70.2	−50.7	−9	18 523	−12	−43 962	+36
Macquarie Island . . .	−54.5	159.0	+27.7	+8	12 847	−24	−63 926	+30
Mawson	−67.6	62.9	−62.5	−7	18 397	+10	−47 269	+111

	Lat.	Long.
North magnetic dip-pole (1980)	+77°.3	258°.2 E
South magnetic dip-pole (1980)	−65°.6	139°.4 E

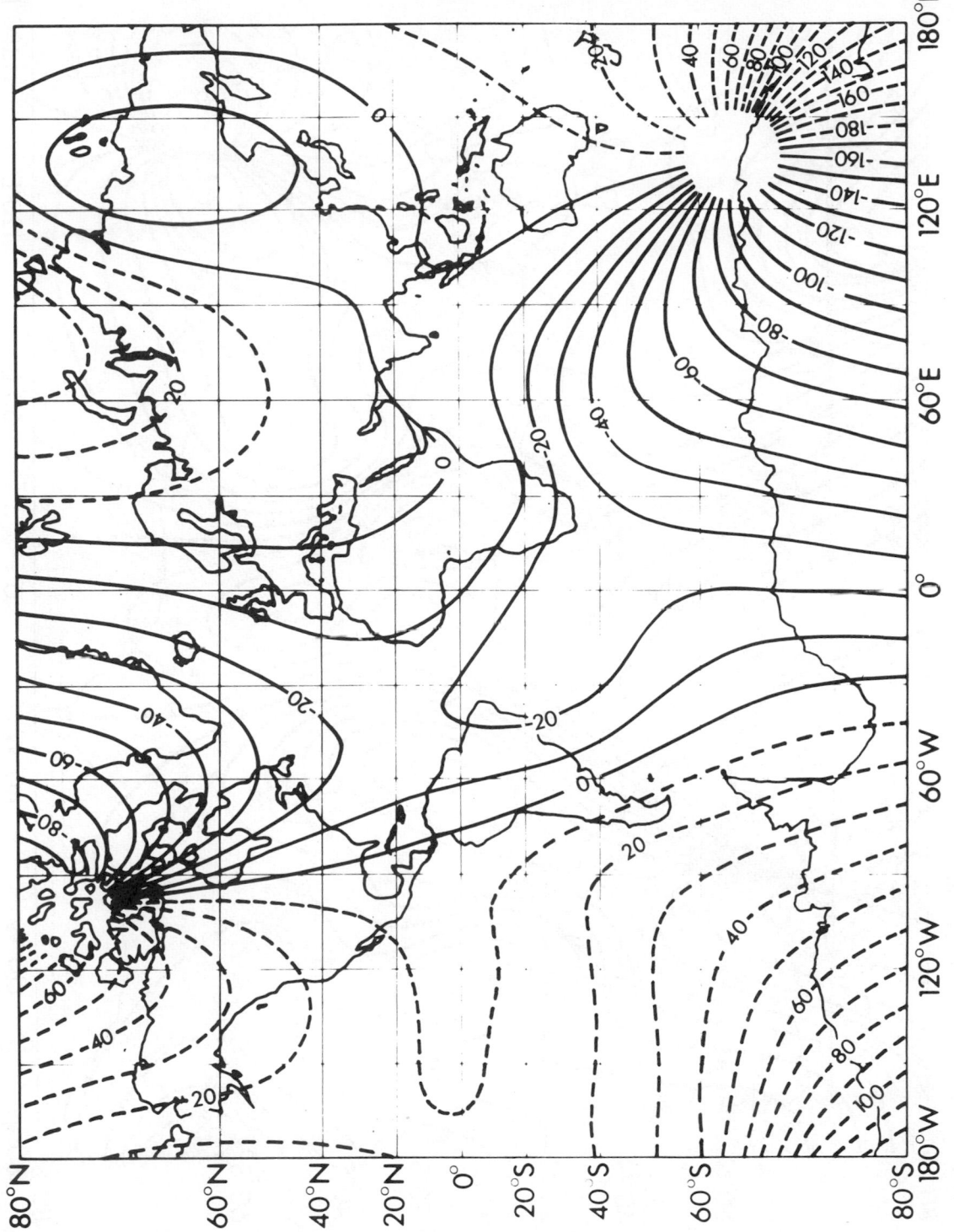

Contours of the magnetic declination at 1980, in degrees
The contour interval is 10°. Positives (or east) declination is indicated by dashed lines; solid lines are used for zero and negative (or west) declination.

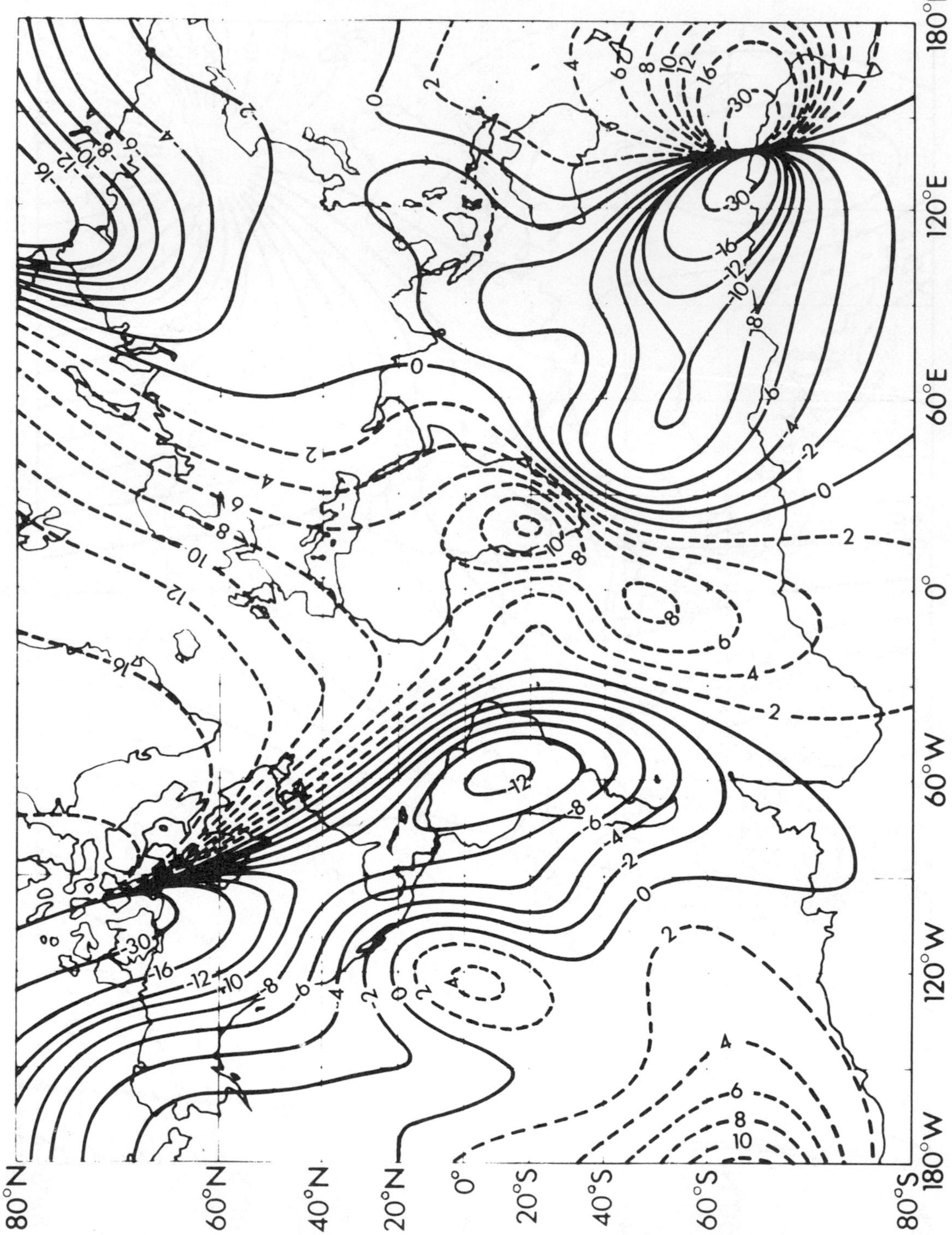

Contours of the secular variation of the magnetic declination for the interval 1980–1985, in arcmin/yr.

The contour interval is 2 arcmin/yr. Positive (or east) values are indicated by dashed lines; solid lines are used for zero and negative (or west) values.

D.R.B.

1.9.6 Cosmic rays

There is in nearby interstellar space a flux of particles—mostly protons and atomic nuclei—travelling at almost the speed of light, having kinetic energies from below 10^8 eV to above 10^{20} eV. The flux reaching the solar system is virtually isotropic (to within 0.1% below 10^{14} eV) and unchanging, but the flux observed at the Earth varies somewhat because: (a) at energies below a few GeV, particles are affected by interplanetary magnetic fields, which cause intensity variations with an irregular 11-year period; and (b) the geomagnetic field deflects low-energy particles away from low-latitude regions. The particles incident at the top of the atmosphere are mainly protons and bare atomic nuclei: encounters with nuclei in the air generate secondary particles, and the less energetic particles are attenuated on traversing the atmosphere, so that near sea level the dominant particles are the penetrating secondary muons and the rapidly generated secondary electrons, positrons and photons.

Occasional bursts of particles originating in solar flares can reach a few GeV; there is also a variable anomalous component of nuclei below 0.1 GeV per nucleon originating in the solar system. These are not tabulated.

Cosmic rays at the top of the atmosphere

The flux of particles per unit time, area and solid angle may be expressed over wide ranges of energy as

$$\mathcal{J}(E)\,dE = A.E^{-\gamma}dE \text{ or, better, } \alpha(E + 2GeV)^{-\gamma}dE$$

where E is the particle's kinetic energy (usually in GeV). There is a wide range of energy in which $\gamma \sim 2.7$, but there are exceptions. The integral flux, $I(E)$, gives the flux of particles of energy $> E$: $\mathcal{J} = -dI/dE$. The table below gives estimates of the particle flux at a time of minimum sunspot activity (when the flux is highest) and also (in *italics*) for a period of near maximum sunspot activity (low flux—though there is no well-defined absolute minimum). $\mathcal{J}$ and I are quoted for protons, and I for all nuclear particles (including protons). As published differential and integral fluxes are somewhat inconsistent, a compromise differential flux has been tabulated: errors of 15–20% are likely to be present. Much larger uncertainties are present in the electron fluxes: some measurements are a factor 2 below those tabulated. I was obtained by integrating the differential flux, except above 1000 GeV. In this latter high-energy range, the flux may be in error by about a factor 2. To interpolate between quoted fluxes, a power law as proposed above is satisfactory.

Fluxes observed where particles are not excluded by geomagnetic field

Differential flux $\mathcal{J}$ in $m^{-2}s^{-1}sr^{-1}GeV^{-1}$, I in $m^{-2}s^{-1}sr^{-1}$.

E(GeV)	$\mathcal{J}_{proton}$		I_{proton}		$I_{all\ nuclei}$		$\mathcal{J}_{elec}$		I_{elec}	
0.1	1100	*92*	2900	*1300*	—	—	—	*~8*	—	—
0.2	1500	*210*	2800	*1300*		—	200	*~5*	90	*26*
0.5	1600	*420*	2300	*1200*	2600	*1400*	70	*10*	60	*24*
1.0	1000	*400*	1700	*1000*	2000	*1100*	30	*9*	38	*20*
2.0	420	*220*	1000	*700*	1200	*830*	11	*6*	20	*12*
5.0	90	*64*	410	*340*	540	*420*	1.8	—	5	*4*
10	24	*20*	180	*160*	240	*210*	0.27	—	1.3	*1.2*
20	5	*4.6*	62	*58*	95	*85*	0.034	—	0.3	*0.3*
100	0.066	*0.065*	3.8	*3.7*	7.6	*7.3*	0.00017	—	—	—
1000	0.00012	—	0.067	—	0.16	—	—	—	—	—
3×10^6	—	—	—	—	3×10^{-7}	—	—	—	—	—
10^{10}	—	—	—	—	2×10^{-14}	—	—	—	—	—
10^{11}	—	—	—	—	4×10^{-16}	~—	—	—	—	—

The 'electron' flux includes positrons—about 30% of the total below 0.3 GeV, but only $\sim 7\%$ above 4 GeV. The photon flux above 0.1 GeV is $0.6\,m^{-2}s^{-1}sr^{-1}$ averaged over the sky, but 50% is from galactic latitudes $< 10°$; about 22% is probably of extragalactic origin.

In the range 0.5–5 GeV per nucleon, comprising most of the particles, nuclei heavier than protons have similar spectra when expressed in terms of magnetic rigidity $R = c.\text{momentum/charge}$. (When $v \sim c$, $R = E/eZ$, Z being the nuclear charge number: thus a 100 GeV He nucleus has a rigidity of 50 GV.) The relative number of nuclei above a rigidity of ~ 4 GV is given below:

H	He	Li	Be	B	C	N	O	F	Ne	Na–Si	P–Cr	Fe	Ni	Cu–Ru
$\sim 24\,000$	3750	16	9	27	100	26	96	2	16	40	14	11	0.5	0.012

The geomagnetic field prevents the arrival in a vertical direction of particles of rigidity less than $15.\cos^4\lambda$ GV, where λ is the geomagnetic latitude: particles with rigidity slightly higher than this are admitted at full intensity.

Cosmic rays near sea level

The flux of particles (in $\text{m}^{-2}\,\text{s}^{-1}\,\text{sr}^{-1}$) arriving nearly vertically above various threshold values of kinetic energy is tabulated below for the more common types of particle (at geomagnetic latitudes $> \sim 40°$). R gives also the median range in lead of muons of the specified energy, allowing for track scattering. Other charged particles are less penetrating.

E(GeV)	I_{muons}	R_{muon}	$I_{\text{electrons}}$*	I_{photons}†	I_{protons}	I_{neutrons}
0.001	100	—	60	130	2.1	—
0.01	100	—	28	60	2.1	~ 30‡
0.02	100	—	20	40	2.1	—
0.1	99	4.8 cm	6.0	8	1.9	~ 10‡
0.2	97	12 cm	3.0	3.5	1.5	—
0.5	86	34 cm	1.0	1.1	0.9	1.5
1	69	69 cm	0.38	0.37	0.51	0.7
2	46	134 cm	0.12	0.11	0.25	—
5	20	3.1 m	0.02	0.02	0.077	—
10	8.6	5.8 m	—	—	0.025	$\sim I_p$
20	3.0	—	—	—	0.008	—
50	0.58	—	—	—	0.0016	—
100	0.14	—	—	—	4.3×10^{-4}	—
200	0.030	—	—	—	1.1×10^{-4}	—
500	3.2×10^{-3}	—	—	—	2×10^{-5}	—
1000	5×10^{-4}	—	—	—	—	—

* 40–50% are positrons, above 0.1 GeV (but only $\sim 5\%$ at 1 MeV).
† Theoretical values, as measurements are inadequate.
‡ Uncertain angular distribution makes vertical flux uncertain.

Above 100 GeV, pions will have fluxes comparable to nucleons: they are less important at lower energies.

Fluxes are averaged over the 11-year cycle. The total muon flux will vary about 3% either way: above 1 GeV the variation is slight. The muon fluxes are the best determined—a few percent at low E (20% say at 100 GeV); the proton flux is uncertain to tens of percent above a few GeV.

Away from the vertical direction, the muon flux per unit solid angle varies with zenith angle θ, approximately as $I \propto \cos^n \theta$, with $n = 2.15$, out to 80° (these refer to the total flux including all energies: muons of tens of GeV vary little with zenith angle). For the protons and neutrons, however, at least above tens of MeV, $n \sim 8$. About 20% of the electrons striking the ground will be distributed like nucleons, the rest like muons. With a zenith-angle variation of this form, the flux passing through unit horizontal surface, integrated over a hemisphere of angles, is $2\pi/(n + 2)$ times the vertical flux per unit solid angle, as tabulated above. Thus a thin, horizontal plane detector will record a flux per m^2 per second of about 150 muons and, if its wall has 1 MeV electron stopping power, 70 electrons, and 1 proton.

Nuclear interactions of cosmic rays in the atmosphere generate about 6×10^4 neutrons s^{-1} per m^2 of the Earth at $45°$ latitude, of which $4 \times 10^4 \, m^{-2} \, s^{-1}$ are absorbed by nitrogen to generate ^{14}C when solar activity is low: the long-term all-Earth average would be about 60% of this. 0.05–0.13 (or 0.2) neutrons s^{-1} are generated per kg of air (or Pb) near sea level, at latitudes $> 40°$ at times of low solar activity.

Primary cosmic rays above $\sim 10^{14}$ eV generate extensive showers of secondary particles in the atmosphere. At sea level, such a shower contains about 1 charged particle per 10 GeV of primary energy (at 10^{14} eV, or 1 per 3 GeV at 10^{17} eV, 1 per 1.6 GeV at 10^{20} eV): 5% of the particles are within 3 m of the centre, 50% are within 40 m, and 90% within ~ 250 m.

Variation with latitude. The number of vertical muons above 0.2 GeV is typically about 13% lower at the equator than at high geomagnetic latitudes; above $50°$ the flux does not change much. The component generating neutrons by nuclear interactions (largely the neutrons below 1 GeV)—long used to monitor cosmic ray variations—falls by about 24% in going from latitude $55°$ to the equator.

Variations with time. During the course of the 11-year sunspot cycle, the flux of neutron-generating particles near sea level varies by about 20%, being highest in years of low solar activity (e.g. 1954, 1965, and a less well-defined minimum in 1976), though the variation is far from smooth. Flux minima occurred in 1947, 1958, 1969.

Cosmic rays as background radiation. Near sea level the dose equivalent rate of cosmic rays, describing their medical (whole body) effect, is about 31 mrem per year (somewhat less at latitudes $< 30°$), but this might be only a third or less of the total natural dose. The dose increases by about 3% per 100 m up in the lower atmosphere.

Cosmic radiation is more penetrating than radioactive emissions. The particles penetrating more than 5 cm of lead are mostly muons (only above 0.7 GeV does an electron produce more than 0.5 particle under such a shield), so the particle flux penetrating a thickness x of absorber normally may be judged from the table above—shielding materials of lighter elements are more effective than lead (for a given mass per unit area) by about a factor 1.4, but lead is much more effective in cutting out electrons and photons.

Cosmic rays at other levels

Some muons have extraordinary penetrating power. The following table gives their flux in the vertical direction (in $m^{-2} \, s^{-1} \, sr^{-1}$) under specified masses h of 'standard rock' (having effective $Z^2/A = 5.5$). The mass is given in tonnes m^{-2}, equivalent in mass to metres of water, and, roughly, to feet of rock. The actual depth of water which is estimated to have the same stopping power is also given, though measurements at great depths of water are not yet available.

$h_{Stand-rock}$	50	100	200	500	1000	2000	4000	7000	10000
Water depth	43 m	87 m	175 m	450 m	0.92 km	2.0 km	4.3 km	8.2 km	12.5 km
Flux ($m^{-2} \, s^{-1} \, sr^{-1}$)	7.7	2.5	0.65	0.081	0.013	1.4×10^{-3}	7.0×10^{-5}	1.9×10^{-6}	1×10^{-7}

At great depths, the chemical composition of the rock (Z^2/A) can have a large effect. The flux at angle θ to the vertical, under rock, may be estimated quite well from the formula

$$I(h,\theta) = \frac{1.7 \times 10^6 \sec\theta}{H + 390 \sec\theta} H^{-1.53} e^{-kH} \, m^{-2} \, s^{-1} \, sr^{-1}$$

H being the mass overlay measured along the slant direction from the top of the atmosphere: $H = h_{rock, \, slant} + (10 \sec\theta)$. For standard rock $k = 7.1 \times 10^{-4}$, but if the rock has $Z^2/A = 6.37$, as at the Kolar Gold Fields, where many of the observations were made, $k = 8 \times 10^{-4}$. The formula fits the observations to about 15% for small angles and for depths of more than a few metres, and probably for θ up to $45°$.

Above ground level, the absorption length, in which the intensity increases by a factor e, is roughly as follows, for the lower half of the atmosphere:

Nucleons above a few GeV, 110 g cm^{-2}; total muon flux, 550 g cm^{-2};
total electronic flux, 180 g cm^{-2}; nuclear disintegrations 165 g cm^{-2}.

The rate of nuclear disintegration reaches a peak near the 100 millibar level, several hundred times the sea-level rate (depending on latitude). At this level, though, large increases due to solar flares occasionally occur, lasting for a few hours.

A.M.H.

1.9.7 The atmosphere

Variation of pressure, density and temperature with altitude

Values given in the table are based on the standard atmosphere of the International Civil Aviation Organization (ICAO). They are representative of average atmospheric conditions in temperate latitudes.

Altitude	Pressure	Density	Temperature	Altitude	Pressure	Density	Temperature
m	kPa	kg m^{-3}	°C	m	kPa	kg m^{-3}	°C
−250	104.4	1.25	17	6 000	47.2	0.66	−24
0	101.3	1.22	15	7 000	41.1	0.59	−30
250	98.4	1.20	13	8 000	35.6	0.53	−37
500	95.5	1.17	12	9 000	30.7	0.47	−44
750	92.6	1.14	10	10 000	26.4	0.41	−50
1000	89.9	1.11	8	12 000	19.3	0.31	−56
1500	84.6	1.06	5	14 000	14.1	0.23	−56
2000	79.5	1.00	2	16 000	10.3	0.17	−56
2500	74.7	0.96	−1	18 000	7.5	0.12	−56
3000	70.1	0.91	−4	20 000	5.5	0.088	−56
3500	65.8	0.86	−8	22 000	4.0	0.064	−54
4000	61.6	0.82	−11	24 000	2.9	0.046	−52
4500	57.7	0.78	−14	26 000	2.2	0.034	−50
5000	54.0	0.74	−18	28 000	1.6	0.025	−48
				30 000	1.2	0.018	−46

Reference

Manual of ICAO Standard Atmosphere, International Civil Aviation Organization Doc. 7488/2, 1964 (ICAO, Montreal, Canada).

K.W.T.E.

Pressure, mean molecular weight and temperature in the upper atmosphere

Altitude/km		Mean molecular weight	Pressure/ Pa	Temp./K
30	. . . . .	28.97	1.20×10^3	230.1
40	. . . . .	28.97	2.94×10^2	250.5
50	. . . . .	28.97	8.10×10^1	271.0
60	. . . . .	28.97	2.19×10^1	243.3
70	. . . . .	28.97	5.12×10^0	216.6
80	. . . . .	28.96	9.75×10^{-1}	186.0
90	. . . . .	28.94	1.63×10^{-1}	186.0
100	. . . . .	28.30	3.10×10^{-2}	208.1
120	. . . . .	27.01	2.73×10^{-3}	355.7
150	. . . . .	25.17	4.70×10^{-4}	652
200	Night . . .	22.7	9.5×10^{-5}	890
	Day	23.0	1.2×10^{-4}	1100
300	Night . . .	18.6	8.9×10^{-6}	980
	Day	19.7	1.8×10^{-5}	1360
500	Night . . .	15.0	2.8×10^{-7}	1000
	Day	16.3	1.3×10^{-6}	1420
800	Night . . .	6.5	1.1×10^{-8}	1000
	Day	12.1	6.6×10^{-8}	1430

The above values are a mean over all conditions. Considerable variation occurs with latitude, season and (above 100 km) with solar activity. Above 100 km the composition of the atmosphere changes due to dissociation and diffusive separation, the main constituents in addition to N_2 and O_2 being atomic oxygen and helium.

Reference

COSPAR International Reference Atmosphere, 1965 (North Holland Publishing Co.).

J.T.H.

1.9.8 Physical properties of sea water

The properties of sea water are a function of temperature, salinity (i.e. total dissolved solids in $g\ kg^{-1}$) and pressure. For the sources of the data see M. Hill (ed.), *The Sea*, vol. 1, chap. 1 and vol. 4, pt. 1, chap, 18 (Wiley) and Cox *et al.*, *Deep Sea Res.*, 1970, **17**, 679. For chemical composition see p. 181.

Density ρ of sea water at atmospheric pressure

Temperature °C	Salinity/(g kg^{-1})				
	20	25	30	35	40
	$(\rho/\text{kg m}^{-3} - 1000)$*				
0	16.04	20.06	24.08	28.10	32.14
5	15.84	19.78	23.73	27.68	31.64
10	15.31	19.18	23.07	26.96	30.86
15	14.48	18.30	22.13	25.97	29.82
20	13.39	17.17	20.96	24.75	28.56
25	12.07	15.82	19.57	23.34	27.12

* In oceanographical work these data are always expressed in terms of $1000(S-1)$ where S is the density relative to water at 4 °C. Values for this can be obtained by adding 0.03 to the values in the table.

Density ρ of sea water at 0 °C and salinity 35 g kg^{-1}

Pressure/MPa	0	20	40	60	80	100
$(\rho/\text{kg m}^{-3} - 1000)$	28.10	37.44	46.37	54.92	63.12	71.02

Mechanical and thermal properties of sea water at salinity 35 g kg^{-1} and atmospheric pressure (unless otherwise stated)

Property	0 °C	20 °C
Dynamic viscosity	1.88×10^{-3} Pa s	1.08×10^{-3} Pa s
Kinematic viscosity, ν	1.83×10^{-6} m^2 s^{-1}	1.05×10^{-6} m^2 s^{-1}
Thermal conductivity	0.563 W m^{-1} K^{-1}	0.596 W m^{-1} K^{-1}
Thermal diffusivity, κ	1.37×10^{-7} m^2 s^{-1}	1.46×10^{-7} m^2 s^{-1}
Prandtl number, ν/κ	13.4	7.2
Specific heat capacity, C_p	3985 J kg^{-1} K^{-1}	3993 J kg^{-1} K^{-1}
Thermal expansion coefficient		
Pressure = 0.1 MN m^{-2}	52×10^{-6} K^{-1}	250×10^{-6} K^{-1}
Pressure = 100 MN m^{-2}	244×10^{-6} K^{-1}	325×10^{-6} K^{-1}
Ratio of specific heat capacities, C_p/C_v	1.0004	1.0106
Velocity of sound*	1449 m s^{-1}	1522 m s^{-1}
Compressibility	4.65×10^{-10} Pa^{-1}	4.28×10^{-10} Pa^{-1}
Freezing point		-1.910 °C
Boiling point		100.56 °C

* See also p. 74.

Electrical conductivity γ of sea water at atmospheric pressure

Temperature °C	Salinity/(g kg^{-1})				
	20	25	30	35	40
	γ/(S m^{-1})				
0	1.745	2.137	2.523	2.906	3.285
5	2.015	2.466	2.909	3.346	3.778
10	2.300	2.811	3.313	3.808	4.297
15	2.595	3.170	3.735	4.290	4.837
20	2.901	3.542	4.171	4.788	5.397
25	3.217	3.926	4.621	5.302	5.974

At a depth of 4000 m the bottom water (about 0 °C, salinity 35 g kg^{-1}) has a conductivity 6% greater than the surface water. The mean conductivity of the oceans (excluding the shallow seas) is 3.27 S m^{-1}.

E.C.B.

1.9.9 The geological timescale

Times of start (before present) and durations of geological periods are given in millions of years (*Quart. J. Geol. Soc. Lond.*, vol. 120S); uncertainties are about 2 Myr for the Tertiary rising to 10 Myr for the Palaeozoic.

Period	Start	Duration
Pleistocene	2	2
Pliocene	7	5
Miocene	26	19
Oligocene	38	12
Eocene	54	16
Palaeocene	65	11
Cretaceous	136	71
Jurassic	192	56
Triassic	225	33
Permian	280	55
Carboniferous	345	65
Devonian	395	50
Silurian	435	40
Ordovician	500	65
Cambrian	570	70
Precambrian	4600	4000

E.C.B.

1.10 Miscellaneous engineering data

1.10.1 Screw threads

The International Organization for Standardization (ISO) has published recommendations for parallel screw threads which recognize the ISO metric thread and the ISO inch thread. The ISO inch thread is the same as the Unified screw thread previously standardized by the USA, Britain and Canada. The ISO metric thread has the same form of thread as that of the Unified thread, namely, a symmetrical vee-form having a 60° included angle between the flanks.

Prior to the issue of the ISO Recommendations it was customary for bolts and machine screws made in Britain to have a thread of Whitworth form (55° included angle) for diameters $\frac{1}{4}$ inch and above, and for sizes less than $\frac{1}{4}$ inch diameter the British Association (BA) thread ($47\frac{1}{2}$° included angle) was most commonly used. The Unified thread was also extensively used in the motor vehicle and aeronautical industries.

In 1965, following the Government announcement of the decision to adopt the metric system of measurement, the British Standards Institution decided to recommend British industry to adopt ISO metric threads for new designs in engineering; this means that BA threads and threads of Whitworth form will eventually become obsolete, but it may be years before this occurs. It is likely also that Unified screw threads will continue to be used for many years in certain industries. Meanwhile ISO metric threads will come into increasing use in Britain.

The major diameter of a screw thread is the diameter over the crests of an external thread or to the roots of an internal thread.

The pitch is the distance between adjoining crests (say) of the same thread, measured parallel to the axis of the screw. It is specified by the reciprocal of the number of turns per inch (t.p.i.) for Whitworth and Unified screws.

For any one thread form various combinations of major diameter and pitch have been standardized to form screw-thread series; in the BA system major diameters and associated pitches are designated by means of numbers. The more common series are given in the table below for sizes up to about 25 mm (1 in).

Micrometer screws are generally made with 40 turns per inch or, in the case of Metric screws, with a pitch of 0.5 mm.

The number designations and nominal sizes of 'woodscrews' are as follows: No. 0 (0.060 inch diameter), No. 1 (0.070), No. 2 (0.082), No. 3 (0.094), each succeeding number adding 0.014 inch to the diameter of the screw for sizes up to No. 10 (0.192): this applies to all lengths. (See BS 1210: 1963.)

ISO inch (Unified) threads				ISO metric threads			
Standard series				Standard series			
Unified Coarse (UNC)		Unified Fine (UNF)		Coarse		Fine	
Major diameter/in	t.p.i.	Major diameter/in	t.p.i.	Major diameter/mm	Pitch/mm	Major diameter/mm	Pitch/mm
—	—	(No. 0) 0.0600	80	1.6	0.35	1.6	0.20
(No. 1) 0.0730	64	(No. 1) 0.0730	72	1.8	0.35	1.8	0.20
(No. 2) 0.0860	56	(No. 2) 0.0860	64	2	0.40	2	0.25
(No. 3) 0.0990	48	(No. 3) 0.0990	56	2.2	0.45	2.2	0.25
(No. 4) 0.1120	40	(No. 4) 0.1120	48	2.5	0.45	2.5	0.35
(No. 5) 0.1250	40	(No. 5) 0.1250	44	3	0.50	3	0.35
(No. 6) 0.1380	32	(No. 6) 0.1380	40	3.5	0.60	3.5	0.35
(No. 8) 0.1640	32	(No. 8) 0.1640	36	4	0.70	4	0.50
(No. 10) 0.1900	24	(No. 10) 0.1900	32	4.5	0.75	4.5	0.50
(No. 12) 0.2160	24	(No. 12) 0.2160	28	5	0.80	5	0.50
$\frac{1}{4}$	20	$\frac{1}{4}$	28	6	1.00	6	0.75
$\frac{5}{16}$	18	$\frac{5}{16}$	24	7	1.00	7	0.75
$\frac{3}{8}$	16	$\frac{3}{8}$	24	8	1.25	8	1.00
$\frac{7}{16}$	14	$\frac{7}{16}$	20	9	1.25	10	1.25
$\frac{1}{2}$	13	$\frac{1}{2}$	20	10	1.50	12	1.25
$\frac{9}{16}$	12	$\frac{9}{16}$	18	11	1.50	14	1.50
$\frac{5}{8}$	11	$\frac{5}{8}$	18	12	1.75	16	1.50
$\frac{3}{4}$	10	$\frac{3}{4}$	16	14	2.00	18	1.50
$\frac{7}{8}$	9	$\frac{7}{8}$	14	16	2.00	20	1.50
1	8	1	12	18	2.50	22	1.50
				20	2.50	24	2.00
				22	2.50		
				24	3.00		

Whitworth form threads				BA threads		
Standard series				Standard series		
British Standard Whitworth (BSW)		British Standard Fine (BSF)		BA		
Major diameter/in	t.p.i.	Major diameter/in	t.p.i.	No.	Major diameter/mm	Pitch/mm
$\frac{1}{4}$	20	$\frac{1}{4}$	26	0	6.0	1.00
—	—	$\frac{9}{32}$	26	1	5.3	0.90
$\frac{5}{16}$	18	$\frac{5}{16}$	22	2	4.7	0.81
$\frac{3}{8}$	16	$\frac{3}{8}$	20	3	4.1	0.73
$\frac{7}{16}$	14	$\frac{7}{16}$	18	4	3.6	0.66
$\frac{1}{2}$	12	$\frac{1}{2}$	16	5	3.2	0.59
$\frac{9}{16}$	12	$\frac{9}{16}$	16	6	2.8	0.53
$\frac{5}{8}$	11	$\frac{5}{8}$	14	7	2.5	0.48
$\frac{11}{16}$	11	$\frac{11}{16}$	14	8	2.2	0.43
$\frac{3}{4}$	10	$\frac{3}{4}$	12	9	1.9	0.39
$\frac{7}{8}$	9	$\frac{7}{8}$	11	10	1.7	0.35
1	8	1	10			

Screw thread standards

Standards are published in Great Britain, for the more common screw threads, by the British Standards Institution, 2 Park Street, London, W1A 2BS. These standards include details of tolerances for different grades of workmanship as well as recommended combinations of major diameter and pitch.

Unified (ISO inch) screw threads   *BS 1580*
Parts 1 and 2, 1962: Diameters $\frac{1}{4}$ inch and larger
Part 3, 1965: Diameters below $\frac{1}{4}$ inch
BS 1580 gives information relating to the following standard series:
    Unified Coarse threads (UNC)
    Unified Fine threads (UNF)

ISO metric threads   *BS 3643*
Part 1, 1981: *Principles and basic data.*
Part 2, 1981: *Specification for selected limits of size.*

Whitworth form screw threads
BS 84: 1956, gives information relating to the following standard series:
    British Standard Whitworth threads (BSW)
    British Standard Fine threads (BSF)
BS 84 is now obsolescent
BS 2779: 1973: Fastening threads of BSP sizes (parallel threads)

British Association (BA) threads   *BS 93: 1951* (now obsolescent)

British Standard Pipe (BSP) threads (taper)   *BS 21: 1973*

L.W.N.

1.10.2 Standard wire sizes

Details of the Standard Wire Gauge (SWG) series once commonly used in Great Britain were given in BS 3737: 1964. This British Standard has now been withdrawn. A more recent British Standard, BS 4391: 1972 *Recommendations for Metric Basic Sizes for Metal Wire*, gives metric diameters from 0.020 mm upwards. The following table gives these recommended diameters in the range of sizes from 0.020 mm to 4.000 mm.

Basic diameters of metal wire (mm)			
0.020	0.080	0.315	1.250
0.025	0.100	0.400	1.600
0.032	0.125	0.500	2.000
0.040	0.160	0.630	2.500
0.050	0.200	0.800	3.150
0.063	0.250	1.000	4.000

L.W.N.

Chemistry

2.1.1 The periodic table of the elements with atomic numbers

Group

Period	IA	IIA	IIIA	IVA	VA	VIA	VIIA	VIII			IB	IIB	IIIB	IVB	VB	VIB	VIIB	
1	1 H																	2 He
2	3 Li	4 Be											5 B	6 C	7 N	8 O	9 F	10 Ne
3	11 Na	12 Mg											13 Al	14 Si	15 P	16 S	17 Cl	18 Ar
4	19 K	20 Ca	21 Sc	22 Ti	23 V	24 Cr	25 Mn	26 Fe	27 Co	28 Ni	29 Cu	30 Zn	31 Ga	32 Ge	33 As	34 Se	35 Br	36 Kr
5	37 Rb	38 Sr	39 Y	40 Zr	41 Nb	42 Mo	43 Tc	44 Ru	45 Rh	46 Pd	47 Ag	48 Cd	49 In	50 Sn	51 Sb	52 Te	53 I	54 Xe
6	55 Cs	56 Ba	57 La	72 Hf	73 Ta	74 W	75 Re	76 Os	77 Ir	78 Pt	79 Au	80 Hg	81 Tl	82 Pb	83 Bi	84 Po	85 At	86 Rn
7	87 Fr	88 Ra	89 Ac	104 Unq	105 Unp	106 Unh	107 Uns											

Lanthanides: 57 La, 58 Ce, 59 Pr, 60 Nd, 61 Pm, 62 Sm, 63 Eu, 64 Gd, 65 Tb, 66 Dy, 67 Ho, 68 Er, 69 Tm, 70 Yb, 71 Lu

Actinides: 89 Ac, 90 Th, 91 Pa, 92 U, 93 Np, 94 Pu, 95 Am, 96 Cm, 97 Bk, 98 Cf, 99 Es, 100 Fm, 101 Md, 102 No, 103 Lr

The Periodic Table listing Groups IA to VIIB has been in use for many years, but the American Chemical Society has now adopted the 18-column format given below (see K. L. Loening, *J. Chem. Ed.*, 1984, **61**, 136).

1	2	3d	4d	5d	6d	7d	8d	9d	10d	11d	12d	13	14	15	16	17	18
H																	He
Li	Be											B	C	N	O	F	Ne
Na	Mg											Al	Si	P	S	Cl	Ar
K	Ca	Sc	Ti	V	Cr	Mn	Fe	Co	Ni	Cu	Zn	Ga	Ge	As	Se	Br	Kr
Rb	Sr	Y	Zr	Nb	Mo	Tc	Ru	Rh	Pd	Ag	Cd	In	Sn	Sb	Te	I	Xe
Cs	Ba	La*	Hf	Ta	W	Re	Os	Ir	Pt	Au	Hg	Tl	Pb	Bi	Po	At	Rn
Fr	Ra	Ac**															

3f {
*Ce Pr, Nd, Pm, Sm, Eu, Gd, Tb, Dy, Ho, Er, Tm, Yb, Lu
**Th Pa, U, Np, Pu, Am, Cm, Bk, Cf, Es, Fm, Md, No, Lr

E.F.G.H.

2.1.2 Properties of the elements

Symbols, atomic weights, atomic numbers, densities, melting points and boiling points (pressure 101.325 kN m^{-2}) of the elements

The values of the relative atomic masses ($A_r(E)$), scaled to the relative mass $A_r(^{12}C) = 12$, are those recommended by the International Union of Pure and Applied Chemistry Commission on Atomic Weights (1983). The footnotes to this table elaborate the types of variations to be expected for individual elements. The values of $A_r(E)$ given here apply to elements as they exist naturally on earth. When used with due regard to the footnotes they are considered reliable to $\pm$ 1 in the last digit unless otherwise noted. An element with an asterisk has no stable isotope and lacks a characteristic terrestrial isotopic composition. A value in parenthesis is the atomic weight of the isotope of the radionuclide of that element which is believed to be the most frequently found and which has the longest half-life.

Element	Symbol	Atomic No.	Atomic weight	Density $\rho/(\text{kg m}^{-3})$	Melting point $\theta/^{\circ}C$	Boiling point $\theta/^{\circ}C$
Actinium* .	Ac	89	(227.027 8)	10 060	1230	3200
Aluminium .	Al	13	26.981 54	2 698	660.46†	2520
Americium*.	Am	95	(243.061 4)	13 670	990	2600
Antimony .	Sb	51	121.75±3	6 692	630.755†	1600
Argon . .	Ar	18	39.948^{g,r}	1656/−233 °C	−189.4	−185.856†
Arsenic . .	As	33	74.921 6	5 776	610 solid sublimes	
Astatine* .	At	85	(209.987 1)		300	350
Barium . .	Ba	56	137.33^g	3 594	730	1640
Berkelium* .	Bk	97	(247.070 3)	14 790	986	
Beryllium .	Be	4	9.012 18	1 846	1280	2480
Bismuth . .	Bi	83	208.980 4	9 803	271.442†	1650
Boron . .	B	5	10.811±5^{m,r}	2 466	2130	3700
Bromine. .	Br	35	79.904	3 120	−7.3	58.9
Cadmium .	Cd	48	112.41^g	8 647	321.108†	770
Caesium. .	Cs	55	132.905 4	1 900	28.4	686
Calcium. .	Ca	20	40.078±4^g	1 530	840	1490
Californium*	Cf	98	(251.079 6)			
Carbon . .	C	6	12.011^r	2 266		
Cerium . .	Ce	58	140.12^g	6 711	800	3000
Chlorine .	Cl	17	35.453	2030/−160 °C	−101	−34.0
Chromium .	Cr	24	51.996 1±6	7 194	1860	2600
Cobalt . .	Co	27	58.933 2	8 800	1495†	2900
Copper . .	Cu	29	63.546±3^r	8 933	1084.88†	2590
Curium* .	Cm	96	(247.070 3)	13 300	1340	
Dysprosium .	Dy	66	162.50±3	8 531	1410	2600
Einsteinium*	Es	99	(252.082 8)			
Erbium . .	Er	68	167.26±3^g	9 044	1520	2600
Europium .	Eu	63	151.96^g	5 248	830	1430
Fermium* .	Fm	100	(257.095 1)			
Fluorine. .	F	9	18.998 403	1140/−200 °C	−219.6	−188.1
Francium* .	Fr	87	(223.019 7)		30	650
Gadolinium.	Gd	64	157.25±3^g	7 870	1330	2900
Gallium . .	Ga	31	69.723±4	5 905	29.8	2200
Germanium.	Ge	32	72.59±3	5 323	959	2830
Gold . . .	Au	79	196.966 5	19 281	1064.43†	2850
Hafnium .	Hf	72	178.49±3	13 276	2230	5300
Helium . .	He	2	4.002 602±2^{g,r}	120/4.22 K	3.5 K‡	4.22 K
Holmium .	Ho	67	164.930 4	8 797	1470	2300
Hydrogen .	H	1	1.007 94±7^{g,m,r}	89/−266.8 °C	−259.2	−252.87†

Properties (*contd*)

Element	Symbol	Atomic No.	Atomic weight	Density $\rho/(\text{kg m}^{-3})$	Melting point $\theta/°C$	Boiling point $\theta/°C$
Indium . .	In	49	114.82^g	7 290	156.634†	2050
Iodine . .	I	53	126.904 5	4 953	113.6	184
Iridium . .	Ir	77	192.22 ± 3	22 550	2447†	4530
Iron . . .	Fe	26	55.847 ± 3	7 873	1540	2760
Krypton . .	Kr	36	83.80^{g,m}	3000/−188 °C	−157.3	−153.5
Lanthanum .	La	57	138.905 5 ± 3^g	6 174	920	3430
Lawrencium*	Lr	103	(260.105 4)			
Lead . . .	Pb	82	207.2^{g,r}	11 343	327.502†	1760
Lithium . .	Li	3	6.941 ± 2^{g,m,r}	533	180.6	1360
Lutetium .	Lu	71	174.967	9 842	1700	3400
Magnesium .	Mg	12	24.305	1 738	650	1100
Manganese .	Mn	25	54.938 0	7 473	1250	2120
Mendelevium*	Md	101	(258.098 6)			
Mercury .	Hg	80	200.59 ± 3	13 546	−38.836†	356.66†
Molybdenum	Mo	42	95.94	10 222	2623†	4630
Neodymium	Nd	60	144.24 ± 3^g	7 000	1024	3100
Neon . .	Ne	10	20.179^{g,m}	1442/−268 °C	−248.6	−246.048†
Neptunium*	Np	93	(237.048 2)	20 450	640	3900
Nickel . .	Ni	28	58.69	8 907	1455†	2900
Niobium .	Nb	41	92.906 4	8 578	2477†	4900
Nitrogen .	N	7	14.006 7^g	1035/−268.8 °C	−210	−195.806†
Nobelium* .	No	102	(259.100 9)			
Osmium . .	Os	76	190.2^g	22 580	3030	5000
Oxygen . .	O	8	15.999 4 ± 3^{g,r}	1460/−252.7 °C	−218.8	−182.962†
Palladium .	Pd	46	106.42^g	11 995	1554†	3000
Phosphorus .	P	15	30.973 76	1820 (yellow)	44.2 (yellow)	280.4
Platinum .	Pt	78	195.08 ± 3	21 450	1769†	3820
Plutonium* .	Pu	94	(244.064 2)	19 814	640	3200
Polonium* .	Po	84	(208.982 4)	9 400	254	960
Potassium .	K	19	39.098 3	862	63.2	770
Praseodymium	Pr	59	140.907 7	6 779	935	3000
Promethium*	Pm	61	(144.912 8)	7 220§	1168§	3300§
Protractinium*	Pa	91	(231.035 9)	15 370	1200	4000
Radium* .	Ra	88	(226.025 4)	5 000	700	1500
Radon* .	Rn	86	(222.017 6)	4400 (liquid −62 °C)	−71	−62
Rhenium .	Re	75	186.207	21 023	3180	5600
Rhodium .	Rh	45	102.905 5	12 420	1963†	3700
Rubidium .	Rb	37	85.467 8 ± 3^g	1 533	38.9	705
Ruthenium .	Ru	44	101.07 ± 2^g	12 360	2330	4100
Samarium .	Sm	62	150.36 ± 3^g	7 536	1050	1600
Scandium .	Sc	21	44.955 91 ± 1	2 992	1500	2800
Selenium .	Se	34	78.96 ± 3	4 808	220	685
Silicon . .	Si	14	28.085 5 ± 3	2 329	1410	3200
Silver . .	Ag	47	107.868 2 ± 3^g	10 500	961.93†	2170
Sodium . .	Na	11	22.989 77	966	97.9	900
Strontium .	Sr	38	87.62^g	2 583	770	1390
Sulphur . .	S	16	32.066 ± 6^r	2 086	115.3	444.674†
Tantalum .	Ta	73	180.947 9	16 670	3000	5500
Technetium*	Tc	43	(97.907 2)	11 496	2200	4600
Tellurium .	Te	52	127.60 ± 3^g	6 247	450	1000
Terbium .	Tb	65	158.925 4	8 267	1360	2500

Properties (*contd*)

Element	Symbol	Atomic No.	Atomic weight	Density $\rho/(\text{kg m}^{-3})$	Melting point $\theta/°C$	Boiling point $\theta/°C$
Thallium .	Tl	81	204.383	11 871	304	1460
Thorium* .	Th	90	232.038 1^{g,x}	11 725	1700	4500
Thulium .	Tm	69	168.934 2	9 325	1550	1900
Tin . . .	Sn	50	118.710 ± 7	7 285	231.9681†	2720
Titanium .	Ti	22	47.88 ± 3	4 508	1670	3300
Tungsten .	W	74	183.85 ± 3	19 254	3422†	5700
Unnilhexium*	Unh	106	(263)	—	—	—
Unnilpentium*	Unp	105	(262)	—	—	—
Unnilquadium*	Unq	104	(261)	—	—	—
Unnilseptium*	Uns	107	(262)	—	—	—
Uranium* .	U	92	238.0289^{g,m,y}	19 050	1135	4000
Vanadium .	V	23	50.941 5	6 090	1920	3400
Xenon . .	Xe	54	131.29 ± 3^{g,m}	3560/(−185 °C)	−111.9	−108.1
Ytterbium .	Yb	70	173.04 ± 3	6 966	824	1500
Yttrium .	Y	39	88.905 9^{g}	4 475	1510	3300
Zinc . .	Zn	30	65.39 ± 2	7 135	419.58†	910
Zirconium .	Zr	40	91.224 ± 2^{g}	6 507	1850	4500

Notes:

^{g} geologically exceptional specimens are known in which the element has an isotopic composition outside the limits for normal material. The difference between the atomic weight of the element in such specimens and that given in the Table may considerably exceed the implied uncertainty.

^{m} modified isotopic compositions may be found in commercially available material because it has been subjected to an undisclosed or inadvertent isotopic separation. Substantial deviations in atomic weight of the element from that given in the Table can occur.

^{r} range in isotopic composition of normal terrestrial material prevents a more precise atomic weight being given; the tabulated A (E) value should be applicable to any normal material.

^{x} Thorium has a well defined (mononuclide) composition in minerals with only rare exceptions. In certain places however (most notably in ocean water), measurable quantities of ^{230}Th (Ionium) can be found.

^{y} Uranium is the only element with no stable isotopes but which has a characteristic terrestrial composition of long-lived isotopes such that a meaningful atomic weight can be given for natural samples.

† Primary and secondary fixed points on the International Practical Temperature Scale of 1968 as amended 1975.

‡ This and other values marked K are in kelvins.

§ This property refers to ^{147}Pm.

E.F.G.H.

2.1.3 Abundances of the elements

This table gives an indication of the abundances of the elements in nature. Parts per million, by mass, are denoted by 'ppm', whereas abundances stated to be in 'atoms' are numbers of atoms per 10^{6} atoms of silicon. The values for sea water and for crustal rocks are from Turekian, *McGraw-Hill Encyclopedia of Science and Technology*, 1970, **4**, 627. Those for stony meteorites are mean values for the commoner varieties of ordinary chondrites taken from Mason, *Handbook of Elemental Abundances in Meteorites*, 1971 (Gordon and Breach, New York) and those for iron meteorites are from Brown, *Rev. Mod. Phys.*, 1949, **21**, 625 (major elements), from Smales *et al.*, in Ahrens (ed.) *Origin and Distribution of the Elements*, 1968, (Pergamon, London) (trace elements) or from Mason, *op. cit.* The values for trace elements in iron meteorites are exceedingly variable and the results given are mean values for medium octahedrites. Values for the Sun are taken from various sources (cf. Ross and Aller, *Science*, 1976, **191**, 1223 and Pagel, in Ahrens (ed.) *Origin and Distribution of the Elements*, 1979 (Pergamon, London), p. 79 and they vary considerably in accuracy; none is better than ± 20% and those that are uncertain by a factor of 2 or more are marked with a colon. Values for the Solar System are taken from Cameron in Barnes,

Clayton and Schramm (eds) *Essays in Nuclear Astrophysics*, 1982, (Cambridge University Press), p. 23 and they refer to initial values at the time of formation of the Solar System. They are mostly based on carbonaceous chondrite meteorites, making an allowance for radioactive decay of thorium and uranium, and using Solar data for the abundant volatile elements relative to silicon, magnesium, etc.

Abundances given for some elements in the Sun are different from those given for the Solar System and in particular this applies to lithium which has been largely destroyed by thermonuclear reactions during the Sun's lifetime. Other differences between abundances for the Sun and for the Solar System are due to uncertainties in the determinations, especially for the Sun.

The Solar System abundances tabulated are similar to those for most stars and for interstellar material in our neighbourhood and for the corresponding parts of other galaxies where, however, minor variations (within a factor of 2 or so either way) may occur in the relative amounts of hydrogen and helium, on the one hand, and carbon and heavier elements on the other. Carbon and heavier elements tend to be relatively more abundant in the central regions of large galaxies (such as our own) than in their outer parts or than in small galaxies; and in stars belonging to the outer spheroidal halo of our Galaxy carbon and heavier elements may be deficient by factors of up to 1000 relative to hydrogen and helium when compared to Solar System values. Peculiar abundances of some elements can also be found in some highly evolved stars as a result of nuclear reactions and also in the surface layers of certain stars where diffusive separation seems to have occurred.

The composition of the atmosphere is from Kuiper (ed.) *Atmospheres of the Earth and Planets*, 1949, (Chicago University Press) with corrections from Glueckauf and Kitt, *Proc. Roy. Soc.*, 1956, **234A**, 557.

Abundances

At. No.	Element	Sea water $\mu g\ dm^{-3}$	Crustal rocks ppm	Meteorites		Sun atoms	Solar system atoms
				Stony ppm	Iron ppm		
1	H	1.1×10^8	—	—	—	2.5×10^{10}	2.7×10^{10}
2	He	0.0072	—	—	—	2×10^9	1.8×10^9
3	Li	1.7×10^2	20	1.7	—	0.2	60
4	Be	0.0006	2.0	0.040	—	0.3	1.2
5	B	4.4×10^3	7.0	2.1	—	5:	9
6	C	2.8×10^4	—	1.0×10^3	1.1×10^3	1.1×10^7	1.1×10^7
7	N	1.6×10^4	20	61	33	2.1×10^6	2.3×10^6
8	O	8.8×10^8	3.7×10^5	3.7×10^5	—	1.9×10^7	1.8×10^7
9	F	1.3×10^3	4.6×10^2	1.2×10^2	—	—	7.8×10^2
10	Ne	0.12	—	—	—	2×10^6	2.6×10^6
11	Na	1.1×10^7	2.3×10^4	6.2×10^3	—	4.9×10^4	6.0×10^4
12	Mg	1.3×10^6	2.8×10^4	1.5×10^5	3.2×10^2	1.0×10^6	1.1×10^6
13	Al	1	8.0×10^4	1.0×10^4	40	6.3×10^4	8.5×10^4
14	Si	2.9×10^3	2.7×10^5	1.8×10^5	40	1.0×10^6	1.0×10^6
15	P	88	1.0×10^3	1.1×10^3	2.2×10^3	8×10^3	6.5×10^3
16	S	9.0×10^5	3.0×10^2	2.1×10^4	3.6×10^2	4.0×10^5	5.0×10^5
17	Cl	1.9×10^7	1.9×10^2	80	—	6×10^3:	4.7×10^3
18	Ar	4.5×10^2	—	—	—	6×10^4:	1.1×10^5
19	K	3.9×10^5	1.7×10^4	8.8×10^2	—	3.5×10^3	3.5×10^3
20	Ca	4.1×10^5	5.1×10^4	1.2×10^4	5.0×10^2	5.0×10^4	6.2×10^4
21	Sc	<0.004	22	7.6	—	25	31
22	Ti	1	8.6×10^3	6.4×10^2	1.0×10^2	1.6×10^3	2.4×10^3
23	V	1.9	1.7×10^2	63	6	2.5×10^2	2.5×10^2
24	Cr	0.2	96	3.6×10^3	15	1.3×10^4	1.3×10^4
25	Mn	1.9	1.0×10^3	2.3×10^3	3.0×10^2	6.3×10^3	9.3×10^3
26	Fe	3.4	5.8×10^4	2.5×10^5	9.1×10^5	8.0×10^5	9.0×10^5

Abundances (*contd*)

At. No.	Element	Sea water $\mu g\ dm^{-3}$	Crustal rocks ppm	Meteorites		Sun atoms	Solar system atoms
				Stony ppm	Iron ppm		
27	Co	0.39	28	7.0×10^2	6.3×10^3	2.5×10^3	2.2×10^3
28	Ni	6.6	72	1.5×10^4	6.7×10^4	5.0×10^4	4.8×10^4
29	Cu	23	58	90	1.3×10^2	3.2×10^2	5.4×10^2
30	Zn	11	82	54	28	1.0×10^3	1.3×10^3
31	Ga	0.03	17	5.1	80	16	38
32	Ge	0.06	1.3	10	37	60	1.2×10^2
33	As	2.6	2.0	1.8	11	—	6.2
34	Se	0.090	0.05	8.0	3	—	67
35	Br	6.7×10^4	4.0	0.4	1	—	9.2
36	Kr	0.21	—	—	—	—	41
37	Rb	1.2×10^2	70	4.0	—	9.5	6.1
38	Sr	8.1×10^3	4.5×10^2	10	—	20	23
39	Y	0.0013	0.35	2.2	—	6	4.8
40	Zr	0.026	1.4×10^2	10	8	14:	12
41	Nb	0.015	20	0.1	0.2	3:	0.9
42	Mo	10	1.2	1.5	7.3	4:	4.0
43	Tc	Unstable with short period					
44	Ru	—	—	0.9	11	2	1.9
45	Rh	—	—	0.23	4.1	1:	0.4
46	Pd	—	0.003	0.84	3.8	1:	1.3
47	Ag	0.28	0.08	0.085	0.035	0.2	0.46
48	Cd	0.11	0.18	0.06	0.02	2	1.6
49	In	—	0.2	0.004	0.010	1	0.19
50	Sn	0.81	1.5	0.65	2	2:	3.7
51	Sb	0.33	0.2	0.10	0.34	0.2:	0.31
52	Te	—	—	1.7	—	—	6.5
53	I	64	0.5	0.036	0.6	—	1.3
54	Xe	0.047	—	—	—	—	5.8
55	Cs	0.30	1.6	0.08	—	—	0.39
56	Ba	21	3.8×10^2	3.5	—	6	4.8
57	La	0.003 4	50	0.32	—	0.3	0.37
58	Ce	0.001 2	83	0.86	—	0.9	1.2
59	Pr	0.000 64	13	0.12	—	0.2	0.18
60	Nd	0.002 8	44	0.59	—	0.4	0.79
61	Pm	Unstable with short period					
62	Sm	0.000 45	7.7	0.19	—	0.13	0.24
63	Eu	0.000 13	2.2	0.07	—	0.1:	0.094
64	Gd	0.000 70	6.3	0.28	—	0.3:	0.42
65	Tb	0.001 4	1.0	0.048	—	—	0.076
66	Dy	0.000 91	8.5	0.31	—	0.2:	0.37
67	Ho	0.000 22	1.6	0.07	—	—	0.092
68	Er	0.008 7	3.6	0.20	—	0.1:	0.23
69	Tm	0.000 17	0.52	0.03	—	0.05	0.035
70	Yb	0.000 82	3.4	0.19	—	0.2:	0.20
71	Lu	0.000 15	0.8	0.033	—	0.15:	0.035
72	Hf	<0.008	4	0.24	—	0.2:	0.17
73	Ta	<0.002 5	2.4	0.022	0.06	—	0.020
74	W	<0.001	1.0	0.15	8.1	0.14:	0.30
75	Re	—	0.000 4	0.058	0.85	—	0.05
76	Os	—	0.000 2	0.77	7.6	0.1:	0.69

Abundances (*contd*)

At. No.	Element	Sea water $\mu g\ dm^{-3}$	Crustal rocks ppm	Meteorites		Sun atoms	Solar system atoms
				Stony atoms	Iron ppm		
77	Ir	—	0.000 2	0.64	3.0	0.2:	0.72
78	Pt	—	—	1.1	19	1.4	1.41
79	Au	0.011	0.002	0.20	1.8	0.14:	0.21
80	Hg	0.15	0.02	0.13	—	—	0.21
81	Tl	—	0.47	0.003	—	0.2:	0.19
82	Pb	0.03	10	0.30	60	2.0	2.6
83	Bi	0.02	0.004	0.012	0.5	—	0.14
84–89	Po–Ac	Unstable with short period					
90	Th	0.001 5	5.8	0.044	0.04	0.04	0.045
91	Pa	Unstable with short period					
92	U	3.3	1.6	0.015	0.007	—	0.027
93–102	Np–No	Unstable with short period					

B.E.J.P.

2.1.4 Composition of the Earth's atmosphere

(Parts in 10^6 of dry air by volume)

N_2	O_2	A	CO_2	Ne	He	CH_4	Kr	N_2O	H_2	O_3	Xe
780 900	209 500	9300	300	18	5.2	1.5	1.14	0.5	0.5	0.4	0.086

E.C.B.

2.2 Properties of inorganic compounds

Formula (including water of crystallization), melting and boiling temperatures, density and solubility

The properties, including melting θ_m, and boiling θ_b temperatures, are for the substances as formulated and at a pressure of $101.325\,kN\,m^{-2}$ (1 atm). Loss of water at a specific temperature $\theta/°C$ is denoted by -1, -2, etc., H_2O/θ, and loss of oxygen similarly. Densities ρ are for 20 °C; values ρ/θ are for temperature $\theta/°C$. The densities of gases and vapours are not given because approximate values can be calculated from the ideal gas laws and more accurate values can be obtained by the use of Table 2.5. Solubilities (S) are for cold water (20 °C) and hot water (100 °C) unless some other temperature (0/°C) is indicated ($S/°C$). The solubility S is 100 times the mass of *anhydrous* compound soluble in a unit mass of water. The solid in equilibrium with the saturated solution may not be the anhydrous compound nor may it be a hydrate listed in this table. For some additional information on the solubilities of solids see Table 2.6.2. For some gases the value of the absorption coefficient (A) is given: the volume of gas, measured at 0 °C and $101.325\,kN\,m^{-2}$ which will dissolve in a unit volume of water at the stated temperature, under a partial pressure of $101.325\,kN\,m^{-2}$ of the gas. For some additional information on the solubilities of gases see Table 2.6.1.

Abbreviations used are: a., acid; abs., absolute; acet., acetone; ac., acetic; al., alcohol; alk., alkali; aq., aqueous; aq. reg., aqua regia; bz., benzene; c., cold; chl., chloroform; conc., concentrated; d., decomposes; dil., dilute; eth., ether; exp., explodes; h., hot; i., insoluble; meth., methyl; min., mineral; pyr., pyridine; s., soluble; sl., slightly; solv., solvents; sb., sublimes; v., very; ∞, completely miscible.

Compound	$\theta_m/°C$	$\theta_b/°C$	$\rho/(kg\,m^{-3})$	Solubility		
				Cold water	Hot water	Other solvents
Aluminium						
Bromide, $AlBr_3$ . . .	97.5	254	2640	d.	d.	s. al., CS_2, acet.
Chloride, $AlCl_3$ . . .	—	sb. 181	2440	d.	d.	s. al., eth., CCl_4; sl. s. bz.
Fluoride, AlF_3	—	sb. 1250	3070	0.50	1.71	i. a., alk., al., acet.
Nitrate, $Al(NO_3)_3 . 9H_2O$	73.5	d. 150	—	74	1.60	s. a., alk., al.
Nitride, AlN	>2200 (in N_2)	d.	3260	d.	d.	d. a., alk.
Oxide, α-Al_2O_3 . . .	2054	2980	3970	i.	i.	v. sl. s. a., alk.
Sulphate, $Al_2(SO_4)_3$.	d. 770	—	2710	36.4	89	s. dil. a.; sl. s. al.
Sulphide, Al_2S_3 . . .	1100	sb. 1500 (in N_2)	2020	d.	—	s. a., alk.; i. acet.
Ammonium						
Ammonia, NH_3 . . .	-77.7	-33.4	—	89.5/0	7.4/100	s. al., eth., org. solv.
Acetate, $NH_4C_2H_3O_2$.	114	d.	1170	148/4	d.	s. al.; sl. s. acet.
Azide, NH_4N_3 . . .	—	sb. 134	1346	25.3	37/40	s. al.; i. eth., bz.
Bromide, NH_4Br . .	—	sb. 395	2429	76.4	145	s. al., eth., acet.
Carbonate, $(NH_4)_2CO_3.H_2O$	d. 58	—	—	22	355 d.	i. al., CS_2, NH_3
Chloride, NH_4Cl . .	—	sb. 340	1527	37.2	77.3	s. al., NH_3
Chloroplatinate, $(NH_4)_2PtCl_6$ . . .	d.	—	2925	0.5	3.4	sl. s. al.; i. eth.
Chromate, $(NH_4)_2CrO_4$	d. 180	—	1910	34.0	d.	sl. s NH_3, acet.; i. al.
Dichromate, $(NH_4)_2Cr_2O_7$ . .	d. 170	—	2150	35.6	156	s. al.; i. acet.
Dihydrogen phosphate, $(NH_4)H_2PO_4$ . . .	190	—	1803	37.4	173	i. acet.
Fluoride, NH_4F . . .	—	sb.	1009/25	100/0	d.	s. al.; i. NH_3
Formate, NH_4CHO_2 .	116	d. 180	1280	143	533/80	s. al.; NH_3
Hydrogen phosphate, $(NH_4)_2HPO_4$ . . .	d. 155	—	1619	68.9	97.2/60	i. al. acet.
Iodate, NH_4IO_3 . .	d. 150	—	3309/21	2.7/16	—	—
Iodide, NH_4I . . .	—	sb. 400	2514/25	172	250	v. s. al., acet., NH_3
Molybdate, $(NH_4)_2MoO_4$	d.	—	2276/25	s.d.	d.	s. a.; i. al., NH_3
Nitrate, NH_4NO_3 . .	170	d. 210	1725/25	192	871	s. al., acet., NH_3
Nitrite, NH_4NO_2 . .	d.	—	1690	v.s.	d.	s. al.; i. eth.
Oxalate, $(NH_4)_2C_2O_4.H_2O$.	d.	—	1500	4.45	34.64	i. NH_3
Sulphate, $(NH_4)_2SO_4$.	d. 235	—	1769/50	75.4	103	i. al., acet., NH_3

Inorganic compounds (contd)

Compound	$\theta_m/°C$	$\theta_b/°C$	$\rho/(kg\ m^{-3})$	Cold water	Hot water	Other solvents
Ammonium (contd)						
Tartrate,						
$(NH_4)_2C_2H_4O_6$ . .	d.	—	1601	63.0	86.9/60	sl. s. al.
Thiocyanate, NH_4CNS.	149.6	d. 170	1305	170	346/60	s. al., acet., NH_3
Vanadate, meta-,						
NH_4VO_3	d. 200	—	2326	0.48	2.42/60	i. al., eth., NH_4Cl
Antimony						
Bromide	96.6	275	4148/23	d.	d.	s. HCl, CS_2, al., NH_3
Chloride, tri-, $SbCl_3$.	73.4	219	3140/25	910	∞/72	s. HCl, a., CS_2
„ , penta-,						
$SbCl_5$	2.8	d.	2336	d.	d.	s. HCl, chl.
Fluoride, tri-, SbF_3 .	292	—	4379/21	445	564/30	i. NH_3
Fluoride, penta-, SbF_5 .	7	149	2990/23	s.	—	s. KF
Hydride, SbH_3 . .	−88	−17	—	0.2/0 (A)	—	v. s. al., CS_2
Iodide, SbI_3 . . .	−167	400	4917/17	d.	d.	s. HCl, KI, al., acet. CS_2
Oxide, tri-, Sb_2O_3 .	655	1425	5250	v. sl. s.	sl. s.	s. HCl, KOH
„ , penta-, Sb_2O_5 .	−O/380	−2O/930	3800	v. sl. s.	sl. s.	s. HCl, KOH
III Oxychloride, SbOCl	d.	—	—	i.	i.	s. HCl, acet., CS_2
III Sulphate $Sb_2(SO_4)_3$	d.	—	3625/4	i.	d.	s. a.
Sulphide, tri-, Sb_2S_3 .	547	—	4640	i.	d.	s. alk., HCl, K_2S
„ , penta-, Sb_2S_5	d.	—	4120	i.	i.	s. alk., HCl
Arsenic						
Bromide, $AsBr_3$ . .	32.8	220	3540/25	d.	d.	s. HCl, CS_2
Chloride, $AsCl_3$ . .	−18	130.2	2163	d.	d.	s. HCl, al., eth.
Fluoride, tri-, AsF_3 .	−6	−56	—	d.	d.	s. al. eth., bz.
Fluoride, penta-, AsF_5 .	−80	−52.8	—	s.	—	s. alk., al., eth.
Hydride, AsH_3 . .	−116.3	−62.5	—	0.2/0 (A)	—	sl. s. alk.; s. chl.
Iodide, AsI_3 . . .	146	403	4390/13	6/25	d.	s. al., eth., chl., bz.
Oxide, tri-, As_2O_3 .	315	457	3738	1.82	8.2	s. alk., HCl; i. eth.
„ , penta-, As_2O_5 .	d. 315	—	4320	65.8	76.4	s. alk., al., a.
Sulphide, tri-, As_2S_3 .	300	707	3430	i.	sl. s.	s. K_2S, $NaHCO_3$
„ , penta-, As_2S_5	—	sb. d. 500	—	i.	d.	s. alk., HNO_3
Barium						
Acetate, $Ba(C_3H_3O_2)_2$	—	—	2468	72	74.8	i. al.
Bromate,						
$Ba(BrO_3)_2.H_2O$ . .	$-H_2O/170$	d. 260	3990/18	0.66	5.7	i. al.; s. acet.
Bromide, $BaBr_2.2H_2O$.	$-H_2O/75$	$-2H_2O/120$	3584/24	101	139	s. al.
„ , $BaBr_2$ . .	857	d.	4781/24	101	139	v. s. al.; v. sl. s. acet.
Carbonate, $BaCO_3$ . .	−	d.	4430	2×10^{-3}	6×10^{-3}	s. a., NH_4Cl; i. al.
Chlorate,						
$Ba(ClO_3)_2.H_2O$ . .	$-H_2O/120$	$-O/250$	3180	34.4	105	sl. s. al., acet.
Chloride, $BaCl_2.2H_2O$.	$-2H_2O/113$	—	3097/24	35.7	58.8	sl. s. HCl, HNO_3
„ , $BaCl_2$.	961	2030	3917	35.7	58.8	v. sl. s. al.
Fluoride, BaF_2 . .	1370	2270	4890	0.16	sl. s.	s. a., NH_4Cl
Hydride, BaH_2 . .	d. 675	—	4210/0	d.	d.	d. a.
Hydroxide,						
$Ba(OH)_2.8H_2O$ . .	78	—	2180/16	3.9	101/80	sl. s. al.; i. acet.
Iodate, $Ba(IO_3)_2$ . .	d.	—	4998	0.035	—	s. HNO_3, HCl
Iodide, $BaI_2.2H_2O$ . .	$-H_2O/100$	$-2H_2O/540$	5150	220	276	s. al., acet.
„ , BaI_2 . . .	740	—	5150/25	220	276	v. s. al.
Nitrate, $Ba(NO_3)_2$ . .	592	d.	3240/23	9.0	34.4	sl. s. meth. al.; i. al.
Nitrite, $Ba(NO_2)_2.H_2O$	d. 115	—	3173	67.5	31.7	—
„ , $Ba(NO_2)_2$.	d. 217	—	3230/23	67.5	31.7	sl. s. al.; i. acet.
Oxide, BaO	2010	3090	5720	3.9	101/80	s. dil. a.
Peroxide . . .	450	$-O/800$	4960	v. sl. s.	d.	s. dil. a.; i. acet.
Sulphate, $BaSO_4$ . .	1580	—	4500/15	2×10^{-4}	4×10^{-4}	sl. s. H_2SO_4
Sulphide, BaS . . .	—	—	4250/15	d.	d.	i. al.
Beryllium						
Bromide, $BeBr_2$ . .	—	sb.475	3465/25	s.	v. s.	s. al. eth.; i. bz.
Chloride, $BeCl_2$ . .	415	532	1899/25	73	v. s. d.	v. s. al., eth., pyr; sl. s. bz., chl.; acet.

Inorganic compounds (*contd*)

Compound	$\theta_m/°C$	$\theta_b/°C$	$\rho/(kg\ m^{-3})$	Solubility		
				Cold water	Hot water	Other solvents
Beryllium (*contd*)						
Fluoride, BeF_2 . . .	550	1170	1986/25	v. s.	v. s.	s. H_2SO_4; sl.s. al.
Nitrate,						
$Be(NO_3)_2.3H_2O$. .	60	142	1557	v. s.	v. s.	v. s. al.
Oxide, BeO	2540	—	3010	$2 \times 10^{-5}/30$	—	s. conc. H_2SO_4
Sulphate, $BeSO_4.4H_2O$	$-2H_2O/100$	$-4H_2O/400$	1713/10.5	38.9	85.1	i. al., acet.
Bismuth						
Bromide, $BiBr_3$ . . .	218	461	5720/25	d.	d.	s. HCl, eth.; i. al.
Chloride, $BiCl_3$ . . .	232	441	4750/25	d.	d.	s. a., al., eth.
Fluoride, BiF_3 . . .	730	—	5320	i.	—	s. a.; i. al.
Iodide, BiI_3	403	500	5778/15	i.	d.	s. HCl., HI, abs. a.
Nitrate, $Bi(NO_3)_3.5H_2O$	d.	—	2830	d.	d.	v. s. HNO_3; s. a., acet.
Oxide, Bi_2O_3 . . .	820	1890	8900	i.	i.	s. a.
Oxychloride, BiOCl .	—	—	7720/15	i.	i.	s. a.
Boron						
Chloride, BCl_3 . . .	-107	12.7	1349/11	d.	d.	d. al.
Fluoride, BF_3 . . .	-126.7	-111	—	1.06/0(A)	d.	s. conc. H_2SO_4
Hydride, B_2H_6 . . .	-165.5	-87	—	sl. s. d.	d.	s. NH_4OH, conc. H_2SO_4
,, , B_4H_{10} . . .	-120.8	16	—	sl. s. d.	d.	d. al.; s. bz.
Nitride, BN	d. 2300	—	2250	i.	d.	sl. s. hot a.
Oxide, B_2O_3 . . .	450	2060	2460	1.1/0	16.4	s. a. al.
Boric acid, H_3BO_3 . .	d. 185	—	1435/15	4.9	38	s. al.; sl. s. acet.
Cadmium						
Acetate, $Cd(C_3H_3O_2)_2$.	256	d.	2341	v. s.	—	s. meth. al.
Bromide, $CdBr_2$ . . .	567	863	5192/25	98.8	166	s. al.; sl. s. eth. acet.
Chloride, $CdCl_2$ . . .	568	967	4047/25	134	147	s. al.; i. acet., eth.
Fluoride, CdF_2 . . .	1110	1758	6640	40	—	s. a., HF; i. al.
Iodide, CdI_2	387	796	5670/30	84.7	125	s. al., acet.; sl. s. NH_3
Nitrate,						
$Cd(NO_3)_2.4H_2O$.	59.4	132	2455/17	158/25	681	s. al., eth., ac., NH_3
Nitrate, $Cd(NO_3)_2$.	350	—	—	150	711	v. s. a.
Oxide, CdO	—	sb. 1559	8150	i.	i.	s. a.; i. alk.
Sulphate, $CdSO_4$ . .	1000	—	4691	76.4	58	i. al., acet.
Caesium						
Carbonate, Cs_2CO_3 .	d. 610	—	—	260/15	v.s.	s. al., eth.
Chloride, CsCl . . .	645	1300	3988	187	271	v. s. al.
Fluoride, CsF . . .	703	1230	4115	362/18	—	s. meth. al.
Hydroxide, CsOH . .	315	—	3675	385/15	—	s. al.
Nitrate, $CsNO_3$ . . .	417	d.	3685	23	197	s. acet.; sl. s. al.
Calcium						
Acetate, $Ca(C_2H_3O_2)_2$.	d.	—	—	34.7	29.7	sl. s. al.
Bromide, $CaBr_2$ . . .	742	1810	3353/25	143	309	s. al., acet.
Carbide, CaC_2 . . .	—	2300	2200	d.	d.	—
Carbonate, $CaCO_3$						
(aragonite) . . .	—	d. 825	2930	i.	i.	s. a., NH_4Cl
Carbonate, $CaCO_3$						
(calcite) . . .	—	d. 899	2710/18	i.	i.	s. a., NH_4Cl
Chloride, $CaCl_2.6H_2O$	$-4H_2O/30$	$-6H_2O/200$	1710/25	74.5	159	s. al.
,, , $CaCl_2$ . .	772	1940	2150/25	74.5	159	s. al., acet.
Fluoride, CaF_2 . . .	1420	2500	3180	2×10^{-3}	—	sl. s. a.
Hydride, CaH_2 . . .	816 (in H_2)	d. 600	1900	d.	d.	d. a.; i. bz.
Hydroxide, $Ca(OH)_2$.	$-H_2O/580$	d.	2240	0.12	0.05	s. NH_4Cl aq., a.; i. al.
Iodide, CaI_2 . . .	778	1700	3956/25	209	426	s. al., acet.
Nitrate, $Ca(NO_3)_2.4H_2O$	40	d. 132	—	129	364	s. al., acet.
,, , $Ca(NO_3)_2$ . .	561	—	2504/18	129	364	s. acet.
Oxide, CaO	2900	—	3300	d.	d.	s. a.
Phosphate, ortho-,						
$Ca_3(PO_4)_2$ . . .	1670	—	3140	d.	d.	s. a.; i. al.
Silicate, meta-, $CaSiO_3$.	1540	—	2905	i.	—	s. HCl
Sulphate, $CaSO_4.2H_2O$	$-1\frac{1}{2}H_2O/130$	$-2H_2O/163$	2320	0.274	0.189	s. a. NH_4 salts.

Inorganic compounds (contd)

Compound	$\theta_m/°C$	$\theta_b/°C$	$\rho/(kg\ m^{-3})$	Solubility Cold water	Hot water	Other solvents
Calcium (contd)						
Sulphate, $CaSO_4$ . .	—	—	2960	0.274	0.189	—
Sulphide, CaS . . .	2525	—	2500	sl. s. d.	—	d. a.
Tungstate, $CaWO_3$ . .	—	—	6060	i.	i.	s. NH_4Cl; i. a., al.
Carbon						
Oxide, mon-, CO . .	−205.06	−191.50	—	$3.5 \times 10^{-2}/o(A)$	$1.5 \times 10^{-2}/6o(A)$	s. al., bz.
,, , di-, CO_2 . .	—	sb. −78.48	—	1.68/o(A)	0.36/6o(A)	s. a., alk., al., acet.
Cyanogen, C_2N_2 . .	−27.84	−20.7	—	4.5/20(A)	—	s. al., eth.
Cerium						
Ammonium nitrate (III)						
$(NH_4)_2Ce(NO_3)_6$.	—	—	—	141/25	227/70	s. H_2SO_4, al.
(III) Chloride, $Ce Cl_3$.	848	1727	3920/0	100/0	d.	s. al., acet.
(III) Fluoride, CeF_3 .	1460	2300	6160	i.	—	—
(IV) ,, , $CeF_4.H_2O$	650	d.	4770	i.	—	s. a.
(III) Nitrate,						
$Ce(NO_3)_3.6H_2O$.	$−3H_2O/150$	d. 200	—	176/25	283/50	s. al.; acet.
(III) Oxide, Ce_2O_3 . .	1692	—	6860	i.	i.	s. H_2SO_4; i. HCl
(IV) ,, , CeO_2 . .	2600	—	7132/23	i.	i.	i. dil. a.; sol. H_2SO_4
(III) Sulphate,						
$Ce_2(SO_4)_3$ . . .	d. 920	—	3912	9.5	0.46	—
(III) Sulphide, Ce_2S_3 .	d. 2100	—	5020/11	i.	d.	s. dil. a.
Chlorine						
Oxide, mon-, Cl_2O . .	−111	2.2	—	2/o (A)	—	s. alk., H_2SO_4
,, , di-, ClO_2 . .	−60	11.1	—	20/4 (A)	d.	s. alk., H_2SO_4
,, , hepta-, Cl_2O_7 .	−91.5	82	—	s. d.	—	s. bz.
Perchloric acid, $HClO_4$.	−112	—	1764/22	39/51	∞	
Chromium						
Ammonium sulphate(III)						
$Cr_2(SO_4)_3(NH_4)_2SO_4$.						
$24H_2O$	94	—	1720	15.8	32.7	s. al., dil. a.
(II) Chloride, $CrCl_2$.	824	—	2878/25	v. s.	v. s.	i. al., eth.
(III) Chloride, $CrCl_3$.	1150	sb. 1300	2760/15	s.	s.	i. a., acet., al.
(III) Nitrate,						
$Cr(NO_3)_3.9H_2O$. .	60	d. 100	—	77.4/25	s.	s. a., alk., al., acet.
Oxide, sesqui-, Cr_2O_3 .	2330	—	5210	i.	i.	i. a., alk.
,, , tri-, CrO_3 . .	196	d.	2700	167	207	s. H_2SO_4, eth., al.
Oxychloride, CrO_2Cl_2 .	−96	117	1911	d.	d.	s. eth., ac. a.; d. al.
Potassium sulphate,						
$Cr_2(SO_4)_3.K_2SO_4$.						
$24H_2O$	89	d.	1826/25	12.5/25	s.	i. al.
Cobalt						
(II) Chloride, $CoCl_2$.	735	1090	3360/25	53	106	s. al. acet.
(II) Hydroxide, $Co(OH)_2$.	d.	—	3597/15	$3 \times 10^{-4}/o$	—	s. a., NH_4 salts.
Nitrate,						
$Co(NO_3)_2.6H_2O$.	$−3H_2O/55$	d.	1870/25	96.8	339/91	s. al., acet.; sl. s. NH_3
(II) Oxide, CoO . .	1800	d.	6450	i.	i.	s. a. NH_3 aq.; i. al.
(III) ,, , Co_3O_4 . .	d. 895	—	6070	i.	i.	v. sl.s. a.
Sulphate, $CoSO_4.7H_2O$	96.8	$−7H_2O/420$	1950/25	35.2	38.5	s. al., meth. al.
Copper						
(I) Acetate,						
$Cu(C_2H_3O_2)_2.H_2O$.	115	d. 240	1882	7.3/25	5	s. al. eth.
(I) Bromide, CuBr .	492	1345	4980	v. sl. s.	d.	s. HCl, HNO_3; i. acet.
(II) ,, , $CuBr_2$ . .	498	—	4770/25	127	132/50	s. al., acet.; i. bz.
(I) Chloride, CuCl .	430	1210	4140	v. sl. s.	—	s. HCl, NH_4OH
(II) ,, , $CuCl_2$ . .	d. 492	d. 990	3386/25	73	120	s. al.
(II) Fluoride,						
$CuF_2.2H_2O$ . . .	d.	—	2930/25	sl. s.	d.	s. al., HCl; i. acet.

Inorganic compounds (*contd*)

Compound	$\theta_m/°C$	$\theta_b/°C$	$\rho/(kg\ m^{-3})$	Solubility		
				Cold water	Hot water	Other solvents
Copper (*contd*)						
(I) Iodide, CuI . . .	605	1336	5620	v. sl. s.	—	s. HCl, KI; i. a., al.
(II) Nitrate,						
$Cu(NO_3)_2.3H_2O$	114.5	d. 170	2320/25	125	247	v. s. al.
(I) Oxide, Cu_2O . .	1235	−O/1800	6000	i.	i.	s. HCl; sl. s. HNO_3; i. al.
(II) ,, , CuO . . .	1120 d.	—	6400	i.	i.	s. a., NH_4Cl
(II) Sulphate,						
$CuSO_4.5H_2O$ . .	$-4H_2O/110$	$-5H_2O/150$	2284	20.5	75.1	sl. s. meth. al.; i. al.
Dysprosium						
Oxide, Dy_2O_3 . . .	2340	—	7810/27	—	—	s. a.
Oxide, $Dy(SO_4)_3.8H_2O$	$-8H_2O/360$	—	—	4.1	2.7/40	—
Erbium						
Oxide, Er_2O_3 . . .	—	—	8640	i.		sl. s. a.
Sulphate,						
$Er_2(SO_4)_3.8H_2O$	$-8H_2O/400$	—	3217	13.0	5.3/40	—
Europium						
Oxide, Eu_2O_3 . . .	—	—	7420	—	—	—
Sulphate,						
$Eu_2(SO_4)_3.8H_2O$.	$-8H_2O/375$	—		2.1	1.8/40	—
Gadolinium						
Nitrate,						
$Gd(NO_3)_3.6H_2O$	91	—	2332	v. s.	v. s.	s. al.
Oxide, Gd_2O_3 . . .	2330	—	7407/15	v. sl. s.	—	s. a.
Sulphate,						
$Gd_2(SO_4)_3.8H_2O$.	—	—	3010/14.6	2.3	1.8/40	—
Gallium						
Bromide, $GaBr_3$ . . .	122	279	3690/25	s.	s.	sl. s. NH_3
Chloride, $GaCl_3$ . . .	77.5	200	2470/25	v. s.	v. s.	s. bz., CCl_4, CS_2
Nitrate, $Ga(NO_3)_3.xH_2O$	d. 110	—	—	v. s.	v. s.	s. abs. al.; i. eth.
Oxide, Ga_2O_3 . . .	1900	—	6440	i.	i.	s. alk.
Sulphate, $Ga_2(SO_4)_3$.	d. 690	—	—	v. s.	v. s.	s. al.; i. eth.
Germanium						
Chloride, $GeCl_4$ . . .	−49.5	84	1840/30	d.	d.	v. s. dil. HCl.; s. al., eth.
Oxide, di-, GeO_2 . .	1086	—	6239	i.	—	s. a., alk.
Sulphide, GeS . . .	530	sb. 430	3310	0.24	i.	s. alk., HCl.; sl. s. NH_4OH
Gold						
(III) Chloride, $AuCl_3$.	d. 254	—	3900	68	v. s.	s. al. eth.; sl. s. NH_3 aq.
Hafnium						
Chloride, $HfCl_4$ . . .	—	sb. 319	—	d.	d.	s. al., eth.
Oxide, di-, HfO_2 . .	2760	—	9680/20	i.	i.	—
Hydrogen						
Bromide, HBr . . .	−86.82	−66.73	—	221/0	130/100	s. al.
Chloride, HCl . . .	−114.22	−85.05	—	82.3/0	56.1/60	s. al., eth., bz.
Cyanide, HCN . . .	−13.24	25.70	699/22	v. s.	v. s.	v. s. al.; s. eth.
Fluoride, HF . . .	−83.7	19.9	987/19.5	v. s.	v. s.	—
Iodide, HI . . .	−50.80	−35.36	—	0.43/0(A)	v. s.	s. al.
Oxide-deuterium, D_2O .	3.8	101.42	1105.6	∞	∞	—
Peroxide, H_2O_2 . . .	−0.41	150.2	1442/25	∞	—	s. al., eth.

Inorganic compounds (contd)

Compound	$\theta_m/°C$	$\theta_b/°C$	$\rho/(kg\ m^{-3})$	Solubility Cold water	Hot water	Other solvents
Hydrogen (contd)						
Phosphide, H_3P	−133.78	−87.74	—	0.26/17(A)	i.	s. al., eth.
Selenide, H_2Se	−65.73	−41.3	—	3.77/4(A)	2.7/22.5(A)	s. CS_2, $COCl_2$
Sulphide, H_2S	−85.53	−60.34	—	4.53/0(A)	1.18/60(A)	sl. s. al.; s. CS_2
Telluride, H_2Te	−51	−2.3	—	s.	—	s. al., alk.
Indium						
Chloride, tri-, $InCl_3$	—	sb. 300	3460/25	v. s.	v. s.	sl. s. al., eth.
Iodide, mono-, InI	351	712	5310	—	s. d.	s. dil. a.; i. al.
Oxide, sesqui-, In_2O_3	—	—	7179	i.	—	—
Sulphide, sesqui-, In_2S_3	1050	—	4900	i.	—	d. a.; sl. s. Na_2S
Iodine						
Bromide, mono-, IBr	—	sb. 50	4416/0	s. d.	—	s. al., eth., chl., CS_2
Chloride, mono-, ICl	27.3	97.4	3182/0	d.	—	s. HCl, al., eth., CS_2
,, , tri-, ICl_3	—	d. 77	3177/15	s. d.	—	s. ac., a., al., bz., eth., CCl_4
Iodic acid, HIO_3	d.	—	4630	309/16	574/101	v. s. al.; sl. s. HNO_3
Oxide, penta, I_2O_5	d. 300	—	4799/25	s. d.	s. d.	sl. s. dil. a.; i. abs. al., eth. chl.
Iridium						
Chloride, tri-, $IrCl_3$	d. 763	—	5300	i.	—	i. a., alk.
Fluoride, hexa-, IrF_6	—	53	6000	d.	d.	—
Oxide, di-, IrO_2	d. 1100	—	3150	i.	i.	i. a., alk.
Iron						
Ammonium sulphate (III) $Fe(NH_4)(SO_4)_2.12H_2O$	40	−12H_2O/230	1710	44.2	v. s.	s. dil. a.; i. al.
Ammonium sulphate (II) $Fe(NH_4)_2(SO_4)_2.6H_2O$	d. 100	—	1860	26.4	52/70	i. al.
Carbonyl, penta, $Fe(CO)_5$	−21	103	1457	i.	—	s. al., eth., bz., alk., conc. H_2SO_4
(III) Chloride, $FeCl_3.6H_2O$	37	280	—	91.9	536	s. al. eth.
(III) Chloride, $FeCl_3$	304	d. 319	2900/25	91.9	536	v. s. al., eth., acet.
(II) Chloride, $FeCl_2$	677	1026	3160/25	62.6	95	v. s. al.; s. acet.; i. eth.
Dicyclopentadiene, $Fe(C_5H_5)_2$	174	249	—	i.	—	s. org. solv.
(III) Nitrate, $Fe(NO_3)_3.9H_2O$	50	d. 125	1684	83	108	s. HCl, H_2SO_4, al., acet.
(III) Oxide, Fe_2O_3	1565	—	5240	i.	i.	s. HCl, H_2SO_4
(III) Oxide, (magnetite), Fe_3O_4	1597	—	5180	i.	i.	sl. s. a.
(II) Oxide, $Fe_{0.95}O$	1377	—	5700	i.	i.	s. a.
(II) Phosphide, FeP	—	—	6070	i.	i.	—
(II) Sulphate, $FeSO_4.7H_2O$	64, −6H_2O/90	−7H_2O/300	1898	26.3	34.2/95	s. l., s. al.
(II) Sulphide, FeS	1190	d.	4740	6×10^{-4}/18	d.	s. a. d.; i. NH_3
Lanthanum						
Chloride, $LaCl_3$	860	>1000	3842/25	97	143/80	v. s. al., pyr.; i. eth.
Nitrate, $La(NO_3)_3.6H_2O$	40	d. 126	—	122/21	215/56	v. s. al.; s. acet.
Oxide, La_2O_3	2300	4200	6510/15	v. sl. s.	d.	s. a.; i. acet.
Sulphate, $La_2(SO_4)_3.9H_2O$	d.	—	2821	2.3/25	0.7	sl. s. HCl; i. acet.
Lead						
Acetate, $Pb(C_2H_3O_2)_2$	280	—	3250	44.3	221/50	v. sl. s. al.
Carbonate, $PbCO_3$	d. 315	—	6600	2×10^{-4}	d.	s. a., alk.; i. al.

Inorganic compounds (*contd*)

Compound	$\theta_m/°C$	$\theta_b/°C$	$\rho/(kg\ m^{-3})$	Solubility		
				Cold water	Hot water	Other solvents
Lead (*contd*)						
Chloride, $PbCl_2$ . .	501	950	5850	1.1	3.34	sl. s. dil. HCl, NH_3; i. al.
Chromate, $PbCrO_4$.	844	d.	6120/15	i.	i.	s. a. alk.
Fluoride, PbF_2 . .	830	1290	8240	0.06/9	0.07/27	s. HNO_3; i. acet.
Iodide, PbI_2 . .	410	830	6160	0.08/25	0.47	s. alk., KI.; i. al.
Nitrate, $Pb(NO_3)_2$.	d. 470	—	9530	56	139	s. aq. al., NH_3, alk.
Oxide, mono-, PbO .	890	—	8000	1×10^{-3}	—	s. HNO_3, alk.
,, (red-lead), Pb_3O_4	d. 500	—	9100	i.	i.	s. h. HCl, ac. a.
,, , di-, PbO_2 .	d. 290	—	9375	i.	i.	s. dil. HCl; sl. s. ac. a.
Sulphate, $PbSO_4$ . .	1170	—	6200	$3 \times 10^{-3}/0$	$6 \times 10^{-3}/50$	s. a., NH_4 salts
Sulphide, PbS . .	1114	d.	7500	i.	i.	s. a.; i. alk.
Tetraethyl, $Pb(C_2H_5)_4$.	−136	d. 200	1659/11	i.	—	s. bz., al. eth.
Lithium						
Aluminium hydride, $LiAlH_4$	d. 125	—	917	d.	d.	s. eth.
Borohydride $LiBH_4$.	d. 275	—	660	d.	d.	s. eth., s. al. (d.)
Bromide, LiBr . .	547	1310	3464/25	168	266	s. al., eth.
Carbonate, Li_2CO_3 .	723	d. 1310	2110	1.33	0.72	s. a.; i. acet., al.
Chloride, LiCl . .	610	1382	2068/25	84.7/25	128	s. al., acet.
Fluoride, LiF . .	848	1720	2635	0.14/25	0.14/35	s. HF; i. al. acet.
Hydride, LiH . .	680	—	820	d.	d.	i. eth.
Iodide, LiI . . .	450	1180	4016	165	481	v. s. al. NH_3
Nitrate, $LiNO_3$ . . .	264	d. 600	2380	73.3	213/90	s. al., acet.
Oxide, Li_2O . .	1570	—	2013/25	d.	d.	—
Phosphate, Li_3PO_4 .	837	—	2537/18	0.03	—	s. a.; i. acet.
Sulphate, Li_2SO_4 . .	845	—	2221	33.5	30.7	i. acet.
Magnesium						
Bromide, $MgBr_2$ . .	710	1280	3720/25	101/25	125	s. HBr., al., acet.; sl. s. NH_3
Carbonate, $MgCO_3$.	d. 350	$-CO_2/900$	2958	sl. s	sl. s.	s. a.
Chloride, $MgCl_2.6H_2O$	d. 117	—	1569	55	73	v. s. al.
,, , $MgCl_2$	708	1412	2320	55	73	v. s. al.
Fluoride, MgF_2 . .	1260	2260	—	0.013/25	—	s. HNO_3; i. al.
Nitrate, $Mg(NO_3)_2.6H_2O$	89	d. 330	1636/25	72.7/25	253	s. al.
Oxide, MgO . . .	2862	—	3580/25	$6 \times 10^{-4}/25$	3×10^{-5}	s. a.; i. al.
Perchlorate, $Mg(ClO_4)_2$	d. 251	—	2210/18	99	v. s.	s. al., sl. s. eth.
Phosphate, $Mg_3(PO_4)_2.22H_2O$.	$-18H_2O/100$	d. 200	1640/15	v. sl. s.	—	d. a.
Sulphate, $MgSO_4$ . .	d. 1124	—	2660	33.7	50.4	s. al., eth.; i. acet.
Manganese						
(II) Carbonate, $MnCO_3$	d.	—	3125	v. sl..s.	v. sl. s.	s. dil. a.; i. al.
(II) Chloride, $MnCl_2$.	650	1190	2977/25	74	115	s. al.; i. eth.
(II) Nitrate, $Mn(NO_3)_2.4H_2O$.	26	129	1820	132	499/75	v. s. al.
(II) Oxide, MnO . .	—	—	5450	i.	i.	s. a.
(III) Oxide, Mn_2O_3 .	$-O/1080$	—	4500	i.	i.	s. a.
,, , di-, MnO_2 .	$-O/535$	—	5026	i.	i.	s. HCl; i. HNO_3
(II) Sulphate, $MnSO_4$.	700	d. 850	3250	62.9	35.7	s. al.; i. eth.
(II) Sulphide, MnS .	d.	—	3990	v. sl. s.	—	s. dil. a.
Mercury						
(II) Bromide, $HgBr_2$.	236	322	6109/25	0.55	4.9	s. al.; sl. s. eth.
(I) ,, , HgBr .	—	sb. 345	7307	v. sl. s.	v. sl. s.	s. a.; i. al., acet.
(II) Chloride, $HgCl_2$.	276	302	5440/25	6.5	57.5	v. s. al., eth.; s. ac. a.
(I) ,, , HgCl .	—	sb. 400	7150	v. sl. s.	v. sl. s.	s. aq. reg.; sl. s. HNO_3, HCl
(II) Cyanide, $Hg(CN)_2$	d.	—	3996	11.2/25	54	s. al., NH_3; i. bz.
(II) Iodide, HgI_2 . .	259	354	6360/25	$5.4 \times 10^{-3}/22$	sl. s.	s. al., eth.
(II) Nitrate, $Hg(NO_3)_2.\frac{1}{2}H_2O$	79	d.	4390	v. s.	d.	s. HNO_3, acet.; i. al.
(II) Oxide, HgO . .	d. 500	—	11100/4	$5 \times 10^{-3}/25$	0.04	s. a.; i. al., eth.
(II) Sulphate, $HgSO_4$.	d.	—	6470	d.	—	s. a.; i. al., acet.
(I) ,, , Hg_2SO_4 .	d.	—	7560	s. d.	s. d.	s. H_2SO_4, HNO_3

Inorganic compounds (*contd*)

Compound	$\theta_m/°C$	$\theta_b/°C$	$\rho/(kg\ m^{-3})$	Solubility		
				Cold water	Hot water	Other solvents
Molybdenum						
Fluoride, hexa-, MoF_6	17	35	2550/18	s. d.	d.	s. NH_4OH, alk.
Oxide, tri-, MoO_3	795	1155	4692/21	0.18/23	0.52/82	s. HF, H_2SO_4, NH_4OH
Silicide, di-, $MoSi_2$	—	—	6310/21	—	—	—
Sulphide, di-, MoS_2	1750	—	4800/14	i.	i.	s. aq. reg., H_2SO_4
Neodymium						
Chloride, $NdCl_3$	784	1600	4134/25	98	140	s. al.; i. eth., chl.
Oxide, Nd_2O_3	1900	—	7240	i.	i.	s. HCl
Sulphate, $Nd_2(SO_4)_3.8H_2O$	1176	—	2850	7.1	1.2	—
Nickel						
Carbonyl, $Ni(CO)_4$	−19.3	42.2	1320/17	0.018/9.8	—	s. HNO_3, aq. reg., al., eth.
Chloride, $NiCl_2$	—	sb. 972	3550	62	88	s. al., NH_4OH
Fluoride, NiF_2	—	sb. 1000 (in HF)	4630	2.56	2.58/90	s.a., alk., eth., NH_3
Nitrate, $NiNO_3.6H_2O$	56.7	136.7	2050	94.2	238	s. al., NH_4OH
Oxide, NiO	1990	—	6670	i.	i.	s. a., NH_4OH
Sulphate, $NiSO_4.6H_2O$	$-6H_2O/103$	—	2070	38	78	v. s. al., NH_4OH
Niobium						
Chloride, $NbCl_5$	205	248	2750	d.	—	s. al., HCl, CCl_4
Fluoride, NbF_5	73	236	3293	d.	—	s. HNO_3, al.; sl. s. chl.
Oxide, mon-, NbO	1940	—	7300	i.	i.	s. HCl, H_2SO_4; i. HNO_3
„ , pent-, Nb_2O_5	1480	—	4470	i.	i.	s. HF, alk.
Nitrogen						
Chloride, tri-, NCl_3	−40	71, exp.	1653	i.	d.	s. chl., bz., CCl_4, CS_2
Oxide, (ous), N_2O	−90.86	−88.48	—	1.3/0(A)	0.57/25(A)	s. al., eth., H_2SO_4
„ , (ic), NO	−163.65	−151.77	—	$7.1 \times 10^{-2}/0(A)$	$2.9 \times 10^{-2}/60(A)$	s. aq. $FeSO_4$, CS_2; sl. s. H_2SO_4
„ , tri-, N_2O_3	−102	d. 3.5	1447/2	s.	d.	s. alk., a., eth.
„ , di-, NO_2	−11.2	21.1	1449/20	s. d.	—	s. alk., CS_2, chl.
„ , penta-, N_2O_5	30	32.4	1642/18	s.	d.	s. chl.
Oxychloride, NOCl	−64.5	−6.4	—	d.	d.	s. fuming H_2SO_4
Hydrazine, N_2H_4	1.5	113.5	1.01/15	∞	∞	s. al.
Hydrazine sulphate, $N_2H_4.H_2SO_4$	254	—	1370	2.9	14.4/80	i. al.
Hydroxylamine, NH_2OH	33.1	56	1200	s.	d.	s. a. al.; v. sl. s. eth.
Hydroxylamine hydrochloride, $NH_2OH.HCl$	151	d.	1670	94.4/25	356/98	s. al.; i. eth.
Osmium						
Oxide, tetr-, OsO_4	39.5	130	4906/22	6.4	s.	v. s. CCl_4; s. al. eth.
Oxygen						
Fluoride, di-, OF_2	−251	−144.8	—	sl. s. d.	—	sl. s. a., alk.
Ozone, O_3	−192.7	−110.51	—	0.49/0(A)	—	s. alk. solns., oils
Palladium						
Chloride, $PdCl_2$	d. 500	d.	4000/18	s.	s.	s. HCl, acet.
Nitrate, $Pd(NO_3)_2$	870	—	—	s. d.	—	s. HNO_3
Oxide, PdO	d. 750	—	8700	i.	i.	i. aq. reg.
Phosphorus						
Bromide, tri-, PBr_3	−40	173	2852/15	d.	d.	s. eth., chl., CCl_4
Chloride, tri-, PCl_3	−112	76	1574/21	d.	d.	s. eth., chl., CS_2, CCl_4

Inorganic compounds (contd)

Compound	$\theta_m/°C$	$\theta_b/°C$	$\rho/(\text{kg m}^{-3})$	Solubility		
				Cold water	Hot water	Other solvents
Phosphorus (contd)						
Chloride, penta-, PCl_5 .	—	sb. 162	1600	d.	d.	s. CCl_4, CS_2
Fluoride, tri-, PF_3 . .	−151.5	−101.5	—	d.	d.	s. al.
,, , penta-, PF_5 . .	−83	−75	—	d.	d.	d. alk.
Oxide, tri-, P_2O_3 . .	23.8	174 (in N_2)	2135/21	d.	d.	s. CS_2, eth., chl.
,, , penta-, P_2O_5 .	—	sb. 358	2390	d.	d.	s. H_2SO_4; i. acet.
Oxychloride, $POCl_3$.	2	105	1675	d.	d.	d. al.
Sulphide, penta-, P_2S_5 .	288	514	2030	d.	d.	s. alk.; sl. s. CS_2
Phosphoric acid, ortho-,						
H_3PO_4	42.4	$\frac{1}{2}H_2O$/213	1834/18	4.2	v. s.	s. al.
Platinum						
Chloride, tetra-, $PtCl_4$.	d. 370 (in Cl_2)	—	4303/25	142/25	572/98	s. acet.; sl. s. al., NH_3; i. eth.
Potassium						
Aluminium sulphate,						
$KAl(SO_4)_2.12H_2O$.	$-9H_2O$/65	$-12H_2O$/200	1757	6.3	v. s.	s. dil. a.; i. al.
Antimonyl tartrate,						
$KSbC_4H_4O_7.\frac{1}{2}H_2O$.	$-\frac{1}{2}H_2O$/100	—	2607	8.1		i. al.
Bromate, $KBrO_3$ . .	d. 370	—	3270/18	6.91	49.9	sl. s. al.
Bromide, KBr . .	730	1383	2750/25	65	104	sl. s. al. eth.
Carbonate, K_2CO_3 .	901	d.	2428	111	156	i. al., acet.
Chlorate, $KClO_3$ . .	356	d. 400	2320	7.3	56.3	sl. s. al., alk.
Chloride, KCl . .	770	1440	1984	34.2	56.3	s. alk.; sl. s. al.
Chromate, K_2CrO_4 . .	968.3	—	2732/18	62.7	80.1	i. al.
Citrate, $K_3C_6H_5O_7.H_2O$	d. 230	—	1980	148	—	sl. s. al.
Cyanide, KCN . . .	622	—	1520/16	71.5/25	122/103	s. meth. al.; sl. s. al.
Dichromate, $K_2Cr_2O_7$.	398	d. 500	2676/25	12.2	97	i. al.
Dihydrogen phosphate,						
KH_2PO_4 . . .	252.6	—	2338	22.4	83.4	i. al.
Ferricyanide,						
$K_3Fe(CN)_6$ . . .	d.	—	1850/25	48.8/25	90.8/99	s. acet.; i. al.
Ferrocyanide,						
$K_4Fe(CN)_6.3H_2O$	$-3H_2O$/70	d.	1850/17	28.2	74.9	s. acet.; i. al.
Fluoride, KF . . .	860	1520	2480	95	150/80	s. HF.; i. al.
Fluoroberyllate, K_2BeF_4	—	—	—	1.54/25	3.48/75	—
Hydrogen phthalate,						
$KHC_8H_4O_4$ . .	—	—	1636	11.4/25	56	—
Hydrogen sulphate						
$KHSO_4$	214	d.	2322	48.6	121.7	i. al.
Hydrogen tartrate,						
$KHC_4H_4O_6$ . .	—	—	1984	0.54	6.9	s. a.; i. al.
Hydroxide, KOH . .	400	1330	2044	112	182	v. s. al.; i. eth.
Iodate, KIO_3 . . .	—	d.	3930/32	8.1	32.3	s. KI.; i. al.
Iodide, KI	680	1340	3130	144	208	v. s. al.; s. NH_3; sl. s. eth.
Metabisulphite,						
$K_2S_2O_5$	d. 190	—	2340	44.5	125/94	sl. s. al.
Nitrate, KNO_3 . . .	334	d. 400	2109/16	31.6	245	i. al., eth.
Nitrite, KNO_2 . . .	440	d.	1915	307	405	v. s NH_3; sl. s. al.
Oxalate, $K_2C_2O_4.H_2O$	$-H_2O$/100	—	2127/4	35.7	80.2	
Oxide, mon-, K_2O . .	d. 350	—	2320/0	v. s. d.	v. s. d.	s. al., eth.
,, , super-, KO_2 . .	380	d.	2140	v. s. d.	—	d. al.
Periodate, KIO_4 . .	−O/300	—	3618/15	0.5/25	6.8/97	v. sl. s. KOH
Permanganate, $KMnO_4$	d. <240	—	2703	6.3	25.3/65	v. sl. s. acet.; s.H_2SO_4
Persulphate, $K_2S_2O_8$.	d. <100	—	2477	4.7		i. al.
Sulphate, K_2SO_4 . .	1069	d. 2290	2662	11.1	24.1	i. al., acet.
Thiocyanate, $KCNS$.	173	d. 500	1886/14	217	687	s. al., acet.
Praseodymium						
Chloride, $PrCl_3$ . . .	786	1700	4020/25	98.6/26	142/80	s. al.; i. eth., chl.
Sulphate,						
$Pr_2(SO_4)_3.8H_2O$.	—	—	2827/13	12.6	3.5/80	—

Inorganic compounds (*contd*)

Compound	$\theta_m/°C$	$\theta_b/°C$	$\rho/(kg\ m^{-3})$	Solubility		
				Cold water	Hot water	Other solvents
Radium						
Bromide, $RaBr_2$ . .	728	sb. 900	5790	70.6	s.	s. al.
Sulphate, $RaSO_4$.	—	—	—	2×10^{-4}	—	i. al.
Rhenium						
Fluoride, hexa-, ReF_6	19	48	6157	s. d.	s. d.	—
Oxide, hepta-, Re_2O_7	297	sb. 250	6103	v. s.	s. d.	v. s. al.; s. a., alk.
Rhodium						
Oxide, Rh_2O_3 . .	d. 1100–1150	—	8200	i.	i.	i. aq. reg., alk.
Rubidium						
Carbonate, Rb_2CO_3 .	d. 740	—	—	450	s.	sl. s. abs. al.
Chloride, RbCl . . .	715	1390	2800	91.1	138.9	sl. s. al., NH_3
Fluoride, RbF . . .	760	1410	3557	130/18	—	s. dil. HF; i. al. eth.
Nitrate, $RbNO_3$. . .	310	—	3110	53.3	452	v. s. HNO_3; s. acet.
Sulphate, Rb_2SO_4 . .	1074	—	3613	48.2	81.8	sl. s. NH_3; i. acet.
Ruthenium						
Chloride, tri-, $RuCl_3$.	>d. 500	—	3110	i.	d.	sl. s. abs. al.; i. a.
Oxide, di-, RuO_2 .	d.	—	6970	i.	i.	i. a.
,, , tetr-, RuO_4 . .	26	—	3290/21	2.0	2.2/74	s. a., alk., al.
Samarium						
Chloride, $SmCl_3$ . .	678	d.	4460/18	93.4	100/50	v. s. al.; s. pyr.
Oxide, Sm_2O_3 . .	—	—	8347	i.	i.	v. s. a., hot HF
Sulphate,						
$Sm_2(SO_4)_3 . 8H_2O$.	$-8H_2O/450$	—	2930	2.1	2.0/40	—
Scandium						
Chloride, $ScCl_3$ . . .	—	sb. 800–850	2390/25	76/0	v. s.	i. al.
Oxide, Sc_2O_3 . . .	2485	—	3864		i.	s. h. a.; sl. s. c. a.
Selenium						
Chloride, mono-, Se_2Cl_2 .	−85	d. 130	2770/25	d.	d.	s. CS_2, CCl_4, bz.
,, , tetra-, $SeCl_4$.	—	d. 192	—	d.	d.	—
Oxide, di-, SeO_2 . .	—	sb. 316	3950/15	39.4/16	s.	s. al., acet., bz., ac. a.
Oxychloride, $SeOCl_2$.	8.5	168	2420/22	d.	d.	s. CS_2, CCl_4, bz.
Selenic acid, H_2SeO_4 .	60	d. 260	2950	v. s.	v. s.	s. H_2SO_4; i. NH_3
Selenous acid, H_2SeO_3 .	d. 70/$-H_2O$	—	3004/15	167	385/90	v. s. al.
Silicon						
Chloride, $SiCl_4$ . . .	−69	57.6	1483	d.	d.	d. al.
Fluoride, SiF_4 . . .	—	sb. −95	—	d.	d.	s. abs. al., HF
Hydride, SiH_4 . . .	−185	−112	—	i.	—	i. al.; d. KOH
Oxide (quartz), SiO_2 .	1726	2230	2650	i.	i.	s. HF; v. sl. s. alk.
Silver						
Bromide, AgBr . . .	432	d. >1300	6473/25	8.4×10^{-6}	3.7×10^{-4}	s. KCN, $Na_2S_2O_3$
Chloride, AgCl . . .	455	1550	5560	1.5×10^{-4}	2×10^{-3}	s. NH_4OH, KCN, $Na_2S_2O_3$
Chromate, Ag_2CrO_4 .	—	—	5625	2.6×10^{-3}	6.4×10^{-2}	s. NH_4OH, KCN
Cyanide, AgCN . . .	d. 320	—	3950	2×10^{-5}	—	s. NH_4OH, KCN, HNO_3
Fluoride, AgF . . .	435	1160	5852/16	172	205	sl. s. NH_4OH
Iodide, AgI . . .	558	1506	6010/15	$2.6 \times 10^{-7}/25$	$2.5 \times 10^{-6}/60$	s. KCN; sl. s. NH_4OH
Nitrate, $AgNO_3$ . . .	212	d. 444	4352/19	218	733	s. eth.; v. sl. s. abs. al.
Oxide, Ag_2O . . .	d. 230	—	7143/17	1.7×10^{-3}	5.5×10^{-2}	s. a., KCN, NH_4OH; i. al.
Sulphate, Ag_2SO_4 . .	652	d. 1085	5450/29	0.80	1.41	s. a., NH_4OH

Inorganic compounds (*contd*)

Compound	θ_m/°C	θ_b/°C	ρ/(kg m^{-3})	Solubility		
				Cold water	Hot water	Other solvents
Sodium						
Acetate, $NaC_2H_3O_2$.	324	—	1528	46.5	170	sl. s. al.
Aluminium fluoride, Na_3AlF_6	1000	—	2900	0.042/25	0.135	d. alk.; i. HCl
Arsenate, Na_3AsO_4 . $12H_2O$.	86.3	—	1760	s.	—	s. al.
Borate, tetra-, $Na_2B_4O_7$. $10H_2O$.	$-8H_2O$/60	$-10H_2O$/320	1730	2.64	39.3	i. al.
Borate, tetra-, $Na_2B_4O_7$	741	d. 1575	2367	2.64	39.3	i. al.
Bromide, NaBr . . .	755	1390	3203/25	90.8	118.8	sl. s. al.
Carbonate, Na_2CO_3 . $10H_2O$.	$-H_2O$/34	—	1440/15	22.2	44.7	i. al.
Carbonate, Na_2CO_3	851	d.	2532	22.2	44.7	sl. s. abs. al.; i. acet.
Chloride, NaCl . . .	801	1465	2165/25	35.8	39.4	sl. s. al.
Citrate, $Na_3C_6H_5O_7$. $2H_2O$.	$-2H_2O$/150	—	—	30.7/25	—	i. al.
Fluoride, NaF . . .	996	1790	2558/41	4.06	5.08	s. HF; v. sl. s. al.
Hydrogen carbonate, $NaHCO_3$	$-CO_2$/270	—	2159	9.5	14.2	sl. s. al.
Hydrogen sulphate, $NaHSO_4$	>315	d.	2435/13	28.7/25	104	sl. s. al.
Hydroxide, NaOH .	318	1375	2130	109	313/80	v. s. al.; i. acet.
Iodide, NaI . . .	651	1304	3667/25	179	302	v. s. al.
Molybdate, Na_2MoO_4 .	687	—	3280/18	65	84	—
Nitrate, $NaNO_3$ . . .	307	d. 380	2261	87	177	v. s. NH_3; sl. s. al.
Nitrite, $NaNO_2$ . . .	271	d. 320	2168/0	62	160	s. meth. al.; sl. s. al., eth.
Nitroprusside, $Na_2Fe(CN)_5NO$. $2H_2O$	—	—	1720	33.5/16	—	s. al.
Oxalate, $Na_2C_2O_4$ . .	d. 250–270	—	2340	3.4	6.5	i. al., eth.
Oxide, mon-, Na_2O . .	1130	1950	2270	d.	d.	d. al.
,, per-, Na_2O_2 .	d. 460	—	2805	s.	d.	d. al.; s. dil. a.
Selenate, Na_2SeO_4 . .	—	—	3213/17	58.5	72.8	—
Sulphate, Na_2SO_4 . .	897	—	2680	19.1	42.2	s. H_2SO_4; i. al.
Sulphide, mono-, Na_2S .	1170	—	1856/14	s.	s.	sl. s. al.; i. eth.
Sulphite, Na_2SO_3 . $7H_2O$.	$-7H_2O$/150	—	1539/15	27.1	26.3	sl. s. al.
Thiosulphate, $Na_2S_2O_3$. $5H_2O$ . .	d. 48	—	1729/17	70.0	191.3	s. NH_3; v. sl. s. al.
Tungstate, Na_2WO_4 .	698	—	4179	73	97	—
Strontium						
Bromide, $SrBr_2$. $6H_2O$.	$-6H_2O$/>180	—	2386/25	102	223	s. al., meth. al., acet.
,, $SrBr_2$ . .	657	2150	4216/24	102	223	s. al.
Carbonate, $SrCO_3$ . .	$-CO_2$/1340	—	3700	8.2×10^{-5}/9	—	s. a.; sl. s. NH_4 salts
Chloride, $SrCl_2$ . . .	873	2060	3052	53	101	v. sl. s. acet. abs. al.
Fluoride, SrF_2 . . .	1480	2490	4240	0.012	—	s. h. HCl; i. HF
Nitrate, $Sr(NO_3)_2$ . .	570	—	2986	85	100/89	s. NH_3; sl. s. acet.
Sulphur						
Chloride, mono-, S_2Cl_2 .	-80	137	1678	d.	d.	s. CS_2, eth., bz.
,, di-, SCl_2 . .	-78	57	1621/15	d.	d.	s. bz., CCl_4; d. al.
Fluoride, hexa-, SF_6 .	-50.2	—	—	v. sl. s	sl. s	s. al., KOH
Oxide, di-, SO_2 . .	-73.2	-10.0	—	s.	s.	s. al., ac., a., H_2SO_4
,, , tri-, SO_3 . .	16.85	43.3	1970	d.	d.	s. al., H_2SO_4, ac. a.
Sulphuric acid, H_2SO_4 .	10.3	33.7	1841	∞	∞	d. al.
Sulphonyl chloride SO_2Cl_2	-54.1	69.1	1667	d.	d.	s. ac. a., bz.
Thionyl chloride, $SOCl_2$	-104.5	75.7	1679	d.	d.	s. bz., chl.; d. a., alk.
Tantalum						
Chloride, $TaCl_5$ . . .	216	234	3680/27	d.	—	s. abs. al.
Oxide, Ta_2O_5 . . .	1780	—	8200	i.	i.	s. fused $KHSO_4$; i. al.

Inorganic compounds (*contd*)

Compound	$\theta_m/°C$	$\theta_b/°C$	$\rho/(kg\ m^{-3})$	Solubility		
				Cold water	Hot water	Other solvents
Tellurium						
Bromide, $TeBr_4$ . .	380	d. 421	4310/15	sl. s. d.	d.	s. eth., a., NaOH
Chloride, $TeCl_4$ . .	224	392	3260/18	s. d.	s. d.	s. dil. HCl, bz., al.; i. CS_2
Oxide, TeO_2 . . .	733	1245	5670	i.	i.	s. HCl, alk.
Terbium						
Sulphate, $Tb_2(SO_4)_3 . 8H_2O$	$-8H_2O/360$	—	—	2.9	2.0/40	—
Thallium						
Chloride, TlCl . . .	430	807	7004/30	0.34	2.5	sl. s. HCl; i. al.
Nitrate, $TlNO_3$ . . .	206	430	—	9.6	414	s. acet.; i. al.
Oxide, Tl_2O . . .	300	$-O/1865$	9520/16	v. s. d.	v. s. d.	s. a., al.
Sulphate, Tl_2SO_4 . .	632	d.	6770	4.9	19.1	—
Thorium						
Chloride, $ThCl_4$ . .	770	d. 928	4590	124/25	v. s.	s. KCl, al., eth.
Oxide, di-, ThO_2 . .	3220	4400	9860	i.	i.	i. alk.
Sulphate, $Th(SO_4)_2 . 9H_2O$ . .	$-9H_2O/400$	—	2770	1.3	0.7/95	s. HCl, HNO_3
Tin						
(IV) Chloride, $SnCl_4$.	-33	114.1	2226	s.	d.	s. CS_2, eth.
(II) ,, , $SnCl_2$.	246	652	3950/25	270/15	—	s. al., eth.
(IV) Oxide, SnO_2 . .	1630	sb. 1800–1900	6950	i.	i.	s. fus. alk.
(II) ,, , SnO .	d. 1080	—	6446/0	i.	i.	sl. s. NH_4Cl
(II) Sulphate, $SnSO_4$.	d.	—	—	18.8/19	18.1	s. H_2SO_4
(IV) Sulphide, SnS_2 .	d. 600	—	4500	i.	—	d. many solvents
(II) ,, , SnS .	882	1230	5220/25	i.	—	d. alk., HCl
Titanium						
Chloride, tetra-, $TiCl_4$.	-25	136.4	1726	d.	d.	s. dil. HCl, al.
Oxide, di-, TiO_2 . .	1843	—	4260	i.	i.	s. H_2SO_4, alk.
Tungsten						
Carbide, WC . . .	2870	6000	15 630/18	i.	i.	s. $HNO_3 + HF$
Chloride, WCl_6 . . .	282	340	3520/25	—	d.	v. s. CS_2; s. al. bz. CCl_4
Oxide, tri-, WO_3 . .	1473	1840	7160	i.	i.	s. h. alk., HF; i. a.
Uranium						
Fluoride, hexa-, UF_6 .	69	sb. 56	4680/21	s.	—	s. CCl_4, chl.; i. CS_2
Oxide, di-, UO_2 . .	2880	—	10 960	i.	i.	s. HNO_3, conc. H_2SO_4
,, , tri-, UO_3 . .	d.	—	7290	i.	i.	s. HNO_3, H_2SO_4
,, , U_3O_8 . . .	d.	—	8300	i.	i.	s. HNO_3, H_2SO_4
Uranyl acetate, $UO_2(C_2H_3O_2)_2 . 2H_2O$	$-2H_2O/110$	—	2893/15	8/17	d.	v. s. al.
Uranyl chloride, UO_2Cl_2	578	d.	—	320/18	v. s.	s. al., eth.
Uranyl nitrate, $UO_2(NO_3)_2 . 6H_2O$.	d. 100	—	2807/13	119	473	v. s. al., eth., ac. a.
Uranyl sulphate, $UO_2SO_4 . 3H_2O$ . .	d. 100	—	3280/16	19/13	22.2/100	s. H_2SO_4
Vanadium						
Chloride, tetra-, VCl_4 .	-28	149	3000/18	s. d.	—	s. abs. al., eth., chl.
Oxide, penta-, V_2O_5 .	670	d. 1690	3357/18	sl. s.	sl. s.	s. a., alk.
Xenon						
Fluoride, di-, XeF_2 . .	120–140	—	4320	d.	d.	v. s. HF
,, , tetra-, XeF_4 .	100	—	4070	d.	d.	s. HF, C_7F_{16}

Inorganic compounds (*contd*)

Compound	$\theta_m/°C$	$\theta_b/°C$	$\rho/(kg\ m^{-3})$	Solubility		
				Cold water	Hot water	Other solvents
Ytterbium						
Chloride, $YbCl_3.6H_2O$	$-6H_2O/180$	—	—	v. s.	v. s.	s. al.
Oxide, Yb_2O_3 . . .	—	—	9170	i.	i.	s. h. dil. a.; i. HF
Sulphate,						
$\quad Yb_2(SO_4)_3.8H_2O$.	—	—	3286	28	4.7	—
Yttrium						
Nitrate, $Y(NO_3)_3.6H_2O$	$-3H_2O/100$	—	2680	135/23	211/67	v. s. al., eth., HNO_3
Oxide, Y_2O_3 . . .	2489	—	5010	i.	i.	s. a.; i. alk., HF
Sulphate,						
$\quad Y_2(SO_4)_3.8H_2O$ . .	$-8H_2O/120$	d. 700	2558	7.5	2/95	v. sl. s. H_2SO_4
Zinc						
Acetate,						
$\quad Zn(C_2H_3O_2)_2.2H_2O$	$-2H_2O/100$	—	1735	34.7/25	—	s. al.
Carbonate, $ZnCO_3$ . .	$-CO_2/300$	—	4398	sl. s.	—	s. a., alk.; i. acet.
Chloride, $ZnCl_2$ . .	283	732	2910/2	368	615	v. s. al., eth.
Nitrate,						
$\quad Zn(NO_3)_2.6H_2O$.	$-6H_2O/105-13$	—	2065/14	139/30	900/70	v. s. al.
Oxide, ZnO	1975	—	5606	i.	—	s. a., alk.
Sulphate,						
$\quad ZnSO_4.7H_2O$ . .	100	$-7H_2O/280$	1957/25	59	60.5	sl. s. al.
Zirconium						
Chloride, tetra-, $ZrCl_4$.	—	sb. 336	2803/15	s.	d.	s. HCl, al., eth.
Oxide, ZrO_2 . . .	2710	—	5600	i.	i.	s. H_2SO_4, HF
Sulphate,						
$\quad Zr(SO_4)_2.4H_2O$ . .	$-3H_2O/135-15$	—	3220/16	79/25	82/40	s. H_2SO_4; i. al.
Zirconyl chloride,						
$\quad ZrOCl_2.8H_2O$ . .	$-6H_2O/150$	$-8H_2O/210$	—	s.	d.	s. al., eth.; sl. s. HCl

E.F.G.H.

2.3 Properties of organic compounds

Formula, melting point, boiling point, density and refractive index

Hydrated forms are indicated thus: Alloxan + $4H_2O$ and the molecules of water are included in the formula.

Melting, θ_m, and boiling, θ_b, points are for a pressure of 101.325 kN m^{-2} unless other conditions are indicated, e.g. 235/2.4 is the boiling point at 2.4 kN m^{-2}. For boiling points at other pressures see section 2.4.4.

Densities and refractive indices (D lines) are for 20 °C unless some other temperature is indicated, e.g. 788/16 is the density at 16 °C; 1.6048/98.8 is the refractive index at 98.8 °C.

Abbreviations: d., decomposes; exp., explodes; sub., sublimes.

Organic compounds

Substance	Formula	Melting point θ_m/°C	Boiling point θ_b/°C	Density $\rho/(\text{kg m}^{-3})$	Refractive index
Acenaphthene	$C_{12}H_{10}$	95.4	277.6	906.4/12	1.6048/98.8
Acetaldehyde	C_2H_4O	−123	20.4	778	1.3311
Acetamide	C_2H_5ON	82	221	1159	1.4274/78
Acetanilide . . .	C_8H_9ON	114	304	1210/4	
Acetic acid . . .	$C_2H_4O_2$	16.7	118	1049	1.3718
Acetic anhydride . .	$C_4H_6O_3$	−73	140	1087/15	1.3904
Acetone	C_3H_6O	−95	56.3	790.0	1.3587
Acetonitrile . . .	C_2H_3N	−43.87	81.56	776.6/25	1.3415/25
Acetophenone . . .	C_8H_8O	20.5	202	1028	1.53423
Acetyl chloride . . .	C_2H_3OCl	−112	51	1105	1.3898
Acetylene	C_2H_2	—	sub. −84	610/−80	
Adipic acid . . .	$C_6H_{10}O_4$	153	—	1366	
Alloxan + $4H_2O$ . . .	$C_4H_{10}O_8N_2$	170 d.			
Allyl alcohol . . .	C_3H_6O	−129	97	855/15	1.4135
4-Aminobenzoic acid . .	$C_7H_7O_2N$	186			
2-Aminopyridine . . .	$C_5H_6N_2$	56	204 sub.		
Aniline	C_6H_7N	−6.0	184.4	1021.8	1.5862
Anilinium hydrochloride .	C_6H_8NCl	198	245	1220	
Anisole	C_7H_8O	−37.5	154	994.0	1.5170
Anthracene	$C_{14}H_{10}$	215.8	341.4	1250	
Anthraquinone . . .	$C_{14}H_8O$	286	379	1419	
Azobenzene . . .	$C_{12}H_{10}N_2$	68	293	1200	
Benzaldehyde . . .	C_7H_6O	−57.1	179	1044.7	1.5455
Benzene	C_6H_6	5.53	80.10	878.9	1.5011
Benzoic acid . . .	$C_7H_6O_2$	122	249	1266/15	1.5397/15
Benzoic anhydride . .	$C_{14}H_{10}O_3$	42	360	1199/15	1.5767/15
Benzoin	$C_{14}H_{12}O_2$	137	343		
Benzonitrile . . .	C_7H_5N	−12.8	191.1	1009/15	1.5282
Benzophenone (α) . .	$C_{13}H_{10}O$	49	306	1085/50	
p-Benzoquinone . . .	$C_6H_4O_2$	115.7	sub.	1318	
Benzoyl chloride . . .	C_7H_5OCl	−1	197	1212	1.5537
Benzoyl peroxide . . .	$C_{14}H_{10}O_4$	108 d.	exp.		
Benzyl alcohol . . .	C_7H_8O	−15.3	205	1049/15	1.5404
Benzyl benzoate . . .	$C_{14}H_{12}O_2$	21	323	1114/18	1.5681/21
Benzyl chloride . . .	C_7H_7Cl	−39	179	1098	1.5415/15
Benzyl cinnamate . . .	$C_{16}H_{14}O_2$	39	350 d.	1109/15	
Biphenyl	$C_{12}H_{10}$	70	255	1180/0	1.5852/79

Organic compounds (*contd*)

Substance	Formula	Melting point $\theta_m/°C$	Boiling point $\theta_b/°C$	Density $\rho/(kg\ m^{-3})$	Refractive index
($\pm$)-Borneol	$C_{10}H_{18}O$	208	sub.	1010	
Bromobenzene . . .	C_6H_5Br	-31	156	1502/15	1.5625/15
Bromoethane	C_2H_5Br	-119	38.4	1456	1.4239
Bromoform	$CHBr_3$	8	150	2888.9	1.5976
Bromomethane . . .	CH_3Br	-93	3.6	1.732/0	
1-Bromonaphthalene . .	$C_{10}H_7Br$	6.2	281	1483.4	1.6580
Butane	C_4H_{10}	-138.4	-0.5		
iso-Butane	C_4H_{10}	-159.6	-11.7		
Butan-1-ol	$C_4H_{10}O$	-88.6	117.7	809.7	1.3993
Butan-2-ol	$C_4H_{10}O$	-114.7	99.6	806.9	1.3972
Butanone	C_4H_8O	-86.4	79.6	804.9	1.3788
Butyric acid . . .	$C_4H_8O_2$	-5	163	958.2	1.3980
iso-Butyric acid . .	$C_4H_8O_2$	-46	155	968.2	1.3930
($\pm$)-Camphene . .	$C_{10}H_{16}$	50	161	879	1.4402/80
(+)-Camphor . . .	$C_{10}H_{16}O$	176	208	992/10	
Carbon disulphide . .	CS_2	-111.9	46.5	1263.2	1.6276
Carbon tetrabromide . .	CBr_4	90	190	2911/99.5	
Carbon tetrachloride . .	CCl_4	-23	76.7	1604/15	1.4607
Catechol	$C_6H_6O_2$	105	245.6	1344	
Chloral hydrate . .	$C_2H_3O_2Cl_3$	51.7	98 d.	1908	
Chloroacetic acid . .	$C_2H_3O_2Cl$	63	190	1390/75	1.4297/65
Chlorobenzene . . .	C_6H_5Cl	-45.6	131.7	1106	1.5248
1-Chlorobutane . . .	C_4H_9Cl	-123	78.4	886.2	1.4021
Chlorodifluoromethane .	$CHClF_2$	-146	-40.8		
1-Chloro-2,3-epoxy propane	C_3H_5OCl	-25.6	118	1180	1.4420/11.6
Chloroethane	C_2H_5Cl	-136	12.3	924/0	1.3790/0
Chloroform	$CHCl_3$	-63.5	61.2	1.498/15	1.4467
Chloromethane . . .	CH_3Cl	-93	-24.0	991/-25	
1-Chloro-2,4,6-trinitrobenzene . . .	$C_6H_2O_6N_3Cl$	83	d.	1797	
Cholesterol	$C_{27}H_{46}O$	148.5	360 d.	1067	
Cineole	$C_{10}H_{18}O$	1.5	176	923.7	1.4575
trans-Cinnamic acid . .	$C_9H_8O_2$	133	300	1247/4	
Cinnamyl alcohol . .	$C_9H_{10}O$	33	258	1044	1.5819
Citric acid	$C_6H_8O_7$	153	d.	1542/18	
o-Cresol	C_7H_8O	31.05	191.00	1135/25	1.5372/40
m-Cresol	C_7H_8O	12.25	202.23	1034.1	1.5406
p-Cresol	C_7H_8O	34.79	201.94	1154/25	1.5316/40
Cumene	C_9H_{12}	-96.5	152.4	8618	1.4915
Cyclohexane . . .	C_6H_{12}	6.5	80.73	778.6	1.4262
Cyclohexanol . . .	$C_6H_{12}O$	25.2	161.1	941.6/30	1.4629/30
Cyclohexanone . . .	$C_6H_{10}O$	-32.1	155.7	951.0/15	1.4510
Cyclohexene . . .	C_6H_{10}	-103.5	83.0	811	1.4466
p-Cymene	$C_{10}H_{14}$	-67.9	177.1	857	1.4909
cis-Decahydronaphthlene	$C_{10}H_{18}$	-43.06	195.8	896.7	1.48098
trans-Decahydro-naphthlene . . .	$C_{10}H_{18}$	-30.4	187.3	869.7	1.46932
1,2-Diaminoethane . .	$C_2H_8N_2$	11.3	117.3	900/15	1.4568
1,2-Dibromoethane . .	$C_2H_4Br_2$	9.8	131.4	2179	1.5387
Dibromomethane . .	CH_2Br_2	-52.7	99	2810/15	1.5420
Dibutyl phthlate . . .	$C_{16}H_{22}O_4$	-35	340	1047	1.4926

Organic compounds (*contd*)

Substance	Formula	Melting point $\theta_m/°C$	Boiling point $\theta_b/°C$	Density $\rho/(kg\ m^{-3})$	Refractive index
Dichlorodifluoromethane	CCl_2F_2	−158	−29.8		
1,2-Dichloroethane . .	$C_2H_4Cl_2$	−35.7	83.5	1252	1.4448
Dichloromethane . . .	CH_2Cl_2	−95.1	39.8	1335/15	1.4242
Diethanolamine . . .	$C_4H_{11}O_2N$	28	268 d.	1097	1.4720/40
Diethylamine	$C_4H_{11}N$	−50	55.5	1083/40	1.3854
Diethyl ether	$C_4H_{10}O$	−116	34.6	707	1.3524
Diethyl oxalate . . .	$C_6H_{10}O_4$	−40.6	185.4	713.4	1.4102
Diethyl sulphate . . .	$C_4H_{10}O_4S$	−24.5	209	1079	1.4010/18
Dimethylamine . . .	C_2H_7N	−96	7.4	1180/18	
N,N-Dimethylaniline . .	$C_8H_{11}N$	2.5	193	956	1.5582
Dimethyl carbonate . .	$C_3H_6O_3$	0.5	90	1069	1.3687
Dimethyl sulphate . .	$C_2H_6O_4S$	−27	188	1335/15	1.3874
1,4-Dioxan . . .	$C_4H_8O_2$	11.8	101.3	1034	1.4224
Diphenylamine . . .	$C_{12}H_{11}N$	54	302	1159	
1,2-Diphenylethane . .	$C_{14}H_{14}$	52	284	995	1.5478/60
Ethane	C_2H_6	−183.3	−88.6		
Ethanediol.	$C_2H_6O_2$	−16	197	1114	1.4318
Ethanethiol	C_2H_6S	−121	35	832/25	1.4351
Ethanol	C_2H_6O	−114	78.3	789.4	1.3614
Ethanolamine	C_2H_7ON	10.5	171	1011/25	1.4539
2-Ethoxyethanol . . .	$C_4H_{10}O_2$		135.1	929	1.4077
Ethyl acetate	$C_4H_8O_2$	−83.6	77.1	900.6	1.3724
Ethyl acetoacetate . .	$C_6H_{10}O_3$	−45	181	1028	1.4192
Ethylamine	C_2H_7N	−80.6	16.6	706/0	
Ethylbenzene	C_8H_{10}	−94.98	136.19	866.96	1.49588
Ethylbenzoate. . . .	$C_9H_{10}O_2$	−35	212	1051/15	1.5057
Ethylene	C_2H_4	−169.2	−103.7		
Ethylene oxide . . .	C_2H_4O	−111.3	10.7	877/7	1.3597/7
Ethyl formate	$C_3H_6O_2$	−79	54.2	917	1.3599
Ethyl nitrate	$C_2H_5O_3N$	−95	87.5	1109	1.3853
Ethyl nitrite	$C_2H_5O_2N$		17	900/15	1.3418/10
Ethyl salicylate . . .	$C_9H_{10}O_3$	1.3	232	1131	1.5296
Eugenol	$C_{10}H_{12}O_2$	10.3	254	1062/25	1.5439/19
Fluorescein	$C_{20}H_{12}O_5$	314 d.			
Fluorobenzene . . .	C_6H_5F	−41.2	84.7	1024	1.4684
Formaldehyde. . . .	CH_2O	−92	−19.1	815/−20	
Formamide	CH_3ON	2.6	210 d.	1133	1.4475
Formic acid	CH_2O_2	8.4	100.6	1220	1.3714
Fructose	$C_6H_{12}O_6$	102 d.		1598	
Fumaric acid	$C_4H_4O_4$	300	sub.	1635	
Furan	C_4H_4O	−86	31.4	937.8	1.4214
Furfuraldehyde . . .	$C_5H_4O_2$	−36.5	162	1160	1.5261
Furfuryl alcohol . . .	$C_5H_6O_2$	−15	170	1129	1.4868
Glucose	$C_6H_{12}O_6$	142		1544/25	
Glycerol	$C_3H_8O_3$	18	290	1261	1.4746
Glycerol trioleate . .	$C_{57}H_{104}O_6$	−4	235/2.4	899/50	1.4561/60
Glycerol tripalmitate .	$C_{51}H_{98}O_6$	65.5	315	875/70	1.4381/80
Glycerol tristearate . .	$C_{57}H_{110}O_6$	72		856/90	1.4385/80
Glycine	$C_2H_5O_2N$	236 d.	289 d.		
Guaiacol	$C_7H_8O_2$	28.2	205	1129/21.4	
Heptane	C_7H_{16}	−90.6	98.4	684	1.3876

Organic compounds (*contd*)

Substance	Formula	Melting point $\theta_m/°C$	Boiling point $\theta_b/°C$	Density $\rho/(kg\ m^{-3})$	Refractive index
Hexachloroethane . .	C_2Cl_6	187 (sealed tube)	sub.	2091	
Hexamine	$C_6H_{12}N_4$	263	sub.		
Hexane	C_6H_{14}	−95.3	68.7	659	1.3749
Hippuric acid . . .	$C_9H_9O_3N$	187	d.	1371	
Indene	C_9H_8	−1.5	182.4	996	1.5764
Iodoethane . . .	C_2H_5I	−111	72.3	1936	1.5133
Iodoform . . .	CHI_3	119	sub. 210 exp.	4008	
Iodomethane . . .	CH_3I	−66.5	42.4	2280	1.5308
Isoprene	C_5H_8	−120	32.6	680	1.4194
Lactic acid . . .	$C_3H_6O_3$	25	d.	1249	1.4414
Lactose + H_2O . . .	$C_{12}H_{24}O_{12}$	(H_2O lost at 130 °C) 230 d.		1525	
Maleic acid . . .	$C_4H_4O_4$	130	d.	1592	
Maleic anhydride . .	$C_4H_2O_3$	60	202		
Malonic acid . . .	$C_3H_4O_4$	135.6	d.	1631/15	
Maltose + H_2O . .	$C_{12}H_{24}O_{12}$	(H_2O lost at 100 °C)		1540	
(−)-Menthol . . .	$C_{10}H_{20}O$	Four forms 44, 35, 33, 31	212	890/15	
Mesitylene	C_9H_{12}	−44.7	164.7	865.2	1.4994
Metaldehyde . . .	$(C_2H_4O)_n$	246.2 (sealed tube)	sub.		
Methane	CH_4	−182.5	−161.5		
Methanol	CH_4O	−97.7	64.7	791.3	1.3284
Methyl acetate . .	$C_3H_6O_2$	−98.1	56.3	934	1.3614
Methylamine . . .	CH_5N	−92.5	−6.3	699/−10.8	
N-Methylaniline . .	C_7H_9N	−57	196	989	1.5684
Methyl anthranilate .	$C_8H_9O_2N$	24	135.5/2.0	1168/18.6	
Methyl benzoate . .	$C_8H_8O_2$	−12.1	199.5	1093/15	1.5170
2-Methylbutane . .	C_5H_{12}	−159.9	27.9	620	1.3537
2-Methyl-butan-1-ol .	$C_5H_{12}O$	−117	129	819	1.4107
2-Methyl-butan-2-ol .	$C_5H_{12}O$	−9	102	810	1.4049
Methyl formate . .	$C_2H_4O_2$	−99	31.5	974	1.3433
2 Methyl furan . .	C_5H_6O		63	916	
Methyl methacrylate .	$C_5H_8O_2$	−48	100	943	1.4146
2-Methyl propan-1-ol .	$C_4H_{10}O$	−108	107.7	802	1.3977
2-Methyl propan-2-ol .	$C_4H_{10}O$	25.8	82.4	789	1.3877
Methyl salicylate . .	$C_8H_8O_3$	−8.6	233	1183	1.5365
Morpholine . . .	C_4H_9ON	−3.1	129	999	1.4542
Naphthalene . . .	$C_{10}H_8$	80.29	218.0	1175	1.5822/100
1-Naphthol . . .	$C_{10}H_8O$	95.8	283	1096/99	1.6206/98.7
2-Naphthol . . .	$C_{10}H_8O$	123	288	1272	1.6011/143
1-Naphthylamine . .	$C_{10}H_9N$	50	301	1120/25	1.6703/51
2-Naphthylamine . .	$C_{10}H_9N$	113	306	1061/98	1.6493/98
(−)-Nicotine . . .	$C_{10}H_{14}N_2$	−79	247	1010	1.5280
Nitrobenzene . . .	$C_6H_5O_2N$	5.8	211	1198/25	1.5530
Nitroethane . . .	$C_2H_5O_2N$	−90	114	1051	1.3919
Nitromethane . . .	CH_3O_2N	−29	101	1138	1.3812
1-Nitropropane . .	$C_3H_7O_2N$	−104	131	1001	1.4016
2-Nitropropane . .	$C_3H_7O_2N$	−91	120	958	1.3944

Organic compounds (*contd*)

Substance	Formula	Melting point $\theta_m/°C$	Boiling point $\theta_b/°C$	Density $\rho/(\text{kg m}^{-3})$	Refractive index
Octane	C_8H_{18}	-56.8	125.7	703	1.3974
Octane-1-ol	$C_8H_{18}O$	-15.0	195	826	1.4295
Oleic acid	$C_{18}H_{34}O_2$	16	360 d.	887/25	1.4599
Oxalic acid	$C_2H_2O_4$	189.5		1900/17	
Palmitic acid	$C_{16}H_{32}O_2$	63	351	853/62	1.4339/60
Paraldehyde	$C_6H_{12}O_3$	12.6	128	994	1.4049
Pentane	C_5H_{12}	-129.7	36.3	626	1.3575
Pentan-1-ol	$C_5H_{12}O$	-78	138	815	1.4100
Pentan-2-ol	$C_5H_{12}O$		119	809	1.4064
Phenanthrene	$C_{14}H_{10}$	99.5	338	1172	1.6567/129
Phenol	C_6H_6O	40.92	181.84	1132/25	1.5425/40.6
Phosgene	$COCl_2$	-104	8.3		
Phthalic acid	$C_8H_6O_4$	d. >191	—	1593	
Phthalic anhydride	$C_8H_4O_3$	131.6	285	1527/4	
Phthalimide	$C_8H_5O_2N$	238	sub.		
2-Picoline	C_6H_7N	-66.70	129.41	944.3	1.50101
3-Picoline	C_6H_7N	-18.14	144.14	956.6	1.50682
4-Picoline	C_6H_7N	3.65	145.36	954.8	1.50584
Picric acid	$C_6H_3O_7N_3$	122.5	exp. >300	1763	
α-Pinene	$C_{10}H_{16}$	-64	156	858	1.4658
Piperidine	$C_5H_{11}N$	-9	106	861	1.4525
Propane	C_3H_8	-187.7	-42.1		
Propan-1-ol	C_3H_8O	-126	97.2	804	1.3856
Propan-2-ol	C_3H_8O	-88	82.3	785	1.3776
Propene	C_3H_6	-185.3	-47.7		
Propyl acetate	$C_5H_{10}O_2$	-92.5	101.5	887	1.3844
Pyridine	C_5H_5N	-41.55	115.26	983.1	1.51015
Pyrogallol	$C_6H_6O_3$	133	309	1453/4	1.561/134
Quinhydrone	$C_{12}H_{10}O_4$	171	sub.	1401	
Quinol	$C_6H_6O_2$	172.5	286	1330	1.630
Quinoline	C_9H_7N	-14.9	237.6	1095	1.6269
iso-Quinoline	C_9H_7N	26.5	241	1098.6	1.6148
Resorcinol	$C_6H_6O_2$	110	276.6	1278	
Salicylic acid	$C_7H_6O_3$	159	255 sub.	1443	
Stearic acid	$C_{18}H_{36}O_2$	71.5	360 d.	941	
Styrene	C_8H_8	-30.6	145 d.	906.2	1.5468
Succinic acid	$C_4H_6O_4$	185	235 gives anhydride	1564/15	
Succinic anhydride	$C_4H_4O_3$	119	261	1234	
Sucrose	$C_{12}H_{22}O_4$	184 d.		1588/15	
meso-Tartaric acid	$C_4H_6O_6$	147		1666	
$(\pm)$-Tartaric acid $+ H_2O$	$C_4H_8O_7$	loses H_2O at 100 °C		1697	
$(+)$-Tartaric acid	$C_4H_6O_6$	170		1760	
$(-)$-Tartaric acid	$C_4H_6O_6$	170		1760	
1,2,3,4-Tetrahydro-naphthalene	$C_{10}H_{12}$	-35.8	207.6	970	1.5414
Thiophen	C_4H_4S	-38.2	84.2	1064.8	1.5287
Thiourea	CH_4N_2S	180	d.	1405	
Thymol	$C_{10}H_{14}O$	51.5	232	969	
Toluene	C_7H_8	-95.0	110.6	868.8	1.4969
o-Toluidine	C_7H_9N	-16	200	99.8	1.5725

Organic compounds (*contd*)

Substance	Formula	Melting point $\theta_m/°C$	Boiling point $\theta_b/°C$	Density $\rho/(kg\ m^{-3})$	Refractive index
m-Toluidine	C_7H_9N	−30	203	987/25	1.5681
p-Toluidine	C_7H_9N	44	201	962/50	1.5535/50
Trichloroethylene . . .	C_2HCl_3	−86	87	1476/15	1.4782
Trichlorofluoromethane .	CCl_3F	−111	23.7	1.494/17	
Triethanolamine . . .	$C_6H_{15}O_3N$	21.6	335	1124	1.4852
2,2,4-Trimethylpentane .	C_8H_{18}	−107.4	99.2	692	1.3915
2,4,6-Trinitrotoluene . .	$C_7H_5O_6N_3$	80.8	exp.	1654	
Triphenylmethane . .	$C_{19}H_{16}$	94	200/1.3		
Urea	CH_4ON_2	132	d.	1323	
Uric acid	$C_5H_4O_3N_4$	d.		1893	
Valeric acid . . .	$C_5H_{10}O_2$	−34	186	939	1.4080
iso-Valeric acid . .	$C_5H_{10}O_2$	−29	177	931/15	1.4063
Vanillin	$C_8H_6O_3$	80	285	1056	
o-Xylene	C_8H_{10}	−25.2	144.5	880.14	1.5055
m-Xylene	C_8H_{10}	−47.9	139.1	864.36	1.4972
p-Xylene	C_8H_{10}	13.3	138.4	860.98	1.4958

E.F.G.H.

2.4 Vapour pressures and boiling points

The plot of the logarithm of vapour pressure (log p) against the reciprocal of absolute temperature $(1/T)$ for a single substance is usually nearly linear and this relation may be used for interpolation and, with caution, for extrapolation. The formula to obtain a value of log p_2 for a temperature T_2 from values of log p_1 and log p_3 for temperatures T_1 and T_3 is

$$\log p_2 = \log p_1 + (\log p_3 - \log p_1)(T_2 - T_1)T_3/(T_3 - T_1)T_2$$

2.4.1 Vapour pressure of water at temperatures between 0 and 360 °C

Pressures are tabulated in kN m^{-2} for values of °C (IPTS-1968); for example at 14 °C the pressure is 1.5983 kN m^{-2} and at 76 °C it is 40.206 kN m^{-2}.

$\theta/°C$	0	1	2	3	4	5	6	7	8	9
0	0.6112	0.6570	0.7059	0.7579	0.8134	0.8723	0.9351	1.0018	1.0726	1.1478
10	1.2277	1.3124	1.4023	1.4975	1.5983	1.7050	1.8180	1.9375	2.0638	2.1973
20	2.3383	2.4871	2.6442	2.8099	2.9845	3.1686	3.3624	3.5666	3.7814	4.0074
30	4.2451	4.4950	4.7575	5.0332	5.3228	5.6265	5.9452	6.2794	6.6297	6.9968

$\theta/°C$	0	2	4	6	8	10	12	14	16	18
40	7.3814	8.2055	9.1077	10.094	11.171	12.345	13.623	15.013	16.522	18.159
60	19.933	21.852	23.925	26.164	28.577	31.177	33.973	36.979	40.206	43.666
80	47.374	51.343	55.587	60.121	64.960	70.119	75.617	81.468	87.692	94.305
100	101.325	108.78	116.67	125.03	133.89	143.25	153.14	163.59	174.61	186.24
120	198.49	211.40	224.97	239.25	254.26	270.03	286.58	303.94	322.15	341.23
140	361.20	382.12	403.99	426.86	450.76	475.72	501.78	528.97	557.32	586.87
160	617.66	649.72	683.10	717.82	753.93	791.46	830.46	870.96	913.01	956.65
180	1001.9	1048.8	1097.5	1147.9	1200.1	1254.2	1310.1	1368.0	1427.8	1489.7
200	1553.6	1619.7	1688.0	1758.4	1831.2	1906.1	1983.6	2063.4	2145.7	2230.5

$\theta/°C$	220	240	260	280	300	320	340	360
Pressure/(kN m^{-2})	2317.9	3344.7	4689.2	6412.7	8583.2	11280	14593	18656

D.A.

2.4.2 Boiling point of water at pressures from 90 to 106.9 kN m^{-2}

The manner of reading the table is indicated by the following examples: for a pressure of 94.3 kN m^{-2} the temperature is 97.999 °C and for a pressure of 94.4 kN m^{-2} the temperature is 98.028 °C.

p/kN m^{-2}	0	0.1	0.2	0.3	0.4	0.5	0.6	0.7	0.8	0.9
90	96.712	96.743	96.773	96.803	96.834	96.864	96.895	96.925	96.955	96.986
91	97.016	97.046	97.076	97.106	97.136	97.167	97.197	97.227	97.257	97.287
92	97.317	97.347	97.377	97.407	97.436	97.466	97.496	97.526	97.556	97.585
93	97.615	97.645	97.674	97.704	97.734	97.763	97.793	97.822	97.852	97.881
94	97.911	97.940	97.969	97.999	98.028	98.057	98.087	98.116	98.145	98.174
95	98.204	98.233	98.262	98.291	98.320	98.349	98.378	98.407	98.436	98.465
96	98.494	98.523	98.552	98.581	98.610	98.638	98.667	98.696	98.725	98.753
97	98.782	98.811	98.839	98.868	98.897	98.925	98.954	98.982	99.011	99.039
98	99.068	99.096	99.124	99.153	99.181	99.209	99.238	99.266	99.294	99.323

Boiling point of water (*contd*)

$p/\text{kN m}^{-2}$	0	0.1	0.2	0.3	0.4	0.5	0.6	0.7	0.8	0.9
99	99.351	99.379	99.407	99.435	99.463	99.491	99.519	99.547	99.576	99.604
100	99.631	99.659	99.687	99.715	99.743	99.771	99.799	99.827	99.854	99.882
101	99.910	99.938	99.965	99.993	100.021	100.048	100.076	100.104	100.131	100.159
102	100.186	100.214	100.241	100.269	100.296	100.323	100.351	100.378	100.406	100.433
103	100.460	100.487	100.515	100.542	100.569	100.596	100.623	100.651	100.678	100.705
104	100.732	100.759	100.786	100.813	100.840	100.867	100.894	100.921	100.948	100.975
105	101.001	101.028	101.055	101.082	101.109	101.135	101.162	101.189	101.216	101.242
106	101.269	101.296	101.322	101.349	101.375	101.402	101.428	101.455	101.481	101.508

D.A.

2.4.3 Vapour pressures of some liquids of low volatility

Element or compound	p/Pa		
	20°C	100°C	150°C
Mercury	0.2	37	370
Dibutyl phthalate	2×10^{-3}	6	200
Tritolyl phosphate	3×10^{-5}	0.25	10
Apiezon oil A	1×10^{-4}	0.4	16
Apiezon oil B	5×10^{-6}	4.6×10^{-2}	2.4
Apiezon oil C	4×10^{-7}	5×10^{-3}	0.26
Silicone fluid DC 702	6.5×10^{-5}	0.13	< 10
Silicone fluid DC 704	1.3×10^{-6}	2.6×10^{-2}	1.3
Silicone fluid DC 705	2.6×10^{-8}	1.3×10^{-3}	< 0.1

D.A.

2.4.4 Vapour pressures from 0.2 to 101.325 kN m^{-2}

Kelvin temperatures are listed for pressures p from 0.2 to 101.325 kN m^{-2}. Values of $\mathrm{d}T/\mathrm{d}p$ in the right-hand column are those applicable at the normal boiling point, i.e. when $p = 101.325$ kN m^{-2}. The letter 's' indicates a solid, 'd' decomposition and 'p' polymerization.

Elements and inorganic compounds

Element or compound	Pressure, $p/(\text{kN m}^{-2})$								$\mathrm{d}T/\mathrm{d}p$ $\text{K}(\text{kN m}^{-2})^{-1}$
	0.2	1	2	5	10	20	50	101.325	
Aluminium	1885	2060	2140	2260	2360	2430	2640	2790	2.1
Aluminium borohydride	s	226	236	251	264	278	299	318	0.28
Aluminium bromide .	356 s	385	400	422	441	463	497	529	0.50
Aluminium chloride .	377 s	394 s	402 s	413 s	422 s	431 s	444 s	454 s	0.14
Aluminium fluoride .	1245 s	1305 s	1335 s	1375 s	1405 s	1440 s	1490 s	1530 s	0.58
Aluminium oxide . .	2460	2630	2710	2820	2910	3000	3140	3260	2
Ammonia	166 s	178.6 s	184.7 s	193.5 s	201.8 s	211.8	226.6	239.8	0.198
Ammonium azide . .	307 s	328 s	338 s	352 s	364 s	376 s	393 s	407 s	0.21
Ammonium bromide .	483 s	519 s	537 s	562 s	584 s	606 s	640 s	669 s	0.42
Ammonium chloride .	445 s	478 s	494 s	517 s	536 s	557 s	587 s	613 s	0.38
Ammonium cyanide .	226 s	242 s	249 s	260 s	269 s	278 s	293 s	305 s	0.18
Ammonium iodide . .	488 s	526 s	544 s	570 s	592 s	615 s	650 s	679 s	0.43
Antimony	1065	1220	1295	1405	1495	1595	1740	1890	1.8
Antimony pentachloride	302	329	343	363	380	d	—	—	—

Elements and inorganic compounds (*contd*)

Element or compound	Pressure, $p/(\text{kN m}^{-2})$								dT/dp $\text{K}(\text{kN m}^{-2})^{-1}$
	0.2	1	2	5	10	20	50	101.325	
Antimony tribromide .	375	408	425	448	468	489	521	548	0.40
Antimony trichloride .	326 s	354 s	367	389	408	430	462	492	0.44
Antimony triiodide .	450 s	488	508	539	565	594	638	674	0.53
Antimony trioxide .	792	911	976	1080	1180	1300	1505	1700	3.0
Argon	55.0 s	60.7 s	63.6 s	67.8 s	71.4 s	75.4 s	81.4 s	87.3	0.093
Arsenic	656 s	701 s	723 s	754 s	780 s	808 s	849 s	883 s	0.50
Arsenic hydride . .	133 s	146 s	152 s	162	171	181	197	211	0.22
Arsenic pentafluoride .	157 s	168 s	173 s	180 s	186 s	193	208	220	0.19
Arsenic tribromide .	322	352	368	390	410	431	464	493	0.43
Arsenic trichloride .	267	291	304	322	337	354	381	404	0.34
Arsenic trifluoride . .	s	s	s	270	281	294	313	329	0.24
Arsenic trioxide .	493 s	526 s	543 s	570 s	592	627	682	730	0.73
Barium	1165	1290	1360	1455	1540	1635	1780	1910	2.0
Beryllium chloride .	584 s	618 s	634 s	656 s	674 s	696	728	754	0.39
Bismuth	1225	1350	1410	1505	1580	1670	1800	1920	1.8
Bismuth tribromide .	503	546	567	598	623	651	694	734	0.62
Bismuth trichloride .	478	528	550	581	606	633	674	714	0.65
Boron tribromide . .	237	259	270	286	300	316	341	365	0.37
Boron trichloride . .	186	203	211	224	236	248	268	286	0.27
Boron trifluoride . .	120 s	130 s	134 s	141 s	147	154	164	172	0.13
Bromine	227 s	244.1 s	252.1 s	263.6 s	275.7	290.0	312.0	332.0	0.301
Bromine pentafluoride .	s	227	236	251	263	276	296	314	0.27
Cadmium	683	748	781	829	870	915	983	1045	0.90
Cadmium chloride .	s	913	951	1006	1050	1100	1175	1240	1.0
Cadmium fluoride . .	1410	1525	1580	1665	1735	1810	1925	2030	1.5
Cadmium iodide . .	704	772	805	854	895	941	1010	1070	0.89
Cadmium oxide . .	1300	1400	1455	1530	1590	1655	1750	1835	1.2
Caesium	559	624	657	708	752	802	883	959	1.2
Caesium bromide . .	1045	1140	1190	1260	1320	1390	1490	1575	1.3
Caesium chloride . .	1040	1140	1185	1260	1315	1385	1485	1575	1.4
Caesium fluoride . .	1005	1100	1145	1215	1275	1340	1440	1525	1.3
Caesium iodide . .	1035	1130	1175	1245	1305	1370	1470	1555	1.3
Calcium	1105	1220	1280	1365	1440	1525	1650	1765	1.7
Carbon dioxide . .	140 s	151.2 s	156.4 s	163.8 s	170.0 s	176.7 s	186.4 s	194.7 s	0.120
Carbon disulphide .	204	224.2	234.4	249.5	262.4	276.9	299.2	319.6	0.310
Carbon monoxide . .	52 s	57 s	59.3 s	62.8 s	66.0 s	69.7	76.0	81.7	0.086
Carbon selenosulphide	231	253	264	280	294	310	335	358	0.34
Carbonyl chloride (phosgene) . .	184	201	209	222	233	245	264	281	0.26
Carbonyl selenide . .	158	175	183	196	206	218	236	252	0.22
Carbonyl sulphide . .	144	157	164	174	183	193	209	223	0.21
Chlorine	158 s	170 s	176.3	187.4	197.0	207.8	224.3	239.2	0.22
Chlorine dioxide . .	s	s	218	229	239	250	267	284	0.25
Chlorine fluoride . .	s	129	134	143	150	159	171	183	0.17
Chlorine monoxide .	182	198	206	219	229	241	259	275	0.24
Chlorine trifluoride .	193	209	218	230	240	252	269	285	0.23
Chromium	2020	2180	2260	2370	2470	2580	2730	2870	2.0
Chromium carbonyl .	314	337	348	363	376	389	408	424	0.23
Chromium oxychloride	260	282	294	310	325	341	366	390	0.37
Cobalt (II) chloride .	945 s	1000 s	1030	1085	1135	1185	1260	1330	1.0
Copper	1940	2110	2200	2320	2420	2540	2710	2860	2.2
Copper (I) bromide .	867	970	1025	1110	1190	1285	1450	1630	3.1
Copper (I) chloride .	842	952	1010	1105	1190	1300	1500	1765	5.5
Copper (I) iodide . .	s	910	965	1055	1140	1240	1420	1610	3.0
Cyanogen	180 s	194 s	201 s	210 s	218 s	227 s	240 s	252	0.22
Cyanogen bromide .	240 s	259 s	268 s	280 s	291 s	302 s	318 s	332 s	0.20
Cyanogen chloride . .	200 s	216 s	224 s	235 s	244 s	254 s	269	288	0.17

Elements and inorganic compounds (*contd*)

Element or compound	Pressure, $p/(\text{kN m}^{-2})$								dT/dp $\text{K}(\text{kN m}^{-2})^{-1}$
	0.2	1	2	5	10	20	50	101.325	
Cyanogen fluoride . .	140 s	152 s	158 s	166 s	173 s	180 s	191 s	200 s	0.14
Cyanogen iodide . .	304 s	326 s	337 s	352 s	364 s	378 s	397 s	413 s	0.23
Deuterium oxide . .	s	282.8	293.2	308.4	321.1	335.2	356.1	374.6	0.273
Dinitrogen oxide . .	132 s	142.1 s	147.0 s	154.2 s	160.2 s	166.6 s	176.0 s	184.7	0.16
Fluorine	53	58	61	65.3	68.9	73.0	79.3	85.0	0.086
Fluorine monoxide .	80	89	93	99	104	110	120	128	0.13
Gallium	1665	1820	1900	2000	2100	2200	2350	2480	1.9
Gallium trichloride .	320 s	346 s	358	379	397	417	447	473	0.38
Germanium bromide .	298 s	324	338	360	378	399	432	462	0.46
Germanium chloride .	233	254	264	280	294	310	335	357	0.34
Germanium hydride .	113	125	132	141	149	158	172	184	0.18
Gold	2100	2290	2390	2520	2640	2770	2960	3120	3.0
Helium	1.3	1.66	1.85	2.17	2.48	2.87	3.54	4.22	0.010
Hydrogen	10 s	11.4 s	12.2 s	13.4 s	14.5	16.0	18.2	20.3	0.033
Hydrogen bromide .	137 s	149.3 s	155.3 s	164.1 s	171.6 s	179.9 s	193.3	206.4	0.20
Hydrogen chloride .	123 s	134.8 s	140.4 s	148.6 s	155.5 s	163.6	176.5	188.1	0.17
Hydrogen cyanide .	206 s	222 s	230 s	242 s	251 s	260.8	276.1	289.2	0.19
Hydrogen fluoride . .	183 s	202.0	212.1	226.9	239.4	253.3	274.2	292.7	0.28
Hydrogen iodide .	159 s	172.4 s	179.2 s	189.3 s	197.9 s	207.6 s	222.2 s	237.6	0.24
Hydrogen peroxide .	295	318.5	330.3	347.6	362.2	378.2	402.3	423.3	0.31
Hydrogen selenide .	158 s	170 s	177 s	186 s	194 s	203 s	218	231	0.20
Hydrogen sulphide .	142 s	154.3 s	160.4 s	169.5 s	177.2 s	185.7 s	199.8	212.8	0.20
Hydrogen telluride .	180 s	195 s	203 s	213 s	227	238	255	271	0.24
Hydroxylamine . .	s	317	325	337	346	357	371	383	0.17
Iodine	318 s	341.8 s	353.1 s	369.3 s	382.7 s	400.8	430.6	457.5	0.40
Iodine heptafluoride .	188 s	205 s	214 s	226 s	236 s	247 s	264 s	279 s	0.22
Iodine pentafluoride .	263 s	280 s	289	305	318	333	356	376	0.30
Iron	2110	2290	2380	2500	2610	2720	2890	3030	2.1
Iron (III) chloride .	475 s	503 s	514 s	529 s	541 s	553 s	570 s	584 s	0.22
Iron (II) chloride .	s	959	995	1048	1095	1148	1229	1300	1.1
Iron pentacarbonyl .	s	273	285	302	317	334	358	378	0.30
Krypton	77 s	84.3 s	88.1 s	93.8 s	98.6 s	103.9 s	112.0 s	119.7	0.127
Lead	1285	1420	1485	1585	1670	1760	1905	2030	1.9
Lead bromide . . .	801	870	904	955	999	1050	1125	1190	1.0
Lead chloride . . .	836	908	943	994	1040	1090	1160	1230	1.0
Lead fluoride . . .	s	1160	1205	1270	1330	1390	1485	1570	1.2
Lead iodide . . .	766	832	864	913	955	1000	1075	1145	1.1
Lead oxide . . .	1240	1340	1390	1460	1515	1580	1670	1750	1.1
Lead sulphide . . .	1145 s	1230 s	1265 s	1320 s	1365 s	1410	1480	1540	0.9
Lithium	1040	1145	1200	1275	1345	1415	1530	1630	1.5
Lithium bromide .	1045	1140	1190	1260	1325	1390	1495	1585	1.4
Lithium chloride .	1080	1185	1235	1310	1375	1445	1555	1660	1.6
Lithium fluoride .	1345	1465	1520	1600	1670	1745	1860	1955	1.5
Lithium iodide . . .	1015	1100	1140	1200	1245	1300	1380	1445	1.0
Magnesium . . .	891	979	1025	1085	1140	1205	1295	1375	1.2
Magnesium chloride .	1075	1185	1235	1315	1385	1465	1585	1695	1.7
Manganese	1560	1710	1780	1890	1980	2080	2240	2390	2.3
Manganese (II) chloride	1010	1100	1145	1210	1265	1330	1420	1505	1.2
Mercury	408	448.8	468.8	498.4	523.4	551.2	592.9	629.8	0.555
Mercury (II) bromide .	417 s	448 s	463 s	485 s	502 s	524	561	593	0.49
Mercury (II) chloride .	415	446	461	482	500	519	546	570	0.35
Mercury (II) iodide .	437 s	469 s	485 s	509 s	525	551	591	627	0.54
Molybdenum . . .	3420	3720	3860	4060	4220	4410	4680	4900	4.0
Molybdenum hexafluoride . . .	212 s	229 s	237 s	250 s	260 s	272 s	289 s	309	0.21
Molybdenum trioxide .	1020 s	1075 s	1095	1155	1205	1265	1350	1430	1.2
Neon	16.3 s	18.1 s	19.0 s	20.4 s	21.6 s	22.9 s	24.9	27.1	0.033

Elements and inorganic compounds (*contd*)

Element or compound	Pressure, p/(kN m^{-2})								dT/dp
	0.2	1	2	5	10	20	50	101.325	K(kN m^{-2})$^{-1}$
Nickel	2230	2410	2500	2630	2740	2860	3030	3180	2.2
Nickel chloride. . .	972	1030	1060	1095	1125	1160	1205	1245	0.56
Nickel carbonyl . .	s	s	s	249	261	276	297	316	0.26
Nitrogen	48.1 s	53.0 s	55.4 s	59.0 s	62.1 s	65.8	71.8	77.4	0.084
Nitrogen dioxide . .	218 s	232.1 s	238.9 s	248.6 s	256.6 s	266.6	285.4	302.2	0.25
Nitrogen oxide . . .	87 s	93.8 s	96.9 s	100.1	104.3	108.9	115.6	121.4	0.085
Nitrogen trifluoride .	91	100.4	104.9	111.8	117.7	124.4	134.9	144.1	0.14
Nitrogen pentoxide .	239 s	253 s	260 s	269 s	277 s	285 s	297 s	307 s	0.14
Nitrosyl chloride . .	179 s	194.4 s	202 s	213	223	235	252	268	0.23
Nitrosyl fluoride . .	146	157	163	171	179	187.5	200.9	213.2	0.19
Nitryl fluoride . . .	s	145	151	160	167	176	189	201	0.18
Osmium tetroxide (white) . . .	273 s	294 s	305 s						
Osmium tetroxide (yellow) . . .	281 s	301 s	310 s	321	336	354	380	402	0.34
Oxygen	55.4	61.3	64.3	68.8	72.7	77.1	83.9	90.2	0.094
Ozone	105	114.8	119.7	127.1	133.5	140.7	151.8	161.8	0.15
Phosphonium bromide .	233 s	249 s	256 s	268 s	277 s	286 s	300 s	312 d	0.16
Phosphonium chloride .	185 s	197 s	203 s	211 s	218 s	226 s	237 s	246 s	0.13
Phosphonium iodide .	252 s	269 s	277 s	289 s	298 s	309 s	323 s	335 s	0.17
Phosphorus (yellow) .	365	398	414	439	460	484	520.5	553.6	0.50
Phosphorus (violet) .	542 s	576 s	592 s	614 s	633 s	652 s	680 s	704 s	0.34
Phosphorus (black) .	572 s	605 s	620 s	642 s	660 s	678 s	705 s	727 s	0.32
Phosphorus hydride .	117 s	128.2 s	133.9 s	142.6	150.4	159.3	172.9	185.4	0.19
Phosphorus oxychloride	s	s	281.1	298.3	313.1	329.8	355.4	378.7	0.35
Phosphorus pentachloride . .	326 s	347 s	358 s	372 s	384.6 s	397.7 s	416.5 s	432.4 s	0.23
Phosphorus pentoxide (stable) form. .	662 s	705 s	725 s	753 s	776 s	801 s	837 s	878	0.80
Phosphorus thiobromide	328	352	364	380	394	408	429	447	0.25
Phosphorus thiochloride	260	284	297	315	330	347	374	397	0.35
Phosphorus tribromide	288	316	329	350	368	388	418	446	0.42
Phosphorus trichloride	226	247	258	274	288	303	327	349	0.33
Phosphorus trioxide .	296	322	335	355	372	391	421	448	0.41
Platinum	2910	3150	3260	3420	3560	3700	3910	4090	2.6
Potassium	633	704	740	794	841	894	975	1050	1.1
Potassium bromide .	1100	1220	1270	1335	1390	1455	1550	1660	2.0
Potassium chloride. .	1120	1220	1270	1350	1410	1480	1590	1680	1.4
Potassium fluoride .	1185	1290	1345	1425	1490	1565	1675	1775	1.5
Potassium hydroxide .	1015	1115	1190	1245	1310	1385	1500	1600	1.6
Potassium iodide .	1040	1140	1190	1265	1325	1395	1505	1600	1.4
Radon	139 s	152 s	158 s	168 s	176 s	184 s	198 s	211	0.2
Rhenium heptoxide .	494	518	529	545	557	570	589	604	0.2
Rubidium	583	649	684	735	779	829	907	978	1.1
Rubidium bromide .	1075	1175	1225	1300	1360	1430	1535	1625	1.4
Rubidium chloride .	1090	1190	1240	1315	1380	1450	1560	1655	1.4
Rubidium fluoride .	1210	1275	1310	1365	1420	1480	1590	1685	1.3
Rubidium iodide .	1045	1140	1190	1260	1320	1385	1490	1580	1.3
Selenium	636	695	724	767	803	844	904	958	0.80
Selenium dioxide . .	469 s	498 s	512 s	532 s	548 s	565 s	590 s	610 s	0.30
Selenium hexafluoride .	160 s	172 s	179 s	188 s	195 s	204 s	216 s	229 s	0.16
Selenium oxychloride .	321	342	352	368.5	383	399	425	449	0.36
Selenium tetrachloride	353 s	376 s	386 s	402 s	414 s	428 s	448 d	464 d	0.24
Silane	96	107.8	113.6	122.1	129.4	137.7	150.3	161.7	0.17
Silicon	2380	2590	2690	2840	2970	3100	3310	3490	2.7
Silicon dioxide . .	1850	1985	2050	2140	2210	2290	2410	2500	1.3
Silicon tetrachloride .	214	234.5	244.7	259.8	272.8	287.4	309.8	329.9	0.30

Elements and inorganic compounds (*contd*)

Element or compound	Pressure, p/(kN m^{-2})								dT/dp
	0.2	1	2	5	10	20	50	101.325	K(kN m^{-2})$^{-1}$
Silicon tetrafluoride .	131 s	141 s	145 s	152 s	158 s	163 s	172 s	178 s	0.10
Silver	1640	1790	1865	1970	2060	2160	2310	2440	1.9
Silver chloride . . .	1210	1325	1385	1470	1540	1620	1735	1840	1.5
Silver iodide . . .	1120	1235	1290	1380	1450	1535	1665	1780	1.8
Sodium	729	807	846	904	954	1010	1095	1175	1.2
Sodium bromide . .	1105	1205	1255	1330	1395	1465	1570	1665	1.5
Sodium chloride . .	1165	1270	1325	1400	1465	1540	1645	1740	1.4
Sodium cyanide . .	1115	1235	1295	1380	1455	1540	1665	1770	1.6
Sodium fluoride . .	1375	1495	1550	1630	1700	1775	1885	1980	1.4
Sodium hydroxide . .	1035	1150	1205	1285	1355	1435	1550	1655	1.5
Sodium iodide . . .	1060	1160	1205	1275	1330	1395	1495	1580	1.3
Strontium	1040 s	1150	1205	1285	1355	1430	1550	1660	1.6
Sulphur	462	505.7	528.0	561.3	590.1	622.5	672.4	717.8	0.68
Sulphur dioxide . .	180 s	193.1 s	199.3 s	211.2	221.0	231.9	248.4	263.2	0.22
Sulphur hexafluoride .	144 s	156.6 s	162.6 s	171.4 s	178.8 s	186.9 s	198.8 s	209.2 s	0.15
Sulphuric acid . . .	427	461	477	501	521	543	576	604 d	0.41
Sulphur monochloride	270	294	306	325	340	358	386	411	0.38
Sulphur trioxide (α) .	237 s	252.8 s	260.3 s	271.0 s	279.8 s	289.2 s	304.6	317.9	0.20
Sulphur trioxide (β) .	243 s	257.9 s	265.1 s	275.3 s	283.5 s	292.3 s	304.8 s	317.9	0.20
Sulphur trioxide (γ) .	262 s	276.1 s	282.4 s	291.3 s	298.4 s	305.8 s	316.2 s	324.7 s	0.12
Sulphuryl chloride . .	226	246	257	272	285	300	322	342	0.30
Tantalum fluoride . .	s	s	s	s	s	420	464	503	0.59
Tellurium	806	888	929	990	1040	1100	1190	1270	1.2
Tellurium tetrachloride	s	510	527	552	572	596	631	662	0.46
Tellurium hexafluoride	164	178	184	194	202	211	224	235	0.17
Thallium	1125	1235	1290	1370	1440	1515	1630	1730	1.5
Thallium bromide . .	s	782	816	867	909	955	1025	1095	1.0
Thallium chloride . .	713	777	810	858	899	946	1015	1080	1.0
Thallium iodide . .	726	793	826	875	917	964	1035	1095	0.94
Thionyl bromide . .	277	301	313	331	346	363	390	413	0.36
Thionyl chloride . .	224	246	257	274	288	303.5	327.4	348.8	0.32
Tin	1930	2120	2210	2350	2470	2600	2800	2990	2.8
Tin (IV) bromide . .	309	340	355	377	396	416	448	478	0.45
Tin (IV) chloride . .	256	278	290	307	322	338	363	386	0.35
Tin (II) chloride . .	604	660	688	730	765	805	866	922	0.84
Tin (IV) hydride . .	136	151	159	170	180	190	207	221	0.21
Tin (IV) iodide . .	403 s	446	466	497	523	552	595	634	0.57
Titanium tetrachloride	265	289	302	320	337	355	384	409	0.38
Tungsten	4300	4630	4790	5020	5200	5400	5690	5940	3.5
Tungsten hexafluoride	205 s	220 s	228 s	239 s	248 s	258 s	272 s	291	0.18
Uranium hexafluoride .	238 s	255.8 s	264.3 s	296.6 s	286.9 s	298.1 s	314.6 s	328.8 s	0.21
Vanadyl trichloride .	255	280	293	312	328	347	375	400	0.38
Water	260 s	280.1	290.6	306.0	319.0	333.2	354.5	373.15	0.276
Xenon	107 s	117.3 s	122.5 s	130.1 s	136.6 s	143.8 s	154.7 s	165.0	0.171
Zinc	780	854	891	945	991	1040	1120	1185	1.01
Zinc chloride . . .	714	770	797	836	869	905	958	1006	0.70
Zinc fluoride . . .	1220	1325	1375	1450	1515	1585	1690	1785	1.4
Zirconium tetrabromide	488 s	518 s	532 s	552 s	568 s	584 s	609 s	630 s	0.32
Zirconium tetrachloride	470 s	498 s	511 s	530 s	545 s	562 s	585 s	604 s	0.28
Zirconium tetraiodide .	545 s	579 s	594 s	616 s	634 s	653 s	681 s	704 s	0.34

Organic compounds

Refrigerant numbers (British Standard 4580) are given in parentheses for some halogenated alkanes.

Compound	Pressure, $p/(kN\ m^{-2})$								dT/dp $K(kN\ m^{-2})^{-1}$
	0.2	1.0	2.0	5.0	10	20	50	101.325	
Acenaphthene	s	397.3	414.4	439.7	461.2	485.0	520.3	550.7	0.444
Acetaldehyde	193	210.3	219.2	232.4	243.7	256.5	276.1	293.5	0.259
Acetamide	s	372.8	386.8	407.2	424.3	443.1	470.9	495.2	0.360
Acetic acid	s	s	297.5	314.4	328.9	345.0	369.4	391.0	0.322
Acetic anhydride	280	304.3	316.2	333.9	348.9	365.6	390.7	412.8	0.326
Acetone	217	237.3	247.2	261.9	274.4	288.5	309.9	329.2	0.289
Acetonitrile	s	252.4	263.3	279.6	293.8	309.7	333.8	354.9	0.311
Acetophenone	316	345.2	359.5	380.7	398.7	418.7	448.9	475.6	0.395
Acetylene	131 s	142.4 s	147.9 s	155.8 s	162.4 s	169.7 s	180.2 s	189.3 s	0.134
Adipic acid	440	472.6	488.4	511.3	530.5	551.6	583.0	610.7	0.414
Allyl alcohol	258	279.5	289.9	305.1	317.7	331.5	351.9	369.8	0.264
Aniline	313	337.3	349.7	368.7	385.3	404.2	433.0	457.6	0.354
Anthracene	s	s	s	s	512.8	539.7	580.6	615.2	0.495
Benzaldehyde	301	327.5	340.9	360.7	377.7	396.7	425.8	451.9	0.392
Benzene	240 s	258.1 s	266.8 s	279.5	293.1	308.4	331.9	353.2	0.321
Benzoic acid	s	399.4	413.7	434.4	451.6	470.5	498.3	522.4	0.354
Benzoic anhydride	425	463.6	482.6	510.4	534.0	560.1	599.1	633.2	0.504
Benzoin	417	453.7	472.0	498.8	521.5	546.5	583.8	616.2	0.476
Benzonitrile	307	335.7	350.0	371.0	388.8	408.6	438.1	463.8	0.378
Benzophenone	389	424.1	441.4	466.8	488.3	512.1	547.6	578.6	0.458
Benzoyl chloride	312	340.2	354.4	375.5	393.5	413.5	443.7	470.4	0.396
Benzyl alcohol	336	360.8	373.1	391.6	407.6	425.7	453.4	477.9	0.362
Benzyl chloride	301	328.5	342.1	362.3	379.4	398.5	427.2	452.6	0.375
Benzyl cinnamate	453	486.6	503.0	527.0	546.8	568.1	598.5	623.2	0.352
Biphenyl	351	383.7	400.1	424.3	444.8	467.3	500.3	528.1	0.401
Bromobenzene	282	308.1	321.1	340.2	356.6	375.0	403.3	429.4	0.398
Bromoethane	203	222.2	231.7	245.9	258.0	271.6	292.6	311.5	0.285
Bromoform	s	302.9	315.9	335.0	351.4	369.6	397.3	422.1	0.372
Bromomethane	s	196.5	205.1	217.9	228.8	241.1	259.8	276.7	0.254
1-Bromonaphthalene	365	399.6	416.6	441.7	463.0	486.6	522.4	554.3	0.478
Bromotrifluoromethane	138	151.1	157.9	168.0	176.7	186.6	201.7	215.4	0.207
Butane	174	191.8	200.4	213.2	224.0	236.4	255.4	272.7	0.261
isoButane	167	183.3	191.5	203.8	214.3	226.3	244.7	261.4	0.252
Butan-1-ol	279	299.8	309.9	324.8	337.3	351.3	372.2	390.9	0.278
Butan-2-ol	267	286.7	296.2	310.2	322.0	335.1	354.9	372.7	0.266
Butanone	235	254.4	265.1	280.8	294.2	309.2	332.1	352.7	0.309
But-1-ene	172	188.3	196.6	208.9	219.6	231.6	250.1	266.9	0.253
isoButene	171	187.7	196.1	208.5	219.2	231.2	249.7	266.4	0.252
cis-But-2-ene	179	196.3	204.8	217.5	228.4	240.8	259.7	276.9	0.259
trans-But-2-ene	176	193.3	201.8	214.6	225.5	237.9	256.8	274.0	0.259
Butyric acid	307	330.9	342.7	356.0	374.6	390.8	415.2	436.9	0.324
isoButyric acid	294	319.2	331.6	349.6	364.7	381.2	405.9	427.7	0.326
Camphene	s	s	327.8	347.2	363.8	382.2	409.8	433.7	0.348
(+)-Camphor	320 s	349.9 s	364.5 s	385.9 s	403.8 s	423.4 s	452.4 s	480.7	0.427
Carbon tetrabromide	s	s	s	368.0	385.1	404.3	434.2	462.7	0.447
Carbon tetrachloride	s	s	259.4	275.3	289.0	304.5	328.2	349.8	0.325
Carbon tetrafluoride (R14)	93	101.6	106.2	113.0	118.9	125.6	135.8	145.1	0.140
Catechol	s	385.4	400.3	422.4	441.0	461.6	492.3	518.7	0.387
Chloracetic acid	s	348.7	361.8	380.8	396.8	414.3	440.2	462.7	0.330
Chloral hydrate	s	s	s	s	s	335.6	353.9	369.4 d	0.225
Chlorobenzene	265	289.8	302.1	320.4	336.0	353.6	380.6	404.9	0.365
Chlorodifluoromethane (R22)	152	166.0	173.1	183.6	192.7	202.8	218.3	232.3	0.211
1-Chloro-2,3-epoxypropane	262	285.1	296.8	314.0	328.6	344.9	369.5	391.1	0.319
Chloroethane	185	202.9	211.7	224.8	236.0	248.6	267.9	285.4	0.262

Organic compounds (*contd*)

Compound	Pressure, $p/(\text{kN m}^{-2})$								dT/dp
	0.2	1.0	2.0	5.0	10	20	50	101.325	$\text{K}(\text{kN m}^{-2})^{-1}$
Chloroform	219	239.4	249.5	264.5	277.6	292.0	314.2	334.3	0.302
Chloromethane . . .	s	176.9	184.6	196.0	205.2	216.9	233.7	249.0	0.229
Chlorotrifluoromethane (R13)	122	134.4	140.5	149.5	157.2	166.0	179.5	191.7	0.194
Cineole (Eucalyptol) .	294	321.6	335.4	355.9	373.4	393.0	422.7	449.2	0.394
trans-Cinnamic acid . .	407	439.8	455.7	478.5	497.3	517.7	547.5	573.2	0.379
Cinnamyl alcohol . .	353	384.3	399.9	422.7	442.1	463.4	495.3	523.2	0.413
o-Cresol	319	344.7	357.5	376.6	392.9	411.2	439.1	464.2	0.377
m-Cresol	329	355.4	368.4	387.7	404.1	422.4	450.3	475.4	0.376
p-Cresol	330	355.9	368.8	387.9	404.2	422.4	450.2	475.1	0.374
Cumene	281	306.4	319.0	337.8	354.0	372.2	400.2	425.6	0.381
Cyclohexane . . .	232 s	253.2 s	263.6 s	278.7 s	292.4	308.0	332.1	353.8	0.326
Cyclohexanol . . .	299	324.0	336.4	354.7	370.2	387.2	412.5	434.2	0.317
Cyclohexanone . . .	280	306.6	319.8	339.4	356.2	375.0	403.5	428.8	0.376
Cyclopentane . . .	209	228.8	238.6	253.4	266.1	280.4	302.5	322.4	0.300
Cyclopropane . . .	154	169.0	176.5	187.7	197.4	208.3	225.1	240.3	0.228
p-Cymene . . .	298	324.0	337.5	357.4	374.6	393.8	423.4	450.3	0.404
cis-Decahydro-naphthalene . . .	308	335.5	349.5	370.5	388.6	409.1	440.5	468.9	0.422
trans-Decahydro-naphthalene . . .	301	328.5	342.4	363.1	381.0	401.2	432.3	460.4	0.422
1,2-Diaminoethane . .	s	289.9	301.0	317.3	331.2	346.6	369.8	390.4	0.304
1,2-Dibromoethane . .	s	290.5	302.7	320.9	336.4	353.9	380.5	404.5	0.361
Dibromomethane . .	243	266.0	277.3	294.2	308.4	324.3	348.2	369.5	0.318
Dibutyl phthalate . .	429	464.7	481.9	506.9	527.7	550.5	584.1	613.2	0.429
Dichlorodifluoro-methane (R12) . .	156	171.1	178.8	190.2	200.0	211.0	228.0	243.3	0.232
1,2-Dichloroethane . .	s	256.7	267.5	283.4	297.0	312.3	335.6	356.6	0.293
Dichlorofluoromethane (R21)	184	201.5	210.1	223.0	234.0	246.3	265.1	282.0	0.255
Dichloromethane . .	206	225.3	234.8	248.7	260.7	274.0	294.5	313.0	0.277
1,2-Dichlorotetra-fluoroethane (R114)	180	196.4	204.8	217.5	228.4	240.8	259.8	277.1	0.259
Diethylamine . . .	s	237.6	247.3	261.8	274.2	288.1	309.4	328.6	0.289
Diethyl ether . . .	202	220.3	229.6	243.4	255.3	268.6	289.1	307.6	0.278
Diethyl oxalate . .	326	352.0	364.4	382.4	397.4	413.8	437.9	458.9	0.309
Diethyl sulphate . .	326	354.9	369.1	389.9	407.5	427.1	456.5	482.2 d	0.381
Dimethylamine . . .	190	206.1	214.1	225.9	236.0	247.2	264.5	280.0	0.234
Dimethylaniline . . .	309	337.4	351.7	372.9	390.9	410.8	440.5	466.3	0.377
Dimethyl ether . . .	161	176.3	184.1	195.7	205.6	216.6	233.3	248.4	0.226
2,2-Dimethylpropane (neopentane) . . .	188 s	204.4 s	212.5 s	224.5 s	234.7 s	246.2 s	264.6	282.6	0.273
1,4-Dioxan	s	s	s	297.1	311.7	328.2	352.9	374.2	0.310
Diphenylamine . . .	389	423.8	440.8	465.8	486.8	510.1	544.8	575.2	0.449
1,2-Diphenylethane . .	368	402.3	419.5	444.8	466.3	490.2	525.9	557.2	0.464
Ethane	116	127.8	133.7	142.6	150.3	159.0	172.4	184.6	0.183
Ethanediol . . .	332	358.6	371.6	390.4	406.2	423.4	448.7	470.5	0.319
Ethanethiol . . .	200	219.1	228.4	242.5	254.6	268.3	289.4	308.1	0.280
Ethanol	247	266.8	276.4	290.5	302.3	315.3	334.6	351.4	0.250
Ethyl acetate . . .	235	255.2	265.4	280.5	293.5	308.0	330.2	350.3	0.300
Ethyl acetoacetate . .	308	334.9	348.3	367.9	384.4	402.7	430.0	454.0	0.357
Ethylamine	196	213.0	221.3	233.5	244.0	255.7	273.6	289.8	0.243
Ethylbenzene . . .	270	294.0	306.2	324.5	340.2	357.8	384.8	409.4	0.368
Ethyl benzoate . .	324	353.2	368.0	389.7	408.2	428.7	459.5	486.6	0.401
Ethylene	107	117.6	123.1	131.2	138.2	146.1	158.3	169.4	0.168
Ethylene oxide . . .	187	204.1	212.6	225.2	236.1	248.2	266.9	283.8	0.255
Ethyl formate . . .	217	236.3	246.0	260.5	273.0	287.0	308.4	327.4	0.284

Organic compounds (*contd*)

Compound	Pressure, $p/(\text{kN m}^{-2})$								dT/dp
	0.2	1.0	2.0	5.0	10	20	50	101.325	$\text{K(kN m}^{-2})^{-1}$
Ethyl methyl ether . .	188	204.2	212.3	224.4	234.7	246.4	264.3	280.6	0.246
Ethyl salicylate . . .	341	371.2	386.1	408.2	426.8	447.4	478.0	504.7	0.393
Eugenol	359	389.6	405.1	427.7	446.9	468.0	499.4	526.7	0.401
Fluorobenzene . . .	236	256.8	267.5	283.4	297.2	312.6	336.4	357.9	0.323
Fluoromethane . .	s	138.2	144.2	153.3	161.0	169.7	182.9	194.8	0.179
Formaldehyde . . .	s	182.6	190.2	201.6	211.4	222.4	239.1	254.1	0.223
Formic acid . . .	s	s	s	295.7	310.5	326.9	351.7	373.8	0.330
Furfuryl alcohol . . .	311	336.1	348.7	367.1	382.4	399.1	423.1	443.2	0.289
Glycerol	405	435.1	449.1	469.3	486.2	504.9	534.2	563.2	0.473
Heptane	244	266.6	277.7	294.2	308.5	324.6	349.2	371.6	0.336
Hept-1-ene . . .	241	262.7	273.6	290.0	304.2	320.1	344.6	366.8	0.344
Hexane	223	243.9	254.2	269.7	283.0	297.9	321.0	341.9	0.315
Hex-1-ene	220	239.8	250.0	265.2	278.3	293.2	316.0	336.6	0.311
Indene	303	329.3	342.6	362.5	379.7	399.0	428.8	455.7	0.402
Iodobenzene . . .	303	331.0	345.0	365.9	383.7	403.6	433.9	461.0	0.406
Iodoethane . . .	224	244.5	256.1	271.8	285.2	300.3	323.5	344.5	0.316
Iodomethane . . .	s	223.8	233.6	248.2	260.7	274.7	296.2	315.6	0.291
Isoprene	197	216.0	225.4	239.5	251.8	265.6	286.8	305.7	0.281
Maleic anhydride . .	s	345.4	361.1	383.2	401.3	420.8	449.7	475.2	0.383
(−)-Menthol . . .	336	363.7	377.5	397.8	414.8	433.5	461.2	485.2	0.354
Mesitylene . . .	291	316.7	329.7	348.9	365.5	384.0	412.4	437.9	0.383
Methane	69 s	76.1 s	79.7 s	85.0 s	89.6 s	95.1	103.7	111.6	0.120
Methanol	233	252.8	262.5	276.6	288.4	301.4	320.7	337.7	0.251
Methyl acetate . . .	221	240.0	249.8	264.2	276.4	290.2	311.2	330.6	0.282
Methylamine . . .	s	196.3	204.0	215.4	225.1	235.8	252.2	266.8	0.219
Methylaniline . . .	315	343.6	357.4	377.8	395.0	414.0	442.9	468.7	0.387
Methyl benzoate . .	318	345.2	358.8	379.0	396.3	415.8	445.7	472.7	0.404
2-Methylbutane (*iso*pentane) . . .	194	212.4	221.7	235.6	247.6	261.2	282.1	301.0	0.286
3-Methylbutan-1-ol .	290	309.9	320.0	335.0	347.9	362.5	384.6	404.2	0.289
Methylcyclohexane .	243	265.4	276.8	293.8	308.6	325.2	350.8	374.1	0.351
Methylcyclopentane .	224	245.2	255.7	271.4	284.9	300.2	323.7	345.0	0.321
Methyl formate . . .	203	220.9	230.0	243.5	255.1	268.0	287.7	305.1	0.258
2-Methylpentane . .	216	236.7	246.9	262.2	275.3	290.0	312.8	333.4	0.310
Methyl phenyl ether (anisole) . . .	285	310.4	322.9	341.5	357.3	375.1	402.3	426.8	0.367
2-Methylpropan-1-ol .	273	292.7	302.5	316.9	329.0	342.6	362.9	381.0	0.271
2-Methylpropan-2-ol .	s	s	s	s	307.6	320.0	338.7	355.5	0.251
1-Methyl-4-*iso*propyl-benzene (*p*-cymene) .	298	324.0	337.5	357.4	374.6	393.8	423.4	450.3	0.404
Methyl salicylate . .	334	362.4	376.7	397.8	415.9	436.2	467.6	496.4	0.439
Naphthalene . . .	331 s	353.4	368.2	390.1	408.8	429.9	462.1	491.2	0.437
1-Naphthol . . .	374	407.9	424.5	449.1	469.8	492.7	526.6	555.7	0.426
2-Naphthol . . .	s	410.6	427.6	452.6	473.7	497.0	531.5	561.2	0.436
1-Naphthylamine . .	385	420.2	437.4	462.6	483.9	507.5	542.8	574.0	0.464
2-Naphthylamine . .	389	424.0	441.3	466.8	488.3	512.1	547.8	579.3	0.467
Nicotine	342	373.8	389.8	413.6	434.0	456.6	490.7	520.5	0.439
Nitrobenzene . . .	324	352.2	366.2	387.2	405.1	425.3	456.0	483.8	0.416
Nitroethane . . .	257	281.0	292.7	309.9	324.5	340.7	365.2	387.2	0.328
Nitromethane . . .	249	271.3	282.5	299.2	313.4	329.4	353.4	374.4	0.309
1-Nitropropane . .	269	293.4	305.6	323.5	338.8	355.8	381.6	404.8	0.347
2-Nitropropane . .	260	283.9	295.9	313.6	328.7	345.4	370.8	393.5	0.338
Octafluorocyclobutane	178	193.2	200.9	212.5	222.5	233.9	251.3	267.2	0.239
Octane.	264	287.7	299.5	317.0	332.1	349.1	375.2	398.8	0.355
Octan-1-ol . . .	333	356.6	368.5	386.2	401.3	418.3	444.4	468.4	0.361
Oct-1-ene	261	283.8	295.5	313.0	328.0	344.9	370.9	394.4	0.353
Oleic acid	457	489.8	505.9	529.7	549.9	572.2	605.4	634.2 d	0.422

Organic compounds (*contd*)

Compound	Pressure, $p/(\text{kN m}^{-2})$								dT/dp $\text{K}(\text{kN m}^{-2})^{-1}$
	0.2	1.0	2.0	5.0	10	20	50	101.325	
Palmitic acid	449	478.2	493.0	515.4	534.9	557.3	592.0	623.9	0.472
Pentane	200	218.9	228.5	242.8	255.0	268.8	290.0	309.2	0.289
Pentan-1-ol	293	314.9	325.6	341.2	354.4	369.0	391.2	411.2	0.299
Pentan-2-ol	280	301.0	311.3	326.2	338.7	352.5	373.5	392.9	0.296
Pent-1-ene	196	214.4	223.8	237.8	249.8	263.4	284.2	303.1	0.285
Phenanthrene . .	400	438.5	457.9	486.7	511.2	538.3	578.6	613.4	0.509
Phenol	315	340.3	352.8	371.2	386.8	404.4	431.0	455.0	0.359
Phthalic anhydride . .	s	s	417.4	443.2	465.6	490.4	527.1	557.7	0.439
Pinene	278	305.0	318.5	338.5	355.7	374.7	403.3	428.2	0.366
Piperidine	s	274.8	285.8	302.3	316.4	332.4	357.0	379.4	0.338
Propadiene (allene) ..	160	171.7	178.0	187.7	196.4	206.6	223.0	238.8	0.245
Propane	147	161.6	168.9	179.8	189.2	199.8	216.2	231.0	0.224
Propan-1-ol . . .	263	283.2	293.2	307.5	319.6	332.9	352.8	370.3	0.261
Propan-2-ol . . .	254	272.8	282.2	295.8	307.4	319.9	338.8	355.4	0.247
Propene	144	158.1	165.2	175.8	184.9	195.2	211.0	225.5	0.217
Prop-2-en-1-ol (allyl alcohol) . . .	258	279.5	289.9	305.1	317.7	331.5	351.9	369.8	0.264
Propyl acetate . . .	251	273.1	284.0	300.1	314.0	329.5	353.3	374.7	0.322
*iso*Propylbenzene (cumene)	281	306.4	319.0	337.8	354.0	372.2	400.2	425.6	0.381
Propyne	166	180.1	187.4	198.3	207.8	218.5	234.9	249.9	0.226
Pyridine	258	280.7	292.2	309.4	324.1	340.5	365.7	388.4	0.340
Pyrogallol . . .	s	434.1	451.1	475.6	496.1	518.6	552.2	581.7 d	0.441
Quinol	s	s	s	463.4	481.9	502.4	532.9	559.4	0.391
*iso*Quinoline . .	345	374.5	390.0	412.7	432.0	453.4	485.5	513.7	0.417
Resorcinol . . .	388	419.4	434.6	456.6	475.0	495.1	524.5	549.7	0.368
Salicylic acid . . .	s	s	s	442.4	459.6	478.2	505.2	528.2	0.336
Stearic acid . . .	467	496.4	511.4	534.4	554.6	578.0	614.8	648.3 d	0.498
Styrene	272	298.4	311.8	331.5	348.2	366.2 p	392.1 p	413.3 p	0.301
Succinic anhydride . .	s	395.3	411.0	434.2	453.9	475.6	507.4	534.2	0.388
1,2,3,4-Tetrahydro-naphthalene . .	318	346.2	360.6	382.0	400.4	421.0	452.4	480.4	0.417
Tetramethylsilane . .	201	221.5	232.0	247.7	261.5	277.1	301.4	323.7	0.339
Thiophen	237	258.0	268.6	284.3	297.7	312.8	336.1	357.6	0.327
Thymol	344	374.2	389.1	411.0	429.5	449.7	479.6	505.0	0.371
Toluene	252	274.7	286.3	303.6	318.4	335.1	360.6	383.8	0.348
o-Toluidine . . .	323	349.2	362.4	381.9	398.7	417.6	446.6	472.9	0.394
m-Toluidine . . .	320	349.2	363.6	384.7	402.5	422.2	451.4	476.5	0.368
p-Toluidine . . .	321	349.2	362.9	382.9	399.9	418.8	447.5	473.6	0.394
Trichloroethylene . .	234	256.2	267.3	283.8	298.0	314.0	338.2	359.8	0.322
Trichlorofluoromethane (R11)	191	209.4	218.7	232.6	244.5	257.9	278.3	296.7	0.277
Triethylamine . . .	237	259.0	270.0	286.5	300.6	316.5	340.8	362.7	0.329
Trimethylamine. . .	178	195.3	203.8	216.6	227.5	239.8	258.8	276.0	0.259
1,3,5-Trimethylbenzene	291	316.7	329.7	348.9	365.5	384.0	412.4	437.9	0.383
2.2.4-Trimethylpentane	242	264.1	275.5	292.5	307.2	323.8	349.2	372.4	0.348
Valeric acid . . .	324	348.6	360.8	378.7	393.9	410.8	436.2	458.7	0.334
*iso*Valeric acid . . .	314	339.4	351.9	370.1	385.2	401.9	426.6	448.3	0.322
Vanillin	388	420.6	436.7	460.0	479.4	500.6	531.6	558.2	0.391
o-Xylene	275	300.1	312.6	331.3	347.3	365.2	392.7	417.6	0.373
m-Xylene . . .	272	296.3	308.8	327.1	342.9	360.6	387.7	412.3	0.368
p-Xylene	s	295.4	307.7	326.2	342.0	359.7	386.9	411.5	0.369

D.A.

2.4.5 Vapour pressures from 0.2 to 6 MN m^{-2}

Kelvin temperatures are listed for pressures p from 0.2 to 6 MN m^{-2}. Values of the critical temperatures and pressures are given in Section 2.5. The letter 's' indicates a solid.

Elements and inorganic compounds

Element or compound	Pressure, $p/(\text{MN m}^{-2})$							
	0.2	0.5	1	2	3	4	5	6
Ammonia	254.3	277.3	298.0	322.5	338.9	351.6	362.0	371.0
Argon	94.3	105.8	116.6	129.8	138.7	145.7		
Bromine	354	389	421	459	485	506	523	537
Carbon dioxide	203.3 s	216.0 s	233.0	253.6	267.6	278.4	287.4	295.1
Carbon disulphide	342	377	409	448	474	495	512	528
Carbon monoxide	88.0	98.8	108.9	121.2	129.5			
Carbonyl chloride (phosgene)	299	330	359	393	415	433	447	
Chlorine	256	283	308	338	358	374	387	398
Cyanogen	268	294	317	345	364	379	391	401
Dinitrogen oxide	197	217	235	257	272	283	293	301
Hydrogen	22.8	27.1	31.2					
Hydrogen bromide	221	244	265	291	309	322	334	343
Hydrogen chloride	201	222	241	264	279	291	301	309
Hydrogen cyanide	318	350	378	410	430	445		
Hydrogen fluoride	312	344	371	402	421	436	448	457
Hydrogen iodide	255	282	307	338	359	376	390	402
Hydrogen sulphide	227	251	273	299	316	330	341	350
Krypton	129.2	144.7	159.2	176.8	188.8	198.2	206.0	
Neon	29.6	33.7	37.5	42.2				
Nitrogen	83.6	94.0	103.8	115.6	123.6			
Nitrogen dioxide	310	332	352	373	387	396	404	411
Nitrogen oxide	128	138	148	159	166	171	175	179
Oxygen	97.2	108.8	119.6	132.7	141.7	148.7	154.4	
Silicon tetrafluoride . . .	185	201	217	238	252			
Sulphur dioxide	279	306	330	358	377	392	404	414
Sulphur hexafluoride . . .	219.6	243.4	265.2	291.1	308.4			
Sulphur trioxide	333	358	381	411	431	447	460	471
Tin (IV) chloride	414	456	495	543	574			
Xenon	177.8	198.9	218.6	242.5	258.8	271.6	282.2	

Organic compounds

Refrigerant numbers (British Standard 4580) are given in parentheses for some halogenated alkanes.

Compound	Pressure, $p/(\text{MN m}^{-2})$							
	0.2	0.5	1	2	3	4	5	6
Acetic acid	415	452	485	523	548	567	582	
Acetone	351	385	416	453	478	497		
Acetylene	201	221	240	263	278	289	299	307
Aniline	485	527	566	614	647	673	694	
Benzene	377	416	452	495	524	546		
Bromobenzene . . .	459	505	547	598	633	659		

Organic compounds (*contd*)

Compound	Pressure, $p/(MN\ m^{-2})$							
	0.2	0.5	1	2	3	4	5	6
Bromoethane . . .	333	367	398	434	458	475	490	501
Bromotrifluoromethane (R13B1)	231	256	279	307	326			
Buta-1,3-diene . . .	289	319	348	388	410	422		
Butane	292	324	352	388	411			
Carbon tetrachloride .	374	414	450	494	524	546		
Chlorobenzene . . .	432	476	516	564	596	621		
Chlorodifluoromethane (R22)	248	273	297	324	343	358		
Chloroethane . . .	305	337	366	401	424	442	457	
Chloroform	357	393	427	466	493	514	531	
Chloromethane . . .	266	294	319	349	370	385	398	410
Chlorotrifluoromethane (R13)	205	228	249	274	290			
Dichlorodifluoromethane (R12)	261	289	315	347	367	383		
1,2-Dichlorotetrafluoroethane (R114)	296	328	357	391	414			
Diethyl ether . . .	328	362	393	430	454			
Ethane	198	220	241	266	283	296		
Ethanol	370	398	424	454	474	489	502	513
Ethyl acetate . . .	373	408	441	480	506			
Ethylene	182	202	221	244	260	272	282	
Hexane	365	403	438	480	507			
Methane	121	135	149	166	177	186		
Methanol	356	385	410	439	458	473	485	495
Methyl acetate . . .	351	385	415	452	476	495		
Pentane	331	366	398	436	462			
Phenol	480	521	557	600	630	653	674	692
Toluene	410	451	490	536	566	590		
Trichlorofluoromethane (R11) . .	317	351	382	420	443	465		

D.A.

2.5 Critical constants and second virial coefficients of gases

The critical pressure p^c, critical molar volume V_m^c, and critical temperature T^c, are the values of the pressure p, molar volume V_m, and thermodynamic temperature T, at which the densities of coexisting liquid and gaseous phases become identical. At the critical point:

$$(\partial p/\partial V)_{T=T^c}=0, \qquad (\partial^2 p/\partial V^2)_{T=T^c}=0 \tag{1}$$

At temperatures above the critical temperature a gas cannot be liquefied.

The state of a gas can be accurately expressed by an equation of the form:

$$Z=pV_m/RT=1+B/V_m+C/V_m^2+\ldots \tag{2}$$

where R is the gas constant: $R=(8.31441\pm0.00026)\,\mathrm{J\,K^{-1}\,mol^{-1}}$; $B, C, \ldots,$ are the second, third, $\ldots,$ virial coefficients which for a given substance are functions of temperature only; and Z is called the compression factor. As a rough rule, neglect of the third and higher virial coefficients in equation (2) leads to percentage errors $10^2\,|\delta Z/Z|<x$ provided that $p/p^c<0.2x^{\frac{1}{2}}$ at $T/T^c=0.6$; or $p/p^c<0.6x^{\frac{1}{2}}$ at $T/T^c=1$; or $p/p^c<1.3x^{\frac{1}{2}}$ at $T/T^c=2$; or $p/p^c<3x^{\frac{1}{2}}$ at $T/T^c=4$.

The second virial coefficient B of a gas can be fitted, usually within experimental errors over the range of temperatures at which measurements have been made, by an equation of the form:

$$B/(\mathrm{cm^3\,mol^{-1}}]=a-b\,\exp\{c(\mathrm{K}/T)\}. \tag{3}$$

Values of a, b, and c are given, where possible, for the range of temperatures from T_L to T_U.

For Ar, Kr, Xe, CH_4, N_2, O_2, and CO accurately, and with useful but lower accuracy for other substances, the critical compression factor $Z^c=0.29$, and B is given by

$$Bp^c/RT^c=0.599-0.467\,\exp(0.694\,T^c/T). \tag{4}$$

An estimate of B for a substance with a value of Z^c appreciably different from 0.29 can be obtained by the method described in *The Properties of Gases and Liquids* by Reid, Prausnitz, and Sherwood, 3rd edn, 1977, p. 53 (McGraw Hill, New York). Experimental values of B can be found in *The Virial Coefficients of Pure Gases and Mixtures* by Dymond and Smith, 2nd edn, 1980 (Clarendon Press, Oxford).

Elements and inorganic compounds

Element or Compound	p^c MPa	V_m^c cm³ mol⁻¹	T^c K	$p^c V_m^c$ RT^c	coefficients of equation (3) a	b	c	T_L K	T_U K
Ammonia	11.35	73	405.5	0.244	44.3	23.6	766.6	273	573
Argon	4.87	75	150.8	0.291	154.2	119.3	105.1	80	1024
Bromine	10.3	127	588	0.268	—	—	—	—	—
Carbon dioxide . . .	7.38	94	304.1	0.274	137.6	87.7	325.7	220	1100
Carbon disulphide . . .	7.9	173	552	0.298	211.0	167.1	538.7	280	430
Carbon monoxide . . .	3.5	93	132.9	0.295	202.6	154.2	94.2	90	573
Carbonyl chloride (phosgene)	5.67	190	455	0.285	—	—	—	—	—
Chlorine	7.98	124	416.9	0.285	—	—	—	—	—
Cyanogen	6.0	—	400	—	746.9	508.2	238.0	308	423
Deuterium	1.650	60	38.26	0.312	155.1	132.5	18.9	18	423
Deuterium oxide . . .	21.66	56	644.0	0.225	−16.8	3.37	1928.2	423	773
Dinitrogen oxide . . .	7.24	97	309.6	0.274	180.7	114.8	305.4	200	423
Fluorine	5.22	66	144.3	0.288	71.4	48.0	165.0	80	300
Helium	0.227	57	5.2	0.302	114.1	98.7	3.245	7	150
Hydrogen	1.297	65	33.2	0.306	315.0	289.7	9.47	14	400
Hydrogen bromide . . .	8.55	—	363.2	—	—	—	—	—	—

Elements and inorganic compounds

Element or Compound	$\dfrac{p^c}{MPa}$	$\dfrac{V_m^c}{cm^3\,mol^{-1}}$	$\dfrac{T^c}{K}$	$\dfrac{p^c V_m^c}{RT^c}$	coefficients of equation (3)			$\dfrac{T_L}{K}$	$\dfrac{T_U}{K}$
					a	b	c		
Hydrogen chloride . . .	8.31	81	324.7	0.249	57.7	37.8	495.9	190	480
Hydrogen cyanide . . .	5.39	139	456.7	0.197	—	—	—	—	—
Hydrogen deuteride . .	1.484	63	35.9	0.312	—	—	—	—	—
Hydrogen fluoride . . .	6.48	69	461	0.117	—	—	—	—	—
Hydrogen iodide . . .	8.31	—	424.0	—	—	—	—	—	—
Hydrogen sulphide . . .	8.94	98	373.2	0.284	47.7	30.3	632.9	278	493
Iodine	—	155	819	—	—	—	—	—	—
Krypton	5.50	91	209.4	0.288	189.6	148.0	145.3	110	700
Mercury	151	43	1765	0.439	—	—	—	—	—
Neon	2.76	42	44.4	0.311	81.0	63.6	30.7	44	973
Nitrogen	3.39	89	126.2	0.290	185.4	141.8	88.7	75	700
Nitrogen dioxide . . .	10.1	84	431	0.240	—	—	—	—	—
Nitrogen oxide . . .	6.48	58	180	0.250	15.9	11.0	372.3	122	311
Oxygen	5.04	73	154.6	0.288	152.8	117.0	108.8	90	400
Ozone	5.57	89	261.1	0.228	—	—	—	—	—
Phosphine	6.5	—	324.5	—	248.6	164.9	276.4	190	297
Radon	6.3	—	377	—	—	—	—	—	—
Silane	4.84	—	269.7	—	—	—	—	—	—
Silicon tetrachloride . .	3.59	326	508.1	0.277	—	—	—	—	—
Silicon tetrafluoride . .	3.72	—	259.0	—	346.7	257.2	185.5	206	475
Sulphur	20.7	—	1314	—	—	—	—	—	—
Sulphur dioxide . . .	7.88	122	430.8	0.269	134.4	72.5	606.5	265	473
Sulphur hexafluoride . .	3.76	198	318.7	0.282	422.1	281.3	273.5	200	525
Sulphur trioxide . . .	8.21	127	491.0	0.256	—	—	—	—	—
Tin (IV) chloride . . .	3.75	351	591.9	0.267	—	—	—	—	—
Uranium hexafluoride . .	4.66	250	505.8	0.277	540.5	380.9	445.0	321	469
Water	22.05	56	647.1	0.231	33.0	15.2	1300.7	293	1248
Xenon	5.84	118	289.7	0.287	247.0	192.9	199.8	160	650
Organic compounds									
Acetaldehyde	—	—	461	—	−71.8	12.3	1351	288	476
Acetic acid	5.79	171	592.7	0.201	—	—	—	—	—
Acetone	4.70	209	508.1	0.232	—	—	—	—	—
Acetonitrile	4.83	173	545.5	0.184	—	—	—	—	—
Acetylene	6.14	113	308.3	0.270	49.2	42.9	354.9	200	273
Aniline	5.31	287	699	0.262	—	—	—	—	—
Benzene	4.90	259	562.2	0.271	571.4	348.4	526.6	290	600
Brombenzene	4.52	324	670	0.263	—	—	—	—	—
Bromoethane	6.23	215	503.9	0.320	—	—	—	—	—
Bromomethane	—	—	464	—	51.1	42.4	792.2	244	383
Bromotrifluoromethane (R13B1)	3.97	196	340.2	0.275	—	—	—	—	—
Buta-1,3-diene	4.33	221	425	0.270	—	—	—	—	—
Butane	3.80	255	425.2	0.274	557.0	344.7	398.0	250	560
iso-Butane	3.65	263	408.2	0.283	556.7	357.0	374.4	273	511
Butanone	4.21	267	536.8	0.252	948.6	314.6	702.4	315	371
But-1-ene	4.02	240	419.6	0.277	522.9	289.6	421.2	243	420
iso-Butene	4.00	239	417.9	0.275	722.1	354.4	405.8	243	333
cis-But-2-ene	4.20	234	435.6	0.271	1369.5	778.2	297.4	251	343

Organic compounds (*contd*)

Element or Compound	p^c MPa	V_m^c cm³ mol⁻¹	T^c K	$\dfrac{p^c V_m^c}{RT^c}$	coefficients of equation (3) a	b	c	T_L K	T_U K
trans-But-2-ene	3.98	238	428.6	0.266	1245.0	763.3	284.3	243	333
Carbon tetrachloride . .	4.56	276	556.4	0.272	607.6	463.0	461.0	316	419
Carbon tetrafluoride (R14)	3.74	140	227.6	0.276	302.9	213.4	181.5	203	773
Chlorobenzene	4.52	308	632.4	0.265	—				
Chlorodifluoromethane (R22)	4.97	166	369.3	0.268	374.7	270.2	287.8	298	473
Chloroethane	5.27	—	460.4	—	182.4	88.4	717.9	273	605
Chloroform	5.47	239	536.4	0.293	485.7	272.5	542.7	316	400
Chloromethane . . .	6.68	139	416.2	0.268	396.7	259.2	337.3	280	600
Chlorotrifluoromethane (R13)	3.87	180	302.0	0.278	388.6	282.9	229.5	192	533
Cyclohexane	4.07	308	553.5	0.273	677.3	393.2	531.4	300	560
Cyclopentane	4.51	260	511.7	0.275	—				
Cyclopropane	5.58	163	398.3	0.274	355.7	255.2	316.7	303	403
Dichlorodifluoromethane (R12)	4.14	217	385.0	0.280	463.1	341.7	291.2	239	478
Dichlorofluoromethane (R21)	5.18	196	451.6	0.271	435.0	401.9	266.7	239	450
1,2-Dichlorotetra-fluoroethane (R114)	3.26	294	418.9	0.275	639.9	431.5	361.3	296	504
Diethylamine	3.71	301	496.6	0.270	673.0	378.0	517.7	314	405
Diethyl ether	3.64	280	466.7	0.262	968.3	346.6	552.3	280	400
Dimethylamine	5.34	—	437.2	—	407.7	234.2	456.0	312	405
Dimethyl ether	5.37	190	400.0	0.307	371.0	267.4	351.3	273	328
2,2-Dimethylpropane . .	3.20	363	433.8	0.269	675.1	439.7	383.3	300	550
1,4-Dioxan	5.21	238	587	0.254	—				
Ethane	4.88	148	305.4	0.285	311.7	230.6	227.8	200	600
Ethanol	6.14	167	513.9	0.240	864.6	45.4	1345	313	399
Ethyl acetate	3.83	286	523.2	0.252	982.9	407.5	598.1	330	399
Ethylamine	5.62	182	456	0.270	533.6	289.0	452.3	293	405
Ethylene	5.04	130	282.3	0.280	278.9	193.5	231.2	181	448
Ethyl formate	4.74	229	508.5	0.257	763.9	341.0	542.6	323	395
Ethyl methyl ether . . .	4.40	221	437.8	0.267	—				
Fluorobenzene	4.55	269	560.1	0.263	613.0	360.0	555.4	318	623
Fluoromethane	5.88	113	317.8	0.252	269.2	172.8	303.0	273	423
Heptane	2.74	432	540.3	0.263	982.0	523.9	582.3	300	700
Hept-1-ene	—	—	537.3	—	1245.0	635.1	530.9	334	411
Hexane	3.01	370	507.5	0.264	839.6	404.1	572.6	300	450
Hex-1-ene	—	—	504.0	—	836.6	460.4	507.1	313	410
Iodobenzene	4.52	351	721	0.265	—				
Iodomethane	—	—	554	—	459.8	248.0	476.0	313	383
Methane	4.60	99	190.6	0.288	206.4	159.5	133.0	110	600
Methanol	8.09	118	512.6	0.224	123.9	17.5	1428	298	573
Methyl acetate	4.69	228	506.8	0.254	762.9	336.7	577.9	323	391
Methylamine	7.61	—	430.7	—	85.3	57.1	685.9	293	550
2-Methylbutane	3.38	306	460.4	0.270	952.1	620.4	355.8	280	450
Methylcyclohexane . . .	3.47	368	572.2	0.268	—				
Methylcyclopentane . .	3.78	319	532.7	0.272	—				
Methyl formate	6.00	172	487.2	0.255	581.0	269.3	525.5	319	397
2-Methylpentane . . .	3.01	367	497.5	0.267	—				

Organic compounds (*contd*)

Element or Compound	$\dfrac{p^c}{MPa}$	$\dfrac{V_m^c}{cm^3 mol^{-1}}$	$\dfrac{T^c}{K}$	$\dfrac{p^c V_m^c}{RT^c}$	coefficients of equation (3)			$\dfrac{T_L}{K}$	$\dfrac{T_U}{K}$
					a	b	c		
Naphthalene	4.05	413	748.4	0.269	—	—	—	—	—
Octafluorocyclobutane .	2.78	324	388.5	0.279	895.7	616.4	285.6	373	623
Octane	2.49	492	568.8	0.259	1138.0	536.1	672.5	300	700
Oct-1-ene	—	—	566.7	—	1328.0	631.5	614.2	359	412
Pentane	3.37	304	469.7	0.263	694.5	407.5	462.7	300	550
Pent-1-ene	3.53	—	464.8	—	655.7	344.2	499.9	294	410
Phenol	6.1	—	694.2	—	—	—	—	—	—
Propadiene	—	—	393	—	391.4	272.0	310.5	223	353
Propane	4.25	203	369.8	0.281	433.6	301.7	301.3	240	550
Propan-1-ol	5.17	219	536.8	0.253	861.9	297.3	669.2	378	423
Propene	4.60	181	364.8	0.274	433.0	310.7	274.2	280	500
Propyne	5.63	164	402.4	0.275	444.6	314.6	292.2	348	473
Pyridine	5.63	254	620.0	0.277	796.1	398.8	576.3	347	438
Tetramethylsilane . . .	2.82	362	448.6	0.273	792.2	502.2	403.4	295	539
Toluene	4.10	316	591.8	0.263	719.1	373.3	646.4	341	583
Trichlorofluoromethane (R11)	4.41	248	471.2	0.279	532.1	394.6	351.2	239	478
Triethylamine	3.03	389	535.6	0.265	1261.6	739.1	442.1	323	405
Trimethylamine . . .	4.09	254	432.8	0.288	747.0	466.5	347.3	298	375
2,2,4-Trimethylpentane .	2.57	468	544.0	0.266	—	—	—	—	—
m-Xylene	3.54	376	617.0	0.259	—	—	—	—	—
o-Xylene	3.73	369	630.3	0.262	—	—	—	—	—
p-Xylene	3.51	379	616.2	0.260	—	—	—	—	—

D.A.
M.L.M.

2.6 Properties of solutions

2.6.1 Solubilities of gases in water

Air in water

A kilogram of water saturated with air at a pressure of 101.325 kN m^{-2} contains the following volumes of dissolved oxygen, etc., in cm^3 at 0 °C and 101.325 kN m^{-2}.

Gas	Temperature of water, $\theta/°C$						
	0	5	10	15	20	25	30
Oxygen	10.19	8.9	7.9	7.0	6.4	5.8	5.3
Nitrogen, argon, etc.	19.0	16.8	15.0	13.5	12.3	11.3	10.4
Sum of above	29.2	25.7	22.9	20.5	18.7	17.1	15.7
% of oxygen in dissolved air (by vol.)	34.9	34.7	34.5	34.2	34.0	33.8	33.6

Gases in water

A indicates the volume of gas measured at 0 °C and 101.325 kN m^{-2} which will dissolve in a unit mass of water at the stated temperature when the gas is at a pressure of 101.325 kN m^{-2}; a indicates the same quantity except the pressure of the gas plus that of the water vapour is 101.325 kN m^{-2}.

Gas		Temperature of water, $\theta/°C$						
		0	10	15	20	30	40	50	60
Ammonia . . .	a	1130	870	770	680	530	400	290	200
Argon . . .	A	0.054	0.041	0.035	0.032	0.028	0.025	0.024	0.023
Carbon dioxide .	A	1.676	1.163	0.988	0.848	0.652	0.518	0.424	0.360
Carbon monoxide .	A	0.035	0.028	0.025	0.023	0.020	0.018	0.016	0.015
Chlorine . . .	a	4.61	3.09	2.63	2.26	1.77	1.41	1.20	1.01
Helium . . .	A	0.0098	0.0091	0.0089	0.0086	0.0084	0.0084	0.0086	0.0090
Hydrogen . .	A	0.0214	0.0195	0.0188	0.0182	0.0170	0.0164	0.0161	0.0160
Hydrogen sulphide .	A	4.53	3.28	2.86	2.51	1.97	1.62	1.37	1.18
Hydrochloric acid .	a	512	475	458	442	412	385	362	339
Nitrogen . .	A	0.0230	0.0183	0.0165	0.0152	0.0133	0.0119	0.0108	0.0100
Nitrous oxide .	A	—	0.88	0.74	0.63	—	—		
Nitric oxide . . .	A	0.071	0.055	0.049	0.046	0.039	0.034	0.031	0.029
Oxygen . . .	A	0.047	0.037	0.033	0.030	0.026	0.022	0.020	0.019
Sulphur dioxide	a	79.8	56.6	47.3	39.4	27.2	18.8	—	—

Values of A for 20 °C for other rare gases are: Ne, 0.0101; Kr, 0.0594; Xe, 0.126.

E.F.G.H.

2.6.2 Solubilities of solids in water

Solubilities are expressed as the number of g of *anhydrous* substance which when dissolved in 100 g of water make a saturated solution at the temperature stated.

The formula given is that of the solid phase which is in equilibrium with the solution. (For more complete data see Seidell's *Solubilities*, Vol. I, 1958; Vol. II, 1965; 4th edn (American Chemical Society, Washington DC). For other solutions see Table 2.2.)

Substance	Temperature of water, $\theta/°C$							
	0	10	15	20	40	60	80	100
Ammonium chloride, NH_4Cl .	29.4	33.3	35.2	37.2	45.8	55.2	65.6	77.3
Barium chloride, $BaCl_2.2H_2O$.	31.6	33.3	34.4	35.7	40.7	46.4	52.4	58.8
Barium hydroxide, $Ba(OH)_2.8H_2O$	1.67	2.48	3.23	3.89	8.22	20.94	101.4	—
Cadmium sulphate, $3CdSO_4.8H_2O$ . . .	75.6	75.9	76.2	76.4	78.4	73.9†	68.8†	58.0†
Calcium hydroxide, $Ca(OH)_2$.	0.137	0.132	0.128	0.118	0.104	0.087	0.063	0.051
Copper sulphate, $CuSO_4.5H_2O$.	14.3	17.4	18.8	20.7	28.5	40.0	55.0	75.4
Lithium carbonate, Li_2CO_3 . .	1.54	1.43	1.38	1.33	1.17	1.01	0.85	0.72
Mercuric chloride, $HgCl_2$ . .	3.6	4.8	5.6	6.5	10.3	16.3	30.0	61.3
Potassium chloride, KCl . . .	28.07	31.23	32.8	34.2	40.0	45.8	51.3	56.3
Potassium bromide, KBr . . .	53.5	59.5	62.5	65.2	75.5	85.5	95.0	104
Potassium iodide, KI . . .	127.5	136	140	144	160	176	192	208
Potassium hydroxide, $KOH.2H_2O$	95.7	103	107	119	138‡	—	161‡	183‡
Potassium nitrate, KNO_3 . .	13.3	20.9	25.8	31.6	63.9	110	169	246
Silver nitrate, $AgNO_3$ . . .	117	162	190	218	312	441	588	733
Sodium carbonate, $Na_2CO_3.10H_2O$ . . .	7.0	12.5	16.4	21.5	48.5§	46.4§	45.8§	45.5§
Sodium chloride, $NaCl$ . . .	35.7	35.8	35.9	36.0	36.6	37.3	38.4	39.8
Sodium sulphate, $Na_2SO_4.10H_2O$ . . .	5.0	9.0	13.4	19.4	48.8‖	45.3‖	43.7‖	42.5‖
Strontium chloride, $SrCl_2.6H_2O$	43.5	47.7		52.9	65.3	81.8	90.5**	100.8**
Succinic acid, $(CH_2)_2(COOH)_2$	2.80	4.51	5.7	6.9	16.2	35.8	70.8	125
Sugar (Cane), $C_{12}H_{22}O_{11}$ . .	179.2	190.5	197.0	203.9	238.1	287.3	362.1	487.2

† Solid phase becomes $CdSO_4.H_2O$ at 43.6°C.
‡ Becomes $KOH.H_2O$ at 32.5°C.
§ Becomes $Na_2CO_3.H_2O$ at 35°C.
‖ Becomes Na_2SO_4 at 32.38°C.
** Becomes $SrCl_2.2H_2O$ at 62.5°C.

E.F.G.H.

2.6.3 Densities of aqueous solutions

The values listed are ρ (kg m^{-3}) for a temperature of 20 °C unless otherwise stated and the indicated % is the number of g of substance in 100 g of aqueous solution.

Density of ethyl alcohol aqueous solutions

As an example of the values listed the density of a 24% solution is 963.1 kg m^{-3}.

%	0	1	2	3	4	5	6	7	8	9
0	998.2	996.4	994.5	992.8	991.0	989.4	987.8	986.3	984.8	983.3
10	981.9	980.5	979.1	977.8	976.4	975.1	973.9	972.6	971.3	970.0
20	968.6	967.3	965.9	964.3	963.1	961.7	960.2	958.7	957.1	955.5
30	953.8	952.1	950.4	948.6	946.8	944.9	943.1	941.1	939.2	937.2
40	935.2	933.1	931.1	929.0	926.9	924.7	922.6	920.4	918.2	916.0
50	913.8	911.6	909.4	907.1	904.9	902.6	900.3	898.0	895.7	893.4
60	891.1	888.8	886.5	884.2	881.8	879.5	877.1	874.8	872.4	870.0
70	867.7	865.3	862.9	860.5	858.1	855.6	853.2	850.8	848.4	845.9
80	843.4	841.0	838.5	836.0	833.5	831.0	828.4	825.8	823.2	820.6
90	818.0	815.3	812.6	809.8	807.1	804.2	801.4	798.5	795.5	792.4
100	789.3									

For other temperatures, interpolate from the above and the following:

At 15 °C: 0%, 999.13; 10%, 983.0; 20%, 970.7; 30%, 956.8; 40%, 938.8; 50%, 917.8; 60%, 895.2; 70%, 871.9; 80%, 847.8; 90%, 822.3; 100%, 793.5.

At 25 °C: 0%, 997.1; 10%, 980.4; 20%, 966.4; 30%, 950.7; 40%, 931.5; 50%, 909.9; 60%, 887.0; 70%, 863.4; 80%, 839.1; 90%, 813.6; 100%, 785.1.

Density of hydrochloric acid aqueous solutions

ρ/(kg m^{-3})	% HCl	ρ/(kg m^{-3})	% HCl	ρ/(kg m^{-3})	% HCl	ρ/(kg m^{-3})	% HCl	ρ/(kg m^{-3})	% HCl
1010	2.40	1050	10.57	1090	18.65	1130	26.60	1170	34.53
1020	4.44	1060	12.60	1100	20.65	1140	28.56	1180	36.58
1030	6.48	1070	14.63	1110	22.65	1150	30.54	1190	38.63
1040	8.53	1080	16.64	1120	24.63	1160	32.54		

Density of nitric acid aqueous solutions

ρ/(kg m^{-3})	% HNO$_3$	ρ/(kg m^{-3})	% HNO$_3$	ρ/(kg m^{-3})	% HNO$_3$	ρ/(kg m^{-3})	% HNO$_3$	ρ/(kg m^{-3})	% HNO$_3$
1010	2.162	1120	20.80	1240	39.02	1360	58.74	1480	90.00
1020	3.983	1140	23.94	1260	42.14	1380	62.64	1490	93.40
1040	7.536	1160	27.00	1280	45.27	1400	66.90	1495	95.90
1060	10.98	1180	30.00	1300	48.38	1420	71.57	1500	97.76
1080	14.32	1200	32.98	1320	51.68	1440	76.66	1505	98.88
1100	17.67	1220	35.92	1340	55.13	1460	82.33	1510	99.60

Density of sulphuric acid aqueous solutions

$\rho/(kg\ m^{-3})$	% H_2SO_4	$\rho/(kg\ m^{-3})$	% H_2SO_4	$\rho/(kg\ m^{-3})$	% H_2SO_4	$\rho/(kg\ m^{-3})$	% H_2SO_4	$\rho/(kg\ m^{-3})$	% H_2SO_4
1010	1.731	1260	35.01	1520	62.00	1780	85.16	1836.5	97.25
1020	3.242	1280	37.36	1540	63.81	1800	87.69	1836.0	98.06
1040	6.221	1300	39.68	1560	65.59	1810	89.23	1835.5	98.38
1060	9.129	1320	41.94	1580	67.35	1820	91.11	1835.0	98.65
1080	11.97	1340	44.17	1600	69.09	1822	91.54	1834.5	98.87
1100	14.73	1360	46.33	1620	70.82	1824	92.00	1834.0	99.04
1120	17.43	1380	48.44	1640	72.53	1826	92.51	1833.5	99.21
1140	20.08	1400	50.50	1660	74.22	1828	93.03	1833.0	99.35
1160	22.67	1420	52.52	1680	75.91	1830	93.64	1832.5	99.50
1180	25.22	1440	54.49	1700	77.63	1832	94.32	1832.0	99.63
1200	27.72	1460	56.41	1720	79.37	1834	95.17	1831.5	99.76
1220	30.18	1480	58.30	1740	81.16	1835	95.72	1831.0	99.88
1240	32.61	1500	60.16	1760	83.06	1836	96.56	1830.5	100

Density of ammonia aqueous solutions

$\rho/(kg\ m^{-3})$	% NH_3	$\rho/(kg\ m^{-3})$	% NH_3	$\rho/(kg\ m^{-3})$	% NH_3	$\rho/(kg\ m^{-3})$	% NH_3	$\rho/(kg\ m^{-3})$	% NH_3
994	0.98	966	7.77	938	15.47	910	24.03	882	33.16
990	1.89	962	8.82	934	16.66	906	25.34	878	34.43
986	2.83	958	9.87	930	17.85	902	26.67	874	35.71
982	3.79	954	10.95	926	19.06	898	28.00	870	37.00
978	4.77	950	12.03	922	20.28	894	29.33	866	38.29
974	5.75	946	13.15	918	21.51	890	30.63		
970	6.76	942	14.29	914	22.76	886	31.89		

Density of sodium carbonate aqueous solutions

$\rho/(kg\ m^{-3})$	% Na_2CO_3	$\rho/(kg\ m^{-3})$	% Na_2CO_3	$\rho/(kg\ m^{-3})$	% Na_2CO_3	$\rho/(kg\ m^{-3})$	% Na_2CO_3	$\rho/(kg\ m^{-3})$	% Na_2CO_3
1010	1.135	1040	4.019	1070	6.895	1100	9.728	1130	12.51
1020	2.096	1050	4.980	1080	7.848	1110	10.66	1140	13.43
1030	3.058	1060	5.942	1090	8.788	1120	11.59	1145	13.88

Density of sodium hydroxide aqueous solutions

$\rho/(kg\ m^{-3})$	% NaOH	$\rho/(kg\ m^{-3})$	% NaOH	$\rho/(kg\ m^{-3})$	% NaOH	$\rho/(kg\ m^{-3})$	% NaOH	$\rho/(kg\ m^{-3})$	% NaOH
1020	1.938	1130	11.92	1240	21.90	1350	32.10	1460	43.11
1030	2.838	1140	12.82	1250	22.82	1360	33.07	1470	44.16
1040	3.746	1150	13.73	1260	23.73	1370	34.04	1480	45.22
1050	4.655	1160	14.63	1270	24.65	1380	35.02	1490	46.28
1060	5.564	1170	15.54	1280	25.56	1390	36.00	1500	47.32
1070	6.473	1180	16.44	1290	26.48	1400	37.00	1510	48.37
1080	7.378	1190	17.35	1300	27.41	1410	37.99	1520	49.44
1090	8.282	1200	18.26	1310	28.34	1420	39.00		
1100	9.191	1210	19.17	1320	29.27	1430	40.00		
1110	10.10	1220	20.08	1330	30.20	1440	41.03		
1120	11.01	1230	20.99	1340	31.15	1450	42.06		

Density of phosphoric acid aqueous solutions

$\rho/(kg\ m^{-3})$	% H_3PO_4	$\rho/(kg\ m^{-3})$	% H_3PO_4	$\rho/(kg\ m^{-3})$	% H_3PO_4	$\rho/(kg\ m^{-3})$	% H_3PO_4	$\rho/(kg\ m^{-3})$	% H_3PO_4	$\rho/(kg\ m^{-3})$	% H_3PO_4
1010	2.15	1090	16.26	1240	38.16	1400	57.23	1720	87.72		
1020	4.00	1100	17.87	1260	40.77	1440	61.43	1760	91.17		
1030	5.84	1120	21.02	1280	43.33	1480	65.49	1780	92.83		
1040	7.64	1140	24.07	1300	45.83	1520	69.41	1800	94.48		
1050	9.43	1160	27.04	1320	48.21	1560	73.21	1820	96.08		
1060	11.18	1180	29.93	1340	50.57	1600	76.94	1840	97.68		
1070	12.91	1200	32.75	1360	52.84	1640	80.63	1860	99.23		
1080	14.60	1220	35.53	1380	55.11	1680	84.20				

Densities of some other aqueous solutions

Values of $\rho/(kg\ m^{-3})$ at 20 °C excepting where a different temperature is given. The indicated % is the number of g of anhydrous substance in 100 g of solution.

Substance	5%	10%	15%	20%	25%	Substance	5%	10%	15%	20%
$BaCl_2$.	1043	1092	1145	1203	1266	NH_4Cl .	1014	1029	1043	1057
$CdSO_4$.	1049	1102	1161	1224	1295	$CuSO_4$.	1051	1107	1167	1230
$FeCl_3$.	1041	1085	1132	1182	1234	$MgSO_4$.	1050	1103	1160	1220
$MgCl_2$.	1040	1082	1125	1171	1218	KCl .	1030	1063	1098	1133
NaA .	1024	1050	1075	1102	1130	$K_2Cr_2O_7$.	1034	1070	—	—
$NaCl$	1034	1071	1109	1148	1189	$K_3Fe(CN)_6$	1026	1054	1083	1113
$NaNO_3$.	1032	1067	1104	1143	1183	$K_4Fe(CN)_6$	1033	1068	1105	—
$SrCl_2$.	1044	1093	1145	1201	1260	KNO_3 . .	1030	1063	1097	1133
$ZnSO_4$.	1051	1107	1168	—	—	K_2SO_4 .	1039	1082	—	—

Substance	5%	10%	15%	20%	25%	30%	35%	40%	45%	50%
NH_4NO_3 . .	1019	1040	1061	1083	1105	1128	1151	1175	1200	1226
$CaCl_2$ . . .	1040	1084	1129	1178	1228	1282	1337	1396	—	—
H_2O_2† (18 °C)	1017	1035	1054	1073	1092	1112	1133	1154	1175	1197
PbA_2 (18 °C)	1037	1077	1120	1166	1217	1271	1330	1399	—	—
$LiCl$ . . .	1027	1056	1085	1115	1146	1179	—	—	—	—
$HClO_4$‡ (15 °C)	1029	1060	1093	1128	1166	1207	1251	1299	1352	1410
KBr . . .	1035	1074	1116	1160	1208	1259	1315	1375	—	—
K_2CO_3 . .	1044	1090	1139	1190	1243	1298	1355	1414	1476	1540
KI . . .	1036	1076	1119	1166	1216	1271	1331	1396	1467	1546
$AgNO_3$ . .	1042	1088	1139	1194	1255	1321	1393	1474	1565	1668
$Na_2S_2O_3$ . .	1040	1083	1127	1174	1223	1274	1327	1383	—	—
Sugar§ . .	1018	1038	1059	1081	1104	1127	1151	1176	1203	1230

† H_2O_2: 60%, 1242; 70%, 1290; 80%, 1341; 90%, 1393; 100%, 1447.
‡ $HClO_4$: 60%, 1539; 70%, 1674.
§ Sugar: 60%, 1287; 70%, 1347.

E.F.G.H.

2.7 Properties of chemical bonds

2.7.1 Dipole moments and dipole lengths

Most of the following data were calculated from an extensive collection given by A. L. McClellan, *Tables of Experimental Dipole Moments*, San Francisco, Copyright by W. H. Freeman and Co., 1963. The values listed represent average values for some common chemicals. The conditions under which the measurements were made are indicated by the following letters, b=substance in benzene solution, g=substance as a gas, l=substance as a liquid.

The SI unit of dipole moment, p, is the coulomb metre (C m) and the dipole length, l_p is equal to p/e where e is the charge of a proton. The unit previously in use for dipole moments was the Debye, symbol D; $1D = 3.335\ 640 \times 10^{-30}$ C m.

Substance	$10^{30}p/(C\,m)$	l_p/pm	Substance	$10^{30}p/(C\,m)$	l_p/pm
Acetaldehyde	8.3b	52	Ethylamine	4.3b	27
Acetamide	12.3b	77	Ethylbenzene	1.2l	7
Acetic acid	3.3–5.0b	21–31	Ethylene carbonate . .	16.0b	100
Acetone	10.0l	62	Ethylene glycol	6.7b	42
Acetonitrile	11.7b	73	Ethylene oxide	6.3l	40
Acetophenone	9.7b	60	Ethyl ether	4.2b	26
Acetylacetone	9.3b	58	Formamide	11.3b	71
Acetyl chloride	8.0b	50	Formic acid	5.0b	31
Acrolein	9.7b	60	Furan	2.2b	14
Acrylonitrile	11.7b	73	Hexane	0 l	0
Ammonia	3.0l	19	Hydrazine	10.0l	62
Aniline	5.2b	32	Hydrogen cyanide . . .	9.8g	61
Anisole	4.0l	25	Hydrogen sulphide . . .	3.2g	20
Benzamide	12.2b	76	Iodoethane	5.8b	36
Benzene	0 l	0	Methanol	5.5b	34
Benzonitrile	10.7l	67	Methyl benzoate . . .	6.2b	39
Benzophenone	9.0b	56	Methyl ether	4.2b	26
Bromobenzene	5.0l	31	Morpholine	5.0b	31
Bromoethane	6.0b	37	Nitric oxide	0.5g	3
Bromoform	3.3b	21	Nitrobenzene	13.3b	83
Carbon dioxide	0 g	0	Nitrogen dioxide . . .	1.3g	8
Carbon disulphide . . .	0 l	0	Nitromethane	10.3b	65
Carbon tetrahalides . . .	0 l	0	Oxygen	0 g	0
Carbonyl sulphide . . .	2.4g	15	Ozone	2.2g	14
Chloroacetic acid . . .	7.7b	48	Phenol	5.3b	33
Chlorobenzene	5.2l	32	Propylene oxide	6.5b	41
Chloroethane	6.7b	42	Pyridine	7.7l	48
Chloroform	3.7b	23	Pyrrole	6.0b	37
Cyclohexanol	6.0b	37	Quinoline	7.0l	44
1,2-Dibromoethane . . .	4.0b	25	Styrene	0.3l	2
Dichloromethane	6.0b	37	Sulphur hexafluoride . . .	0 g	0
Diethylamine	3.7l	23	Tetrahydrofuran	5.7b	35
Dimethylaniline	5.2b	32	Thiophen	1.6l	10
Dimethylformamide . . .	12.7l	79	Toluene	1.3l	8
Dimethylsulphoxide . . .	13.0b	81	Triethylamine	2.8l	18
Dioxan	1.3b	8	s-Trioxane	7.3b	46
Ethanol	5.7b	35	Water	6.7–10.0l	42–62
Ethyl acetate	6.2b	39	Water	6.2g	39

E.F.G.H.

2.7.2 Bond lengths and dissociation energies of diatomic molecules

The tabulated values are for the internuclear separation r_e in the absence of vibration; those marked * are for the vibrational ground state. They are taken from the compilations: *Tables of Interatomic Distances and Configuration of Molecules and Ions*, Special Publication No 11; Supplement 1956–1959, Special Publication No 18, Chemical Society, London, 1958, 1965. Bond Dissociation Energies, D, are taken from A. G. Gaydon, *Dissociation Energies and Spectra of Diatomic Molecules*, 3rd edn, 1968 (Chapman and Hall, London).

Molecule	r_e/pm	D/kJ	Molecule	r_e/pm	D/kJ
H—H	74.130	432.00	C—H	112.0	335
H—D	74.140	435.39	N—H	103.8	310
D—D	74.143	439.53	P—H	143.3*	340
C—C	131.2	603	O—H	97.1	425
N—N	109.76	942	F—H	91.7	564
P—P	189.3	485	Cl—H	127.5	428
O—O	120.741	493.6	Br—H	140.8	362
S—S	188.7	423	I—H	160.0	295
F—F	141.8*	155	C—N	117.7	745
Cl—Cl	198.8	239	C—O	113.1	1069
Br—Br	228.4	190	N—O	115.02	627
I—I	266.7	147	Cl—F	162.8	247

J.H.S.G.

2.7.3 Bond lengths and angles in polyatomic molecules

The values are selected from those given in *Tables of Interatomic Distances and Configuration of Molecules and Ions*, Special Publication No. 11; Supplement 1956–1959, Special Publication No 18, Chemical Society, London 1958, 1965.

Molecule	Bond	length/pm	Bond	angle/deg.	Molecule	Bond	length/pm	Bond	angle/deg.
CO_2 . .	C—O	115.98	O—C—O	180	C_2H_2 . .	C—H	105.9	H—C—C	180
CS_2 . .	C—S	155.30	S—C—S	180		C—C	120.5		
CSe_2 . .	C—Se	198	Se—C—Se	180	C_2H_4 . .	C—H	108.4	H—C—H	115.5
SO_2 . .	S—O	143.21	O—S—O	119.5		C—C	133.2		
SO_3 . .	S—O	143	O—S—O	120	C_2H_6 . .	C—H	109.3	H—C—H	109.75
H_2O . .	O—H	95.8	H—O—H	104.45		C—C	153.4		
H_2O_2 .	O—O	148	O—O—H	100	C_6H_6 . .	C—H	108.4	H—C—C	120
ClO_2 .	Cl—O	149	O—Cl—O	118.5		C—C	139.7	C—C—C	120
H_2S . .	S—H	134.55	H—S—H	93.3	CH_3OH .	C—H	109.5	C—O—H	109
NH_3 . .	N—H	100.8	H—N—H	107.3		C—O	142.8		
PH_3 . .	P—H	143.7	H—P—H	93.3		O—H	96.0		
AsH_3 .	As—H	151.9	H—As—H	91.83	$(CH_3)_2O$.	C—H	109.4	C—O—C	111.5
$AsCl_3$.	As—Cl	216.1	Cl—As—Cl	98.4		C—O	141.6		
$SbCl_3$.	Sb—Cl	235.2	Cl—Sb—Cl	99.5	$(CH_3)_3N$.	C—H	109	H—C—H	107.1
$BiCl_3$.	Bi—Cl	248	Cl—Bi—Cl	100		C—N	147.2	C—N—C	108.7
SiH_3F .	Si—H	146.0	H—Si—H	109.3	C_2H_5Cl .	C—C	159.5	H—C—H	110
	Si—F	159.5				C—Cl	177.9	C—C—Cl	110.5
GeH_3Cl .	Ge—H	152	H—Ge—H	110.9	$(CH_3)_2CO$	C—C	151.5	C—C—C	116.22
$POCl_3$.	P—Cl	199	Cl—P—Cl	103.5		C—O	121.5	C—C—O	121.9
CH_4 . .	C—H	109.3	H—C—H	109.5	CH_3SH .	C—S	181.9	C—S—H	96.5
CCl_4 . .	C—Cl	176.6	Cl—C—Cl	109.5	$(CH_3)_3As$.	C—As	195.9	C—As—C	96

J.H.S.G.

2.7.4 Bond dissociation energies in polyatomic molecules

The values have been selected from those listed by T. L. Cottrell, *The Strengths of Chemical Bonds*, 2nd edn, 1958 (Butterworths, London); V. I. Vedeneyev, L. V. Gurvich, V. N. Kondrat'yev, V. A. Medvedev and Y. L. Frankevich, *Bond Energies, Ionization Potentials and Electron Affinities*, 1966 (Edward Arnold, London); and the references cited in J. A. Kerr and A. F. Trotman-Dickenson, *Handbook of Chemistry and Physics*, 50th edn, 1969/70 (The Chemical Rubber Co., Cleveland, Ohio).

$X-Y$	$D(X-Y)/\text{kJ}$	$X-Y$	$D(X-Y)/\text{kJ}$	$X-Y$	$D(X-Y)/\text{kJ}$
$CH-H$	448	CF_3-CF_3	406	tC_3H_7-Cl	339
CH_2-H	439	CH_3-CF_3	418	CF_3-Cl	356
CH_3-H	435	CH_3-CHO	314	CCl_3-Cl	305
C_2H_5-H	435	CH_3-CN	431	CH_3-Br	293
tC_3H_7-H	395	$NC-CN$	607	tC_3H_7-Br	285
C_6H_5-H	469	CH_3-NH_2	335	C_6H_5-Br	335
$C_6H_5CH_2-H$	356	$C_6H_5-NH_2$	418	CF_3-Br	293
CCl_3-H	401	CH_3-OH	377	CCl_3-Br	226
CF_3-H	444	C_6H_5-OH	469	CH_3-I	234
CH_3NH-H	385	CH_3-OCH_3	335	tC_3H_7-I	222
$(CH_3)_2N-H$	360	$C_6H_5-OCH_3$	423	C_6H_5-I	272
C_6H_5NH-H	335	$tC_3H_7-OCH_3$	347	O_2N-NO	40
NH_2-H	427	CH_3-SH	368	O_2N-NO_2	57
$OH-H$	492	CH_3-F	452	$CO=O$	531
$SH-H$	377	tC_3H_7-F	439	$HO-OH$	213
$HC\equiv CH$	962	C_6H_5-F	523	CH_3O-OCH_3	151
$H_2C=CH_2$	699	CF_3-F	540	$F_3P=O$	544
$F_2C=CF_2$	319	CCl_3-F	444	$Cl_3P=O$	510
CH_3-CH_3	368	CH_3-Cl	352	$Br_3P=O$	498

J.H.S.G.

2.7.5 Atomic radii

Atomic radii, also called single-bond metallic radii or metallic radii, have been calculated by Pauling using interatomic distances and an equation relating such distances with bond number. Metallic radii (12) have been obtained similarly with an assumed number of nearest neighbours equal to 12.

Atomic and metallic (12) radii in pm

Element	Atomic radius	Metallic (12) radius	Element	Atomic radius	Metallic (12) radius
Aluminium	125	143	Boron	80	98
Antimony	139	166	Bromine	—	117
Arsenic	121	148	Cadmium	138	151
Barium	198	222	Caesium	235	267
Beryllium	89	112	Calcium	174	197
Bismuth	151	178	Carbon	—	86

Atomic and metallic (12) radii in pm (*contd*)

Element	Atomic radius	Metallic (12) radius	Element	Atomic radius	Metallic (12) radius
Chlorine	—	91	Phosphorus	110	128
Chromium	119	128	Platinum.	130	139
Cobalt	116	125	Potassium	203	235
Copper	118	128	Rhenium	128	137
Gallium	125	140	Rhodium	125	134
Germanium	124	144	Rubidium	216	248
Gold	134	144	Ruthenium	125	134
Hafnium	144	159	Scandium	144	162
Helium	—	122	Selenium	117	140
Hydrogen.	—	78	Silicon.	117	138
Indium	142	158	Silver	134	144
Iodine	—	139	Sodium	157	190
Iridium	127	136	Strontium	191	215
Iron	117	126	Sulphur	104	127
Lanthanium	169	187	Tantalum	134	146
Lead	150	170	Technetium	127	136
Lithium	123	155	Tellurium	137	160
Magnesium	136	160	Thallium	144	160
Manganese	118	127	Tin.	142	163
Mercury	139	151	Titanium	132	147
Molybdenum . . .	130	139	Tungsten	130	139
Nickel	115	124	Vanadium	122	134
Niobium	134	146	Yttrium	162	180
Nitrogen	—	53	Zinc	121	134
Osmium	126	135	Zirconium	145	160
Palladium	128	137			

E.F.G.H.

2.7.6 Ionic radii

Empirical values for ionic radii are used for predicting interatomic distances and packing in crystals, even when the forces are far from being purely ionic. The ions are regarded as hard spheres, whose size is, to a first approximation, constant in all environments. To a second approximation, it has long been recognized that there is a dependence on *coordination number*, and, more recently, for transition-metal ions, a dependence also on *spin state*. Separate values are listed accordingly in the table.

The *coordination number* of an atom is the number of its first-nearest neighbours of opposite sign. It may sometimes be ambiguous when the neighbours are not all at exactly the same distance; judgement and experience, rather than formal rules, must then decide which of them are to be counted as first-nearest. Different decisions on this point will lead to small differences in estimates of the mean cation–anion distance within the polyhedron (the figure whose vertices are the first-neighbour anions round the cation).

The *spin state* of a transition-metal ion refers to its spin angular momentum. Normally this depends on the ground state of the free atom, and is given by Hund's rule; this is the *high-spin state*. In a particular crystal structure, however, ions may be situated in strong ligand fields which reduce the spin angular momentum; this is the *low-spin state*.

In using the table, no *a priori* knowledge of the ionic character of the bonds, i.e. of the degree of polarization or covalent character, is needed. An estimate of the degree of polarization is given by the *Pauling electrostatic valence*, whose value (measured in v.u.) is the valency of the cation divided by its coordination number. The radii tabulated for the different coordination numbers have been so derived as to allow for this, and will predict the mean cation–anion distance within a polyhedron to an accuracy, generally, of about ± 3 pm. Individual distances may, however, show a rather larger scatter. The tabulated values also predict the shortest anion–anion polyhedron edges and packing distances for polyhedra where the Pauling valence is not greater than 1 v.u.; for greater Pauling valences, the polyhedron edges are considerably shorter than the anion diameter.

The radii tabulated are taken, with some simplification, from those listed by R. D. Shannon and C. T. Prewitt, *Acta Cryst.*, 1969, **B25**, 925, with corrections *ibid.*, 1970, **B26**, 1046. Methods of derivation and references to earlier work will be found there. For the sort of corrections to be made for more precise work, see, for example, W. H. Baur, *Trans. Am. Cryst. Assoc.*, 1970, **6**, 129.

Ionic Radii, in pm

Cation	Coordination no.				Cation	Coordination no.				Cation	Coordination no.			
	4	6	8	12		4	6	8	12		4	6	8	12
Li^+	59	74			Pb^{2+}	94	118	129	149	Si^{4+}	26	40		
Na^+		102	116							Ge^{4+}	40	54		
K^+		138	151	160	B^{3+}	12				Sn^{4+}		69		
Rb^+		149	160	173	Al^{3+}	39	53			Pb^{4+}		78	94	
Cs^+		170	182	188	Ga^{3+}	47	62							
					In^{3+}		79	92		Ti^{4+}		60		
Ag^+		115	130		Tl^{3+}		88	100		Zr^{4+}		72	84	
Tl^+		150	160	176						Hf^{4+}		71	83	
					Sc^{3+}		73	87		Th^{4+}		100	106	
Be^{2+}	27				Y^{3+}		90	101		U^{4+}		97	100	
Mg^{2+}	58	72			La^{3+}		106	118	132					
Ca^{2+}		100	112	135						P^{5+}	17	35		
Sr^{2+}		116	125	144	Cr^{3+}		61			As^{5+}	34	50		
Ba^{2+}		136	142	160	Mn^{3+} LS		58			Sb^{5+}		61		
					Mn^{3+} HS		65			Bi^{5+}		74		
Zn^{2+}	60	75			Fe^{3+} LS		55			V^{5+}	36	54		
Cd^{2+}	80	95	107	131	Fe^{3+} HS	49	65			Nb^{5+}	32	64		
					Co^{3+} LS		53			Ta^{5+}		64	69	
Mn^{2+} LS		67			Co^{3+} HS		61							
Mn^{2+} HS		82								S^{6+}	12			
Fe^{2+} LS		61			Bi^{3+}		102	111						
Fe^{2+} HS	63	78								Cr^{6+}	30	52		
Co^{2+} LS		57			Pu^{3+}		100			Mo^{6+}	42	60		
Co^{2+} HS		74								W^{6+}	41	58		
Ni^{2+}		70								U^{6+}	48	75		
Cu^{2+}		73												

Anion	Coordination no.			Anion	Coordination no.				
	2	4	6		2	4	6		
F^-	128	131	133	O^{2-}	135	138	140		
Cl^-		181†		S^{2-}		184†			
Br^-		196†							
I^-		220†							

LS = low-spin state; HS = high-spin state.
† Independent of coordination number.

H.D.M.

2.7.7 Crystal structures

A perfect infinite crystal possesses a *lattice*, an infinite set of points generated by three non-parallel vectors, such that each point is identical in itself and its surroundings.

With each lattice point may be associated a number of atoms. If their coordinates relative to the lattice point are given, together with the lengths and directions of the lattice vectors chosen to define the axes of reference, the complete structure is defined.

Position coordinates x, y, z are commonly expressed as fractions of the *lattice parameters* a, b, c; the parallelepiped defined by the lattice vectors a, b, c, is the *unit cell*.

Symmetry elements may be present imposing certain relations (i) between lattice parameters, (ii) between position coordinates of different atoms. Axes of reference are generally chosen in accordance with the symmetry. As a result of (i), and with a conventional choice of axes, crystals are classified into systems as follows:

Cubic:	$a = b = c,$	$\alpha = \beta = \gamma = 90°$
Tetragonal:	$a = b \neq c,$	$\alpha = \beta = \gamma = 90°$
Orthorhombic:	$a \neq b \neq c,$	$\alpha = \beta = \gamma = 90°$
Hexagonal:	$a = b \neq c,$	$\alpha = \beta = 90°, \gamma = 120°$
Monoclinic:	$a \neq b \neq c,$	$\alpha = \gamma = 90°, \beta \neq 90°$
Triclinic:	$a \neq b \neq c,$	$\alpha \neq \beta \neq \gamma \neq 90°$

where α is the angle between b and c, and similarly for β and γ. Accidental equality of unit cell edges, and special values of interaxial angles not required by the symmetry, are disregarded in making this classification.

Atoms may be in *general positions*, or in *special positions* on point-symmetry elements. In the latter case, some or all of the position coordinates x, y, z are simple fractions; in the former, they are variable parameters whose values may change with temperature, pressure, or composition. It often happens, however, that atoms in general positions have, accidentally, parameters which are to a good approximation simple fractions.

If the translation vectors chosen as axes of reference generate all the points of the lattice, it is said to be *primitive* (P); otherwise it is *centred*. For centred lattices, if there is an atom at x, y, z, all translation-repeats of it within one unit cell can be derived by adding to x, y, z, components which are fractions of the lattice parameters. The kind of centring is summarized in the lattice centring operator: if, for example, this is written $(0, 0, 0; 0, \frac{1}{2}, \frac{1}{2})+$ it means that for every point at $0+x$, $0+y$, $0+z$, there is another identical point at $0+x$, $\frac{1}{2}+y$, $\frac{1}{2}+z$. The complete set of possibilities is as follows.

For any symmetry except hexagonal:

One-face-centred lattice A [2 points],	$(0, 0, 0; \quad 0, \frac{1}{2}, \frac{1}{2})+$
Body-centred lattice I [2 points],	$(0, 0, 0; \quad \frac{1}{2}, \frac{1}{2}, \frac{1}{2})+$
All-face-centred lattice F [4 points]	$(0, 0, 0; \quad 0, \frac{1}{2}, \frac{1}{2}; \quad \frac{1}{2}, 0, \frac{1}{2}; \quad \frac{1}{2}, \frac{1}{2}, 0)+$

For hexagonal symmetry only:

Rhombohedrally centred lattice R [3 points],	$(0, 0, 0; \quad \frac{2}{3}, \frac{1}{3}, \frac{1}{3}; \quad \frac{1}{3}, \frac{2}{3}, \frac{2}{3})+$

There are expressions similar to A for different choices of axes, giving B- and C-face-centring. A- or B-face-centring is impossible in the tetragonal system; A-, B- or C- in the cubic. The same lattice could be named either P or C in the tetragonal and monoclinic systems, according to the choice of axes; similarly for I and F. A rhombohedrally centred lattice rotated through 180° from the conventional (obverse) orientation listed above is obtained by the operations $(0, 0, 0; \quad \frac{1}{3}, \frac{2}{3}, \frac{1}{3}; \quad \frac{2}{3}, \frac{1}{3}, \frac{2}{3})+$.

To describe a structure fully, we specify its system, its independent lattice parameters (thus choosing our axes of reference), its lattice type, and the position parameters of a set of atoms so chosen that (a) no two are separated by a lattice vector, (b) no atom not excluded by (a) is omitted. This description is complete and general, applicable to crystals of any system and any degree of complexity. Note that, if a letter or number specifying a position parameter is negative, it is conventional to write a bar over it instead of a minus sign in front of it: thus $\bar{x}, \bar{x}, \frac{1}{4}$ stands for $-x, -x, \frac{1}{4}$.

If the space group is known, the description can be abbreviated by listing position parameters for the *asymmetric unit* only, i.e. the set of atoms not related to each other by any symmetry element. This kind of abbreviation will not be used for the structures described below (see *International Tables for X-ray Crystallography*, Vol. 1, for details of its use).

Some simple structures

The structures described below are particularly simple and important types. Most of them are cubic, and most of them have all their atoms in special positions. The structure type is commonly named after the particular material described, except where otherwise specified. Lattice parameters and variable position parameters (if any) apply to this material. Other isostructural (isomorphous) materials (of which examples are given in the table) have slightly different numerical values for these parameters. A short qualitative comment is added to the formal description, which, however, is sufficient in itself to allow scale diagrams to be drawn and interatomic distances calculated.

Note that atoms are not separately listed if their coordinates can be derived from those listed by the use of lattice centring operators, as explained above. As a check, the total number of atoms of each kind in the unit cell is indicated.

1. *Copper ('Monatomic face-centred cubic')*
   Cubic; all-face-centred lattice F
   $a = 361.5$ pm
   4 Cu at 0, 0, 0
       Each atom has 12 equidistant neighbours, in directions parallel to the face diagonal of the cube, at distances $a/\sqrt{2}$. The structure is a *cubic close packing* of equal spheres.
       An alternative description uses hexagonal axes of reference, with c_H along the diagonal of the cube and a_H along a face diagonal perpendicular to it; this unit cell is rhombohedrally centred and contains 3 Cu.

2. *Iron ('Monatomic body-centred cubic')*
   Cubic; body-centred lattice I
   $a = 286.6$ pm
   2 Fe at 0, 0, 0
       Each atom has 8 equidistant neighbours in directions parallel to the cube body diagonal, at distances $\sqrt{(3)}a/2$.

3. *Magnesium ('Hexagonal close-packed')*
   Hexagonal; primitive lattice P
   $a = 320.9$ pm, $c = 521.0$ pm
   2 Mg at $\pm(\frac{2}{3}, \frac{1}{3}, \frac{1}{4})$
   With an alternative choice of origin, Mg atoms are at 0, 0, 0 and $\frac{2}{3}, \frac{1}{3}, \frac{1}{2}$.
       Each atom has 6 equidistant neighbours in its own plane at a distance a, and two sets of 3 above and below it. If c/a has the ideal value of 1.633, the whole array is in *hexagonal close packing*. The unit cell contains two close-packed layers, not related by a lattice vector (in contrast to cubic close packing, with three layers related by lattice vectors).

4. *Diamond*
   Cubic; all-face-centred lattice F
   $a = 356$ pm
   8 C at $\pm(\frac{1}{8}, \frac{1}{8}, \frac{1}{8})$
   With an alternative choice of origin, C atoms are at 0, 0, 0 and $\frac{1}{4}, \frac{1}{4}, \frac{1}{4}$.

Each atom has 4 equidistant neighbours in directions parallel to the body diagonals of the cube, at distances $\sqrt{(3)}a/4$; but adjacent atoms have all their nearest-neighbour bonds oppositely directed. The structure could be derived by taking 8 body-centred cells, of side $a/2$, and leaving out half the atoms systematically. The C—C bonds are at the tetrahedral angle of $109\frac{1}{2}°$.

5. *Rock Salt*, NaCl
Cubic; all face-centred lattice F
$a = 563$ pm
4 Na at 0, 0, 0
4 Cl at $\frac{1}{2}$, 0, 0
With an alternative choice of origin, the position coordinates of Na and Cl can be interchanged.

Each atom has 6 neighbours of the opposite kind, in directions parallel to the cube edges, at a distance $a/2$.

6. *Caesium Chloride*, CsCl
Cubic; primitive lattice P
$a = 411$ pm
1 Cs at 0, 0, 0
1 Cl at $\frac{1}{2}$, $\frac{1}{2}$, $\frac{1}{2}$
With an alternative choice of origin, the position coordinates of Cs and Cl can be interchanged.

Each atom has 8 neighbours of the opposite kind, in directions parallel to the cube body diagonal, at a distance of $\sqrt{(3)}a/2$.

The structure is closely related to that of iron, to which it would reduce if Cs and Cl were replaced by indistinguishable atoms.

7. *Fluorite*, CaF$_2$
Cubic; all-face-centred lattice F
$a = 545$ pm
4 Ca at 0, 0, 0
8 F at $\pm(\frac{1}{4}, \frac{1}{4}, \frac{1}{4})$
Each Ca atom has 8 equidistant F neighbours, at the corners of a cube; each F atom has 4 equidistant Ca neighbours, at the vertices of a regular tetrahedron. The Ca–F distance is $\sqrt{(3)}a/4$.

8. *Zinc blende*, ZnS
Cubic; all-face-centred lattice F
$a = 542$ pm
4 Zn at 0, 0, 0
4 S at $\frac{1}{4}$, $\frac{1}{4}$, $\frac{1}{4}$
Each Zn atom is surrounded by 4 equidistant S atoms, at the corners of a regular tetrahedron, at a distance $\sqrt{(3)}a/4$; similarly, each S atom is surrounded tetrahedrally by 4 Zn atoms. All the Zn—S bonds lying parallel to a given body diagonal of the cube point in the same direction.

The structure is closely related to that of diamond, to which it would reduce if Zn and S were replaced by indistinguishable atoms.

Alternatively, the structure may be described using hexagonal axes of reference, chosen as for copper (q.v.); the unit cell is rhombohedrally centred, and contains 3 atoms of each kind, with Zn at 0, 0, 0 and S at 0, 0, $\frac{1}{4}$ (or by reversing the sense of the c-axis and changing the origin these coordinates may be interchanged).

Unlike any of the structures previously described, this has no centre of symmetry.

9. *Wurtzite*, ZnS
Hexagonal; primitive lattice P
$a = 382$ pm, $c = 626$ pm
2 Zn at $\frac{2}{3}$, $\frac{1}{3}$, 0; $\frac{1}{3}$, $\frac{2}{3}$, $\frac{1}{2}$
2 S at $\frac{2}{3}$, $\frac{1}{3}$, u; $\frac{1}{3}$, $\frac{2}{3}$, $\frac{1}{2}+u$; with $u \simeq \frac{3}{8}$

As in the zinc-blende structure, each Zn atom is surrounded by 4 S atoms at the corners of a tetrahedron, and each S similarly by 4 Zn atoms. The tetrahedra are, however, only regular if c/a and u have the ideal values of 1.633 and 0.375 respectively. The Zn—S bonds parallel to the c-axis all point in the same direction; in contrast to the zinc-blende structure, this is a unique axis, and the symmetry is therefore *polar*.

The relation between the structures of wurtzite and zinc blende is the same as that between the structures of magnesium and copper, i.e. between the operations of hexagonal close packing and cubic close packing.

Intermediates between the wurtzite and zinc-blende structure, sometimes of considerable complexity ('polytypes'), are found in a number of materials, including ZnS itself and SiC (carborundum).

10. *Rutile*, TiO_2

Tetragonal; primitive lattice P

$a = 459.4$ pm, $c = 296.2$ pm

2 Ti at $0, 0, 0$ and $\frac{1}{2}, \frac{1}{2}, \frac{1}{2}$

4 O at $\pm(u, u, 0)$ and $\pm(\frac{1}{2}+u, \frac{1}{2}-u, \frac{1}{2})$, with $u = 0.31$

Each Ti atom has as neighbours 6 O atoms, forming a nearly regular octahedron. Edges of the octahedra parallel to face-diagonals of the square base are shared with similar octahedra, forming chains parallel to the c-axis; octahedra in neighbouring chains share corners, making each O atom neighbour to 3 Ti atoms.

11. *Ideal perovskite: example*, $SrTiO_3$

('Perovskite' is the mineral name of $CaTiO_3$, whose actual structure, though it approximates to that of $SrTiO_3$, is more complicated and of lower symmetry.)

Cubic; primitive lattice P

$a = 390.5$ pm

1 Ti at $0, 0, 0$

1 Sr at $\frac{1}{2}, \frac{1}{2}, \frac{1}{2}$

3 O at $\frac{1}{2}, 0, 0;\ \ 0, \frac{1}{2}, 0;\ \ 0, 0, \frac{1}{2}$

Each Ti atom has 6 neighbouring O atoms, at a distance $a/2$, forming a regular octahedron. Each octahedron shares each corner with one similar octahedron to form a three-dimensional framework, with cavities holding the larger Sr atoms. Each Sr atom has 12 neighbouring O atoms at a distance $\sqrt{(2)}a/2$. Each O atom is linked to 2 Ti atoms and 4 Sr atoms.

The ideal perovskite structure imposes a particular relation between ionic radii as a condition of cation–anion contact for both cations, and only occurs when this is nearly satisfied, as in $SrTiO_3$. With moderate misfit, related structures of lower symmetry are found, collectively described as 'structures of the perovskite family'. Many materials possessing such structures at room temperature have a high-temperature form with the ideal perovskite structure.

12. *Calcite*, $CaCO_3$

Rhombohedrally centred hexagonal lattice, R

$a_H = 499.0$ pm, $c_H = 1700$ pm

6 Ca at $0, 0, 0$ and $0, 0, \frac{1}{2}$

6 C at $0, 0, \frac{1}{4}$ and $0, 0, \frac{3}{4}$

18 O at $x, 0, \frac{1}{4};\ \ 0, x, \frac{1}{4};\ \ \bar{x}, \bar{x}, \frac{1}{4};\ \ \bar{x}, 0, \frac{3}{4};\ \ 0, \bar{x}, \frac{3}{4};\ \ x, x, \frac{3}{4};$ with $x = 0.257$

Each Ca atom has 6 equidistant O atoms at the vertices of an octahedron. Each O atom is shared between two octahedra, and also forms one corner of an equilateral triangle, perpendicular to the c-axis, with the carbon atom at its centre. The edges of the CO_3 triangle are shorter than any of the octahedron edges. CO_3 groups in successive layers perpendicular to the c-axis are rotated through 180°.

An alternative description uses the edges of the primitive rhombohedron as axes of reference; then $a_r = 636$ pm, $\alpha = 46°\ 05'$.

If the CO_3 groups were replaced by cylindrical discs, the unit cell would be half the height; the primitive rhombohedron would have $a_r' = 404$ pm, $\alpha' = 76°\ 10'$.

References

(1) For a full account of crystallographic symmetry and notation, see *International Tables for X-ray Crystallography*, vol. I (ed. N. F. M. Henry and K. Lonsdale), Kynoch Press (Birmingham, England) for the International Union of Crystallography.

(2) For a classified reference book of crystal structures, with lattice parameters and position parameters, see R. W. G. Wyckoff, *Crystal Structures*, 2nd edn (Interscience, New York).

(3) For a description and discussion of numerous important structures and their symmetry, see H. D. Megaw, *Crystal Structures—A Working Approach* (Saunders, Philadelphia, Pa.).

Table of Materials

Name of structure type	Some representative examples†
1. Copper (A1)‡	Cu, Ag, Au, Al, Pt, Ni, Pb
2. Iron (A2)	Fe, Mo, W, Li, Na, K, Ba
3. Magnesium (A3)	Mg, Be, Gd, Rh, Zr; Zn§, Cd§
4. Diamond (A4)	C (diamond), Si, Ge, Sn (grey)
5. Rock salt (sodium chloride) (B1)	NaCl, and most alkali halides except Cs compounds; MgO, CaO, SrO, FeO, CaS, CaSe, AsSn, ThC
6. Caesium chloride (B2)	CsCl, CsBr, CsI, TlCl; many ordered intermetallic compounds such as AgMg, AuMg, BaCd
7. Fluorite (calcium fluoride) (C1)	CaF_2, BaF_2, UO_2, ThO_2, K_2O, UN_2, $AuAl_2$
8. Zinc blende (B3)	ZnS, ZnSe, BeS, CdS, GaAs, BN
9. Wurtzite (B4)	ZnS, ZnSe, ZnO, BeO, CdS, GaN
10. Rutile (C4)	TiO_2, SnO_2, MnO_2, CrO_2, MgF_2, FeF_2, MnF_2
11. Ideal perovskite	$SrTiO_3$, $KTaO_3$, $BaZrO_3$, $BaSnO_3$, $DyMnO_3$, $KMgF_3$, $KFeF_3$; also high-temperature forms of $BaTiO_3$, $KNbO_3$, $NaNbO_3$, $LaAlO_3$; also, by omission of the 12-coordinated cation, ReO_3, RhF_3, MoF_3
12. Calcite	$CaCO_3$, $MgCO_3$, $FeCO_3$, $NaNO_3$

† Unless otherwise specified, the structures listed are those found at room temperature.

‡ The type names (A1), (A2), etc. are those given to the structures in Vol. 1 of *Strukturbericht* (1920) and are still sometimes found in the literature.

§ These elements have axial ratios c/a much greater than the ideal value characteristic of close packing.

H.D.M.

2.8 Nuclear magnetic resonance and Mössbauer effect

2.8.1 Nuclear moments and magnetic resonance

The table includes all naturally occurring isotopes having non-zero spin (excluding ^{180}Ta). The magnetic moment, μ, is given in units of the nuclear magneton μ_n. The resonant frequency of an isotope is given by

$$v = \mu H / hI$$

where H is the magnetic field at the nucleus, h is Planck's constant and I is the nuclear spin. Since radio frequencies are among the most accurately measurable physical quantities known, nuclear magnetic resonance (nmr) measurements should provide highly accurate values of the nuclear moments. However, it was found that the resonance frequency of a given isotope is dependent of the chemical nature of the compound containing that isotope. This 'chemical shift' effect, while limiting the accuracy of nuclear moment measurements, has proved to be a most important tool for the determination of molecular structures (see 2.8.2).

The values of the nuclear moments given are corrected for diamagnetism, arrived at from considering values obtained by various experimental techniques including nmr, atomic beam resonance, electron spin resonance, optical spectroscopy and electron-nuclear double resonance (ENDOR). For a few nuclei the moment values obtained by two different methods are consistent and when this occurs those values are quoted but when discrepancies exist the nmr values have been listed. See G. H. Fuller *J. Phys. Chem. Ref. Data*, 1976, **5**, 835, for further details, from which reference most of the quoted nuclear moment and nuclear electronic quadrupole moment values were selected.

Nuclear electric quadrupole moments are conveniently written $e.Q$ where e is the charge on the electron and the parameter Q is in units of 10^{-28} m^2 (barns). The quadrupole moment values reported are averages of the values reported in the literature (or derived from these), since some uncertainty attaches to the accuracy of any particular value due to the difficulties in estimating the electric field inhomogeneity at the nucleus arising from its molecular and electronic environment. Only those values marked * contain any correction for polarization shielding effects. Such corrections are difficult to quantify and may amount to 30 per cent of the Q value for some nuclei. Spin $\frac{1}{2}$ nuclei have no quadrupole moment.

The nmr frequency values tabulated are those for an externally applied field of 1 tesla (10 kilogauss).

The chemical reference standards listed in the table are either those universally employed (e.g. (Me_4Si) or (H_3PO_4)) or those frequently used or in a few instances are the only compound of the nuclide to have been observed by nmr techniques.

Some abbreviations used in the table are aq. for aqueous solution, Et for ethyl, Me for methyl, neat for undiluted compound.

The sign of μ/μ_n and of $Q/10^{-28}$m^2 is uncertain for those nuclides for which no sign is given.

Nuclear spins, moments and resonant frequencies

Nuclide	I	μ/μ_N	$Q/10^{-28}$ m^2	v/MHz	Chemical shift reference standard
^{0}n	$\frac{1}{2}$	-1.91312	—	29.167	—
^{1}H	$\frac{1}{2}$	$+2.79278$	—	42.5760	Me$_4$Si
^{2}H	1	$+0.8572$	$+0.0028$	6.5357	—
^{3}H	$\frac{1}{2}$	$+2.9789$	—	45.4131	Me$_4$Si-tritiated
^{3}He	$\frac{1}{2}$	-2.1276	—	32.4338	—
^{6}Li	1	$+0.82203$	-0.0008	6.2655	Li$^+$aq.
^{7}Li	$\frac{3}{2}$	$+3.25636$	-0.04	16.5465	Li$^+$aq.
^{9}Be	$\frac{3}{2}$	-1.17745	$+0.05$	5.9827	Be^{2+}aq.

Nuclear spins, moments and resonant frequencies (*contd*)

Nuclide	I	μ/μ_N	$Q/10^{-28}\ m^2$	ν/MHz	Chemical shift reference standard
^{10}B	3	$+1.8006$	$+0.085*$	4.574	$Et_2O.BF_3$ or BH_4^-
^{11}B	$\frac{3}{2}$	$+2.6885$	$+0.041*$	13.6595	$Et_2O.BF_3$ or BH_4^-
^{13}C	$\frac{1}{2}$	$+0.7024$		10.7054	Me_4Si
^{14}N	1	$+0.40375$	$+0.01$	3.076	$MeNO_2$ or NO_3^-
^{15}N	$\frac{1}{2}$	-0.2831	—	4.315	$MeNO_2$ or NO_3^-
^{17}O	$\frac{5}{2}$	-1.8937	$-0.026*$	5.7719	H_2O
^{19}F	$\frac{1}{2}$	$+2.6288$		40.0543	CCl_3F
^{21}Ne	$\frac{3}{2}$	-0.66176	$+0.09$	3.3611	—
^{23}Na	$\frac{3}{2}$	$+2.21740$	$+0.10*$	11.2621	Na^+aq.
^{25}Mg	$\frac{5}{2}$	-0.8554	$+0.22$	2.606	Mg^{2+}aq.
^{27}Al	$\frac{5}{2}$	$+3.6413$	$+0.15^+$	11.0940	$[Al(H_2O)_6]^{3+}$
^{29}Si	$\frac{1}{2}$	-0.55526		8.458	Me_4Si
^{31}P	$\frac{1}{2}$	$+1.1317$	—	17.238	$85\%\ H_3PO_4$
^{33}S	$\frac{3}{2}$	$+0.6435$	-0.055	3.266	CS_2 or SO_4^{2-}
^{35}Cl	$\frac{3}{2}$	$+0.82181$	$-0.10*$	4.1717	Cl^-aq.
^{36}Cl	2	$+1.2853$	$-0.021*$	4.8931	—
^{37}Cl	$\frac{3}{2}$	$+0.68407$	$-0.079*$	3.472	Cl^-aq.
^{39}K	$\frac{3}{2}$	$+0.39147$	$+0.049*$	1.9864	K^+aq.
^{40}K	4	-1.2981	$-0.061*$	2.470	—
^{41}K	$\frac{3}{2}$	$+0.21487$	$+0.060*$	1.0903	K^+aq.
^{43}Ca	$\frac{7}{2}$	-1.3172	—	2.8654	Ca^{2+}aq.
^{45}Sc	$\frac{7}{2}$	$+4.7559$	-0.22	10.3434	$Sc(ClO_4)_3$aq.
^{47}Ti	$\frac{5}{2}$	-0.78846	$+0.29$	2.3997	$TiF_6^{2-}/48\%HF$
^{49}Ti	$\frac{7}{2}$	-1.10414	$+0.24$	2.4004	$TiF_6^{2-}/48\%HF$
^{50}V	6	$+3.3470$	0.06	4.243	—
^{51}V	$\frac{7}{2}$	$+5.1485$	-0.05	11.1922	$VOCl_3$ neat
^{53}Cr	$\frac{3}{2}$	-0.4744	$+0.03$	2.4063	$[CrO_4]^{2-}$
^{55}Mn	$\frac{5}{2}$	$+3.4680$	$+0.4$	10.5542	$KMnO_4$aq.
^{57}Fe	$\frac{1}{2}$	$+0.09060$	—	1.38	$Fe(CO)_5$
^{59}Co	$\frac{7}{2}$	$+4.616$	$+0.38$	10.072	$[Co(CN)_6]^{3-}$aq.
^{61}Ni	$\frac{3}{2}$	-0.7498	$+0.16*$	3.8048	$Ni(CO)_4$ neat
^{63}Cu	$\frac{3}{2}$	$+2.2262$	$-0.211*$	11.285	$[Cu(CN)_4]^{3-}$
^{65}Cu	$\frac{3}{2}$	$+2.3849$	$-0.195*$	12.090	$[Cu(CN)_4]^{3-}$
^{67}Zn	$\frac{5}{2}$	$+0.8756$	$+0.16$	2.663	Zn^{2+}aq.
^{69}Ga	$\frac{3}{2}$	$+2.0161$	$+0.19$	10.2188	$[Ga(H_2O)_6]^{3+}$
^{71}Ga	$\frac{3}{2}$	$+2.5617$	$+0.12$	12.9840	$[Ga(H_2O)_6]^{3+}$
^{73}Ge	$\frac{9}{2}$	-0.87918	-0.18	1.485	$GeCl_4$ neat
^{75}As	$\frac{3}{2}$	$+1.439$	$+0.29$	7.292	$KAsF_6$
^{77}Se	$\frac{1}{2}$	$+0.534$	—	8.118	Me_2Se
^{79}Br	$\frac{3}{2}$	$+2.1055$	$+0.37*$	10.669	Br^-aq.
^{81}Br	$\frac{3}{2}$	$+2.2696$	$+0.31*$	11.498	Br^-aq.
^{83}Kr	$\frac{9}{2}$	-0.9703	$+0.26$	1.6380	—
^{85}Rb	$\frac{5}{2}$	$+1.3527$	$+0.26*$	4.111	Rb^+aq.
^{87}Rb	$\frac{3}{2}$	$+2.7506$	$+0.13*$	13.932	Rb^+aq.
^{87}Sr	$\frac{9}{2}$	-1.093	$+0.3$	1.8451	Sr^{2+}aq.
^{89}Y	$\frac{1}{2}$	-0.13733	—	2.0860	$Y(NO_3)_3$aq.
^{91}Zr	$\frac{5}{2}$	-1.3028	—	3.9578	
^{93}Nb	$\frac{9}{2}$	$+6.167$	-0.22	10.4048	$NbF_6^-/48\%HF.$

Nuclear spins, moments and resonant frequencies (contd)

Nuclide	I	μ/μ_N	$Q/10^{-28}\ m^2$	ν/MHz	Chemical shift reference standard
^{95}Mo	$\frac{5}{2}$	-0.9135	0.12	2.774	$[MoO_4]^{2-}$
^{97}Mo	$\frac{5}{2}$	-0.9327	1.1	2.832	$[MoO_4]^{2-}$
^{99}Tc	$\frac{9}{2}$	$+5.681$	$+0.3$	9.5832	$[TcO_4]^-$ aq.
^{99}Ru	$\frac{5}{2}$	-0.64	—	1.9	$K_4[Ru(CN)_6]$ aq.
^{101}Ru	$\frac{5}{2}$	-0.72	—	2.1	
^{103}Rh	$\frac{1}{2}$	-0.0883	—	1.3434	mer$[RhCl_3(SMe_2)_3]$
^{105}Pd	$\frac{5}{2}$	-0.642	$+0.8$	1.8921	
^{107}Ag	$\frac{1}{2}$	-0.1135	—	1.7230	
^{109}Ag	$\frac{1}{2}$	-0.1305	—	1.9808	Ag^+ aq.
^{111}Cd	$\frac{1}{2}$	-0.59500	—	9.0283	$CdMe_2$ neat
^{113}Cd	$\frac{1}{2}$	-0.62245	—	9.4441	$CdMe_2$ neat
^{113}In	$\frac{9}{2}$	$+5.5229$	$+0.82$	9.312	$[In(H_2O)_6]^{3+}$
^{115}In	$\frac{9}{2}$	$+5.5348$	$+0.83$	9.331	—
^{115}Sn	$\frac{1}{2}$	-0.9178	—	13.922	
^{117}Sn	$\frac{1}{2}$	-0.9999	—	15.168	
^{119}Sn	$\frac{1}{2}$	-1.0461	—	15.868	Me_4Sn
^{121}Sb	$\frac{5}{2}$	$+3.3592$	-0.28	10.192	$SbCl_6^-$
^{123}Sb	$\frac{7}{2}$	$+2.5466$	-0.36	5.519	$SbCl_6^-$
^{123}Te	$\frac{1}{2}$	-0.7359	—	11.159	
^{125}Te	$\frac{1}{2}$	-0.8872	—	13.453	Me_2Te
^{127}I	$\frac{5}{2}$	$+2.8091$	-0.79	8.517	I^- aq.
^{129}I	$\frac{7}{2}$	$+2.6174$	-0.55	5.6694	
^{129}Xe	$\frac{1}{2}$	-0.7768	—	11.777	$XeOF_4$
^{131}Xe	$\frac{3}{2}$	$+0.6908$	-0.12	3.4902	
^{133}Cs	$\frac{7}{2}$	$+2.5788$	-0.0030*	5.58469	Cs^+ aq.
^{135}Ba	$\frac{3}{2}$	$+0.8372$	$+0.18$	4.2295	Ba^{2+} aq.
^{137}Ba	$\frac{3}{2}$	$+0.9365$	$+0.28$	4.7314	Ba^{2+} aq.
^{138}La	5	$+3.704$	0.51*	5.6171	—
^{139}La	$\frac{7}{2}$	$+2.778$	$+0.22$*	6.0144	—
^{141}Pr	$\frac{5}{2}$	$+4.16$	-0.058	13.0	—
^{143}Nd	$\frac{7}{2}$	-1.063	-0.48	2.303	—
^{145}Nd	$\frac{7}{2}$	-0.654	-0.25	1.414	—
^{147}Sm	$\frac{7}{2}$	-0.813	-0.18	1.72	—
^{149}Sm	$\frac{7}{2}$	-0.670	$+0.052$	1.39	—
^{151}Eu	$\frac{5}{2}$	$+3.466$	$+1.1$	10.35	—
^{153}Eu	$\frac{5}{2}$	$+1.530$	$+2.8$	4.56	—
^{155}Gd	$\frac{3}{2}$	-0.2584	$+1.6$	1.4	—
^{157}Gd	$\frac{3}{2}$	-0.3388	$+1.7$	1.9	—
^{159}Tb	$\frac{3}{2}$	2.008	$+1.3$*	9.7	—
^{161}Dy	$\frac{5}{2}$	-0.482	$+2.4$*	1.39	—
^{163}Dy	$\frac{5}{2}$	$+0.676$	$+2.5$*	1.94	—
^{165}Ho	$\frac{7}{2}$	$+4.12$	$+2.7$	8.73	—
^{167}Er	$\frac{7}{2}$	-0.569	$+2.83$	1.23	—
^{169}Tm	$\frac{1}{2}$	-0.231	—	3.47	—
^{171}Yb	$\frac{1}{2}$	$+0.4930$	—	7.456	—
^{173}Yb	$\frac{5}{2}$	-0.6791	$+3.0$	2.050	—
^{175}Lu	$\frac{7}{2}$	$+2.230$	$+5.6$	4.8189	—
^{176}Lu	7	$+3.18$	$+8.0$	3.39	—
^{177}Hf	$\frac{7}{2}$	$+0.7902$	$+4.5$*	1.3	—

Nuclear spins, moments and resonant frequencies (*contd*)

Nuclide	I	μ/μ_N	$Q/10^{-28}\ m^2$	ν/MHz	Chemical shift reference standard
^{179}Hf	$\frac{9}{2}$	-0.638	$+5.1*$	0.8	—
^{181}Ta	$\frac{7}{2}$	$+2.35$	$+3*$	5.096	TaF_6^- in HF/HNO_3
^{183}W	$\frac{1}{2}$	$+0.1169$	—	1.7716	WF_6 neat
^{185}Re	$\frac{5}{2}$	$+3.172$	$+2.3$	9.586	ReO_4^- aq.
^{187}Re	$\frac{5}{2}$	$+3.204$	$+2.2$	9.684	ReO_4^- aq.
^{187}Os	$\frac{1}{2}$	$+0.0643$	—	1.0	OsO_4
^{189}Os	$\frac{3}{2}$	$+0.6565$	$+0.8*$	3.304	OsO_4
^{191}Ir	$\frac{3}{2}$	$+0.1454$	$+1.1$	0.9	—
^{193}Ir	$\frac{3}{2}$	$+0.1583$	$+1.0$	0.9	—
^{195}Pt	$\frac{1}{2}$	$+0.6022$	—	9.1523	$[Pt(CN)_6]^{2-}$
^{197}Au	$\frac{3}{2}$	$+0.14486$	$+0.59$	0.7412	—
^{199}Hg	$\frac{1}{2}$	$+0.50415$	—	7.612	Me_2Hg neat
^{201}Hg	$\frac{3}{2}$	-0.55830	$+0.44$	2.8781	Me_2Hg neat
^{203}Tl	$\frac{1}{2}$	$+1.6115$	—	24.332	—
^{205}Tl	$\frac{1}{2}$	$+1.627$	—	24.567	$TlNO_3$aq.
^{207}Pb	$\frac{1}{2}$	$+0.5881$	—	8.898	Me_4Pb
^{209}Bi	$\frac{9}{2}$	$+4.080$	-0.38	6.842	—
^{235}U	$\frac{7}{2}$	-0.43	$+4.9$	0.75	—

References

(1) G. H. Fuller, *J. Phys. Chem. Ref. Data*, 1976, **5**, 835.
(2) R. K. Harris and B. E. Mann (eds), *NMR and the Periodic Table*, 1978 (Academic Press).

C.P.R.

2.8.2 NMR chemical shifts in diamagnetic molecules

Chemical shift values (δ) are defined to be $\delta = 10^6\ (\nu_R - \nu)/\nu_R$ where ν and ν_R are the resonance frequencies of a nuclei in a sample and of the same nuclei in a reference standard. The chemical shift value is thus dimensionless and is expressed in parts per million (ppm). A material is chosen as the most convenient reference standard and its resonance frequency is assigned a chemical shift value of zero. The sign convention recommended by IUPAC for all nuclei is that the signals of frequencies greater than that of the reference standard are given positive chemical shift values. A contrary sign convention was used in early nmr studies of certain nuclei, for example ^{19}F and ^{31}P, and care needs to be exercised, even today, to ascertain the sign convention authors are using, especially for nuclei other than ^{1}H and ^{13}C. A most convenient presentation of the relationship between molecular structure and chemical shift is the bar chart. Such charts exist for many active nmr elements and bar charts for ^{1}H, ^{13}C and ^{15}N are given here. The charts for ^{1}H and ^{13}C are widely used because of their importance in organic, organo-metallic and metal complex chemistry. The nmr spectroscopy of ^{15}N has not hitherto been widely studied because of the experimental difficulties associated with the small magnetic moment and low natural abundance of that nuclide, but recent improvements in nmr spectrometers have largely overcome the problem of low sensitivity. Consequently the nmr spectroscopy of ^{15}N promises to become important because of the widespread occurrence and range of bond types of nitrogen. The column marked No. CPDS on the ^{1}H bar charts shows the number of compounds from which the diagrams were constructed and indicates the degree of confidence to be placed in the shift ranges depicted.

The ^{13}C nmr chart indicates the multiplicity of the ^{13}C nmr signal if recorded under conditions of proton-coupling or proton single frequency off-resonance decoupling:

$$S = \text{singlet}, \quad D = \text{doublet}, \quad T = \text{triplet and } Q = \text{quadruplet}.$$

Such internuclear spin-spin couplings are of great importance in molecular structure analyses.

References

(1) N. F. Chamberlain, *The Practice of NMR Spectroscopy—with Spectra-Structure Correlations for Hydrogen-1*, 1974 (Plenum Press).

(2) F. W. Wehrli and T. Wirthlin, *Interpretation of Carbon-13 NMR Spectra*, 1978 (Heyden).

(3) T. Clerc, E. Pretsch, J. Seibl and W. Simon, *Tabellen zur Strukturaufklärung organischer Verbindungen mit spektroskopischen Methoden*, 1981 (Springer-Verlag).

(4) G. C. Levy and R. L. Lichter, ^{15}N *nmr Spectroscopy*, 1979 (Wiley).

(5) J. Mason, *Chemistry in Britain*, 1983, **19**, 654.

Ranges of ^1H chemical shifts in functional groups (see also next chart)

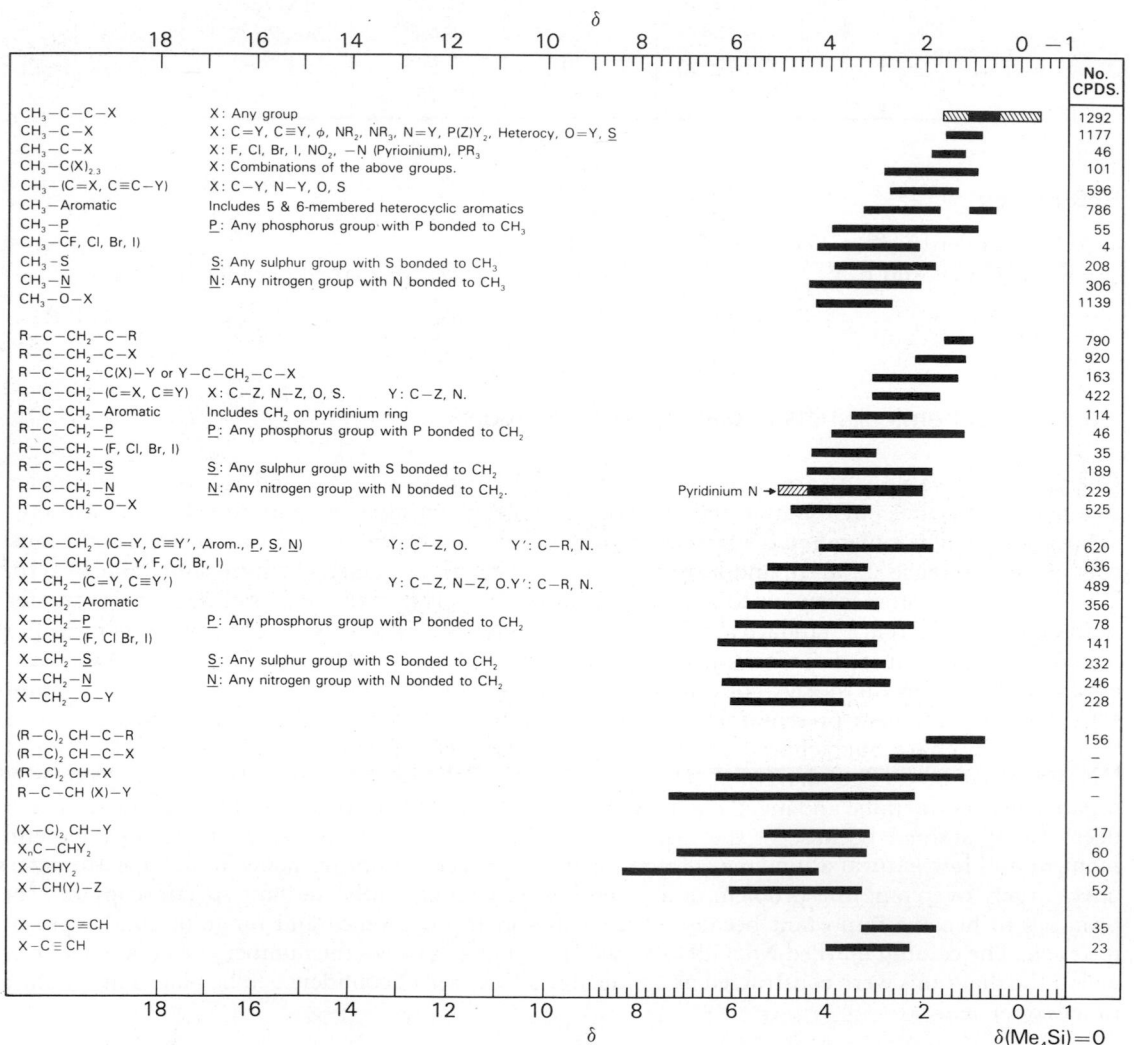

Ranges of ¹H chemical shifts in functional groups *(contd)*

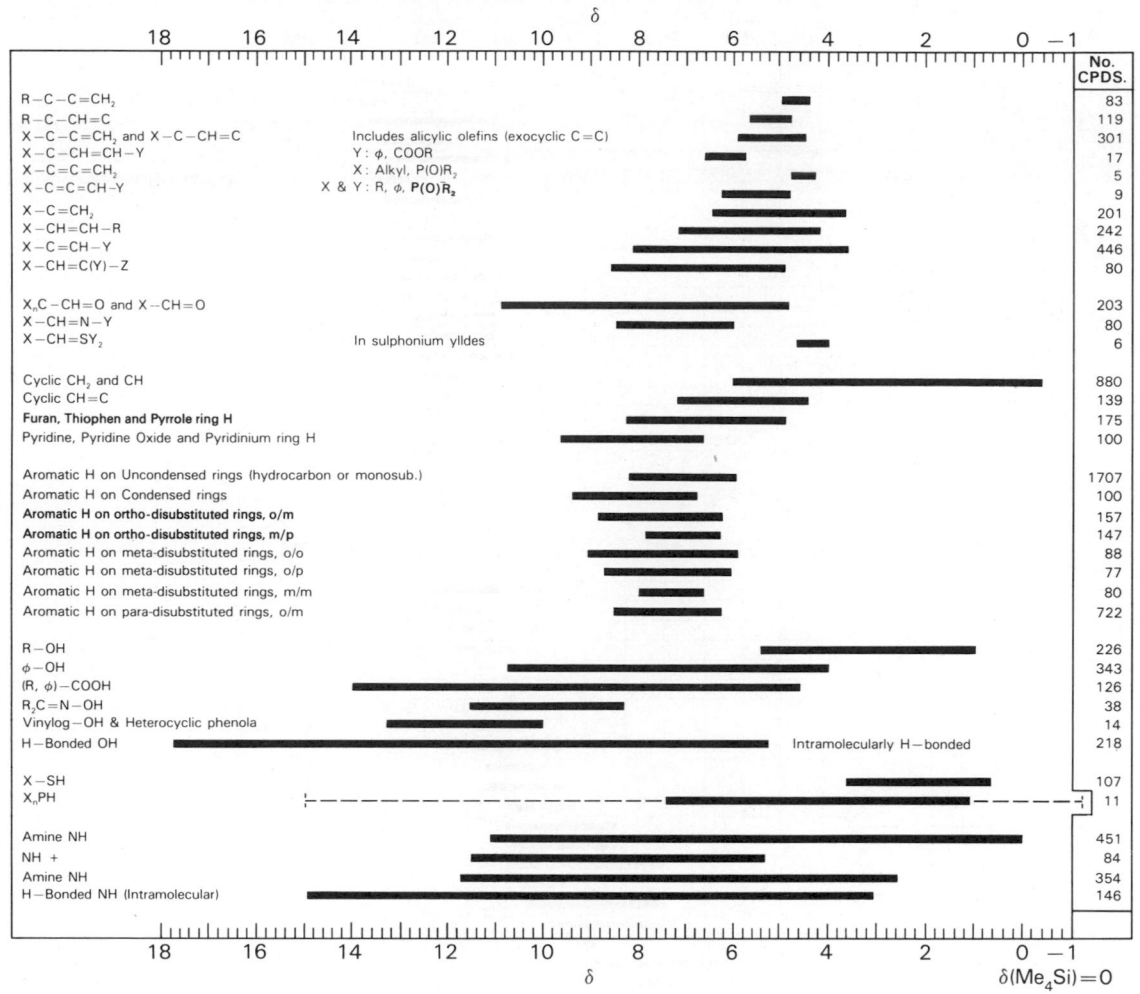

Ranges of ^{13}C chemical shifts in functional groups

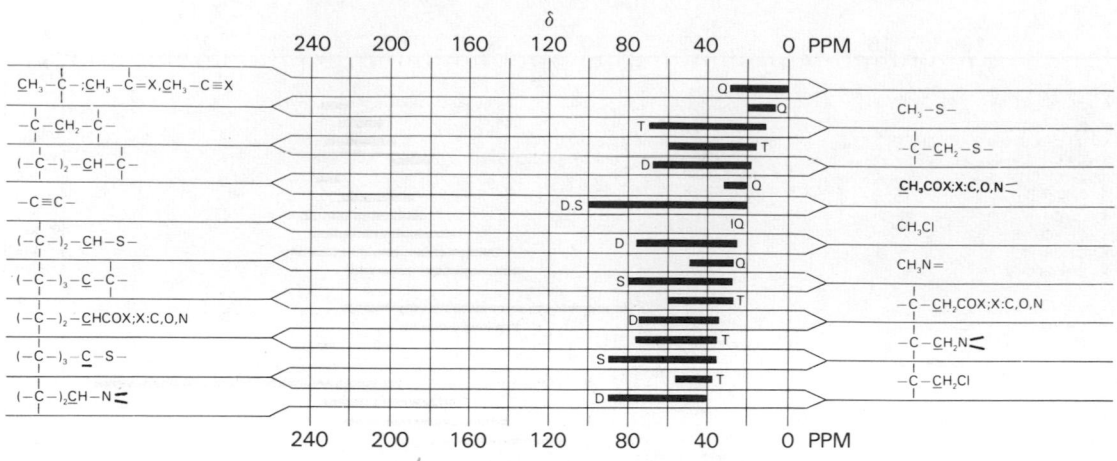

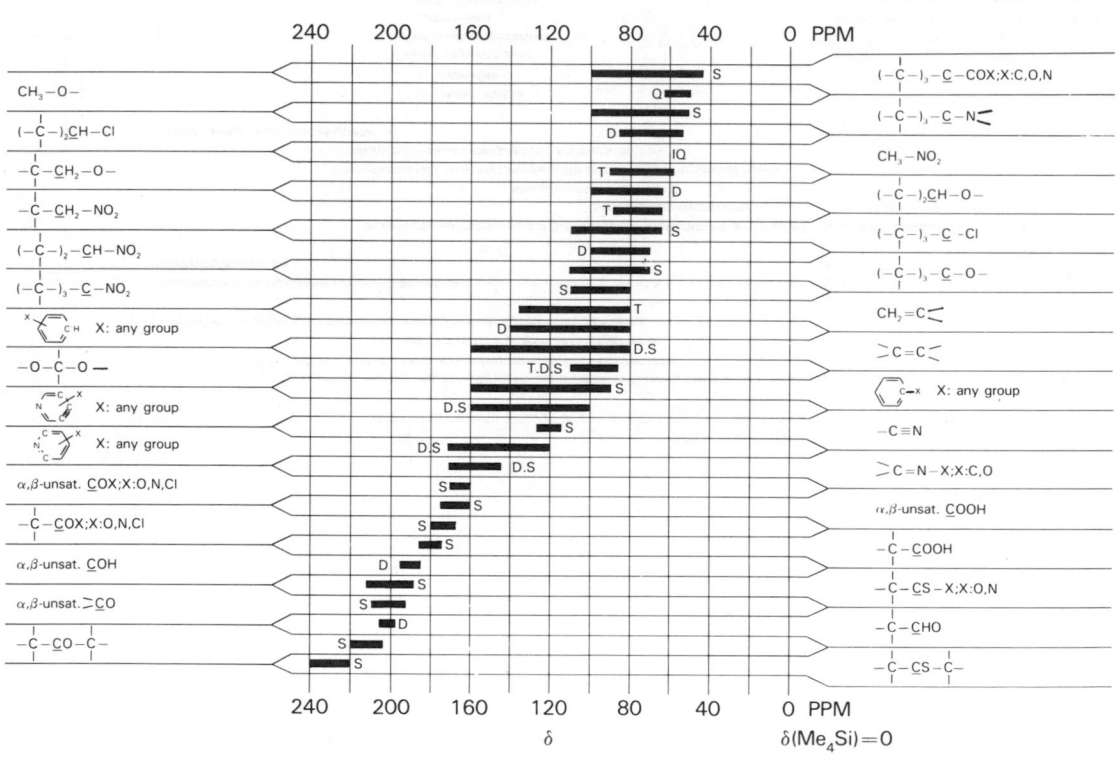

$\delta(Me_4Si) = 0$

Ranges of ^{15}N chemical shifts in functional groups

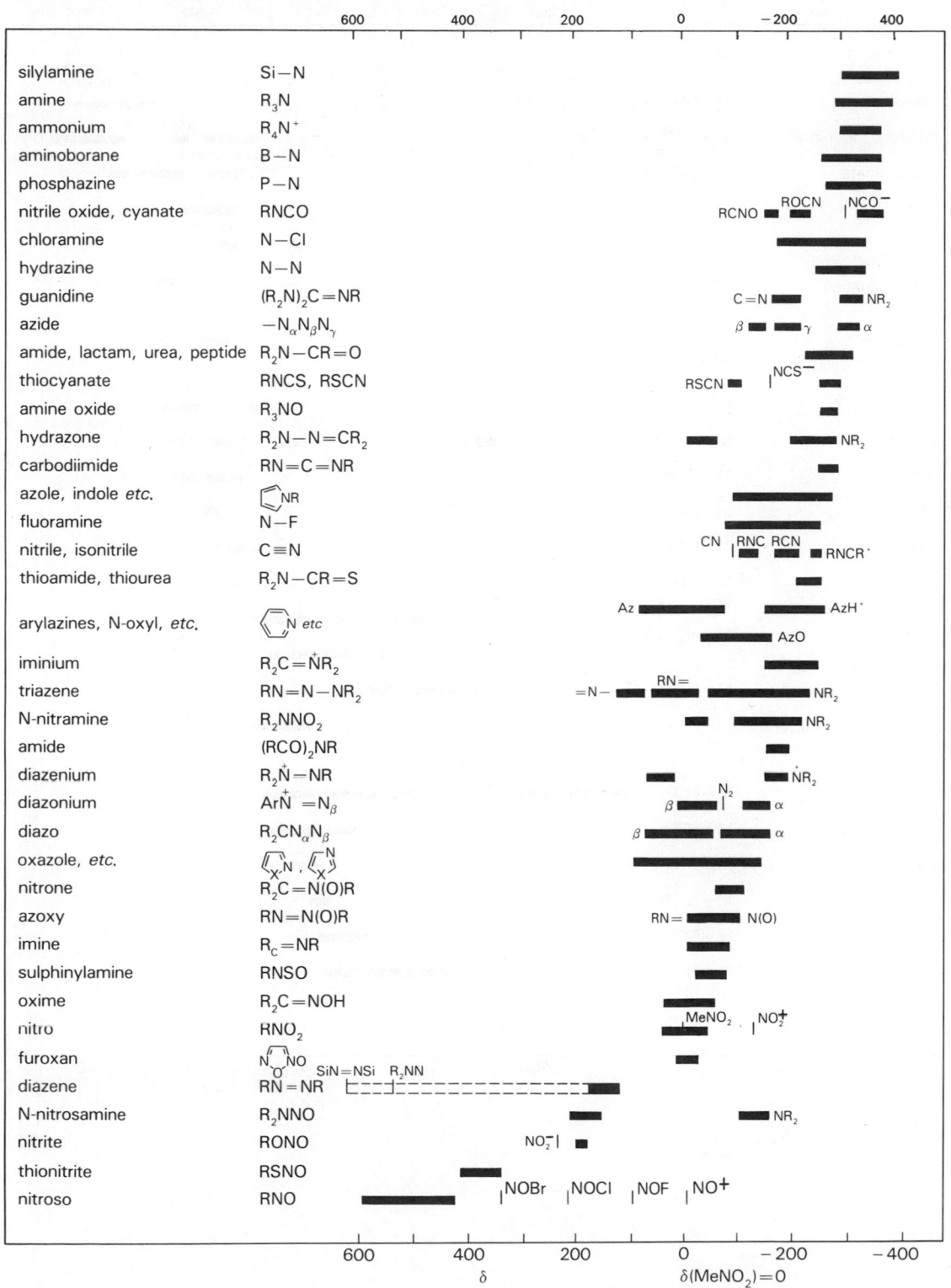

Ranges of ^{15}N chemical shifts in ligands attached to metals

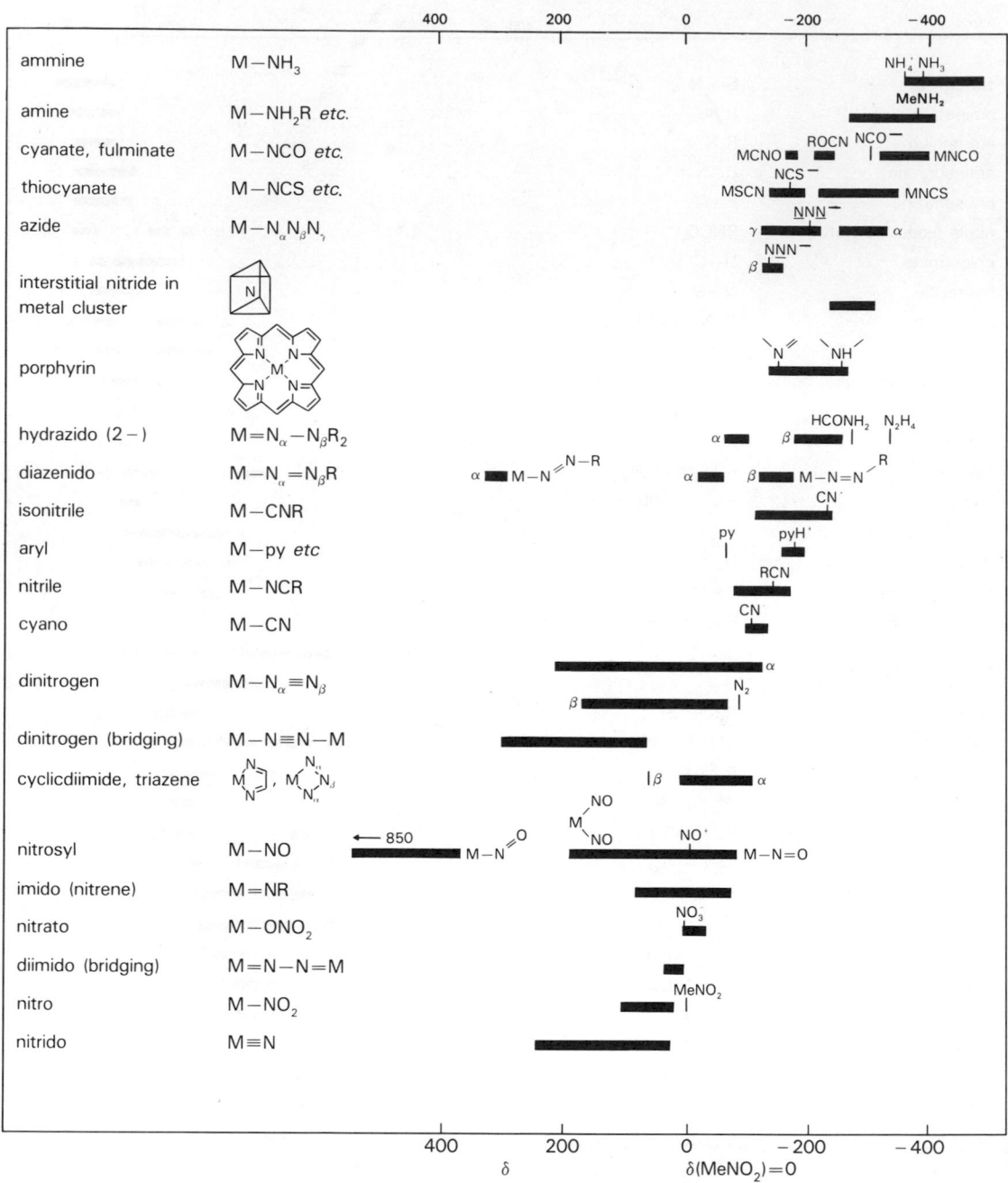

2.8.3 Nuclear spin relaxation time

When a material is in thermal equilibrium, the nuclear spins are either randomly oriented or have a distribution of orientations related to the crystal structure. This equilibrium distribution of nuclear spins may be altered by various re-orienting processes, but after such a process is removed the spin orientations (and hence the nuclear magnetization of the sample) will revert to their original equilibrium distribution exponentially with time. The time constant for this reversion is known as the spin-lattice relaxation time, τ. Relaxation times are usually temperature sensitive but are not very dependent on magnetic field. In the table a magnetic field of about 1 tesla (10 kilogauss) is assumed and the temperature T, or temperature range of the observations, is listed. The relaxation times in insulating diamagnetic materials are often very much reduced if paramagnetic impurities are present.

Nuclide	Material	τ/s	T/K
^1H	H_2 (gas, 10 atm)	0.015	293
^1H	H_2 (liquid, 75% ortho)	0.017	20
^1H	H_2O (liquid)	2.8	293
^1H	H_2O (solid)	20	263
^1H	C_6H_6 (benzene, liquid)	15	293
^1H	Polyethylene (solid)	0.9	295
^2D	D_2O (liquid)	0.36	293
^3He	^3He (liquid)	550	3
^7Li	LiF (solid)	300	293
^7Li	Li (solid)	$45/(T/K)$	1–300
^{13}C	C_6H_6 (benzene, liquid)	23	310
^{13}C	Polyethylene (solid)	0.24	298
^{17}O	H_2O (liquid)	0.0059	293
^{19}F	LiF (solid)	120	293
^{23}Na	NaCl (solid)	12	293
^{23}Na	Na (solid)	$4.9/(T/K)$	1–300
^{27}Al	Al (solid)	$1.82/(T/K)$	1–900
^{29}Si	Si (solid)	2×10^4	293
^{31}P	P (white)	20	300
^{31}P	P (red or black)	300	300
^{35}Cl	NaCl (solid)	6	293
^{51}V	V (solid)	$0.79/(T/K)$	20–290
^{63}Cu	Cu (solid)	$1.27/(T/K)$	1–300
^{69}Ga	GaAs (solid)	0.27	293
^{109}Ag	Ag (solid)	$9/(T/K)$	1–4
^{115}In	InSb (solid)	0.05	298
^{127}I	KI (solid)	0.04	293
^{195}Pt	Pt (solid)	$0.032/(T/K)$	0.5–290
^{197}Au	Au (solid)	$4.0/(T/K)$	1–4

Reference: A. Abragam, *The Principles of Nuclear Magnetism*, 1961 (Oxford University Press).

L.E.D.

2.8.4 The Mössbauer effect

The recoil-free resonance absorption and emission of γ-rays arising from nuclear excited states (with half-lives typically in the range 10^{-6} to 10^{-10} second) are known as the Mössbauer effect. In the optical spectrum, resonance absorption, for example the absorption of sodium light by sodium vapour, is well known, but in the γ-ray case it is not normally observed because the frequency shift and broadening of the line by the Doppler effect are much greater than the line width. However, if the emitting and absorbing atoms are tightly bound in a crystal so that the recoil energy is less than the binding energy, a fraction f of the γ-rays (known as the Lamb–Mössbauer fraction) is emitted without suffering these broadening effects.

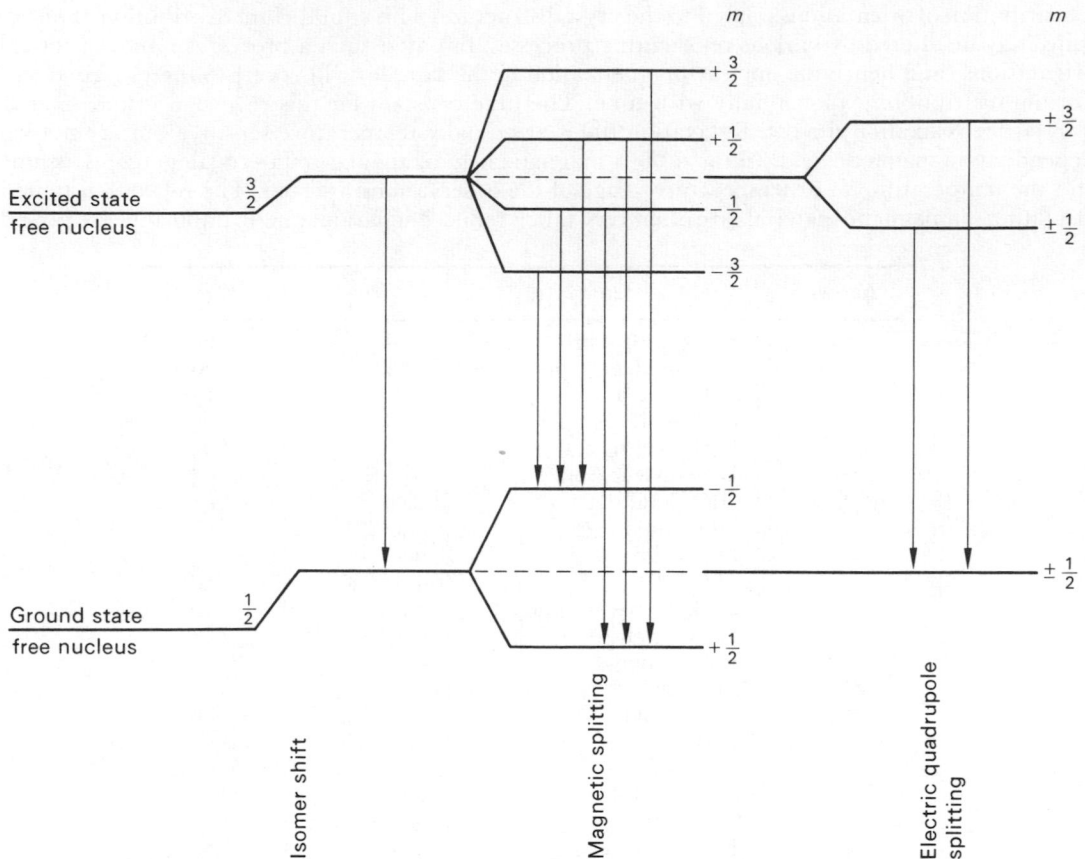

The usefulness of the Mössbauer effect results from the fact that transition energies and, hence, absorption spectra are affected by interaction between the nucleus and its surroundings—local electrons or applied magnetic fields. Mössbauer spectra are obtained by measuring the resonance absorption as a function of γ-ray energy. The variation of energy is provided by moving the source or absorber and using the Doppler effect. It is then convenient to express the effect—chemical shift or quadrupole splitting—in mm s^{-1}. The relation between Mössbauer spectrum and hyperfine inter-actions can be seen from the figure, which shows the allowed transitions in the case of ^{57}Fe, the most widely studied nucleus.

A change in the electronic environment of a nucleus causes changes in the energy of the ground and excited states. This results in a change in the Mössbauer spectrum known as the *chemical shift* or *isomer shift*. The shift is proportional to the electron density in the vicinity of the nucleus and can give useful information on valency states.

Interaction between the nucleus and a magnetic field can give rise to *magnetic splitting* or *hyperfine Zeeman splitting* which can be used to detect magnetic exchange interactions and to determine the magnitude and sign of the magnetic field at a selected site in a crystal. This has some similarity to the splitting observed in n.m.r., but in this case the transition is always from a magnetic sublevel of the excited state to a magnetic sublevel of the ground state, and not between levels of one state as in n.m.r.

Hyperfine interactions between nuclear quadrupoles and electric field gradients give rise to *electric quadrupole splitting* in Mössbauer spectra. Such splitting can be used to obtain information about site symmetries and field gradients in crystals.

Isotopes commonly used in Mössbauer spectroscopy

I_π and I_π^* are the spins and parities of the ground and excited states, E_γ is the γ-ray energy, $t_{1/2}$ is the half-life of the excited state, W_0 is the width of the resonance and σ_0 is the cross-section for resonance absorption. More exactly, statistical factors and branching into alternative modes of decay, for example internal conversion, have to be taken into account giving the equation

$$\sigma_0 = 2.44 \times 10^{-13} E_\gamma^{-2} (1 + 2I^*)(1 + 2I)^{-1}(1 + \alpha)^{-1}$$

where α is the internal conversion coefficient.

Nuclide	I_π	I_π^*	E_γ/keV	$t_{1/2}$/ns	$W_0/(\text{mm s}^{-1})$	σ_0/mm^2
^{197}Au	$\frac{3}{2}+$	$\frac{1}{2}+$	77.3	1.8	2.0	4.4×10^{-18}
^{151}Eu	$\frac{5}{2}+$	$\frac{7}{2}+$	21.6	8.8	1.4	2.3×10^{-17}
^{57}Fe	$\frac{1}{2}-$	$\frac{3}{2}-$	14.4	97.7	0.19	2.4×10^{-16}
^{127}I	$\frac{5}{2}+$	$\frac{7}{2}+$	57.6	1.9	2.5	2.1×10^{-17}
^{121}Sb	$\frac{5}{2}+$	$\frac{7}{2}+$	37.2	3.5	2.1	2.1×10^{-17}
^{119}Sn	$\frac{1}{2}+$	$\frac{3}{2}+$	23.9	18.4	0.63	1.3×10^{-16}

To obtain the probability of absorption at resonance account must be taken of any splitting of the line by hyperfine interactions and of the Lamb–Mössbauer fraction, f. The value of f depends on the probability of no change of phonon occupation numbers during the transition, and in the Debye approximation is given by the equation

$$f = \exp\{-6E_R/k\theta_D[1/4 + (T/\theta_D)^2 \int_0^{\theta_D/T} x\,(e^x + 1)^{-1}\mathrm{d}x]\}$$

where E_R is the recoil energy, $E_\gamma^2/2Mc^2$, and θ_D is the Debye temperature. When the value of f is not near unity its value is clearly strongly dependent on the values of E_γ and T. A typical value for ^{57}Fe γ-rays is 0.7 at room temperature, but for ^{197}Au it is about 0.1 at 4 K. Values of f much less than 0.1 make observation of the resonance difficult or impossible.

Mössbauer spectroscopy

Chemical (isomer) shifts, internal magnetic field (H_{int}) in ferromagnetic materials and quadrupole splittings for the isotopes listed in the preceding table.

Nuclide	Absorbing substance	$\dfrac{Temp}{\text{K}}$	$\dfrac{Chemical\ shift}{\text{mm s}^{-1}}$	$\dfrac{H_{\text{int}}}{\text{T}}$	$\dfrac{Quadrupole\ splitting}{\text{mm s}^{-1}}$
^{197}Au	Pt Au		0		0
	Fe Au dilute alloy	4.2	4.4	128	0
	Au$_2$Mn	4.2	1.5	163	1.45
	Ni Au dilute alloy	4.2	3.8	29	0
	AuCl	4.2	-1.42	0	4.65
^{151}Eu	EuF$_3$		0		
	Eu	4.2	-7.6	-25.7	
	EuS	4.2	-11.8	-31.4	
	EuSO$_4$	0.054	0	32.4	
		300	-14.0	0	$\}$ †
	EuCl$_2$	0.054	0	31.5	
		300	-13.0	0	
	EuF$_2$	300	-15.0	0	
	Eu$_2$(SO$_4$)$_3$	300	0.35	0	

Mössbauer spectroscropy (*contd*)

Nuclide	Absorbing substance	Temp K	Chemical shift mm s^{-1}	H_{int} T	Quadrupole splitting mm s^{-1}
^{57}Fe	Fe metal	293	0	−33.0	0
		4.2		−33.8	0
	αFe_2O_3	293	0.36	−51.7	†
	$Na_2(Fe(CN)_3NO)2H_2O$	293	0.257	0	1.712
	$K_4Fe(CN)_63H_2O$	293	0.045	0	0
	FeF_2	298	1.36	0	2.79
		4.2	1.47	33	2.85
	FeF_3	298	0.49	0	−0.04
		4.2	0.56	62	−0.04
^{127}I	ZnTe(I)		0		
	$KICl_4$	4.2	−1.4	0	
	ICl	4.2	−0.6	0	†
	KIO_4	4.2	0.7	0	
	KI	4.2	0.14	0	
^{121}Sb	InSb		0		
	$KSbF_6$	77	12.3	0	
	Sb_2O_5	77	9.9	0	†
	Sb_2O_3	77	−3.02	0	
	SbF_3	77	−6.0	0	
^{119}Sn	βSn		0		
	SnO_2	77	−2.8	0	0.45
	SnO	77	0.0	0	1.4
	SnF_2	77	0.8	0	1.67
	SnF_4	77	−3.1	0	1.7
	$SnCl_2$	77	1.8	0	0
	$(CH_3)_4Sn$	77	−1.6	0	0
	$(CH_2)_3SnCH_2COCH_3$	77	−1.7	0	2.1
	FeSn dilute alloy	77	−1.1	−8.5	0

† When either of the spins is greater than $\frac{3}{2}$, or there is a magnetic field plus an electric field gradient, the line splitting is more complicated and cannot be represented by a single splitting parameter.

References

(1) N. N. Greenwood and T. C. Gibb, *Mössbauer Spectroscopy*, 1971 (Chapman and Hall).
(2) J. G. and V. E. Stevens, *Mössbauer Effect Data Index*, 1970 (Adam Hilger Ltd, London).

T.E.C.

2.9 Electrochemistry

2.9.1 Standard solutions for calibrating conductivity vessels

The following table is from Jones and Bradshaw, *J. Am. Chem. Soc.*, 1933, **55**, 1799. The unit of conductivity κ is Ω^{-1} cm^{-1}. The original data have been converted from (int. ohm)$^{-1}$ cm^{-1} where 1 int. ohm = 1.000 49 ohm. It is still common practice not to do this, and conductivity results are therefore usually quoted in (int. ohm)$^{-1}$ cm^{-1} units. The experimental values of conductivity have been corrected for that of the water used.

Potassium chloride solutions

g KCl per kg solution in vacuo	$\kappa/(\Omega^{-1}$ cm$^{-1})$		
	0 °C	18 °C	25 °C
71.135 2	0.065 14$_4$	0.097 79$_0$	0.111 28$_7$
7.419 13	0.007 134$_4$	0.011 161$_2$	0.012 849$_7$
0.745 263	0.000 773 2$_6$	0.001 219 9$_2$	0.001 408 0$_8$

For valves recommended by I UPAC see *Pure & Appl. Chem.*, 1981, **53**, 1841.

The following equation, which gives the molar conductivity in Ω^{-1} cm^{-1} of aqueous potassium chloride solution at 25 °C in the range 0.01 mol dm$^{-3} < c <$ 0.10 mol dm^{-3}, is also useful for cell calibration (Ying-Chech Chiu and Fuoss, *J. Phys. Chem.*, 1968, **72**, 4123):

$$\Lambda = 149.86 - 94.83c + 25.47c \ln c + 220.9c - 229c^{3/2}$$

J.E.P.

2.9.2 Conductivities

Molar conductivities of electrolytes in aqueous solution at 25 °C

The unit of Λ in the table is Ω^{-1} cm^2 mol^{-1}. In the following table the entities to which the mole relates are given in the first column, e.g. $\frac{1}{2}MgCl_2$ (not $MgCl_2$). The values are taken by permission from Harned and Owen, *Physical Chemistry of Electrolytic Solutions*, 2nd edn, 1950, Appendix A, p. 537 (Reinhold, New York).

Electrolyte	$c/(\text{mol dm}^{-3})$							
	0	0.0005	0.001	0.005	0.01	0.02	0.05	0.1
HCl . . .	425.95	422.53	421.15	415.60	411.80	407.04	398.89	391.13
LiCl . . .	114.97	113.09	112.34	109.34	107.27	104.60	100.06	95.81
NaCl . . .	126.39	124.44	123.68	120.59	118.45	115.70	111.01	106.69
KCl . . .	149.79	147.74	146.88	143.48	141.20	138.27	133.30	128.90
NH_4Cl . .	149.6	—	—	—	141.21	138.26	133.26	128.69
KBr . . .	151.8	—	—	146.02	143.36	140.41	135.61	131.32
NaI . . .	126.88	125.30	124.19	121.19	119.17	116.64	112.73	108.73
KI . . .	150.31	—	—	144.30	142.11	139.38	134.90	131.05
KNO_3 . .	144.89	142.70	141.77	138.41	135.75	132.34	126.25	120.34
$KHCO_3$ . .	117.94	116.04	115.28	112.18	110.03	107.17	—	—
$NaOOC.CH_3$	91.0	89.2	88.5	85.68	83.72	81.20	76.88	72.76
NaOH . .	247.7	245.5	244.6	240.7	237.9	—	—	—
$AgNO_3$ . .	133.29	131.30	130.45	127.14	124.68	121.35	115.18	109.09

Molar conductivities of electrolytes in aqueous solution at 25 °C (*contd*)

Electrolyte	$c/(\text{mol dm}^{-3})$							
	0	0.0005	0.001	0.005	0.01	0.02	0.05	0.1
$\frac{1}{2}MgCl_2$ . .	129.36	125.55	124.05	118.25	114.49	109.99	103.03	97.05
$\frac{1}{2}CaCl_2$ . .	135.77	131.87	130.30	124.19	120.30	115.59	108.42	102.41
$\frac{1}{2}SrCl_2$ . .	135.73	131.84	130.27	124.18	120.23	115.48	108.20	102.14
$\frac{1}{2}BaCl_2$ . .	139.91	135.89	134.27	127.96	123.88	119.03	111.42	105.14
$\frac{1}{2}Na_2SO_4$ . .	129.8	125.68	124.09	117.09	112.38	106.73	97.70	89.94
$\frac{1}{2}CuSO_4$ . .	133.5	121.5	115.20	94.02	83.08	72.16	59.02	50.55
$\frac{1}{2}ZnSO_4$ . .	132.7	121.3	115.47	95.44	84.87	72.20	61.17	52.61
$\frac{1}{3}LaCl_3$ . .	145.7	139.5	136.9	127.4	121.7	115.2	106.1	99.1
$\frac{1}{3}K_3Fe(CN)_6$.	174.4	166.3	163.0	150.6	—	—	—	—
$\frac{1}{4}K_4Fe(CN)_6$.	184.4	—	167.16	146.02	134.76	122.76	107.64	97.82

There are extensive tables of conductivity data for both aqueous and non-aqueous solutions in Landolt–Börnstein, 6th edn, 1960, Vol. II, Pt 7, pp. 27–726 (Springer-Verlag, Berlin).

Ionic conductivities at infinite dilution at 25 °C

The unit of ionic conductance in the table is $\Omega^{-1}\,\text{cm}^2\,\text{mol}^{-1}$. The values in the table are taken by permission from Robinson and Stokes, *Electrolyte Solutions*, 2nd edn, 1959, p. 463 (Butterworths, London).

Cation	Λ^∞	Cation	Λ^∞	Anion	Λ^∞	Anion	Λ^∞
H^+ . . .	349.6	$\frac{1}{2}Mg^{2+}$ . .	53.0	OH^- . . .	199.1†	HCO_2^- . .	54.6
Li^+ . . .	38.7	$\frac{1}{2}Ca^{2+}$ . .	59.5	Cl^- . . .	76.3	$CH_3CO_2^-$ . .	40.9
Na^+ . . .	50.1	$\frac{1}{2}Sr^{2+}$ . .	59.4	Br^- . . .	78.1	$\frac{1}{2}CO_3^{2-}$ . .	69.3
K^+ . . .	73.5	$\frac{1}{2}Ba^{2+}$ . .	63.6	I^- . . .	76.8	$\frac{1}{2}C_2O_4^{2-}$ . .	74.1
NH_4^+ . .	73.6	$\frac{1}{2}Cu^{2+}$ . .	53.6	NO_3^- . . .	71.4	$\frac{1}{2}SO_4^{2-}$ . .	80.0
Ag^+ . . .	61.9	$\frac{1}{2}Zn^{2+}$ . .	52.8	ClO_4^- . . .	67.3	$\frac{1}{3}Fe(CN)_6^{3-}$.	100.9
Tl^+ . . .	74.7	$\frac{1}{3}La^{3+}$ . . .	69.7	HCO_3^- . .	44.5	$\frac{1}{4}Fe(CN)_6^{4-}$.	110.5

† Marsh and Stokes, *Austral. J. Chem.*, 1964, **17**, 740.

J.E.P.

2.9.3 Standard half-cell EMFs at 25 °C

Each conventional half-cell process implies the real process obtained by adding the half-cell process $H^+ + e^- \rightarrow \frac{1}{2}H_2(g)$. The word 'standard' implies that each solute species is present at an activity equal to one mole kg^{-1} and each gaseous species at a partial pressure of 101.325 kN m^{-2}. All values relate to aqueous solutions at 25 °C and are taken by permission from Latimer, *The Oxidation States of the Elements and their Potentials in Aqueous Solutions*, 1952, pp. 340–345 (Prentice-Hall, New York). If the sign is reversed, the quantities may be called standard electrode potentials (IUPAC *Manual of Symbols and Terminology for Physicochemical Quantities and Units*, 1970, p. 32 (Butterworths, London)).

Standard half-cell EMFs at 25 °C

Process	$E^{\bullet}$ volt	Process	$E^{\bullet}$ volt
$Li \rightarrow Li^{+} + e^{-}$	3.04	$Ag + I^{-} \rightarrow AgI + e^{-}$	0.15
$K \rightarrow K^{+} + e^{-}$	2.92	$\frac{1}{2}Sn \rightarrow \frac{1}{2}Sn^{2+} + e^{-}$	0.14
$Rb \rightarrow Rb^{+} + e^{-}$	2.92	$\frac{1}{2}Pb \rightarrow \frac{1}{2}Pb^{2+} + e^{-}$	0.13
$Cs \rightarrow Cs^{+} + e^{-}$	2.92	$\frac{1}{2}H_2 \rightarrow H^{+} + e^{-}$	o exactly
$\frac{1}{2}Ba \rightarrow \frac{1}{2}Ba^{2+} + e^{-}$	2.90	$Ag + Br^{-} \rightarrow AgBr + e^{-}$	−0.10
$\frac{1}{2}Sr \rightarrow \frac{1}{2}Sr^{2+} + e^{-}$	2.89	$\frac{1}{2}Sn^{2+} \rightarrow \frac{1}{2}Sn^{4+} + e^{-}$	−0.15
$\frac{1}{2}Ca \rightarrow \frac{1}{2}Ca^{2+} + e^{-}$	2.87	$Cu^{+} \rightarrow Cu^{2+} + e^{-}$	−0.15
$Na \rightarrow Na^{+} + e^{-}$	2.71	$Ag + Cl^{-} \rightarrow AgCl + e^{-}$	−0.22
$\frac{1}{3}La \rightarrow \frac{1}{3}La^{3+} + e^{-}$	2.52	$Hg + Cl^{-} \rightarrow HgCl + e^{-}$	−0.27
$\frac{1}{2}Mg \rightarrow \frac{1}{2}Mg^{2+} + e^{-}$	2.37	$\frac{1}{2}Cu \rightarrow \frac{1}{2}Cu^{2+} + e^{-}$	−0.34
$\frac{1}{3}Sc \rightarrow \frac{1}{3}Sc^{3+} + e^{-}$	2.08	$Fe(CN)_6^{4-} \rightarrow Fe(CN)_6^{3-} + e^{-}$	−0.36
$\frac{1}{4}Th \rightarrow \frac{1}{4}Th^{4+} + e^{-}$	1.90	$Cu \rightarrow Cu^{+} + e^{-}$	−0.52
$\frac{1}{2}Be \rightarrow \frac{1}{2}Be^{2+} + e^{-}$	1.85	$I^{-} \rightarrow \frac{1}{2}I_2 + e^{-}$	−0.54
$\frac{1}{3}Al \rightarrow \frac{1}{3}Al^{3+} + e^{-}$	1.66	$\frac{3}{2}I^{-} \rightarrow \frac{1}{2}I_3^{-} + e^{-}$	−0.54
$\frac{1}{2}Ti \rightarrow \frac{1}{2}Ti^{2+} + e^{-}$	1.63	$Hg + \frac{1}{2}SO_4^{2-} \rightarrow \frac{1}{2}Hg_2SO_4 + e^{-}$	−0.62
$\frac{1}{2}Mn \rightarrow \frac{1}{2}Mn^{2+} + e^{-}$	1.18	$\frac{1}{3}Au + \frac{4}{3}SCN^{-} \rightarrow \frac{1}{3}(AuSCN)_4^{-} + e^{-}$	−0.66
$\frac{1}{2}Zn \rightarrow \frac{1}{2}Zn^{2+} + e^{-}$	0.76	$Fe^{2+} \rightarrow Fe^{3+} + e^{-}$	−0.77
$\frac{1}{3}Cr \rightarrow \frac{1}{3}Cr^{3+} + e^{-}$	0.74	$Hg \rightarrow \frac{1}{2}Hg_2^{2+} + e^{-}$	−0.79
$\frac{1}{3}Ga \rightarrow \frac{1}{3}Ga^{3+} + e^{-}$	0.53	$Ag \rightarrow Ag^{+} + e^{-}$	−0.80
$\frac{1}{2}Fe \rightarrow \frac{1}{2}Fe^{2+} + e^{-}$	0.44	$\frac{1}{2}Hg_2^{2+} \rightarrow Hg^{2+} + e^{-}$	−0.92
$Cr^{2+} \rightarrow Cr^{3+} + e^{-}$	0.41	$Pu^{3+} \rightarrow Pu^{4+} + e^{-}$	−0.97
$\frac{1}{2}Cd \rightarrow \frac{1}{2}Cd^{2+} + e^{-}$	0.40	$\frac{1}{2}Pd \rightarrow \frac{1}{2}Pd^{2+} + e^{-}$	−0.99
$Ti^{2+} \rightarrow Ti^{3+} + e^{-}$	0.37	$\frac{1}{3}Au + \frac{4}{3}Cl^{-} \rightarrow \frac{1}{3}AuCl_4^{-} + e^{-}$	−1.00
$\frac{1}{2}Pb + \frac{1}{2}SO_4^{2-} \rightarrow \frac{1}{2}PbSO_4 + e^{-}$	0.36	$Br^{-} \rightarrow \frac{1}{2}Br_2(l) + e^{-}$	−1.06
$\frac{1}{3}In \rightarrow \frac{1}{3}In^{3+} + e^{-}$	0.34	$\frac{1}{2}H_2O \rightarrow \frac{1}{4}O_2 + H^{+} + e^{-}$	−1.23
$Tl \rightarrow Tl^{+} + e^{-}$	0.34	$\frac{1}{2}Tl^{+} \rightarrow \frac{1}{2}Tl^{3+} + e^{-}$	−1.25
$\frac{1}{2}Co \rightarrow \frac{1}{2}Co^{2+} + e^{-}$	0.28	$Cl^{-} \rightarrow \frac{1}{2}Cl_2 + e^{-}$	−1.36
$V^{2+} \rightarrow V^{3+} + e^{-}$	0.26	$\frac{1}{3}Au \rightarrow \frac{1}{3}Au^{3+} + e^{-}$	−1.50
$\frac{1}{2}Ni \rightarrow \frac{1}{2}Ni^{2+} + e^{-}$	0.25		

J.E.P.

2.9.4 Ionization constant of water

Values taken by permission from Harned and Owen, *Physical Chemistry of Electrolytic Solutions*, 2nd edn, 1950, p. 485 (Reinhold, New York). The unit of K_w in the table is mol² (kg water)$^{-2}$.

°C	$10^{14}K_w$	°C	$10^{14}K_w$	°C	$10^{14}K_w$
0	0.114	25	1.008	45	4.02
5	0.185	30	1.47	50	5.47
10	0.292	35	2.09	55	7.30
15	0.450	40	2.92	60	9.61
20	0.681				

J.E.P.

2.9.5 pH values

Standardization of pH

According to BS 1647: 1961 and Amendment PD 4565, 1962 the pH of a solution X, denoted by pH(X), is related to the pH of a standard solution, denoted by pH(S) by

$$pH(X) - pH(S) = E/2.3026RTF^{-1}$$

where R denotes the gas constant, T the absolute temperature, F the Faraday constant and E the electromotive force of the cell

$$Pt, H_2 \,|\, Solution \; X \,|\, 3.5M \; KCl \,|\, Solution \; S \,|\, H_2, Pt$$

The primary standard solution S is a one-twentieth molar solution of pure potassium hydrogen phthalate, of which the pH is defined as having the value 4 exactly at 15 °C and at other temperatures t °C between 0 °C and 95 °C values defined by the formulae:

$$pH(S) = 4.000 + \frac{1}{2}\left(\frac{t-15}{100}\right)^2 \qquad \text{if } t \text{ is between 0 and 55;}$$

$$pH(S) = 4.000 + \frac{1}{2}\left(\frac{t-15}{100}\right) - \frac{t-55}{500} \quad \text{if } t \text{ is between 55 and 95.}$$

The following table is from the British Standard:

°C	$2.3026RT/F$ mV	$F/2.3026RT$ V^{-1}	$pH(S)$	°C	$2.3026RT/F$ mV	$F/2.3026RT$ V^{-1}	$pH(S)$
0	54.20	18.45	4.011	50	64.12	15.60	4.061
5	55.19	18.12	4.005	55	65.11	15.36	4.080
10	56.18	17.80	4.001	60	66.10	15.13	4.091
15	57.17	17.49	4.000	65	67.09	14.90	4.105
20	58.17	17.19	4.001	70	68.08	14.69	4.121
25	59.16	16.90	4.005	75	69.08	14.48	4.140
30	60.15	16.63	4.011	80	70.07	14.27	4.161
35	61.14	16.36	4.020	85	71.06	14.07	4.185
40	62.13	16.09	4.031	90	72.06	13.88	4.211
45	63.13	15.84	4.045	95	73.04	13.69	4.240

The following table gives solutions of which the pH, as defined in the British Standard, has been accurately determined and which are recommended for calibration of glass electrodes.

pH Values of aqueous solutions recommended for calibration of glass electrodes

Composition/(mol dm^{-3} soln)	pH values		
	12 °C	25 °C	38 °C
0.1 $KH_3(C_2O_4)_2 . 2H_2O$	—	1.48	1.50
0.01 $HCl + 0.09$ KCl	—	2.07	2.08
0.05 $HOOC.C_6H_4.COOK$	4.000	4.005	4.026
(Primary Standard)			
0.1 $CH_3.COOH + 0.1$ $CH_3.COONa$ . .	4.65	4.64	4.65
0.01 $CH_3.COOH + 0.01$ $CH_3.COONa$ . .	4.71	4.70	4.72
0.025 $KH_2PO_4 + 0.025$ $Na_2HPO_4 . 2H_2O$ . .	—	6.85	6.84
0.05 $Na_2B_4O_7 . 10H_2O$	—	9.18	9.07
0.025 $NaHCO_3 + 0.025$ Na_2CO_3	—	10.00	—

For other national pH scales see *Pure & Appl. Chem.*, 1978, **50**, 1485.

Buffer solutions

The following table of pH values covering a wider range of temperatures is taken from the IUPAC *Manual of Symbols and Terminology for Physico-chemical Quantities and Units*, 1970, p. 34 (Butterworths, London). The values are based on a convention slightly different from the British Standard, but the values agree almost certainly to within better than ± 0.01.

Values of pH(S) for five standard solutions

t/°C	A	B	C	D	E
0 . . .		4.003	6.984	7.534	9.464
5 . . .		3.999	6.951	7.500	9.395
10 . . .		3.998	6.923	7.472	9.332
15 . . .		3.999	6.900	7.448	9.276
20 . . .		4.002	6.881	7.429	9.225
25 . . .	3.557	4.008	6.865	7.413	9.180
30 . . .	3.552	4.015	6.853	7.400	9.139
35 . . .	3.549	4.024	6.844	7.389	9.102
38 . . .	3.548	4.030	6.840	7.384	9.081
40 . . .	3.547	4.035	6.838	7.380	9.068
45 . . .	3.547	4.047	6.834	7.373	9.038
50 . . .	3.549	4.060	6.833	7.367	9.011
55 . . .	3.554	4.075	6.834		8.985
60 . . .	3.560	4.091	6.836		8.962
70 . . .	3.580	4.126	6.845		8.921
80 . . .	3.609	4.164	6.859		8.885
90 . . .	3.650	4.205	6.877		8.850
95 . . .	3.674	4.227	6.886		8.833

The compositions of the standard solutions are:

A: KH tartrate (saturated at 25 °C)
B: KH phthalate, $m = 0.05$ mol kg^{-1}
C: KH_2PO_4, $m = 0.025$ mol kg^{-1};
   Na_2HPO_4, $m = 0.025$ mol kg^{-1}

D: KH_2PO_4, $m = 0.008$ 695 mol kg^{-1};
   Na_2HPO_4, $m = 0.030$ 43 mol kg^{-1}
E: $Na_2B_4O_7$, $m = 0.01$ mol kg^{-1}

where m denotes molality and the solvent is water.

Acid-base indicators

An acid-base indicator is a conjugate acid-base pair such that the acid form and the base form have different colours. The following table is from A. I. Vogel, *Quantitative Inorganic Analysis*, 3rd edn, 1961, p. 55 (Longman, Green & Co.).

Indicator	pH range	Acid	Base
Brilliant cresyl blue (acid)	0.0–1.0	Red-orange	Blue
Cresol red (acid) . . .	0.2–1.8	Red	Yellow
Quinaldine red . . .	1.0–2.0	Colourless	Red
Thymol blue (acid) . .	1.2–2.8	Red	Yellow
m-Cresol purple . . .	1.2–2.8	Red	Yellow
Pentamethoxy red . . .	1.2–3.2	Red-violet	Colourless
Tropaeolin OO . . .	1.3–3.0	Red	Yellow
Meta-cresol purple . .	1.2–2.8	Red	Yellow
Bromo-phenol blue . .	3.0–4.6	Yellow	Blue
Methyl yellow	2.9–4.0	Red	Yellow
Ethyl orange	3.0–4.5	Red	Orange
Methyl orange	3.1–4.4	Red	Orange
Congo red	3.0–5.0	Violet	Red
α-Naphthyl red . . .	3.7–5.0	Red	Yellow
Bromo-cresol green . .	3.8–5.4	Yellow	Blue
Methyl red	4.2–6.3	Red	Yellow
Ethyl red	4.5–6.5	Red	Orange
Propyl red	4.6–6.6	Red	Yellow
Chlorophenol red . . .	4.8–6.4	Yellow	Red
p-Nitrophenol	5.6–7.6	Colourless	Yellow
Bromocresol purple . .	5.2–6.8	Yellow	Purple
Bromophenol red . . .	5.2–6.8	Yellow	Red
Azolitmin (litmus) . .	5.0–8.0	Red	Blue
Bromo-thymol blue . .	6.0–7.6	Yellow	Blue
Neutral red	6.8–8.0	Red	Orange
Phenol red	6.8–8.4	Yellow	Red
Cresol red (base) . . .	7.2–8.8	Yellow	Red
α-Naphtholphthalein . .	7.3–8.7	Yellow	Blue
m-Cresol purple . . .	7.6–9.2	Yellow	Purple
Thymol blue (base) . .	8.0–9.6	Yellow	Blue
o-Cresolphthalein . . .	8.2–9.8	Colourless	Red
Phenolphthalein . . .	8.3–10.0	Colourless	Red
Turmeric	8.0–10.0	Yellow	Orange
Thymolphthalein . . .	8.3–10.5	Colourless	Blue
Alizarin yellow R . . .	10.1–12.0	Yellow	Orange red
Brilliant cresyl blue (base) .	10.8–12.0	Blue	Yellow
Tropaeolin O	11.1–12.7	Yellow	Orange
Nitramine	10.8–13.0	Colourless	Orange-brown

J.E.P.

2.9.6 Half-wave potentials of metals

Half-wave potentials $E_{1/2}$ are referred to the saturated calomel electrode at 25 °C for which $E^{\ominus} = -0.242$ V. The subjoined data are taken by permission from Kolthoff and Lingane, *Polarography*, 2nd edn, 1952 (Interscience, New York).

Half-wave potentials of metals

Element	Supporting electrolyte	$E_{1/2}/V$	Reaction
Al	0.025M $BaCl_2$	-1.75	
Ba	Tetramethylammonium iodide in 80% ethanol	-1.82	
Bi	Bi^{3+} in 0.5M tartrate (pH = 4.5)	-0.29	
Ca	Tetramethylammonium iodide in 80% ethanol	-2.13	
Cd	0.1M KCl or HCl	-0.599	
	0.5M tartrate (pH = 4.5)	-0.64	
	1M NH_3 + 1M NH_4Cl	-0.81	
Co	Co^{2+} in 0.1M KCl or NaCl	-1.4	
	Co^{3+} in 1M NH_3 + 1M NH_4Cl	-0.5	$Co^{3+} \rightarrow Co^{2+}$
		-1.3	$Co^{2+} \rightarrow Co$
	Co^{2+} in 1M KSCN	-1.03	
	Co^{2+} in 0.1M pyridine + 0.1M pyridinium chloride	-1.07	
Cr	Cr^{3+} in 0.1M KCl	-0.91	$Cr^{3+} \rightarrow Cr^{2+}$
		-1.47	$Cr^{2+} \rightarrow Cr$
	CrO_4^{2-} in 1M NaOH	-0.85	$Cr^{6+} \rightarrow Cr^{3+}$
	CrO_4^{2-} in 0.1M KCl	-0.3	
		-1.0	$Cr^{6+} \rightarrow Cr^{3+}$
		-1.5	$Cr^{3+} \rightarrow Cr^{2+}$
		-1.7	$Cr^{2+} \rightarrow Cr$
Cu	0.1M KCl or HCl	$+0.04$	$Cu^{2+} \rightarrow Cu$
	1M NH_3 + 1M NH_4Cl	-0.24	$Cu^{2+} \rightarrow Cu^+$
		-0.50	$Cu^+ \rightarrow Cu$
	0.1M KSCN	-0.02	$Cu^{2+} \rightarrow Cu^+$
		-0.39	$Cu^+ \rightarrow Cu$
	Cu^{2+} in 0.5M tartrate (pH = 4.5)	-0.09	
Fe	Fe^{3+} in KCl, HCl, $HClO_4$, etc.	0.0†	
	Fe^{2+} in 0.1M KCl or 0.05M $BaCl_2$	-1.3	$Fe^{2+} \rightarrow Fe$
	Fe^{3+} and/or Fe^{2+} in 0.5M tartrate (pH = 5.8)	$-0.17‡$	$Fe^{3+} \rightleftharpoons Fe^{2+}$
	Fe^{3+} and/or Fe^{2+} in 0.2M oxalate (pH = 5.25)	$-0.245‡$	$Fe^{3+} \rightleftharpoons Fe^{2+}$
	Fe^{3+} in 1M KOH + 8% mannitol	-0.9	$Fe^{3+} \rightarrow Fe^{2+}$
		-1.5	$Fe^{2+} \rightarrow Fe$
K	Tetraethylammonium hydroxide in 50% ethanol	-2.10	
Li	Tetraethylammonium hydroxide in 50% ethanol	-2.31	
Mn	Mn^{2+} in 1M KCl	-1.51	$Mn^{2+} \rightarrow Mn$
	Mn^{2+} in 1M NH_3 + 1M NH_4Cl	-1.65	
	Mn^{2+} in 1M KSCN	-1.55	
Na	Tetraethylammonium hydroxide in 50% ethanol	-2.07	
Ni	1M KCl	-1.1	
	1M NH_3 + 0.2M NH_4Cl	-1.06	
	1M KSCN	-0.70	
	1M KCl + 0.5M pyridine	-0.78	
Pb	1M KNO_3 or HNO_3	-0.405	
	0.1M KCl or HCl	-0.396	
	$HPbO_2^-$ in 1M NaOH	-0.755	
	0.5M tartrate (pH = 4.5)	-0.48	
Sb	Sb^{3+} in 1M HCl	-0.15	$Sb^{3+} \rightarrow Sb$
	Sb^{3+} in 1M NaOH	-1.26	
Sn	Sn^{2+} in 1M HCl	-0.47	
	SnO_2^{2-} in 1M NaOH	-1.22	$SnO_2^{2-} \rightarrow Sn$
		$-0.73§$	$SnO_2^{2-} \rightarrow SnO_3^{2-}$
Ti	Ti^{4+} in 0.1M HCl or HNO_3	-0.81	$Ti^{4+} \rightarrow Ti^{3+}$
Tl	Tl^+ in 0.1M KCl, KNO_3, HCl or NaOH	-0.460	
U	UO_2^{2+} in 0.02M HCl	-0.18	$U^{6+} \rightarrow U^{5+}$
		-0.92	$U^{5+} \rightarrow U^{3+}$

Half-wave potentials of metals (*contd*)

Element	Supporting electrolyte	$E_{1/2}$/V	Reaction
Zn	0.1M KCl or KNO$_3$	-0.995	
	ZnO$_2^{2-}$ in 1M NaOH	-1.53	
	0.1M KSCN	-1.01	
	0.5M tartrate (pH $= 4.5$)	-1.23	

† Diffusion current of $Fe^{3+} \rightarrow Fe^{2+}$ at zero applied potential.   ‡ Reversible.   § Anodic wave.

J.E.P.

2.9.7 Osmotic coefficients

Osmotic coefficients at 25 °C in aqueous solutions used as standards for isopiestic measurements

The practical osmotic coefficient ϕ in a solution of an electrolyte consisting of v ions is given by $vr\phi = \ln (p^0/p)$, where r is the mole ratio of electrolyte to solvent, p is the partial vapour pressure of the solvent over the solution and p^0 is the vapour pressure of the pure solvent. In aqueous solutions the mole ratio r is related to the molality m, expressed in mol (kg water)$^{-1}$ by $m = 55.51r$. The values of ϕ in the following table are taken by permission from Robinson and Stokes, *Electrolyte Solutions*, 2nd edn, 1959, Appendices 8.3 and 8.5 (Butterworths, London).

$\dfrac{m}{\text{mol kg}^{-1}}$	NaCl	KCl	CaCl$_2$	$\dfrac{m}{\text{mol kg}^{-1}}$	NaCl	KCl	CaCl$_2$
	ϕ	ϕ	ϕ		ϕ	ϕ	ϕ
0.1	0.9324	0.9266	0.854	3.6	1.0867	0.9531	
0.2	0.9245	0.9130	0.862	3.8	1.1013	0.9588	
0.3	0.9215	0.9063	0.876	4.0	1.1158	0.9647	2.182
0.4	0.9203	0.9017	0.894	4.2	1.1306	0.9707	
0.5	0.9209	0.8989	0.917	4.4	1.1456	0.9766	
0.6	0.9230	0.8976	0.940	4.5			2.383
0.7	0.9257	0.8970	0.963	4.6	1.1608	0.9824	
0.8	0.9288	0.8970	0.988	4.8	1.1761	0.9883	
0.9	0.9320	0.8971	1.017	5.0	1.1916		2.574
1.0	0.9355	0.8974	1.046	5.2	1.2072		
1.2	0.9428	0.8986	1.107	5.4	1.2229		
1.4	0.9513	0.9010	1.171	5.5			2.743
1.6	0.9616	0.9042	1.237	5.6	1.2389		
1.8	0.9723	0.9081	1.305	5.8	1.2548		
2.0	0.9833	0.9124	1.376	6.0	1.2706		2.891
2.2	0.9948	0.9168		6.5			3.003
2.4	1.0068	0.9214		7.0			3.081
2.5			1.568	7.5			3.127
2.6	1.0192	0.9264		8.0			3.151
2.8	1.0321	0.9315		8.5			3.165
3.0	1.0453	0.9367	1.779	9.0			3.171
3.2	1.0587	0.9421		9.5			3.171
3.4	1.0725	0.9477		10.0			3.169
3.5			1.981				

J.E.P.

2.9.8 Activity coefficients

Activity coefficients of typical strong electrolytes in aqueous solution at 25 °C

The mean activity coefficient γ of an electrolyte, consisting of v_+ cations R and v_- anions X, is related to the chemical potential μ of one mole of electrolyte by

$$\mu = \mu^{\ominus} + v_+ RT \ln m_R + v_- RT \ln m_X + (v_+ + v_-) RT \ln \gamma$$

where m_R and m_X denote the molalities of the cation R and anion X respectively while $\mu^{\ominus}$ is independent of the composition of the solution and is so chosen that $\gamma \to 1$ at infinite dilution.

In the following table of γ for typical 1–1 electrolytes at 25 °C the values at molalities 0.1 upwards are from Robinson and Stokes, *Trans. Faraday Soc.*, 1949, **45**, 612. The values at lower molalities for HCl, NaCl and KCl have been obtained from e.m.f. measurements and for the other electrolytes have been interpolated by a Debye–Hückel type of formula. All γ values are considered to be accurate to within ± 0.002.

Typical 1–1 electrolytes at 25 °C

$m/(\text{mol kg}^{-1})$	HCl	LiCl	NaCl	KCl	CsCl	LiNO$_3$	NaNO$_3$	KNO$_3$	CsNO$_3$
0.01	0.904	0.903	0.902	0.901	0.899	0.903	0.900	0.897	0.896
0.02	0.875	0.873	0.870	0.868	0.865	0.872	0.866	0.861	0.860
0.05	0.830	0.825	0.820	0.816	0.807	0.825	0.811	0.799	0.795
0.10	0.796	0.790	0.778	0.770	0.756	0.788	0.762	0.739	0.733
0.2	0.767	0.757	0.735	0.718	0.694	0.752	0.703	0.663	0.655
0.4	0.755	0.740	0.693	0.666	0.628	0.728	0.638	0.576	0.561
0.6	0.763	0.743	0.673	0.637	0.589	0.727	0.599	0.519	0.501
0.8	0.783	0.755	0.662	0.618	0.563	0.733	0.570	0.476	0.458
1.0	0.809	0.774	0.657	0.604	0.544	0.743	0.548	0.443	0.422
1.2	0.840	0.796	0.654	0.593	0.529	0.757	0.530	0.414	0.393
1.4	0.876	0.823	0.655	0.586	0.518	0.774	0.514	0.390	0.368
1.6	0.916	0.853	0.657	0.580	0.509	0.792	0.501	0.369	
1.8	0.960	0.885	0.662	0.576	0.501	0.812	0.489	0.350	
2.0	1.009	0.921	0.668	0.573	0.495	0.835	0.478	0.333	
2.5	1.147	1.026	0.688	0.569	0.484	0.896	0.455	0.297	
3.0	1.316	1.156	0.714	0.569	0.478	0.966	0.437	0.269	
3.5	1.518	1.317	0.746	0.572	0.474	1.044	0.422	0.246	
4.0	1.762	1.510	0.783	0.577	0.473	1.125	0.408		
4.5	2.04	1.741	0.826	0.583	0.473	1.215	0.396		
5.0	2.38	2.02	0.874		0.474	1.310	0.386		

The values of γ for typical 2–1 and 1–2 electrolytes in the following table are from Stokes, *Trans. Faraday Soc.*, 1948, **44**, 295 as revised by Guggenheim and Stokes, *Trans. Faraday Soc.*, 1958, **54**, 1646. They are derived from experimental values of the osmotic coefficient ϕ by use of the Gibbs–Duhem relation.

Typical 2–1 and 1–2 electrolytes at 25 °C

$m/(\text{mol kg}^{-1})$	0.1	0.2	0.3	0.4	0.5	0.6	0.7	0.8	0.9	1.0
MgCl$_2$	0.528	0.488	0.476	0.474	0.480	0.490	0.505	0.521	0.543	0.569
CaCl$_2$	0.518	0.472	0.455	0.488	0.448	0.453	0.460	0.470	0.484	0.500
SrCl$_2$	0.515	0.466	0.446	0.436	0.433	0.434	0.437	0.445	0.453	0.465
BaCl$_2$	0.508	0.451	0.426	0.411	0.403	0.397	0.397	0.397	0.398	0.401
MnCl$_2$	0.518	0.471	0.452	0.444	0.442	0.445	0.450	0.457	0.468	0.481
FeCl$_2$	0.520	0.475	0.456	0.450	0.452	0.456	0.465	0.475	0.490	0.508

Typical 2–1 and 1–2 electrolytes at 25 °C   (contd)

$m/(\text{mol kg}^{-1})$	0.1	0.2	0.3	0.4	0.5	0.6	0.7	0.8	0.9	1.0
$CoCl_2$	0.523	0.480	0.464	0.460	0.463	0.471	0.480	0.493	0.512	0.532
$NiCl_2$	0.523	0.480	0.464	0.461	0.465	0.472	0.483	0.497	0.516	0.537
$CuCl_2$	0.510	0.457	0.431	0.419	0.413	0.411	0.411	0.412	0.415	0.419
$MgBr_2$	0.542	0.511	0.510	0.519	0.537	0.562	0.590	0.625	0.667	0.712
$CaBr_2$	0.532	0.492	0.482	0.483	0.491	0.505	0.522	0.543	0.568	0.597
$SrBr_2$	0.527	0.484	0.469	0.466	0.468	0.474	0.485	0.498	0.516	0.536
$BaBr_2$	0.517	0.469	0.450	0.440	0.438	0.442	0.446	0.452	0.462	0.473
$Mg(NO_3)_2$	0.522	0.480	0.467	0.465	0.469	0.478	0.488	0.501	0.518	0.536
$Ca(NO_3)_2$	0.488	0.429	0.397	0.378	0.365	0.356	0.349	0.344	0.340	0.338
$Sr(NO_3)_2$	0.478	0.410	0.373	0.348	0.329	0.314	0.302	0.292	0.283	0.275
$Co(NO_3)_2$	0.521	0.474	0.455	0.448	0.448	0.451	0.458	0.468	0.480	0.493
$Cu(NO_3)_2$	0.512	0.461	0.440	0.430	0.427	0.428	0.432	0.438	0.446	0.456
Li_2SO_4	0.478	0.406	0.369	0.344	0.326	0.313	0.303	0.295	0.288	0.283
Na_2SO_4	0.452	0.371	0.325	0.294	0.270	0.252	0.237	0.225	0.213	0.204
K_2SO_4	0.436	0.356	0.313	0.283	0.261	0.243	0.229	—	—	—
Rb_2SO_4	0.460	0.381	0.338	0.307	0.285	0.268	0.254	0.243	0.233	0.223
Cs_2SO_4	0.464	0.389	0.344	0.317	0.296	0.279	0.267	0.256	0.246	0.239

J.E.P.

2.9.9 Acidity constants

Acidity constants in water at 25 °C

The acidity constant K of an acid-base equilibrium $A \rightleftharpoons B + H^+$ is defined as the limiting value at infinite dilution of $[B][H^+]/[A]$ and pK is an abbreviation for $-\log_{10}(K/\text{mol kg}^{-1})$. The values in the following table relate to aqueous solutions at 25 °C and are taken from the sources quoted.

Name	Formula	pK	Formula	pK	Reference
Oxalic	$H_2C_2O_4$	1.271	$HC_2O_4^-$	4.266	1
Phosphoric	H_3PO_4	2.148	$H_2PO_4^-$	7.198	1
Glycine	$^+H_3N.CH_2.CO_2H$	2.350	$^+H_3N.CH_2.CO_2^-$	9.780	1
p-Aminobenzoic . .	$^+H_3N.C_6H_4.CO_2H$	2.45	$^+H_3N.C_6H_4.CO_2^-$	4.85	1
Cyanoacetic . . .	$CH_2CN.CO_2H$	2.469			1
Malonic	$CH_2.(CO_2H)_2$	2.855	$(CO_2H)CH_2.CO_2^-$	5.696	1
Chloroacetic . . .	$CH_2Cl.CO_2H$	2.865			1
Bromoacetic . . .	$CH_2Br.CO_2H$	2.901			1
o-Phthalic	$C_6H_4.(CO_2H)_2$	2.950	$(CO_2H)C_6H_4.CO_2^-$	5.408	1
o-Hydroxybenzoic . .	$HO.C_6H_4.CO_2H$	2.996			1
Tartaric	$C_2H_4O_2.(CO_2H)_2$	3.033	$(CO_2H)C_2H_4O_2.CO_2^-$	4.366	1
Citric	$C_3H_5O.(CO_2H)_3$	3.128	$(CO_2H)_2C_3H_5O.CO_2^-$	4.761	1
Iodoacetic	$CH_2I.CO_2H$	3.174			1
2,6-Dinitrophenol . .	$(O_2N)_2.C_6H_4.OH$	3.72			2
Formic	$H.CO_2H$	3.739			1
Glycollic	$CH_2OH.CO_2H$	3.831			1
Lactic	$CH_3.CHOH.CO_2H$	3.860			1
2,4-Dinitrophenol . .	$(O_2N)_2.C_6H_4.OH$	4.09			2
Benzoic	$C_6H_5.CO_2H$	4.201			1
Succinic	$C_2H_4.(CO_2H)_2$	4.207	$(CO_2H)C_2H_4.CO_2^-$	5.638	1
Anilinium	$C_6H_5.NH_3^+$	4.60			3
Acetic	$CH_3.CO_2H$	4.756			1
Butyric	$C_3H_7.CO_2H$	4.820			1
Propionic	$C_2H_5.CO_2H$	4.874			1

Acidity constants in water at 25 °C (*contd*)

Name	Formula	pK	Formula	pK	Reference
2,3-Dinitrophenol . . .	$(O_2N)_2.C_6H_4.OH$	4.96			5
2,5-Dinitrophenol . . .	$(O_2N)_2.C_6H_4.OH$	5.21			2
3,4-Dinitrophenol . . .	$(O_2N)_2.C_6H_4.OH$	5.42			2
Hydroxylammonium . .	NH_3OH^+	5.96			5
Carbonic	H_2CO_3	6.352	HCO_3^-	10.329	1
3,5-Dinitrophenol . . .	$(O_2N)_2.C_6H_4.OH$	6.69			5
p-Nitrophenol . . .	$O_2N.C_6H_4.OH$	7.156			4
o-Nitrophenol . . .	$O_2N.C_6H_4.OH$	7.230			6
Triethanolammonium . .	$(C_2H_4OH)_3.NH^+$	7.762			3
Tris-hydroxymethyl-methyl-ammonium .	$(CH_2OH)_3.CNH_3^+$	8.075			1
Hydrazinium	$N_2H_5^+$	8.11			1
m-Nitrophenol . . .	$O_2N.C_6H_4.OH$	8.355			7
Diethanolammonium . .	$(C_2H_4OH)_2.NH_2^+$	8.883			3
Phenol-sulphonate . .	$HO.C_6H_4.SO_3^-$	9.053			1
Hydrogen cyanide . . .	HCN	9.216			4
Boric	H_3BO_3	9.234			1
Ammonium	NH_4^+	9.245			1
Ethanol-ammonium . .	$C_2H_4OH.NH_3^+$	9.498			1
Trimethyl-ammonium . .	$(CH_3)_3.NH^+$	9.800			1
Phenol	C_6H_5OH	9.998			1
m-Cresol	$CH_3.C_6H_4.OH$	10.09			1
p-Cresol	$CH_3.C_6H_4.OH$	10.26			1
o-Cresol	$CH_3.C_6H_4.OH$	10.29			1
Methyl-ammonium . .	$CH_3.NH_3^+$	10.624			1
Ethyl-ammonium . . .	$C_2H_5.NH_3^+$	10.631			1
Triethyl-ammonium . .	$(C_2H_5)_3.NH^+$	10.715			1
Dimethyl-ammonium . .	$(CH_3)_2.NH_2^+$	10.774			1
Diethyl-ammonium . .	$(C_2H_5)_2.NH_2^+$	10.933			1
Piperidinium	$C_5H_{10}NH_2^+$	11.123			1

References: (1) Robinson and Stokes, *Electrolyte Solutions*, 2nd edn, 1959, pp. 517–535 (Butterworths, London). (2) Kortüm, Vogel and Andrussow, *Dissociation Constants of Organic Acids in Aqueous Solution*, 1961 (Butterworths, London). (3) Perrin, *Dissociation Constants of Organic Bases in Aqueous Solution*, 1965, *Supplement*, 1972 (Butterworths, London). Perrin, *Dissociation Constants of Inorganic Acids and Bases in Aqueous Solution*, 1969 (Butterworths, London). (5) Robinson, Davis, Paabo and Bower, *J. Res. Nat. Bur. Stand.*, 1960, **64A**, 347. (6) Robinson and Peiperl, *J. Phys. Chem.*, 1963, **67**, 1723. (7) Robinson and Peiperl, *J. Phys. Chem.*, 1963, **67**, 2860. (7) Serjeant and Demsey, *Ionization Constants of Organic Acids in Aqueous Solution*, IUPAC Chemical Data Series, No. 23, 1979 (Pergamon). (8) Perrin, *Ionization Constants of Inorganic Acids and Bases in Aqueous Solution*, IUPAC Chemical Series, No. 29, 1982 (Pergamon).

Extensive tables of acidity constants will also be found in Landolt–Börnstein, 6th edn, 1960, Vol. II, Pt 7, pp. 839–953 (Springer-Verlag, Berlin).

J.E.P.

2.9.10 Stability constants and solubility products

Stability constants in aqueous solutions

The stability constants K_n of step-wise complex ion equilibria $ML_{n-1} + L \rightleftharpoons ML_n$ are defined as the limiting values at infinite dilution of $[ML_n]/[ML_{n-1}]$ $[L]$. pK_n is an abbreviation for

$-\log_{10}(K_n/\text{dm}^3\,\text{mol}^{-1})$. Q_n is used to denote the corresponding stability quotient in a swamping concentration of a specified inert salt. The values in the following tables are selected from Sillén and Martell, *Stability Constants*, Chem. Soc. Special Publ., 1964, No. 17.

$\text{L} = \text{NH}_3$; *temperature* 30 °C.

n	Co^{2+} $-pK_n$	Ni^{2+} $-pK_n$	Cu^{2+} $-pK_n$	Zn^{2+} $-pK_n$	Cd^{2+} $-pK_n$	Hg^{2+}† $-pQ_n$	Ag^{+}‡ $-pK_n$
1	1.99	2.67	4.25	2.18	2.51	8.8	3.32
2	1.51	2.12	3.61	2.25	1.96	8.7	3.92
3	0.93	1.61	2.98	2.31	1.30	1.0	
4	0.64	1.07	2.24	1.96	0.79	0.8	
5	0.06	0.63	0.52				
6	−0.74	−0.09					

† pQ_n in 2M ammonium nitrate at ∼22 °C.
‡ 25 °C.

$\text{L} = $ *ethylenediaminetetraacetate ion*; *temperature* 20 °C.

	Mg^{2+}	Ca^{2+}	Ba^{2+}	Mn^{2+}	Fe^{2+}	Co^{2+}	Ni^{2+}	Cu^{2+}	Fe^{3+}	La^{3+}
$-pQ_1$† . .	8.69	10.70	7.76	13.58	14.33	16.21	18.56	18.79	25.1	15.1
	Zn^{2+}	Cd^{2+}	Hg^{2+}	Pb^{2+}	Ag^{+}					
$-pQ_1$‡ . .	16.50	16.46	21.80	18.04	7.32					

† In 0.1M potassium chloride.
‡ In 0.1M potassium nitrate.

Solubility products in water

The solubility product P of a salt M_nX_m is defined by the value extrapolated to infinite dilution of $[M]_n[X]_m$ in a saturated solution of the salt. pP is an abbreviation for $-\log_{10}(P/\text{mol}^{m+n}\,\text{dm}^{-3(m+n)})$. The values in the following table relate to aqueous solutions at 25 °C and are selected from Sillén and Martell, *Stability Constants*, Chem. Soc. Special Publ., 1964, No. 17.

	AgCl	AgBr	AgI	$AgIO_3$	AgSCN	TlCl	$TlIO_3$
pP . .	9.75	12.28	16.08	7.51	12.0	3.7	5.51
	$BaSO_4$	$PbSO_4$	$PbCl_2$	Hg_2Cl_2	$Ca(IO_3)_2$	$La(IO_3)_3$	Ag_2SO_4
pP . .	10.0	7.80	4.7	17.88	6.15	11.21	4.8

J.E.P.

2.10 Chemical thermodynamics

2.10.1 Molar heat capacities, heats of fusion and vaporization of the elements

Values for the molar heat capacities of the elements at 298.15 K are given in Section 2.10.5. In the following table heat capacities are expressed in $J\,mol^{-1}\,K^{-1}$ and heats of transition are expressed in $kJ\,mol^{-1}$. The horizontal lines indicate the positions of melting and boiling points. The elements are arranged in alphabetical order of name but are listed by chemical symbols. The tables have been compiled mainly from the JANAF Thermochemical Tables (for a summary index see *J. Phys. Chem. Ref. Data*, 1982, **11**, 695) and K. K. Kelley and E. G. King, 'Contributions to the data on theoretical metallurgy, XIV–Entropies of the elements and inorganic compounds', *US Bureau of Mines, Bulletin*, 592.

† indicates rounded values for primary and secondary fixed points on the *International Practical Temperature Scale* of 1968 amended 1975; for exact values see Table 2.1.2.

T/K	Al 26.98	Sb 121.75	Ar 39.948	As 74.92	Ba 137.34	Be 9.012	Bi 208.98	B 10.81
50	3.8	13.0	24.8	7.9		0.2	17.7	0.1
100	13.0	20.6	20.8	16.7	24.3	1.8	23.0	1.1
150	18.5	23.2	20.8	20.7		5.7	24.4	3.2
200	21.6	24.4	20.8	22.7	26.4	10.0	25.0	6.1
400	25.8	26.0	20.8	25.6	33.2	20.0	27.8	15.6
600	28.1	27.4	20.8	27.4	33.9	23.3	31.4	20.9
800	30.8	28.9	20.8	29.3	39.1	25.4	31.4	23.3
1000	31.7	31.4	20.8		39.1	27.3	31.4	24.9
1500	31.7	31.4	20.8		38.9	32.3	31.4	27.9
2000	31.7	18.7	20.8		39.7	29.7	21.0	29.8
2500	31.7	18.7	20.8		39.9	30.8	21.4	30.5
m.p. (T/K)	933.6†	903.9†	83.75		1000	1556	544.6†	2450
$\Delta H_m/(kJ\,mol^{-1})$	10.7	19.83	1.18		8.01	11.71		22.6
b.p. (T/K)	2790	1870	87.29†		1910	2757	1920	4000
$\Delta H_v/(kJ\,mol^{-1})$	290.8	67.91	6.52		140.2	297.6		507.8
Molecular formula of gas	Al	Sb_2	Ar		Ba		Bi	B

Molar heat capacities, heats of fusion and vaporization of the elements (contd)

T/K	Element and formula weight							
	Br_2 159.808	Cd 112.40	Cs 132.905	Ca 40.08	C 12.011	Ce 140.12	Cl_2 70.906	Cr 51.996
50	33.3	16.2	24.3	10.9	0.5	20.8	29.2	2.0
100	43.6	22.2	25.8	19.5	1.7	27.0	42.3	10.0
150	49.2	24.0	26.8	23.0	3.2	28.1	51.0	16.5
200	53.8	24.9	27.8	24.5	5.0	28.9	54.2	19.9
400	36.7	27.2	31.5	26.9	11.8	30.5	35.3	25.2
600	37.3	29.7	31.0	27.3	16.8	33.9	36.6	27.7
800	37.5	29.7	30.9	34.5	19.8	37.2	37.1	29.4
1000	37.7	29.7	20.8	39.0	21.6	40.6	37.5	31.9
1500	38.0	20.8	20.8	30.1	23.9	33.5	37.9	35.2
2000	38.2	20.8	21.3	21.0	25.1	33.5	38.3	41.2
2500	38.5	20.8	23.0	21.8	25.9	33.5	38.7	39.3
m.p. (T/K)	265.9	594.3†	301.55	1112		1073	172.15	2130
$\Delta H_m/(kJ\ mol^{-1})$	10.57	6.07	2.09	8.53		9.20	6.41	20
b.p. (T/K)	332.6	1045	959	1765		3270	239.2	2870
$\Delta H_v/(kJ\ mol^{-1})$	29.56	99.87	67.77	154.67		313.80	20.41	339.5
Molecular formula of gas . . .	Br_2	Cd	Cs	Ca		Ce	Cl_2	Cr

T/K	Element and formula weight							
	Co 58.93	Cu 63.54	Er 167.26	Eu 151.96	F_2 38.00	Gd 157.25	Ga 69.72	Ge 72.59
50	4.4	6.3	28.4			19.4	10.3	6.2
100	13.9	16.2	24.6		29.1	28.9	18.5	13.8
150	19.5	20.5	26.0			32.7	22.0	18.5
200	22.3	22.6	27.0		29.7	36.1	23.8	20.9
400	26.5	25.3	28.7	27.6	33.0	29.3	27.8	24.9
600	29.7	26.5	30.0	29.3	35.1	30.7	27.8	26.4
800	32.4	27.5	31.3	31.0	36.3	32.1	27.8	27.4
1000	37.0	28.7	32.5	32.6	36.9	33.5	27.8	28.3
1500	39.7	32.8	35.6	33.5	37.8	37.0	27.8	29.3
2000	40.5	32.8	33.5	21.3	38.4	33.5	27.8	29.3
2500	40.5	32.8	33.5	23.5	38.8	33.5	27.8	29.3
m.p. (T/K)	1768†	1358.0†	1790	1100	53.55	1600	303	1232
$\Delta H_m/(kJ\ mol^{-1})$	16.2	13.14	17.15	10.46	5.10	15.48	5.59	31.80
b.p. (T/K)	3170	2860	2900	1700	85.05	3200	2480	3100
$\Delta H_v/(kJ\ mol^{-1})$	373.3	300.5	292.88	175.73	6.54	311.71	256.06	334.30
Molecular formula of gas . . .	Co	Cu	Er	Eu	F_2	Gd	Ga	Ge

Molar heat capacities, heats of fusion and vaporization of the elements (*contd*)

T/K	Element and formula weight							
	Au 196.97	Hf 178.49	He 4.003	Ho 164.93	H_2 2.0158	In 114.82	I_2 253.81	Ir 192.22
50	14.3		═	24.5	═	18.6	35.8	7.3
100	21.4			39.2	28.2	23.3	45.7	17.3
150	23.5			26.6		25.0	49.6	21.5
200	24.4			26.5	27.4	25.9	51.6	23.4
400	25.9	26.0	20.8	27.9	29.2	28.9	80.7	25.6
600	26.8	27.0	20.8	29.4	29.3	29.7	37.6	26.8
800	27.8	28.0	20.8	30.8	29.6	29.7	37.8	28.0
1000	28.9	29.0	20.8	32.2	30.2	29.7	37.9	29.2
1500	29.3	31.6	20.8	35.8	32.3	29.7	38.2	32.2
2000	29.3	34.1	20.8	33.5	34.3	29.7	38.5	35.1
2500	29.3	33.5	20.8	33.5	35.8	23.9	38.8	38.1
m.p. (T/K)	1337.6†	2500	3.5	1740	13.95	429.8†	386.8	2720†
$\Delta H_m/(kJ\ mol^{-1})$	12.36	21.76	0.02	17.15	0.11	3.26	15.52	26.36
b.p. (T/K)	3120	5600	4.22	2600	20.28†	2320	457.5	4800
$\Delta H_v/(kJ\ mol^{-1})$	324.43	661.07	0.08	251.04	0.90	226.35	41.94	563.58
Molecular formula of gas	Au	Hf	He	Ho	H_2	In	I_2	Ir

T/K	Element and formula weight							
	Fe 55.85	Kr 83.30	La 138.91	Pb 207.2	Li 6.941	Mg 24.31	Mn 54.94	Hg 200.59
50	3.0	25.1	18.5	21.4	4.2	5.8		19.9
100	12.1	31.6	23.6	24.4	13.4	15.8		24.3
150	18.1	20.8	25.4	25.4	17.8	20.4		25.9
200	21.6	20.8	26.3	25.9	21.6	22.8		27.3
400	27.4	20.8	6.8	27.7	27.6	26.3	28.2	27.4
600	32.0	20.8	7.1	29.4	29.5	28.5	31.5	27.1
800	37.9	20.8	7.5	30.0	28.9	31.0	34.4	20.8
1000	54.4	20.8	7.8	29.4	28.8	33.0	38.9	20.8
1500	36.5	20.8	8.0	28.7	28.5	20.8	47.3	20.8
2000	46.0	20.8	8.0	29.5	20.8	20.8	46.0	20.8
2500	46.0	20.8	8.0	29.1	21.1	20.8	21.1	20.8
m.p. (T/K)	1810	115.9	1195	600.65†	453.7	922	1520	234.31†
$\Delta H_m/(kJ\ mol^{-1})$	13.8	1.64	11.30	4.77	3.00	8.95	14.64	2.30
b.p. (T/K)	3030	119.75	3700	2030	1630	1375	2390	629.81†
$\Delta H_v/(kJ\ mol^{-1})$	349.5	9.03	399.57	177.9	147.1	127.6	219.74	59.14
Molecular formula of gas	Fe	Kr	Fe	Pb	Li	Mg	Mn	Hg

Molar heat capacities, heats of fusion and vaporization of the elements (*contd*)

T/K	Element and formula weight							
	Mo 95.94	Nd 144.24	Ne 20.179	Ni 58.71	Nb 92.91	N_2 28.013	Os 190.2	O_2 31.999
50	3.7	21.6	═══	4.0	8.5	41.5		46.1
100	13.5	26.6		13.6	17.4	29.1		29.1
150	18.8	28.3		19.3	21.6	29.1		29.1
200	21.5	29.1		22.5	23.1	29.1		29.1
400	25.1	32.4	20.8	28.5	25.4	29.3	25.3	30.1
600	26.5	36.9	20.8	34.9	26.3	30.1	26.0	32.1
800	27.4	41.3	20.8	31.0	27.2	31.4	26.7	33.7
1000	28.4	45.8	20.8	32.2	28.0	32.7	27.5	34.9
1500	31.7	33.5	20.8	36.3	30.1	34.8	29.3	36.5
2000	36.7	33.5	20.8	38.9	33.4	36.0	31.2	37.7
2500	43.9	33.5	20.8	38.9	38.6	36.6	33.0	38.8
m.p. (T/K)	2896†	1297	24.56	1728†	2750†	63.2	3300	54.36
$\Delta H_m/(kJ\ mol^{-1})$	36.0	10.88	0.34	17.2	26.9	0.72	29.29	0.44
b.p. (T/K)	4900	3370	27.10†	3180	5140	77.34†	5300	90.19†
$\Delta H_v/(kJ\ mol^{-1})$	590.4	283.68	1.77	377.5	690.1	5.59	627.60	6.82
Molecular formula of gas	Mo	Nd	Ne	Ni	Nb	N_2	Os	O_2

T/K	Element and formula weight							
	Pd 106.4	P (red) 30.974	Pt 195.09	K 39.098	Pr 140.91	Ra 226.03	Rn 222	Re 186.2
50	8.2		10.7	20.9				7.9
100	17.9	8.7	19.6	24.6				18.0
150	22.2		23.0	26.0				22.3
200	24.2	17.0	24.6	27.0			═══	24.3
400	26.6	23.2	26.5	31.5	28.4	29.3	20.8	25.9
600	27.7	25.8	27.5	30.1	31.0	33.5	20.8	26.9
800	28.9		28.6	29.8	33.7	37.7	20.8	28.0
1000	30.0		29.7	30.4	36.4	31.4	20.8	29.1
1500	33.0		32.4	20.8	33.5	31.4	20.8	31.8
2000	34.7		34.5	21.0	33.5	21.5	20.8	34.6
2500	34.7		34.7	21.6	33.5	24.4	20.8	37.3
m.p. (T/K)	1827†	870	2042†	336.4	1208	973	202	3453
$\Delta H_m/(kJ\ mol^{-1})$	16.74	18.8	19.66	2.33	10.04	8.37	2.90	33.05
b.p. (T/K)	3300	704	4090	1050	3270	1800	211	5900
$\Delta H_v/(kJ\ mol^{-1})$	393.30	(sublimes)	510.45	76.9	332.63	136.82	16.40	707.10
Molecular formula of gas	Pd		Pt	K	Pr	Ra	Rn	Re

Molar heat capacities, heats of fusion and vaporization of the elements (contd)

T/K	Element and formula weight							
	Rh 102.91	Rb 85.47	Ru 101.1	Sm 150.4	Sc 44.96	Se 78.96	Si 28.09	Ag 107.868
50	4.9	23.4	3.7			11.0	2.2	11.6
100	15.1	25.5	13.5			18.2	7.3	20.2
150	20.2	26.5	18.9			21.7	11.9	23.0
200	22.6	27.4	21.6			23.3	15.6	24.3
400	26.4	31.4	24.5	28.4	25.6	27.8	22.1	25.9
600	28.2	31.4	25.7	30.7	26.5	35.1	24.2	27.1
800	29.9	31.4	27.0	33.1	27.4	35.1	25.4	28.5
1000	31.6	20.8	28.2	35.4	28.3	18.8	26.3	29.8
1500	35.9	20.8	30.1	33.5	30.6	19.2	28.5	31.4
2000	40.2	21.0	31.4	26.4	33.5	19.6	27.2	31.4
2500	41.8	21.8	31.4	25.4	33.5	20.0	27.2	20.8
m.p. (T/K)	2236†	312.0	2600	1320	1770	493	1685	1235.1†
$\Delta H_m/(\text{kJ mol}^{-1})$	21.76	2.34	25.52	11.09	16.11	5.44	50.2	11.30
b.p. (T/K)	4000	978	4400	1870	3070	958	3490	2440
$\Delta H_v/(\text{kJ mol}^{-1})$	495.39	69.20	567.77	191.63	304.80	26.32	359.0	255.06
Molecular formula of gas	Rh	Rb	Rn	Sm	Sc	Se$_2$		Ag

T/K	Element and formula weight							
	Na 22.990	Sr 87.62	S 32.06	Ta 180.95	Te 127.60	Tl 204.37	Th 232.04	Sn 118.69
50	15.8		7.4	10.9	14.9	21.0	16.9	10.9
100	22.5	22.2	12.8	19.7	21.6	24.4	22.9	19.9
150	24.7		16.6	22.8	23.7	25.4	25.0	22.8
200	26.0	25.5	19.4	24.1	24.7	25.7	26.0	24.1
400	31.5	28.4	32.2	25.8	27.9	27.8	29.3	29.0
600	29.8	32.0	34.3	26.8	32.3	30.5	33.1	30.5
800	28.9	36.8	18.3	27.5	37.7	30.5	36.9	30.5
1000	28.9	37.7	18.6	27.9	37.7	30.5	40.7	30.5
1500	20.8	35.1	19.4	29.3	18.7	30.5	50.2	30.5
2000	20.8	21.0	20.1	31.2	18.7	22.7	46.0	30.5
2500	20.9	22.4	20.5	34.2	18.7	24.4	46.0	30.5
m.p. (T/K)	371	1041	388.4	3260	723	577	1970	505.1†
$\Delta H_m/(\text{kJ mol}^{-1})$	2.60	8.2	1.73	36	17.49	4.27	15.65	7.20
b.p. (T/K)	1175	1660	717.8†	5790	1270	1730	4800	2990
$\Delta H_v/(\text{kJ mol}^{-1})$	97.4	136.9		737.0	50.63	162.09	543.92	290.37
Molecular formula of gas	Na	Sr	S$_2$	Ta	Te$_2$	Tl	Th	Sn

Molar heat capacities, heats of fusion and vaporization of the elements (*contd*)

T/K	Ti 47.90	W 183.85	U 238.03	V 50.94	Xe 131.30	Y 88.91	Zn 65.38	Zr 91.22
50	4.7	5.8	15.5	4.0	25.1		11.1	9.2
100	14.3	16.0	22.2	13.5	28.2		19.2	18.7
150	19.5	20.5	24.7	19.1	33.6		22.5	22.3
200	22.3	22.5	26.1	21.9	20.8		23.8	24.1
400	26.5	24.9	29.6	26.2	20.8	25.6	26.4	26.6
600	28.3	25.8	34.5	27.5	20.8	26.4	28.4	28.9
800	30.2	26.7	41.3	28.7	20.8	27.3	31.4	31.3
1000	32.5	27.0	42.5	30.1	20.8	28.1	31.4	33.6
1500	31.5	29.9	38.3	34.8	20.8	30.3	20.8	30.2
2000	35.6	32.3	38.3	40.9	20.8	33.5	20.8	32.5
2500	35.6	34.7	38.3	46.2	20.8	33.5	20.8	33.5
m.p. (T/K)	1940	3695†	1408	2190	161.3	1780	692.7†	2123
$\Delta H_m/(\text{kJ mol}^{-1})$	18.6	35.4	15.48	22.8	2.30	17.15	7.38	21
b.p. (T/K)	3590	5940	4300	3690	165.0	3600	1185	4780
$\Delta H_v/(\text{kJ mol}^{-1})$	425.2	805.9	422.58	446.7	12.64	393.30	115.31	590.5
Molecular formula of gas	Ti	W	U	V	Xe	Y	Zn	Zr

E.F.G.H.

2.10.2 Molar heat capacities of some solids at various temperatures

The values for the molar heat capacities ($\text{J mol}^{-1}\text{K}^{-1}$) were calculated from data given by K. K. Kelley, *US Bureau of Mines Bulletin 584*, 1960 or by C. E. Wicks and F. E. Block, *US Bureau of Mines Bulletin 605*, 1963. Some values for solid elements and for other substances are given in Section 2.10.5.

T/K	BN 24.82	CaO 56.08	CaCO₃ calcite 100.09	CaAl₂Si₂O₈ glass 278.21	Fe₃O₄ 231.54	MgO 40.30	SiO₂ quartz 60.08	SiC 40.10
298	19.72	42.12	81.84	206.99	147.23	37.11	44.59	26.84
400	26.28	46.63	97.07	250.24	171.13	42.56	53.43	34.10
600	35.23	50.48	110.47	286.62	212.55	47.43	64.42	41.79
800	40.46	52.40	118.00	305.79	252.88	49.74	73.70	45.88
1000	44.35	53.74	123.85	319.98	200.83	51.21	68.96	48.42
1200	46.86	54.83	129.02		200.83	52.33		50.16
1500	48.70	56.27			200.83	53.69		51.94
2000	48.95	58.41				55.65		53.80

Molar heat capacities of some solids at various temperatures (*contd*)

T/K	Substance and formula weight							
	NaCl	ThO$_2$	TiO$_2$ *rutile*	WC	UO$_2$	ZnO	ZnS	ZrO$_2$
	58.44	264.04	79.90	195.86	254.03	81.83	97.44	123.22
298 . . .	50.51	62.33	55.10	36.09	63.70	40.24	46.02	56.19
400 . . .	52.35	66.91	62.84	37.02	72.69	45.34	49.40	63.85
600 . . .	55.48	71.94	69.93	38.84	79.80	49.52	52.41	70.24
800 . . .	59.31	74.87	73.08	40.65	83.17	51.65	54.14	73.45
1000 . . .	64.86	77.66	74.85	42.47	85.45	53.19	55.50	75.75
1200 . . .		80.27	76.02	44.28	87.32	54.49	56.71	77.68
1500 . . .		84.05	77.32	47.01	89.76	56.25		74.48
2000 . . .			78.87	51.55				74.48

E.F.G.H.

2.10.3 Molar heat capacities of gases at constant pressure at various temperatures

The values listed are the molar heat capacities, $C_p^{\ominus}$, in J mol^{-1} K^{-1} for the substances as ideal gases. Useful sources of information are *JANAF Thermochemical Tables*, 2nd edn, 1971, Project Directors D. R. Stull and H. Prophet, National Standard Reference Data Series, Nat. Bur. Stand. (US), and D. R. Stull, E. F. Westrum and G. C. Sinke, *The Chemical Thermodynamics of Organic Compounds*, 1969 (J. Wiley & Sons, Inc.).

T/K	Substance and formula weight							
	Acetone	Acetylene	Argon	Ammonia	Benzene	Bromine	Butane	Carbon dioxide
	58.0798	26.0378	39.948	17.0304	78.1134	159.808	58.123	44.0098
298 . . .	74.89	44.10	20.79	35.65	81.67	36.05	97.45	37.13
400 . . .	92.05	50.48	20.79	38.71	111.88	36.71	123.85	41.33
600 . . .	122.76	58.29	20.79	45.29	157.90	37.27	168.62	47.32
800 . . .	146.15	63.76	20.79	51.23	188.53	37.53	201.79	51.43
1000 . . .	163.80	68.27	20.79	56.49	209.87	37.70	226.86	54.31
1200 . . .		72.05	20.79	61.05		37.83		56.34

T/K	Substance and formula weight							
	Carbon disulphide	Carbon monoxide	Carbon tetrachloride	Chlorine	Chloroform	Deuterium	Diethyl ether	Ethane
	76.131	28.0104	153.823	70.906	119.3779	4.0282	74.1224	30.0694
298 . . .	45.66	29.14	83.40	33.94	65.38	29.20	112.51	52.63
400 . . .	49.68	29.34	91.70	35.30	74.25	29.24	138.11	65.61
600 . . .	54.61	30.44	99.68	36.57	85.26	29.62	183.76	89.33
800 . . .	57.50	31.90	103.09	37.15	91.51	30.50	218.66	108.07
1000 . . .	59.28	33.18	104.80	37.47	95.52	31.63	244.81	122.72
1200 . . .	60.45	34.17	105.77	37.70	98.30	32.76		

Molar heat capacities of gases at constant pressure at various temperatures (*contd*)

T/K	Substance and formula weight							
	Ethanol 46.0688	Ethylene 28.0536	Hydrochloric acid 36.4609	Hydrogen 2.0158	Hydrogen sulphide 34.0758	Methane 16.0426	Methanol 32.042	Nitric oxide 30.0061
298 . . .	65.44	42.89	29.14	28.84	34.19	35.64	43.89	29.84
400 . . .	81.00	53.05	29.18	29.18	35.58	40.50	51.42	29.94
600 . . .	107.49	70.66	29.58	29.33	38.94	52.23	67.03	31.24
800 . . .	126.90	83.84	30.50	29.62	42.52	62.93	79.66	32.77
1000 . . .	141.54	93.90	31.63	30.20	45.79	71.80	89.45	33.99
1200 . . .		101.63	32.71	30.99	48.47	78.83		34.88

T/K	Substance and formula weight							
	Nitrogen 28.0134	Nitrogen peroxide 46.0055	Nitrous oxide 44.0128	Oxygen 31.9988	Propane 44.0962	Sulphur dioxide 64.058	Sulphur trioxide 80.0582	Water 18.0152
298 . . .	29.12	36.97	38.62	29.38	73.51	39.87	50.66	33.58
400 . . .	29.25	40.17	42.62	30.11	94.31	43.49	57.67	34.25
600 . . .	30.11	45.84	48.39	32.09	129.20	49.05	67.26	36.30
800 . . .	31.43	49.71	52.24	33.73	155.14	52.43	72.76	38.69
1000 . . .	32.70	52.17	54.86	34.87	175.02	54.48	75.97	41.22
1200 . . .	33.72	53.75	56.67	35.67		55.79	77.94	43.70

E.F.G.H.

2.10.4 Cryoscopic and ebullioscopic constants, heats of fusion and vaporization for some common solvents

The cryoscopic constant of a solvent, k_f, would be the depression of freezing point in kelvin if one mole of any substance which does not dissociate or associate were dissolved in one kilogram of solvent and provided the laws of dilute solutions held at such a concentration. The analogous constant for boiling-point elevation is k_b. These constants k_f and k_b are connected with the heats of fusion, l_f, and vaporization l_v expressed in J g^{-1} by the equations

$$k_f = \boldsymbol{R}T_f^2/1000\,l_f \text{ and } k_b = \boldsymbol{R}T_b^2/1000\,l_v$$

where T_f and T_b are the freezing points and boiling points of the pure solvents and $\boldsymbol{R}$ is the gas constant $(8.314\,\mathrm{J\,mol^{-1}\,K^{-1}})$.

Cryoscopic and ebullioscopic constants, heats of fusion and vaporization for some common solvents (*contd*)

Solvent	$k_f/(\text{K kg mol}^{-1})$	$k_b/(\text{K kg mol}^{-1})$	$l_f/(\text{J g}^{-1})$	$l_v/(\text{J g}^{-1})$
Acetic acid	3.90	3.07	195	405
Acetone	2.40	1.71	98	524
Aniline	5.87	3.22	88	460
Benzene	5.12	2.53	126	394
Camphor	40	—	45	—
Carbon disulphide . . .	3.8	2.37	58	351
Carbon tetrachloride . .	30	4.95	16	195
Chloroform	4.90	3.66	80	246
Cyclohexane	20.1	2.79	32	356
Diethyl ether	1.79	1.82	97	358
Naphthalene	6.94	5.8	152	316
Nitrobenzene	6.90	5.26	94	331
Phenol	7.27	3.04	122	485
Pyridine	4.75	—	94	460
Water	1.86	0.51	333	2257

Heats of vaporization at various temperatures

Approximate values of the heat of vaporization can be computed from vapour pressure–temperature tables (see Section 2.4.4.) by the use of the Clapeyron equation. For conditions where the molar volume of the liquid is small compared with the molar volume of the vapour and where the vapour can be treated as an ideal gas the molar heat of vaporization H_v in J mol^{-1} can be calculated from the expression

$$H_v = \boldsymbol{R}T^2 p^{-1}(\mathrm{d}T/\mathrm{d}p)^{-1}$$

Values of $\mathrm{d}T/\mathrm{d}p$ for $p = 101.325$ kN m^{-2} for many substances are given in Section 2.4.4.

The following tables give values for the heats of vaporization (J g^{-1}) for several substances at various temperatures (°C).

Substance	$\theta/°C$						
	0	20	40	60	80	120	140
Acetone	585	563	541	519	498	458	438
Carbon disulphide	375	367	357	345	331	300	286
Carbon tetrachloride . . .	—	214	208	206	191	171	163
Chloroform	280	273	265	257	248	229	219
Diethyl ether	393	385	374	361	344	342	277
Sulphur dioxide	384	352	—	—	—	—	—

Some values for the heats of vaporization (J g^{-1}) at the normal boiling points are given in the preceding table.

Heats of vaporization for water at various temperatures

The subjoined highly accurate data for *water* which refer to ITS-68 are recalculated from those due to Osborne, Simpson and Ginnings, *J. Res. Bur. Stand.*, 1939, **23**, 197, 261.

$\theta/°C$	Heat vaporization/$(J\ g^{-1})$	$\theta/°C$	Heat vaporization/$(J\ g^{-1})$	$\theta/°C$	Heat vaporization/$(J\ g^{-1})$
0	2500.58	30	2429.95	90	2282.72
5	2488.87	40	2406.14	100	2256.67
10	2477.14	50	2382.10	130	2174.17
15	2465.39	60	2357.82	150	2114.39
20	2453.59	70	2333.20	170	2049.60
25	2441.78	80	2308.19	200	1940.56

E.F.G.H.

2.10.5 Standard free energies, heats of formation, entropies and molar heat capacities

The chemical species for which data are included in the following table are either of common occurrence or they have properties of particular interest. Most of the numerical values quoted are derived from the very comprehensive collection of data in *Selected Values of Chemical Thermodynamic Properties*, National Bureau of Standards Circular 500, Technical Notes 270–3, 270–4 and 270–5.

The National Bureau of Standards has now published revised tables of selected values for inorganic and C1 and C2 organic substances (*J. Phys. Chem. Ref. Data*, 1982, **11**, Suppl. 2) which gives SI values but employs a standard pressure of 100 000 Pa instead of 101 325 Pa as used hitherto. Adoption of this new convention does not change values for the standard heat of formation or molar heat capacity, but in some instances changes values for the standard free energy of formation and entropy.

Extensive tables giving values for thermodynamic quantities for a wide temperature range (Supplement JANAF Thermochemical Tables, *J. Phys. Chem. Ref. Data*, 1982, **11**, 695) use a standard pressure of 101 325 Pa and the thermochemical calorie. The new National Bureau of Standards tables are not consistent with extant tables and so data from Circular 500 have mainly been used in Table 2.10.5. Great care is necessary if data from different sources are to be combined.

Under 'State' s, sI and sII refer to different crystalline forms; allomorphic forms are also indicated: s. solid; l, liquid; g, gas; a, amorphous; and aq., aqueous solution. The entry hyp. m = 1 in this table means a hypothetical concentration of 1 mole in 1 kg water.

$\Delta G_f^\ominus$ is the standard free energy of formation and $\Delta H_f^\ominus$ is the standard heat of formation of the compound. These are incremental values associated with the formation of the compound from its elements when the product and reactants are taken in their appropriate standard states at 298.15 K. The standard state for a liquid or solid is that of the component under 101.325 kN m^{-2} (1 standard atmosphere) pressure and for a gas that of the ideal gas at a pressure of 101.325 kN m^{-2}. By definition the free energy of formation and heat of formation of an element in the standard state is zero, and by convention the corresponding quantities for the hydrogen ion in the standard state of unit activity in aqueous solution (i.e. hyp. m = 1) are taken as equal to zero.

For entities other than ions, $S^\ominus$ is the entropy in the standard state as above defined taking as zero the entropy at 0 K of the crystalline form in which all the molecules are orientated regularly. Thus nuclear contributions and those arising from isotopic mixing are excluded. For ions the entropy (as well as $\Delta G_f^\ominus$ and $\Delta H_f^\ominus$) is taken by convention with respect to the hydrogen ion in the standard state, as zero.

$C_p^\ominus$ is the molar heat capacity of the substance in the standard state at constant pressure.

As a consequence of the sign convention a negative value for a heat corresponds to heat evolution. Thus, for example, the entry in the table under $\Delta H_f^\ominus/kJ\ mol^{-1}$ for Hydrogen, H_2O, l is -285.8, which means that 285.8 kJ of heat are evolved when one mole of hydrogen gas combines with half a mole of oxygen to give one mole of liquid water with the reactants and products at 298.15 K.

Inorganic compounds

Substance and formula	State	$\Delta G_f^\ominus$ kJ mol^{-1}	$\Delta H_f^\ominus$ kJ mol^{-1}	$S^\ominus$ JK^{-1} mol^{-1}	$C_p^\ominus$ JK^{-1} mol^{-1}
Aluminium					
Al	s	0	0	28.3	24.4
Al_2O_3	s(α, corundum)	−1582.4	−1675.7	50.9	79.0
Al_2O_3	s, γ		−1652.7		
AlF_3	s	−1425.1	−1504.2	66.4	75.2
$AlCl_3$	s	−628.9	−704.2	110.7	91.8
$AlBr_3$	s		−527.2		101.7
$Al(NO_3)_3 . 9H_2O$ . . .	s		−3757		
$Al_2(SO_4)_3$	s	−3100.1	−3440.8	239.3	259.4
$NH_4Al(SO_4)_2 . 12H_2O$ . .	s	−4938.0	−5942.4	697.1	683.2
Al^{3+}	hyp. m = 1	−485.3	−531.4	−321.7	
Antimony					
Sb	sIII	0	0	45.7	25.2
Sb_2O_4	s	−795.8	−907.5	127.2	114.6
Sb_2O_5	s	−829.3	−971.9	125.1	
SbF_3	s		−915.5		
$SbCl_3$	s	−323.7	−382.2	184.1	108.0
$SbCl_5$	l	−350.2	−440.2	301.3	
Sb_2S_3	s, black	−173.6	−174.8	182.0	119.9
Argon					
Ar	g	0	0	154.7	20.8
Arsenic					
As	s, α, grey metallic	0	0	35.1	24.6
As_2O_5	s	−782.4	−924.9	105.4	116.5
As_4O_6	s, octahedral	−1152.5	−1313.9	214.2	191.3
AsH_3	g	68.9	66.4	222.7	38.1
AsF_3	l	−909.1	−956.3	181.2	126.6
$AsCl_3$	l	−256.7	−305.0	207.5	
As_2S_3	s	−168.6	−169.0	163.6	116.3
H_3AsO_3	hyp. m = 1	−639.9	−742.2	195.0	
H_3AsO_4	hyp. m = 1	−766.1	−902.5	184.1	
$H_2AsO_3^-$	hyp. m = 1	−587.2	−714.8	110.5	
$H_2AsO_4^-$	hyp. m = 1	−753.3	−909.6	117.2	
$HAsO_4^{2-}$	hyp. m = 1	−714.7	−906.3	−1.7	
Barium					
Ba	sII	0	0	66.9	26.4
BaO	s	−528.4	−558.1	70.3	47.4
BaF_2	s	−1148.5	−1200.4	96.2	71.2
$BaCl_2 . 2H_2O$	s	−1295.8	−1461.7	202.9	162.0
$Ba(OH)_2 . 8H_2O$ . . .	s		−3345.1		
$BaSO_4$	s	−1353.1	−1465.2	132.2	101.8
$BaCO_3$	sII, witherite	−1138.9	−1218.8	112.1	85.4
Ba^{2+}	hyp. m = 1	−560.7	−538.4	12.6	

Inorganic compounds (*contd*)

Substance and formula	State	$\Delta G_f^{\ominus}$ kJ mol^{-1}	$\Delta H_f^{\ominus}$ kJ mol^{-1}	$S^{\ominus}$ JK^{-1} mol^{-1}	$C_p^{\ominus}$ JK^{-1} mol^{-1}
Beryllium					
Be .	s	0	0	9.5	16.4
BeO .	s	-581.6	-610.9	14.1	25.4
Be^{2+} .	aq. acid soln		-389.1		
Bismuth					
Bi .	s	0	0	56.7	25.5
Bi$_2$O$_3$.	s	-493.7	-573.9	151.5	113.5
BiCl$_3$.	s	-315.1	-379.1	177.0	104.6
BiOCl	s	-322.2	-366.9	120.5	
Bi$_2$S$_3$.	s	-140.6	-143.1	200.4	122.2
Boron					
B .	s, β	0	0	5.9	11.1
B$_2$O$_3$.	s	-1193.7	-1272.8	54.0	62.9
BF$_3$.	g	-1120.3	-1137.0	254.0	50.5
BCl$_3$.	l	-387.4	-427.2	206.3	106.7
B$_2$H$_6$.	g	86.6	35.6	232.0	56.9
BN	s	-228.4	-254.4	14.8	19.7
H$_3$BO$_3$.	s	-969.0	-1094.3	88.8	81.4
BF$_4^-$.	hyp. m = 1	-1487.0	-1574.9	179.9	
Bromine					
Br$_2$.	l	0	0	152.2	75.7
Br$_2$.	g	3.1	30.9	245.4	36.0
HBr .	g	-53.4	-36.4	198.6	29.1
Br$^-$.	hyp. m = 1	-104.0	-121.5	82.4	-141.8
BrO$_3^-$.	hyp. m = 1	1.7	-83.7	163.2	
Cadmium					
Cd .	s, γ	0	0	51.8	26.0
CdO .	s	-228.4	-258.2	54.8	43.4
Cd(OH)$_2$.	s, precipitated	-473.6	-560.7	96.2	
CdCl$_2$.$2\frac{1}{2}$H$_2$O .	s	-944.1	-1131.9	227.2	
CdBr$_2$.4H$_2$O .	s	-1248.0	-1492.6	316.3	
CdSO$_4$.	s	-822.8	-933.3	123.0	99.6
CdS .	s	-156.5	-161.9	64.9	
Cd^{2+} .	hyp. m = 1	-77.6	-75.9	-73.2	
Caesium					
Cs .	s	0	0	82.8	31.0
CsCl .	sII		-433.0		52.5
Cs$^+$.	hyp. m = 1	-282.0	-247.7	133.1	
Calcium					
Ca .	sII	0	0	41.6	26.3
CaO .	s	-604.2	-635.5	39.7	42.8
Ca(OH)$_2$.	s	-896.8	-986.6	76.1	84.5

Inorganic compounds (*contd*)

Substance and formula	State	$\Delta G_f^{\ominus}$ kJ mol^{-1}	$\Delta H_f^{\ominus}$ kJ mol^{-1}	$S^{\ominus}$ JK^{-1} mol^{-1}	$C_p^{\ominus}$ JK^{-1} mol^{-1}
CaH$_2$	s	−149.8	−188.7	41.8	
CaC$_2$	s	−67.8	−62.8	70.3	62.7
CaCO$_3$	s, calcite	−1128.8	−1206.9	92.9	81.9
CaCO$_3$	s, aragonite	−1127.7	−1207.0	88.7	81.3
CaF$_2$	s	−1161.9	−1214.6	68.9	67.0
CaCl$_2$	s	−750.2	−795.0	113.8	72.6
Ca$_3$(PO$_4$)$_2$	s, α	−3889.9	−4126.3	241.0	231.6
Ca$_3$(PO$_4$)$_2$	s, β	−3899.5	−4137.6	236.0	227.8
CaSO$_4$	s, anhydrite	−1320.3	−1432.7	106.7	99.6
CaSO$_4$	s, soluble α	−1311.8	−1423.7	108.4	100.0
CaSO$_4$	s, soluble β	−1307.3	−1419.3	108.4	99.6
CaSO$_4$.$\frac{1}{2}$H$_2$O . . .	s, α	−1435.2	−1575.2	130.5	119.7
CaSO$_4$.$\frac{1}{2}$H$_2$O . . .	s, β	−1434.2	−1573.1	134.3	123.8
CaSO$_4$.2H$_2$O	s	−1795.7	−2021.1	194.0	186.2
Ca(NH$_2$)$_2$	s		−383.2		
Ca(NO$_3$)$_2$.4H$_2$O . . .	sII	−1700.8	−2131.2	338.9	
Ca^{2+}	hyp. m = 1	−553.0	−543.0	−55.2	
Carbon					
C	s, graphite	0	0	5.7	8.5
C	s, diamond	2.9	1.9	2.4	6.1
CO	g	−137.2	−110.5	197.6	29.1
CO$_2$	g	−394.4	−393.5	213.6	37.1
CO$_2$	hyp. m = 1	−386.0	−413.8	117.6	
CF$_4$	g	−879.6	−924.7	261.5	61.1
CCl$_4$	l	−65.3	−135.4	216.4	131.8
COCl$_2$	g	−204.6	−218.8	283.4	57.7
CS$_2$	l	65.3	89.7	151.3	75.7
CNI	s	185.0	166.1	96.2	
HCN	g	124.7	135.1	201.7	35.9
HCN	l	124.9	108.9	112.8	70.6
C$_2$N$_2$	g	297.4	308.9	241.8	56.8
CO$_3^{2-}$	hyp. m = 1	−527.9	−677.1	−56.9	
HCO$_3^-$	hyp. m = 1	−586.8	−692.0	91.2	
CN$^-$	hyp. m = 1	172.4	150.6	94.1	
CNS$^-$	aq.	92.7	76.4	144.3	−40.2
H.COO$^-$	hyp. m = 1 formate	−351.0	−426.6	92.0	−87.9
C$_2$H$_3$O$_2^-$	hyp. m = 1 acetate	−369.4	−486.0	86.6	−6.3
C$_2$O$_4^{2-}$	hyp. m = 1 oxalate	−674.0	−825.1	45.6	
HC$_2$O$_4^-$	hyp. m = 1 bioxalate	−698.4	−818.4	149.4	
Chlorine					
Cl$_2$	g	0	0	223.0	33.9
Cl$_2$O	g	97.9	80.3	266.1	45.4
ClO$_2$	g	120.5	102.5	256.7	42.0
Cl$_2$O$_7$	g		272.0		
HCl	g	−95.3	−92.3	186.8	29.1
Cl$^-$	hyp. m = 1	−131.3	−167.2	56.5	−136.4
ClO$_3^-$	hyp. m = 1	−3.3	−99.2	162.3	
ClO$_4^-$	hyp. m = 1	−8.6	−129.3	182.0	

Inorganic compounds (*contd*)

Substance and formula	State	ΔG_f° kJ mol^{-1}	ΔH_f° kJ mol^{-1}	S° JK^{-1} mol^{-1}	C_p° JK^{-1} mol^{-1}
Chromium					
Cr	s	0	0	23.8	23.3
Cr_2O_3	s	−1058.1	−1139.7	81.2	118.7
CrO_3	s		−589.5		
$CrCl_2$	s	−356.1	−395.4	115.3	71.2
$CrCl_3$	s	−486.2	−556.5	123.0	91.8
CrO_2Cl_2	l	−510.9	−579.5	221.8	
Cr^{2+}	aq.		−143.5		
$[Cr.6H_2O]^{3+}$ . . .	aq. violet		−1999.1		
$HCrO_4^-$	hyp. m = 1	−764.8	−878.2	184.1	
CrO_4^{2-}	hyp. m = 1	−727.8	−881.2	50.2	
$Cr_2O_7^{2-}$	hyp. m = 1	−1300.2	−1490.3	261.9	
Cobalt					
Co	sIII	0	0	30.0	24.8
$CoCl_2$	s	−269.9	−312.5	109.2	78.5
$CoCl_2.6H_2O$	s	−1725.5	−2115.4	343.1	
$CoSO_4.7H_2O$	s	−2473.8	−2979.9	406.1	390.5
Co^{2+}	hyp. m = 1	−54.4	−58.2	−113.0	
Co^{3+}	hyp. m = 1	133.9	92.0	−305.4	
Copper					
Cu	s	0	0	33.1	24.4
Cu_2O	s	−146.0	−168.6	93.1	63.6
CuCl	s	−119.9	−137.2	86.2	48.5
$CuCl_2$	s	−175.7	−220.1	108.1	71.9
$CuCl_2.2H_2O$	s	−656.1	−821.3	167.4	
$Cu(NO_3)_2.3H_2O$ . . .	s		−1217.1		
$CuSO_4.5H_2O$	sII	−1880.1	−2279.7	300.4	280.3
CuI	s	−69.5	−67.8	96.7	54.1
Cu_2S	s, α	−86.2	−79.5	120.9	76.3
CuS	s	−53.6	−53.1	66.5	47.8
Cu^+	hyp. m = 1	50.0	71.7	40.6	
Cu^{2+}	hyp. m = 1	65.5	64.8	−99.6	
Deuterium					
D_2	g	0	0	144.9	29.2
HD	g	−1.5	0.3	143.7	29.2
D_2O	g	−234.6	−249.2	198.2	34.3
D_2O	l	−243.5	−294.6	75.9	84.3
HDO	g	−233.1	−245.3	199.4	33.8
Europium					
Eu	s	0	0		26.7
Eu^{3+}	hyp. m = 1	−690.8	−708.3		
Fluorine					
F_2	g	0	0	202.7	31.3
HF	g	−273.2	−271.1	173.7	29.1
F^-	hyp. m = 1	−278.8	−332.6	−13.8	−106.7
HF_2^-	aq.	−578.1	−649.9	92.5	

Inorganic compounds (*contd*)

Substance and formula	State	$\Delta G_f^{\bullet}$ kJ mol^{-1}	$\Delta H_f^{\bullet}$ kJ mol^{-1}	$S^{\bullet}$ JK^{-1} mol^{-1}	$C_p^{\bullet}$ JK^{-1} mol^{-1}
Gallium					
Ga	sI	0	0	40.9	25.9
Ga^{3+}	hyp. m = 1	−159.0	−211.7	−330.5	
GaO$_3^{3-}$	hyp. m = 1		−619.2		
Germanium					
Ge	s	0	0	31.1	23.3
GeO$_2$	s, hexagonal	−497.1	−551.0	55.3	52.1
GeH$_4$	g	113.4	90.8	217.0	45.0
Gold					
Au	s	0	0	47.4	25.4
Au(OH)$_3$	s, precipitated	−317.0	−424.7	189.5	
AuCl	s		−34.7		
AuCl$_3$	s		−117.6		
AuCl$_4^-$	hyp. m = 1	−235.2	−322.2	266.9	
Au(CN)$_2^-$	hyp. m = 1	285.8	242.3	171.5	
Hafnium					
Hf	s, α hexagonal	0	0	43.6	25.7
HfO$_2$	s	−1088.3	−1144.7	59.3	60.3
Helium					
He	g	0	0	126.0	20.8
Hydrogen					
H$_2$	g	0	0	130.6	28.8
H	g	203.3	218.0	114.6	20.8
H$_2$O	g	−228.6	−241.8	188.7	33.6
H$_2$O	l	−237.2	−285.8	69.9	75.3
H$_2$O$_2$	g	−105.6	−136.3	232.6	43.1
H$_2$O$_2$	l	−120.4	−187.8	109.6	89.1
OH	g	34.2	39.0	183.6	29.0
H$^+$	hyp. m = 1	0	0	0	0
OH$^-$	hyp. m = 1	−157.3	−230.0	−10.8	−148.5
Indium					
In	s	0	0	57.8	26.7
InCl$_3$	s		−537.2		
In$_2$(SO$_4$)$_3$	s	−2439.3	−2786.5	272.0	280.3
In^{3+}	hyp. m = 1	−97.9	−104.6	−150.6	
Iodine					
I$_2$	s	0	0	116.1	54.4
I$_2$	g	19.4	62.4	260.6	36.9
I$_2$O$_5$	s		−158.1		
ICl	g	−5.4	17.8	247.4	35.6
ICl$_3$	s	−22.3	−89.5	167.4	
HI	g	1.7	26.5	206.5	29.2
HIO$_3$	s		−230.1		
I$^-$	hyp. m = 1	−51.6	−55.2	111.3	−142.3
I$_3^-$	hyp. m = 1	−51.5	−51.5	239.3	
IO$_3^-$	hyp. m = 1	−128.0	−221.3	118.4	

Inorganic compounds (*contd*)

Substance and formula	State	$\Delta G_f^{\bullet}$ kJ mol^{-1}	$\Delta H_f^{\bullet}$ kJ mol^{-1}	$S^{\bullet}$ JK^{-1} mol^{-1}	$C_p^{\bullet}$ JK^{-1} mol^{-1}
Iridium					
Ir .	s	0	0	35.5	25.1
IrCl	s		−81.6		
IrCl$_3$	s		−245.6		
IrCl$_6^{2-}$	aq.		−619.7		
IrCl$_6^{3-}$	aq.		−751.0		
Iron					
Fe .	s, α	0	0	27.3	25.1
Fe$_{0.947}$O	wustite	−245.1	−266.3	57.5	48.1
Fe$_2$O$_3$	hematite	−742.2	−824.2	87.4	103.8
Fe$_3$O$_4$	magnetite	−1015.5	−1118.4	146.4	143.4
FeCl$_2$.	s	−302.3	−341.8	117.9	76.7
FeCl$_3$.	s	−334.1	−399.5	142.3	96.7
FeCl$_3$.6H$_2$O .	s		−2223.8		
FeSO$_4$.7H$_2$O	s	−2510.3	−3004.6	409.2	394.5
Fe$_{1.000}$S	Iron-rich pyrrhotite, α	−100.4	−100.0	60.3	50.5
FeS$_2$.	s, pyrites	−166.9	−178.2	52.9	62.2
FeS$_2$	s, magkasite		−154.8		
Fe(NO$_3$)$_3$.9H$_2$O	s		−3285.3		
Fe(CO)$_5$.	l	−705.4	−774.0	338.1	240.6
Fe$_3$C .	s, cementite	20.1	25.1	104.6	105.9
FeCO$_3$.	s, siderite	−666.7	−740.6	92.9	82.1
Fe^{2+}	hyp. m = 1	−78.9	−89.1	−137.7	
Fe^{3+}	hyp. m = 1	−4.6	−48.5	−315.9	
Krypton					
Kr	g	0	0	164.0	20.8
KrF$_2$.	g		58.6		
Lead					
Pb.	s	0	0	64.8	26.4
PbO	s, red	−188.9	−219.0	66.5	45.8
PbO	s, yellow	−187.9	−217.3	68.7	46.8
PbO$_2$	s	−217.4	−277.4	68.6	64.6
Pb$_3$O$_4$	s	−601.2	−718.4	211.3	146.9
PbF$_2$	s	−617.1	−664.0	110.5	
PbCl$_2$.	s	−314.1	−359.4	136.0	
PbI$_2$.	s	−173.6	−175.5	174.8	77.4
Pb(NO$_3$)$_2$	s		−451.9		
PbSO$_4$	s	−813.2	−919.9	148.6	103.2
PbS	s	−98.7	−100.4	91.2	49.5
Pb(N$_3$)$_2$	s, monoclinic	624.7	478.2	148.1	
Pb(C$_2$H$_5$)$_4$	l		52.7		
Pb(CH$_3$COO)$_2$.3H$_2$O .	s		−1851.5		
Pb^{2+}	hyp. m = 1	−24.4	−1.7	10.5	
Lithium					
Li .	s	0	0	28.0	24.8
LiCl .	s		−408.8		48.0
LiAlH$_4$.	s		−101.3		83.2
LiBH$_4$.	s		−186.6		82.6

Inorganic compounds (*contd*)

Substance and formula	State	$\Delta G_f^{\ominus}$ kJ mol^{-1}	$\Delta H_f^{\ominus}$ kJ mol^{-1}	$S^{\ominus}$ JK^{-1} mol^{-1}	$C_p^{\ominus}$ JK^{-1} mol^{-1}
Li$^+$	hyp. m = 1	-293.8	-278.5	14.2	
Magnesium					
Mg	s	0	0	32.5	23.9
MgO	s	-569.6	-601.8	26.8	37.4
MgO	s, finely divided	-566.1	-598.1	27.9	37.8
MgCl$_2$.6H$_2$O . . .	s	-2115.6	-2499.6	366.1	315.7
MgF$_2$.	s	-1049.3	-1102.5	57.2	61.6
Mg(NO$_3$)$_2$.6H$_2$O . .	s		-2612.3		
MgSO$_4$.7H$_2$O . . .	s		-3383.6		
Mg(ClO$_4$)$_2$.2H$_2$O . .	s		-1216.3		
Mg(ClO$_4$)$_2$	s		-588.3		
MgCO$_3$	s	-1029.3	-1112.9	65.7	75.5
Mg^{2+}	hyp. m = 1	-456.0	-462.0	-118.0	
Manganese					
Mn	s, α	0	0	32.0	26.3
Mn	s, γ	1.4	1.5	32.4	27.6
MnO	s	-362.9	-385.2	59.7	45.4
MnO$_2$	s	-465.2	-520.0	53.1	54.1
MnCl$_2$	s	-440.5	-481.3	118.2	72.9
MnSO$_4$.4H$_2$O . . .	s		-2258.1		
Mn^{2+}	hyp. m = 1	-228.0	-220.7	-73.6	50.2
MnO$_4^-$	aq.	-447.3	-541.4	191.2	
Mercury					
Hg	l	0	0	76.0	28.0
Hg	g	31.9	61.3	174.8	20.8
HgO	s, red, orthorhombic	-58.6	-90.8	70.3	44.1
HgO	s, yellow	-58.4	-90.5	71.1	
Hg$_2$Cl$_2$	s	-210.8	-265.2	192.5	
HgCl$_2$	s	-178.7	-224.3	146.0	
Hg$_2$I$_2$.	s	-111.0	-121.3	233.5	
HgI$_2$	s, red	-101.7	-105.4	179.9	
HgI$_2$	s, yellow		-102.9		
HgS	s, red	-50.6	-58.1	82.4	48.4
HgS	s, black	-47.7	-53.6	88.3	
Hg(ONC)$_2$	s, fulminate		267.8		
Hg$_2^{2+}$	hyp. m = 1	153.6	172.4	84.5	
Hg^{2+}	hyp. m = 1	164.4	171.1	-32.2	
Molybdenum					
Mo	s	0	0	28.7	24.1
H$_2$MoO$_4$.H$_2$O . . .	s, yellow		-1359.8		
MoO$_3$	s	-668.0	-745.1	77.7	75.0
Neon					
Ne	g	0	0	146.2	20.8
Nickel					
Ni	s	0	0	29.9	26.1
NiO	s	-211.7	-239.7	38.0	44.3
NiCl$_2$.6H$_2$O. . . .	s	-1713.5	-2103.2	344.3	
Ni(NO$_3$)$_2$.6H$_2$O . . .	s		-2211.7		464.4

Inorganic compounds (*contd*)

Substance and formula	State	$\Delta G_f^\ominus$ kJ mol^{-1}	$\Delta H_f^\ominus$ kJ mol^{-1}	$S^\ominus$ JK^{-1} mol^{-1}	$C_p^\ominus$ JK^{-1} mol^{-1}
NiSO$_4$.6H$_2$O	s, α tetragonal, green	−2225.0	−2682.8	334.5	327.9
Ni(CO)$_4$	l	−588.3	−633.0	313.4	204.6
Ni^{2+}	hyp. m = 1	−45.6	−54.0	−128.9	
Ni(CN)$_4^{2-}$	hyp. m = 1	472.0	367.8	217.6	
Niobium					
Nb	s	0	0	36.4	24.6
Nb$_2$O$_5$	s, high temp. form	−1766.1	−1899.5	137.2	132.1
Nitrogen					
N$_2$	g	0	0	191.5	29.1
NO	g	86.6	90.2	210.7	29.8
NO$_2$	g	51.3	33.2	240.0	37.2
N$_2$O$_4$	g	97.8	9.2	304.2	77.3
N$_2$O	g	104.2	82.0	219.7	38.5
N$_2$O$_5$	g	115.1	11.3	355.6	84.5
NH$_3$	g	−16.5	−46.1	192.3	35.1
NH$_3$	hyp. m = 1	−26.6	−80.3	111.3	
N$_2$H$_4$	l	149.2	50.6	121.2	98.9
N$_2$H$_4$.HCl	s		−196.7		
NH$_2$OH.HCl	s		−317.6		
HNO$_3$	l	−80.8	−174.1	155.6	109.9
HNO$_3$	hyp. m = 1	−111.3	−207.4	146.4	−86.6
NH$_4$NO$_3$	s	−184.0	−365.6	151.1	139.3
NH$_4$NO$_2$	s		−256.5		
NOCl	g	66.1	51.7	261.6	44.7
NH$_4$Cl	s	−203.0	−314.4	94.6	84.1
NH$_4$HSO$_4$	s		−1027.0		
(NH$_4$)$_2$SO$_4$	s	−901.9	−1180.9	220.1	187.5
HNO$_2$	hyp. m = 1	−55.6	−119.2	152.7	
NH$_4$HS	s	−50.6	−156.9	97.5	
(NH$_4$)$_2$S	hyp. m = 1	−72.8	−231.8	212.1	
(NH$_4$)$_2$S$_2$	hyp. m = 1	−79.1	−234.7	255.2	
NH$_4^+$	hyp. m = 1	−79.4	−132.5	113.4	79.9
NO$_3^-$	hyp. m = 1	−111.3	−207.4	146.4	−86.6
NO$_2^-$	hyp. m = 1	−37.2	−104.6	140.2	−97.5
Osmium					
Os	s	0	0	32.6	24.7
OsO$_4$	s, white	−303.8	−385.8	167.8	
OsO$_4$	s, yellow	−305.0	−394.1	143.9	
Oxygen					
O$_2$	g	0	0	205.0	29.4
O$_3$	g	163.2	142.7	238.8	39.2
Palladium					
Pd	s	0	0	37.6	26.0
PdCl$_2$	s	−125.1	−171.5	104.6	
PdCl$_4^{2-}$	m = 1 in 1N HCl	−416.7	−522.2	259.4	
PdCl$_6^{2-}$	m = 1 in 1N HCl	−430.1	−598.3	272.0	
Phosphorus					
P	s, α, white	0	0	41.1	23.8

Inorganic compounds (*contd*)

Substance and formula	State	$\Delta G_f^{\ominus}$ kJ mol^{-1}	$\Delta H_f^{\ominus}$ kJ mol^{-1}	$S^{\ominus}$ JK^{-1} mol^{-1}	$C_p^{\ominus}$ JK^{-1} mol^{-1}
P	s, red, triclinic	-12.1	-17.6	22.8	21.2
P	s, black		-39.3		
P_4O_{10}	s, hexagonal	-2697.8	-2984.0	228.9	211.7
PH_3	g	13.4	5.4	210.1	37.1
PCl_3	g	-267.8	-287.0	311.7	71.8
PCl_3	l	-272.4	-319.7	217.1	
PCl_5	s		-443.5		
PCl_5	g	-305.0	-374.9	364.5	112.8
PBr_3	g	-162.8	-139.3	348.0	76.0
PBr_3	l	-175.7	-184.5	240.2	
$POCl_3$	g	-513.0	-558.5	325.3	84.9
$POCl_3$	l	-520.9	-597.1	222.5	138.8
PBr_5	s		-269.9		
H_3PO_4	s	-1119.22	-1279.0	110.5	106.1
$(NH_4)_2HPO_4$	s		-1566.9		
PO_4^{3-}	hyp. m = 1	-1018.8	-1277.4	-221.8	
HPO_4^{2-}	hyp. m = 1	-1089.3	-1292.1	-33.5	
$H_2PO_4^-$	hyp. m = 1	-1130.4	-1296.3	90.4	
Platinum					
Pt	s	0	0	41.63	25.85
$PtCl_2$	s		-109.4		
$PtCl_4^{2-}$	hyp. m = 1	-368.6	-503.3	167.4	
$PtCl_6^{2-}$	hyp. m = 1	-489.5	-673.6	220.1	
Potassium					
K	s	0	0	63.6	29.2
K_2O	s		-361.5		
KF	s	-533.1	-562.6	66.6	49.1
KCl	s	-408.3	-435.9	82.7	51.5
KBr	s	-379.2	-392.2	96.4	53.6
KI	s	-322.3	-327.6	104.3	52.9
KHF_2	s	-852.4	-920.4	104.3	76.9
KNO_3	s	-393.1	-492.7	132.9	96.3
KNO_2	s		-370.3		107.4
K_2SO_4	s	-1316.4	-1433.7	175.7	130.1
$KClO_3$	s	-289.9	-391.2	143.0	100.2
$KClO_4$	s	-304.2	-433.5	151.0	110.2
$KBrO_3$	s	-243.5	-332.2	149.2	104.9
KIO_3	s	-423.4	-508.4	151.5	106.4
K_2CO_3	s		-1146.1		114.4
$KHCO_3$	s		-959.4		
KCN	s		-112.5		66.3
KCNO	s		-412.1		
KCNS	s		-203.4		88.5
$KMnO_4$	s		-813.4		
$K_2Cr_2O_7$	s		-2033.0		
KH_2PO_4	s		-1568.6		
K^+	hyp. m = 1	-282.3	-251.2	102.5	
Radium					
Ra	s	0	0	71.1	27.0
Ra^{2+}	hyp. m = 1	-562.7	-527.2	54.4	

Inorganic compounds (*contd*)

Substance and formula	State	ΔG_f° kJ mol^{-1}	ΔH_f° kJ mol^{-1}	S° JK^{-1} mol^{-1}	C_p° JK^{-1} mol^{-1}
Radon					
Rn	g	0	0	176.1	20.8
Rhenium					
Re	s	0	0	36.9	25.5
ReO_4^-	aq.	−694.5	−787.4	201.3	−13.4
Rhodium					
Rh	s	0	0	31.5	25.0
$RhCl_3$	s		−299.2		
Rubidium					
Rb	s	0	0	69.5	30.4
RbCl	s		−430.6		52.4
Rb^+	hyp. m = 1	−282.2	−246.4	124.3	
Ruthenium					
Ru	s	0	0	28.5	24.1
$RuCl_3$	s, black		−205.0		
Scandium					
Sc	s	0	0	34.6	25.5
Sc^{3+}	hyp. m = 1	−586.6	−614.2	−255.2	
Selenium					
Se	s, black hexagonal	0	0	42.4	25.4
SeO_2	s		−225.4		
H_2Se	g	15.9	29.7	218.9	34.7
SeO_3^{2-}	hyp. m = 1	−369.9	−509.2	12.6	
SeO_4^{2-}	hyp. m = 1	−441.4	−599.1	54.0	
Silicon					
Si	s	0	0	18.8	20.0
SiO_2	s, α, quartz	−856.7	−910.9	41.8	44.4
SiO_2	s, α, cryostobalite	−855.9	−909.5	42.7	44.2
SiO_2	s, α, tridymite	−855.3	−909.1	43.5	44.6
SiO_2	a	−850.7	−903.5	48.9	44.4
SiH_4	g	56.9	34.3	204.5	42.8
SiF_4	g	−1572.7	−1614.9	282.4	73.6
$SiCl_4$	g	−617.0	−657.0	330.6	90.2
$SiCl_4$	l	−619.9	−687.0	239.7	145.3
SiC	s, β, cubic	−62.8	−65.3	16.6	26.9
SiF_6^{2-}	aq.	−2199.5	−2389.1	122.2	
Silver					
Ag	s	0	0	42.6	25.4
Ag_2O	s	−11.2	−31.0	121.3	65.9
AgCl	s	−109.8	−127.1	96.2	50.8
AgBr	s	−96.9	−100.4	107.1	52.4
AgI	s	−66.2	−61.8	115.5	56.8
$AgNO_3$	s	−33.5	−124.4	140.9	93.1
Ag_2S	s, α, orthorhombic	−40.7	−32.6	144.0	76.5

Inorganic compounds (*contd*)

Substance and formula	State	$\Delta G_f^{\ominus}$ kJ mol^{-1}	$\Delta H_f^{\ominus}$ kJ mol^{-1}	$S^{\ominus}$ JK^{-1} mol^{-1}	$C_p^{\ominus}$ JK^{-1} mol^{-1}
Ag$_2$S	s, β	-39.5	-29.4	150.6	
Ag$_2$SO$_4$	s	-618.5	-715.9	200.4	131.4
Ag$^+$	hyp. m $=$ 1	77.1	105.6	72.7	21.8
Ag(CN$_2$)$^-$	hyp. m $=$ 1	305.4	270.3	192.5	
Sodium					
Na	s	0	0	51.0	28.4
Na$_2$O	s	-376.6	-415.9	72.8	68.2
NaF	s	-541.0	-569.0	58.6	46.0
NaCl	s	-384.0	-411.0	72.4	49.7
NaBr	s		-359.9		51.4
NaI	s		-288.0		52.1
NaOH	s		-426.7		59.5
NaNO$_3$	s	-365.9	-466.7	116.3	93.1
Na$_2$SO$_4$	s	-1266.8	-1384.5	149.5	127.6
Na$_2$CO$_3$	s	-1047.7	-1130.9	136.0	112.3
Na$_2$CO$_3$.10H$_2$O	s		-4081.9		550.3
NaHCO$_3$	s	-851.9	-947.7	102.1	87.6
Na$_2$O$_2$	s		-504.6		89.2
NaClO$_3$	s		-358.7		
NaClO$_4$	s		-385.7		
Na$_2$SO$_3$.7H$_2$O	s		-3152.2		
Na$_2$SO$_4$.10H$_2$O	s	-3644.0	-4324.1	592.9	587.4
Na$_2$S$_2$O$_3$.5H$_2$O	s		-2602.0		
NaNO$_2$	s		-359.4		
Na$_2$HPO$_4$.12H$_2$O	s		-5298.6		
Na$_3$PO$_4$	s		-1997.9		
NaC$_2$H$_3$O$_2$	s, acetate		-710.4		79.9
NaCN	sIII		-89.8		70.4
NaCNS	s		-174.6		
Na$_2$SiO$_3$	s	-1426.7	-1518.8	113.8	111.8
Na$_2$CrO$_4$	s		-1328.8		142.1
Na$_2$B$_4$O$_7$.10H$_2$O	s		-6264.3		615
NaBH$_4$	sI	-119.5	-183.3	104.7	86.6
Na$^+$	hyp. m $=$ 1	-261.9	-239.7	60.2	
Strontium					
Sr	s	0	0	54.4	25.1
Sr(NO$_3$)$_2$.4H$_2$O	s		-2152.7		
Sr(OH)$_2$.8H$_2$O	s		-3352.2		
Sr^{2+}	hyp. m $=$ 1	-557.3	-547.3	39.3	
Sulphur					
S	s, rhombic	0	0	31.8	22.6
S	s, monoclinic		0.3		
S	g	238.3	278.8	167.7	23.7
SO$_2$	g	-300.2	-296.8	248.1	39.9
SO$_3$	g	-371.1	-395.7	256.6	50.7
H$_2$S	g	-33.6	-20.6	205.7	34.2
H$_2$SO$_4$	l	-690.1	-814.0	156.9	138.9
S$_2$Cl$_2$	g	-31.8	-18.4	331.4	73.6
SO$_2$Cl$_2$	l		-394.1		133.9

Inorganic compounds (*contd*)

Substance and formula	State	$\Delta G_f^{\oplus}$ kJ mol^{-1}	$\Delta H_f^{\oplus}$ kJ mol^{-1}	$S^{\oplus}$ JK^{-1} mol^{-1}	$C_p^{\oplus}$ JK^{-1} mol^{-1}
S^{2-}	hyp. m = 1	85.8	33.1	−14.6	
HS$^-$	hyp. m = 1	12.0	−17.6	62.8	
SO$_3^{2-}$	hyp. m = 1	−486.6	−635.5	−29.3	
SO$_4^{2-}$	hyp. m = 1	−744.6	−909.3	20.1	−292.9
Tantalum					
Ta	s	0	0	41.5	25.4
Ta$_2$O$_5$	s, β	−1911.3	−2046.0	143.1	135.1
Tellurium					
Te	s	0	0	49.7	25.7
TeO$_2$	s	−270.3	−322.6	79.5	
H$_6$TeO$_6$	s		−1298.7		
H$_2$Te	g		99.6		
TeCl$_4$	s		−326.4		138.5
Thallium					
Tl	s	0	0	64.2	26.3
Tl$^+$	hyp. m = 1	−32.4	5.4	125.5	
Tl^{3+}	hyp. m = 1	214.6	196.6	−192.5	
Thorium					
Th	s	0	0	56.9	27.4
ThO$_2$	s		−1221.7		61.8
Th(SO$_4$)$_2$	s		−2542.6		
Tin					
Sn	sI, white	0	0	51.5	27.0
Sn	sII, grey	0.1	−2.1	44.1	25.8
SnO	s	−256.9	−285.8	56.5	44.3
SnO$_2$	s	−519.7	−580.7	52.3	52.6
SnCl$_4$	l	−440.2	−511.3	258.6	165.3
SnCl$_2$.2H$_2$O	s		−921.3		
Sn^{2+}	in aq. HCl	−27.2	−8.8	−16.7	
Sn^{4+}	in aq. HCl	2.5	30.5	−117.2	
Titanium					
Ti	sII	0	0	30.6	25.0
TiO$_2$	s, rutile	−889.5	−944.7	50.3	55.0
TiCl$_3$	s	−653.5	−720.9	139.7	97.2
TiCl$_4$	l	−737.2	−804.2	252.3	145.2
Tungsten					
W	s	0	0	32.6	24.3
WO$_3$	s	−764.1	−842.9	75.9	73.8
Uranium					
U	sIII	0	0	50.3	27.5
UO$_2$	s	−1075.3	−1129.7	77.8	
UO$_3$	s	−1184.1	−1263.6	98.6	81.7
UF$_4$	s	−1761.5	−1853.5	151.0	117.7
UF$_6$	g	−2029.2	−2112.9	379.7	129.6

Inorganic compounds (*contd*)

Substance and formula	State	$\Delta G_f^{\ominus}$ kJ mol^{-1}	$\Delta H_f^{\ominus}$ kJ mol^{-1}	$S^{\ominus}$ JK^{-1} mol^{-1}	$C_p^{\ominus}$ JK^{-1} mol^{-1}
$UO_2(NO_3)_2.6H_2O$	s	−2615.0	−3197.8	505.6	466.9
$UO_2SO_4.3H_2O$	s	−2451.8	−2789.9	263.6	282.8
U^{3+}	hyp. m = 1	−520.5	−514.6	−125.5	
U^{4+}	hyp. m = 1	−579.1	−613.8	−326.4	
UO_2^{2+}	hyp. m = 1	−989.1	−1047.7	−71.1	
Vanadium					
V	s	0	0	28.9	24.9
V_2O_5	s	−1419.6	−1550.6	131.0	127.7
Xenon					
Xe	g	0	0	169.6	20.8
XeO_3	s		401.7		
XeO_3	g		527.2		
XeO_4	g		642.2		
XeF_2	g	−74.0	−107.6	259.5	54.1
XeF_4	g	−132.8	−206.5	327.8	90.4
XeF_6	g		−279.0		
Yttrium					
Y	s	0	0	44.4	26.5
Y_2O_3	s	−1816.7	−1905.3	99.1	102.5
Zinc					
Zn	s	0	0	41.6	25.4
ZnO	s	−318.3	−348.3	43.6	40.3
$ZnCl_2$	s	−369.4	−415.1	111.5	71.3
$ZnBr_2$	s	−312.1	−328.7	138.5	
ZnI_2	s	−208.9	−208.0	161.1	
$ZnSO_4.7H_2O$	s	−2563.1	−3077.8	388.7	383.4
ZnS	s, sphalerite	−201.3	−206.0	57.7	46.0
ZnS	s, wurtzite		−192.6		
$Zn(NO_3)_2.6H_2O$	s	−1773.1	−2306.6	456.9	323.0
Zn^{2+}	hyp. m = 1	−147.0	−153.9	−112.1	46.0
Zirconium					
Zr	s, α, hexagonal	0	0	39.0	25.4
$ZrOCl_2.8H_2O$	s		−3471.9		
$Zr(SO_4)_2.H_2O$	s		−2553.9		

Organic compounds

Some additional sources of data are D. R. Stull, E. F. Westrum and G. C. Sinke, *The Chemical Thermodynamics of Organic Compounds*, 1969 (Wiley & Sons, Inc., New York) and J. D. Cox and G. Pilcher, *Thermochemistry of Organic and Organometallic Compounds*, 1970 (Academic Press, New York).

See inorganic compounds for data on other carbon compounds.

Organic compounds (*contd*)

Substance	State	$\Delta G_f^{\ominus}$ kJ mol^{-1}	$\Delta H_f^{\ominus}$ kJ mol^{-1}	$S^{\ominus}$ JK^{-1} mol^{-1}	$C_p^{\ominus}$ JK^{-1} mol^{-1}
Acetaldehyde	g	−128.9	−166.2	250.2	57.3
Acetamide	s		−317.6		
Acetic acid	l	−389.9	−484.5	159.8	124.3
Acetone	l	−155.4	−248.1	200.4	124.7
Acetonitrile	l	99.2	53.6	149.6	91.5
Acetyl chloride	l	−208.1	−273.8	200.8	117.2
Acetylene	g	209.2	226.7	200.8	43.9
Benzene	g	129.7	82.9	269.2	81.7
Benzene	l	124.3	49.0	173.3	136.1
Benzoic acid	s	−245.3	−385.1	167.6	146.8
Bromoethane	l	−27.8	−92.0	198.7	100.8
Bromoethane	g	−25.9	−35.2	246.3	42.4
Butane	g	−17.2	−126.1	310.1	97.4
Buta-1,3-diene	g	150.7	110.2	278.7	79.5
Chloroacetic acid . . .	sI		−513.0		
Chloroethane	g	−60.5	−112.2	275.9	62.8
Chloroform	l	−73.7	−134.5	201.7	113.8
Chloromethane	g	−57.4	−80.8	234.5	40.8
Cyclohexane	l	6.4	−37.3	204.3	156.5
Diazomethane	g	217.8	192.5	242.8	52.5
1,2-Dichloroethane . . .	l	−79.6	−165.2	208.5	129.3
1,1-Dichloroethylene . .	l	24.5	−24.3	201.5	111.3
Dichloromethane . . .	l	−67.3	−121.5	177.8	100.0
Diethyl ether	g	−122.3	−252.2	342.7	112.5
Dimethylamine	g	68.4	−18.4	273.0	70.7
Diphenyl ether	s	144.2	−31.7	233.9	216.6
Ethane	g	−32.9	−84.7	229.5	52.6
Ethanediol	l	−323.2	−454.8	166.9	149.8
Ethanol	l	−174.1	−277.0	160.7	113.4
Ethyl acetate	l	−332.7	−479.0	259.4	170.1
Ethylbenzene	g	130.6	29.8	360.5	128.4
Ethylene	g	68.1	52.3	219.5	43.6
Ethylene oxide	g	−13.1	−52.6	242.4	47.9
Ethyl hydrogen peroxide . .	l		−245.2		
Ethyl nitrate	l	−43.1	−190.4	247.2	170.3
Ethyl nitrite	g		−103.8		
Formaldehyde	g	−113.0	−117.2	218.7	35.4
Formamide	l		−257.7		107.6
Formic acid	l	−361.4	−424.7	129.0	99.0
Glucose	s	−910.5	−1274.4	212.1	218.9
Glycine	s	−373.4	−532.9	103.5	99.2
Heptane	l	1.0	−224.4	328.6	224.3
Iodoform	g	177.9	210.9	355.6	74.9
Iodomethane	g	14.6	13.0	254.0	44.1
Ketene	g	−61.9	−61.1	247.5	51.8
Methane	g	−50.8	−74.8	186.2	35.3
Methanethiol	g	−9.3	−22.3	255.1	50.2
Methanol	l	−166.4	−238.7	126.8	81.6
Methylamine	g	32.1	−23.0	243.3	53.1
Methyl formate	g	−297.2	−347.5	301.2	66.5
Nitroethane	l		−140.2		134.2
Nitromethane	l	−14.5	−113.1	171.8	106.0
Oxalic acid	s	−701.2	−830.0	120.1	117.2
Pentane	g	−8.4	−146.4	348.9	120.2

Organic compounds (*contd*)

Substance	State	$\dfrac{\Delta G_f^\circ}{\text{kJ mol}^{-1}}$	$\dfrac{\Delta H_f^\circ}{\text{kJ mol}^{-1}}$	$\dfrac{S^\circ}{\text{JK}^{-1}\,\text{mol}^{-1}}$	$\dfrac{C_p^\circ}{\text{JK}^{-1}\,\text{mol}^{-1}}$
Phenol	s	-50.4	-165.0	144.0	127.4
Propane	g	-23.5	-103.8	269.9	73.5
Propene	g	62.7	20.4	266.9	63.9
Pyridine	g	190.2	140.2	282.8	78.1
Styrene	g	213.8	147.4	345.1	122.1
Sucrose	s	-1544.6	-2222.1	360.2	425.0
Thiophen	g	126.8	115.7	278.9	72.9
Thiourea	s	21.7	-90.0	115.8	
Toluene	g	122.0	50.0	320.7	103.6
Trichloroethylene	g	18.0	-7.8	324.7	80.3
Trimethylamine	g	98.9	-23.8	288.8	91.8
Urea	s	-197.2	-333.2	104.6	93.1
Vinyl chloride	g	51.9	35.6	263.9	53.7
o-Xylene	g	122.1	19.0	352.8	133.3
m-Xylene	g	118.9	17.2	357.7	127.6
p-Xylene	g	121.1	17.9	352.4	126.9

E.F.G.H.

2.10.6 CODATA key values for thermodynamics, 1977

The values are taken from CODATA Bulletin 28, 1977 (see also *J. Chem. Thermodynamics*, 1978, **10**, 903). Where values for the same quantities are given in the self-consistent set in Section 2.10.5 the CODATA values are more accurate.

The reference state for each element at 298.15 K is the thermodynamically stable state except for phosphorus, for which the 'white' crystal modification has been selected; for tin the 'white form' (thermodynamically stable at 298.15 K) is taken as the reference state at all temperatures down to zero, even although that form is known to be metastable below 286 K. For crystalline solids (cr) and liquids (l) the standard state is that of the pure substance under a pressure of 101 325 Pa. For gases the standard state is that of the ideal gas at a pressure of 101 325 Pa. For species in aqueous solution (aq) the standard state is the hypothetical ideal solution at unit activity (molality scale); the properties of ideal aqueous ionic solutions are equal to the sum of the properties of the individual ions.

The values of H° (298.15 K) $- H^\circ$(o) for gases relate to the hypothetical ideal-gas state at zero temperature, whilst values of H° (298.15 K) $- H^\circ$(o) for both liquids and crystalline solids relate to crystalline solids at zero temperature.

The following values were used in the calculations: IUPAC Atomic Weights, 1970; gas constant $8.31433 \pm 0.0080\,\text{J K}^{-1}\,\text{mol}^{-1}$; Faraday constant $96\,487.0 \pm 1.0\,\text{J V}^{-1}\,\text{mol}^{-1}$; constant relating wave number and energy $0.119\,625\,6 \pm 0.000\,000\,26\,\text{J m mol}^{-1}$.

CODATA recommended key values for thermodynamics, 1977

Substance	State	$\dfrac{\Delta_f H^\circ (298.15\,\text{K})}{\text{kJ mol}^{-1}}$	$\dfrac{S^\circ (298.15\,\text{K})}{\text{J K}^{-1}\,\text{mol}^{-1}}$	$\dfrac{H^\circ (298.15\,\text{K}) - H^\circ (\text{o})}{\text{kJ mol}^{-1}}$
O	g	249.17 ± 0.10	160.946 ± 0.020	6.728 ± 0.003
O_2	g	0	205.037 ± 0.033	8.682 ± 0.004
H	g	217.997 ± 0.006	114.604 ± 0.015	6.197 ± 0.002
H^+	aq	0	0	—
H_2	g	0	130.570 ± 0.033	8.468 ± 0.003
OH^-	aq	-230.025 ± 0.045	-10.71 ± 0.20	—

CODATA recommended key values for thermodynamics, 1977 (*contd*)

Substance	*State*	$\Delta_f H°(298.15 \text{ K})$ kJ mol^{-1}	$S°(298.15 \text{ K})$ J K^{-1} mol^{-1}	$H°(298.15 \text{ K}) - H°(0)$ kJ mol^{-1}
H_2O	l	-285.830 ± 0.042	69.950 ± 0.080	13.293 ± 0.020
H_2O	g	-241.814 ± 0.042	188.724 ± 0.040	9.908 ± 0.008
He	g	0	126.039 ± 0.012	6.197 ± 0.002
Ne	g	0	146.214 ± 0.016	6.197 ± 0.002
Ar	g	0	154.732 ± 0.020	6.197 ± 0.002
Kr	g	0	163.971 ± 0.020	6.197 ± 0.002
Xe	g	0	169.573 ± 0.020	6.197 ± 0.002
F	g	79.39 ± 0.30	158.640 ± 0.020	6.518 ± 0.004
F$^-$	aq	-335.35 ± 0.65	-13.18 ± 0.54	—
F_2	g	0	202.685 ± 0.040	8.825 ± 0.004
HF	g	-273.30 ± 0.70	173.665 ± 0.035	8.599 ± 0.004
Cl	g	121.302 ± 0.008	165.076 ± 0.020	6.272 ± 0.003
Cl$^-$	aq	-167.080 ± 0.088	56.73 ± 0.16	—
Cl_2	g	0	222.965 ± 0.040	9.180 ± 0.008
HCl	g	-92.31 ± 0.13	186.786 ± 0.033	8.640 ± 0.004
Br	g	111.86 ± 0.12	174.904 ± 0.020	6.197 ± 0.020
Br$^-$	aq	-121.50 ± 0.15	82.84 ± 0.20	
Br_2	l	0	152.210 ± 0.040	24.52 ± 0.13
Br_2	g	30.91 ± 0.11	245.350 ± 0.054	9.724 ± 0.012
HBr	g	-36.38 ± 0.17	198.585 ± 0.033	8.648 ± 0.004
I	g	106.762 ± 0.040	180.673 ± 0.020	6.197 ± 0.002
I$^-$	aq	-56.90 ± 0.84	106.70 ± 0.20	—
I_2	cr	0	116.139 ± 0.080	13.196 ± 0.040
I_2	g	62.421 ± 0.080	260.567 ± 0.063	10.117 ± 0.021
HI	g	26.36 ± 0.80	206.480 ± 0.040	8.657 ± 0.006
S	cr, rhombic	0	32.054 ± 0.050	4.412 ± 0.060
S	g	276.98 ± 0.25	167.715 ± 0.035	6.657 ± 0.004
S_2	g	128.49 ± 0.30	228.055 ± 0.050	9.131 ± 0.008
SO_2	g	-296.81 ± 0.20	248.11 ± 0.06	10.548 ± 0.013
SO_4^{2-}	aq	-909.60 ± 0.40	18.83 ± 0.50	—
N	g	472.68 ± 0.40	153.189 ± 0.020	6.197 ± 0.002
N_2	g	0	191.502 ± 0.025	8.669 ± 0.003
NO_3^-	aq	—	146.94 ± 0.85	—
NH_3	g	-45.94 ± 0.35	192.67 ± 0.08	10.046 ± 0.008
NH_4^+	aq	-133.26 ± 0.25	111.17 ± 0.75	—
P	cr, white	0	41.09 ± 0.25	5.360 ± 0.015
P	g	316.5 ± 1.0	163.085 ± 0.020	6.197 ± 0.002
P_2	g	144.0 ± 2.0	218.01 ± 0.04	8.903 ± 0.008
P_4	g	58.9 ± 0.3	279.9 ± 0.5	14.10 ± 0.24
C	cr	0	5.74 ± 0.12	1.050 ± 0.020
C	g	716.67 ± 0.44	157.988 ± 0.020	6.535 ± 0.006
CO	g	-110.53 ± 0.17	197.556 ± 0.032	8.673 ± 0.008
CO_2	g	-393.51 ± 0.13	213.677 ± 0.040	9.364 ± 0.008
Si	cr	0	18.81 ± 0.08	3.217 ± 0.008
Si	g	450 ± 8	167.870 ± 0.035	7.550 ± 0.004
SiO_2	cr, α-quartz	-910.7 ± 1.0	41.46 ± 0.20	6.916 ± 0.020
SiF_4	g	-1614.95 ± 0.85	282.65 ± 0.40	15.36 ± 0.05

CODATA recommended key values for thermodynamics, 1977 (*contd*)

Substance	State	$\Delta_f H°(298.15\ K)$ kJ mol^{-1}	$S°(298.15\ K)$ J K^{-1} mol^{-1}	$H°(298.15\ K) - H°(0)$ kJ mol^{-1}
Ge	cr, cubic	0	31.09 ±0.13	4.636±0.015
GeO$_2$	cr, tetragonal	−580.2 ±1.2	39.71 ±0.15	7.230±0.020
GeF$_4$	g	−1190.15 ±0.50	301.8 ±1.0	17.30 ±0.080
Sn	cr, white	0	51.18 ±0.08	6.323±0.008
Sn	g	301.2 ±1.7	168.380±0.020	6.215±0.002
Sn^{2+}	aq	−8.9 ±0.8	−15.8 ±4.0	—
SnO	cr	−285.93 ±0.70	57.17 ±0.30	8.736±0.022
SnO$_2$	cr	−580.78 ±0.40	52.3 ±1.2	8.76 ±0.08
Pb	cr	0	64.80 ±0.30	6.870±0.020
Pb	g	195.20 ±0.80	175.270±0.020	6.197±0.002
Pb^{2+}	aq	0.92 ±0.25	17.7 ±0.8	—
PbSO$_4$	cr	−919.94 ±0.90	148.49 ±0.40	20.050±0.040
B	cr	0	5.90 ±0.08	1.222±0.008
B	g	560 ±12	153.325±0.035	6.315±0.004
B$_2$O$_3$	cr	−1273.5 ±1.4	53.97 ±0.30	9.301±0.040
BF$_3$	g	−1135.95 ±0.80	254.31 ±0.10	11.650±0.020
Al	cr	0	28.35 ±0.08	4.565±0.010
Al	g	329.7 ±4.0	164.440±0.030	6.919±0.004
Al$_2$O$_3$	cr, α-corundum	−1675.7 ±1.3	50.92 ±0.10	10.016±0.020
AlF$_3$	cr	−1510.4 ±1.3	66.5±0.4	11.62 ±0.04
Zn	cr	0	41.63 ±0.13	5.657±0.020
Zn	g	130.42 ±0.20	160.875±0.025	6.197±0.020
Zn^{2+}	aq	−153.39 ±0.20	−109.6 ±0.7	—
ZnO	cr	−350.46 ±0.27	43.64±0.40	6.933±0.040
Cd	cr	0	51.80 ±0.15	6.247±0.004
Cd^{2+}	aq	−75.88 ±0.60	−72.8 ±1.2	—
CdO	cr	−258.1 ±0.8	54.8 ±1.7	8.41 ±0.08
CdSO$_4$.$\frac{8}{3}$H$_2$O . .	cr	−1729.55 ±0.80	229.66 ±0.40	35.56 ±0.04
Hg	l	0	75.90 ±0.12	9.342±0.008
Hg	g	61.38 ±0.04	174.860±0.020	6.197±0.002
Hg^{2+}	aq	170.16 ±0.20	−36.32 ±0.80	—
Hg$_2^{2+}$	aq	166.82 ±0.20	65.52 ±0.80	—
HgO	cr, red	−90.83 ±0.12	70.25±0.30	9.117±0.025
Hg$_2$Cl$_2$	cr	−265.45 ±0.30	191.6 ±1.5	23.35 ±0.20
Hg$_2$SO$_4$	cr	−743.41 ±0.50	200.71 ±0.20	26.070±0.030
Cu	cr	0	33.15 ±0.08	5.004±0.008
Cu	g	337.6 ±1.2	166.285±0.025	6.197±0.002
Cu^{2+}	aq	65.69 ±0.80	−97.1 ±1.2	—
CuSO$_4$	cr	−771.1 ±1.2	109.2 ±0.4	16.86 ±0.08
Ag	cr	0	42.55± 0.21	5.745±0.020
Ag	g	248.9 ±0.8	172.883±0.025	6.197±0.002
Ag$^+$	aq	105.750±0.085	73.38 ±0.40	—
AgCl	cr	−127.070±0.085	96.23 ±0.20	12.033±0.040
Th	cr	0	53.39 ±0.40	6.510±0.020
Th	g	598 ±6	190.06 ±0.40	6.197±0.002
ThO$_2$	cr	−1226.4 ±3.5	65.23 ±0.20	10.560±0.020
U	cr	0	50.20 ±0.20	6.364±0.020
UO$_2$	cr	−1085.0 ±1.0	77.03 ±0.20	11.280±0.020

CODATA recommended key values for thermodynamics, 1977 *(contd)*

Substance	State	$\Delta_f H^\circ (298.15\,K)$ $kJ\,mol^{-1}$	$S^\circ (298.15\,K)$ $J\,K^{-1}\,mol^{-1}$	$H^\circ (298.15\,K) - H^\circ (0)$ $kJ\,mol^{-1}$
UO_2^{2+}	aq	-1019.2 ± 2.5	-98.3 ± 4.0	—
UO_3	cr, gamma	-1223.8 ± 2.0	96.11 ± 0.40	14.585 ± 0.080
U_3O_8	cr	-3574.8 ± 2.5	282.55 ± 0.50	42.74 ± 0.10
Be	cr	0	9.50 ± 0.08	1.950 ± 0.020
Be	g	324 ± 5	136.165 ± 0.020	6.197 ± 0.002
BeO	cr	-609.4 ± 2.5	13.77 ± 0.04	2.837 ± 0.008
Mg	cr	0	32.68 ± 0.10	5.000 ± 0.030
Mg	g	147.10 ± 0.80	148.535 ± 0.020	6.197 ± 0.002
MgO	cr	-601.5 ± 0.3	26.95 ± 0.15	5.160 ± 0.020
MgF_2	cr	-1124.2 ± 1.2	57.2 ± 0.4	9.92 ± 0.06
Ca	cr	0	41.6 ± 0.4	5.73 ± 0.04
Ca	g	177.8 ± 0.8	154.775 ± 0.020	6.197 ± 0.002
Ca^{2+}	aq	-543.10 ± 0.80	-56.4 ± 0.4	—
CaO	cr	-635.09 ± 0.90	38.1 ± 0.4	6.75 ± 0.06
Li	cr	0	29.12 ± 0.20	4.632 ± 0.040
Li^+	aq	-278.455 ± 0.090	11.30 ± 0.35	—
Na	cr	$0\;000$	51.30 ± 0.20	6.460 ± 0.020
Na^+	aq	-240.300 ± 0.065	58.41 ± 0.20	—
K	cr	0	64.68 ± 0.20	7.088 ± 0.020
K^+	aq	-252.17 ± 0.10	101.04 ± 0.25	—
Rb	cr	0	76.78 ± 0.30	7.489 ± 0.020
Rb^+	aq	-251.12 ± 0.13	120.46 ± 0.40	—
Cs	cr	0	85.23 ± 0.40	7.711 ± 0.020
Cs^+	aq	-258.04 ± 0.13	132.84 ± 0.40	

E.F.G.H.

2.11 Miscellaneous data

2.11.1 Properties of polymers

As the physical properties of polymers vary from sample to sample and are dependent on composition and pretreatment of the specimen, only ranges of values are recorded. There are now available a large number of co-polymers and filled polymers concerning which information can be found in J. Brandrup and E. H. Immergut (eds), *Polymer Handbook*, 1966 (Interscience, New York) and in *Modern Plastics, 1970–71 Encyclopedia* (McGraw-Hill, Inc., New York).

Polymer	ρ / kg m^{-3}	Tensile strength / MN m^{-2}	10^5 Coeff. linear expansion / °C	Heat capacity / J g^{-1} K^{-1}	Thermal conductivity / W m^{-1} K^{-1}	n_D	Resistivity / Ω cm	Dielectric constant	Stability
Acetals	1420	65	8	1.46	0.23	1.48	10^{15}	—	Attacked by some acids and alkalies, good resistance to organic solvents.
Cellulose (cotton, wood pulp) and regenerated	1480–1530	80–240	—	1.3–1.5	0.23	1.53–1.56	10^7–10^{14}	6.7–7.5	Decomposed by oxid. agents and strong acids or by heating to 270 °C. Absorbs water.
Cellulose acetate: Moulded	1220–1340	12–58	8–18	1.26–1.8	0.17–0.33	1.46–1.50	10^{10}–10^{14}	3.5–7.5	Softens at ~130 °C, decomposed by strong acids and alkalies, soluble in ketones and esters, softened or dissolved by chlorinated or aromatic hydrocarbons.
Sheet	1280–1320	30–52	10–15	1.26–2.1	0.17–0.33	1.49–1.50	10^{11}–10^{15}	3.5–7.5	
Cellulose nitrate (Celluloid)	1350–1400	50	8–12	1.3–1.7	0.12–0.21	1.49–1.51	10^{10}	7.0–7.5	Softens at ~60 °C and ignites ~160 °C. Decomposed by strong acids and alkalies. Soluble in alcohol, ethers and esters.
Chlorinated polyether	1400	39	8	—	0.13	—	10^{15}	3.1	Unattacked by acids or alkalies except by oxid. acids. Resists most solvents.
Epoxy cast resins	1110–1400	26–85	4.5–6.5	1.0	0.17–0.21	1.55–1.61	10^{12}–10^{17}	3.5–5.0	Unattacked by weak acids and alkalies, some attack by strong acids and alkalies. Generally resistant to organic solvents.
Nylon-6 (Poly-ϵ-caprolactam)	1120–1170	45–90	28	1.6	0.25	—	10^{12}–10^{15}	3.7–5.5	Stable at 100 °C. Softens at ~230 °C. Attacked by strong acids but resistant to alkalies. Resistant to common organic solvents but dissolves in phenol and formic acid.

Properties of polymers (*contd*)

Polymer	ρ / kg m⁻³	Tensile strength / MN m⁻²	10⁵ Coeff. linear expansion / °C	Heat capacity / J g⁻¹ K⁻¹	Thermal conductivity / W m⁻¹ K⁻¹	n_D	Resistivity / Ω cm	Dielectric constant	Stability
Nylon-66 (Polyhexamethylene-adipamide)	1130–1150	60–80	8	1.7	0.25	1.53	10^{14}–10^{15}	4.0–4.6	Stable at 100 °C. Softens at ~230 °C. Attacked by strong acids but resistant to alkalies. Resistant to common organic solvents but dissolves in phenol and formic acid.
Phenolic cast resins	1240–1320	33–59	6–8	1.25–1.67	0.15	1.58–1.66	10^{12}–10^{13}	6.5–17.5	Not affected by acids, attacked by strong alkalies, little affected by organic solvents.
Phenol–formaldehyde moulding compounds	1250–1300	45–52	2.5–6.0	1.59–1.76	0.13–0.25	1.5–1.7	10^{11}–10^{12}	5.0–6.5	Slightly attacked by acids and alkalies. Depending on composition soluble or insoluble in organic solvents.
Polyacrylonitrile	1160–1180	200 (fibres)	7	—	—	1.52	10^{14}	6.5	Fair resistance to weak alkalies but soluble in strong nitric and sulphuric acids.
Polycarbonates	1200	52–62	6.6	1.17–1.25	0.19	1.59	10^{16}	2.97–3.17	Unattacked by weak acids, some attack by strong acids and alkalies. Resistant to paraffinic solvents but soluble in aromatic and chlorinated hydrocarbons.
Poly-2-chloro-1:3 butadiene (Neoprene)	1230–1250	—	20	2.18	0.19	1.56	10^{10}–10^{13}	—	Soluble in many organic solvents but resistance to heat and oil greater than that of natural rubber.
Polychlorotrifluoro-ethylene	2100–2200	29–39	4.5–7.0	0.9	0.20–0.22	1.43	10^{18}	2.24–2.8	Not affected by acids or alkalies but halogenated organic solvents cause slight swelling.
Polyethylene Low density	910–925	4–15	10–20	—	0.33	1.51	10^{15}–10^{18}	2.25–2.35	All grades melt ~133 °C, insoluble in common solvents below 60 °C, resistant to acids and alkalies but attacked by oxid. acids.
Medium density	926–940	8–22	10–20	—	0.33–0.42	1.52	10^{15}–10^{18}	2.25–2.35	
High density	940–965	20–36	10–20	—	0.45–0.52	1.54	10^{15}–10^{18}	2.30–2.35	

Properties of polymers (*contd*)

Polymer	ρ kg m^{-3}	Tensile strength MN m^{-2}	10^5 Coeff. linear expansion °C	Heat capacity J g^{-1} K^{-1}	Thermal conductivity W m^{-1} K^{-1}	n_D	Resistivity Ω cm	Dielectric constant	Stability
Polyethylene terephthalate	1370–1380	66	6	1.25	0.29	1.54	10^{16}	3.25	Softens ~250 °C. Resistant to dilute acids and alkalies but attacked by oxid. acids. Insoluble in many common solvents but soluble in hot aromatic and chlorinated hydrocarbons.
Polyimide aromatics	1430	68	4–5	1.12	—	opaque	$>10^{16}$	3.4	Thermally very stable and does not burn. Resistant to acids but attacked by alkalies. Insoluble in organic solvents.
Polyisoprene Natural rubber	906–913	—	22	1.88	0.13	1.52	10^6	—	Swells in many organic solvents. More resistive to attack by chemicals than natural rubber.
Hard rubber	1130–1180	39	6	1.38	0.16	1.6	10^{16}	—	
Polymethylmethacrylate	1170–1200	45–72	5–9	1.5	0.17–0.25	1.48–1.50	$>10^{14}$	3.3–4.5	Softens at ~180 °C. Attacked by high concentrations of oxid. acids. Soluble in ketones, esters, aromatic and chlorinated hydrocarbons.
Polypropylene	902–906	28–36	6–10	1.92	0.12	1.49	$>10^{16}$	2.2–2.6	Softens at ~160 °C. Resistant to acids and alkalies but attacked by oxid. acids. Insoluble in common organic solvents below 80 °C.
Polystyrene	1040–1090	30–100	3.4–21	1.3–1.5	0.04–0.14	1.59–1.60	$>10^{16}$	2.45–2.65	Softens at ~98 °C and burns. Unaffected by acids or alkalies, attacked by strong oxid. acids. Soluble in aromatic and chlorinated hydrocarbons.
Polytetrafluoroethylene moulding compound	2140–2200	13–33	10	1.0	0.25	1.35	$>10^{18}$	—	Not affected by acids, alkalies or organic solvents.
Polyurethane Cast liquid	1100–1500	1–65	10–20	1.8	0.21	1.50–1.60	10^{11}–10^{15}	4.0–7.5	Slightly attacked by weak acids and alkalies. Strongly attacked by strong acids and alkalies. Only slightly affected by organic solvents.

Properties of polymers (*contd*)

Polymer	ρ kg m^{-3}	Tensile strength MN m^{-2}	10^5 Coeff. linear expansion °C	Heat capacity J g^{-1} K^{-1}	Thermal conductivity W m^{-1} K^{-1}	n_D	Resistivity Ω cm	Dielectric constant	Stability
Polyurethane Elastomer	1110–1250	29–55	10–20	1.8	0.07–0.31	—	10^{11}–10^{13}	5.4–7.6	Attacked by acids and alkalies and may dissolve. Resistant to most organic solvents.
Polyvinylchloride (PVC)	1300–1400	~50	7–8	0.84–1.17	0.12–0.17	1.53–1.55	10^{16}	3.2–3.6	Softens at 80 °C. Soluble in organic solvents.
Polyvinylidene chloride moulding compounds	1650–1720	20–33	19	1.34	0.13	1.60–1.63	10^{14}–10^{16}	4.5–6.0	Resistant to acids and alkalies, little or only slightly affected by organic solvents.
Silicone cast resin	1300	—	8–30	—	0.15–0.32	1.43	10^{14}–10^{15}	2.7–4.2	Excellent stability to heat and oxidation, water repellent, insoluble in many organic solvents.

Permeability constants for polymers

The permeability constant P is defined as

$$P = \text{(amount of permeant)}/\text{(area)}\,\text{(time)}\,\text{(driving force across the film)}$$

The unit employed here for P is the amount of vapour expressed in cm^3 measured at $0\,°C$ and at 1 standard atmosphere ($101.325\,kN\,m^{-2}$) passed in 1 second for an area of $1\,cm^2$ of film when the pressure across the film of $1\,cm$ thickness is $101.325\,kN\,m^{-2}$.

Additional data can be found in an article by H. Yasuda, *Polymer Handbook* (eds J. Brandrup and E. H. Immergut), 1966 (Interscience, New York).

Polymer	$10^{10}P$								$10^6 P$	
	H_2		N_2		O_2		CO_2		H_2O	
	$\theta/°C$		$\theta/°C$		$\theta/°C$		$\theta/°C$		$\theta/°C$	
Cellulose (Cellophane)	25	0.494	25	0.243	25	0.160	25	0.357	—	—
Cellulose acetate	20	266	30	21.3†	30	59.3†	30	181†	25	41.8
Cellulose nitrate	20	152	25	9.1	25	148	25	161	25	47.8
Polychlorotrifluoroethylene (Amorphous)	20	71	25	0.380	25	3.0	40	16.0	25	0.0022
Polyethylene:										
Low density	30	912	30	103	30	300	30	1269	25	0.684
High density	20	228	30	13.7	30	38.8	30	160	25	0.091
Polyethylene terephthalate	20	281	30	0.836	30	3.42	30	1.14	25	0.988
Polyisoprene (Natural rubber)	25	3952	25	623	25	1809	25	10260	—	—
Polypropylene	20	312	30	33.4	30	175	30	700	25	0.388
Polystyrene	—	—	30	22.0	30	83.6	30	669	25	9.12
Polytetrafluoroethylene	20	152	—	—	—	—	20	357	—	—
Polyvinylchloride	20	182	30	8.36	30	22.8	30	114	25	1.19
Polyvinylidenechloride	—	—	30	0.071	30	0.403	30	2.28	25	0.0038
Nylon-6	—	—	30	0.722	30	2.89	20	6.69	—	—

† Plasticized polymer was used for this measurement.   E.F.G.H.

2.11.2 Some characteristics of common glasses

Glass	Approximate composition	Annealing temperature/°C	$10^6 \times$ linear coefficient of expansion
Soda (S95)	Sodium calcium silicate	515	9.2†
Lead (L92)	Lead silicate	435	9.0†
Pyrex	Sodium aluminium silicate	530–600	3.2
Silica	Silica	—	0.45
GEC B37	Silica, boron, sodium	560	3.7‡
Monax	Silica (small amounts of sodium potassium, zinc and aluminium)	560	4.4§

† Suitable for sealing to platinum.
‡ Suitable for sealing to tungsten.
§ Suitable for sealing to platinum, tungsten and molybdenum, also impermeable to liquid helium.

For further information see W. E. S. Turner, *The Elements of Glass Technology for Scientific Glass Blowers (Lampworkers)*, 1954, The Glass Delegacy of the University of Sheffield, Sheffield, and J. H. Partridge, *Glass-to-Metal Seals*, 1949, The Society of Glass Technology, Sheffield. For information on the properties of American glasses see W. E. Barr and V. J. Anhorn, *Scientific and Industrial Glass Blowing and Laboratory Techniques*, 1959 (Instruments Publishing Company, Pittsburgh).

E.F.G.H.

2.11.3 Cooling agents

Composition	$\theta/°C$	Composition	$\theta/°C$
Crushed ice 2 parts, sodium chloride 1 part	-18	Liquid oxygen	-183
Crushed ice 3 parts, cryst. calcium chloride 4 parts	-48	Liquid nitrogen	-196
		Liquid helium	-269
Solid CO_2 added to equal parts CCl_4 and $CHCl_3$	-80		

Ethylene glycol-water antifreeze

The values listed are ρ (kg m^{-3}) and freezing temperature θ(°C) and % is the number of g of ethylene glycol in 100 g of solution.

%	ρ	θ	%	ρ	θ
10	1010.8	-3.4	40	1051.4	-23.5
20	1024.1	-7.9	50	1063.9	-36.8
30	1037.5	-13.7	60	1076.5	-52.8

E.F.G.H.

2.11.4 Mohs's scale of mineral hardness

The numbers are not quantitative but merely indicate the sequence of hardness.

Mineral	Hardness	Mineral	Hardness	Mineral	Hardness
Talc	1	Apatite	5	Corundum . . .	9
Rock salt	2	Felspar	6	Diamond . . .	10
Calcspar	3	Quartz	7		
Fluorspar	4	Topaz	8	Finger-nail . . .	~2.5
				Penknife . . .	~6.5

E.F.G.H.

2.11.5 Calorific values of solid, liquid and gaseous fuels

By custom the basic calorific value for solid and liquid fuels is the gross calorific value at constant volume and for gaseous fuels it is the gross calorific value at constant pressure. The word 'gross' here signifies that the water formed and liberated during combustion is in the liquid phase. The values given are approximate because many of the substances listed are not well defined. The calorific values of pure substances can be calculated from information in Section 2.10. More detailed information on technical fuels can be found in J. W. Rose (ed.) and J. R. Cooper (ed.), *Technical Data on Fuel*, 7th edn, 1977 (British National Committee, World Energy Conference, London).

Calorific values of solid, liquid and gaseous fuels

Solid and liquid fuels	Gross calorific value/MJ kg^{-1}
Alcohols	
Ethanol	30
Methanol	23
Coal and coal products	
Anthracite (4% water)	36
Coal tar fuels	36–41
Domestic coke (8–12% water)	28
General purpose coal (5–10% water)	32–42
High-volatile coking coals (4% water)	35
Low temperature coke (15% water)	26
Medium-volatile coking coal (1% water)	37
Steam coal (1% water)	36
Peat	
Peat (20% water)	16
Petroleum and petroleum products	
Diesel fuel	46
Gas oil	46
Heavy fuel oil	43
Kerosine	47
Light distillate	48
Light fuel oil	44
Medium fuel oil	43
Petrol	44.8–46.9
Wood	
Wood (15% water)	16

Gaseous fuels at 15 °C, 101.325 kPa, dry	Gross calorific value/MJ m^{-3}
Coal gas coke oven (debenzolized)	20
Coal gas continuous vertical retort (steaming)	18
Coal gas low temperature	34
Commerical butane	118
Commerical propane	94
North Sea gas natural	39
Producer gas coal	6
Producer gas coke	5
Water gas carburetted	19
Water gas blue	11

E.F.G.H.

Atomic and
Nuclear Physics

3.1 Free electrons and ions in gases

3.1.1 Ionic mobility

The **ionic mobility** (k) is defined as the velocity attained by an ion moving through a gas under unit electric field. A suitable unit is therefore $m^2 \, s^{-1} \, volt^{-1}$. In general the mobility is a function of E/ρ, where E is the field strength and ρ the gas pressure, but is constant for low values of E/ρ, the departure from constancy occurring when the ions attain drift velocities comparable with the agitation velocity of the gas molecules. The mobility is critically dependent on gas purity, as polar impurity molecules tend to cluster the ion and so reduce its mobility. Early measurements were frequently rendered quite valueless by inattention to this matter. The mobility depends little on the charge of the ion, a doubly charged ion having practically the same mobility as a singly charged ion (of either sign). By theory, $k \propto (\text{density})^{-1}$ so at constant temperature $k \propto (\text{pressure})^{-1}$. This is observed to hold from 10 to $6 \times 10^6 \, N \, m^{-2}$. The reduced mobility $K_0 \; (= 273 \, kp / 1.01 \times 10^5 \, T)$ is the mobility at 273 K and $1.01 \times 10^5 \, N \, m^{-2}$ pressure in units of $m^2 \, s^{-1} \, volt^{-1}$. K_0 is thus the value corresponding to a gas number density of $2.69 \times 10^{25} \, m^{-3}$. The ratio of the electric field strength to the neutral gas number density is often used in preference to E/ρ. The unit is the Townsend (Td), $1 \, Td = 10^{-21} \, Vm^2$. All values in the table below are for K_0. A dash indicates that either the particular ion does not exist or has not been examined. Where the mobility is not independent of E/ρ it is usual to take the mobility corresponding to the limit $E = 0$.

It should be noted that considerable confusion existed in the early measurements due to the production of molecular ions in the noble gases. Modern techniques have shown that the number of ion species produced is much greater and more complex than was thought earlier. (See McDaniel, *Collision Phenomena in Ionized Gases*, 1964, Wiley, London.)

Ionic mobilities for ions in their parent gas

$K_0/(10^{-4} \, m^2 \, s^{-1} \, volt^{-1})$
at 273 K and $1.01 \times 10^5 \, N \, m^{-2}$

Ionic species	H$_2$	He	N$_2$	O$_2$	Ne	Ar	Kr	Xe	CO	CH$_4$
Atomic positive ion . . .	16.0	10.2	3.3	—	4.2	1.53	0.9	0.6	—	—
Molecular positive ion . .	11.3(H_3^+)	20.3	1.8	2.2	6.5	2.3	1.2	0.8	1.6	3.5
Molecular negative ion . .	—	—	—	3.3	—	—	—	—	—	—

Ionic mobilities for ions in various gases

$k/(10^{-4} \, m^2 \, s^{-1} \, volt^{-1})$

Ion	*Gas*									
	He	Ne	Ar	Kr	Xe	H$_2$	N$_2$	CO	CO$_2$	H$_2$O Vap.
Li$^+$	25.6	12	4.99	4.0	3.04	13.3	4.21	2.63		0.725
Na$^+$	23.4	8.70	3.23	2.34	1.80	15.0	3.40	2.44	1.63	0.715
K$^+$	22.7	8.0	2.81	1.98	1.44	14	2.7	2.32	1.45	0.705
Rb$^+$	21.2	7.18	2.40	1.59	1.10	13.4	2.39	2.08	1.23	0.700
Cs$^+$	19.1	6.50	2.23	1.42	0.99	13.4	2.25	1.98		0.695

These values, from Tyndall, *Mobility of Positive Ions in Gases*, 1938 (Cambridge University Press), are for gases at 293 K and $1.01 \times 10^5 \, N \, m^{-2}$, and from Ellis *et al. Atomic Data and Nuclear Data Tables*, **17**, 177–210, 1976.   J.W.L.

3.1.2 Electron mobility

As an electron loses little energy on colliding with gas atoms or molecules, it rapidly attains energies in excess of the thermal agitation energy of the gas atoms and so a significant mobility value cannot be given (see ionic mobility). Instead the drift velocity as a function of E/ρ is normally found in the literature. For low E/ρ, the drift velocity is given by $v = \mathrm{const}\,(E/\rho)^{1/2}$. At larger E/ρ values this relation is no longer valid, v increasing more rapidly than given by this relation. The breakdown is usually associated with the onset of inelastic collisions. Impurities, especially with the noble gases, are again of importance (see D. H. Wilkinson, *Ionization Chambers and Counters*, 1950, p. 31 (Cambridge University Press)).

The table below is compiled mainly from Pack, Voshall and Phelps, *Phys. Rev.*, 1962, **127**, 2084–2089, and McDaniel, *Collision Phenomena in Ionized Gases*, 1964 (Wiley, London).

Revised data for Ar/CH_4 and data for Ar/CF_4, Ar/C_2H_2 is from Christophorou L. G., McCorkle, D. L., Maxey, D. V., and Carter, J. G. 'Fast gas mixtures for gas-filled particle detectors', *Nucl. Inst. & Methods*, 1979, **163**, 141–9.

Drift velocity/$(10^3 \mathrm{~m~s}^{-1})$

Gas	Field strength per unit pressure $(E/\rho)/(\mathrm{V~mN}^{-1})$											
	0.075	0.15	0.3	0.45	0.6	0.75	1.5	3	4.5	6	7.5	15
H_2	3.1	5	6.8	8.0	9.2	10	15	23	30	36	42	70
He	2.7	4	6	7	8.5	9.6	15	36				
N_2	3.1	3.9	5	6	7	8	14	22	30	36	43	80
Ne	4	5	7	10	12	16	26	52				
Ar	2.3	2.7	3.2	3.5	3.9	4.1	6.2	13	30			
Kr	1.5	1.8	2.1	2.4	2.5	3.1						
Xe	1.0	1.2	1.4	1.6	1.7	1.8						
CO_2	0.56	1.1	2.3	3.4	4.6	5.7	11	33	66	94	109	137
CO	4.8	6.5	9.4	11	13	14	17	20	25	28	29	
BF_3	0.13	0.25	0.5	0.75	1.0	1.3	2.5	5.0	7.5	10	12.5	25
Air	3.5	5	8	9.3	11	12	17	26	34	43	51	88
CH_4	8	24	60	80	95	100	100					
Ar/CH_4	49	55	45	36	32	30	26	24				
Ar/N_2	4	6	9	13	16.8	19.3	27.3					
Ar/C_2H_2	14.3	27.3	44.3	49.3	48.3	45.8	45.2					
Ar/CF_4	34	63	100	120	120	111	66					

All binary mixtures are 90%/10% by volume.   J.W.L.

3.1.3 Ionic recombination

The neutralization of electrical carriers of opposite sign in a gas is termed recombination. The positive carrier is a positive ion. The negative carrier may be a negative ion or an electron depending on the nature of the gas. This table deals only with the former type of recombination, and only with the situation in which the ions are randomly distributed throughout the gas (see Loeb, *Basic Processes of Gaseous Electronics*, 1955, p. 477 *et seq*. California University Press). Then $\mathrm{d}n_+/\mathrm{d}t = \mathrm{d}n_-/\mathrm{d}t = -\alpha n_- \cdot n_+$, where n_+ and n_- are the ion densities and α is the recombination coefficient. α is commonly taken as a constant. Experimentally, however, measurements show that α changes with ion age, presumably due to a change in the nature of the ions owing to the presence of small quantities of impurity. Further there is a small dependence on ion density, α decreasing with increasing ion density. α is dependent on pressure and temperature and from $10^5 \mathrm{~N~m}^{-2}$ to a few hundred $\mathrm{N~m}^{-2}$ is adequately described by the theory of Thomson. For pressures above $10^6 \mathrm{~N~m}^{-2}$ the Langevin theory is employed. Three-body and mutual neutralization processes are by far the most important for ionic recombination.

See the summary by Massey, *Advances in Physics*, 1952, Vol. I, p. 395. For general theoretical discussion, see Loeb (above); Jaffe, *Phys. Rev.*, 1940, **58**, 968; 1941, **59**, 652.

All values given below are at 1.01×10^5 N m^{-2} pressure unless otherwise stated.

Gas	H$_2$	N$_2$	O$_2$	Air	Ar (*impure*)
$\alpha/(10^{-12}\,\mathrm{m}^3\,\mathrm{s}^{-1})$	0.28 (22 °C)	1.06 (22 °C)	1.32 (22 °C) 2.08 (25 °C)	1.65 (18 °C)	1.06 (22 °C)

Dependence on ion density, pressure and temperature

n = No. of ions/m^3; ρ = pressure; T = temperature.

Air at	$n/(10^{12}\,\mathrm{m}^{-3})$	1.5	2.7	3.1	4.1	4.4
15 °C	$\alpha/(10^{-12}\,\mathrm{m}^3\,\mathrm{s}^{-1})$	2.65	2.5	2.5	2.3	2.3

$\rho/(10^3\,\mathrm{N\,m}^{-2})$	5	10	20	30	40	50	60	70	80	90	100	110
Air 18 °C $\alpha/(10^{-12}\,\mathrm{m}^3\,\mathrm{s}^{-1})$ . .	0.12	0.21	0.37	0.55	0.7	0.88	1.05	1.22	1.4	1.55	1.72	
O$_2$ 20 °C $\alpha/(10^{-12}\,\mathrm{m}^3\,\mathrm{s}^{-1})$ . .		0.7	1.3	1.6	1.75	1.88	1.96	2.01	2.05	2.08	2.08	2.08

$\rho/(10^5\,\mathrm{N\,m}^{-2})$	4.6	6.4	7.9	8.0	9.7	12.5	15.1	17.4	17.8	21.1	22.0	23.6	25.7
Air 18 °C $\alpha/(10^{-13}\,\mathrm{m}^3\,\mathrm{s}^{-1})$	10.58	8.44		6.98	5.94	4.80	4.07		3.44	2.90		2.59	2.37
CO$_2$ 18 °C $\alpha/(10^{-13}\,\mathrm{m}^3\,\mathrm{s}^{-1})$	5.76		3.58			2.24		1.56			1.24		

O$_2$ T/K	200	250	300	350	400
$\alpha/(10^{-12}\,\mathrm{m}^3\,\mathrm{s}^{-1})$ . . .	3.7	2.7	1.9	1.5	1.2

J.W.L

3.1.4 Ionic diffusion

The coefficient of diffusion D is given by the equation $dn/dt = D(\nabla^2 n)$, where n is the ion density and ∇^2 is the Laplacian operator. Actual measurements of D are seldom undertaken as the coefficient can be calculated from the equation $k/D = Ne/\rho$, where k is the mobility, N is the number of molecules per m^3, ρ is the pressure in Newtons per square metre and e is the electronic charge. Thus, numerically, $D = 0.0234k$ m^2 s^{-1} at s.t.p. Values of k are given in the section on Ionic Mobility.

J.W.L.

3.1.5 Electron diffusion

The coefficient of electron diffusion D_e may be computed from the equation $k_e/D_e = Ne/\eta\rho$, where k_e is the electron mobility, N is the number of gas molecules per m^3 at s.t.p., e is the electronic charge, ρ is the pressure in $N\,m^{-2}$ and η is the ratio of the mean agitation energy of an electron to that of a gas molecule. η is also known as the Townsend energy factor. Experimental values of η are given below as a function of E/ρ.

Field strength per unit pressure $(E/\rho)/(V\,mN^{-1})$

Gas	0.5	1	2	3	4	5	10	20	30	40
H_2 . . .	6	10	18	23	28	32	55	100	130	150
He . . .	37	74	126	152	176					
N_2 . . .	16	25	34	39	42	44	52	68	89	114
O_2 . . .	11	18	32	41	47	52	68	96	133	
Air . . .	7	15	26	34	39	42	50	68	90	106
CO . . .	5	8	14	19	23	27	38	48	68	
NO . . .	7	8	10	11	12	14	38			
Ar . . .	220	320	316	312	310	314	324			
CO_2 . . .	1.3	1.6	2	4	9	17	58	89	117	
Cl_2 . . .							53	71	77	82
Br_2 . . .			40	52	59		74	82	89	91
I_2							27	76	91	99

Most of the values in the above table were obtained in the 1920s and are from Healey and Reed, *The Behaviour of Slow Electrons in Gases*, 1941 (Amalgamated Wireless, Sydney). Recent determinations have shown only small changes for H_2, He, N_2 and Air but the values for O_2 over a limited range are higher than those in the above table (see Huxley *et al.*, *Aus. J. Phys.*, 1959, **12**, 303).

J.W.L.

3.1.6 Electron collisions

Observed total collision cross-section of electrons with atoms and molecules

Q is the total collision cross-section in units of πa_0^2, where a_0 is the radius of the first Bohr orbit of the hydrogen atom; $a_0 = 0.53 \times 10^{-10}$ m. Hence the unit of Q in the tables below is 0.88×10^{-20} m^2. The values are obtained by interpolation from graphs. General references: *Landolt-Bornstein Tables*, 1950, Vol. I (Springer-Verlag); Massey and Burhop, *Electronic and Ionic Impact Phenomena*, 1952 (Oxford University Press). For molecules, the data given relate to the Ramsauer method.

Q in units of πa_0^2

Gas	$(Energy)^{1/2}/V^{1/2}$											
	0.5	1	1.5	2	3	4	5	6	7	8·	9	10
H . .	—	—	—	—	—	6	7.6	8.6	9.0	—	—	—
H_2 . .	—	15	17	16	10.5	7	5.5	4	3.5	—	—	—
He . .	6.3	6.3	6.2	5.6	4.6	3.5	2.6	2.0	1.7	1.5	1.3	1.2
N_2 . .	17	10	29	13	12	12.5	14	13	11.5	—	—	—
O_2 . .	6	7	7.5	8	11	12	12	12	—	—	—	—
CO . .	15	14	38	18	14	14	14	—	—	—	—	—
Ne . .	1.0	1.9	2.5	2.8	3.5	3.7	3.7	3.6	3.5	3.4	3.0	2.9

Q in units of πa_0^2 (contd)

Gas	$(Energy)^{1/2}/V^{1/2}$											
	0.5	1	1.5	2	3	4	5	6	7	8	9	10
Na . .	—	280	380	180	130	110	85	65	50	48	46	35
Ar . .	0.6	0.6	4.0	8.5	21	22	15	11	8	7.5	7	6.5
K . .	—	450	420	270	220	160	125	80	60	50	50	40
Zn . .	—	—	—	80	45	35	27	24	23	23	23	21.5
Kr . .	2.0	1.0	6.0	15	30	29	20	14	11.5	9	8	7.5
Rb . .	—	374	390	329	220	170	142	85	68	51	45	43
Cd . .	—	—	—	—	54	43	40	41.5	39	34	31	29
Xe . .	7.5	2.0	15	33	43	32.5	22.5	16	—	—	—	—
Cs . .	—	470	675	350	300	230	160	135	100	80	67	46
Hg . .	—	80	70	55	23	18	18	20	20	18.5	18	17.5
Tl . .	—	6	10	16	12.5	11	9.5	9.0	8.8	8.0	7.0	6.2

Cross-section (Q) for ionization by electron impact

Q is expressed in units of πa_0^2 (i.e. 0.88×10^{-20} m^2) for removal of the first electron.

Gas	Energy/V					
	50	100	200	300	400	500
H_2	1.1	1.1	0.8	0.66	0.56	0.48
He	0.23	0.4	0.34	0.28	0.25	0.23
N_2	2.5	3.3	2.9	2.4	2.1	1.8
O_2	2.5	3.3	3.0	2.5	2.1	1.8
CO	2.9	3.5	3.0	2.4	2.0	1.8
NO	2.9	3.7	3.2	2.7	2.5	2.0
Ne	0.46	0.85	0.89	0.77	0.68	0.61
Ar	3.6	3.3	2.7	2.4	1.9	1.7
Hg	6.6	5.7	4.0	3.15	2.6	—

J.W.L.

3.2 Work function

Energy required to extract an electron from a solid

The work function ϕ is the energy required to remove an electron from the highest filled level in the Fermi distribution of a solid to a point a great distance from the solid, at absolute zero. An estimate of the work function can be obtained *thermionically* from Richardson's equation

$$I = AT^2 \exp(-\phi/kT)$$

where I is the thermionic current, T the absolute temperature, k is Boltzmann's constant and A is a constant having the theoretical value 120 amp cm^{-2} deg^{-2}. Since A contains the reflection coefficient, which varies with temperature, observed A values usually differ from the theoretical value. ϕ too varies very slightly with temperature, but for most metals the difference between ϕ at absolute zero and at room temperature is less than experimental errors.

ϕ may also be estimated *photoelectrically* for metals. Einstein's expression for the photoelectric effect is $h\nu = e\phi + E$, where E is the kinetic energy of the ejected photoelectron. The photoelectric current J released when light of energy $h\nu$ falls on the surface of a metal, for which the threshold frequency is given by $h\nu_0 = e\phi$ (for then $E = 0$), is given by the Fowler equation

$$J = B(kT^2) . f\{(h\nu - h\nu_0)/kT\}$$

where f is a universal function of $(h\nu - h\nu_0)/kT$ and B is constant provided that $h\nu$ is near to $h\nu_0$.

Work functions

Metal	Work function φ/ev		Metal	Work function φ/eV		
	Photoelectric	C.P.D.		Thermionic	Photoelectric	C.P.D.
Li	—	2.32	Nb	4.30	—	4.37
Na	2.36	2.46	Mo	4.33	4.49	4.21
K	2.30	2.01	Ta	4.33	4.30	4.22
Rb	2.05	—	W	4.55	4.55	4.55
Cs	1.95	1.82	Re	4.72	—	—
Be	—	3.91	Ti	4.10	4.33	4.20
Mg. . . .	—	3.61	Cr	4.60	4.44	—
Ca	2.87	—	Mn	—	4.08	—
Ba	2.52	2.35	Fe	—	4.60	4.16
			Co	—	4.97	—
Zn	3.63	4.11	Ni	5.24	5.15	5.25
Cd	—	4.22				
			Zr	4.00	—	—
Al	4.28	4.19	Hf	3.65	—	—
Ga	4.35	—				
In	4.08	—	Ru	—	4.71	4.73
			Rh	4.72	—	—
Sn	4.28	4.43	Pd	—	5.40	—
Pb	4.25	3.83	Ir	4.57	—	—
			Pt	5.36	5.63	—
Cu	4.65	4.51				
Ag	4.26	4.29	Th	—	—	3.71
Au	5.10	5.28	U	3.47	3.47	3.63
As	4.79	—	C (dag) . .	—	—	4.65–5.0
Sb	4.56	—	Si	—	4.95	4.75
Bi	4.34	—	Ge	—	5.15	4.83

Note: Values selected from the articles by J. C. Rivière, *Solid State Surface Science* (ed. Mino Green), Vol. 1, 1969 (Marcel Dekker, New York), and by J. Hölzl and F. K. Schulte, *Springer Tracts in Modern Physics*, **85**, 1 (1979).

The third common method of measuring ϕ is by the *contact potential difference* (c.p.d.) V_{AB} that exists between the surfaces of two solids A and B of work functions ϕ_A and ϕ_B, when connected electrically, since

$$\phi_B - \phi_A = eV_{AB}$$

for the two solids at the same temperature. The method involves a prior knowledge of the work function of one of the solids if that of the other is to be measured absolutely.

Typical errors of the tabulated quantities for the metals are 0.02 eV.

J.C.R.

3.3 Electrons in atoms

3.3.1 Arrangement of electrons in atoms

In the following table (p. 305) the electrons in an atom are shown arranged in shells and sub-shells. The charge of the nucleus is $+Ze$ ($Z=$ atomic number), and it is surrounded by Z electrons (charge $-e$) in a neutral atom.

Shells

The electrons with the same principal quantum number n are said to form a shell. Proceeding from the nucleus outwards the shells are called K, L, etc. Thus if $n=1, 2, 3, 4$, the shell is the K, L, M or N shell respectively. This nomenclature had its origin in X-ray spectroscopy, in which a K-series line is due to an electron transition from an outer to the K shell.

Sub-shells. The electrons in each shell are arranged in sub-shells (often also called shells), specified by the value of l ($l=0, 1, \ldots, n-1$). In Bohr's theory, n and l determine the size and shape of the orbit. The nl sub-shell is complete or closed when it contains $2(2l+1)$ electrons.

Quantum numbers for one electron

These are a development of those introduced by Bohr and Sommerfeld.

Principal quantum number n, in Bohr's theory of the H atom, specifies the major axis of the orbit and determines the energy $E=-Rhc/n^2$. With other atoms, the orbits with any given value of l are still numbered serially by n, starting with $n=l+1$ (circular orbit), but the energy now depends on l too. In wave mechanics, $n-l-1$ is the number of nodes (zeros) in the radial factor in the wave function.

Azimuthal quantum number l specifies the angular momentum $lh/2\pi$ of the orbital motion, and is connected with the kind of symmetry possessed by the wave function ($l=0$ implies spherical symmetry). Code letters for the value of l:

$l=0$	1	2	3	4	5	6	7
Code letter s	p	d	f	g	h	i	k

and so on, omitting j (not used) and s, p (used out of turn). The value of the principal quantum number n is put in front of the code letter, e.g. 3d meaning $n=3$, $l=2$.

Magnetic quantum number m_l specifies one component of the orbital angular momentum: $m_l h/2\pi$ is the moment of momentum about some given line through the nucleus, e.g. drawn parallel to an external magnetic field. The $2l+1$ possible values of m_l are $l, (l-1), \ldots, -l$.

Spin quantum number m_s specifies the component of the intrinsic angular momentum (spin) of the electron; $m_s=\pm\frac{1}{2}$. The third and fourth quantum numbers are often of less direct significance owing to coupling effects. **Pauli's Exclusion Principle** states that no two electrons can have all their four quantum numbers the same.

Several electrons: spectral terms

In atoms with many electrons, it is usually a good approximation to assign values of (n, l) to each electron, corresponding to motion in an effective central field. Thus $1s^2 2s^2 2p^5$ means a **configuration** with two electrons in the K-shell, two with $n=2$ and $l=0$, and five with $n=2$ and $l=1$.

LS coupling. With the lighter atoms and with outer electrons, the energy of the various possible terms in any configuration depends on the extent to which the electrons keep apart (in other words on the probability of two electrons being found close together); and this in turn depends on the vector sums $L = \sum l$ and $S = \sum s$. The magnitude of L is specified by a capital letter (S, P, D...), and the resultant spin S is specified by writing the number $2S+1$ as a superscript: the relative orientation of L and S merely affects the magnetic energy and splits the terms into $2S+1$ (or fewer) levels. Thus a term with $L=2$ and $S=1$ is written ^{3}D (read 'triplet D'). Note that S is used in two senses, both as a code letter and as the resultant of spins.

Inner quantum number J specifies the total angular momentum $J = L + S$, and is written as a subscript (^{3}D$_3$, ^{3}D$_2$, ^{3}D$_1$).

X-ray spectra and jj coupling. With inner electrons or with heavy atoms, the coupling of the spin and orbital magnetic moments commonly makes it necessary to divide the nl 'orbits' into two groups according to the value of $j = |l+s| = l \pm \frac{1}{2}$, with a maximum number $2j+1$ of electrons in each (total $(2l+2) + 2l = 2(2l+1)$ as before), and L and S cease to be useful.

Electron configurations of the elements
Ground states of free neutral atoms

Element	K	L		M			N				O					Lowest state
	1s	2s	2p	3s	3p	3d	4s	4p	4d	4f	5s	5p	5d	5f	5g	
1. H	1															^{2}S$_{1/2}$
2. He	2															^{1}S$_0$
3. Li	2	1														^{2}S$_{1/2}$
4. Be	2	2														^{1}S$_0$
5. B	2	2	1													^{2}P$_{1/2}$
6. C	2	2	2													^{3}P$_0$
7. N	2	2	3													^{4}S$_{3/2}$
8. O	2	2	4													^{3}P$_2$
9. F	2	2	5													^{2}P$_{3/2}$
10. Ne	2	2	6													^{1}S$_0$
11. Na	2	2	6	1												^{2}S$_{1/2}$
12. Mg	2	2	6	2												^{1}S$_0$
13. Al	2	2	6	2	1											^{2}P$_{1/2}$
14. Si	2	2	6	2	2											^{3}P$_0$
15. P	2	2	6	2	3											^{4}S$_{3/2}$
16. S	2	2	6	2	4											^{3}P$_2$
17. Cl	2	2	6	2	5											^{2}P$_{3/2}$
18. Ar	2	2	6	2	6											^{1}S$_0$
19. K	2	2	6	2	6		1									^{2}S$_{1/2}$
20. Ca	2	2	6	2	6		2									^{1}S$_0$
21. Sc	2	2	6	2	6	1	2									^{2}D$_{3/2}$
22. Ti	2	2	6	2	6	2	2									^{3}F$_2$
23. V	2	2	6	2	6	3	2									^{4}F$_{3/2}$
24. Cr	2	2	6	2	6	5	1									^{7}S$_3$
25. Mn	2	2	6	2	6	5	2									^{6}S$_{5/2}$
26. Fe	2	2	6	2	6	6	2									^{5}D$_4$
27. Co	2	2	6	2	6	7	2									^{4}F$_{9/2}$
28. Ni	2	2	6	2	6	8	2									^{3}F$_4$

Electron configurations of the elements (*contd*)

Element	K	L		M			N				O					Lowest state
	1s	2s	2p	3s	3p	3d	4s	4p	4d	4f	5s	5p	5d	5f	5g	
29. Cu	2	2	6	2	6	10	1									$^2S_{1/2}$
30. Zn	2	2	6	2	6	10	2									$^1S_0$
31. Ga	2	2	6	2	6	10	2	1								$^2P_{1/2}$
32. Ge	2	2	6	2	6	10	2	2								$^3P_0$
33. As	2	2	6	2	6	10	2	3								$^4S_{3/2}$
34. Se	2	2	6	2	6	10	2	4								$^3P_2$
35. Br	2	2	6	2	6	10	2	5								$^2P_{3/2}$
36. Kr	2	2	6	2	6	10	2	6								$^1S_0$
37. Rb	2	2	6	2	6	10	2	6			1					$^2S_{1/2}$
38. Sr	2	2	6	2	6	10	2	6			2					$^1S_0$
39. Y	2	2	6	2	6	10	2	6	1		2					$^2D_{3/2}$
40. Zr	2	2	6	2	6	10	2	6	2		2					$^3F_2$
41. Nb	2	2	6	2	6	10	2	6	4		1					$^6D_{1/2}$
42. Mo	2	2	6	2	6	10	2	6	5		1					$^7S_3$
43. Tc	2	2	6	2	6	10	2	6	5		2					$^6S_{5/2}$
44. Ru	2	2	6	2	6	10	2	6	7		1					$^5F_5$
45. Rh	2	2	6	2	6	10	2	6	8		1					$^4F_{9/2}$
46. Pd	2	2	6	2	6	10	2	6	10							$^1S_0$
47. Ag	2	2	6	2	6	10	2	6	10		1					$^2S_{1/2}$
48. Cd	2	2	6	2	6	10	2	6	10		2					$^1S_0$
49. In	2	2	6	2	6	10	2	6	10		2	1				$^2P_{1/2}$
50. Sn	2	2	6	2	6	10	2	6	10		2	2				$^3P_0$
51. Sb	2	2	6	2	6	10	2	6	10		2	3				$^4S_{3/2}$
52. Te	2	2	6	2	6	10	2	6	10		2	4				$^3P_2$
53. I	2	2	6	2	6	10	2	6	10		2	5				$^2P_{3/2}$
54. Xe	2	2	6	2	6	10	2	6	10		2	6				$^1S_0$

{ 2   8   18 }

Element	K	L	M	N				O					P						Q
				4s	4p	4d	4f	5s	5p	5d	5f	5g	6s	6p	6d	6f	6g	6h	7s
55. Cs . .	2	8	18	2	6	10		2	6				1						
56. Ba . .	2	8	18	2	6	10		2	6				2						
57. La . .	2	8	18	2	6	10		2	6	1			2						
58. Ce . .	2	8	18	2	6	10	2	2	6				2						
59. Pr . .	2	8	18	2	6	10	3	2	6				2						
60. Nd . .	2	8	18	2	6	10	4	2	6				2						
61. Pm . .	2	8	18	2	6	10	5	2	6				2						
62. Sm . .	2	8	18	2	6	10	6	2	6				2						
63. Eu . .	2	8	18	2	6	10	7	2	6				2						
64. Gd . .	2	8	18	2	6	10	7	2	6	1			2						
65. Tb . .	2	8	18	2	6	10	9	2	6				2						
66. Dy . .	2	8	18	2	6	10	10	2	6				2						
67. Ho . .	2	8	18	2	6	10	11	2	6				2						

Electron configurations of the elements (*contd*)

Element	K	L	M	N				O					P						Q
				4s	4p	4d	4f	5s	5p	5d	5f	5g	6s	6p	6d	6f	6g	6h	7s
68. Er . .	2	8	18	2	6	10	12	2	6				2						
69. Tm .	2	8	18	2	6	10	13	2	6				2						
70. Yb . .	2	8	18	2	6	10	14	2	6				2						
71. Lu . .	2	8	18	2	6	10	14	2	6	1			2						
72. Hf . .	2	8	18	2	6	10	14	2	6	2			2						
73. Ta . .	2	8	18	2	6	10	14	2	6	3			2						
74. W . .	2	8	18	2	6	10	14	2	6	4			2						
75. Re . .	2	8	18	2	6	10	14	2	6	5			2						
76. Os . .	2	8	18	2	6	10	14	2	6	6			2						
77. Ir . .	2	8	18	2	6	10	14	2	6	7			2						
78. Pt . .	2	8	18	2	6	10	14	2	6	9			1						
79. Au . .	2	8	18	2	6	10	14	2	6	10			1						
80. Hg .	2	8	18	2	6	10	14	2	6	10			2						
81. Tl . .	2	8	18	2	6	10	14	2	6	10			2	1					
82. Pb . .	2	8	18	2	6	10	14	2	6	10			2	2					
83. Bi . .	2	8	18	2	6	10	14	2	6	10			2	3					
84. Po . .	2	8	18	2	6	10	14	2	6	10			2	4					
85. At . .	2	8	18	2	6	10	14	2	6	10			2	5					
86. Em . .	2	8	18	2	6	10	14	2	6	10			2	6					
87. Fr . .	2	8	18	2	6	10	14	2	6	10			2	6					1
88. Ra . .	2	8	18	2	6	10	14	2	6	10			2	6					2
89. Ac . .	2	8	18	2	6	10	14	2	6	10			2	6	1				2
90. Th . .	2	8	18	2	6	10	14	2	6	10			2	6	2				2
91. Pa . .	2	8	18	2	6	10	14	2	6	10	2		2	6	1				2
92. U . .	2	8	18	2	6	10	14	2	6	10	3		2	6	1				2
93. Np . .	2	8	18	2	6	10	14	2	6	10	4		2	6	1				2
94. Pu . .	2	8	18	2	6	10	14	2	6	10	6		2	6					2
95. Am .	2	8	18	2	6	10	14	2	6	10	7		2	6					2
96. Cm .	2	8	18	2	6	10	14	2	6	10	7		2	6	1				2
97. Bk . .	2	8	18	2	6	10	14	2	6	10	9		2	6					2
98. Cf . .	2	8	18	2	6	10	14	2	6	10	10		2	6					2
99. Es . .	2	8	18	2	6	10	14	2	6	10	11		2	6					2
100. Fm . .	2	8	18	2	6	10	14	2	6	10	12		2	6					2
101. Md . .	2	8	18	2	6	10	14	2	6	10	13		2	6					2
102. No . .	2	8	18	2	6	10	14	2	6	10	14		2	6					2
103. Lr . .	2	8	18	2	6	10	14	2	6	10	14		2	6	1				2
104. Ku . .	2	8	18	2	6	10	14	2	6	10	14		2	6	2				2
105. Ha . .	2	8	18	2	6	10	14	2	6	10	14		2	6	3				2

J.C.R.

3.3.2 Ionization potentials

The ionization potential is the least energy, in electron volts, which is necessary to remove an electron from a free unexcited neutral atom (or additional electron from an ionized atom). It can be determined either directly from the speed of the slowest electrons which will ionize the atom by collision, or more accurately from the limit of a spectroscopic series.

H	13.598	Ne	21.56	K	4.34	Ni+	18.15	Tc	7.28	Ba+	10.00
He	24.587	Ne+	41.08	K+	31.71	Ni2+	35.20	Ru	7.36	La	5.61
He+	54.42	Ne2+	63.45	Ca	6.11	Cu	7.72	Rh	7.46	Ta	7.88
Li	5.39	Na	5.14	Ca+	11.87	Zn	9.39	Pd	8.33	W	7.98
Li+	75.60	Na+	47.30	Sc	6.54	Ga	6.00	Ag	7.57	Re	7.87
Be	9.32	Mg	7.65	Ti	6.82	Ge	7.88	Cd	8.99	Os	8.70
B	8.30	Mg+	15.04	V	6.74	As	9.81	In	5.79	Pt	9.00
C	11.26	Al	5.99	Cr	6.77	Se	9.75	Sn	7.34	Au	9.22
C+	24.38	Al+	18.83	Mn	7.43	Br	10.56	Sb	8.64	Hg	10.43
C2+	47.88	Si	8.15	Fe	7.87	Kr	14.00	Te	9.01	Tl	6.11
N	14.53	P	10.49	Fe+	16.18	Rb	4.18	I	10.45	Pb	7.42
N+	29.60	S	10.36	Fe2+	30.65	Sr	5.69	Xe	12.13	Ra	5.28
N2+	47.44	S+	23.41	Co	7.86	Y	6.25	Xe+	23.39	Th	7.50
O	13.62	Cl	12.97	Co+	17.06	Zr	6.95	Cs	3.89	U	6.19
O+	35.12	A	15.75	Co2+	33.50	Nb	6.77	Cs+	25.10	Np	6.16
O2+	54.90	A+	27.63	Ni	7.64	Mo	7.06	Ba	5.21	Pu	5.71
F	17.42										

References

(1) W. Lotz, *J. Opt. Soc. Am.*, 1967, **57**, 873.
(2) *Handbook of Physics*, 2nd edn, 1967.
(3) *J. Chem. Phys.*, 1969, **51**, 3105.

J.C.R.

3.3.3 Auger spectroscopy

Impact of electrons with a solid causes ionization of atomic energy levels of atoms in the solid. An ionized atom can relax by the ejection either of an X-ray photon or of an Auger electron; the latter process is far more probable at low primary energies, e.g. less than $10 \, keV$. In the Auger process the ionized level, of binding energy E_A, is filled by an electron from an outer level, of binding energy E_B, and the excess energy $(E_A - E_B)$ is given to another electron either in the same level E_B or in a still more shallow one of binding energy E_C. The energy E_{ABC} of the Auger transition ABC, i.e. the kinetic energy of the ejected electron, is then

$$E_{ABC} = E_A - E_B - E_C'$$ ...(1)

where E_C' is primed to show that the binding energy of the electron in C has changed due to the fact that the electron is being ejected from an already ionized atom. To a good approximation for most purposes, the Auger energy E_{ABC} may be estimated from the Chung and Jenkins' expression,[1] for an atom of atomic number Z,

$$E_{ABC} = E_A(Z) - \tfrac{1}{2}\{E_B(Z) + E_B(Z+1)\} - \tfrac{1}{2}\{E_C(Z) + E_C(Z+1)\}$$ ...(2)

where $E_B(Z+1)$, $E_C(Z+1)$ are the binding energies of electrons in the same levels in the next element up the Periodic Table. Since equation (2) can be used in conjunction with tables[2] of atomic energy levels to calculate the expected energies of Auger transitions, analysis of the observed Auger spectrum from a solid enables its atomic composition to be established.

In electron-excited Auger spectroscopy, the range of energies normally used is $1000-10000 \, eV$, and the energies of the resultant Auger electrons are typically $20-1000 \, eV$. At these low energies, the inelastic mean free paths of electrons in solids (see section 3.5.2) are very short, and therefore the Auger electrons must originate at, or very close to, the true surface if they are to escape and be observed externally. Auger spectroscopy is thus very surface-specific.

Auger electrons may also be produced by initial excitation with X-rays, ions, and other particles.

The table sets out the experimentally observed energies of the most intense Auger transitions of the more common elements in solid elemental or compound form. Allocation of the transitions has been made on the basis of equation (2), using the tables of binding energies published by Bearden and Burr.[2]

For more comprehensive reading about Auger spectroscopy see the article by Rivière.[3]

References

(1) M. F. Chung and J. H. Jenkins, *Surface Science*, 1970, **21**, 253.
(2) J. A. Bearden and A. F. Burr, *Rev. Mod. Phys.*, 1967, **39**, 125.
(3) J. C. Rivière, *The Analyst*, 1983, **108**, 649.

Energies of the major Auger lines for the more common elements

Element	Z	E/eV	Transition†	Element	Z	E/eV	Transition†
C . . .	6	270	KVV	Zn . . .	30	58	M_3VV
N . . .	7	380	KVV			830	$L_3M_{2,3}M_{2,3}$
O . . .	8	510	KVV			910	$L_3M_{2,3}V$
F . . .	9	650	KVV			990	L_3VV
Na . . .	11	30	$L_{2,3}VV$	Ge . . .	32	24	$M_{4,5}VV$
		990	$KL_{2,3}L_{2,3}$			44	$M_3M_{4,5}M_{4,5}$
Mg . . .	12	47	$L_{2,3}VV$			48	$M_2M_{4,5}M_{4,5}$
		1180	$KL_{2,3}L_{2,3}$			84	$M_{2,3}M_{4,5}V$
Al . . .	13	66	$L_{2,3}VV$			1140	$L_3M_{4,5}M_{4,5}$
		1380	$KL_{2,3}L_{2,3}$			1173	$L_2M_{4,5}M_{4,5}$
Si . . .	14	91	$L_{2,3}VV$	As . . .	33	36	$M_{4,5}VV$
		1610	$KL_{2,3}L_{2,3}$			42	$M_3M_{4,5}M_{4,5}$
P . . .	15	116	$L_{2,3}VV$			47	$M_2M_{4,5}M_{4,5}$
S . . .	16	150	$L_{2,3}VV$			91	$M_{2,3}M_{4,5}V$
Cl . . .	17	180	$L_{2,3}VV$	Zr . . .	40	20	$N_{2,3}VV$
K . . .	19	250	$L_{2,3}VV$			90	$M_{4,5}N_1N_{2,3}$
Ca . . .	20	290	$L_{2,3}M_{2,3}M_{2,3}$			115	$M_{4,5}N_{2,3}N_{2,3}$
Cr . . .	24	35	$M_{2,3}VV$			145	$M_{4,5}N_{2,3}V$
		486	$L_3M_{2,3}M_{2,3}$	Nb . . .	41	22	$N_{2,3}VV$
		527	$L_3M_{2,3}V$			102	$M_{4,5}N_1N_{2,3}$
Mn . . .	25	40	$M_{2,3}VV$			164	$M_{4,5}N_{2,3}N_{2,3}$
		540	$L_3M_{2,3}M_{2,3}$			194	$M_{4,5}N_{2,3}V$
		588	$L_3M_{2,3}V$	Mo . . .	42	27	$N_{2,3}VV$
		637	L_3VV			120	$M_{4,5}N_1N_{2,3}$
Fe . . .	26	44	$M_{2,3}VV$			185	$M_{4,5}N_{2,3}N_{2,3}$
		594	$L_3M_{2,3}M_{2,3}$			220	$M_{4,5}N_{2,3}V$
		647	$L_3M_{2,3}V$	Ag . . .	47	48	N_3VV
		700	L_3VV			56	N_2VV
Co . . .	27	52	$M_{2,3}VV$			270	$M_{4,5}N_1V$
		651	$L_3M_{2,3}M_{2,3}$			308	$M_{4,5}N_{2,3}V$
		711	$L_3M_{2,3}V$			362	$M_{4,5}VV$
		771	L_3VV	Sn . . .	50	22	$N_{4,5}VV$
Ni . . .	28	60	$M_{2,3}VV$			64	$N_{2,3}N_{4,5}V$
		716	$L_3M_{2,3}M_{2,3}$			310	$M_5N_1N_{4,5}$
		781	$L_3M_{2,3}V$			360	$M_5N_{2,3}N_{4,5}$
		847	L_3VV			423	$M_5N_{4,5}N_{4,5}$
Cu . . .	29	60	M_3VV	Ta . . .	73	26	$N_{6,7}VV$
		772	$L_3M_{2,3}M_{2,3}$			163	$N_5N_{6,7}N_{6,7}$
		844	$L_3M_{2,3}V$			173	$N_4N_{6,7}N_{6,7}$
		917	L_3VV			199	$N_5N_{6,7}V$
						209	$N_4N_{6,7}V$
						341	$N_3N_{6,7}N_{6,7}$

Energies of the major Auger lines for the more common elements (*contd*)

Element	Z	E/eV	Transition†	Element	Z	E/eV	Transition†
W . . .	74	20	$N_{6,7}VV$	Pb . . .	82	46	$O_3O_{4,5}O_{4,5}$
		164	$N_5N_{6,7}N_{6,7}$			57	$O_2O_{4,5}O_{4,5}$
		176	$N_4N_{6,7}N_{6,7}$			92	$N_7O_{4,5}O_{4,5}$
		205	$N_5N_{6,7}V$			116	$N_7O_{4,5}V$
		216	$N_4N_{6,7}V$			246	$N_5N_7O_{4,5}$
		347	$N_3N_{6,7}N_{6,7}$			265	$N_4N_7O_{4,5}$
Pt . . .	78	13	N_7O_3V	Bi . . .	83	60	$O_2O_{4,5}O_{4,5}$
		45	O_3VV			102	$N_7O_{4,5}O_{4,5}$
		68	N_7VV			128	$N_7O_{4,5}V$
		160	$N_5N_{6,7}N_{6,7}$			250	$N_5N_7O_{4,5}$
		171	$N_4N_{6,7}N_{6,7}$			270	$N_4N_7O_{4,5}$
		238	$N_5N_{6,7}V$				
		254	$N_4N_{6,7}V$				
Au . . .	79	42	O_3VV				
		71	N_7VV				
		151	$N_5N_{6,7}N_{6,7}$				
		165	$N_4N_{6,7}N_{6,7}$				
		244	$N_5N_{6,7}V$				
		261	$N_4N_{6,7}V$				

† The symbols K, L, M, N, and O are in the usual X-ray notation for atomic energy levels; the symbol V refers to the valence band of the solid.

J.C.R.

3.3.4 X-ray photoemission spectroscopy

Interaction of a photon of energy hv with an electron in an atomic energy level of binding energy E_B occurs by transfer of all the photon energy to the electron. If hv is greater than E_B, a photoelectron is ejected from the atom with a kinetic energy given by

$$E_{Kin} = hv - E_B \qquad \qquad ...(1)$$

When the atom forms part of a solid, there is a work function term to be subtracted from the right-hand side of (1), but it is comparatively small and nearly constant. Measured E_{Kin} values thus reflect the E_B values directly, so that measurement of E_{kin} provides elemental analysis.

In X-ray photoemission spectroscopy the two sources in most common use are magnesium and aluminium, the principal line energies used being the $K\alpha_{1,2}$, 1253.6 eV for magnesium and 1486.6 eV for aluminium. Since the kinetic energies of the photoelectrons will therefore be in the range 0–1500 eV, the inelastic mean free paths of the electrons (see section 0.00) will be very short, and the photoelectrons must originate at, or very close to, the true surface if they are to escape and be measured. Like Auger spectroscopy, X-ray photoelectron spectroscopy is thus a surface-specific technique.

The tables list the photoelectron kinetic energies expected for all elements from H to U, for Mg $K\alpha$, and for Al $K\alpha$, excitation, respectively, to the nearest electron-volt. Where the element exists as a pure elemental non-volatile solid, the values will correspond to that state with reasonable accuracy, but for other elements, particularly the non-metallic ones, the values must be regarded as approximate since shifts of several eV can occur between different states of chemical combination of an element. The reference point for all the values is a binding energy for the Au $4f_{7/2}$ level assumed to be 84.0 eV.

For further reading the reader is referred to the articles by Roberts[1] and by Menzel.[2]

References

(1) M. W. Roberts, *Sci. Prog., Oxford*, 1982, **68**, 65
(2) D. Menzel, *Crit. Rev. Solid State Mat. Sci.*, 1978, **7**, 357.

J.C.R.

Table 1

Photoelectron kinetic energies, to nearest electron-volt and referred to the binding energy $Au\ 4f_{7/2} = 84.0\ eV$, using Mg K_α excitation. The energy of the most intense photoelectron peak for each element is underlined.

	1s	2s	$2p_{1/2}$	$2p_{3/2}$	3s	$3p_{1/2}$	$3p_{3/2}$	$3d_{3/2}$	$3d_{5/2}$	4s	$4p_{1/2}$	$4p_{3/2}$	$4d_{3/2}$	$4d_{5/2}$	$4f_{5/2}$	$4f_{7/2}$	5s	$5p_{1/2}$	$5p_{3/2}$	$5d_{3/2}$	$5d_{5/2}$	$5f_{5/2}$	$5f_{7/2}$	6s	$6p_{1/2}$	$6p_{3/2}$	$6d_{3/2}$	$6d_{5/2}$
1 H	**1239**																											
2 He	**1228**																											
3 Li	**1198**																											
4 Be	**1141**																											
5 B	**1064**		1248																									
6 C	**969**		1246																									
7 N	**854**		1243																									
8 O	**720**	1229	1245																									
9 F	**567**	1221	1244																									
10 Ne	**386**	1207	1234																									
11 Na	**180**	1189	1221		1251																							
12 Mg		1163	1201		1250																							
13 Al		**1135**	1179	1180	1251	1250																						
14 Si		1104	1152	1153	1244	1249																						
15 P		1063	1116	1117	1236	1242																						
16 S		1023	1087	1088	1236	1244																						
17 Cl		982	1051	1052	1235	1246																						
18 Ar		932	1005	1007	1227	1240																						
19 K		875	956	**959**	1219	1235				1250																		
20 Ca		815	902	906	1209	1227				1247																		
21 Sc		752	846	850	1199	1220		1245		1250																		
22 Ti		689	791	797	1192	1218		1249		1249																		
23 V		624	732	740	1186	1215		1250		1249																		
24 Cr		558	669	678	1178	1210		1250		1249																		
25 Mn		483	600	612	1169	1204		1249		1249																		
26 Fe		406	531	544	1160	1198		1249		1249																		
27 Co		327	459	474	1152	1193		1250		1249																		
28 Ni		244	381	398	1141	1184		1250		1248																		
29 Cu		156	301	321	1133	1179		1251		1250																		
30 Zn		59	210	233	1117	1166		1244		1248																		
31 Ga			110	**137**	1094	1146	1150	1235			1251																	
32 Ge			5	**36**	1072	1125	1132	1224			1249																	
33 As					1049	1106	1112	1211			1250																	
34 Se					1021	1084	1091	1196			1247																	
35 Br					996	1063	1071	1182	1183	1225	1247	1248																
36 Kr					963	1030	1039	1164	1164	1228	1242																	
37 Rb					930	1005	1014	1141	1142	1223	1238	1239																
38 Sr					895	973	983	1117	1119	1215	1233																	

Table 1 (contd)

	6d$_{5/2}$	6d$_{3/2}$	6p$_{3/2}$	6p$_{1/2}$	6s	5f$_{7/2}$	5f$_{5/2}$	5d$_{5/2}$	5d$_{3/2}$	5p$_{3/2}$	5p$_{1/2}$	5s	4f$_{7/2}$	4f$_{5/2}$	4d$_{5/2}$	4d$_{3/2}$	4p$_{3/2}$	4p$_{1/2}$	4s	3d$_{5/2}$	3d$_{3/2}$	3p$_{3/2}$	3p$_{1/2}$	3s	2p$_{3/2}$	2p$_{1/2}$	2s	1s
39 Y																1249		1227	1207	1095	1093	952	940	859				
40 Zr																1249		1224	1201	1072	1070	922	908	822				
41 Nb																1249		1219	1194	1048	1045	889	874	784				
42 Mo																1251		1218	1191	1025	1022	860	843	748				
43 Tc																1250		1214	1184	1000	996	827	808	708				
44 Ru																1250		1209	1178	973	969	792	770	667				
45 Rh																1250		1205	1171	945	941	756	731	625				
46 Pd																1249		1201	1166	918	912	721	693	583				
47 Ag																1243	1197	1190	1157	886	880	681	650	535				
48 Cd																1236	1186	1175	1145	849	842	636	602	482				
49 In											1252					1229	1175	1164	1131	809	802	588	550	427				
50 Sn											1251					1221	1164	1154	1116	768	759	538	496	369				
51 Sb											1250	1252				1213	1154	1142	1100	725	716	487	441	309				
52 Te											1250	1246				1203	1142	1130	1084	680	670	434	383	246				
53 I											1249	1241				1189	1130	1106	1066	633	621	378	322	180				
54 Xe											1245	1239					1106	1080	1044	580	567	315	253	107				
55 Cs										1239	1241	1234			1176	1174	1091	1061	1022	527	513	255	187	35				
56 Ba										1236	1238	1230			1163	1160	1073	1047	999	472	456	190	116					
57 La										1238		1213				1154	1061	1029	982	421	404	129	48					
58 Ce										1233		1220				1142	1045	1016	963	369	351	67						
59 Pr										1230		1215				1139	1035	1009	948	321	301	10						
60 Nd										1231		1215	1252	1252		1135	1028	997	937	275	253							
61 Pm										1231		1215	1250	1250		1132	1015	987	921	226	201							
62 Sm										1230		1214	1251	1251		1123	1005	969	907	172	146							
63 Eu										1232		1215	1248	1248		1119	996	964	892	122	92							
64 Gd										1227		1221	1247	1247		1112	982	942	877	67	35							
65 Tb										1226		1216	1252	1252		1105	967	921	855	11								
66 Dy										1223		1213	1248	1252		1098	960	909	836									
67 Ho										1229		1190	1252	1250		1091	946	886	817									
68 Er										1229		1193	1250	1249	1085	1076	932	867	803									
69 Tm										1224		1199	1248	1248		1073	916	856	781									
70 Yb											1214	1198	1247	1247	1068	1054	909	842	765									
71 Lu									1247		1208	1196	1246	1246	1057	1048	893	815	746									
72 Hf									1245	1222	1214	1188	1235	1234	1039	1029	872	788	714									
73 Ta									1247	1216	1208	1181	1227	1226	1023	1011	848	761	687									
74 W									1246	1217	1206	1175	1219	1216	1007	994	827	735	657									
75 Re									1248	1218	1207	1170	1212	1210	992	979	808	706	627									
76 Os									1252	1207	1194	1169	1202	1200	980	963	784	675	598									
77 Ir									1249	1202	1189	1157	1192	1189	958	941	758	643	562									
78 Pt									1250	1201	1187	1151	1181	1178	939	922	733	609	530									
79 Au									1250	1199	1181	1145	1170	1166	919	900	707	576	494									
80 Hg									1246	1195	1172	1132	1154	1150	893	874	681		452									

Table 1 (contd)

	1s	2s	2p1/2	2p3/2	3s	3p1/2	3p3/2	3d3/2	3d5/2	4s	4p1/2	4p3/2	4d3/2	4d5/2	4f5/2	4f7/2	5s	5p1/2	5p3/2	5d3/2	5d5/2	5f5/2	5f7/2	6s	6p1/2	6p3/2	6d3/2	6d5/2
81 Tl										407	531	643	846	866	1130	1134	1116	1153	1177	1237	1239			1249				
82 Pb										359	489	608	817	840	1110	1114	1105	1148	1166	1231	1233			1249	1252			
83 Bi										314	447	574	789	812	1091	1095	1093	1136	1160	1226	1228			1244	1250			
84 Po										257	401	547	752	779	1066	1071	1075	1120	1148	1221				1240	1247			
85 At										210	366	512	719	745	1042		1057	1104	1137	1212				1234	1244			
86 Rn										155	323	484	686	711	1014		1038	1088	1125	1204				1226	1241			
87 Fr										99	272	442	649	675	984		1018	1070	1112	1194				1218	1237			
88 Ra										44	195	373	617	650	954		998	1052	1100	1185				1209	1234			
89 Ac											172	362	578	613	933		980	1037	1085	1172				1200	1220		1250	
90 Th											84	285	538	576	908	917	962	1023	1071	1158	1165			1193	1203	1209	1250	
91 Pa											28	246	509	544	881	893	943	1029		1158		1219	1219	1187	1210		1249	
92 U												208	472	515	861	872	929	993	1057	1147	1156	1215	1215	1182	1210	1220	1248	

Table 2

Photoelectron kinetic energies, to nearest electron-volt and referred to the binding energy $Au\ 4f_{7/2} = 84.0$ eV, using $Al\ K_\alpha$ excitation. The energy of the most intense photoelectron peak for each element is underlined.

	$1s$	$2s$	$2p_{1/2}$	$2p_{3/2}$	$3s$	$3p_{1/2}$	$3p_{3/2}$	$3d_{3/2}$	$3d_{5/2}$	$4s$	$4p_{1/2}$	$4p_{3/2}$	$4d_{3/2}$	$4d_{5/2}$	$4f_{5/2}$	$4f_{7/2}$	$5s$	$5p_{1/2}$	$5p_{3/2}$	$5d_{3/2}$	$5d_{5/2}$	$5f_{5/2}$	$5f_{7/2}$	$6s$	$6p_{1/2}$	$6p_{3/2}$	$6d_{3/2}$	$6d_{5/2}$
1 H	1472																											
2 He	1461																											
3 Li	<u>1431</u>																											
4 Be	<u>1374</u>																											
5 B	1297		1481																									
6 C	1202		1479																									
7 N	1087		1476																									
8 O	<u>953</u>	1462	1478																									
9 F	800	1454	1477																									
10 Ne	619	1440	1467																									
11 Na	<u>413</u>	1422	1454		1484																							
12 Mg	<u>180</u>	1396	1434		1483																							
13 Al		<u>1368</u>	1412	1413	1484	1483																						
14 Si		1337	1385	<u>1386</u>	1477	1482																						
15 P		1296	1349	<u>1350</u>	1469	1475																						
16 S		1256	1320	<u>1321</u>	1468	1477																						
17 Cl		1215	1284	<u>1285</u>	1460	1479																						
18 Ar		1165	1238	<u>1240</u>	1452	1473																						
19 K		1108	1189	<u>1192</u>	1442	1468				1483																		
20 Ca		1048	1135	<u>1139</u>	1432	1460				1480																		
21 Sc		985	1079	<u>1083</u>	1425	1453		1478		1483																		
22 Ti		922	1024	<u>1030</u>	1419	1451		1482		1482																		
23 V		857	965	<u>973</u>	1411	1448		1483		1482																		
24 Cr		791	902	<u>911</u>	1402	1443		1483		1482																		
25 Mn		716	833	<u>845</u>	1393	1437		1482		1482																		
26 Fe		639	764	<u>777</u>	1385	1431		1482		1482																		
27 Co		560	692	<u>707</u>	1374	1426		1483		1482																		
28 Ni		477	614	<u>631</u>	1366	1417		1482		1481																		
29 Cu		389	534	<u>554</u>	1350	1412		1484		1481																		
30 Zn		292	443	<u>466</u>	1327	1399		1477		1483																		
31 Ga		188	343	<u>370</u>	1305	1379	1383	1468		1481	1484																	
32 Ge		71	238	<u>269</u>	1282	1358	1365	1457		1481	1482																	
33 As			127	<u>162</u>	1254	1339	1345	1444			1483																	
34 Se			9	<u>50</u>	1229	1317	1324	1429			1480																	
35 Br					1196	1296	1304	1415	<u>1416</u>	1458	1480	1481																
36 Kr					1163	1263	1272	<u>1397</u>		1461	1475																	
37 Rb					1128	1238	1247	1374	<u>1375</u>	1456	1471	1472																
38 Sr						1206	1216	1350	<u>1352</u>	1448	1466																	

Table 2 (*Contd*)

	$1s$	$2s$	$2p_{1/2}$	$2p_{3/2}$	$3s$	$3p_{1/2}$	$3p_{3/2}$	$3d_{3/2}$	$3d_{5/2}$	$4s$	$4p_{1/2}$	$4p_{3/2}$	$4d_{3/2}$	$4d_{5/2}$	$4f_{5/2}$	$4f_{7/2}$	$5s$	$5p_{1/2}$	$5p_{3/2}$	$5d_{3/2}$	$5d_{5/2}$	$5f_{5/2}$	$5f_{7/2}$	$6s$	$6p_{1/2}$	$6p_{3/2}$	$6d_{3/2}$	$6d_{5/2}$
39 Y					1092	1173	1185	1326	1328	1440	1460	1460	1482	1482														
40 Zr					1055	1141	1155	1303	1305	1434	1457	1457	1482	1482														
41 Nb					1017	1107	1122	1278	1281	1427	1452	1452	1482	1482														
42 Mo					981	1076	1093	1255	1258	1424	1451	1451	1484	1484														
43 Tc					941	1041	1060	1229	1233	1417	1447	1447	1483	1483														
44 Ru					900	1003	1025	1202	1206	1411	1442	1442	1483	1483														
45 Rh					858	964	989	1174	1178	1404	1438	1438	1483	1483														
46 Pd					816	926	954	1145	1151	1399	1434	1434	1484	1484														
47 Ag					768	883	914	1113	1119	1390	1423	1430	1482	1482														
48 Cd					715	835	869	1075	1082	1378	1419	1419	1476	1476														
49 In					660	783	821	1035	1042	1364	1408	1408	1469	1469			1485											
50 Sn					602	729	771	992	1001	1349	1397	1397	1462	1462			1479	1485	1485									
51 Sb					542	674	720	949	958	1333	1387	1387	1454	1454			1474	1484	1484									
52 Te					479	616	667	903	913	1317	1375	1375	1446	1446			1472	1483	1483									
53 I					413	555	611	854	866	1299	1363	1363	1436	1436			1467	1483	1483									
54 Xe					340	486	548	800	813	1277	1339	1339	1422	1422			1463	1482	1482									
55 Cs					268	420	488	746	760	1255	1313	1324	1407	1409	1485	1485	1446	1472	1474									
56 Ba					193	349	423	689	705	1232	1294	1306	1393	1396	1483	1483	1453	1469	1471									
57 La					124	281	362	637	654	1215	1280	1294	1387	1387	1484	1484	1448	1471	1471									
58 Ce					51	213	300	584	602	1196	1262	1278	1375	1375	1481	1481	1448	1466	1466									
59 Pr						148	243	534	554	1181	1249	1268	1372	1372	1480	1480	1447	1463	1463									
60 Nd						83	188	486	508	1170	1242	1261	1368	1368	1485	1485	1448	1464	1464									
61 Pm						14	129	434	459	1154	1230	1248	1365	1365	1485	1485	1454	1463	1463									
62 Sm							66	379	405	1140	1220	1238	1356	1356	1483	1483	1449	1464	1464									
63 Eu							5	325	355	1125	1202	1229	1352	1352	1481	1481	1446	1463	1463									
64 Gd								268	300	1110	1197	1215	1345	1345	1482	1482	1446	1465	1465									
65 Tb								210	244	1088	1175	1200	1338	1338	1481	1481	1423	1460	1460									
66 Dy								153	191	1069	1154	1193	1331	1331	1480	1480	1434	1459	1459									
67 Ho								94	134	1050	1142	1179	1324	1324	1479	1479	1426	1465	1465									
68 Er								32	76	1036	1119	1165	1309	1318	1479	1479	1432	1456	1456									
69 Tm									18	1014	1089	1149	1306	1306	1467	1468	1431	1462	1462									
70 Yb										998	1075	1126	1287	1301	1459	1460	1429	1457	1457									
71 Lu										979	1048	1105	1281	1290	1449	1452	1421	1447	1455	1480								
72 Hf										947	1021	1081	1262	1272	1443	1445	1414	1441	1449	1478								
73 Ta										920	994	1060	1244	1256	1433	1435	1408	1439	1450	1480								
74 W										890	968	1041	1227	1240	1422	1425	1403	1440	1451	1479								
75 Re										860	941	1017	1212	1225	1411	1414	1402	1427	1440	1481								
76 Os										831	908	991	1196	1213	1399	1403	1390	1422	1435	1485								
77 Ir										795	876	966	1174	1191	1383	1387	1384	1420	1434	1482								
78 Pt										763	842	940	1155	1172			1378	1414	1432	1483								
79 Au										727	809	914	1133	1152			1365	1405	1428	1483								
80 Hg										685			1107	1126						1479								

Table 2 (*Contd*)

	1s	2s	2p₁/₂	2p₃/₂	3s	3p₁/₂	3p₃/₂	3d₃/₂	3d₅/₂	4s	4p₁/₂	4p₃/₂	4d₃/₂	4d₅/₂	4f₅/₂	4f₇/₂	5s	5p₁/₂	5p₃/₂	5d₃/₂	5d₅/₂	5f₅/₂	5f₇/₂	6s	6p₁/₂	6p₃/₂	6d₃/₂	6d₅/₂
81 Tl										640	764	876	1079	1099	1363	1367	1349	1386	1410	1470	1472			1482				
82 Pb										592	722	841	1050	1073	1353	1357	1338	1381	1399	1464	1466			1482				
83 Bi										547	680	807	1022	1045	1324	1328	1326	1369	1393	1459	1461			1477	1485			
84 Po										490	634	780	985	1012	1299	1304	1308	1353	1381	1454				1473	1483			
85 At										443	599	745	952	978	1275		1290	1337	1370	1445				1467	1480			
86 Rn										388	556	717	919	944	1247		1271	1321	1358	1437				1459	1477			
87 Fr										332	505	675	882	908	1217		1251	1303	1345	1427				1451	1474			
88 Ra										277	428	606	850	883	1187		1231	1285	1333	1418				1442	1470			
89 Ac										216	405	595	811	846	1166		1213	1270	1318	1405				1433	1467		1483	
90 Th										156	317	518	771	809	1141	1150	1195	1256	1304	1391	1398			1426	1453		1483	
91 Pa										98	261	479	742	777	1114	1126	1176	1262		1391		1452		1420	1436	1442	1482	
92 U										45	213	441	705	748	1094	1105	1162	1226	1290	1380	1389	1448		1415	1443	1453	1481	

3.4 X-rays

3.4.1 X-ray absorption edges, characteristic X-ray lines and fluorescence yields

Characteristic X-ray lines are generated by the movement of an electron from an 'initial' level or edge to a 'terminal' level in which a vacancy exists as a result of excitation of some sort (e.g. X-ray, electron, etc.); the energy of the line is equal to the difference in energy of the terminal and initial levels. Depending on atomic number, the X-ray spectra from the elements can include lines from the K, L, M, N and O series corresponding to excitation of the K, L, M, N or O levels; the table lists the energies—in keV—of the principal lines of the common K, L and M series along with the corresponding terminal edge, or excitation, energy. Lines are identified both by the common labels—e.g. $K\alpha_1$, $K\alpha_2$, etc—and the term labels giving, first, the terminal and, secondly, the initial edges—e.g. KL_{III}, KL_{II}, etc. With the exception of the elements 92–103 the table has been prepared by calculation of line energies from the compilation of smoothed edge energies given by Dewey, Mapes and Reynolds, which in turn drew extensively on the compilation by Bearden; the edge data for elements 92–103 are taken directly from Bearden. The energies of the softer radiations may be affected by the chemical state of the elements concerned; generally the shifts do not exceed a few electron volts. The wavelength λ, in pm, can be derived from the tabulated energy E, in keV, by the relationship $\lambda = 1239.81/E$.

Approximate K and L line intensities are given at the head of the columns in the table, relative to the line in the series which is normally the strongest. Where a range of values is indicated, the first number represents the value for lowest Z, the second number that for the highest Z, in the section of the table to which it refers. The values given are based on the compilation of experimental relative intensities of Salem, Panossian and Krause and the intensities calculated by Scofield (Salem *et al.*, pp. 121–37). The values are intended only as a rough guide and the original references should be consulted for details. The accompanying figure gives a plot of the relative intensities for the commonly encountered $K\beta$ line.

In addition to the lines given in the table, satellite, or non-diagram, lines also occur; these are generally only of significance in the case of the K-satellites of the lighter elements Al, Mg, Si, etc. where their intensities may be a few per cent of that of the $K\alpha$ line (see Clark or Sandstrom).

The transition of an electron to fill, for example, a vacancy in the K-shell may be accompanied by either the emission of an X-ray photon or the transfer of energy to another electron which is then emitted (an Auger electron); the probability that a vacancy in a given shell will result in emission of an X-ray is the fluorescence yield of that shell.

The accompanying figures give plots of the fluorescence yield ω_K for the K shell and an effective yield $\gamma_{L_{II}}$ for the L_{II} shell versus Z; the figures are taken from the extensive survey by Krause. The effective yield contains contributions arising from inter-shell transitions. The variations in fluorescence yields of the L_I and L_{III} shells are similar to that shown for the L_{II} shell, but the original reference should be consulted for details. A few values for the average M shell fluorescence yield taken from *The Handbook of Spectroscopy* (ed. J. W. Robinson) are also given in the figure.

References

(1) Dewey, Mapes and Reynolds, *Progress in Nuclear Energy*, 1969, Series IX, *Analytical Chemistry*, eds Elion and Stewart (Pergamon Press).
(2) Bearden, *Rev. Mod. Phys.*, 1967, **39** (1), 78.
(3) Salem, Panossian and Krause, *Atomic and Nuclear Data Tables* 1A, 1974, 91–109.
(4) Clark, *Encyclopaedia of X-rays and Gamma Rays*, 1963 (Reinhold).
(5) Sandström, *Handbuch der Physik*, Vol. XXX, *X-rays*, 1957 (Springer-Verlag).
(6) Krause, *J. Physical and Chemical Reference Data*, 1979, **8**, 307–27.
(7) Robinson (ed.), *Handbook of Spectroscopy*, Vol. 1, p. 228, 1974, (CRC Press, Cleveland).

X-ray absorption edges, characteristic X-ray lines and fluorescence yields

Atomic number and element	K edge	KN_{III} $K\beta_2$	KM_{III} $K\beta_1$	KM_{II} $K\beta_3$	KL_{III} $K\alpha_1$	KL_{II} $K\alpha_2$	L_I edge	L_IN_{III} $L\gamma_3$	L_IM_{III} $L\beta_3$	L_IM_{II} $L\beta_4$	L_{II} edge	$L_{II}N_{IV}$ $L\gamma_1$
Intensity	—	2–5	~20	~10	100	50–53	—	~5	50–35	20	—	~5
4 Be	0.115				0.109							
5 B	0.188				0.183							
6 C	0.282				0.277							
7 N	0.397				0.393							
8 O	0.533				0.525							
9 F	0.692				0.677							
10 Ne	0.874		0.858		0.848							
11 Na	1.080		1.071		1.041							
12 Mg	1.309		1.302		1.253		0.062					
13 Al	1.562		1.557		1.487	1.486	0.087				0.076	
14 Si	1.840		1.836		1.740	1.739	0.118				0.101	
15 P	2.143		2.139		2.014	2.013	0.153				0.130	
16 S	2.471		2.464		2.308	2.307	0.193				0.164	
17 Cl	2.824		2.816		2.622	2.620	0.237				0.204	
18 Ar	3.203		3.190		2.958	2.956	0.286				0.247	
19 K	3.607		3.590		3.314	3.311	0.340				0.296	
20 Ca	4.034		4.013		3.692	3.688	0.403				0.346	
21 Sc	4.486		4.461		4.090	4.086	0.462				0.400	
22 Ti	4.965		4.932		4.511	4.505	0.529				0.460	
23 V	5.463		5.427		4.952	4.944	0.626		0.585		0.519	
24 Cr	5.987		5.947		5.415	5.405	0.694		0.654		0.582	
25 Mn	6.537		6.490		5.899	5.888	0.768		0.721		0.649	
26 Fe	7.112		7.058		6.404	6.391	0.846		0.792		0.721	
27 Co	7.712		7.649		6.930	6.915	0.929		0.870		0.797	
28 Ni	8.339		8.265		7.478	7.461	1.016		0.941		0.878	
29 Cu	8.993		8.905	8.903	8.048	8.028	1.109		1.023	1.019	0.965	
30 Zn	9.673	9.658[1]	9.572	9.567	8.639	8.616	1.208		1.107	1.102	1.057	
31 Ga	10.386	10.366[1]	10.271	10.261	9.252	9.231	1.316		1.197	1.191	1.155	
32 Ge	11.115	11.101[1]	10.983	10.978	9.887	9.856	1.426		1.294	1.289	1.259	
33 As	11.877	11.864[1]	11.727	11.721	10.544	10.509	1.536		1.386	1.380	1.368	
34 Se	12.666	12.652[1]	12.496	12.489	11.222	11.181	1.662		1.492	1.485	1.485	
35 Br	13.483	13.470[1]	13.292	13.285	11.924	11.878	1.791		1.600	1.593	1.605	
36 Kr	14.330	14.315[1]	14.113	14.105	12.650	12.598	1.923		1.706	1.698	1.732	
37 Rb	15.202	15.185[1]	14.962	14.952	13.396	13.336	2.067	2.051[2]	1.827	1.817	1.866	
38 Sr	16.106	16.085[1]	15.836	15.826	14.166	14.098	2.217	2.197[2]	1.947	1.937	2.008	
39 Y	17.037	17.015[1]	16.737	16.725	14.958	14.882	2.372	2.347[2]	2.072	2.060	2.155	
40 Zr	17.997	17.963[1]	17.662	17.649	15.770	15.692	2.535	2.503[2]	2.200	2.187	2.305	2.292
41 Nb	18.985	18.947[1]	18.623	18.606	16.615	16.521	2.698	2.660[2]	2.336	2.319	2.464	2.449
42 Mo	20.002	19.960	19.608	19.590	17.479	17.374	2.867	2.825[2]	2.473	2.455	2.628	2.611
43 Tc	21.048	21.002	20.619	20.599	18.367	18.251	3.047	3.001[2]	2.618	2.598	2.797	2.778
44 Ru	22.123	22.072	21.656	21.637	19.279	19.150	3.230	3.179[2]	2.763	2.744	2.973	2.952
45 Rh	23.229	23.173	22.723	22.698	20.216	20.073	3.421	3.365[2]	2.915	2.890	3.156	3.132
46 Pd	24.365	24.303	23.819	23.792	21.178	21.021	3.619	3.557	3.073	3.046	3.344	3.318
47 Ag	25.531	25.463	24.943	24.912	22.163	21.991	3.822	3.754	3.234	3.203	3.540	3.511
48 Cd	26.727	26.653	26.095	26.061	23.173	22.985	4.034	3.960	3.402	3.368	3.742	3.710
49 In	27.953	27.872	27.275	27.237	24.209	24.002	4.250	4.169	3.572	3.534	3.951	3.915
50 Sn	29.211	29.122	28.491	28.439	25.272	25.044	4.475	4.377	3.750	3.703	4.167	4.127

X-ray absorption edges, characteristic X-ray lines and fluorescence yields (*contd*)

Z	L_II M_IV Lβ₁	L_III edge	L_III N_V Lβ₂	L_III M_V Lα₁	L_III M_IV Lα₂	M_III edge	M_III N_V Mγ	M_IV edge	M_IV N_VI Mβ	M_V edge	M_V N_VII Mα₁	M_V N_VI Mα₂
	L-series (contd)					*M-series*						
Int.	100	—	~5	~90	10	—		—		—		
4		0.006										
5		0.005										
6		0.005										
7		0.004										
8		0.008										
9		0.015										
10		0.026										
11		0.039										
12		0.056										
13		0.075										
14		0.100										
15		0.129										
16		0.163										
17		0.202										
18		0.245										
19		0.293										
20		0.342										
21	0.400	0.396		0.395								
22	0.458	0.454		0.452								
23	0.519	0.511		0.511								
24	0.583	0.572		0.573								
25	0.649	0.638		0.637								
26	0.719	0.708		0.705								
27	0.791	0.782		0.776								
28	0.869	0.861		0.852								
29	0.950	0.945		0.930				0.015				
30	1.035	1.034		1.012				0.022				
31	1.125	1.134		1.098		0.115		0.030				
32	1.218	1.228		1.188		0.132		0.041				
33	1.316	1.333		1.282		0.150		0.052				
34	1.419	1.444		1.379		0.170		0.066				
35	1.523	1.559		1.480		0.191		0.082				
36	1.637	1.680		1.586		0.217		0.095				
37	1.752	1.806		1.694	1.692	0.240		0.114		0.112		
38	1.872	1.940		1.806	1.804	0.270		0.136		0.134		
39	1.996	2.079		1.923	1.920	0.300		0.159		0.156		
40	2.118	2.227	2.215	2.043	2.040	0.335	0.323	0.187		0.184		
41	2.257	2.370	2.357	2.166	2.163	0.362	0.349	0.207		0.204		
42	2.396	2.523	2.508	2.295	2.291	0.394	0.379	0.232		0.228		
43	2.537	2.681	2.664	2.424	2.421	0.429	0.412	0.260		0.257		
44	2.683	2.844	2.825	2.556	2.554	0.467	0.448	0.290		0.288		
45	2.835	3.013	2.992	2.698	2.692	0.506	0.485	0.321		0.315		
46	2.990	3.187	3.163	2.838	2.833	0.546	0.522	0.354		0.349		
47	3.151	3.368	3.342	2.985	2.979	0.588	0.562	0.389		0.383		
48	3.319	3.554	3.525	3.134	3.131	0.632	0.603	0.423		0.420		
49	3.487	3.744	3.712	3.288	3.280	0.678	0.646	0.464		0.456		
50	3.661	3.939	3.903	3.442	3.433	0.720	0.684	0.506		0.497		

X-ray absorption edges, characteristic X-ray lines and fluorescence yields (contd)

Atomic number and element	K-series						L-series					
	K edge	KN$_{III}$	KM$_{III}$	KM$_{II}$	KL$_{III}$	KL$_{II}$	L$_I$ edge	L$_I$N$_{III}$	L$_I$M$_{III}$	L$_I$M$_{II}$	L$_{II}$ edge	L$_{II}$N$_{IV}$
		Kβ$_2$	Kβ$_1$	Kβ$_3$	Kα$_1$	Kα$_2$		Lγ$_3$	Lβ$_3$	Lβ$_4$		Lγ$_1$
Intensity	—	5-15	~20	~10	100	53-65	—	~5	35-20	20	—	5-25
51 Sb	30.499	30.402	29.725	29.677	26.359	26.110	4.706	4.609	3.932	3.884	4.389	4.345
52 Te	31.817	31.712	30.995	30.944	27.472	27.201	4.942	4.837	4.120	4.069	4.616	4.568
53 I	33.168	33.054	32.295	32.239	28.612	28.317	5.186	5.072	4.313	4.257	4.851	4.799
54 Xe	34.551	34.428	33.625	33.562	29.779	29.459	5.442	5.319	4.516	4.453	5.092	5.035
55 Cs	35.966	35.833	34.985	34.918	30.973	30.625	5.700	5.567	4.719	4.652	5.341	5.278
56 Ba	37.414	37.270	36.378	36.303	32.194	31.817	5.964	5.820	4.928	4.853	5.597	5.529
57 La	38.894	38.739	37.802	37.721	33.442	33.034	6.235	6.080	5.143	5.062	5.860	5.786
58 Ce	40.410	40.243	39.258	39.170	34.720	34.279	6.516	6.349	5.364	5.276	6.131	6.051
59 Pr	41.958	41.778	40.748	40.653	36.026	35.550	6.802	6.622	5.592	5.497	6.408	6.321
60 Nd	43.538	43.345	42.272	42.166	37.361	36.847	7.095	6.902	5.829	5.723	6.691	6.597
61 Pm	45.152	44.947	43.825	43.713	38.725	38.171	7.398	7.193	6.071	5.959	6.981	6.880
62 Sm	46.801	46.584	45.413	45.289	40.118	39.523	7.707	7.490	6.319	6.195	7.278	7.169
63 Eu	48.486	48.256	47.036	46.902	41.542	40.902	8.024	7.794	6.574	6.440	7.584	7.467
64 Gd	50.207	49.964	48.696	48.554	42.996	42.309	8.343	8.100	6.832	6.690	7.898	7.772
65 Tb	51.965	51.709	50.382	50.228	44.481	43.744	8.679	8.423	7.096	6.942	8.221	8.086
66 Dy	53.761	53.491	52.119	51.956	45.999	45.208	9.013	8.743	7.371	7.208	8.553	8.409
67 Ho	55.593	55.308	53.878	53.707	47.547	46.699	9.365	9.080	7.650	7.479	8.894	8.740
68 Er	57.464	57.164	55.681	55.491	49.128	48.221	9.725	9.425	7.942	7.752	9.243	9.078
69 Tm	59.374	59.059	57.513	57.303	50.742	49.773	10.097	9.782	8.236	8.026	9.601	9.426
70 Yb	61.322	60.991	59.374	59.157	52.389	51.354	10.479	10.148	8.531	8.314	9.968	9.781
71 Lu	63.311	62.960	61.286	61.049	54.070	52.965	10.869	10.518	8.844	8.607	10.346	10.144
72 Hf	65.345	64.973	63.236	62.979	55.790	54.611	11.262	10.890	9.153	8.896	10.734	10.517
73 Ta	67.405	67.011	65.221	64.946	57.533	56.277	11.672	11.278	9.488	9.213	11.128	10.894
74 W	69.517	69.100	67.244	66.951	59.318	57.982	12.092	11.675	9.819	9.526	11.535	11.284
75 Re	71.670	71.230	69.309	68.994	61.140	59.718	12.522	12.082	10.161	9.846	11.952	11.682
76 Os	73.869	73.404	71.416	71.077	63.001	61.487	12.968	12.503	10.515	10.176	12.382	12.092
77 Ir	76.111	75.620	73.560	73.203	64.896	63.287	13.416	12.925	10.865	10.508	12.824	12.514
78 Pt	78.400	77.883	75.751	75.364	66.832	65.123	13.880	13.363	11.231	10.844	13.277	12.944
79 Au	80.729	80.182	77.985	77.580	68.804	66.990	14.353	13.806	11.609	11.204	13.739	13.383
80 Hg	83.109	82.532	80.261	79.822	70.819	68.894	14.835	14.258	11.987	11.548	14.215	13.834
81 Tl	85.532	84.924	82.575	82.384	72.872	70.832	15.344	14.736	12.387	12.196	14.700	14.293
82 Pb	88.008	87.367	84.936	84.450	74.969	72.804	15.863	15.222	12.791	12.305	15.204	14.769
83 Bi	90.540	89.866	87.354	86.831	77.118	74.815	16.391	15.717	13.205	12.682	15.725	15.261
84 Po	93.113	92.403	89.801	89.250	79.301	76.863	16.940	16.230	13.628	13.077	16.250	15.756
85 At	95.730	94.983	92.302	91.722	81.523	78.943	17.495	16.748	14.067	13.487	16.787	16.262
86 Rn	98.402	97.617	94.866	94.246	83.793	81.065	18.047	17.262	14.511	13.891	17.337	16.777
87 Fr	101.131	100.306	97.477	96.807	86.114	83.231	18.630	17.805	14.976	14.306	17.900	17.307
88 Ra	103.909	103.039	100.130	99.432	88.476	85.434	19.222	18.352	15.443	14.745	18.475	17.848
89 Ac	106.738	105.837	102.846	102.101	90.884	87.675	19.823	18.922	15.931	15.186	19.063	18.402
90 Th	109.641	108.690	105.611	104.831	93.358	89.952	20.449	19.498	16.419	15.639	19.689	18.993
91 Pa	112.599	111.606	108.435	107.606	95.883	92.287	21.088	20.095	16.924	16.095	20.312	19.581
92 U	115.606	114.561	111.303	110.424	98.440	94.659	21.757	20.712	17.454	16.575	20.947	20.167
93 Np	118.678	117.591	114.243	113.312	101.068	97.077	22.427	21.340	17.992	17.061	21.601	20.785
94 Pu	121.818	120.703	117.261	116.277	103.761	99.552	23.097	21.982	18.540	17.556	22.266	21.417
95 Am	125.027	123.891	120.360	119.317	106.523	102.083	23.773	22.637	19.106	18.063	22.944	22.065

X-ray absorption edges, characteristic X-ray lines and fluorescence yields (contd)

Z	$L_{II}M_{IV}$ $L\beta_1$	L_{III} edge	$L_{III}N_V$ $L\beta_2$	$L_{III}M_V$ $L\alpha_1$	$L_{III}M_{IV}$ $L\alpha_2$	M_{III} edge	$M_{III}N_V$ $M\gamma$	M_{IV} edge	$M_{IV}N_{VI}$ $M\beta$	M_V edge	M_VN_{VII} $M\alpha_1$	M_VN_{VI} $M\alpha_2$
Int.	100	—	5–20	~90	10	—		—		—		
51	3.843	4.140	4.101	3.604	3.594	0.774	0.735	0.546		0.536		
52	4.030	4.345	4.302	3.770	3.759	0.822	0.779	0.586		0.575		
53	4.221	4.556	4.509	3.938	3.926	0.873	0.826	0.630		0.618		
54	4.415	4.772	4.720	4.110	4.095	0.926	0.874	0.677		0.662		
55	4.619	4.993	4.936	4.289	4.271	0.981	0.924	0.722		0.704		
56	4.827	5.220	5.158	4.470	4.450	1.036	0.974	0.770		0.750		
57	5.037	5.452	5.385	4.651	4.629	1.092	1.025	0.823	0.854	0.801	0.833	
58	5.261	5.690	5.617	4.839	4.820	1.152	1.079	0.870	0.902	0.851	0.883	
59	5.485	5.932	5.853	5.034	5.009	1.210	1.131	0.923	0.950	0.898	0.929	
60	5.722	6.177	6.091	5.231	5.208	1.266	1.180	0.969	0.997	0.946	0.978	
61	5.962	6.427	6.334	5.433	5.408	1.327	1.234	1.019		0.994		
62	6.205	6.683	6.582	5.635	5.610	1.388	1.287	1.073	1.100	1.048	1.081	
63	6.455	6.944	6.835	5.843	5.815	1.450	1.341	1.129	1.153	1.101	1.131	
64	6.713	7.211	7.034	6.058	6.026	1.511	1.334	1.185	1.209	1.153	1.185	
65	6.976	7.484	7.358	6.273	6.239	1.583	1.457	1.245	1.266	1.211	1.240	
66	7.249	7.762	7.627	6.496	6.458	1.642	1.507	1.304	1.325	1.266	1.293	
67	7.529	8.046	7.901	6.719	6.681	1.715	1.570	1.365	1.383	1.327	1.348	
68	7.813	8.336	8.180	6.951	6.906	1.783	1.627	1.430	1.443	1.385	1.406	
69	8.103	8.632	8.465	7.181	7.134	1.861	1.694	1.498	1.503	1.451	1.462	
70	8.402	8.933	8.755	7.415	7.367	1.948	1.770	1.566	1.568	1.518	1.521	1.507
71	8.709	9.241	9.049	7.655	7.604	2.025	1.833	1.637	1.623	1.586		1.572
72	9.016	9.555	9.348	7.891	7.837	2.109	1.902	1.718	1.700	1.664		1.646
73	9.345	9.872	9.649	8.147	8.089	2.184	1.961	1.783	1.760	1.725		1.702
74	9.671	10.199	9.959	8.396	8.335	2.273	2.033	1.864	1.835	1.803	1.776	1.774
75	10.006	10.530	10.273	8.651	8.584	2.361	2.104	1.946	1.910	1.879	1.845	1.843
76	10.349	10.868	10.592	8.905	8.835	2.453	2.177	2.033	1.988	1.963	1.921	1.918
77	10.705	11.215	10.919	9.175	9.096	2.551	2.255	2.119	2.062	2.040	1.988	1.983
78	11.073	11.568	11.251	9.439	9.364	2.649	2.332	2.204	2.134	2.129	2.065	2.059
79	11.432	11.925	11.585	9.705	9.618	2.744	2.404	2.307	2.220	2.220	2.142	2.133
80	11.823	12.290	11.927	9.999	9.898	2.848	2.485	2.392	2.285	2.291	2.195	2.184
81	12.217	12.660	12.272	10.271	10.177	2.957	2.569	2.483	2.360	2.389	2.270	2.266
82	12.618	13.039	12.625	10.555	10.453	3.072	2.658	2.586	2.442	2.484	2.345	2.340
83	13.031	13.422	12.981	10.836	10.728	3.186	2.745	2.694	2.534	2.586	2.422	2.426
84	13.452	13.812	13.342	11.131	11.014	3.312	2.842	2.798	2.620	2.681	2.501	2.503
85	13.882	14.207	13.708	11.427	11.302	3.428	2.929	2.905	2.707	2.780	2.581	2.582
86	14.323	14.609	14.079	11.727	11.595	3.536	3.006	3.014	2.794	2.882	2.663	2.662
87	14.775	15.017	14.456	12.031	11.892	3.654	3.093	3.125	2.881	2.986	2.746	2.742
88	15.238	15.433	14.839	12.340	12.196	3.779	3.185	3.237	2.967	3.093	2.829	2.823
89	15.711	15.854	15.227	12.652	12.502	3.892	3.265	3.352	3.054	3.202	2.913	2.904
90	16.215	16.283	15.622	12.970	12.809	4.030	3.369	3.474	3.145	3.313	2.996	2.984
91	16.715	16.716	16.022	13.300	13.119	4.164	3.470	3.597	3.251	3.416	3.083	3.070
92	17.219	17.166	16.429	13.614	13.438	4.303	3.566	3.728	3.337	3.552	3.171	3.161
93	17.751	17.610	16.840	13.944	13.760	4.435	3.665	3.850	3.435	3.666	3.262	3.251
94	18.293	18.057	17.256	14.279	14.084	4.557	3.756	3.973	3.527	3.778	3.346	3.332
95	18.852	18.504	17.676	14.617	14.412	4.667	3.839	4.092	0.0	3.887	0.0	0.0

X-ray absorption edges, characteristic X-ray lines and fluorescence yields (contd)

Atomic number and element	K-series						L-series					
	K edge	KN_{III}	KM_{III}	KM_{II}	KL_{III}	KL_{II}	L_I edge	$L_I N_{III}$	$L_I M_{III}$	$L_I M_{II}$	L_{II} edge	$L_{II}N_{IV}$
		$K\beta_2$	$K\beta_1$	$K\beta_3$	$K\alpha_1$	$K\alpha_2$		$L\gamma_3$	$L\beta_3$	$L\beta_4$		$L\gamma_1$
Intensity	—	~15	~20	~10	100	~65	—	~5	~20	20	—	~25
96 Cm	128.220	127.066	123.423	122.325	109.290	104.441	24.460	23.306	19.663	18.565	23.779	
97 Bk	131.590	130.355	126.663	125.443	112.138	107.205	25.275	24.040	20.348	19.128	24.385	
98 Cf	135.960	134.681	130.851	129.601	116.030	110.710	26.110	24.831	21.001	19.751	25.250	
99 Es	139.490	138.169	134.238	132.916	119.080	113.470	26.900	25.579	21.648	20.326	26.020	
100 Fm	143.090	141.724	137.693	136.347	122.190	116.280	27.700	26.334	22.303	20.957	26.810	
101 Md	146.780	145.370	141.234	139.761	125.390	119.170	28.530	27.120	22.984	21.511	27.610	
102 No	150.540	149.092	144.852	143.295	128.660	122.100	29.380	27.932	23.692	22.135	28.440	
103 Lw	154.380	152.900	148.670	146.920	132.020	125.100	30.240	28.760	24.530	22.780	29.280	

Unresolved lines:

1—K $N_{II, III}$ ($K\beta_2$); 2—$L_I N_{II, III}$ ($L\gamma_{2, 3}$)

Depending on the resolving power of the dispersing system used (e.g. crystal spectrometer, solid state energy dispersive detector) line pairs shown separately in the table may not be resolved and the effective energy of the doublet will be close to the mean value weighted by the relative intensity of the components.

Z	L-series (contd)					M-series						
	$L_{II}M_{IV}$	L_{III}	$L_{III}N_V$	$L_{III}M_V$	$L_{III}M_{IV}$	M_{III}	$M_{IV}N_V$	M_{IV}	$M_{IV}N_{VI}$	M_V	$M_V N_{VII}$	$M_V N_{VI}$
	$L\beta_1$	edge	$L\beta_2$	$L\alpha_1$	$L\alpha_2$	edge	$M\gamma$	edge	$M\beta$	edge	$M\alpha_1$	$M\alpha_2$
Int.	~100	—	~20	~90	10	—		—		—		
96	19.552	18.930		14.959	14.703	4.797		4.227		3.971		
97	20.019	19.452		15.320	15.086	4.927		4.366		4.132		
98	20.763	19.930		15.677	15.443	5.109		4.487		4.253		
99	21.390	20.410		16.036	15.780	5.252		4.630		4.374		
100	22.044	20.900		16.402	16.134	5.397		4.766		4.498		
101	22.707	21.390		16.768	16.487	5.546		4.903		4.622		
102	23.403	21.880		17.139	16.843	5.688		5.037		4.741		
103	24.130	22.360		17.500	17.210	5.710		5.150		4.860		

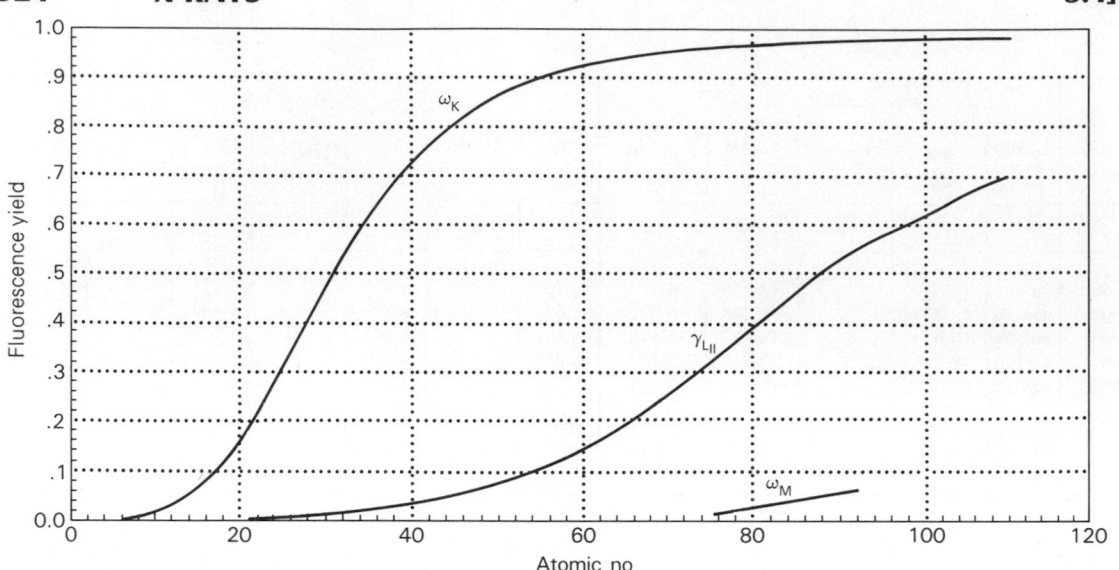

Variations of the fluorescence yields ω_K and ω_M for the K-shell and M-shell respectively and of the effective fluorescence yield γ_{LI} for the L_{II} shell with atomic number.

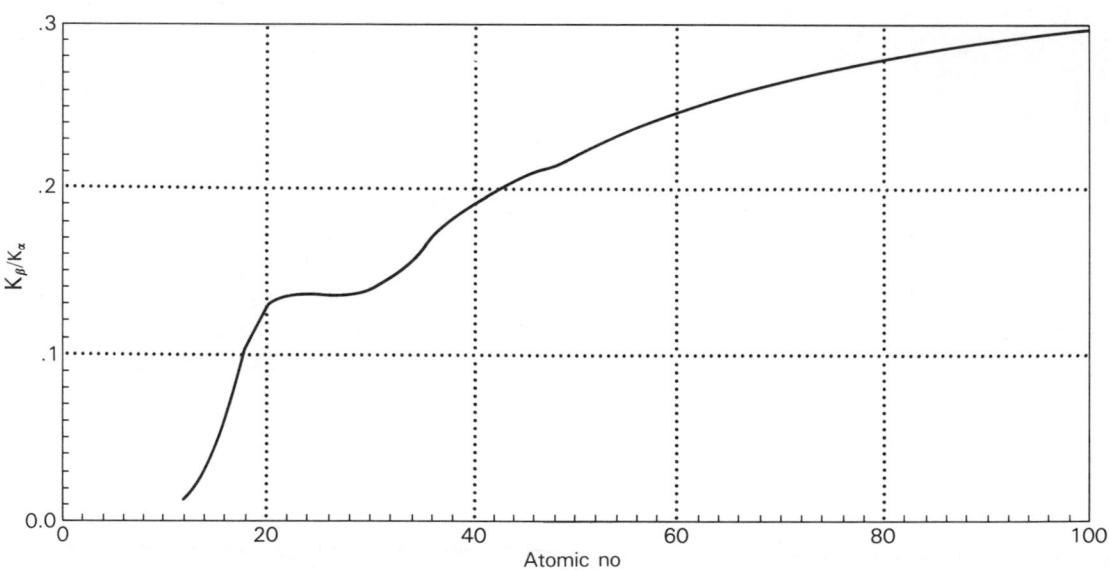

Variation of K_β/K_α with atomic number.

D.M.P.

3.4.2 Attenuation of photons

The intensity $I(t)$ of a photon beam after passing through t kg m^{-2} of matter is given by

$$I(t) = I(o)\exp(-(\mu/\rho)t)$$

where μ/ρ is the **mass absorption coefficient**. The values of μ/ρ given in the accompanying figure and table refer to the attenuation of a collimated beam of photons striking the attenuating material and an emerging beam consisting only of those photons which are undeflected and have suffered no energy loss. Inclusion of secondary products in the transmitted beam reduces this attenuation. The values given here are taken from Hubbell, *Int. J. Appl. Radioat. Isot.*, 1982, **33**, 1269 and Hubbell, Gimm and Overbo, *J. Phys. Chem. Ref. Data*, 1980, **9**, 1023.

The four main processes by which photons interact with matter are photoelectric absorption and coherent (Rayleigh) scattering which dominate at low energies, incoherent (Compton) scattering which is most important between about 0.2 and 5 MeV, and pair production which dominates at high energies. These four components can be obtained separately from the two references above. The photonuclear interaction is excluded from the values given here.

Mass absorption coefficients for composite materials may be calculated by taking the average value of the absorption coefficients for the constituents weighted in proportion to their abundance by weight.

Uncertainties in the mass absorption coefficients are estimated to be about $\pm 5\%$ up to 5 keV and about $\pm 2\%$ above 5 keV. Above 10 MeV the photonuclear interaction can contribute an extra ~ 1–5% to the mass absorption coefficient.

Mass absorption coefficient as a function of element and photon energy in units of $0.01\,\mathrm{m^2\,kg^{-1}}$.

MeV	Element											
	H	Be	C	Al	Fe	Ge	Ag	I	W	Pb	U	Con-crete†
0.2	2.429	1.089	1.229	1.223	1.458	1.658	2.963	3.650	7.844	9.985	12.98	1.27
0.3	2.112	0.946	1.066	1.042	1.098	1.130	1.557	1.768	3.238	4.026	5.191	1.08
0.4	1.893	0.847	0.954	0.928	0.940	0.933	1.130	1.215	1.925	2.323	2.922	0.96
0.5	1.729	0.774	0.871	0.845	0.841	0.821	0.931	0.969	1.378	1.613	1.976	0.88
0.6	1.599	0.716	0.806	0.780	0.770	0.745	0.814	0.830	1.093	1.248	1.490	0.81
0.8	1.405	0.629	0.708	0.684	0.670	0.643	0.676	0.674	0.806	0.887	1.016	0.71
1.0	1.263	0.565	0.636	0.615	0.599	0.573	0.592	0.584	0.662	0.710	0.789	0.64
1.5	1.027	0.460	0.518	0.501	0.488	0.466	0.475	0.465	0.500	0.522	0.559	0.52
2.0	0.877	0.394	0.444	0.432	0.426	0.409	0.421	0.412	0.443	0.461	0.488	0.45
3.0	0.692	0.314	0.356	0.354	0.362	0.352	0.375	0.372	0.408	0.423	0.445	0.37
4.0	0.581	0.266	0.305	0.311	0.331	0.328	0.361	0.361	0.404	0.420	0.439	0.32
5.0	0.505	0.235	0.271	0.284	0.315	0.316	0.358	0.361	0.410	0.427	0.446	0.29
6.0	0.450	0.212	0.247	0.265	0.306	0.311	0.360	0.366	0.421	0.439	0.458	0.27
8.0	0.375	0.182	0.215	0.244	0.299	0.310	0.372	0.382	0.447	0.468	0.488	0.24
10.0	0.325	0.163	0.196	0.232	0.299	0.316	0.388	0.400	0.475	0.497	0.519	0.23
15.0	0.254	0.136	0.170	0.220	0.309	0.334	0.428	0.446	0.538	0.566	0.593	0.22
20.0	0.215	0.123	0.158	0.217	0.322	0.353	0.461	0.482	0.589	0.620	0.651	0.21
30.0	0.175	0.110	0.147	0.220	0.347	0.385	0.513	0.540	0.665	0.702	0.739	0.21
40.0	0.154	0.104	0.144	0.225	0.367	0.410	0.551	0.581	0.720	0.761	0.802	0.21
50.0	0.142	0.102	0.143	0.231	0.383	0.430	0.581	0.613	0.762	0.806	0.849	0.22
60.0	0.134	0.100	0.143	0.236	0.396	0.446	0.604	0.638	0.795	0.841	0.887	0.22
80.0	0.124	0.099	0.144	0.245	0.417	0.471	0.640	0.676	0.844	0.893	0.943	0.23
100.0	0.119	0.099	0.146	0.252	0.433	0.489	0.666	0.704	0.880	0.931	0.983	0.23

† Approximate composition by weight of concrete: 50% oxygen, 31% silicon, 8% calcium, 5% aluminium, 6% other materials.

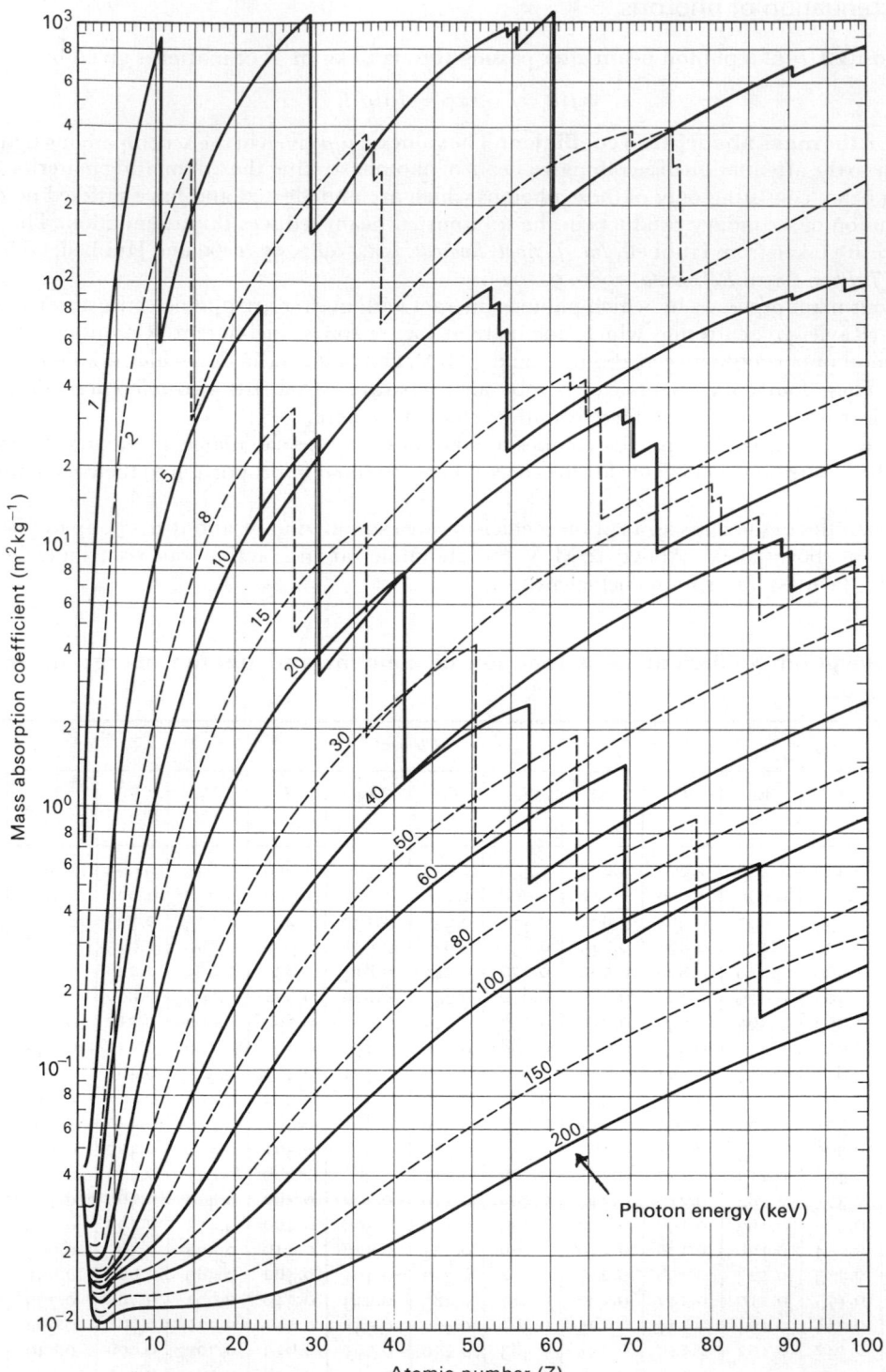

Mass absorption coefficient (μ/ρ) **as a function of element for a set of photon energies between 1 and 200 keV.**

D.J.S.F.

3.5 Absorption of particles and dosimetry

3.5.1 Range and stopping power of ions in various materials

When a fast ion passes through matter, it loses energy principally by scattering from atomic electrons and, more importantly at low energies, by scattering from atomic nuclei.

The **mean range** of an ion is defined as the mean thickness of material, which is usually expressed in terms of mass per unit area, traversed before it comes to rest. The **stopping power** (S) of a material for an ion is defined as the rate of change of the energy $(-dE/dT)$ with material thickness (T).

For very thin absorbers, the energy loss of an ion is approximately ST. For thick absorbers, the energy loss ΔE is better obtained from the range tables (p. 328) as follows. Let R_0 be the mean range of an ion of energy E_0 in a given absorber material. After passing through an absorber of thickness T, it emerges as an ion with a mean range $(R_0 - T)$ and a corresponding energy E_1, from which the energy lost in passing through the absorber is seen to be $\Delta E = E_0 - E_1$.

Statistical fluctuations in the processes of energy loss result in a distribution about the mean energy loss in passing through the absorber and also about the mean range. For light ions, the fractional variation about the mean range is of the order of 10% for low energies, falling to a few per cent at high energies. The fractional variation in energy loss is at least a few per cent and can be 40% for very thin absorbers and high energies. A more detailed treatment of energy loss variations for thick absorbers is given by Tschalär in *Nuclear Instruments and Methods*, Vol. 64, p. 237, 1968.

The stopping power goes through a maximum at an energy of about 0.1 MeV/amu for protons to about 5 MeV/amu for uranium ions, depending also on the atomic number of the absorber. This maximum is evident in the table of stopping power in air of protons and alpha particles as a function of distance from end of range.

The proton range and stopping powers were derived from J. F. Janni, *Atomic Data and Nuclear Data Tables*, 1982, Vol. 27, Nos. 2–5, which tabulate values for 92 elements and 63 compounds. For α-particles, the data were based on *The Stopping and Ranges of Ions in Matter*, ed. J. F. Ziegler, vols. 5 and 6, 1980, (Pergamon Press), where ranges and stopping powers in 92 elements are plotted.

For pi-mesons and muons, the data in the tables were derived from Janni's proton tables by using the following scaling relations. For particles of charge ze, mass M and energy E, the stopping power, S, for a given material is given by

$$[S(E)]_{M,ze} = z^2 [S(EM_p/M)]_{proton}$$

where M_p is the proton mass.

Similarly, their range is given by:

$$[R(E)]_{M,ze} = M/(M_p z^2) [R(EM_p/M)]_{proton}$$

These formulae are based on the assumption that the rate of energy loss for a particle of a given charge in a given medium depends only on its velocity, and are generally accepted as accurate for $z \leqslant 2$, and above 1 MeV/amu. They can be used to extend the above tables to other isotopes of hydrogen and helium and to other light elementary particles.

The range/energy functions have a power law relationship varying approximately between $R \propto E^{1/2}$ and $E^{5/3}$ below and above the stopping power maximum, respectively. Interpolation in energy should therefore be made by interpolating logarithms of the range against logarithms in energy.

Stopping power (keV/mm) as a function of distance from end of range

Distance (mm)	50	30	20	15	10	7	5	3	2	1.5	1.0	0.5
α	81	105	126	146	179	214	254	252	201	168	123	72
p	20	24	29	33	39	47	55	70	82	91	95	77

Ranges of light ions

Energy MeV	$Range/(kg\,m^{-2})$ for protons in:										
	H	C	Al	Ti	Cu	Ag	Pb	U	Water	Air	Multiplier
0.005	0.37	1.76	2.15	2.76	3.14	3.38	4.87	4.75	1.32	1.79	10^{-4}
0.01	0.62	3.02	3.84	5.19	6.00	6.23	9.24	8.65	2.30	3.10	10^{-4}
0.05	0.19	0.94	1.32	1.96	2.51	2.58	4.04	3.61	0.71	0.97	10^{-3}
0.10	0.32	1.61	2.41	3.36	4.66	4.93	7.32	6.82	1.22	1.65	10^{-3}
0.5	0.27	0.96	1.50	1.76	2.44	2.84	4.02	4.05	0.80	1.01	10^{-2}
1.0	0.88	2.75	3.93	4.76	6.02	7.26	10.3	10.6	2.40	2.95	10^{-2}
2.0	2.94	8.44	11.4	13.5	16.3	20.3	28.4	29.2	7.60	9.20	10^{-2}
5.0	1.52	4.02	5.13	5.99	6.87	8.39	11.3	11.8	3.68	4.36	10^{-1}
10.0	5.40	13.6	16.8	19.4	21.7	25.8	33.9	35.6	12.5	14.6	10^{-1}
20.0	1.93	4.71	5.68	6.45	7.08	8.23	10.6	11.1	4.31	5.02	10^{0}
50.0	10.4	24.6	29.0	32.4	35.0	39.7	49.5	51.5	22.5	26.0	10^{0}
100	3.66	8.54	9.90	11.0	11.7	13.2	16.1	16.8	7.79	8.97	10^{1}
200	12.5	28.7	33.0	36.4	38.6	42.9	51.8	53.7	26.2	30.0	10^{1}
500	5.69	13.0	14.7	16.1	17.0	18.7	22.3	23.1	11.8	13.5	10^{2}
1000	1.59	3.60	4.08	4.45	4.67	5.12	6.05	6.25	3.28	3.74	10^{3}

Energy MeV	$Range/(kg\,m^{-2})$ for alphas in:										
	H	C	Al	Ti	Cu	Ag	Pb	U	Water	Air	Multiplier
0.01	9.0	18.8	35.1	35.7	47.4	47.1	56.6	61.0	16.9	28.8	10^{-5}
0.05	3.0	6.6	10.9	13.6	20.6	20.9	27.2	28.6	5.95	9.57	10^{-4}
0.10	4.3	11.1	16.7	22.1	34.9	36.6	47.5	49.5	10.0	15.7	10^{-4}
0.5	1.12	3.6	4.9	6.6	9.2	12.0	17.0	15.8	3.24	3.92	10^{-3}
1.0	1.71	5.95	8.77	10.8	17.9	19.4	28.3	27.6	4.95	6.53	10^{-3}
2.0	3.40	12.3	17.8	20.2	32.0	36.4	52.7	54.9	10.3	12.8	10^{-3}
5.0	1.28	4.23	5.85	6.75	9.08	10.9	15.5	16.5	3.68	4.35	10^{-2}
10.0	4.35	12.4	16.3	19.5	23.6	28.9	42.3	43.8	11.3	12.7	10^{-2}
20.0	1.51	3.97	4.97	6.00	6.84	83.9	11.7	12.3	3.73	4.14	10^{-1}
50.0	0.80	2.03	2.37	2.85	3.15	3.79	5.00	5.22	1.86	2.17	10^{0}
100	2.90	7.18	8.10	9.60	10.6	12.4	15.7	16.7	6.45	7.48	10^{0}
200	1.03	2.58	2.80	3.25	3.61	4.11	5.00	5.23	2.25	2.60	10^{1}
500	5.42	13.3	14.2	16.3	17.9	19.5	23.9	24.7	11.6	13.3	10^{1}
1000	1.83	4.28	4.64	5.40	5.70	5.99	7.55	7.85	3.83	4.39	10^{2}

Ranges of light ions (*contd*)

Energy MeV	Range/(kg m^{-2}) for pions in:										
	H	C	Al	Ti	Cu	Ag	Pb	U	Water	Air	Multiplier
1.0	3.88	10.0	12.6	14.7	16.6	20.1	26.7	28.0	9.19	10.8	10^{-2}
2.0	1.38	3.43	4.20	4.81	5.33	6.28	8.15	8.57	3.15	3.67	10^{-1}
5.0	7.45	17.9	21.2	23.9	26.0	29.8	37.5	39.2	16.4	18.9	10^{-1}
10.0	2.65	6.24	7.29	8.13	8.73	9.86	12.2	12.7	5.70	6.57	10^{0}
20.0	9.22	21.4	24.7	27.4	29.1	32.5	39.6	41.1	19.5	22.5	10^{0}
50.0	4.46	10.2	11.7	12.8	13.5	15.0	17.9	18.5	9.30	10.7	10^{1}
100	13.3	30.2	34.3	37.5	39.5	43.4	51.5	53.3	27.5	31.5	10^{1}
200	3.55	7.98	9.03	9.84	10.3	11.3	13.3	13.7	7.28	8.29	10^{2}
500	10.8	24.1	27.3	29.6	30.8	33.7	39.2	40.5	22.1	25.0	10^{2}
1000	22.6	50.2	57.0	61.8	64.0	69.7	80.3	82.8	46.4	51.7	10^{2}

Energy MeV	Range/kg m^{-2}) for muons in:										
	H	C	Al	Ti	Cu	Ag	Pb	U	Water	Air	Multiplier
1.0	4.89	12.4	15.4	17.8	20.0	23.9	31.5	33.1	11.4	13.4	10^{-2}
2.0	1.74	4.28	5.19	5.90	6.50	7.58	9.76	10.2	3.92	4.57	10^{-1}
5.0	9.44	22.5	26.5	29.7	32.1	36.5	45.7	47.7	20.6	23.7	10^{-1}
10.0	3.32	7.78	9.05	10.1	10.8	12.1	14.8	15.4	7.10	8.18	10^{0}
20.0	11.4	26.4	30.3	33.5	35.5	39.5	47.9	49.6	24.0	27.6	10^{0}
50.0	5.32	12.1	13.8	15.2	16.0	17.6	21.0	21.8	11.1	12.7	10^{1}
100	15.2	34.3	38.9	42.5	44.6	48.9	57.9	59.8	31.3	35.7	10^{1}
200	3.85	8.63	9.76	10.6	11.1	12.1	14.3	14.7	7.88	8.95	10^{2}
500	11.1	24.8	28.1	30.4	31.6	34.5	40.0	41.2	22.8	25.6	10^{2}
1000	22.7	50.2	57.0	61.8	63.9	69.5	79.8	82.2	46.6	51.5	10^{2}

Stopping powers of light ions

Energy MeV	Stopping power/(MeV kg^{-1} m^{2}) for protons in:										
	H	C	Al	Ti	Cu	Ag	Pb	U	Water	Air	Multiplier
0.005	17.1	2.71	1.73	0.991	0.700	0.647	0.287	0.373	3.80	2.56	10^{1}
0.01	22.0	4.01	2.54	1.41	1.03	0.967	0.401	0.543	5.13	3.67	10^{1}
0.05	38.9	7.14	4.48	2.98	1.91	1.72	0.963	1.08	9.72	7.07	10^{1}
0.10	34.1	7.34	4.31	3.50	2.16	1.86	1.23	1.23	9.53	7.18	10^{1}
0.5	11.0	3.64	2.51	2.08	1.65	1.33	0.893	0.863	4.15	3.40	10^{1}
1.0	66.3	22.9	17.2	14.1	11.8	9.28	6.39	6.15	25.4	21.0	10^{0}
2.0	38.5	14.3	11.0	9.35	8.03	6.38	4.54	4.33	15.5	13.0	10^{0}
5.0	18.1	7.14	5.74	4.96	4.42	3.67	2.74	2.60	7.78	6.63	10^{0}
10.0	10.1	4.13	3.40	2.98	2.71	2.31	1.77	1.68	4.50	3.87	10^{0}
20.0	5.64	2.35	1.98	1.76	1.62	1.41	1.11	1.06	2.57	2.22	10^{0}
50.0	26.3	11.2	9.66	8.68	8.10	7.19	5.80	5.57	12.3	10.7	10^{-1}
100	15.2	6.58	5.72	5.17	4.86	4.36	3.57	3.44	7.22	6.29	10^{-1}
200	9.28	4.06	3.55	3.23	3.05	2.76	2.29	2.20	4.45	3.89	10^{-1}
500	5.60	2.48	2.19	2.00	1.90	1.73	1.45	1.40	2.72	2.38	10^{-1}
1000	4.48	2.00	1.77	1.62	1.55	1.41	1.20	1.16	2.19	1.93	10^{-1}

Stopping powers of light ions (*contd*)

Energy MeV	Stopping power/(MeV kg^{-1} m^2) *for alphas in:*										
	H	C	Al	Ti	Cu	Ag	Pb	U	Water	Air	Multiplier
0.01	12.7	5.67	2.63	2.09	1.13	1.06	0.64	0.54	5.00	3.48	10^1
0.05	28.1	9.8	6.41	4.40	2.35	2.17	1.40	1.31	10.5	7.4	10^1
0.10	41.9	12.5	9.04	8.35	3.31	2.99	1.97	1.93	16.3	10.4	10^1
0.50	80.6	19.1	13.0	11.0	6.51	5.75	3.81	4.20	25.1	19.7	10^1
1.0	71.1	18.1	12.3	10.6	7.06	6.25	4.01	4.45	22.2	19.2	10^1
2.0	46.1	13.6	9.87	8.28	6.28	5.29	3.49	3.57	16.0	13.9	10^1
5.0	22.6	7.72	6.05	5.08	4.28	3.43	2.40	2.27	8.81	7.62	10^1
10.0	13.0	4.79	3.79	3.21	2.81	2.28	1.61	1.54	5.27	4.69	10^1
20.0	7.41	2.85	2.28	1.92	1.71	1.42	1.10	0.99	3.11	2.81	10^1
50.0	33.8	13.6	11.4	10.0	9.19	7.87	6.10	5.82	15.0	13.4	10^0
100	18.8	7.77	6.61	5.90	5.46	4.75	3.81	3.62	8.59	7.62	10^0
200	10.5	4.45	3.82	3.45	3.21	2.87	2.29	2.25	4.92	4.36	10^0
500	5.13	2.18	1.91	1.76	1.63	1.48	1.23	1.19	2.45	2.14	10^0
1000	3.22	1.41	1.23	1.12	1.06	0.96	0.80	0.78	1.55	1.38	10^0

Energy MeV	Stopping power/(MeV kg^{-1} m^2) *for pions in:*										
	H	C	Al	Ti	Cu	Ag	Pb	U	Water	Air	Multiplier
1.0	14.1	5.66	4.61	4.00	3.61	3.03	2.29	2.17	6.17	5.28	10^0
2.0	7.88	3.25	2.71	2.39	2.18	1.88	1.46	1.38	3.55	3.06	10^0
5.0	36.5	15.5	13.2	11.8	10.9	9.63	7.70	7.37	16.9	14.7	10^{-1}
10.0	20.7	8.91	7.69	6.93	6.50	5.79	4.70	4.52	9.76	8.64	10^{-1}
20.0	12.2	5.33	4.63	4.20	3.96	3.56	2.93	2.82	5.83	5.08	10^{-1}
50.0	6.80	2.99	2.63	2.40	2.28	2.06	1.72	1.66	3.28	2.87	10^{-1}
100	5.00	2.22	1.96	1.80	1.71	1.56	1.31	1.27	2.44	2.14	10^{-1}
200	4.26	1.91	1.69	1.55	1.49	1.36	1.16	1.12	2.08	1.84	10^{-1}
500	4.12	1.87	1.64	1.51	1.47	1.34	1.16	1.13	2.02	1.81	10^{-1}
1000	4.34	1.96	1.72	1.59	1.54	1.42	1.25	1.21	2.09	1.93	10^{-1}

Energy MeV	Stopping power/(MeV kg^{-1} m^2) *for muons in:*										
	H	C	Al	Ti	Cu	Ag	Pb	U	Water	Air	Multiplier
1.0	11.2	4.54	3.73	3.26	2.95	2.51	1.92	1.81	4.94	4.24	10^0
2.0	6.23	2.59	2.18	1.93	1.77	1.54	1.21	1.15	2.83	2.44	10^0
5.0	28.9	12.3	10.6	9.48	8.85	7.83	6.30	6.04	13.5	11.7	10^{-1}
10.0	16.7	7.19	6.24	5.63	5.29	4.74	3.87	3.72	7.88	6.86	10^{-1}
20.0	10.0	4.38	3.83	3.48	3.29	2.97	2.46	2.37	4.81	4.20	10^{-1}
50.0	5.90	2.60	2.30	2.10	1.99	1.81	1.52	1.47	2.86	2.50	10^{-1}
100	4.61	2.05	1.81	1.66	1.59	1.45	1.23	1.18	2.25	1.98	10^{-1}
200	4.14	1.86	1.64	1.51	1.46	1.33	1.14	1.10	2.03	1.80	10^{-1}
500	4.19	1.90	1.67	1.54	1.49	1.37	1.19	1.16	2.04	1.85	10^{-1}
1000	4.47	2.01	1.76	1.63	1.58	1.47	1.29	1.26	2.13	2.00	10^{-1}

For heavy ions, the range data were derived from the semi-empirical calculations in U. Littmark and J. F. Ziegler, *The Range and Energy Loss of Particles in Matter*, vol. 6, 1980, ed. J. F. Ziegler, (Pergamon Press), where ranges of ions from protons to uranium in all elemental substrates are plotted graphically from 0.2 MeV/amu up to 1000 MeV. The authors account for shell effects in the atomic number of the substrate by scaling to detailed α-particle stopping powers in which those effects are experimentally observed.

Where the tables extend beyond the range of Littmark and Ziegler, extrapolation is made using, as a guide, L. C. Northcliffe and R. F. Schilling, *Nuclear Data Tables*, vol. 7A, 1970, Nos. 3–4, where calculated ranges and stopping powers for 103 species in 24 materials are tabulated. These tables do not include atomic shell effects in the substrate and thus were used only to provide the shape of the range curves in the extrapolation region.

There are limited experimental data, mainly for the lighter heavy ions, in a few substrates; agreement with the curves is generally within 10%, which is comparable with scatter in the data. There is little, if any, experimental data for heavier ions at the higher energies.

Ranges of heavy ions

Energy per mass unit	Ranges of heavy ions						
MeV/amu	$Range/(\text{kg m}^{-2} \times 10^{-2})$ *for* ^4_2He *ions in*:						
	C	Al	Ti	Ge	Ag	Au	U
0.0125	0.084	0.120	0.150	0.25	0.30	0.58	0.47
0.050	0.20	0.28	0.36	0.56	0.68	1.20	1.05
0.200	0.50	0.72	0.85	1.35	1.60	2.55	2.35
0.800	2.20	3.10	3.46	5.2	6.0	9.4	9.3
3.20	18.4	24.5	28.7	35.2	42.2	58.5	61.5
12.00	189	233	262	305	357	459	490
MeV/amu	$Range/(\text{kg m}^{-2} \times 10^{-2})$ *for* $^{12}_6\text{C}$ *ions in*:						
	C	Al	Ti	Ge	Ag	Au	U
0.0125	0.065	0.096	0.11	0.19	0.24	0.35	0.36
0.050	0.17	0.25	0.27	0.47	0.58	0.82	0.80
0.200	0.52	0.73	0.90	1.15	1.32	1.90	1.80
0.800	1.48	2.10	2.35	3.55	3.95	6.25	6.15
3.20	8.3	10.8	12.9	16.0	19.2	27.0	28.1
12.00	65.5	81.0	91.2	109	128	163	179
MeV/amu	$Range/(\text{kg m}^{-2} \times 10^{-2})$ *for* $^{27}_{13}\text{Al}$ *ions in*:						
	C	Al	Ti	Ge	Ag	Au	U
0.0125	0.087	0.12	0.15	0.21	0.24	0.36	0.34
0.050	0.24	0.34	0.41	0.62	0.64	1.00	0.99
0.200	0.59	0.85	0.94	1.47	1.58	2.40	2.28
0.800	1.53	2.17	2.43	3.65	4.02	6.27	6.12
3.20	6.33	8.2	9.51	12.2	14.5	20.5	21.0
12.00	38.3	48.0	54.1	66.0	72.2	97.0	104

Ranges of heavy ions (contd)

Energy per mass unit	Ranges of heavy ions						
MeV/amu	$Range/(kg\ m^{-2} \times 10^{-2})$ for $^{40}_{18}Ar$ ions in:						
	C	Al	Ti	Ge	Ag	Au	U
0.0125	0.090	0.11	0.13	0.21	0.26	0.39	0.38
0.050	0.28	0.33	0.43	0.63	0.73	1.18	1.05
0.200	0.67	0.97	1.08	1.68	1.83	2.76	2.63
0.800	1.68	2.40	2.65	4.05	4.40	7.0	6.80
3.20	5.99	8.0	9.20	12.0	14.3	20.5	21.0
12.00	34.6	42.0	47.0	55.0	62.8	83.0	89.5
MeV/amu	$Range/(kg\ m^{-2} \times 10^{-2})$ for $^{58}_{28}Ni$ ions in:						
	C	Al	Ti	Ge	Ag	Au	U
0.0125	0.11	0.15	0.14	0.23	0.25	0.41	0.39
0.050	0.36	0.46	0.49	0.78	0.84	1.32	1.25
0.200	0.88	1.15	1.38	2.10	2.27	3.35	3.22
0.800	1.80	2.71	3.05	4.55	5.05	7.65	7.50
3.20	5.6	7.6	8.8	11.6	13.2	19.3	19.9
12.00	27.2	31.5	36.5	43.5	49.0	65.5	70.5
MeV/amu	$Range/(kg\ m^{-2} \times 10^{-2})$ for $^{74}_{32}Ge$ ions in:						
	C	Al	Ti	Ge	Ag	Au	U
0.0125	0.13	0.16	0.15	0.25	0.28	0.42	0.38
0.050	0.43	0.54	0.56	0.88	0.96	1.38	1.35
0.200	1.02	1.43	1.57	2.39	2.57	3.79	3.65
0.800	2.18	3.08	3.44	5.09	5.65	8.6	8.4
3.20	6.15	8.3	9.7	12.7	14.8	21.2	21.7
12.00	27.2	33.5	38.3	46.5	53.5	69.8	74.5
MeV/amu	$Range/(kg\ m^{-2} \times 10^{-2})$ for $^{90}_{40}Zr$ ions in:						
	C	Al	Ti	Ge	Ag	Au	U
0.0125	0.11	0.14	0.15	0.21	0.23	0.36	0.35
0.050	0.36	0.49	0.53	0.79	0.87	1.31	1.24
0.200	0.93	1.32	1.46	2.23	2.42	3.63	3.52
0.800	2.16	3.07	3.40	5.10	5.64	8.75	8.65
3.20	6.1	8.1	9.5	12.5	14.5	20.8	21.6
12.00	25.0	30.5	36.0	41.5	49.8	63.0	67.5
MeV/amu	$Range/(kg\ m^{-2} \times 10^{-2})$ for $^{107}_{47}Ag$ ions in:						
	C	Al	Ti	Ge	Ag	Au	U
0.0125	0.12	0.16	0.15	0.24	0.27	0.38	0.38
0.050	0.42	0.55	0.60	0.91	1.00	1.47	1.55
0.200	1.12	1.56	1.73	2.60	2.82	4.17	4.04
0.800	2.44	3.56	3.86	5.75	6.37	9.70	9.60
3.20	6.18	8.30	9.70	12.7	14.8	21.4	20.3
12.00	26.0	31.8	33.8	41.9	49.1	65.0	69.5

Ranges of heavy ions (*contd*)

Energy per mass unit	Ranges of heavy ions						
MeV/amu	$Range/(\text{kg m}^{-2} \times 10^{-2})$ for $^{153}_{63}$Eu ions in:						
	C	Al	Ti	Ge	Ag	Au	U
0.0125	0.13	0.16	0.16	0.24	0.26	0.36	0.34
0.050	0.45	0.60	0.65	0.96	1.00	1.52	1.45
0.200	1.23	1.72	1.92	2.86	3.11	4.66	4.49
0.800	2.81	3.94	4.39	6.55	7.20	11.1	11.0
3.20	7.25	9.68	11.2	14.8	17.1	24.7	25.5
12.00	25.8	33.1	35.4	43.7	51.9	67.2	72.5
MeV/amu	$Range/(\text{kg m}^{-2} \times 10^{-2})$ for $^{181}_{73}$Ta ions in:						
	C	Al	Ti	Ge	Ag	Au	U
0.0125	0.14	0.18	0.17	0.28	0.28	0.37	0.36
0.050	0.53	0.68	0.70	1.13	1.12	1.65	1.59
0.200	1.42	1.96	2.16	3.43	3.54	5.17	5.12
0.800	3.26	4.51	5.02	7.32	8.08	12.3	12.2
3.20	7.81	10.4	12.0	15.8	18.4	26.5	27.3
12.00	26.0	32.0	36.9	43.2	53.1	69.0	75.2
MeV/amu	$Range/(\text{kg m}^{-2} \times 10^{-2})$ for $^{197}_{79}$Au ions in:						
	C	Al	Ti	Ge	Ag	Au	U
0.0125	0.15	0.19	0.20	0.27	0.29	0.38	0.36
0.050	0.54	0.72	0.79	1.09	1.18	1.67	1.64
0.200	1.47	2.16	2.39	3.47	3.76	5.53	5.40
0.800	3.37	4.67	5.20	7.62	8.34	12.7	12.8
3.20	8.15	10.8	12.4	16.3	19.0	27.4	28.3
12.00	27.0	33.2	36.5	45.4	53.4	70.3	76.0
MeV/amu	$Range/(\text{kg m}^{-2} \times 10^{-2})$ for $^{238}_{92}$U ions in:						
	C	Al	Ti	Ge	Ag	Au	U
0.0125	0.16	0.19	0.20	0.27	0.29	0.36	0.35
0.050	0.60	0.75	0.80	1.12	1.21	1.66	1.65
0.200	1.68	2.28	2.53	3.64	3.97	5.73	5.67
0.800	3.67	5.10	5.68	8.27	9.11	13.7	13.7
3.20	8.65	11.4	13.3	17.5	20.2	29.0	29.9
12.00	27.3	33.5	36.6	46.3	53.9	70.8	76.5

J.A.

3.5.2 Inelastic mean free path of electrons in solids

The electron inelastic mean free path (IMFP), λ, is defined as the thickness of material through which electrons may pass with a probability e^{-1} that they survive without inelastic scattering. Conventionally the inelastic scattering is considered to be significant only for energy losses ≥ 1 eV, phonon excitations being ignored. (The IMFP should not be confused with the total electron range which may be 10 to 100 times greater.) The value of λ depends on both the material and electron energy. For electrons emitted at an angle θ to the normal from a solid surface, the unscattered intensity from a source at a depth z below the surface follows the approximate relation

$$I = I_0 \exp(-z/\lambda\cos\theta)$$

Thus, in Auger and X-ray photoelectron spectra from solid surfaces, the information depth is characterized by $\lambda\cos\theta$.

For elements, an empirical and approximate description of λ in nm is given by

$$\lambda = 538\, a E^{-2} + 0.41\, a^{3/2}\, E^{1/2}$$

where $E\,eV$ is the electron energy and $a^3\,nm^3$ is the volume of one atom of the element in the solid state. This description fits the experimental data on average to a standard deviation of 30%.

A selection of recent measurements of λ is given in the table, either as the values at certain energies $E\,eV$ or, when sufficient data exists in the energy range 200–2000 eV, as the value at 1 keV and the power of the energy dependence where n is defined in the equation,

$$\lambda = \lambda(1\,\text{keV})(E/1000)^n$$

References

M. P. Seah and W. A. Dench, *Surface and Interface Analysis*, 1979, **1**, 2–11.
C. D. Wagner, L. E. Davis and W. M. Riggs, *Surface and Interface Analysis*, 1980, **2**, 53–55.

Inelastic mean free path of electrons

Element	$\lambda/$(nm) at 1 keV	n	$\lambda/$(nm), $E/$(eV)	Inorganic compound	$\lambda/$(nm) at 1 keV	n	$\lambda/$(nm), $E/$(eV)
Ag	1.5	0.50		Al_2O_3	1.4	0.54	
Al	1.9	0.74		Cr_2O_3	1.9	0.52	
Au	1.7	0.50		Fe_3O_4			2.3, 519
Be	1.6	0.50		GeO_2			0.6, 250
C(amorphous)							
			1.4, 1000	KI	6.3	0.59	
C(diamond)			2.0, 1000	NaCl	3.8	0.56	
C(graphite)			3.7, 1000	NaF	4.2	0.56	
Co			1.2, 1194	NiO			1.0, 511
Cr			1.1, 1211	SiO_2	2.2	0.70	
Cs			2.6, 1260	WO_3			2.6, 1450
Cu	1.4	0.75		*Organic compound*			
Fe			1.3, 1200	alpha-iodostearic acid			14.0, 867
Ge	2.3	0.60		barium stearate	5.0	0.62	
In			1.0, 408	butylamine	4.2	0.48	
K			2.1, 173	cadmium arachidate	3.4	0.50	
Mo	1.6	0.71		graphite fluoride			4.4, 969

Inelastic mean free path of electrons (contd)

Element	λ/(nm) at 1 keV	n	λ/(nm), E/(eV)	Organic compound	λ/(nm) at 1 keV	n	λ/(nm), E/(eV)
Na			1.3, 173	polybromoparaxylylene			1.9, 1065
Ni			1.3, 1186	polychloroparaxylylene			1.9, 1065
Pt			0.9, 1168	polyethylene			8.4, 969
Rb			2.0, 173	polyethylene terephthalate			6.8, 969
Si	2.9	0.62		polyparaxylene	1.5	2.0	
Sn			1.8, 520	polystyrene			7.4, 969
Ti			1.0, 422	polytetrafluoroethylene			7.2, 969
V			1.1, 1216	polyvinylchloride			8.4, 969
W	1.3	0.36		polyvinylidene fluoride	1.4	1.4	
Zn			1.8 1167				

M.P.S.

3.5.3 Range of electrons and beta rays

The slowing down and stopping of electrons (except perhaps in H and He) is so much influenced by multiple scattering that the process almost resembles diffusion. The practical maximum range tabulated below is roughly that thickness of aluminium through which very few particles will penetrate. It is nearly the same for beta rays of maximum energy E and for homogeneous electrons of energy E, and depends little on counter geometry and on the absorbing material (except for H and heavy elements). For detailed definitions and methods, see Katz and Penfold, *Rev. Mod. Phys.*, 1952, **24**, 28; Glendenin, *Nucleonics*, 1948, **2**, 12; and Feather, *Proc. Camb. Phil. Soc.*, 1938, **34**, 599.

Energy E keV	Range in Al mg cm^{-2}	Energy E keV	Range in Al mg cm^{-2}	Energy E MeV	Range in Al mg cm^{-2}	Energy E MeV	Range in Al g cm^{-2}	Energy E MeV	Range in Al g cm^{-2}
		30	1.5	0.15	26.5	0.8	0.31	4	2.02
		40	2.6	0.20	42	1.0	0.42	5	2.54
10	0.16	50	4.0	0.25	59	1.2	0.52	6.5	3.3
12	0.24	65	6.4	0.30	78	1.5	0.68	8	4.1
15	0.38	80	9.2	0.40	120	2.0	0.95	10	5.2
20	0.68	100	13.5	0.50	165	2.5	1.21	12	6.2
25	1.06	120	18.5	0.65	235	3.0	1.48	15	7.8

J.G.C.

3.5.4 Radiation quantities and units

(a) Physical quantities

The quantity of particulate radiations, whether the particles do or do not possess electric charge, may be defined in terms of **fluence**, which is the number of particles that enter a sphere of unit cross-sectional area. For a unidirectional beam the fluence so defined is equal to the number of particles that cross a unit area placed perpendicular to the beam. When the radiation is distributed isotropically the number of particles that cross a unit area of any plane is one half the fluence. The **fluence rate** is known as the **flux density** or **flux**.

The quantity of X- or gamma-radiation to which an object is exposed can be specified in terms of **exposure**, which is the ionization that the radiation would produce in air. Exposure is defined as the quotient of ΔQ by Δm, where ΔQ is the sum of the electrical charges on all ions of one sign produced in air when all the electrons (negatrons and positrons) liberated by photons in a volume of air whose mass is Δm are completely stopped in air. The ionization arising from the absorption of bremsstrahlung or annihilation radiation emitted by the electrons is not to be included in ΔQ. Owing to the difficulty of measurement, exposure is not normally used when the photon energy exceeds 3 MeV. The unit of exposure is the coulomb per kilogram of air. The older unit is the **roentgen (R)**, which is equal to 2.58×10^{-4} coulomb per kilogram of air.

The energy imparted to matter by charged or uncharged ionizing particles is known as the **absorbed dose** for which the unit is the **gray (Gy)**, which is equivalent to 1 joule per kilogram or 6.242×10^{12} MeV per kilogram. The older unit is the **rad**, which is 0.01 gray.

The sum of the kinetic energies of all charged particles produced per unit mass of a medium by interaction of uncharged particles, such as photons and neutrons, is known as **kerma**. The kinetic energy includes the energy that the charged particles may lose subsequently as bremsstrahlung or annihilation radiation and also the kinetic energy of charged particles, such as Auger electrons, produced by other secondary processes. Since the charged particles do not dissipate their energy at the point at which they are produced, kerma is not identical to absorbed dose unless complete secondary equilibrium exists in the medium, i.e. those charged particles that leave their point of production are exactly balanced by others produced elsewhere that arrive at the point. Exposure for X- and gamma-radiation is the ionization equivalent of kerma in air, except at very high energies when a difference arises due to bremsstrahlung and annihilation radiation effects. In the dosimetry of X- and gamma-radiation, ionization chambers are often used for the dual purposes of measuring the exposure and the absorbed dose. In order to avoid the difficulties of different units and instrument scales for these two quantities, some authorities now recommend that the quantity of X- and gamma-radiation should be expressed in terms of air kerma, since the unit of kerma is identical to that of absorbed dose, i.e. the gray. An exposure of one roentgen is equivalent to an air kerma of 0.8734 rad or $8.734 . 10^{-3}$ gray.

The **Linear energy transfer** or **restricted linear collison stopping power** ($L\Delta$) of charged particles is the differential mean energy loss per unit distance traversed due to collisions with energy transfers less than some specified value Δ in electron volts. L_{100}, for example, designates the linear energy transfer when $\Delta = 100$ electron volts. The symbol L_{∞} is used when all possible energy transfers are possible. L_{∞} is sometimes referred to as the **unrestricted linear collision stopping power**. Linear energy transfer is usually expressed in units of kiloelectron volt per micrometre. In a medium with a density of 1000 kilograms per metre^3, a L_{∞} of 1 keV per micrometre is equivalent to a specific energy loss (or stopping power (p. 327)) of 1 MeV metre^2 per kilogram.

The expectation value of the rate of spontaneous nuclear transitions in a quantity of material is known as its **activity**. The special name for the unit of activity is the **becquerel (Bq)** and is equal to one transition per second. The older unit is the **curie (Ci)**, which is equivalent to 3.7×10^{10} Bq. In its original definition the curie was that quantity of radon that is in radioactive equilibrium with 1 gram of radium. As the total number of nuclear transitions in 1 gram of radium and hence in the radon is 3.7×10^{10} per second, this rate was applied at first unofficially as the unit of measure of the quantity of any radioactive material.

The **decay constant**, λ, of a radioactive nuclide is the probability per unit time that a given nucleus will undergo a spontaneous nuclear transition. The quantity $\lambda^{-1} . \ln 2$ is commonly called the **half-life**, $T_{1/2}$, and is the time required for the activity of an amount of a radioactive nuclide to decrease to half its initial value.

The **air kerma-rate constant** (Γ_δ) of a radionuclide emitting photons enables the air kerma-rate (K_δ) due to photons with energies greater than δ to be determined at a distance (l) from a point source of the nuclide with activity (A). The unit is metre^2 (gray per second) per becquerel. In older units it would be metre^2 (rad per second) per curie, which is 2.703×10^{-13} metre^2 (gray per second) per becquerel. Without allowance for absorption and scattering of the radiation, the air kerma-rate is given by:

$$K_\delta = A\Gamma_\delta \, l^{-2} \text{ gray per second}$$

Table 1 gives values for the air kerma-rate constant for some common radionuclides.

(b) Protection quantities

For protection purposes the quantity **dose equivalent** is employed to express on a common scale the risk to exposed persons from all ionizing radiations. The unit of dose equivalent is the **sievert (Sv)**, which is equal to the absorbed dose to tissue in gray multiplied by appropriate modifying factors. Although there is provision in the definition for a variety of modifying factors, e.g. one based on dose rate, the only factor used at present is the **quality factor**, which depends on the unrestricted linear energy transfer (L_∞) in water of the charged particles responsible for the absorbed dose in tissue in a manner defined by the International Commission on Radiological Protection (ICRP). The quality factor is to some extent representative of the observed differences in the absorbed doses of different radiations required to produce the same level of effect. The minimum value of the quality factor of 1 applies to charged particles with unrestricted linear energy transfer values of less than 3.5 keV per micrometre, such as those produced in tissue by X- and gamma-radiation. Recoiling protons produced by fast neutron collisions in tissue have quality factors that range up to about 12 and alpha particles are usually assigned the maximum value of 20, which applies to charged particles with unrestricted linear energy transfer values of greater than 175 keV per micrometre. The older unit of dose equivalent is the **rem**, which is 0.01 sievert. Both physical and protection quantities are summarized in Table 2.

The most significant effects expected to be produced by low (< 50 milligray) or moderate (< 2 gray) amounts of radiation are cancers in some of the exposed persons and/or of hereditary effects in some of their descendants. The probability of inducing cancer in some organs is higher than in others, and ICRP have recommended weighting factors to indicate their relative sensitivities (Table 3). The **effective dose equivalent** is obtained by summing the products of the weighting factors and dose equivalents in the different tissues. This quantity provides an approximate method of comparing the hazards of different exposures to radiation. In particular it enables the hazard of uniformly irradiating the whole body to be compared with non-uniform irradiation, such as would result from a radioactive material incorporated in the body tissues. The dose equivalent to the tissues from such material will be spread out in time and delivered gradually as the material decays. The total dose equivalent to a tissue during the period of 50 years following the incorporation of a quantity of a radionuclide, after allowing for the metabolic processes of excretion, is known as the **committed dose equivalent** arising at the time of incorporation.

An effective dose equivalent of one sievert to the average adult implies a fatal cancer risk of about 1 in 100, and a non-fatal cancer risk of about the same magnitude. The risk that the children born subsequently to a person exposed to this dose will have serious hereditary defect as a result is also about 1 in 100, with diminishing risks for succeeding generations. These estimates of cancer and hereditary risks have fairly large uncertainties, which are difficult to quantify but may be a factor of two in either direction. Depending upon age, women may be a factor of two at greater risk than men because of their greater sensitivity to breast and thyroid cancers.

To keep the risks from radiation to acceptable levels, ICRP recommend that workers should not be exposed in any year to more than 50 millisieverts and members of the public to not more than 5 millisieverts of effective dose equivalent and committed effective dose equivalent. (There are additional restrictions on the maximum dose equivalent and committed dose equivalent allowed to specific tissues.) It is not regarded as acceptable for any person to be exposed at these levels for more than a few years without due justification. A further ICRP recommendation is that all exposures to radiation should be kept as low as reasonably achievable, economic and social factors having been taken into account.

(c) Natural exposures

In areas of normal background the average annual effective dose equivalent from radiation of natural origin is about 2.0 millisievert. This is comprised of 0.30 millisievert from cosmic radiation, 0.35 millisievert from terrestrial gamma radiation, 0.38 millisievert from internal radioactivity (e.g. [14]C and ^{40}K) and 0.97 millisievert from inhaled radon and thoron daughter products. The cosmic radiation component is comparatively independent of latitude at sea level, but becomes increasingly latitude-dependent with increasing altitude and is highest in polar regions. At a latitude of 50° it increases by a factor of about 60 between sea-level and a height of 10 kilometres and by even larger factors at times of intense solar flares. The contribution to effective dose equivalent from terrestrial gamma radiation and to airborne radon and thoron daughter products depends on the local geology. Values that are factors of two or three above or below the normal average values are not uncommon and in a very few limited areas can be up to 100 times greater.

(d) Man-made exposures

In addition to the natural exposures, there are some sources of radiation due to man's activities. In the United Kingdom the average annual effective dose equivalent due to medical procedures is 250 μSv, and 30 μSv for the sum of all other sources (weapons fall-out, discharges to environment, occupational exposure and miscellaneous sources).

References

(i) *Radiation Quantities and Units*, ICRP Report 33, 1980.
(2) *Recommendations of the International Commission on Radiological Protection*, ICRP Publication 26, 1977.
(3) *United Nations Scientific Committee on the Effects of Atomic Radiation. Sources and effects of ionising radiation*, 1977 Report to the General Assembly.
(4) *United Nations Scientific Committee on the Effects of Atomic Radiation. Ionising Radiation: Sources and Biological Effects*, 1982 Report to the General Assembly.
(5) *Living with Radiation*, NRPB booklet, 2nd edn (1981).

Table 1. Air kerma-rate constants for some common radionuclides ($\delta = 50\,\mathrm{keV}$)

Nuclide Z	A		Half-life	Intensity of major photon emissions (in MeV, %)	Kerma-rate constant $\mathrm{m^2\,Gy\,s^{-1}\,Bq^{-1}}$
11	Na	24	15.03 h	1.369, 100%; 2.754, 100%	$12.0.10^{-17}$
26	Fe	59	44.56 d	1.099, 56.5%; 1.292, 43.5%	$4.1.10^{-17}$
27	Co	60	5.272 y	1.173, 100%; 1.333, 100%	$8.5.10^{-17}$
53	I	131	8.040 d	0.365, 81%	$1.4.10^{-17}$
55 Cs + 56 Ba	Cs Ba	137 137m	30.17 y	0.662; 85.1%	$2.1.10^{-17}$
57	La	140	40.3 h	0.329, 18.5%; 0.487, 43.0% 0.816, 22.4%; 1.597, 95.5%	$7.4.10^{-17}$
69	Tm	170	128.6 d	0.051, 1.2%; 0,052, 2.1%; 0.059, 0.7%; 0.061, 0.2%; 0.084, 3.2%	$0.018.10^{-17}$
73	Ta	182	115 d	0.068, 41.4%; 0.100, 14.1%; 0.152, 7.2%; 0.222, 7.6%; 1.121, 35.1%; 1.189, 16.5%; 1.221, 27.5%; 1.231, 11.6%	$4.4.10^{-17}$
77	Ir	192	74.2 d	0.296, 28.7%; 0.308, 29,7%; 0.316, 82.9%; 0.468, 58.0%; 0.604, 8.3%	$2.9.10^{-17}$
Radium plus decay products in 0.5 mm platinum			1600 y	0.075, 15%; 0.295, 18.9%; 0.352, 36%; 0.609, 41.2%; 1.120, 13.6%; 1.764, 15.8%	$5.5.10^{-17}$

Note: In practice the air kerma-rate constant may depend on the details of the source construction, and may include a contribution from bremsstrahlung and annihilation radiation generated in the source by beta particles and positrons.

Table 2. Radiation quantities and units

Quantity	Symbol	SI unit	Older unit	Comment
Fluence	Φ	m^{-2}	cm^{-2}	
Exposure	X	$C\,kg^{-1}$	roentgen $(2.58.10^{-4}\,C\,kg^{-1})$	Quantity of X- or gamma radiation
Absorbed dose	D	$Gy\ (1\,J\,kg^{-1})$ $(6.242\ 10^{12}\,MeV\ kg^{-1})$	rad $(0.01\,Gy)$	Energy absorbed as a result of interactions by charged or uncharged particles
Kerma	K	$Gy\ (1\,J\,kg^{-1})$	rad $(0.01\,Gy)$	Energy released by uncharged particles in the form of charged particles
Linear energy transfer	L_{Δ}	$keV\,\mu m^{-1}$ $(1\,MeV\,m^2\,kg^{-1})$	$keV\,\mu m^{-1}$	Linear collision stopping power
Dose equivalent	H	$Sv\ (1\,J\,kg^{-1})$	rem $(0.01\,Sv)$	Expresses on a common scale the risks from different radiations
Activity	A	$Bq\ (1\,s^{-1})$	curie $(3.7.10^{10}Bq)$	Rate of spontaneous nuclear transformations in a quantity of radioactive material

Table 3. Risk factors and risk weighting factors recommended by the ICRP for fatal cancers and for serious hereditary defects

Tissue or organ	Risk factor[†] per sievert	Risk weighting factor
Breast	$25\ 10^{-4}$	0.15
Red bone marrow	$20\ 10^{-4}$	0.12
Lung	$20\ 10^{-4}$	0.12
Thyroid	$5\ 10^{-4}$	0.03
Bone surfaces	$5\ 10^{-4}$	0.03
Any other tissue[‡]	$10\ 10^{-4}$	0.06
Skin	$1.7\ 10^{-4}$	0.01
Gonads . . . . . (hereditary risk)	$40\ 10^{-4}$	0.25
Whole body	$165\ 10^{-4}$	1.00

† These risk factors apply to a population of adult workers of all ages and composed equally of both sexes. Risks to individuals depend on age, sex and child expectancy.

‡ The other tissues considered in the calculation of dose equivalent are the five receiving the highest doses, but excluding the skin and lens of the eye.

§ The skin weighting factor is not used in controlling and limiting doses to workers, since for these purposes separate controls are applied to skin exposures. However, in estimating effects on populations it may be appropriate to embody this factor in the estimate.

J.A.D.

3.6 Radioactive elements

3.6.1 Table of nuclides

The nuclei listed in this table are those for which the identification of both atomic number and mass are considered probably correct (categories A, B and C of Reference 1, from which the information has been extracted). Long-lived isomeric states are included and are indicated by the symbols m (for a conventional isomer) or f (for a fission isomer) following the mass number. In some cases energy measurements are insufficiently precise to determine which of the entries for a particular mass number should be considered as the ground state.

Following the identification of the nuclide, the second column gives the half-life or the natural abundance in atoms per cent; in certain cases both quantities are relevant. It should be noted that the natural isotopic abundance of elements in the Earth's crust can vary according to the source, particularly in the case of the lighter elements.

The third column attempts to summarize the decay modes and principal radiations emitted by a radioactive nuclide. Branching ratios (where known) are expressed as a percentage of the total decay events and are quoted to the nearest 1%; branches weaker than 0.5% are excluded. The energy of emitted radiations is expressed in MeV. The possible decay modes are β^+; β^-; $\beta^-\beta^-$ (double beta decay); α; p (proton emission); n (neutron emission); SF (spontaneous fission); IT (isomeric transition–electromagnetic transition from an isomeric state to the ground state of the same nuclide); K and L. K is used to denote electron capture processes to which capture from the K shell contributes (capture from higher shells may be important but is not distinguished); L is used for those rare cases of electron capture where capture from the K shell is energetically impossible. Generally the dominant decay mode (though not necessarily the strongest branch) is listed first; however, direct particle emission modes (α, p, n) are always listed first to distinguish them from emissions following $\beta\pm$ decay to particle-unstable daughter states, which are listed after the other decay (thus β^-; 10% n indicates that 10% of the decays consist of β^- decay to particle-unstable states which emit neutrons).

In the case of β^+ or β^- decay, the branch with the highest end point energy is listed first; if this is not the dominant branch it is followed by information on other important branches. The maximum possible energy is the difference in masses of the parent and daughter nuclides for β^- decay, and 1.02 MeV less than this for β^+ decay. For α decay, up to four branches are listed in order of energy; if there are more than four significant α branches, the three strongest are listed and the remaining α strength denoted by (for example) 13% α (dist). Where the individual α branching ratios are uncertain, the form 11% α (3.32) or α 4.45 is used to indicate the energy of the principal branch, where the total α decay probability is respectively known and unknown. If the listed decay branches add up to less than 100% this indicates that there are also β^+ or β^- branches weaker than those listed. Cases where some of the decays lead to an isomeric state of the daughter are generally indicated: (88% m) denotes that 88% of all decay events lead to an isomeric state.

After the decay modes, the three strongest γ rays (if any) emitted during the decay of the daughter(s) to their ground state(s) are listed in order of strength. In one case (Mg 29) only relative γ ray intensities are given. Annihilation radiation, which always follows β^+ emission, is not included. Gamma rays emitted in the IT decay of a daughter which is itself an isomeric state are distinguished by the symbol (m); in such cases the branching ratio is still expressed as a percentage of the total decay events of the parent—note that if the half-life of the isomeric state is comparable with or longer than that of the parent the instantaneous decay rate will differ from this branching ratio. The single entry γ indicates that γ-rays are emitted but the branching ratios are unknown. Details of conversion electrons, X-rays and Auger electrons have not been included.

The fourth column gives the mass excess expressed in MeV, i.e. the difference between the measured atomic mass and A atomic mass units on the ^{12}C scale (1 a.m.u. = 931.478 MeV).

The fifth column gives the thermal neutron cross-section in barns for various processes indicated as follows: (γ) for (n,γ); (γm) for (n,γ) leading to an isomeric state; (α) for (n,α); (f) for neutron-induced fission; and (t) for the sum of all absorption processes. The cross-sections were measured for a neutron velocity of 2200 ms^{-1} (an energy of 0.0253 eV) except where indicated by the symbols (r) or (s), which denote respectively neutrons with a thermal reactor spectrum or 'subcadmium' neutrons with a spectrum extending up to an energy of approximately 0.5 eV.

The sixth column gives the spin (I) and parity (π).

It is impossible in so short a space to do justice to the available information on the nuclides, which is continually being extended. For more details the reader is referred to Reference 1.

Reference

(1) C. M. Lederer and V. S. Shirley, *Table of Isotopes*, (7th edn), 1978.

Table of the nuclides

Nuclide Z	A	Natural abundance or half-life	Decay modes and major radiations	Mass excess MeV	σ/b	I$^\pi$
0 n	1	10.6 min	100% β^- 0.7824	8.0714		$\frac{1}{2}^+$
1 H	1	99.985%		7.2890	0.332 (γ)	$\frac{1}{2}^+$
H	2	0.015%		13.1358	0.0005 (γ)	1^+
H	3	12.35 yr	100% β^- 0.01862	14.9499		$\frac{1}{2}^+$
2 He	3	0.00014%		14.9313	5330 (p)	$\frac{1}{2}^+$
He	4	99.9999%		2.4249		0^+
He	6	0.81 s	100% β^- 3.51	17.597		0^+
He	8	0.122 s	88% β^- 9.7; 12% n; 88% γ 0.981	31.61		0^+
3 Li	6	7.5%		14.086	942 (α)	1^+
Li	7	92.5%		14.908	0.045 (γ) (r)	$\frac{3}{2}^-$
Li	8	0.844 s	100% β^- 13.1; 100% 2 α	20.947		2^+
Li	9	0.178 s	65% β^- 13.6; 35% n, 2 α	24.955		
Li	11	0.0085 s	39% β^- 20.4; 61% n	40.94		
4 Be	7	53.3 d	100% K; 10% γ 0.477	15.770	49 000 (γ) (r)	$\frac{3}{2}^-$
Be	8	7×10^{-17} s	100% 2 α	4.942		0^+
Be	9	100%		11.348	0.0076 (γ)	$\frac{3}{2}^-$
Be	10	1.6×10^6 yr	100% β^- 0.556	12.608		0^+
Be	11	13.8 s	57% β^- 11.5; 33% γ 2.12; 5% γ 6.79; 2% γ 5.85	20.18		$\frac{1}{2}^+$
Be	12	0.011 s	β^-; n	25.0		0^+
5 B	8	0.77 s	93% β^+ 13.7; 100% 2 α	22.922		2^+
B	10	19.8%		12.052	3838 (α)	3^+
B	11	80.2%		8.668	0.005 (γ) (r)	$\frac{3}{2}^-$
B	12	0.020 s	97% β^- 13.37; 2% α; 1% γ 4.439	13.370		1^+
B	13	0.017 s	92% β^- 13.44; 8% γ 3.68	16.562		$\frac{3}{2}^-$
B	14	0.016 s	87% β^- 14.5; 90% γ 6.09; 9% γ 6.73	23.66		2^-
6 C	9	0.127 s	β^+; 100% p	28.91		
C	10	19.15 s	99% β^+ 1.87; 99% γ 0.718; 1% γ 1.022	15.703		0^+
C	11	20.38 min	100% β^+ 0.96	10.650		$\frac{3}{2}^-$
C	12	98.89%		0.0	0.0034 (γ)	0^+
C	13	1.11%		3.1250	0.0009 (γ)	$\frac{1}{2}^-$
C	14	5730 yr	100% β^- 0.156	3.0199		0^+
C	15	2.45 s	32% β^- 9.77; 68% β^- 4.47; 68% γ 5.299	9.873		$\frac{1}{2}^+$
C	16	0.75 s	84% β^- 5.4; 100% n	13.69		0^+

Table of the nuclides (*contd*)

Nuclide		Natural abundance or half-life	Decay modes and major radiations	Mass excess MeV	σ/b	I^{π}
Z	A					
7 N	12	0.0110 s	94% β^+ 16.3; 4% α; 2% γ 4.439	17.338		1^+
N	13	9.96 min	100% β^+ 1.19	5.346		$\frac{1}{2}^-$
N	14	99.63%		2.8634	1.8 (p)	1^+
N	15	0.366%		0.1015	4×10^{-5} (γ) (r)	$\frac{1}{2}^-$
N	16	7.1 s	26% β^- 10.42; 68% β^- 4.29; 69% γ 6.129; 5% γ 7.115; 1% γ 2.75	5.682		2^-
N	17	4.17 s	2% β^- 8.7; 50% β^- 3.3; 95% n; 3% γ 0.871	7.87		$\frac{1}{2}^-$
N	18	0.63 s	100% β^- 9.6; 100% 1.982; 72% γ 0.82; 72% γ 1.65	13.3		
8 O	13	0.009 s	β^+ 16.7; 12% p	23.11		$\frac{3}{2}^-$
O	14	70.6 s	1% β^+ 4.12; 99% β^+ 1.81; 99% γ 2.313	8.008		0^+
O	15	122.1 s	100% β^+ 1.73	2.855		$\frac{1}{2}^-$
O	16	99.76%		-4.7370	0.00018 (γ) (r)	0^+
O	17	0.038%		-0.810	0.24 (α) (s)	$\frac{5}{2}^+$
O	18	0.202%		-0.783	0.00016 (γ)	0^+
O	19	26.8 s	40% β^- 4.62; 60% β^- 3.27; 91% γ 0.197; 55% γ 1.357; 4% γ 0.110	3.331		$\frac{5}{2}^+$
O	20	13.5 s	100% β^- 2.75; 100% γ 1.057	3.80		0^+
9 F	17	64.5 s	100% β^+ 1.74	1.9517		$\frac{5}{2}^+$
F	18	109.7 min	97% β^+ 0.64; 3% K	0.873		1^+
F	19	100%		-1.4874	0.0096 (γ) (r)	$\frac{1}{2}^+$
F	20	11.00 s	100% β^- 5.392; 100% γ 1.634	-0.017		2^+
F	21	4.35 s	29% β^- 5.69; 63% β^- 5.34; 71% γ 0.351; 8% γ 1.395	-0.05		$\frac{5}{2}^+$
F	22	4.23 s	3% β^- 7.50; 54% β^- 5.33; 16% β^- 5.21; 100% γ 1.275; 82% γ 2.083; 62% γ 2.166	2.83		4^+
F	23	2.2 s	14% β^- 6.7; 10% β^- 6.2; 22% β^- 5.1; 35% β^- 4.7; 48% γ 1.701; 33% γ 2.129; 23% γ 1.822	3.4		
10 Ne	17	0.109 s	1% β^+ 13.5; 54% β^+ 8.0; 98% p	16.48		$\frac{1}{2}^-$
Ne	18	1.67 s	92% β^+ 3.42; 8% γ 1.041	5.32		0^+
Ne	19	17.22 s	99% β^+ 2.22	1.751		$\frac{1}{2}^+$
Ne	20	90.51%		-7.043	0.038 (γ) (r)	0^+
Ne	21	0.27%		-5.733	0.7 (γ) (r)	$\frac{3}{2}^+$
Ne	22	9.22%		-8.026	0.05 (γ) (r)	0^+
Ne	23	37.6 s	67% β^- 4.37; 32% β^- 3.94; 33% γ 0.440	-5.155		$\frac{5}{2}^+$
Ne	24	3.38 min	92% β^- 1.99; (100% m); 100% γ (m) 0.472; 8% γ 0.874	-5.95		0^+
Ne	25	0.60 s	77% β^- 7.1; 96% γ 0.090; 18% γ 0.980; 2% γ 1.069	-2.2		

Table of the nuclides (*contd*)

Nuclide Z	A	Natural abundance or half-life	Decay modes and major radiations	Mass excess MeV	σ/b	Iπ
11 Na	20	0.448 s	79% β^+ 11.23; 20% α; 79% γ 1.634	6.84		2^+
Na	21	22.5 s	95% β^+ 2.53; 5% γ 0.351	−2.186		$\frac{3}{2}^+$
Na	22	2.602 yr	90% β^+ 0.545; 10% K; 100% γ 1.275	−5.184	0.0003 (γ)	3^+
Na	23	100%		−9.530	0.53 (γ)	$\frac{3}{2}^+$
Na	24	15.03 hr	100% β^- 1.39; 100% γ 1.368; 100% γ 2.754	−8.418		4^+
Na	24m	0.020 s	IT; β^- 5.99 (weak); ~100% γ 0.472	−7.945		1^+
Na	25	60 s	63% β^-3.83; 27% β^- 2.86; 15% γ 0.975; 13% γ 0.390; 13% γ 0.585	−9.36		$\frac{5}{2}^+$
Na	26	1.1 s	88% β^- 7.52; 99% γ 1.809; 6% γ 1.130; 3% γ 1.412	−6.89		3^+
Na	27	0.30 s	86% β^- 8.0; 14% β^- 7.3; 86% γ 0.985; 14% γ 1.699	−5.6		
Na	28	0.031 s	69% β^- 13.9; 30% γ 1.474; 16% γ 2.380	−1.1		1^+
Na	29	0.043 s	β^-; 15% n; 12% γ 2.570; 7% γ 1.510	2.66		
Na	30	0.055 s	β^-; 33% n	8.4		
Na	31	0.017 s	β^-; 30% n	10.6		
Na	32	0.015 s	β^-	16.4		
Na	33	0.02 s	β^-			
12 Mg	21	0.123 s	16% β^+ 12.08; 41% β^+ 11.75; 11% β^+ 10.36; 10% β^+ 7.61; 32% p	10.91		$\frac{5}{2}^+$
Mg	22	3.86 s	40% β^+ 3.19; 54% β^+ 3.11; 100% γ 0.583; 60% γ 0.074; 5% γ 1.280	−0.394		0^+
Mg	23	11.33 s	91% β^+ 3.04; 9% γ 0.440	−5.471		$\frac{3}{2}^+$
Mg	24	78.99%		−13.931	0.053 (γ) (r)	0^+
Mg	25	10.00%		−13.191	0.18 (γ) (r)	$\frac{5}{2}^+$
Mg	26	11.01%		−16.212	0.038 (γ)	0^+
Mg	27	9.46 min	72% β^- 1.77; 28% β^- 1.60; 73% γ 0.844; 27% γ 1.014	−14.585	0.15 (γ) (r)	$\frac{1}{2}^+$
Mg	28	20.9 hr	95% β^- 0.46; 95% γ 0.031; 54% γ 1.342; 36% γ 0.401; 36% γ 0.942	−15.016		0^+
Mg	29	1.5 s	β^-; (50%) γ 2.224; (25%) γ 0.960; (25%) γ 1.398	−10.8		
Mg	30	1 s	β^-	−9.8		
13 Al	23	0.47 s	β^+; p	6.77		
Al	24	2.05 s	7% β^+ 8.7; 48% β^+ 4.4; 39% β^+ 3.4; 96% γ 1.368; 43% γ 2.753; 41% γ 7.066	−0.05		4^+
Al	24m	0.13 s	93% IT; 4% β^+ 12.9; 93% γ 0.439	0.39		1^+
Al	25	7.17 s	99% β^+ 3.26; 1% γ 1.612	−8.913		$\frac{5}{2}^+$

Table of the nuclides (*contd*)

Nuclide Z	A	Natural abundance or half-life	Decay modes and major radiations	Mass excess MeV	σ/b	I$^\pi$
Al	26	7.2×10^5 yr	82% β^+ 1.17; 18% K; 100% γ 1.809; 2% γ 1.130	-12.208		5$^+$
Al	26m	6.35 s	100% β^+ 3.2	-11.979		0$^+$
Al	27	100%		-17.194	0.23 (γ)	$\frac{5}{2}^+$
Al	28	2.241 min	100% β^- 2.87; 100% γ 1.779	-16.848		3$^+$
Al	29	6.6 min	89% β^- 2.41; 89% γ 1.273; 7% γ 2.426; 4% γ 2.028	-18.21		$\frac{5}{2}^+$
Al	30	3.7 s	18% β^- 6.3; 67% β^- 5.0; 65% γ 2.235; 41% γ 1.263; 33% γ 3.498	-15.9		
Al	31	0.64 s	β^-; γ	-15.1		
14 Si	25	0.22 s	β^+; p	3.82		
Si	26	2.21 s	75% β^+ 3.83; 22% β^+ 3.00; (100% m)	-7.143		0$^+$
Si	27	4.11 s	100% β^+ 3.79	-12.385		$\frac{5}{2}^+$
Si	28	92.23%		-21.491	0.17 (γ) (r)	0$^+$
Si	29	4.67%		-21.894	0.10 (γ) (r)	$\frac{1}{2}^+$
Si	30	3.10%		-24.432	0.11 (γ)	0$^+$
Si	31	2.62 hr	100% β^- 1.49	-22.949	0.5 (γ) (r)	$\frac{3}{2}^+$
Si	32	650 yr	100% β^- 0.21	-24.09		0$^+$
Si	33	6.2 s	β^-; γ	-20.6		
Si	34	2.8 s	β^-; γ	-19.9		0$^+$
15 P	28	0.270 s	68% β^+ 11.5; 98% γ 1.779; 12% γ 4.497; 9% γ 7.536	-7.16		3$^+$
P	29	4.15 s	99% β^+ 3.92; 1% γ 1.273	-16.949		$\frac{1}{2}^+$
P	30	2.50 min	β^+ 3.21; K	-20.205		1$^+$
P	31	100%		-24.440	0.18 (γ) (r)	$\frac{1}{2}^+$
P	32	14.28 d	100% β^- 1.71	-24.305		1$^+$
P	33	25.3 d	100% β^- 0.249	-26.337		$\frac{1}{2}^+$
P	34	12.5 s	85% β^- 5.4; 15% γ 2.128	-24.6		1$^+$
P	35	47 s	β^-; 99% γ 1.572	-24.9		
16 S	29	0.20 s	β^+; p	-3.2		$\frac{5}{2}^+$
S	30	1.22 s	19% β^+ 5.1; 78% β^+ 4.4; 80% γ 0.678; 3% γ 2.342	-14.062		0$^+$
S	31	2.61 s	99% β^+ 4.4; 1% γ 1.266	-19.044		$\frac{1}{2}^+$
S	32	95.02%		-26.015	0.53 (γ) (r)	0$^+$
S	33	0.75%		-26.586	0.09 (α) (r)	$\frac{3}{2}^+$
S	34	4.21%		-29.931	0.24 (γ) (r)	0$^+$
S	35	87.4 d	100% β^- 0.167	-28.846		$\frac{3}{2}^+$
S	36	0.017%		-30.666	0.15 (γ) (r)	0$^+$
S	37	4.99 min	6% β^- 4.8; 94% β^- 1.7; 94% γ 3.103	-26.91		
S	38	170 min	12% β^- 2.94; 84% β^- 1.00; 84% γ 1.942; 2% γ 1.746	-26.86		0$^+$
17 Cl	32	0.298 s	1% β^+ 11.7; 60% β^+ 9.5; 92% γ 2.230; 21% γ 4.770; 4% γ 2.464	-13.33		1$^+$
Cl	33	2.51 s	98% β^+ 4.56; 1% γ 0.841; 1% γ 1.966; 1% γ 2.867	-21.003		$\frac{3}{2}^+$
Cl	34	1.526 s	100% β^+ 4.47	-24.438		0$^+$

Table of the nuclides (contd)

Nuclide Z	A	Natural abundance or half-life	Decay modes and major radiations	Mass excess MeV	σ/b	Iπ
Cl	34m	32.0 min	28% β^+ 2.5; 24% β^+ 1.3; 47% IT; 42% γ 0.146; 42% γ 2.127; 13% γ 1.176	−24.292		3^+
Cl	35	75.77%		−29.014	43 (γ)	$\frac{3}{2}^+$
Cl	36	3.00×10^5 yr	98% β^- 0.710; 2% K	−29.522		2^+
Cl	37	24.23%		−31.762	0.43 (γ)	$\frac{3}{2}^+$
Cl	38	37.3 min	58% β^- 4.92; 31% β^- 1.11; 42% γ 2.167; 31% γ 1.642	−29.798		2^-
Cl	38m	0.715 s	100% IT; 100% γ 0.671	−29.127		5^-
Cl	39	56.2 min	7% β^- 3.44; 83% β^- 1.92; 54% γ 1.267; 47% γ 0.250; 37% γ 1.517	−29.80		$\frac{3}{2}^+$
Cl	40	1.32 min	9% β^- 7.5; 28% β^- 3.2; 20% β^- 2.9; 77% γ 1.461; 30% γ 2.840; 15% γ 2.622	−27.5		2^-
Cl	41	34 s	β^-; γ	−27.4		
18 Ar	33	0.18 s	18% β^+ 10.6; 48% β^+ 9.8; 27% β^+ 5.1; 34% p; 48% γ 0.810	−9.39		$\frac{1}{2}^+$
Ar	34	0.84 s	94% β^+ 5.0; 3% γ 0.666; 1% γ 0.461; 1% γ 3.129	−18.379		0^+
Ar	35	1.77 s	98% β^+ 4.94; 1% γ 1.219	−23.049		$\frac{3}{2}^+$
Ar	36	0.337%		−30.231	5 (γ) (r)	0^+
Ar	37	35.02 d	100% K	−30.948		$\frac{3}{2}^+$
Ar	38	0.063%		−34.715	0.8 (γ) (r)	0^+
Ar	39	269 yr	100% β^- 0.57	−33.24	600 (γ) (r)	$\frac{7}{2}^-$
Ar	40	99.60%		−35.040	0.6 (γ)	0^+
Ar	41	1.83 h	1% β^- 2.49; 99% β^- 1.20; 99% γ 1.294	−33.068		$\frac{7}{2}^-$
Ar	42	33 yr	100% β^- 0.60	−34.42		0^+
Ar	43	5.4 min	β^-; γ	−32.0		
Ar	44	11.9 min	β^-; 66% γ 0.182; 66% γ 1.705; 33% γ 1.887	−32.3		0^+
19 K	36	0.344 s	44% β^+ 9.8; 42% β^+ 5.2; 82% γ 1.970; 32% γ 2.433; 30% γ 2.208	−17.43		2^+
K	37	1.23 s	98% β^+ 5.13; 2% γ 2.794	−24.799		$\frac{3}{2}^+$
K	38	7.61 min	100% β^+ 2.72; 100% γ 2.168	−28.802		3^+
K	38m	0.926 s	100% β^+ 5.0	−28.671		0^+
K	39	93.26%		−33.806	2.1 (γ) (r)	$\frac{3}{2}^+$
K	40	1.28×10^9 yr 0.0117%	89% β^- 1.31; 11% K; 11% γ 1.461	−33.535	70 (γ) (r)	4^-
K	41	6.73%		−35.560	1.5 (γ)	$\frac{3}{2}^+$
K	42	12.361 hr	81% β^- 3.52; 19% γ 1.525	−35.023		2^-
K	43	22.2 hr	2% β^- 1.8; 89% β^- 0.8; 88% γ 0.373; 80% γ 0.618; 11% γ 0.397; 11% γ 0.594	−36.59		$\frac{3}{2}^+$
K	44	22.2 min	34% β^- 5.66; 29% β^- 2.36; 58% γ 1.157; 23% γ 2.151; 10% γ 2.519	−35.81		2^-
K	45	20 min	β^-; γ	−36.61		$\frac{3}{2}^+$

Table of the nuclides (*contd*)

Z	A	Natural abundance or half-life	Decay modes and major radiations	Mass excess MeV	σ/b	I^π
K	46	115 s	50% β^- 6.3; 28% β^- 2.7; 91% γ 1.347; 28% γ 3.700; 9% γ 3.015	−35.42		
K	47	17.5 s	β^-; 100% γ 2.013; 85% γ 0.586; 15% γ 0.565	−35.70		$\frac{1}{2}^+$
K	48	6.8 s	β^- 8.2; 80% γ 3.832; 32% γ 0.780; 17% γ 0.675	−32.2		
20 Ca	37	0.173 s	16% β^+ 10.6; 47% β^+ 5.5; 76% p; γ	−13.16		$\frac{3}{2}^+$
Ca	38	0.44 s	74% β^+ 5.6; 25% β^+ 4.0; (100% m); 25% γ 1.568	−22.06		0^+
Ca	39	0.860 s	100% β^+ 5.49	−27.282		$\frac{3}{2}^+$
Ca	40	96.94%		−34.847	0.4 (γ) (r)	0^+
Ca	41	1.03×10^5 yr	100% K	−35.139		$\frac{7}{2}^-$
Ca	42	0.647%		−38.544	0.7 (γ) (r)	0^+
Ca	43	0.135%		−38.405	6 (γ) (r)	$\frac{7}{2}^-$
Ca	44	2.09%		−41.466	0.88 (γ)	0^+
Ca	45	165 d	100% β^- 0.26	−40.810		$\frac{7}{2}^-$
Ca	46	0.0035%		−43.138	0.7 (γ) (r)	0^+
Ca	47	4.540 d	16% β^- 1.99; 84% β^- 0.68; 77% γ 1.297; 7% γ 0.489; 7% γ 0.808	−42.343		$\frac{7}{2}^-$
Ca	48	0.187%		−44.216	1.1 (γ) (r)	0^+
Ca	49	8.72 min	92% β^- 2.2; 92% γ 3.084; 7% γ 4.072	−41.286		$\frac{3}{2}^-$
Ca	50	14 s	100% β^- 3.1; (100% m) 97% γ (m) 0.257; 58% γ 0.072; 58% γ 1.519	−39.57		0^+
21 Sc	40	0.183 s	19% β^+ 9.6; 50% β^+ 5.6; 100% γ 3.736; 41% γ 0.756; 25% γ 2.046	−20.527		4^-
Sc	41	0.596 s	100% β^+ 5.48	−28.644		$\frac{7}{2}^-$
Sc	42	0.681 s	100% β^+ 5.40	−32.121		0^+
Sc	42m	62.0 s	100% β^+ 2.8; 100% γ 0.438; 100% γ 1.226; 100% γ 1.525	−31.503		7^+
Sc	43	3.89 hr	β^+ 1.2; K; 22% γ 0.373	−36.185		$\frac{7}{2}^-$
Sc	44	3.93 hr	95% β^+ 1.4; 5% K; 100% γ 1.157; 1% γ 1.499	−37.811		2^+
Sc	44m	2.442 d	99% IT; 1% K; 87% γ 0.271; 1% γ 1.002; 1% γ 1.126	−37.540		6^+
Sc	45	100%		−41.067	17 (γ); 9 (γm) (r)	$\frac{7}{2}^-$
Sc	45m	0.32 s	100% IT	−41.054		$\frac{3}{2}^+$
Sc	46	83.80 d	100% β^- 0.36; 100% γ 0.889; 100% γ 1.121	−41.756	8 (γ) (s)	4^+
Sc	46m	18.7 s	100% IT; 62% 0.143	−41.613		1^-
Sc	47	3.422 d	31% β^- 0.60; 69% β^- 0.44; 69% γ 0.159	−44.331		$\frac{7}{2}^-$
Sc	48	43.7 hr	91% β^- 0.65; 100% γ 0.984; 100% γ 1.312; 98% γ 1.037	−44.50		6^+
Sc	49	57.0 min	100% β^- 2.00	−46.555		

Table of the nuclides (*contd*)

Nuclide		Natural abundance or half-life	Decay modes and major radiations	Mass excess MeV	σ/b	Iπ
Z	A					
Sc	50	1.71 min	14% β^- 4.2; 86% β^- 3.6; 100% γ 1.121; 100% γ 1.554; 86% γ 0.524	−44.54		
Sc	50m	0.35 s	100% IT; 97% γ 0.257	−44.28		
Sc	51	12.4 s	28% β^- 5.1; 34% β^- 4.4; 52% γ 1.437; 32% γ 2.144; 15% γ 1.568	−43.22		
22 Ti	41	0.080 s	9% β^+ 9.7; 15% β^+ 7.6; 24% β^+ 6.0; 100% p	−15.78		$\frac{3}{2}^+$
Ti	42	0.202 s	43% β^+ 6.0; 56% β^+ 5.4; 56% γ 0.611	−25.12		0^+
Ti	43	0.49 s	100% β^+ 5.84	−29.32		$\frac{7}{2}^-$
Ti	44	48 yr	100% K; 94% γ 0.078; 88% γ 0.068	−37.546		0^+
Ti	45	3.08 hr	86% β^+ 1.04; 14% K	−39.004		$\frac{7}{2}^-$
Ti	46	8.2%		−44.123	0.6 (γ) (r)	0^+
Ti	47	7.4%		−44.931	1.7 (γ) (r)	$\frac{5}{2}^-$
Ti	48	73.7%		−48.488	7.9 (γ) (r)	0^+
Ti	49	5.4%		−48.559	2.1 (γ) (r)	$\frac{7}{2}^-$
Ti	50	5.2%		−51.432	0.18 (γ)	0^+
Ti	51	5.80 min	92% β^- 2.15; 93% γ 0.320; 7% γ 0.929	−49.733		$\frac{3}{2}^-$
Ti	52	1.7 min	100% β^- 1.83; 100% γ 0.124	−49.47		0^+
Ti	53	33 s	β^-; 45% γ 0.128; 39% γ 0.228; 25% γ 1.676	−46.8		
23 V	46	0.4223 s	100% β^+ 6.0	−37.071		0^+
V	47	32.6 min	β^+ 1.9; K	−42.001		$\frac{3}{2}^-$
V	48	15.976 d	50% β^+ 0.70; 50% K; 100% γ 0.984; 98% γ 1.312; 8% γ 0.944	−44.473		4^+
V	49	330 d	100% K	−47.957		$\frac{7}{2}^-$
V	50	0.250%		−49.219	50 (γ)	6^+
V	51	99.750%		−52.199	4.9 (γ)	$\frac{7}{2}^-$
V	52	3.75 min	99% β^- 2.55; 100% γ 1.434	−51.439		3^+
V	53	1.6 min	89% β^- 2.4; 90% γ 1.006; 10% γ 1.289	−51.86		$\frac{7}{2}^-$
V	54	43 s	11% β^- 5.2; 45% β^- 2.9; 100% γ 0.835; 82% γ 0.986; 50% γ 2.255	−49.9		
24 Cr	45	0.05 s	β^+; p	−19.5		
Cr	46	0.26 s	100% β^+ 6.6	−29.46		0^+
Cr	48	21.56 hr	100% K; 100% γ 0.116; 100% γ 0.308	−42.82		0^+
Cr	49	41.9 min	β^+ 1.6; K; 51% γ 0.091; 29% γ 0.153; 17% γ 0.062	−45.329		$\frac{5}{2}^-$
Cr	50	4.35%		−50.258	15.9 (γ)	0^+
Cr	51	27.70 d	100% K; 10% γ 0.320	−51.448		$\frac{7}{2}^-$
Cr	52	83.79%		−55.415	0.8 (γ) (r)	0^+
Cr	53	9.50%		−55.284	18 (γ) (r)	$\frac{3}{2}^-$
Cr	54	2.36%		−56.931	0.38 (γ) (r)	0^+
Cr	55	3.52 min	100% β^- 2.60	−55.106		$\frac{3}{2}^-$

Table of the nuclides (*contd*)

Nuclide		Natural abundance or half-life	Decay modes and major radiations	Mass excess MeV	σ/b	Iπ
Z	A					
Cr	56	5.9 min	100% β^- 1.53; 100% γ 0.026; 100% γ 0.083	−55.27		0+
25 Mn	50	0.2827 s	100% β^+ 6.61	−42.626		0+
Mn	50m	1.76 min	8% β^+ 3.7; 69% β^+ 3.5; 28% β^+ 3.0; 100% γ 0.783; 100% γ 1.098; 69% γ 1.443	−42.40		5+
Mn	51	46.2 min	β^+ 2.2; K	−48.240		$\frac{5}{2}^-$
Mn	52	5.591 d	72% K; 28% β^+ 0.58; 100% γ 1.434; 95% γ 0.936; 90% γ 0.744	−50.704		6+
Mn	52m	21.1 min	98% β^+ 3.7; 2% IT; 98% γ 1.434; 2% γ 0.378	−50.327		2+
Mn	53	3.74×10^6 yr	100% K	−54.687	70 (γ) (r)	$\frac{7}{2}^-$
Mn	54	312.20 d	100% K; 100% γ 0.835	−55.554		3+
Mn	55	100%		−57.710	13.2 (γ)	$\frac{5}{2}^-$
Mn	56	2.579 hr	56% β^- 2.84; 28% β^- 1.04; 99% γ 0.847; 27% 1.811; 14% 2.113	−56.909		3+
Mn	57	1.54 min	81% β^- 2.68; 10% γ 0.122; 4% γ 0.692; 1% γ 0.352	−57.49		$\frac{5}{2}^-$
Mn	58	65 s	β^-; 82% γ 0.811; 55% γ 1.323; 20% γ 0.459	−55.80		3+
Mn	58	3.0 s	β^- 6.3	−55.83		
26 Fe	49	0.08 s	β^+; p	−24.5		
Fe	52	8.28 hr	57% β^+ 0.80; 43% K; (100% m); 99% γ 0.169	−48.33		0+
Fe	53	8.51 min	β^+ 2.7; K; 42% γ 0.378	−50.944		$\frac{7}{2}^-$
Fe	53m	2.51 min	100% IT; 99% γ 0.701; 86% γ 1.012; 86% γ 1.328	−47.904		
Fe	54	5.8%		−56.251	2.2 (γ) (r)	0+
Fe	55	2.69 yr	100% K	−57.479		$\frac{3}{2}^-$
Fe	56	91.8%		−60.604	2.6 (γ) (r)	0+
Fe	57	2.15%		−60.179	2.4 (γ) (r)	$\frac{1}{2}^-$
Fe	58	0.29%		−62.152	1.14 (γ)	0+
Fe	59	44.56 d	53% β^- 0.47; 45% β^- 0.27; 57% γ 1.099; 43% γ 1.292; 3% γ 0.192	−60.661		$\frac{3}{2}^-$
Fe	60	3×10^5 yr	100% β^- 0.15 (100% m); 2% γ (m) 0.059	−61.44		0+
Fe	61	5.98 min	15% β^- 2.8; 37% β^- 2.6; 27% β^- 2.5; 44% γ 1.205; 43% γ 1.027; 22% γ 0.298	−59.0		
Fe	62	68 s	100% β^- 2.1; 100% γ 0.506	−58.86		0+
27 Co	53m	0.25 s	1% p; β^+	−39.45		
Co	54	0.193 s	100% β^+ 7.22	−48.010		0+
Co	54m	1.43 min	100% β^+ 4.5; 100% γ 0.411; 100% γ 1.130; 100% γ 1.407	−47.811		
Co	55	17.54 hr	β^+ 1.5; 23% K; 75% γ 0.931; 20% 0.477; 16% γ 1.409	−54.024		$\frac{7}{2}^-$

Table of the nuclides (*contd*)

Nuclide			Natural abundance or half-life	Decay modes and major radiations	Mass excess MeV	σ/b	I^π
	Z	A					
	Co	56	78.8 d	81% K; 19% β^+ 1.5; 100% γ 0.847; 67% γ 1.238; 17% γ 2.599	−56.037		4^+
	Co	57	271.7 d	100% K; 86% γ 0.122; 11% γ 0.136; 10% γ 0.014	−59.342		$\frac{7}{2}^-$
	Co	58	70.8 d	85% K; 15% β^+ 0.47; 99% γ 0.811; 1% γ 0.864	−59.844	1900 (γ)	2^+
	Co	58m	9.2 hr	100% IT	−59.819	1.4×10^5 (γ)	5^+
	Co	59	100%		−62.226	18 (γ) 19 (γm)	$\frac{7}{2}^-$
	Co	60	5.272 yr	100% β^- 0.32; 100% γ 1.173; 100% γ 1.332	−61.647	2.0 (γ) (s)	5^+
	Co	60m	10.47 min	100% IT; 2% γ 0.059	−61.588	58 (γ) (s)	2^+
	Co	61	1.65 hr	96% β^- 1.25; 86% γ 0.067; 3% γ 0.909	−62.897		$\frac{7}{2}^-$
	Co	62 (g)	1.50 min	64% β^- 4.1; 28% β^- 3.0; 83% γ 1.173; 16% γ 2.302; 13% γ 1.129	−61.43		
	Co	62(m)	13.9 min	62% β^- 3.0; 20% β^- 2.2; 98% γ 1.173; 69% γ 1.164; 19% γ 2.004	−61.41		
	Co	63	27.5 s	94% β^- 3.6; 49% γ 0.087; 3% γ 0.982; 2% γ 0.156	−61.85		
	Co	64	0.30 s	90% β^- 7.3; 10% γ 1.346; 5% γ 0.931	−59.79		
28	Ni	53	0.05 s	β^+; p	−29.4		
	Ni	56	6.10 d	100% K; 99% γ 0.158; 75% γ 0.812; 48% γ 0.750	−53.90		0^+
	Ni	57	36.0 hr	60% K; 34% β^+ 0.86; 78% γ 1.378; 15% γ 1.919; 13% γ 0.127	−56.077		$\frac{3}{2}^-$
	Ni	58	68.3%		−60.224	4.6 (γ) (r)	0^+
	Ni	59	8×10^4 yr	100% K	−61.153	92 (t)	$\frac{3}{2}^-$
	Ni	60	26.1%		−64.470	2.8 (γ) (r)	0^+
	Ni	61	1.13%		−64.219	2.5 (γ) (r)	$\frac{3}{2}^-$
	Ni	62	3.59%		−66.745	14.2 (γ)	0^+
	Ni	63	100 yr	100% β^- 0.066	−65.513	23 (γ) (r)	$\frac{1}{2}^-$
	Ni	64	0.91%		−67.098	1.49 (γ)	0^+
	Ni	65	2.520 hr	61% β^- 2.14; 28% β^- 0.66; 24% γ 1.482; 16% γ 1.115; 5% γ 0.366	−65.125	24 (γ) (s)	$\frac{5}{2}^-$
	Ni	66	54.8 hr	100% β^- 0.24	−66.02		0^+
	Ni	67	18 s	β^- 3.8; γ	−63.5		
29	Cu	58	3.20 s	82% β^+ 7.5; 16% γ 1.455; 12% γ 1.448; 5% γ 0.040	−51.662		1^+
	Cu	59	81.5 s	58% β^+ 3.8; 15% γ 1.302; 12% γ 0.878; 8% γ 0.339	−56.352		$\frac{3}{2}^-$
	Cu	60	23.4 min	β^+ 3.8; 7% K; 88% γ 1.332; 45% γ 1.792; 22% γ 0.826	−58.343		2^+
	Cu	61	3.41 hr	β^+ 1.2; 38% K; 13% γ 0.284; 12% γ 0.656; 7% γ 0.067	−61.981		$\frac{3}{2}^-$
	Cu	62	9.73 min	98% β^+ 2.9; 2% K	−62.796		1^+

Table of the nuclides (*contd*)

Nuclide		Natural abundance or half-life	Decay modes and major radiations	Mass excess MeV	σ/b	Iπ
Z	A					
Cu	63	69.2%		−65.579	4.4 (γ)	$\frac{3}{2}^-$
Cu	64	12.70 hr	41% K; 19% β^+ 0.66; 40% β^- 0.58; 1% γ 1.346	−65.423		1^+
Cu	65	30.8%		−67.262	2.17 (γ)	$\frac{3}{2}^-$
Cu	66	5.10 min	92% β^- 2.6; 8% γ 1.039	−66.257	140 (γ) (s)	1^+
Cu	67	62.0 hr	20% β^- 0.58; 23% β^- 0.49; 56% β^- 0.40; 47% γ 0.185; 17% γ 0.093; 7% γ 0.091	−67.31		$\frac{3}{2}^-$
Cu	68	30 s	31% β^- 4.6; 40% β^- 3.5; 17% β^- 2.3; γ	−65.39		1^+
Cu	68m	3.75 min	86% IT; 2% β^- 2.9; 74% γ 0.526; 71% γ 0.084; 16% γ 0.111	−64.66		
Cu	69	3.0 min	79% β^- 2.5; 10% γ 1.006; 6% γ 0.834; 3% γ 0.530	−65.9		
Cu	70	5 s	46% β^- 6.2; 54% β^- 5.3; 54% γ 0.885	−63.4		1^+
Cu	70	42 s	30% β^- 4.5; 60% β^- 3.3; 100% γ 0.885; 87% γ 0.902; 57% γ 1.252	−63.3		
30 Zn	57	0.04 s	β^+; p	−32.6		
Zn	60	2.42 min	β^+ 3.1; 3% K; 68% γ 0.670; 23% γ 0.062; 10% γ 0.273	−54.18		0^+
Zn	61	89.1 s	68% β^+ 4.4; 16% γ 0.475; 7% γ 1.661; 2% γ 0.970	−56.2		$\frac{3}{2}^-$
Zn	62	9.13 hr	93% K; 7% β^+ 0.61; 25% γ 0.041; 24% γ 0.597; 14% γ 0.548	−61.17		0^+
Zn	63	38.0 min	β^+ 2.3; 7% K; 8% γ 0.670; 6% γ 0.962; 1% γ 1.412	−62.211		$\frac{3}{2}^-$
Zn	64	48.6%		−66.001	0.78 (γ)	0^+
Zn	65	244.0 d	99% K; 1% β^+ 0.33; 50% γ 1.116	−65.910		$\frac{5}{2}^-$
Zn	66	27.9%		−68.898	1 (γ) (r)	0^+
Zn	67	4.10%		−67.880	7 (γ) (r)	$\frac{5}{2}^-$
Zn	68	18.8%		−70.006	0.81 (γ); 0.07 (γm)	0^+
Zn	69	56 min	100% β^- 0.90	−68.417		$\frac{1}{2}^-$
Zn	69m	13.76 hr	100% IT; 95% γ 0.439	−67.978		$\frac{9}{2}^+$
Zn	70	0.62%		−69.560	0.09 (γ); 0.008 (γm)	0^+
Zn	71	2.5 min	57% β^- 2.82; 30% β^- 2.31; 30% γ 0.512; 7% γ 0.910; 3% γ 0.390	−67.32		$\frac{1}{2}^-$
Zn	71m	3.92 hr	β^-; 92% γ 0.386; 62% γ 0.487; 56% γ 0.620	−67.17		
Zn	72	46.5 hr	86% β^- 0.30; 83% γ 0.145; 9% γ 0.192; 8% γ 0.016	−68.13		0^+
Zn	73	24 s	β^-; γ	−65.0		
Zn	74	95 s	39% β^- 2.3; 57% β^- 2.1; (90% m); 80% γ 0.057; 34% γ 0.140; 23% γ 0.190	−65.7		0^+
Zn	75	10.2 s	β^-	−62.5		

Table of the nuclides (*contd*)

Z	Nuclide	A	Natural abundance or half-life	Decay modes and major radiations	Mass excess MeV	σ/b	Iπ
	Zn	76	5.7 s	β^-	−62.6		0^+
31	Ga	63	32.4 s	55% β^+ 4.5; 11% γ 0.637; 10% γ 0.627; 6% γ 0.193	−56.7		
	Ga	64	2.62 min	β^+ 6.1; 43% γ 0.992; 14% γ 0.808; 13% γ 3.366	−58.84		0^+
	Ga	65	15.2 min	β^+ 2.2; 14% K; 55% γ 0.115; 12% γ 0.061; 9% γ 0.153	−62.654		$\frac{3}{2}^-$
	Ga	66	9.45 hr	51% β^+ 4.2; 44% K; 38% γ 1.039; 23% γ 2.752; 6% γ 0.834	−63.723		0^+
	Ga	67	78.3 hr	100% K; 38% γ 0.093; 24% γ 0.185; 19% γ 0.300	−66.879		$\frac{3}{2}^-$
	Ga	68	68.3 min	89% β^+ 1.9; 10% K; 3% γ 1.077	−67.085		1^+
	Ga	69	60.1%		−69.322	1.7 (γ)	$\frac{3}{2}^-$
	Ga	70	21.1 min	99% β^- 1.66	−68.905		1^+
	Ga	71	39.9%		−70.142	4.6 (γ)	$\frac{3}{2}^-$
	Ga	72	14.12 hr	11% β^- 3.2; 28% β^- 1.8; 21% β^- 0.7; 96% γ 0.834; 26% γ 2.202; 24% γ 0.630	−68.591		3^-
	Ga	73	4.86 hr	44% β^- 1.5; 47% β^- 1.2; (99% m); 47% γ 0.297; 10% γ (m) 0.053; 7% γ 0.326	−69.73		
	Ga	74	8.25 min	5% β^- 4.8; 51% β^- 2.5; 91% γ 0.596; 45% γ 2.354; 15% γ 0.608	−68.0		
	Ga	74m	10 s	100% IT; 75% γ 0.057	−68.0		1^+
	Ga	75	2.10 min	β^-; γ	−68.6		
	Ga	76	27.1 s	14% β^- 6.2; 10% β^- 5.7; 10% β^- 5.2; 66% γ 0.563; 26% γ 0.546; 16% γ 1.108	−66.4		
	Ga	77	13.0 s	β^-	−66.4		
	Ga	78	5.09 s	β^-	−63.7		
	Ga	79	3.0 s	β^-; γ	−62.8		
	Ga	80	1.66 s	β^-; n	−59.5		
	Ga	81	1.23 s	β^-; n; γ			
	Ga	83	0.31 s	β^-; n			
32	Ge	64	64 s	β^+; 37% γ 0.427; 17% γ 0.667; 11% γ 0.128	−54.4		0^+
	Ge	65	31 s	β^+; 33% γ 0.650; 27% γ 0.062; 21% γ 0.809	−56.4		
	Ge	66	2.27 hr	72% K; β^+; 29% γ 0.044; 28% γ 0.382; 11% γ 0.109, 0.273	−61.62		0^+
	Ge	67	19.0 min	β^+ 3.2; 4% K; 84% γ 0.167; 5% γ 1.473; 3% γ 0.911	−62.45		
	Ge	68	288 d	100% K	−66.97		0^+
	Ge	69	39.1 hr	64% K; β^+ 1.2; 27% γ 1.106; 12% γ 0.574; 10% γ 0.872	−67.096		$\frac{5}{2}^-$
	Ge	70	20.5%		−70.561	3.3 (γ) (r)	0^+

Table of the nuclides (*contd*)

Nuclide			Natural abundance or half-life	Decay modes and major radiations	Mass excess MeV	σ/b	I^π
Z		A					
	Ge	71	11.2 d	100% K	−69.906		$\frac{1}{2}-$
	Ge	72	27.4%		−72.583	1.0 (γ) (r)	$0+$
	Ge	73	7.8%		−71.294	15 (γ) (r)	$\frac{9}{2}+$
	Ge	73m	0.50 s	100% IT; 11% γ 0.053	−71.227		$\frac{1}{2}-$
	Ge	74	36.5%		−73.422	0.36 (γ) (r); 0.16 (γm) (r)	$0+$
	Ge	75	82.78 min	87% β^- 1.18: 11% γ 0.265; 1% γ 0.199	−71.856		$\frac{1}{2}-$
	Ge	75m	47.7 s	100% IT; 39% γ 0.140	−71.716		$\frac{7}{2}+$
	Ge	76	7.8%		−73.214	0.06 (γ) (r) 0.10 (γm) (r)	$0+$
	Ge	77	11.30 hr	3% β^- 2.5; 21% β^- 1.5; 53% γ 0.264; 30% γ 0.211; 28% γ 0.216	−71.214		$\frac{7}{2}$
	Ge	77m	53 s	58% β^- 2.8; 21% β^- 2.6; 20% IT; 21% γ 0.216; 11% γ 0.160	−71.055		$\frac{1}{2}-$
	Ge	78	1.47 hr	96% β^- 0.71; 96% γ 0.277; 4% γ 0.294	−71.8		$0+$
	Ge	79	42 s	75% β^- 4.2; 25% γ 0.230; 15% γ 0.543	−69.6		
	Ge	80	29.5 s	β^-; 25% γ 0.266; 6% γ 0.110; 5% γ 1.564	−69.4		$0+$
	Ge	81	10 s	β^- 6.3; γ	−66.3		
	Ge	82	4.6 s	β^-; γ	−66.0		$0+$
	Ge	83	1.9 s	β^-	−62.5		
	Ge	84	1.2 s	β^-			$0+$
33	As	68	2.7 min	9% β^+ 7.2; 28% β^+ 6.2; 29% β^+ 4.8; 66% γ 1.016; 24% γ 0.651; 23% γ 0.763	−58.8		
	As	69	15 min	90% β^+ 2.95; 2% K; 5% γ 0.233; 2% γ 0.146; 1% γ 0.087	−63.12		
	As	70	53 min	β^+ 2.8; 16% K; 82% γ 1.040; 21% γ 0.668; 21% γ 1.114	−64.34		4
	As	71	61 hr	68% K; β^+ 0.81; 84% γ 0.175; 4% γ 1.096; 3% γ 0.500	−67.893		$\frac{5}{2}-$
	As	72	26.0 hr	β^+ 3.3; 23% K; 80% γ 0.834; 8% γ 0.630; 1% γ 1.051	−68.23		$2-$
	As	73	80.3 d	100% K (100% m); 11% γ (m) 0.053	−70.949		$\frac{3}{2}-$
	As	74	17.79 d	17% β^- 1.35; 15% β^- 0.72; 31% β^+; 37% K; 60% γ 0.596; 15% γ 0.635	−70.860		$2-$
	As	75	100%		−73.034	4.4 (γ)	$\frac{3}{2}-$
	As	76	26.3 hr	51% β^- 2.97; 36% β^- 2.41; 45% γ 0.559; 6% γ 0.657; 3% γ 1.216	−72.291		$2-$
	As	77	38.83 hr	98% β^- 0.69; 2% γ 0.239	−73.916		$\frac{3}{2}-$

Table of the nuclides (*contd*)

Nuclide		Natural abundance or half-life	Decay modes and major radiations	Mass excess MeV	σ/b	I^π
Z	A					
As	78	90.7 min	34% β^- 4.3; 17% β^- 3.7; 54% γ 0.614; 18% γ 0.695; 11% γ 1.309	−72.7		
As	79	9.0 min	95% β^- 2.1; (98% m); 9% γ (m) 0.096; 2% γ 0.365; 2% γ 0.432	−73.71		$\frac{3}{2}^-$
As	80	15.2 s	56% β^- 5.7; 27% β^- 5.0; 42% γ 0.662; 7% γ 1.645; 4% γ 1.207	−72.1		1
As	81	33 s	67% β^- 3.8; (3% m); 20% γ 0.468; 8% γ 0.491; 1% γ 0.521	−72.6		
As	82	14 s	15% β^- 5.5; 44% β^- 4.3; 19% β^- 3.7; 72% γ 0.654; 39% γ 1.895; 27% γ 0.819, 1.731			
As	82	19 s	80% β^- 7.2; 15% γ 0.654; 4% γ 1.731; 2% γ 2.353	−70.4		
As	83	13 s	β^-; (70% m); γ	−69.9		
As	84	0.7 s	β^-	−66.2		
As	84	5.3 s	β^-; 49% γ 1.455; 21% γ 0.667; 5% γ 2.087	−66.2		
As	85	2.03 s	β^-; 23% n; 16% γ 1.455; 7% γ 0.667; 2% γ 1.112	−63.5		
As	86	0.9 s	β^-; 4% n; γ	−59.7		
As	87	0.6 s	β^-	−56		
34　Se	69	27.4 s	β^+ 5.8; 63% γ 0.098; 27% γ 0.066; 14% γ 0.691	−56.30		
Se	70	41.1 min	β^+; K; 35% γ 0.050; 29% γ 0.426; 9% γ 0.377	−61.7		0+
Se	71	4.9 min	β^+ 3.4; K; 47% γ 0.147; 13% γ 0.831; 10% γ 1.096	−63.5		
Se	72	8.4 d	100% K; 57% γ 0.046	−67.89		0+
Se	73	7.2 hr	β^+ 1.7; 35% K; 97% γ 0.361; 71% γ 0.067	−68.21		$\frac{7}{2}^+$
Se	73m	41 min	73% IT; β^+ 1.7; K; 3% γ 0.254; 2% γ 0.085; 2% γ 0.393	−68.18		$\frac{1}{2}^-$
Se	74	0.87%		−72.213	52 (γ)	0+
Se	75	118.5 d	100% K; 58% γ 0.265; 54% γ 0.136; 24% γ 0.280	−72.169		$\frac{5}{2}^+$
Se	76	9.0%		−75.259	64 (γ); 21 (γm) (r)	0+
Se	77	7.6%		−74.606	42 (γ) (r)	$\frac{1}{2}^-$
Se	77m	17.4 s	100% IT; 53% γ 0.162	−74.444		$\frac{7}{2}^+$
Se	78	23.5%		−77.032	0.41 (γ); 0.3 (γm) (r)	0+
Se	79	≤6.5 × 10⁴ yr	100% β^- 0.16	−75.91		$\frac{7}{2}^+$
Se	79m	3.91 min	100% IT; 10% γ 0.096	−75.82		$\frac{1}{2}^-$
Se	80	49.8%		−77.761	0.6 (γ); 0.07 (γm)	0+
Se	81	18.2 min	99% β^- 1.6	−76.391		
Se	81 m	57.3 min	100% IT; 10% γ 0.103	−76.288		
Se	82	9.2% 10²⁰ yr	$\beta^- \beta^-$	−77.59	0.006 (γ); 0.04 (γm)	0+

Table of the nuclides (*contd*)

Nuclide		Natural abundance or half-life	Decay modes and major radiations	Mass excess MeV	σ/b	I^π
Z	A					
Se	83	22.4 min	β^- 2.9; 69% γ 0.357; 44% 0.510; 32% γ 0.225	−75.33		
Se	83m	70.4 s	31% β^- 3.9; 33% β^- 2.9; 21% γ 1.031; 17% γ 0.357; 15% γ 0.674, 0.988	−75.11		
Se	84	3.5 min	100% β^- 1.4; 100% γ 0.409	−75.94		0^+
Se	85	31 s	60% β^- 6.1; 22% γ 0.345; 4% γ 0.609; 3% γ 3.396	−72.6		
Se	86	16.7 s	β^-; γ	−70.9		0^+
Se	87	5.9 s	β^-; γ	−66		
Se	88	1.5 s	β^-; 1% n; γ	−64.1		0^+
Se	91	0.27 s	β^-; 21% n			
35 Br	72	1.31 min	23% β^+ 7.1; 20% β^+ 6.6; 70% γ 0.862; 17% γ 1.317; 13% γ 0.455	−58.9		
Br	73	3.3 min	β^+ 3.5; K; γ	−63.7		
Br	74	25.3 min	β^+; K; 84% γ 0.635 (double); 19% γ 0.282; 8% γ 2.615	−65.30		
Br	74m	42 min	β^+; K; 117% γ 0.635 (double); 37% γ 0.728; 9% γ 1.269	−65.1		
Br	75	96 min	β^+ 1.7; 25% K; 92% γ 0.286; 7% γ 0.141; 4% γ 0.377, 0.428, 0.432	−69.16		
Br	76	16.1 hr	β^+ 3.9; 43% K; 74% γ 0.559; 16% γ 0.657; 14% γ 1.854	−70.30		1^-
Br	77	57.0 hr	99% K; 1% β^+ 0.34; 23% γ 0.239; 22% γ 0.521; 4% γ 0.297	−73.242		$\frac{3}{2}^-$
Br	77m	4.3 min	100% IT; 14% γ 0.106	−73.136		$\frac{9}{2}^+$
Br	78	6.46 min	β^+ 2.5; 8% K; 14% γ 0.614	−73.458		1^+
Br	79	50.69%		−76.070	10.8 (γ); 2.4 (γm)	$\frac{3}{2}^-$
Br	79m	4.86 s	100% IT; 76% γ 0.207	−75.863		$\frac{9}{2}^+$
Br	80	17.68 min	85% β^- 2.0; 6% β^- 1.4; 6% K; 3% β^+ 0.86; 7% γ 0.616; 1% γ 0.666	−75.891		1^+
Br	80m	4.42 hr	100% IT; 39% γ 0.037	−75.805		5^-
Br	81	49.31%		−77.976	2.7 (γ)	$\frac{3}{2}^-$
Br	82	35.34 hr	99% β^- 0.44; 83% γ 0.776; 70% γ 0.554; 43% γ 0.619	−77.498		5^-
Br	82m	6.05 min	98% IT; 2% β^- 3.1	−77.452		2^-
Br	83	2.39 hr	99% β^- 0.92; (100% m); 5% γ (m) 0.009; 1% γ 0.530	−79.03		
Br	84	31.8 min	33% β^- 4.7; 13% β^- 3.8; 42% γ 0.882; 15% γ 1.898; 7% γ 2.484, 3.928	−77.76		2^-
Br	84m	6.0 min	100% β^- 2.2; 100% γ 0.424; 98% γ 0.882; 97% γ 1.463	−77.5		
Br	85	2.87 min	96% β^- 2.5; (100% m); 14% γ (m) 0.305; 3% γ 0.802; 2% γ 0.925	−78.7		$\frac{3}{2}^-$

Table of the nuclides (*contd*)

Nuclide Z	A	Natural abundance or half-life	Decay modes and major radiations	Mass excess MeV	σ/b	Iπ
Br	86	55.7 s	15% β^- 7.3; 51% β^- 3.0; 62% γ 1.565; 19% γ 2.751; 10% γ 1.362	−76.0		
Br	87	55.6 s	β^-; 2% n; 32% γ 1.420; 12% γ 1.476; 9% γ 1.578	−74.2		
Br	88	16.7 s	β^-; 6% n; 77% γ 0.775; 16% γ 0.802; 5% γ 1.441, 3.932	−71.1		
Br	89	4.37 s	β^-; 13% n; γ	−69.1		
Br	90	1.96 s	β^-; 23% n; γ	−65.2		
Br	91	0.541 s	β^-; 10% n			
Br	92	0.37 s	β^-; 16% n	−58		
36 Kr	72	17.4 s	β^+; K; 19% γ 0.415; 15% γ 0.310; 8% γ 0.163	−53.9		0^+
Kr	73	26 s	44% β^+ 5.5; 1% p; 66% γ 0.178; 13% γ 0.151; 11% γ 0.474	−57.0		
Kr	74	11.5 min	β^+ 2.2; K; 31% γ 0.090; 20% γ 0.203; 11% γ 0.297	−62.0		0^+
Kr	75	4.2 min	β^+; K; 75% γ 0.132; 22% γ 0.155; 6% γ 0.153	−64.2		
Kr	76	14.82 hr	100% K; 38% γ 0.316; 20% γ 0.270; 18% γ 0.046	−69.1		0^+
Kr	77	75 min	β^+ 1.9; 20% K; (8% m); 84% γ 0.130; 39% γ 0.147; 4% γ 0.312	−70.24		
Kr	78	0.356%		−74.15	4.7 (γ); 0.21 (γm)	0^+
Kr	79	35.0 hr	93% K; 7% β^+ 0.61; 13% γ 0.261; 10% γ 0.398; 8% γ 0.606	−74.44		$\frac{1}{2}^-$
Kr	79m	50 s	100% IT; 27% γ 0.130	−74.31		$\frac{7}{2}^+$
Kr	80	2.27%		−77.90	11.5 (γ); 4.6 (γm)	0^+
Kr	81	2.1 × 10⁵ yr	100% K; 4% γ 0.276	−77.65		$\frac{7}{2}^+$
Kr	81m	13 s	100% IT; 67% γ 0.190	−77.46		$\frac{1}{2}^-$
Kr	82	11.6%		−80.59	23 (γ); 20 (γm)	0^+
Kr	83	11.5%		−79.985	200 (γ)	$\frac{9}{2}^+$
Kr	83m	1.83 hr	100% IT; 5% γ 0.009	−79.943		$\frac{1}{2}^-$
Kr	84	57.0%		−82.432	0.042 (γ); 0.09 (γm)	0^+
Kr	85	10.70 yr	100% β^- 0.69	−81.472	1.7 (γ)	$\frac{9}{2}^+$
Kr	85m	4.48 hr	79% β^- 0.84; 21% IT; 79% γ 0.151; 14% γ 0.305	−81.167		$\frac{1}{2}^-$
Kr	86	17.3%		−83.263	0.06 (γ) (r)	0^+
Kr	87	76.3 min	31% β^- 3.9; 40% β^- 3.5; 50% γ 0.403; 9% γ 2.555; 8% γ 0.845	−80.707		
Kr	88	2.86 hr	14% β^- 2.9; 67% β^- 0.5; 35% γ 2.392; 26% γ 0.196; 13% γ 0.835, 2.196	−79.69		0^+
Kr	89	3.18 min	23% β^- 4.9; 15% β^- 2.3; 20% γ 0.221; 17% γ 0.586; 7% γ 0.498, 0.904, 1.473	−76.8		

Table of the nuclides *(contd)*

Nuclide		Natural abundance or half-life	Decay modes and major radiations	Mass excess MeV	σ/b	Iπ
Z	A					
Kr	90	32.3 s	29% β^- 4.4; 63% β^- 2.6; (12% m); 38% γ 1.119; 33% γ 0.122; 30% γ 0.539	−75.2		0⁺
Kr	91	8.57 s	10% β^- 6.1; 42% γ 0.109; 18% γ 0.507; 7% γ 0.613	−71.8		
Kr	92	1.84 s	2% β^- 6.0; 88% β^- 4.6; 64% γ 0.142; 60% γ 1.219; 15% γ 0.813	−69.2		0⁺
Kr	93	1.29 s	5% β^- 7.3; 27% β^- 6.8; 2% n; 36% γ 0.254; 24% γ 0.252; 24% γ 0.324	−65.6		
Kr	94	0.20 s	β^-; 6% n; γ	−61.3		0⁺
37 Rb	74	0.065 s	100% β^+ 9.6	−51.4		
Rb	75	18 s	β^+; γ	−57.5		
Rb	76	39 s	β^+; γ	−60.6		
Rb	77	3.9 min	β^+; K; 66% γ 0.066; 26% γ 0.179; 12% γ 0.393	−65.1		
Rb	78	18 min	β^+; K; γ	−68.8		
Rb	78m	6.0 min	β^+; K; IT; γ	−68.7		
Rb	79	23.0 min	β^+; 16% K; (37% m) 24% γ 0.688; 16% γ 0.183; 13% γ 0.505	−70.9		
Rb	80	34 s	74% β^+ 4.7; 22% β^+ 4.1; 25% γ 0.616; 2% γ 0.640; 2% γ 0.704	−72.19		1⁺
Rb	81	4.58 hr	73% K; β^+ 1.1; (97% m); 66% γ (m) 0.190; 19% γ 0.446; 2% γ 0.457	−75.39		$\frac{3}{2}^-$
Rb	81m	32 min	β^+; γ	−75.31		$\frac{9}{2}^+$
Rb	82	1.25 min	β^+ 3.4; 4% K; 14% γ 0.776	−76.21		1⁺
Rb	82m	6.2 hr	74% K; β^+; 83% γ 0.776; 63% γ 0.554; 37% γ 0.619	−76.1		5⁻
Rb	83	86.2 d	100% K; (76% m); 46% γ 0.520; 30% γ 0.530; 14% γ 0.553	−78.91		$\frac{5}{2}^-$
Rb	84	32.8 d	75% K; 11% β^+ 1.7; 11% β^+ 0.8; 3% β^- 0.9; 74% γ 0.882; 1% γ 1.900	−79.752	12 (p) (r)	2⁻
Rb	84m	20.5 min	100% IT; 65% γ 0.248; 33% γ 0.216; 32% γ 0.464	−79.288		
Rb	85	72.17%		−82.159	0.40 (γ); 0.047 (γm)	$\frac{5}{2}^-$
Rb	86	18.82 d	91% β^- 1.77; 9% γ 1.077	−82.738		2⁻
Rb	86m	1.020 min	100% IT; 98% γ 0.556	−82.182		6⁻
Rb	87	4.72 × 10¹⁰ yr 27.83%	100% β^- 0.27	−84.596	0.12 (γ) (r)	$\frac{3}{2}^-$
Rb	88	17.8 min	77% β^- 5.3; 21% γ 1.836; 14% γ 0.898; 2% γ 2.678	−82.60	1.0 (γ) (r)	2⁻
Rb	89	15.15 min	25% β^- 4.5; 33% β^- 2.2; 33% β^- 1.3; 58% γ 1.032; 43% γ 1.248; 13% γ 2.196	−81.72		
Rb	90	153 s	37% β^- 6.4; γ	−79.6		

Table of the nuclides (*contd*)

Z	Nuclide	A	Natural abundance or half-life	Decay modes and major radiations	Mass excess MeV	σ/b	Iπ
	Rb	90m	258 s	β^-; IT; γ	−79.5		
	Rb	91	58.7 s	β^- 5.7; 32% γ 0.094; 12% γ 2.564; 10% γ 3.600	−77.97		
	Rb	92	4.54 s	94% β^- 7.8; 4% γ 0.815; 1% γ 0.570, 1.385, 2.821	−75.1		
	Rb	93	5.85 s	β^-; 1% n; γ	−72.9		
	Rb	94	2.73 s	β^-; 10% n; γ	−68.8		
	Rb	95	0.38 s	β^-; 8% n; γ	−66.6		
	Rb	96	0.203 s	β^-; 13% n; γ	−62.8		
	Rb	97	0.170 s	β^-; 27% n; γ			
	Rb	98	0.12 s	β^-; 13% n			
	Rb	99	0.076 s	β^-			
38	Sr	77	9 s	β^+; γ	−58.0		
	Sr	78	31 min	β^+; K	−65.5		0⁺
	Sr	79	8.1 min	β^+; K; γ	−65.5		
	Sr	80	106 min	β^+; K; 40% γ 0.589; 10% γ 0.175; 7% γ 0.553	−70.4		0⁺
	Sr	81	26 min	β^+; 13% K; 36% γ 0.153; 30% γ 0.148; 21% γ 0.188	−71.40		
	Sr	82	25.0 d	100% K	−76.00		0⁺
	Sr	83	32.4 hr	76% K; β^+ 1.23; 30% γ 0.763; 20% γ 0.382 (double); 5% γ 0.418	−76.66		7/2⁺
	Sr	83m	5.0 s	100% IT; 88% γ 0.259	−76.41		1/2⁻
	Sr	84	0.56%		−80.641	0.3 (γ); 0.6 (γm) (s)	0⁺
	Sr	85	64.85 d	100% K; 99% γ 0.514	−81.10		9/2⁺
	Sr	85m	68.0 min	87% IT; 13% K; 85% γ 0.232; 11% γ 0.151	−80.86		1/2⁻
	Sr	86	9.8%		−84.512	0.84 (γm) (s)	0⁺
	Sr	87	7.0%		−84.869		9/2⁺
	Sr	87m	2.805 hr	100% IT; 82% γ 0.388	−84.481		1/2⁻
	Sr	88	82.6%		−87.911	0.0057 (γ) (r)	0⁺
	Sr	89	50.6 d	100% β^- 1.49	−86.203	0.42 (γ) (r)	5/2⁺
	Sr	90	28.8 yr	100% β^- 0.55	−85.935	0.8 (γ) (r)	0⁺
	Sr	91	9.48 hr	31% β^- 2.7; 24% β^- 1.4; 35% β^- 1.1; (57% m); 54% γ (m) 0.556; 33% γ 1.024; 23% γ 0.750; 8% γ 0.653	−83.666		5/2⁺
	Sr	92	2.71 hr	3% β^- 1.93; 96% β^- 0.55; 90% γ 1.384; 4% γ 0.953; 3% γ 0.242, 0.431, 1.142	−82.89		0⁺
	Sr	93	7.43 min	11% β^- 3.4; 23% β^- 2.4; (39% m); 73% γ 0.590; 25% γ 0.876; 24% γ 0.888	−80.3		
	Sr	94	74.1 s	99% β^- 2.0; 95% γ 1.428; 3% γ 0.724; 2% γ 0.622, 0.704	−79.0		0⁺
	Sr	95	24.4 s	53% β^- 6.1; 24% γ 0.686; 5% γ 2.717; 4% γ 2.933	−75.1		
	Sr	96	1.06 s	β^-; 75% γ 0.809; 66% γ 0.122; 15% γ 0.932	−73.1		0⁺

Table of the nuclides (*contd*)

Nuclide		Natural abundance or half-life	Decay modes and major radiations	Mass excess MeV	σ/b	Iπ
Z	A					
Sr	97	0.40 s	20% β^- 7.2; 42% β^- 5.0; 24% γ 1.905; 23% γ 0.954; 12% γ 0.307, 0.652	−69.1		
Sr	98	0.65 s	65% β^- 5.8; 29% β^- 5.2; 29% γ 0.119; 10% γ 0.445; 9% γ 0.429	−67.4		0⁺
Sr	99	0.6 s	β^-; 3% n			
39 Y	81	5.0 min	β^+; K; γ			
Y	83	7.1 min	β^+ 3.3; 5% K; (2% m); 20% γ 0.036; 6% γ 0.490; 6% γ 0.882 (double)	−72.4		
Y	83	2.85 min	β^+; K; (100% m); 88% γ (m) 0.259; 33% γ 0.422; 13% γ 0.495	−72.4		
Y	84	39 min	β^+; K; 98% γ 0.793; 78% γ 0.974; 57% γ 1.040	−73.69		
Y	84	4.6 s	β^+; K; 35% γ 0.793			
Y	85(g)	2.7 hr	β^+; 45% K; (100% m); 85% γ (m) 0.232; 64% γ 0.504; 7% γ 0.914	−77.86		
Y	85(m)	4.7 hr	β^+ 2.2; 30% K; (4% m); 23% γ 0.232; 5% γ 2.124; 5% 0.768 (double)	−77.84		
Y	86	14.74 hr	66% K; β^+ 3.2; 83% γ 1.077; 33% γ 0.628; 31% γ 1.153	−79.24		4⁻
Y	86m	48 min	99% IT; 1% K; 94% γ 0.208	−79.02		8⁺
Y	87	80.3 hr	100% K; (100% m); 92% γ 0.485; 82% γ (m) 0.388	−83.007		½⁻
Y	87m	13.2 hr	98% IT; 1% β^+ 1.2; 1% K; 78% γ 0.381	−82.626		9/2⁺
Y	88	106.61 d	100% K; 99% γ 1.836; 91% γ 0.898; 1% γ 2.734	−84.298		4⁻
Y	89	100%		−87.695	1.2 (γ)	½⁻
Y	89m	16.06 s	100% IT; 99% γ 0.909	−86.786		9/2⁺
Y	90	64.1 hr	100% β^- 2.28	−86.481		2⁻
Y	90m	3.19 hr	100% IT; 96% γ 0.203; 91% γ 0.480	−85.799		7⁺
Y	91	58.5 d	100% β^- 1.54	−86.350	1.4 (γ) (r)	½⁻
Y	91m	49.71 min	100% IT; 95% γ 0.556	−85.794		9/2⁺
Y	92	3.54 hr	86% β^- 3.63; 14% γ 0.934; 5% γ 1.405; 2% γ 0.449, 0.561	−84.82		2⁻
Y	93	10.25 hr	90% β^- 2.89; 7% γ 0.267; 2% γ 0.947; 1% γ 1.918	−84.23		½⁻
Y	93m	0.82 s	100% IT; γ	−83.47		9/2⁺
Y	94	18.7 min	41% β^- 4.88; 40% β^- 3.96; 56% γ 0.919; 6% γ 1.139; 5% γ 0.551	−82.38		2⁻
Y	95	10.3 min	58% β^- 4.43; 19% γ 0.954; 8% γ 2.176; 8% γ 3.577	−81.23		

Table of the nuclides (*contd*)

Nuclide		Natural abundance or half-life	Decay modes and major radiations	Mass excess MeV	σ/b	Iπ
Z	A					
Y	96	9.8 s	9% β^- 3.3; 91% β^- 2.7; 89% γ 1.751; 60% γ 0.915; 55% γ 0.617			
Y	96	6.0 s	75% β^- 7.0; 25% β^- 5.4; 25% γ 1.594	−78.4		
Y	97(g)	3.7 s	40% β^- 6.7; 27% β^- 3.4; 18% γ 3.288; 14% γ 3.401; 7% γ 1.997	−76.3		
Y	97(m)	1.21 s	37% β^- 6.1; 45% β^- 5.1; 92% γ 1.103; 71% γ 0.161; 40% γ 0.970	−75.6		
Y	98	0.6 s	51% β^- 8.1; 26% β^- 7.2; 11% γ 1.223; 5% γ 2.941; 4% γ 1.591	−73.2		
Y	98	2.0 s	12% β^- 6.3; 46% β^- 3.8; 96% γ 1.223; 74% γ 0.621; 55% γ 0.647	−73.2		
Y	99	1.5 s	β^-; 1% n; γ	−71.5		
Y	100	0.8 s	β^-; γ	−68.0		
40 Zr	83	0.7 min	β^+; 10% γ 0.106; 9% γ 0.475; 9% γ 1.525	−65		
Zr	84	5 min	K; β^+	−71.4		0$^+$
Zr	85	7.85 min	β^+; K; (98% m); 41% γ 0.454; 24% γ 0.416; 4% γ 1.198	−73.2		
Zr	85m	10.9 s	IT; β^+; K; γ	−72.9		0$^+$
Zr	86	16.5 hr	100% K; 96% γ 0.243; 20% γ 0.028; 5% γ 0.612	−77.9		
Zr	87	1.6 hr	β^+ 2.26; K; (99% m); 78% γ (m) 0.381; 4% γ 1.228; 1% γ 1.210	−79.4		
Zr	87m	14.0 s	100% IT; 97% γ 0.201; 26% γ 0.135	−79.1		
Zr	88	83.4 d	100% K; 97% γ 0.394	−83.62		0$^+$
Zr	89	78.4 hr	78% K; 22% β^+ 0.9; (99% m); 98% γ (m) 0.909; 1% γ 1.713	−84.860		$\frac{9}{2}^+$
Zr	89m	4.18 min	94% IT; 5% K; 1% β^+ 0.9; 90% γ 0.588; 6% γ 1.507	−84.272		$\frac{1}{2}^-$
Zr	90	51.5%		−88.765	0.03 (γ) (r)	0$^+$
Zr	90m	0.809 s	100% IT; 84% γ 2.319; 16% γ 0.426; 4% γ 0.133	−86.446		5$^-$
Zr	91	11.2%		−87.893	1.1 (γ) (r)	$\frac{5}{2}^+$
Zr	92	17.1%		−88.456	0.2 (γ) (r)	0$^+$
Zr	93	1.5 × 10^6 yr	β^- 0.06; (>95% m)	−87.117	1 (γ) (r)	$\frac{5}{2}^+$
Zr	94	17.4%		−87.264	0.06 (γ)	0$^+$
Zr	95	64.0 d	1% β^- 0.88; 44% β^- 0.40; 55% β^- 0.36; (1% m); 55% γ 0.757; 44% γ 0.724	−85.663		$\frac{5}{2}^+$
Zr	96	2.80%		−85.445	0.02 (γ)	0$^+$

Table of the nuclides (*contd*)

Nuclide		Natural abundance or half-life	Decay modes and major radiations	Mass excess MeV	σ/b	Iπ
Z	A					
Zr	97	16.9 hr	86% β⁻ 1.92; (95% m); 93% γ (m) 0.743; 5% γ 0.508; 3% γ 1.148	−82.954		$\frac{1}{2}^+$
Zr	98	30.7 s	100% β⁻ 2.2	−81.29		0⁺
Zr	99	2.1 s	39% β⁻ 3.5; 58% β⁻ 3.4; (36% m); 56% γ 0.469; 45% γ 0.546; 27% γ 0.594	−77.9		
Zr	100	7.1 s	β⁻; γ	−76.6		0⁺
Zr	101	2.0 s	β⁻	−73.1		
Zr	102	2.9 s	β⁻	−72.4		0⁺
41 Nb	86	1.4 min	β⁺; γ	−69.3		
Nb	87	2.6 min	β⁺; K; 26% γ 0.201; 19% γ 0.471; 10% γ 1.066	−74.4		
Nb	87	3.9 min	β⁺; K; (100% m); 97% γ (m) 0.201; 26% γ (m) 0.135	−74.4		
Nb	88	7.8 min	β⁺; K; 90% γ 1.057; 57% γ 1.083; 43% γ 0.400	−76.4		
Nb	88	14.3 min	β⁺; K; 100% γ 1.057; 98% γ 1.083; 80% γ 0.503	−76.4		
Nb	89	2.0 hr	β⁺ 3.2; K; (1% m); 4% γ 1.627; 3% γ 1.833; 3% γ 2.572	−80.62		
Nb	89	66 min	β⁺ 2.6; 26% K; (100% m); 90% γ (m) 0.588; 86% γ 0.507; 7% γ 0.770	−80.62		
Nb	90	14.6 hr	53% β⁺ 1.5; 47% K; (98% m); 93% γ 1.129; 83% γ (m) 2.319; 66% γ 0.141	−82.65		8⁺
Nb	90m	18.8 s	100% IT; 64% γ 0.122	−82.53		4⁻
Nb	91m	62 d	97% IT; 3% K; 3% γ 1.205; 1% γ 0.104	−86.532		$\frac{1}{2}^-$
Nb	92	3 × 10⁷ yr	100% K; γ	−86.448		7⁺
Nb	92m	10.14 d	100% K; 99% γ 0.934; 2% γ 0.913; 1% γ 1.848	−86.313		2⁺
Nb	93	100%		−87.209	1.1 (γ)	$\frac{9}{2}^+$
Nb	93m	13.6 yr	100% IT	−87.179		$\frac{1}{2}^-$
Nb	94	2.0 × 10⁴ yr	100% β⁻ 0.47; 100% γ 0.871; 98% γ 0.703	−86.367	15 (γ) (s); 0.59 (γm) (r)	6⁺
Nb	94m	6.26 min	100% IT	−86.326		3⁺
Nb	95	34.97 d	100% β⁻ 0.16; 100% γ 0.766	−86.787		$\frac{9}{2}^+$
Nb	95m	87 hr	97% IT; 3% β⁻ 1.0; 26% γ 0.236; 2% γ 0.204	−86.552		$\frac{1}{2}^-$
Nb	96	23.4 hr	96% β⁻ 0.75; 97% γ 0.778; 55% γ 0.569; 49% γ 1.091	−85.608		
Nb	97	72 min	98% β⁻ 1.27; 98% γ 0.658; 1% γ 1.025	−85.612		$\frac{9}{2}^+$
Nb	97m	60 s	100% IT; 98% γ 0.743	−84.868		$\frac{1}{2}^-$
Nb	98	2.9 s	90% β⁻ 4.59; 3% γ 0.787; 2% γ 1.024; 1% γ 0.645, 0.972, 1.432	−83.53		1⁺

Table of the nuclides (*contd*)

Nuclide Z	A	Natural abundance or half-life	Decay modes and major radiations	Mass excess MeV	σ/b	I^π
Nb	98m	51.3 min	5% β^- 3.2; 16% β^- 2.5; 28% β^- 2.0; 93% γ 0.787; 73% γ 0.723; 18% γ 1.169	−83.45		
Nb	99	15.0 s	100% β^- 3.4; 90% γ 0.138; 45% γ 0.098	−82.35		
Nb	99m	2.6 min	β^-; γ	−81.98		
Nb	100	1.5 s	β^-; γ	−80.0		
Nb	100	3.1 s	β^-; γ			
Nb	101	7.0 s	β^-; γ	−79.0		
Nb	102	4.3 s	β^-; γ	−76.4		
Nb	102	1.3 s	β^-; γ	−76.4		
Nb	103	1.5 s	β^-; γ	−75.4		
Nb	104	0.8 s	β^-; γ	−72.7		
Nb	104	4.8 s	β^-; γ			
Nb	106	1 s	β^-; γ			
42 Mo	90	5.67 hr	75% K; 25% β^+ 1.1; (94% m); 78% γ 0.257; 64% γ 0.122; 6% γ 0.203	−80.17		0^+
Mo	91	15.49 min	94% β^+ 3.4; 6% K	−82.20		$\frac{9}{2}^+$
Mo	91m	64 s	50% IT; β^+; K; (50% m); 48% γ 0.653; 24% γ 1.508; 19% γ 1.208	−81.55		$\frac{1}{2}^-$
Mo	92	14.8%		−86.807	0.3 (t) (r)	0^+
Mo	93	4×10^3 yr	100% K; (96% m)	−86.803		$\frac{5}{2}^+$
Mo	93m	6.9 hr	100% IT; 100% γ 0.685; 99% γ 1.477; 57% γ 0.263	−84.378		$\frac{21}{2}^+$
Mo	94	9.3%		−88.412		0^+
Mo	95	15.9%		−87.712	14 (γ) (r)	$\frac{5}{2}^+$
Mo	96	16.7%		−88.795	1.0 (γ) (r)	0^+
Mo	97	9.6%		−87.545	2 (γ) (r)	$\frac{5}{2}^+$
Mo	98	24.1%		−88.115	0.13 (γ)	0^+
Mo	99	66.02 hr	81% β^- 1.21; (87% m); 76% γ (m) 0.141; 13% γ 0.739; 6% γ 0.181	−85.970		$\frac{1}{2}^+$
Mo	100	9.6%		−86.19	0.20 (γ) (r)	0^+
Mo	101	14.6 min	12% β^- 2.6; 21% β^- 0.8; 23% γ 0.591 (double); 20% γ 0.192; 13% γ 1.013	−83.52		$\frac{1}{2}^+$
Mo	102	11.0 min	β^-; γ	−83.56		0^+
Mo	103	60 s	β^-	−80.6		
Mo	104	1.00 min	β^-; γ	−81.7		0^+
Mo	105	36 s	β^-; γ	−77.1		
Mo	106	9.5 s	β^-; γ	−76		0^+
Mo	108	0.9 s	β^-	−71		0^+
43 Tc	90	50 s	β^+; γ			
Tc	90	7.9 s	β^+ 7.9; γ	−71		
Tc	91	3.14 min	β^+ 5.2; K; (1% m); 14% γ 2.451; 8% γ 1.605; 7% γ 1.565	−76.0		

Table of the nuclides (*contd*)

Nuclide		Natural abundance or half-life	Decay modes and major radiations	Mass excess MeV	σ/b	Iπ
Z	A					
Tc	91	3.3 min	β^+; K; (98% m); 55% γ 0.503; 47% γ (m) 0.653; 4% γ 0.928			
Tc	92	4.4 min	β^+ 4.1; K; 100% γ 0.773; 100% γ 1.510; 80% γ 0.329	−78.94		
Tc	93	2.75 hr	87% K; β^+ 0.8; 66% γ 1.363; 24% γ 1.520; 10% γ 1.477	−83.61		$\frac{9}{2}^+$
Tc	93m	44 min	80% IT; 20% K; 60% γ 0.392; 16% γ 2.645; 2% γ 3.129	−83.22		$\frac{1}{2}^-$
Tc	94	293 min	89% K; 11% β^+ 0.82; 100% γ 0.871; 100% γ 0.703; 98% γ 0.850	−84.16		7^+
Tc	94m	53 min	β^+ 2.4; 28% K; 94% γ 0.871; 6% γ 1.869; 5% γ 1.522	−84.08		
Tc	95	20.0 hr	100% K; 93% γ 0.766; 4% γ 1.074; 2% γ 0.948	−86.01		$\frac{9}{2}^+$
Tc	95m	61 d	96% K; 4% IT; 66% γ 0.204; 32% γ 0.582; 28% γ 0.835	−85.97		$\frac{1}{2}^-$
Tc	96	4.35 d	100% K; 99% γ 0.778; 97% γ 0.850; 81% γ 0.813	−85.82		7^+
Tc	96m	52 min	98% IT; 2% K; 2% γ 0.778; 1% γ 1.200	−85.79		4^+
Tc	97	2.6×10^6 yr	100% K	−87.224		$\frac{9}{2}^+$
Tc	97m	90 d	100% IT	−87.128		$\frac{1}{2}^-$
Tc	98	4.2×10^6 yr	100% β^- 0.39; 100% γ 0.652; 100% γ 0.745	−86.43	3 (γm) (r)	
Tc	99	2.1×10^5 yr	100% β^- 0.29	−87.326	19 (γ)	$\frac{9}{2}^+$
Tc	99m	6.01 hr	100% IT; 89% γ 0.141	−87.184		$\frac{1}{2}^-$
Tc	100	15.8 s	93% β^- 3.20; 7% γ 0.540; 6% γ 0.591	−86.019		1^+
Tc	101	14.3 min	89% β^- 1.3; 88% γ 0.307; 6% γ 0.545; 3% γ 0.127	−86.33		$\frac{9}{2}^+$
Tc	102	5.3 s	41% β^- 4.5; 30% β^- 4.0; 53% γ 0.475; 11% γ 1.105 (double); 10% γ 0.628	−84.6		1^+
Tc	102m	4.4 min	3% β^- 3.7; 35% β^- 2.1; 2% IT; 85% γ 0.475; 26% γ 0.628; 16% γ 0.630			
Tc	103	50 s	β^-; γ	−84.9		
Tc	104	18.1 min	24% β^- 3.9; 89% γ 0.358; 15% γ 0.531; 14% γ 0.535	−83.9		
Tc	105	7.6 min	41% β^- 3.4 (double); 11% γ 0.143; 10% γ 0.108; 8% γ 0.321	−82.5		
Tc	106	36 s	β^-; γ	−80.0		
Tc	107	21.2 s	β^-; γ	−79.5		
Tc	108	5.1 s	β^-; γ	−76		
Tc	109	1.4 s	β^-			
Tc	110	0.82 s	β^-; γ			
44 Ru	92	3.7 min	β^+; K; 92% γ 0.214; 88% γ 0.259; 63% γ 0.135	−75		0^+

Table of the nuclides (*contd*)

Nuclide		Natural abundance or half-life	Decay modes and major radiations	Mass excess MeV	σ/b	Iπ
Z	A					
Ru	93	60 s	β^+; K; 5% γ 0.681; 1% γ 1.435	−77.3		
Ru	93m	10.8 s	β^+; K; 21% IT; (79% m); 36% γ 1.396; 24% γ 1.111; 20% γ 0.734	−76.6		
Ru	94	52 min	100% K; (100% m); 79% γ 0.367; 21% γ 0.892; 2% γ 0.525	−82.57		0^+
Ru	95	1.65 hr	85% K; β^+ 1.2; (3% m); 71% γ 0.336; 21% γ 1.097; 18% γ 0.627	−83.45		$\frac{5}{2}^+$
Ru	96	5.5%		−86.08	0.25 (γ)	0^+
Ru	97	2.88 d	100% K; 86% γ 0.216; 10% γ 0.325; 1% γ 0.569	−86.07		$\frac{5}{2}^+$
Ru	98	1.86%		−88.226		0^+
Ru	99	12.7%		−87.620	4 (γ)	$\frac{5}{2}^+$
Ru	100	12.6%		−89.222	5.8 (γ)	0^+
Ru	101	17.0%		−87.952	5 (γ)	$\frac{5}{2}^+$
Ru	102	31.6%		−89.101	1.3 (γ)	0^+
Ru	103	39.4 d	6% β^- 0.72; 87% β^- 0.22; (100% m); 86% γ 0.497; 5% γ 0.610	−87.261		$\frac{5}{2}^+$
Ru	104	18.7%		−88.099	0.47 (γ)	0^+
Ru	105	4.44 hr	2% β^- 1.8; 48% β^- 1.2; (28% m); 48% γ 0.724; 18% γ 0.469; 16% γ 0.676	−85.94	0.30 (γ)	
Ru	106	367 d	100% β^- 0.039	−86.33		0^+
Ru	107	4.2 min	75% β^- 3.2; 14% γ 0.194; 7% γ 0.849; 6% γ 0.463	−83.7		
Ru	108	4.5 min	70% β^- 1.2; 30% β^- 1.0; 30% γ 0.165	−83.8		0^+
Ru	109	34 s	β^-; γ	−80.8		
Ru	109	13 s	β^-	−80.8		
Ru	110	16 s	β^-	−80		0^+
Ru	111	1.5 s	β^-			
45 Rh	95	5.0 min	β^+ 3.2; K; 72% γ 0.942; 21% γ 1.352; 6% γ 0.678	−78.3		
Rh	95m	1.96 min	88% IT; β^+; K; 80% γ 0.543; 8% γ 0.784; 2% γ 3.407	−77.8		
Rh	96	9.9 min	β^+; K; 100% γ 0.833; 98% γ 0.685; 80% γ 0.635	−79.63		
Rh	96m	1.51 min	60% IT; β^+; K; 39% γ 0.833; 8% γ 1.099; 6% γ 1.692	−79.58		
Rh	97	31 min	β^+; K; 75% γ 0.421; 12% γ 0.840; 9% γ 0.879	−82.6		
Rh	97m	44 min	β^+; K; 5% IT; 51% γ 0.189; 13% γ 0.422; 13% γ 2.245	−82.3		
Rh	98(g)	8.7 min	β^+ 3.4; K; 94% γ 0.652; 5% γ 0.745; 5% γ 1.817	−83.17		
Rh	98(m)	3.5 min	β^+; K; 96% γ 0.652; 96% γ 1.414; 78% γ 0.745	−83.16		

Table of the nuclides (*contd*)

Nuclide Z	A	Natural abundance or half-life	Decay modes and major radiations	Mass excess MeV	σ/b	Iπ
Rh	99	15.0 d	97% K; β^+ 1.1; 38% γ 0.528; 35% γ 0.353; 33% γ 0.090	−85.52		
Rh	99m	4.7 hr	90% K; β^+; 66% γ 0.341; 16% γ 1.261; 13% γ 0.618	−85.45		$\frac{9}{2}^+$
Rh	100	20.8 hr	95% K; 5% β^+ 2.6; 78% γ 0.540; 35% γ 2.376; 20% γ 0.823, 1.553	−85.59		1^-
Rh	100m	4.7 min	93% IT; K; β^+; 70% γ 0.075; 31% γ 0.265; 11% γ 0.540	−85.25		
Rh	101	3.3 yr	100% K; 70% γ 0.127; 68% γ 0.198; 13% γ 0.325	−87.41		$\frac{1}{2}^-$
Rh	101m	4.34 d	93% K; 7% IT; 87% γ 0.307; 4% γ 0.545	−87.25		$\frac{9}{2}^+$
Rh	102	2.9 yr	100% K; 95% γ 0.475; 56% γ 0.631; 44% γ 0.697			
Rh	102m	206 d	62% K; 10% β^+ 1.3; 4% β^+ 0.8; 17% β^- 1.1; 2% β^- 0.5; 5% IT; 45% γ 0.475; 4% γ 0.628; 3% γ 0.469	−86.78		
Rh	103	100%		−88.024	134 (γ); 11 (γm)	$\frac{1}{2}^-$
Rh	103m	56.12 min	100% IT	−87.984		
Rh	104	42.8 s	98% β^- 2.45; 2% γ 0.556	−86.952	40 (γ) (r)	1^+
Rh	104m	4.4 min	100% IT; 48% γ 0.051; 3% γ 0.097; 2% γ 0.078	−86.823	800 (γ) (r)	5^+
Rh	105	35.5 hr	75% β^- 0.57; 19% γ 0.319; 5% γ 0.306	−87.86	11 000 (γ); 5000 (γm)	
Rh	105m	45 s	100% IT; 20% γ 0.130	−87.73		$\frac{1}{2}^-$
Rh	106	29.8 s	80% β^- 3.5; 19% γ 0.512; 10% γ 0.622; 2% γ 1.051	−86.37		1^+
Rh	106m	130 min	1% β^- 1.7; 84% β^- 0.9; 86% γ 0.512; 31% γ 1.047; 29% γ 0.717	−86.24		
Rh	107	21.7 min	β^-; 66% γ 0.303; 9% γ 0.392; 5% γ 0.312	−86.86		
Rh	108	16.8 s	54% β^- 4.5; 21% β^- 3.4; 43% γ 0.434; 21% γ 0.619; 11% γ 0.511	−85.0		1^+
Rh	108	6.0 min	7% β^- 3.4; 68% β^- 1.6; 91% γ 0.434; 58% γ 0.581; 50% γ 0.947	−85.1		
Rh	109	80 s	β^- 2.5; 62% γ 0.327; 22% γ 0.291; 14% γ 0.249	−85.1		
Rh	110	3.0 s	β^- 5.5; γ	−82.8		
Rh	110	29 s	19% β^- 4.5; 42% β^- 2.6; 92% γ 0.374; 41% γ 0.546; 28% γ 0.440	−82.9		
Rh	112	4.7 s	β^-; γ	−80		
46 Pd	97	3.3 min	β^+; K; γ	−77.8		
Pd	98	17.5 min	K; β^+; γ	−81.3		0^+

Table of the nuclides (*contd*)

Nuclide Z	A	Natural abundance or half-life	Decay modes and major radiations	Mass excess MeV	σ/b	I^π
Pd	99	21.4 min	β^+ 2.2; K; (97% m); 73% γ 0.136; 15% γ 0.264; 7% γ 0.673	−86.11		
Pd	100	3.6 d	100% K; γ	−85.23		0^+
Pd	101	8.5 hr	94% K; 6% β^+ 0.8; (100% m); 18% γ 0.296; 11% γ 0.590; 6% γ 0.270	−85.43		
Pd	102	1.0%		−87.93	5 (γ) (r)	0^+
Pd	103	16.96 d	100% K; (100% m)	−87.48		$\frac{5}{2}^+$
Pd	104	11.0%		−89.400		0^+
Pd	105	22.2%		−88.422		$\frac{5}{2}^+$
Pd	106	27.3%		−89.913	0.28 (γ); 0.013 (γm)	0^+
Pd	107	6.5×10^6 yr	100% β^- 0.03	−88.37		$\frac{5}{2}^+$
Pd	107m	21.3 s	100% IT; 68% γ 0.215	−88.16		$\frac{11}{2}^-$
Pd	108	26.7%		−89.523	11 (γ) (r); 0.19 (γm) (r)	0^+
Pd	109	13.43 hr	100% β^- 1.0; (100% m); 4% γ (m) 0.088	−87.606		$\frac{5}{2}^+$
Pd	109m	4.69 min	100% IT; 56% γ 0.189	−87.417		$\frac{11}{2}^-$
Pd	110	11.8%		−88.34	0.36 (γ) (r); 0.02 (γm) (r)	0^+
Pd	111	22 min	96% β^- 2.1; (99% m); 1% γ 0.071; 1% γ 0.580; 1% γ (m) 0.060	−86.03		
Pd	111m	5.5 hr	71% IT; β^-; (23% m); 32% γ 0.172; 7% γ 0.071; 5% γ 0.391	−85.86		
Pd	112	21.1 hr	100% β^- 0.28; 4% γ 0.019	−86.33		0^+
Pd	113	1.5 min	β^-	−83.6		
Pd	114	2.4 min	β^-	−83.8		0^+
Pd	115	37.4 s	β^-			
Pd	116	14 s	β^-	−80		0^+
Pd	117	5 s	β^-			
Pd	118	3.1 s	β^-	−76.2		0^+
47 Ag	100	2.3 min	β^+; K; γ	−77.9		
Ag	101	10.8 min	β^+; K; 92% γ 0.263; 33% γ 0.668; 33% γ 1.164	−81.3		
Ag	102	13.0 min	β^+; K; 97% γ 0.557; 58% γ 0.719; 17% γ 1.745	−82.33		5^+
Ag	102m	7.7 min	β^+ 4.0; K; 49% IT; 42% γ 0.557; 10% γ 1.835; 7% γ 2.055	−82.32		2^+
Ag	103	1.10 hr	58% K; 21% β^+ 1.7; 22% γ 0.119; 20% γ 0.148; 9% γ 0.267	−84.80		$\frac{7}{2}^+$
Ag	103m	5.7 s	100% IT; 21% γ 0.134	−84.67		
Ag	104	69 min	β^+; K; 92% γ 0.556; 66% γ 0.768; 25% γ 0.942	−85.15		5^+
Ag	104m	33 min	β^+ 2.7; K; 33% IT; 60% γ 0.556; 3% γ 1.239; 1% γ 1.782	−85.14		2^+

Table of the nuclides (*contd*)

Nuclide		Natural abundance or half-life	Decay modes and major radiations	Mass excess MeV	σ/b	Iπ
Z	A					
Ag	105	41.3 d	100% K; 42% γ 0.345; 31% γ 0.280; 12% γ 0.443, 0.645	−87.08		$\frac{1}{2}^-$
Ag	105m	7.2 min	100% IT	−87.05		
Ag	106	23.96 min	K; β⁺ 2.0; 17% γ 0.512	−86.93		1⁺
Ag	106m	8.5 d	100% K; 88% γ 0.512; 29% γ 0.717; 28% γ 0.451	−86.84		6⁺
Ag	107	51.83%		−88.404	37 (γ); 0.3 (γm) (s)	$\frac{1}{2}^-$
Ag	107m	44.3 s	100% IT; 5% γ 0.093	−88.311		$\frac{7}{2}^+$
Ag	108	2.37 min	96% β⁻ 1.6; 2% β⁻ 1.0; 2% K; 2% γ 0.633	−87.60		1⁺
Ag	108m	127 yr	K; β⁺; 9% IT; 91% γ 0.434; 91% γ 0.614; 91% γ 0.723	−87.49		6⁺
Ag	109	48.17%		−88.722	88 (γ); 4 (γm)	$\frac{1}{2}^-$
Ag	109m	39.8 s	100% IT; 4% γ 0.088	−88.634		$\frac{7}{2}^+$
Ag	110	24.4 s	95% β⁻ 2.9; 5% γ 0.658	−87.456		1⁺
Ag	110m	252.2 d	30% β⁻ 0.5; 67% β⁻ 0.09; 1% IT; 94% γ 0.658; 73% γ 0.885; 34% γ 0.937	−87.338	80 (γ)	6⁺
Ag	111	7.45 d	94% β⁻ 1.03; 5% γ 0.342; 1% γ 0.246	−88.226	3 (γ) (r)	$\frac{1}{2}^-$
Ag	111m	65 s	100% IT; 1% γ 0.060	−88.166		
Ag	112	3.14 hr	54% β⁻ 4.0; 21% β⁻ 3.4; 42% γ 0.617; 5% γ 1.388; 3% γ 0.607, 1.614	−86.62		2
Ag	113	1.15 min	β⁻; γ	−86.8		
Ag	113	5.37 hr	85% β⁻ 2.0; (2% m); 9% γ 0.298; 2% γ 0.259; 1% γ 0.316	−87.04		$\frac{1}{2}$
Ag	114	4.5 s	90% β⁻ 4.9; 10% γ 0.558; 1% γ 0.576	−85.2		1⁺
Ag	115	18 s	β⁻; γ			
Ag	115	20.0 min	32% β⁻ 3.2; 24% β⁻ 3.0; (9% m); 32% γ 0.230; 8% γ 0.214; 6% γ 0.473	−84.9		
Ag	116	2.68 min	β⁻; 84% γ 0.513; 13% γ 0.700; 9% γ 1.305	−82.6		
Ag	116m	10.5 s	β⁻; 2% IT; 95% γ 0.513; 42% γ 0.706; 16% γ 1.029	−82.5		
Ag	117	1.21 min	β⁻; γ	−82.2		
Ag	117	5.3 s	β⁻	−82.2		
Ag	118	3.7 s	β⁻; 95% γ 0.488; 40% γ 0.677; 12% γ 3.226	−80.2		
Ag	118m	2.8 s	β⁻; 41% IT; 59% γ 0.488; 58% γ 0.677; 32% γ 1.059	−80.1		
Ag	119	2.1 s	3% β⁻ 4.2; 31% β⁻ 3.9; (22% m); 11% γ 0.626; 10% γ 0.366; 10% γ 0.399 (double)	−79.3		
Ag	120	1.17 s	β⁻; γ	−78		
Ag	120m	0.32 s	β⁻; 37% IT; γ	−78		
Ag	121	<3 s	β⁻; γ			
Ag	122	1.5 s	β⁻; γ	−71		

Table of the nuclides (*contd*)

Nuclide		Natural abundance or half-life	Decay modes and major radiations	Mass excess MeV	σ/b	I^{π}
Z	A					
Ag	123	0.39 s	β^-; n			
48 Cd	100	1.1 min	β^+; K; γ	-73.4		0^+
Cd	101	1.2 min	β^+; K; γ	-75.5		
Cd	102	5.5 min	K; β^+; (95% m); 62% γ 0.481; 13% γ 1.037; 9% γ 0.505	-79.4		0^+
Cd	103	7.3 min	β^+; K; γ	-80.6		
Cd	104	58 min	99% K; 1% β^+ 0.4; (100% m) 47% γ 0.084; 20% γ 0.709; 6% γ 0.559	-83.6		0^+
Cd	105	56 min	K; β^+ 1.7; (85% m); 5% γ 0.962; 4% γ 0.347; 4% γ 1.302	-84.34		$\frac{5}{2}^+$
Cd	106	1.25%		-87.13	1 (γ) (r)	0^+
Cd	107	6.50 hr	100% K; (100% m); 5% γ (m) 0.093	-86.99		$\frac{5}{2}^+$
Cd	108	0.89%		-89.25	1.2 (γ) (r)	0^+
Cd	109	453 d	100% K; (100% m); 4% γ (m) 0.088	-88.540	700 (γ) (r)	$\frac{5}{2}^+$
Cd	110	12.5%		-90.349	11 (γ) (r); 0.10 (γm) (r)	0^+
Cd	111	12.8%		-89.254	24 (γ) (r)	$\frac{1}{2}^+$
Cd	111m	48.6 min	100% IT; 94% γ 0.245; 31% γ 0.150	-88.858		$\frac{11}{2}^-$
Cd	112	24.1%		-90.578	2 (γ) (r)	0^+
Cd	113	9 × 10^{15} yr 12.2%	100% β^- 0.32	-89.050	19 800 (γ)	$\frac{1}{2}^+$
Cd	113m	14 yr	100% β^- 0.58	-88.787		$\frac{11}{2}^-$
Cd	114	28.7%		-90.020	0.30 (γ) (s); 0.04 (γm) (s)	0^+
Cd	115	53.38 hr	63% β^- 1.1; 33% β^- 0.6; (100% m); 46% γ (m) 0.336; 34% γ 0.528; 10% γ 0.492	-88.09		$\frac{1}{2}^+$
Cd	115m	44.8 d	97% β^- 1.6; 2% γ 0.934; 1% γ 1.291	-87.92		$\frac{11}{2}^-$
Cd	116	7.5%		-88.718	0.05 (γ) (s); 0.025 (γm) (s)	0^+
Cd	117	2.4 hr	21% β^- 2.2; 32% β^- 0.6; (94% m); 29% γ 0.273; 18% γ 0.344; 18% γ 1.303	-86.42		$\frac{1}{2}^+$
Cd	117m	3.3 hr	52% β^- 0.6; (1% m); 25% γ 1.997; 23% γ 1.066; 15% γ 0.564, 1.434	-86.29		$\frac{11}{2}^-$
Cd	118	50.3 min	100% β^- 0.7	-86.71		0^+
Cd	119	2.7 min	β^- 3.2; (93% m); 25% γ 0.293; 13% γ 0.343; 8% γ 1.610, 1.764	-84.2		$\frac{1}{2}^+$
Cd	119m	1.9 min	1% β^- 2.2; 30% β^- 1.6; 28% γ 1.025; 24% γ 2.021; 21% γ 0.721	-84.1		$\frac{11}{2}^-$
Cd	120	50.8 s	β^- 1.7	-83.98		0^+

Table of the nuclides (contd)

Nuclide			Natural abundance or half-life	Decay modes and major radiations	Mass excess MeV	σ/b	I^π
Z		A					
	Cd	121	12.8 s	β^-; γ	-81		
	Cd	121	4.8 s	β^-; γ	-81		
	Cd	122	5.8 s	β^-	-80		0^+
	Cd	124	0.9 s	β^-; 50% γ 0.180; 23% γ 0.063; 13% γ 0.143	-76		0^+
49	In	104	1.5 min	β^+; K; 100% γ 0.658; 100% γ 0.833; 32% γ 0.879	-75.6		
	In	105	5.1 min	β^+; K; γ	-79.3		
	In	106	5.32 min	β^+ 4.9; K; 93% γ 0.633; 8% γ 0.861; 6% γ 1.621	-80.59		
	In	106	6.3 min	β^+; K; 100% γ 0.633; 90% γ 0.861; 39% γ 0.998			
	In	107	32.4 min	65% K; β^+ 2.3; 47% γ 0.205; 12% γ 0.506; 10% γ 0.321	-83.5		$\frac{9}{2}^+$
	In	107m	50 s	100% IT; 94% γ 0.678	-82.9		$\frac{1}{2}^-$
	In	108	40 min	K; β^+ 3.5; 76% γ 0.633; 12% γ 1.986; 9% γ 3.452	-84.1		3^+
	In	108	58 min	K; β^+; 100% γ 0.633; 95% γ 0.876; 39% γ 0.243	-84.1		
	In	109	4.2 hr	94% K; β^+ 0.8; 74% γ 0.203; 5% γ 0.624, 1.149	-86.52		$\frac{9}{2}^+$
	In	109m₁	1.3 min	100% IT; 93% γ 0.650	-85.87		$\frac{1}{2}^-$
	In	109m₂	0.20 s	100% IT; 98% γ 0.678; 75% γ 1.428; 20% γ 0.404, 1.044	-84.41		
	In	110	4.9 hr	100% K; 99% γ 0.658; 95% γ 0.885; 69% γ 0.937			7^+
	In	110	69 min	β^+ 2.3; K; 98% γ 0.658; 2% γ 2.129; 2% γ 2.211	-86.41		2^+
	In	111	2.83 d	100% K; 94% γ 0.171; 94% γ 0.245	-88.41		$\frac{9}{2}^+$
	In	111m	7.6 min	100% IT; 87% γ 0.537	-87.87		$\frac{1}{2}^-$
	In	112	14.4 min	44% β^- 0.66; 34% K; 22% β^+ 1.6; 6% γ 0.617; 2% γ 0.607	-88.00		1^+
	In	112m	20.9 min	100% IT; 13% γ 0.155	-87.85		4^+
	In	113	4.3%		-89.372	3 (γ); 8 (γm)	$\frac{9}{2}^+$
	In	113m	99.5 min	100% IT; 64% γ 0.392	-88.980		$\frac{1}{2}^-$
	In	114	71.9 s	98% β^- 1.98; 2% K	-88.576		1^+
	In	114m	49.51 d	97% IT; 3% K; 16% γ 0.190; 3% γ 0.558; 3% γ 0.725	-88.386		5^+
	In	115	5×10^{14} yr 95.7%	100% β^- 0.49	-89.54	41 (γ); 70 (γm₁); 91 (γm₂)	$\frac{9}{2}^+$
	In	115m	4.486 hr	95% IT; 5% β^- 0.83; 46% γ 0.336	-89.21		$\frac{1}{2}^-$
	In	116	14.10 s	99% β^- 3.3; 1% γ 1.294	-88.25		1^+
	In	116m₁	54.1 min	51% β^- 1.1; 36% 0.9; 85% γ 1.294; 56% γ 1.097; 32% γ 0.417	-88.13		5^+
	In	116m₂	2.16 s	100% IT; (100% m); 35% γ 0.162	-87.96		8^-

Table of the nuclides (*contd*)

Nuclide		Natural abundance or half-life	Decay modes and major radiations	Mass excess MeV	σ/b	Iπ
Z	A					
In	117	42 min	100% β^- 0.74; 100% γ 0.553; 87% γ 0.159	−88.94		$\frac{9}{2}^+$
In	117m	1.93 hr	37% β^- 1.8; 16% β^- 1.6; 47% IT; 17% γ 0.315; 14% γ 0.159	−88.63		$\frac{1}{2}^-$
In	118	5.0 s	85% β^- 4.2; 15% γ 1.230; 5% γ 0.528	−87.5		1^+
In	118	4.4 min	24% β^- 2.0; 64% β^- 1.3; 96% γ 1.230; 82% γ 1.051; 55% γ 0.683	−87.4		
In	118	8.5 s	99% IT; 1% β^- 1.9; 22% γ 0.138; 1% γ 0.254	−87.2		
In	119	2.1 min	100% β^- 1.6; 99% γ 0.763	−87.73		$\frac{9}{2}^+$
In	119m	18 min	94% β^- 2.6; 5% IT; 2% γ 0.311	−87.42		$\frac{1}{2}^-$
In	120	44 s	19% β^- 3.1; 42% β^- 2.2; 97% γ 1.172; 60% γ 1.023; 30% γ 0.864	−85.8		
In	120	3.0 s	81% β^- 5.6; 19% γ 1.172; 1% γ 0.704; 1% γ 1.186	−85.5		1^+
In	121	30.0 s	100% β^- 2.4; 87% γ 0.926; 8% γ 0.262; 7% γ 0.657	−85.84		$\frac{9}{2}^+$
In	121m	3.8 min	β^- 3.7; 1% IT; 21% γ 0.060; 1% γ 1.041; 1% γ 1.102	−85.53		$\frac{1}{2}^-$
In	122	9.2 s	100% β^- 4.3; 100% γ 1.141; 56% γ 1.003; 18% γ 1.194	−83.4		
In	122	1.5 s	71% β^- 6.4; 29% γ 1.141	−83.5		
In	123(g)	6.0 s	β^-; (100% m); 63% γ 1.130; 32% γ 1.020; 3% γ 0.619	−83.44		
In	123(m)	48 s	β^- 4.6; (100% m); γ	−83.12		
In	124	2.4 s	β^-; γ			
In	124	3.2 s	β^-; γ	−81.1		
In	125	2.32 s	1% β^- 4.8; 85% β^- 4.0; 76% γ 1.335; 10% γ 1.032; 8% γ 0.618	−80.5		
In	125	12.2 s	β^-; γ			
In	126	1.53 s	β^-	−77.9		
In	127	1.3 s	β^-; γ	−77.4		
In	127	3.7 s	β^-; n; γ	−77.4		
In	129	2.5 s	β^-; n; γ			
In	129	0.99 s	β^-; n; γ	−73.1		
In	130	0.58 s	β^-; n; 100% γ 0.775; 100% γ 1.217; 80% γ 0.127	−70.1		
In	131	0.29 s	β^-; n; γ	−70		
In	132	0.12 s	β^-; n; γ	−65		
50 Sn	106	1.9 min	K; β^+; γ	−77.0		0^+
Sn	107	2.90 min	β^+; K; γ	−78.4		
Sn	108	10.5 min	100% K; 74% γ 0.397; 51% γ 0.273; 24% γ 0.169	−81.9		0^+

Table of the nuclides (*contd*)

Nuclide		Natural abundance or half-life	Decay modes and major radiations	Mass excess MeV	σ/b	I^π
Z	A					
Sn	109	18.0 min	β^+; K; (25% m_1); 35% γ 1.098; 23% γ (m) 0.650; 14% γ 1.322	−82.6		$\frac{7}{2}^+$
Sn	110	4.1 hr	100% K; 98% γ 0.283	−85.83		0^+
Sn	111	35 min	71% K; 29% β^+ 1.5; 1% γ 0.761, 1.152, 1.915	−85.94		$\frac{7}{2}^+$
Sn	112	1.01%		−88.658	0.4 (γ) (r); 0.3 (γm) (r)	0^+
Sn	113	115.1 d	100%K; (100%m); 64%γ (m) 0.392; 2%γ 0.255	−88.332		$\frac{1}{2}^+$
Sn	113m	21.4 m	91% IT; 9% K; 1% γ 0.079	−88.253		$\frac{7}{2}^+$
Sn	114	0.67%		−90.560		0^+
Sn	115	0.38%		−90.035	50 (γ) (r)	$\frac{1}{2}^+$
Sn	116	14.8%		−91.526	0.006 (γm) (r)	0^+
Sn	117	7.75%		−90.399	3 (γ) (r)	$\frac{1}{2}^+$
Sn	117m	14.0 d	100% IT; 86% γ 0.159	−90.084		$\frac{11}{2}^-$
Sn	118	24.3%		−91.654	0.08 (γm) (r)	0^+
Sn	119	8.6%		−90.067	2 (γ)	$\frac{1}{2}^+$
Sn	119m	250 d	100% IT; 16% γ 0.024	−89.977		$\frac{11}{2}^-$
Sn	120	32.4%		−91.102	0.16 (γ); 0.001 (γm)	0^+
Sn	121	27.06 hr	100% β^- 0.39	−89.202		$\frac{3}{2}^+$
Sn	122	4.56%		−89.946	0.001 (γ); 0.15 (γ m)	0^+
Sn	123	129 d	99% β^- 1.40; 1% γ 1.089	−87.821		$\frac{11}{2}^-$
Sn	123m	40.1 min	100% β^- 1.3; 86% γ 0.160	−87.796		
Sn	124	5.64%		−88.240	0.005 (γ); 0.13 (γm)	0^+
Sn	125	9.63 d	82% β^- 2.4; 9% γ 1.066; 4% γ 1.089; 3% γ 0.823	−85.903		$\frac{11}{2}^-$
Sn	125m	9.52 min	98% β^- 2.0; 97% γ 0.332	−85.876		$\frac{3}{2}^+$
Sn	126	10^5 yr	100% β^- 0.25; (100% m); 37% γ 0.088; 10% γ 0.064; 9% γ 0.087	−86.02		0^+
Sn	127	2.16 hr	22% β^- 2.9; 12% β^- 0.3; 38% γ 1.114; 19% γ 1.096; 11% γ 0.823	−83.79		
Sn	127m	4.13 min	β^- 2.4; 99% γ 0.491; 5% γ 1.348; 4% γ 1.564	−83.78		
Sn	128	59.3 min	84% β^- 0.6; (100% (m)); 58% γ 0.482; 27% γ 0.075; 17% γ 0.557	−83.4		0^+
Sn	129	2.23 min	β^-; γ	−80.6		
Sn	129m	7.5 min	β^-; γ	−80.6		
Sn	130	3.7 min	15% β^- 1.3; 85% β^- 1.0; 71% γ 0.192; 59% γ 0.780; 36% γ 0.070	−80.4		0^+
Sn	130m	1.7 min	β^-; γ			
Sn	131	63 s	β^-; γ	−77.5		
Sn	132	41 s	100% β^- 1.8; 49% γ 0.086; 43% γ 0.340; 42% γ 0.247, 0.899	−76.6		0^+
Sn	133	1.47 s	β^- 7.5; n; γ	−71.5		
Sn	134	1.04 s	β^-; 17% n			0^+
51 Sb	108	7.0 s	75% β^+ 7.3; 25% β^+ 6.4; 100% γ 1.206; 25% γ 0.905	−72.4		

Table of the nuclides (*contd*)

Nuclide Z	A	Natural abundance or half-life	Decay modes and major radiations	Mass excess MeV	σ/b	I$^\pi$
Sb	109	18.3 s	β^+; K; γ	−76.1		
Sb	110	23.0 s	β^+; K; 92% γ 1.211; 31% γ 0.985; 13% γ 1.243	−76.8		
Sb	111	75 s	β^+ 3.3; K; 64% γ 0.154; 27% γ 0.489; 8% γ 1.032	−81.5		
Sb	112	54 s	β^+ 4.8; K; 95% γ 1.257; 14% γ 0.991; 3% γ 0.670	−81.6		
Sb	113	6.7 min	β^+ 2.4; K; (16% m); 80% γ 0.498; 10% γ 0.331; 5% γ 0.939	−84.44		
Sb	114	3.5 min	β^+ 4.1; K; 100% γ 1.300; 18% γ 0.888; 6% γ 0.322	−84.1		
Sb	115	31.8 min	33% β^+ 1.5; 67% K; 99% γ 0.497; 3% γ 0.491	−87.01		$\frac{5}{2}^+$
Sb	116	16 min	72% K; β^+ 2.3; 85% γ 1.294; 25% γ 0.932; 14% γ 2.225	−86.93		3$^+$
Sb	116m	60.4 min	81% K; β^+ 1.3; 100% γ 1.294; 72% γ 0.973; 52% γ 0.543	−86.32		8$^-$
Sb	117	2.80 hr	98% K; 2% β^+ 0.57; 86% γ 0.159	−88.65		$\frac{5}{2}^+$
Sb	118	3.5 min	75% β^+ 2.66; K; 3% γ 1.230; 1% γ 1.267	−87.967		1$^+$
Sb	118m	5.00 hr	100% K; 100% γ 0.255; 100% γ 1.229; 95% γ 1.050	−87.75		8$^-$
Sb	119	38.0 hr	100% K; 16% γ 0.024	−89.48		$\frac{5}{2}^+$
Sb	120	15.8 min	56% K; 44% β^+ 1.7; 2% γ 1.172	−88.42		1$^+$
Sb	120	5.76 d	100% K; 100% γ 0.197; 100% γ 1.172; 99% γ 1.023			8$^-$
Sb	121	57.3%		−89.588	6.1 (γ); 0.06 (γm)	$\frac{5}{2}^+$
Sb	122	2.68 d	26% β^- 1.98; 66% β^- 1.42; 5% β^- 0.72; 3% K; 70% γ 0.564; 4% γ 0.693; 1% γ 1.140, 1.257	−88.323		2$^-$
Sb	122m	4.21 min	100% IT; 43% γ 0.061; 18% γ 0.076; 10% γ 0.025	−88.160		
Sb	123	42.7%		−89.218	4.0 (γ); 0.04 (γm)	$\frac{7}{2}^+$
Sb	124	60.20 d	23% β^- 2.3; 52% β^- 0.6; 98% γ 0.603; 49% γ 1.691; 11% γ 0.723	−87.613	7 (γ) (r)	3$^-$
Sb	124m$_1$	93 s	80% IT; 20% β^- 1.2; 20% γ 0.498; 20% γ 0.603; 20% γ 0.646	−87.603		
Sb	124m$_2$	20.2 min	100% IT; (100% m); γ	−87.58		
Sb	125	2.71 yr	14% β^- 0.62; 40% β^- 0.31; (23% m); 30% γ 0.428; 18% γ 0.601; 12% γ 0.636	−88.252		$\frac{7}{2}^+$
Sb	126	12.4 d	16% β^- 1.9; 32% β^- 0.5; 100% γ 0.666; 100% γ 0.695; 88% γ 0.415 (double)	−86.40		

Table of the nuclides (*contd*)

Nuclide		Natural abundance or half-life	Decay modes and major radiations	Mass excess MeV	σ/b	I$^\pi$
Z	A					
Sb	126m	19.0 min	82% β^- 1.9; 14% IT; 86% γ 0.415, 86% γ 0.661; 86% γ 0.695	−86.38		
Sb	127	3.9 d	2% β^- 1.5; 35% β^- 0.9; (17% m); 36% γ 0.686; 25% γ 0.473; 15% γ 0.784	−86.70		$\frac{7}{2}^+$
Sb	128(g)	9.10 hr	20% β^- 2.0; 100% γ 0.743; 100% γ 0.754; 61% γ 0.314	−84.8		8$^-$
Sb	128(m)	10.0 min	78% β^- 2.4; 4% IT; 96% γ 0.743; 96% γ 0.754; 91% γ 0.314	−84.7		5$^+$
Sb	129	4.41 hr	3% β^- 2.3; 27% β^- 0.6; 23% β^- 0.5; (17% m); 46% γ 0.812; 21% γ 0.915; 19% γ 0.545	−84.63		$\frac{7}{2}^+$
Sb	130	40 min	18% β^- 2.9; 100% γ 0.793; 100% γ 0.839; 78% γ 0.331	−82.4		
Sb	130	6.5 min	3% β^- 3.4; 36% β^- 2.2; 100% γ 0.893; 86% γ 0.793; 41% γ 0.182			
Sb	131	23.03 min	7% β^- 2.9; 29% β^- 1.2; (7% m); 46% γ 0.943; 26% γ 0.933; 23% γ 0.642	−82.1		
Sb	132	2.8 min	20% β^- 3.9; 99% γ 0.974; 86% γ 0.697; 15% γ 0.990	−79.7		
Sb	132	4.2 min	44% β^- 3.6; 100% γ 0.697; 100% γ 0.974; 66% γ 0.151			
Sb	133	2.7 min	β^-; γ	−79.0		
Sb	134	10.4 s	43% β^- 7.1; 57% β^- 6.4; 100% γ 1.279; 97% γ 0.297; 57% γ 0.706	−73.9		
Sb	134	0.8 s	β^- 8.8	−73.9		
Sb	135	1.71 s	β^-; 20% n; γ	−70.4		
Sb	136	0.82 s	β^-; 32% n			
52 Te	107	2.2 s	α 3.28			
Te	108	5.3 s	α 3.08	−65.3		0$^+$
Te	109	4.2 s	α 3.08	−67.5		
Te	111	19 s	β^+; K; p	−74.1		
Te	113	2.0 min	β^+; K; γ	−79.0		
Te	114	17 min	β^+; K	−81.5		0$^+$
Te	115	6.0 min	β^+ 2.7; K; 32% γ 0.723; 25% γ 1.381; 24% γ 1.327	−82.58		
Te	115	7.5 min	β^+; K; 61% γ 0.770; 24% γ 1.072; 15% γ 1.032			
Te	116	2.50 hr	K; β^+; 29% γ 0.094; 5% γ 0.103; 1% γ 0.630	−85.4		0$^+$
Te	117	61 min	70% K; β^+ 1.8; 65% γ 0.720; 16% γ 1.716; 11% γ 2.300	−85.16		$\frac{1}{2}^+$
Te	118	6.00 d	100% K	−87.67		0$^+$

Table of the nuclides (*contd*)

Nuclide		Natural abundance or half-life	Decay modes and major radiations	Mass excess MeV	σ/b	Iπ
Z	A					
Te	119	16.0 hr	97% K; 3% β⁺ 0.6; 84% γ 0.644; 10% γ 0.700; 4% γ 1.750	−87.19		$\frac{1}{2}^+$
Te	119m	4.68 d	100% K; 66% γ 0.153; 66% γ 1.213; 27% γ 0.270	−86.9		$\frac{11}{2}^-$
Te	120	0.091%		−89.40	2.0 (γ) (r); 0.3 (γm) (r)	0⁺
Te	121	16.8 d	100% K; 80% γ 0.573; 18% γ 0.508; 1% γ 0.470	−88.49		$\frac{1}{2}^+$
Te	121m	154 d	90% IT; 10% K; 83% γ 0.212; 2% γ 1.102	−88.19		$\frac{11}{2}^-$
Te	122	2.5%		−90.304	3 (γ) (r)	0⁺
Te	123	10¹⁴ yr 0.89%	100% K	−89.166	400 (γ) (r)	$\frac{1}{2}^+$
Te	123m	119.7 d	100% IT; 84% γ 0.159	−88.918		$\frac{11}{2}^-$
Te	124	4.6%		−90.518	6.7 (γ) (r); 0.05 (γm) (r)	0⁺
Te	125	7.0%		−89.019	1.6 (γ) (r)	$\frac{1}{2}^+$
Te	125m	58 d	100% IT	−88.874		$\frac{11}{2}^-$
Te	126	18.7%		−90.066	0.9 (γ) (r); 0.13 (γm) (r)	0⁺
Te	127	9.4 hr	99% β⁻ 0.69; 1% γ 0.418	−88.285		$\frac{3}{2}^+$
Te	127m	109 d	98% IT; 2% β⁻ 0.7; 1% γ 0.058	−88.197		$\frac{11}{2}^-$
Te	128	31.7% 10²⁴ yr	β⁻β⁻	−88.992	0.20 (γ); 0.016 (γm)	0⁺
Te	129	70 min	89% β⁻ 1.47; 16% γ 0.028; 7% γ 0.460; 1% γ 0.487	−87.007		$\frac{3}{2}^+$
Te	129m	33.5 d	63% IT; 33% β⁻ 1.6; γ	−86.901		$\frac{11}{2}^-$
Te	130	34.5% 2.10²¹ yr	β⁻β⁻	−87.348	0.2 (γ) (r); 0.03 (γm) (r)	0⁺
Te	131	25.0 min	60% β⁻ 2.10; 22% β⁻ 1.65; 68% γ 0.150; 18% γ 0.452; 5% γ 1.147	−85.201		$\frac{3}{2}^+$
Te	131m	30 hr	β⁻ 2.4; 22% IT; 38% γ 0.774; 21% γ 0.852; 20% γ 0.150	−85.019		$\frac{11}{2}^-$
Te	132	78 hr	100% β⁻ 0.22; 88% γ 0.228; 14% γ 0.050; 2% γ 0.112, 0.116	−85.21		0⁺
Te	133	12.5 min	26% β⁻ 2.7; 34% β⁻ 2.2; 73% γ 0.312; 33% γ 0.408; 12% γ 1.333	−82.9		
Te	133m	55.4 min	30% β⁻ 2.4; (10% m); 17% IT; 63% γ 0.913; 22% γ 0.648; 12% γ 0.915	−82.6		
Te	134	42 min	42% β⁻ 0.5; 43% β⁻ 0.4; 30% γ 0.767; 22% γ 0.210; 20% γ 0.278	−82.7		0⁺
Te	135	19.2 s	β⁻; γ	−77.6		
Te	136	17.5 s	β⁻; 1% n; γ	−74.8		0⁺
Te	137	3 s	β⁻; 3% n; γ			

Table of the nuclides (*contd*)

Nuclide Z	Nuclide A	Natural abundance or half-life	Decay modes and major radiations	Mass excess MeV	σ/b	Iπ
Te	138	1.4 s	β^-; 6% n			
53 I	115	1.3 min	β^+; K	−76.8		
I	116	2.9 s	β^+ 6.7; K; 8% γ 0.679; 1% γ 0.540	−77.6		1$^+$
I	117	2.20 min	54% K; β^+ 3.3; γ	−80.9		
I	118	14.3 min	β^+ 5.4; 46% K; 95% γ 0.605; 12% γ 0.545; 12% γ 1.338	−80.6		
I	118m	8.5 min	β^+; K; IT; γ	−80.5		
I	119	19.3 min	β^+ 2.4; 49% K; 90% γ 0.257; 3% γ 0.636; 2% γ 0.321	−83.8		
I	120	1.35 hr	54% K; 19% β^+ 4.6; 73% γ 0.560; 11% γ 1.523; 9% γ 0.641	−83.79		2$^-$
I	120m	53 min	β^+ 3.7; K; 100% γ 0.560; 87% γ 0.601; 67% γ 0.615	−82.9		
I	121	2.12 hr	94% K; β^+ 1.2; 84% γ 0.212; 6% γ 0.532; 2% γ 0.599	−86.12		$\frac{5}{2}^+$
I	122	3.6 min	β^+ 3.1; 23% K; 18% γ 0.564; 1% γ 0.693; 1% γ 0.793	−86.16		1$^+$
I	123	13.02 hr	100% K; 83% γ 0.159; 1% γ 0.529	−88.0		$\frac{5}{2}^+$
I	124	4.15 d	75% K; 11% β^+ 2.1; 61% γ 0.603; 10% γ 0.723; 10% γ 1.691	−87.361		2$^-$
I	125	60.2 d	100% K; 7% γ 0.035	−88.841	900 (γ) (s)	$\frac{5}{2}^+$
I	126	13.0 d	53% K; 1% β^+ 1.1; 10% β^- 1.25; 32% β^- 0.86; 4% β^- 0.37; 35% γ 0.389; 34% γ 0.666; 4% γ 0.754	−87.91	6000 (γ) (r)	2$^-$
I	127	100%		−88.980	6.1 (γ)	$\frac{5}{2}^+$
I	128	24.99 min	77% β^- 2.1; 15% β^- 1.7; 2% β^- 1.1; 6% K; 16% γ 0.443; 2% γ 0.527	−87.734		1$^+$
I	129	1.57 × 10^7 yr	100% β^- 0.15; 8% γ 0.040	−88.505	9 (γ); 18 (γm)	$\frac{7}{2}^+$
I	130	12.36 hr	1% β^- 1.2; 48% β^- 1.0; 46% β^- 0.6; 99% γ 0.536; 96% γ 0.669; 82% γ 0.739	−86.90	18 (γ) (r)	5$^+$
I	130m	9.16 min	83% IT; 15% β^- 2.5; 17% γ 0.536 1% γ 0.586	−86.85		2$^+$
I	131	8.040 d	89% β^- 0.61; (1% m); 81% γ 0.364; 7% γ 0.637; 6% γ 0.284	−87.451	0.7 (γ)	$\frac{7}{2}^+$
I	132	2.285 hr	16% β^- 2.1; 19% β^- 1.2; 99% γ 0.668; 76% γ 0.773; 18% γ 0.955	−85.71		4$^+$
I	132m	83 min	86% IT; 8% β^- 1.5; 13% γ 0.600; 13% γ 0.668; 13% γ 0.773	−85.6		
I	133	20.9 hr	1% β^- 1.5; 84% β^- 1.2; (3% m); 86% γ 0.530; 5% γ 0.875; 2% γ 1.298	−85.90		$\frac{7}{2}^+$

Table of the nuclides *(contd)*

Nuclide Z	A	Natural abundance or half-life	Decay modes and major radiations	Mass excess MeV	σ/b	I^π
I	133m	9 s	100% IT; 100% γ 0.647; 100% γ 0.912	−84.27		
I	134	52.5 min	12% β^- 2.4; 32% β^- 1.3; 95% γ 0.847; 65% γ 0.884; 15% γ 1.073	−84.0		
I	134m	3.50 min	98% IT; 2% β^- 2.5; (2% m); 79% γ 0.272; 10% γ 0.044	−83.7		
I	135	6.61 hr	2% β^- 2.2; 23% β^- 1.4; 22% β^- 1.0; (15% m); 29% γ 1.260; 23% γ 1.132; 10% γ 1.678	−83.80		$\frac{7}{2}^+$
I	136	45 s	67% β^- 5.1; 100% γ 0.381; 100% γ 1.313; 78% γ 0.197			
I	136	83.4 s	30% β^- 5.6; 34% β^- 4.4; 68% γ 1.313; 25% γ 1.321; 7% γ 2.415, 2.634	−79.4		
I	137	24.5 s	β^-; 6% n; 13% γ 1.219; 7% γ 0.601; 5% γ 1.303	−76.7		
I	138	6.3 s	58% β^- 7.7; 5% n; 93% γ 0.589; 17% γ 0.875; 7% γ 0.484	−71.9		
I	139	2.3 s	β^-; 10% n; γ	−69		
I	140	0.86 s	β^-; 14% n; γ			
I	141	0.48 s	β^-; n; γ			
54 Xe	113	2.8 s	β^+; K; p			
Xe	115	18 s	β^+; K; γ	−68.9		
Xe	116	57 s	β^+ 3.3; K; γ	−73.3		0$^+$
Xe	117	61 s	65% K; β^+; γ	−74.5		
Xe	118	6 min	86% K; β^+; γ	−77.3		0$^+$
Xe	119	6 min	82% K; β^+; γ	−78.8		
Xe	120	40 min	97% K; 2% β^+ 0.9; 29% γ 0.025; 9% γ 0.073; 7% γ 0.178	−81.8		0$^+$
Xe	121	38.8 min	92% K; β^+ 2.8; γ	−82.3		
Xe	122	20.1 hr	100% K; 8% γ 0.350; 3% γ 0.149; 2% γ 0.417	−85.2		0$^+$
Xe	123	2.08 hr	87% K; β^+ 1.5; 49% γ 0.149; 15% γ 0.178; 9% γ 0.330	−85.3		
Xe	124	0.096%		−87.5	100 (γ); 20 (γm)	0$^+$
Xe	125	17.3 hr	100% K; 55% γ 0.188; 29% γ 0.243; 6% γ 0.055	−87.11		
Xe	125m	57 s	100% IT; 58% γ 0.112; 19% γ 0.142	−86.86		
Xe	126	0.090%		−89.16	3 (γ); 0.4 (γm)	0$^+$
Xe	127	36.41 d	100% K; 68% γ 0.203; 25% γ 0.172; 17% γ 0.375	−88.32		
Xe	127m	69 s	100% IT; 68% γ 0.125; 37% γ 0.173	−88.02		
Xe	128	1.92%		−89.861	0.4 (γm)	0$^+$
Xe	129	26.4%		−88.698	20 (γ) (r)	$\frac{1}{2}^+$
Xe	129m	8.89 d	100% IT; 5% γ 0.197	−88.461		$\frac{11}{2}^-$

Table of the nuclides (*contd*)

Nuclide Z	Nuclide A	Natural abundance or half-life	Decay modes and major radiations	Mass excess MeV	σ/b	Iπ
	Xe 130	4.1%		−89.881	0.4 (γm)	0+
	Xe 131	21.2%		−88.421	90 (γ) (r)	3/2+
	Xe 131m	11.77 d	100% IT; 2% γ 0.164	−88.257		11/2−
	Xe 132	26.9%		−89.286	0.4 (γ); 0.03 (γm)	0+
	Xe 133	5.25 d	99% β⁻ 0.35; 37% γ 0.081	−87.66	190 (γ) (r)	3/2+
	Xe 133m	2.19 d	100% IT; 10% γ 0.233	−87.43		11/2−
	Xe 134	10.4%		−88.13	0.25 (γ); 0.003 (γm)	0+
	Xe 134m	0.29 s	100% IT; 100% γ 0.847; 100% γ 0.884; 68% γ 0.233	−86.16		
	Xe 135	9.10 hr	96% β⁻ 0.91; 90% γ 0.250; 3% γ 0.608	−86.51	2.6 × 10⁶ (γ)	3/2+
	Xe 135m	15.6 min	100% IT; 81% γ 0.527	−85.98		11/2−
	Xe 136	8.9%		−86.43	0.16 (γ)	0+
	Xe 137	3.82 min	67% β⁻ 4.3; 31% γ 0.455; 1% γ 0.849	−82.22		
	Xe 138	14.1 min	β⁻; 30% γ 0.258; 20% γ 0.434; 19% γ 1.768	−80.2		0+
	Xe 139	39.7 s	22% β⁻ 4.9; 21% β⁻ 4.4; 50% γ 0.219; 19% γ 0.297; 18% γ 0.175	−75.8		
	Xe 140	13.6 s	β⁻; 8% γ 0.622; 5% γ 0.118; 4% γ 0.080	−73.2		0+
	Xe 141	1.73 s	β⁻; γ	−69.0		
	Xe 142	1.24 s	β⁻; γ	−66.1		0+
	Xe 143	0.30 s	β⁻			
	Xe 143	0.96 s	β⁻			
	Xe 144	1.2 s	β⁻			0+
	Xe 145	0.9 s	β⁻			
55	Cs 116	57 s	β⁺ 3.3; K	−73.3		
	Cs 117	8 s	β⁺; K	−66.9		
	Cs 118	16 s	β⁺; K	−67.9		
	Cs 119	38 s	β⁺; K	−72.5		
	Cs 120	61 s	β⁺; K; γ	−73.4		
	Cs 121	126 s	β⁺; K; γ	−77.1		
	Cs 122	4.5 min	β⁺; K; γ			
	Cs 122	21 s	β⁺; K; γ	−78.0		
	Cs 123	5.9 min	β⁺; K; 15% γ 0.098; 15% γ 0.598; 9% γ 0.177	−81.2		
	Cs 123m	1.6 s	100% IT; γ			
	Cs 124	31 s	β⁺; K; γ	−81.5		
	Cs 125	45 min	61% K; 27% β⁺ 2.05; 25% γ 0.526; 9% γ 0.112; 5% γ 0.412	−84.04		1/2+
	Cs 126	1.64 min	β⁺ 3.8; 18% K; 42% γ 0.389; 5% γ 0.491; 5% γ 0.925	−84.3		1+
	Cs 127	6.2 hr	97% K; 1% β⁺ 1.1; 62% γ 0.411; 12% γ 0.125; 5% γ 0.462	−86.23		1/2+
	Cs 128	3.62 min	β⁺ 2.9; 39% K; 26% γ 0.443; 2% γ 0.527; 1% γ 1.140	−85.94		1+

Table of the nuclides (*contd*)

Nuclide		Natural abundance or half-life	Decay modes and major radiations	Mass excess MeV	σ/b	Iπ
Z	A					
Cs	129	32.3 hr	100% K; 32% γ 0.372; 23% γ 0.411; 4% γ 0.549	−87.49		$\frac{1}{2}^+$
Cs	130	29.9 min	K; β⁺ 2.0; 2% β⁻ 0.44; 4% γ 0.536	−86.86		1⁺
Cs	131	9.688 d	100% K	−88.07		$\frac{5}{2}^+$
Cs	132	6.47 d	97% K; 2% β⁻ 0.8; 1% β⁺ 0.4; 98% γ 0.668; 2% γ 0.465; 1% γ 0.630	−87.18		2
Cs	133	100%		−88.09	27 (γ); 2.5 (γm)	$\frac{7}{2}^+$
Cs	134	2.06 yr	70% β⁻ 0.66; 98% γ 0.605; 85% γ 0.796; 15% γ 0.569	−86.91	140 (γ) (r)	4⁺
Cs	134m	2.91 hr	100% IT; 13% γ 0.127	−86.77		8⁻
Cs	135	3.0 × 10⁶ yr	100% β⁻ 0.21	−87.67	9 (γ) (s)	$\frac{7}{2}^+$
Cs	135m	53 min	100% IT; 100% γ 0.787; 96% γ 0.840	−86.04		
Cs	136	13.1 d	2% β⁻ 0.68; 94% β⁻ 0.34; (15% m); 100% γ 0.818; 80% γ 1.048; 47% γ 0.341	−86.36		5⁺
Cs	136m	19 s	100% IT			
Cs	137	30.17 yr	5% β⁻ 1.17; 95% β⁻ 0.51; (95% m); 85% γ (m) 0.662	−86.56	0.11 (γ) (r)	$\frac{7}{2}^+$
Cs	138	32.2 min	9% β⁻ 3.9; 41% β⁻ 2.8; 75% γ 1.436; 28% γ 1.010; 27% γ 0.463	−83.0		3⁻
Cs	138m	2.9 min	75% IT; 20% β⁻ 3.3; 25% γ 0.463; 25% γ 1.436; 20% γ 0.192	−82.9		
Cs	139	9.5 min	84% β⁻ 4.3; 7% γ 1.283; 2% γ 0.627; 1% γ 1.421	−80.6		
Cs	140	65 s	14% β⁻ 6.0; 20% β⁻ 5.4; 72% γ 0.602; 12% γ 0.908; 6% γ 1.201	−77.2		
Cs	141	24.9 s	50% β⁻ 5.0; γ	−75.0		
Cs	142	1.69 s	β⁻; γ	−71.0		
Cs	143	1.78 s	β⁻ 5.7; 2% n; γ	−68.4		
Cs	144	1.00 s	β⁻; 3% n; γ	−63.9		
Cs	145	0.58 s	β⁻; 12% n; γ	−61.7		
Cs	146	0.34 s	β⁻; 14% n			
56 Ba	117	1.9 s	β⁺; K; p			
Ba	119	5.4 s	β⁺; K; p	−64.5		
Ba	120	32 s	β⁺; K; γ	−69		0⁺
Ba	121	30 s	β⁺; K	−70.6		
Ba	122	2.0 min	β⁺; K	−74.3		0⁺
Ba	123	2.7 min	β⁺; K; γ	−75.7		
Ba	124	10.5 min	β⁺; K; γ	−78.8		0⁺
Ba	125	3.5 min	β⁺; K; γ	−79.5		
Ba	125	8 min	β⁺; K			
Ba	126	100 min	β⁺; K; 20% γ 0.234; 8% γ 0.258; 6% γ 0.241	−82.6		0⁺
Ba	127	12.7 min	β⁺; K; 11% γ 0.181; 8% γ 0.115; 2% γ 0.066	−82.8		

Table of the nuclides (*contd*)

Nuclide Z	A	Natural abundance or half-life	Decay modes and major radiations	Mass excess MeV	σ/b	I^π
Ba	128	2.4 d	100% K; 14% γ 0.273	−85.48		0$^+$
Ba	129	2.2 hr	K; β^+ 1.4; γ	−85.05		$\frac{1}{2}^+$
Ba	129m	2.1 hr	β^+; K; γ	−84.77		
Ba	130	0.106%		−87.30	8 (γ) (r); 2.5 (γm) (r)	0$^+$
Ba	131	12.0 d	100% K; 42% γ 0.496; 28% γ 0.124; 22% γ 0.216	−86.73		$\frac{1}{2}^+$
Ba	131m	14.6 min	100% IT; 56% γ 0.108; 1% γ 0.078	−86.54		$\frac{9}{2}^-$
Ba	132	0.101%		−88.45	7 (γ) (r); 0.6 (γm) (r)	0$^+$
Ba	133	10.7 yr	100% K; 62% γ 0.356; 34% γ 0.081; 18% γ 0.303	−87.57		$\frac{1}{2}^+$
Ba	133m	38.9 hr	100% IT; 18% γ 0.276	−87.28		$\frac{11}{2}^-$
Ba	134	2.42%		−88.97	2 (t)	0$^+$
Ba	135	6.59%		−87.87	6 (γ) (r); 0.014 (γm) (r)	$\frac{3}{2}^+$
Ba	135m	28.7 hr	100% IT; 16% γ 0.268	−87.60		$\frac{11}{2}^-$
Ba	136	7.85%		−88.91	0.4 (t)	0$^+$
Ba	136m	0.308 s	100% IT; 100% γ 0.819; 100% γ 1.048	−86.88		7$^-$
Ba	137	11.2%		−87.73	5.1 (γ) (r)	$\frac{3}{2}^+$
Ba	137m	2.551 min	100% IT; 90% γ 0.662	−87.07		$\frac{11}{2}^-$
Ba	138	71.7%		−88.27	0.4 (γ) (r)	0$^+$
Ba	139	82.9 min	72% β^- 2.3; 22% γ 0.166	−84.93	6 (γ) (r)	
Ba	140	12.79 d	22% β^- 1.01; 40% β^- 0.99; 24% γ 0.537; 13% γ 0.030; 6% γ 0.163	−83.29	1.6 (γ)	0$^+$
Ba	141	18.3 min	β^-; 46% γ 0.190; 25% γ 0.304; 23% γ 0.277	−80.0		
Ba	142	10.6 min	β^- 2.1; 20% γ 0.255; 15% γ 1.204; 13% γ 0.895	−77.8		0$^+$
Ba	143	13.6 s	β^-; γ	−74.0		
Ba	144	11.9 s	β^-; γ	−72.0		0$^+$
Ba	145	5 s	β^-; γ	−67.8		
Ba	148	0.5 s	β^-			0$^+$
57 La	126	1.0 min	β^+; K; γ			
La	127	3.8 min	β^+; K	−77.8		
La	128	4.9 min	β^+; K; γ	−78.7		
La	129	10 min	β^+; K; γ	−81.1		
La	129m	0.56 s	100% IT; 24% γ 0.068; 4% γ 0.105	−80.9		
La	130	8.7 min	β^+; K; 81% γ 0.357; 27% γ 0.551; 18% γ 0.544, 0.908	−81.6		
La	131	61 min	76% K; β^+ 1.9; 24% γ 0.108; 19% γ 0.418; 17% γ 0.366, 1.910	−83.8		$\frac{3}{2}^+$
La	132	4.8 hr	β^+ 3.7; K; 76% γ 0.465; 16% γ 0.567; 9% γ 0.663	−83.74		2$^-$
La	132m	24.3 min	76% IT; β^+; K; 44% γ 0.135; 22% γ 0.465; 7% γ 0.285	−83.55		6$^-$

Table of the nuclides (*contd*)

Nuclide Z	Nuclide	A	Natural abundance or half-life	Decay modes and major radiations	Mass excess MeV	σ/b	I^π
	La	133	3.91 hr	β^+; K; 2% γ 0.279; 1% γ 0.290; 1% γ 0.302	−85.6		$\frac{5}{2}^+$
	La	134	6.67 min	61% β^+ 2.7; 38% K; 5% γ 0.605	−85.27		1^+
	La	135	19.4 hr	100% K; 2% γ 0.481	−86.67		$\frac{5}{2}^+$
	La	136	9.87 min	64% K; β^+ 1.9; 3% γ 0.818	−86.0		1^+
	La	137	6×10^4 yr	100% K	−87.1		$\frac{7}{2}^+$
	La	138	1.12×10^{11} yr 0.089%	68% K; 32% β^- 0.25; 68% γ 1.436; 32% γ 0.788	−86.52	57 (γ)	5^+
	La	139	99.911%		−87.23	9.2 (γ)	$\frac{7}{2}^+$
	La	140	40.3 hr	10% β^- 2.2; 41% β^- 1.4; 96% γ 1.596; 43% γ 0.487; 23% γ 0.816	−84.32	2.7 (γ) (s)	3^-
	La	141	3.9 hr	97% β^- 2.4; 3% γ 1.354	−83.01		
	La	142	93 min	13% β^- 4.5; 20% β^- 2.1; 49% γ 0.641; 15% γ 2.398; 10% γ 2.543	−80.02		2^-
	La	143	14.0 min	96% β^- 3.3; 2% γ 0.620; 1% γ 0.644	−78.3		
	La	144	39.9 s	β^-; γ	−74.9		
	La	145	29 s	β^-; γ	−72.9		
	La	146	11 s	β^-; γ	−69.5		
	La	148	1.3 s	β^-; γ	−64.0		
58	Ce	129	3.5 min	β^+; K; γ			
	Ce	130	25 min	β^+; K; γ			0^+
	Ce	131	5 min	β^+; K; γ	−79.5		
	Ce	131	10 min	89% K; β^+; 20% γ 0.169; 9% γ 0.396; 9% γ 1.440	−79.5		
	Ce	132	3.5 hr	100% K; 79% γ 0.182; 11% γ 0.155; 5% γ 0.217	−82.3		0^+
	Ce	133	97 min	β^+; K; 45% γ 0.097; 16% γ 0.077; 11% γ 0.558	−82.2		$\frac{1}{2}$
	Ce	133	5.4 hr	β^+; K; 39% γ 0.477; 20% γ 0.510; 19% γ 0.058	−82.2		$\frac{9}{2}^-$
	Ce	134	76 hr	100% K	−84.8		0^+
	Ce	135	17.8 hr	99% K; β^+; 42% γ 0.266; 23% γ 0.300; 19% γ 0.607	−84.6		$\frac{1}{2}$
	Ce	135m	20 s	100% IT; 74% γ 0.214; 21% γ 0.150; 20% γ 0.083, 0.296	−84.1		
	Ce	136	0.190%		−86.50	6 (γ) (r); 1.0 (γm) (r)	0^+
	Ce	137	9.0 hr	100% K; 2% γ 0.447	−85.9		$\frac{3}{2}^+$
	Ce	137m	34.4 hr	99% IT; 1% K; 11% γ 0.254	−85.7		$\frac{11}{2}^-$
	Ce	138	0.254%		−87.57	1.0 (γ) (r); 0.015 (γm) (r)	0^+
	Ce	139	137.2 d	100% K; 80% γ 0.166	−86.97		$\frac{3}{2}^+$
	Ce	139m	56.4 s	100% IT; 93% γ 0.754	−86.21		$\frac{11}{2}^-$
	Ce	140	88.5%		−88.08	0.56 (γ) (r)	0^+
	Ce	141	32.55 d	31% β^- 0.58; 69% β^- 0.44; 48% γ 0.145	−85.44	29 (γ) (r)	$\frac{7}{2}^-$
	Ce	142	11.1%		−84.54	0.95 (γ)	0^+

Table of the nuclides (*contd*)

Nuclide		Natural abundance or half-life	Decay modes and major radiations	Mass excess MeV	σ/b	Iπ
Z	A					
Ce	143	33.0 hr	39% β^- 1.4; 47% β^- 1.1; 42% γ 0.293; 12% γ 0.057; 5% γ 0.665, 0.722	−81.61	6.2 (γ) (r)	$\frac{3}{2}^-$
Ce	144	284 d	76% β^- 0.32; (2% m); 11% γ 0.134; 2% γ 0.080	−80.43	1.0 (γ)	0^+
Ce	145	3.0 min	76% β^- 1.7; 68% γ 0.725; 14% γ 0.063; 10% γ 1.148	−77.1		
Ce	146	14 min	β^- 0.7; 53% γ 0.317; 20% γ 0.218; 8% γ 0.133	−75.8		0^+
Ce	147	56 s	β^-; γ	−72.2		
Ce	148	48 s	β^-; γ	−70.8		0^+
Ce	149	5.0 s	β^-; γ	−67.5		
Ce	150	4 s	β^-; γ	−65		0^+
Ce	151	1.0 s	β^-; γ	−62.7		
59 Pr	129	24 s	β^+; K			
Pr	130	28 s	β^+; K			
Pr	132	1.6 min	β^+; K; γ	−75.3		
Pr	133	6.5 min	β^+; K; γ	−78.0		$\frac{5}{2}^+$
Pr	134	17 min	β^+; K; γ	−78.5		2^+
Pr	134	11 min	β^+; K; γ	−78.5		
Pr	135	25 min	75% K; β^+ 2.5; γ	−81.0		$\frac{3}{2}^+$
Pr	136	13.1 min	β^+; K; 76% γ 0.552; 52% γ 0.540; 18% γ 1.092	−81.4		2^+
Pr	137	1.28 hr	75% K; 25% β^+ 1.7; 2% γ 0.837; 1% γ 0.434; 1% γ 0.514	−83.2		$\frac{5}{2}^+$
Pr	138	1.4 min	β^+ 3.4; K; 2% γ 0.789; 1% γ 0.688	−83.13		1^+
Pr	138m	2.02 hr	77% K; 23% β^+ 1.6; 100% γ 0.789; 100% γ 1.038; 80% γ 0.302	−82.77		7^-
Pr	139	4.41 hr	92% K; β^+ 1.1	−84.85		$\frac{5}{2}^+$
Pr	140	3.39 min	51% K; 48% β^+ 2.4	−84.69		1^+
Pr	141	100%		−86.02	7.6 (γ); 3.9 (γm)	$\frac{5}{2}^+$
Pr	142	19.2 hr	96% β^- 2.2; 4% γ 1.576	−83.79	20 (γ) (r)	2^-
Pr	142m	14.6 min	100% IT	−83.79		5^-
Pr	143	13.57 d	100% β^- 0.93	−83.07	90 (γ)	$\frac{7}{2}^+$
Pr	144	17.3 min	98% β^- 3.0; 1% γ 0.696	−80.75		0^-
Pr	144m	7.2 min	100% IT	−80.69		3^-
Pr	145	5.98 hr	97% β^- 1.8	−79.63		
Pr	146	24.0 min	β^- 4.1; 48% γ 0.454; 18% γ 1.525; 8% γ 0.736, 0.789	−76.8		
Pr	147	13 min	β^-; 22% γ 0.315; 17% γ 0.641; 15% γ 0.578	−75.4		
Pr	148	2.30 min	42% β^- 4.5; 91% γ 0.302; 16% γ 0.450; 12% γ 0.697	−72.6		
Pr	149	2.3 min	55% β^- 3.0; 7% γ 0.139; 6% γ 0.110; 6% γ 0.165	−71.4		
Pr	150	6.2 s	β^- 5.7; γ	−68.0		
Pr	151	4.0 s	β^-; γ	−67.4		
60 Nd	129	6 s	β^+; K; p			

Table of the nuclides (*contd*)

Nuclide		Natural abundance or half-life	Decay modes and major radiations	Mass excess MeV	σ/b	I^π
Z	A					
Nd	130	28 s	β^+; K			0^+
Nd	132	1.8 min	β^+; K			0^+
Nd	133	1.2 min	β^+; K; γ			
Nd	134	9 min	β^+; K; γ			0^+
Nd	135	12 min	β^+; K; 51% γ 0.204; 23% γ 0.042; 15% γ 0.441	−76.3		$\frac{9}{2}$
Nd	135	6 min	β^+; K	−76.3		
Nd	136	50.6 min	94% K; β^+; 33% γ 0.109; 20% γ 0.040; 12% γ 0.575	−79.2		0^+
Nd	137	39 min	β^+; K; 17% γ 0.076; 13% γ 0.581; 10% γ 0.307	−79.4		$\frac{1}{2}^+$
Nd	137m	1.6 s	100% IT; 64% γ 0.234; 58% γ 0.178; 35% γ 0.108	−78.9		$\frac{11}{2}^-$
Nd	138	5.1 hr	100% K; 3% γ 0.326	−82.0		0^+
Nd	139	29.7 min	74% K; 25% β^+ 1.8; 6% γ 0.405; 2% γ 1.074; 1% γ 0.669	−82.05		$\frac{3}{2}^+$
Nd	139m	5.5 hr	87% K; 1% β^+ 1.2; 12% IT; 40% γ 0.114; 35% γ 0.738; 26% γ 0.708, 0.982	−81.82		$\frac{11}{2}^-$
Nd	140	3.37 d	100% K	−84.22		0^+
Nd	141	2.5 hr	97% K; β^+ 0.8; 1% γ 1.127	−84.20		$\frac{3}{2}^+$
Nd	141m	61 s	100% IT; 91% γ 0.756	−83.45		$\frac{11}{2}^-$
Nd	142	27.2%		−85.95	19 (γ) (r)	0^+
Nd	143	12.2%		−84.00	320 (γ)	$\frac{7}{2}^-$
Nd	144	2.1×10^{15} yr 23.8%	α 1.83	−83.75	4 (γ) (r)	0^+
Nd	145	8.3%		−81.43	41 (γ)	$\frac{7}{2}^-$
Nd	146	17.2%		−80.92	1.3 (γ)	0^+
Nd	147	10.98 d	83% β^- 0.81; 27% γ 0.091; 12% γ 0.531; 2% γ 0.319	−78.14	440 (γ) (r)	$\frac{5}{2}^-$
Nd	148	5.7%		−77.41	2.5 (γ)	0^+
Nd	149	1.73 hr	1% β^- 1.58; 26% β^- 1.48; 24% β^- 1.15; 31% γ 0.211; 22% γ 0.114; 20% γ 0.030	−74.37		$\frac{5}{2}^-$
Nd	150	5.6%		−73.68	1.2 (γ)	0^+
Nd	151	12.4 min	β^- 2.3; 47% γ 0.116; 17% γ 0.256; 15% γ 1.181	−70.95		
Nd	152	11.4 min	β^-; 31% γ 0.278; 21% γ 0.250; 8% γ 0.016	−70.15		0^+
Nd	154	40 s	β^-; γ			0^+
61　Pm	132	4 s	β^+; K			
Pm	133	12 s	β^+; K			
Pm	134	24 s	β^+; K; γ			
Pm	135	0.9 min	β^+; K; γ			
Pm	136	107 s	β^+; K; 89% γ 0.374; 49% γ 0.603; 31% γ 0.858	−71.4		
Pm	137	2.4 min	β^+; K; γ	−74.2		
Pm	138	3.5 min	β^+; K; 93% γ 0.521; 37% γ 0.729; 20% γ 0.493	−75.0		

Table of the nuclides (*contd*)

Nuclide Z	A	Natural abundance or half-life	Decay modes and major radiations	Mass excess MeV	σ/b	Iᴨ
Pm	139	4.15 min	β^+; K; 12% γ 0.403; 3% γ 0.368; 3% γ 0.463	−77.6		
Pm	140	9.2 s	β^+ 5.0; K; 5% γ 0.774; 1% γ 0.717; 1% γ 1.490	−78.2		1^+
Pm	140m	5.9 min	58% K; β^+ 3.2; 100% γ 0.774; 100% γ 1.028; 92% γ 0.420	−77.8		
Pm	141	20.9 min	β^+ 2.7; 43% K; 4% γ 1.223; 2% γ 0.886; 1% γ 0.194	−80.47		$\frac{5}{2}^+$
Pm	142	40.5 s	β^+ 3.9; 31% K; 3% γ 1.576	−81.1		1^+
Pm	143	265 d	100% K; 39% γ 0.742	−82.96		$\frac{5}{2}^+$
Pm	144	350 d	100% K; 100% γ 0.696; 99% γ 0.618; 42% γ 0.477	−81.42		5^-
Pm	145	17.7 yr	100% K; 2% γ 0.072; 1% γ 0.067	−81.27		$\frac{5}{2}^+$
Pm	146	5.5 yr	63% K; 34% β^- 0.8; 3% β^- 0.2; 63% γ 0.454; 37% γ 0.747; 23% γ 0.736	−79.44	8000 (γ) (r)	3^-
Pm	147	2.623 yr	100% β^- 0.22	−79.04	97 (γ); 85 (γm)	$\frac{7}{2}^+$
Pm	148	5.37 d	54% β^- 2.46; 35% β^- 0.99; 23% γ 0.550; 22% γ 1.465; 13% γ 0.915	−76.87	< 3000 (γ) (r)	1^-
Pm	148m	41.3 d	1% β^- 1.0; 56% β^- 0.4; 5% IT; 93% γ 0.550; 89% γ 0.630; 32% γ 0.726	−76.73	1.1×10^4 (γ)	6^-
Pm	149	53.1 hr	96% β^- 1.0; 3% γ 0.286	−76.06	1400 (γ) (r)	$\frac{7}{2}^+$
Pm	150	2.68 hr	8% β^- 3.2; 28% β^- 2.3; 73% γ 0.334; 19% γ 1.325; 17% γ 1.166	−73.6		
Pm	151	28.4 hr	11% β^- 1.2; 42% β^- 0.8; 22% γ 0.005; 22% γ 0.340; 8% γ 0.168	−73.39	< 700 (γ) (r)	$\frac{5}{2}^+$
Pm	152	4.1 min	61% β^- 3.5; 16% γ 0.122; 5% γ 0.842; 5% γ 0.961, 0.963	−71.3		
Pm	152	7.5 min	β^-; 54% γ 0.245; 44% γ 0.122; 27% γ 0.340			
Pm	152	15 min	β^-; γ			
Pm	153	5.3 min	55% β^- 1.7; 25% γ 0.036; 14% γ 0.127; 8% γ 0.028	−70.8		
Pm	154	1.7 min	12% β^- 3.1; 24% β^- 2.0; 35% β^- 1.9; 19% γ 2.059; 12% γ 0.082; 12% γ 0.840, 1.394	−68.5		
Pm	154	2.7 min	β^-; 39% γ 0.185; 25% γ 1.440 (double); 19% γ 0.082			
62 Sm	133	32.0 s	β^+; K; p			
Sm	134	12 s	β^+; K			0^+
Sm	135	10 s	β^+; K; p			
Sm	137	44 s	β^+; K			
Sm	138	3.0 min	β^+; K; γ			0^+
Sm	139	2.5 min	β^+; K; γ	−72.4		
Sm	139m	9.5 s	94% IT; β^+; 37% γ 0.190; 34% γ 0.267; 33% γ 0.155	−71.9		

Table of the nuclides (*contd*)

Nuclide		Natural abundance or half-life	Decay modes and major radiations	Mass excess MeV	σ/b	I$^\pi$
Z	A					
Sm	140	14.8 min	β^+; K; γ	−75.5		0^+
Sm	141	10.2 min	53% K; β^+ 3.2; 42% γ 0.404; 37% γ 0.438; 7% γ 1.293	−75.9		$\frac{1}{2}^+$
Sm	141m	22.5 min	β^+; K; 74% γ 0.197; 40% γ 0.432; 20% γ 0.777	−75.7		$\frac{11}{2}^-$
Sm	142	72.5 min	90% K; β^+ 1.0; γ	−78.98		0^+
Sm	143	8.83 min	54% K; β^+ 2.4; 2% γ 1.057; 1% γ 1.515	−79.51		$\frac{3}{2}^+$
Sm	143m	65 s	100% IT; 90% γ 0.754	−78.76		$\frac{11}{2}^-$
Sm	144	3.1%		−81.96	0.7 (γ) (r)	0^+
Sm	145	340 d	100% K; 12% γ 0.061	−80.66	110 (γ) (r)	$\frac{7}{2}^-$
Sm	146	1.03×10^8 yr	α 2.55	−80.98		0^+
Sm	147	1.06×10^{11} yr 15.1%	α 2.23	−79.27	60 (γ)	$\frac{7}{2}^-$
Sm	148	8×10^{15} yr 11.3%	α 1.96	−79.34	4.7 (γ)	0^+
Sm	149	13.9%		−77.14	4.2×10^4 (γ)	$\frac{7}{2}^-$
Sm	150	7.4%		−77.05	104 (γ)	0^+
Sm	151	90 yr	99% β^- 0.076	−74.57	1.5×10^4 (γ)	$\frac{5}{2}^-$
Sm	152	26.6%		−74.76	204 (γ)	0^+
Sm	153	46.8 hr	21% β^- 0.81; 44% β^- 0.71; 28% γ 0.103; 4% γ 0.070; 1% γ 0.097	−72.56		$\frac{3}{2}^+$
Sm	154	22.6%		−72.45	5 (γ)	0^+
Sm	155	22.4 min	93% β^- 1.53; 70% γ 0.104; 4% γ 0.246; 2% γ 0.141	−70.20		$\frac{3}{2}^-$
Sm	156	9.4 hr	52% β^- 0.69; 45% β^- 0.42; 24% γ 0.088; 21% γ 0.204; 11% γ 0.166	−69.37		0^+
Sm	157	8.0 min	β^-; γ	−66.9		
63 Eu	138	1.5 s	β^+			
Eu	138	35 s	β^+			
Eu	139	22 s	β^+; K; γ			
Eu	140	1.3 s	β^+; K; γ			
Eu	141	40 s	β^+; K; (3% m); 14% γ 0.394; 9% γ 0.385; 5% γ 0.383, 0.593	−69.9		
Eu	141m	3.3 s	β^+; K; (66% m); 33% IT; 1% γ 0.096; 1% γ 0.395 (double)	−69.8		
Eu	142	2.4 s	β^+ 7.0; K; 15% γ 0.768; 2% γ 0.890; 2% γ 1.287, 1.658	−71.5		1^+
Eu	142	1.22 min	β^+; K; 100% γ 0.768; 92% γ 1.023; 87% γ 0.557	−71.5		
Eu	143	2.61 min	β^+ 4.1; 28% K; 7% γ 1.107; 3% γ 1.537; 2% γ 0.108, 1.913	−74.41		
Eu	144	10.2 s	β^+ 5.3; K; 9% γ 1.660; 2% γ 0.818; 1% γ 2.423	−75.64		1^+

Table of the nuclides (*contd*)

Nuclide		Natural abundance or half-life	Decay modes and major radiations	Mass excess MeV	σ/b	I$^\pi$
Z	A					
Eu	145	5.93 d	98% K; 2% β^+ 1.7; 65% γ 0.894; 16% γ 1.659; 15% γ 0.654	−77.94		$\frac{5}{2}^+$
Eu	146	4.59 d	96% K; β^+ 1.5; 99% γ 0.747; 44% γ 0.633; 38% γ 0.634	−77.11		4$^-$
Eu	147	22 d	99% K; β^+ 0.7; 22% γ 0.197; 19% γ 0.121; 9% γ 0.678	−77.54		$\frac{5}{2}^+$
Eu	148	54 d	100% K; 99% γ 0.550; 71% γ 0.630; 19% γ 0.414, 0.611	−76.24		5$^-$
Eu	149	93.1 d	100% K; 4% γ 0.328; 3% γ 0.277; 1% γ 0.255	−76.44		$\frac{5}{2}^+$
Eu	150	36 yr	100% K; 94% γ 0.334; 79% γ 0.439; 52% γ 0.584			
Eu	150	12.6 hr	89% β^- 1.01; 11% K; 4% γ 0.334; 2% γ 0.407	−74.76		0
Eu	151	47.9%		−74.65	5800 (γ); 3200 (γm$_1$); 4 (γm$_2$)	$\frac{5}{2}^+$
Eu	152	13.2 yr	73% K; 8% β^- 1.48; 31% γ 0.122; 27% γ 0.344; 21% γ 1.408	−72.88		3$^-$
Eu	152m$_1$	9.3 hr	73% β^- 1.87; 24% K; 13% γ 0.842; 11% γ 0.963; 6% γ 0.122	−72.84		0$^-$
Eu	152m$_2$	96 min	100% IT; 72% γ 0.090; 1% γ 0.018, 0.077	−72.74		8$^-$
Eu	153	52.1%		−73.36	380 (γ)	$\frac{5}{2}^+$
Eu	154	8.5 yr	14% β^- 1.86; 34% β^- 0.61; 40% γ 0.123; 36% γ 1.274; 20% γ 0.723	−71.73		3$^-$
Eu	154m	45.8 min	100% IT; 36% γ 0.068; 28% γ 0.101; 9% γ 0.036	−71.6		
Eu	155	4.96 yr	13% β^- 0.25; 49% β^- 0.14; 34% γ 0.086 (double); 23% γ 0.105; 1% γ 0.045	−71.83	4000 (γ)	$\frac{5}{2}^+$
Eu	156	15.1 d	31% β^- 2.5; 30% β^- 0.5; 10% γ 0.811; 9% γ 0.089; 8% γ 1.231	−70.08		0$^+$
Eu	157	15.1 hr	41% β^- 1.30; 23% γ 0.064; 19% γ 0.411; 11% γ 0.371	−69.47		
Eu	158	45.9 min	5% β^- 3.4; 23% β^- 2.5; 27% γ 2.4; 32% γ 0.944; 17% γ 0.977; 14% γ 0.079	−67.2		
Eu	159	18.1 min	β^- 2.6; 20% γ 0.068; 10% γ 0.079; 7% γ 0.096	−65.9		
Eu	160	50 s	β^-; γ	−63.5		
64 Gd	143	108 s	β^+; K; 95% γ 0.272; 18% γ 0.588; 12% γ 0.799	−68.5		
Gd	144	4.5 min	β^+; K; γ	−71.9		0$^+$
Gd	145	22 min	β^+; K; 35% γ 1.758; 34% γ 1.881; 10% γ 1.042	−72.9		$\frac{1}{2}^+$

Table of the nuclides (*contd*)

Nuclide Z	A	Natural abundance or half-life	Decay modes and major radiations	Mass excess MeV	σ/b	I^π
Gd	145m	85 s	95% IT; β^+; K; 83% γ 0.721; 4% γ 0.387	−72.2		$\frac{11}{2}^-$
Gd	146	48.3 d	100% K; 50% γ 0.115; 50% γ 0.116; 50% γ 0.155	−75.36		0^+
Gd	147	38.1 hr	100% K; 60% γ 0.229; 32% γ 0.396; 20% γ 0.929	−75.21		$\frac{7}{2}^-$
Gd	148	98 yr	α 3.18	−76.27		0^+
Gd	149	9.3 d	100% K; 53% γ 0.150; 31% γ 0.299; 25% γ 0.347	−75.13		$\frac{7}{2}^-$
Gd	150	1.8×10^6 yr	α 2.73	−75.77		0^+
Gd	151	120 d	100% K; γ	−74.17		$\frac{7}{2}^-$
Gd	152	1.1×10^{14} yr 0.20%	α 2.14	−74.70	1100 (γ)	0^+
Gd	153	242 d	100% K; 27% γ 0.097; 19% γ 0.103; 2% γ 0.070	−73.12		$\frac{3}{2}^-$
Gd	154	2.1%		−73.70	90 (γ)	0^+
Gd	155	14.8%		−72.07	6.1×10^4 (γ)	$\frac{3}{2}^-$
Gd	156	20.6%		−72.54	2 (γ)	0^+
Gd	157	15.7%		−70.83	2.6×10^5 (γ)	$\frac{3}{2}^-$
Gd	158	24.8%		−70.69	2.4 (γ)	0^+
Gd	159	18.6 hr	β^- 0.97; 10% γ 0.363; 2% γ 0.058	−68.56		$\frac{3}{2}^-$
Gd	160	21.8%		−67.94	0.77 (γ)	0^+
Gd	161	3.6 min	5% β^- 1.6; 85% β^- 1.5; 61% γ 0.361; 23% γ 0.315; 14% γ 0.102	−65.51	4×10^4 (γ) (r)	$\frac{5}{2}^-$
Gd	162	9 min	100% β^- 1.0; 53% γ 0.442; 46% γ 0.403; 7% γ 0.039	−64.4		0^+
65 Tb	146	23 s	β^+; K; 90% γ 1.580; 46% γ 1.079; 15% γ 1.417	−67.3		
Tb	147	1.6 hr	95% K; β^+; 75% γ 1.152; 32% γ 0.694; 20% γ 0.140	−70.5		$\frac{5}{2}^+$
Tb	147	1.8 min	β^+; K; 85% γ 1.398; 14% γ 1.798; 1% γ 0.998	−70.5		$\frac{11}{2}^-$
Tb	148	2.2 min	β^+; K; γ			
Tb	148	60 min	80% K; β^+ 4.6; 86% γ 0.784; 25% γ 0.489; 12% γ 1.077	−70.6		2^-
Tb	149	4.15 hr	17% α 3.97; 79% K; β^+; 33% γ 0.352; 28% γ 0.165; 21% γ 0.388	−71.43		
Tb	149m	4.16 min	β^+; K; 90% γ 0.796; 7% γ 0.165	−71.39		
Tb	150	3.3 hr	90% K; β^+ 3.7; 72% γ 0.638; 15% γ 0.496; 5% γ 0.792	−71.10		
Tb	150	6.0 min	β^+; K; 100% γ 0.638; 70% γ 0.650; 42% γ 0.438			
Tb	151	17.6 hr	99% K; β^+; 26% γ 0.252; 25% γ 0.108; 25% γ 0.287	−71.61		$\frac{1}{2}$
Tb	152	17.5 hr	87% K; 6% β^+ 2.8; 66% γ 0.344; 10% γ 0.586; 9% γ 0.271 (double)	−70.85		2^-

Table of the nuclides (*contd*)

Nuclide		Natural abundance or half-life	Decay modes and major radiations	Mass excess MeV	σ/b	Iπ
Z	A					
Tb	152m	4.2 min	78% IT; 22% K; 62% γ 0.283; 20% γ 0.344; 20% γ 0.411	−70.35		
Tb	153	2.30 d	100% K; 40% γ 0.212; 9% γ 0.170; 8% γ 0.110	−71.33		$\frac{5}{2}^+$
Tb	154	21 hr	98% K; β+ 2.4; γ	−70.24		0
Tb	154m₁	9 hr	β+; K; 22% IT; γ			3
Tb	154m₂	23 hr	98% K; 2% IT; 87% γ 0.248; 76% γ 0.347; 49% γ 1.420			
Tb	155	5.3 d	100% K; 29% γ 0.087; 23% γ 0.105; 7% γ 0.180	−71.26		$\frac{3}{2}^+$
Tb	156	5.4 d	100% K; 66% γ 0.534; 42% γ 0.199; 32% γ 1.222	−70.10		3⁻
Tb	156	24 hr	100% IT; γ			
Tb	156m	5.0 hr	IT; K; γ	−70.01		
Tb	157	150 yr	100% K	−70.77		$\frac{3}{2}^+$
Tb	158	150 yr	82% K; 17% β⁻ 0.84; 43% γ 0.944; 20% γ 0.962; 11% γ 0.080	−69.48		3⁻
Tb	158m	10.5 s	100% IT; 1% γ 0.110	−69.37		0⁻
Tb	159	100%		−69.54	23 (γ)	$\frac{3}{2}^+$
Tb	160	72.1 d	27% β⁻ 0.87; 47% β⁻ 0.57; 30% γ 0.879; 27% γ 0.299; 25% γ 0.966	−67.84	500 (γ) (r)	3⁻
Tb	161	6.90 d	10% β⁻ 0.59; 66% β⁻ 0.52; 16% γ 0.049; 10% γ 0.075; 2% γ 0.057	−67.47		$\frac{3}{2}^+$
Tb	162	7.8 min	99% β⁻ 1.3; 80% γ 0.260; 43% γ 0.807; 39% γ 0.888	−65.8		
Tb	163	19.5 min	9% β⁻ 1.3; 35% β⁻ 0.8; 26% γ 0.351; 22% γ 0.495; 11% γ 0.422	−64.7		$\frac{3}{2}^+$
Tb	164	3.0 min	β⁻; 24% γ 0.169; 22% γ 0.755; 20% γ 0.215, 0.611	−62.1		
66 Dy	147m	59 s	IT; γ	−63.5		
Dy	148	3.1 min	β+; K; 100% γ 0.620	−67.8		0⁺
Dy	149	4.1 min	β+; K; γ	−67.5		
Dy	150	7.17 min	31% α (4.23); β+ 0.5; K; 68% γ 0.397	−69.1		0⁺
Dy	151	17 min	6% α (4.07); β+; K; 11% γ 0.477; 10% γ 0.547; 9% γ 0.176	−68.60		$\frac{7}{2}^-$
Dy	152	2.37 hr	100% K; 98% γ 0.257	−70.12		0⁺
Dy	153	6.3 hr	β+; K; 7% γ 0.081; 7% γ 0.214; 6% γ 0.100	−69.16		$\frac{7}{2}$
Dy	154	10⁷ yr	α 2.87	−70.39		0⁺
Dy	155	10.0 hr	97% K; β+ 1.08; 68% γ 0.227; 4% γ 0.185; 3% γ 1.090	−69.16		$\frac{3}{2}$
Dy	156	0.057%		−70.53	33 (γ) (s)	0⁺
Dy	157	8.1 hr	100% K; 94% γ 0.326; 2% γ 0.182; 1% γ 0.083	−69.43		$\frac{3}{2}^-$
Dy	158	0.100%		−70.41	70 (γ) (r)	0⁺

Table of the nuclides (*contd*)

Nuclide		Natural abundance or half-life	Decay modes and major radiations	Mass excess MeV	σ/b	I^π
Z	A					
Dy	159	144.4 d	100% K; 2% γ 0.058	−69.17		$\frac{3}{2}^-$
Dy	160	2.3%		−69.67	60 (γ)	0^+
Dy	161	19.0%		−68.06	570 (γ)	$\frac{5}{2}^+$
Dy	162	25.5%		−68.18	160 (γ)	0^+
Dy	163	24.9%		−66.38	130 (γ)	$\frac{5}{2}^-$
Dy	164	28.1%		−65.97	900 (γ); 1800 (γm)	0^+
Dy	165	2.33 hr	83% β^- 1.29; 4% γ 0.095	−63.61	4000 (γ) (r)	$\frac{7}{2}^+$
Dy	165m	1.26 min	98% IT; 2% β^- 0.9; 3% γ 0.108	−63.50	2100 (γ) (r)	$\frac{1}{2}^-$
Dy	166	81.5 hr	5% β^- 0.48; 92% β^- 0.40; 13% γ 0.082	−62.58		0^+
Dy	167	6.2 min	7% β^- 2.0; 85% β^- 1.8; 48% γ 0.570; 28% γ 0.259; 25% γ 0.310	−60.0		
67 Ho	150	40 s	β^+; K; 100% γ 0.391; 100% γ 0.654; 100% γ 0.804	−62.0		
Ho	151	47 s	10% α (4.61); β^+; K; γ	−63.4		
Ho	151	35.6 s	20% α (4.52); β^+; K; γ	−63.4		
Ho	152	52 s	6% α (4.46); β^+; K; 97% γ 0.648; 95% γ 0.684; 93% γ 0.614			
Ho	152	2.4 min	2% α (4.39); β^+; K; 97% γ 0.614; 16% γ 0.648	−63.7		
Ho	153	2.0 min	β^+; K; γ			
Ho	153	9.3 min	β^+; K; γ			
Ho	154	12 min	β^+; K; γ	−64.64		1
Ho	154	3.2 min	β^+; K; 95% γ 0.335; 76% γ 0.412; 51% γ 0.477			
Ho	155	49 min	α 3.94; β^+ 2.1; K; 8% γ 0.240; 4% γ 0.136; 2% γ 0.185, 0.325	−66.06		$\frac{5}{2}$
Ho	156(m)	56 min	β^+; K; IT; γ			1
Ho	157	12.6 min	β^+; K; 20% γ 0.280; 15% γ 0.341; 7% γ 0.193	−66.9		$\frac{7}{2}^-$
Ho	158	11.5 min	β^+ 2.9; K; γ	−66.43		5^+
Ho	158m₁	27 min	65% IT; β^+; K; γ	−66.37		2^-
Ho	158(m₂)	21 min	β^+; K; γ			
Ho	159	33 min	100% K; 34% γ 0.121; 23% γ 0.132; 14% γ 0.253, 0.310	−67.32		$\frac{7}{2}^-$
Ho	159m	8.3 s	100% IT; 40% γ 0.206; 5% γ 0.166	−67.11		$\frac{1}{2}^+$
Ho	160	25.6 min	100% K; γ	−66.39		5^+
Ho	160m	5.0 hr	65% IT; β^+; K; γ	−66.33		2^-
Ho	160	3 s	?			
Ho	161	2.5 hr	100% K; 29% γ 0.026; 3% γ 0.078; 3% γ 0.103	−67.20		$\frac{7}{2}^-$
Ho	161m	6.7 s	100% IT; 45% γ 0.211	−66.99		$\frac{1}{2}^+$
Ho	162	15 min	95% K; β^+ 1.1; 8% γ 0.080; 4% γ 1.319; 1% γ 1.373	−66.05		1^+
Ho	162m	68 min	61% IT; β^+; K; γ	−65.9		6^-
Ho	163	33 yr	100% K	−66.38		$\frac{7}{2}^-$

Table of the nuclides (*contd*)

Nuclide		Natural abundance or half-life	Decay modes and major radiations	Mass excess MeV	σ/b	I^π
Z	A					
Ho	163m	1.09 s	100% IT; 78% γ 0.298	−66.08		$\frac{1}{2}^+$
Ho	164	29 min	58% K; 29% $β^-$ 1.00; 13% $β^-$ 0.91; 3% γ 0.091	−64.94		1^+
Ho	164m	37 min	100% IT; 11% γ 0.037; 7% γ 0.057	−64.80		6
Ho	165	100%		−64.90	62 (γ); 3 (γm)	$\frac{7}{2}^-$
Ho	166	26.8 hr	51% $β^-$ 1.85; 48% $β^-$ 1.77; 6% γ 0.081; 1% γ 1.379	−63.07		0^-
Ho	166m	1200 yr	80% $β^-$ 0.06; 73% γ 0.184; 63% γ 0.810; 59% γ 0.712	−63.06		
Ho	167	3.1 hr	15% $β^-$ 0.97; 43% $β^-$ 0.30; (13%m); 57% γ 0.347; 24% γ 0.321; 5% γ 0.208, 0.238	−62.32		
Ho	168	3.0 min	71% $β^-$ 1.9; 36% γ 0.741; 34% γ 0.821; 18% γ 0.816	−60.3		3^+
Ho	169	4.8 min	$β^-$; 22% γ 0.788; 12% γ 0.853; 11% γ 0.761, 0.778	−58.79		
Ho	170	43 s	$β^-$ 4.0; γ	−56.1		
Ho	170	2.8 min	23% $β^-$ 3.0; 28% $β^-$ 2.1; 35% γ 0.932; 25% γ 0.182; 19% γ 0.890, 1.139	−56.1		
68 Er	151	23 s	$β^+$; K	−58.2		
Er	152	9.8 s	90% α (4.80); $β^+$; K	−60.4		0^+
Er	153	36 s	38% α (4.67); $β^+$; K	−60.3		
Er	154	3.8 min	$β^+$; K	−62.4		0^+
Er	155	5.3 min	$β^+$; K	−62.06		
Er	156	20 min	$β^+$; K; 37% γ 0.035; 6% γ 0.030	−63.9		0^+
Er	157	24 min	$β^+$; K; γ	−63.1		$\frac{3}{2}^-$
Er	158	2.4 hr	K; $β^+$ 0.7; 11% γ 0.072; 5% γ 0.387; 2% γ 0.249	−65.0		0^+
Er	159	36 min	$β^+$; K; (21%m); 37% γ 0.624; 28% γ 0.649; 8% γ(m) 0.206	−64.4		$\frac{3}{2}^-$
Er	160	28.6 hr	100% K	−66.05		0^+
Er	161	3.24 hr	100% K; (27%m); 61% γ 0.827; 12% γ (m) 0.211; 3% γ 0.592	−65.20		$\frac{3}{2}^-$
Er	162	0.14%		−66.34	19 (γ)	0^+
Er	163	75.1 min	100% K	−65.17		$\frac{5}{2}^-$
Er	164	1.56%		−65.94	13 (γ)	0^+
Er	165	10.4 hr	100% K	−64.52		$\frac{5}{2}^-$
Er	166	33.4%		−64.92	5 (γ) (r); 15 (γm) (r)	0^+
Er	167	22.9%		−63.29	650 (γ) (r)	$\frac{7}{2}^+$
Er	167m	2.28 s	100% IT; 42% γ 0.208	−63.08		$\frac{1}{2}^-$
Er	168	27.1%		−62.99	2.0 (γ)	0^+
Er	169	9.40 d	55% $β^-$ 0.35; 45% $β^-$ 0.34	−60.92		$\frac{1}{2}^-$
Er	170	14.9%		−60.10	5.7 (γ)	0^+
Er	171	7.52 hr	2% $β^-$ 1.49; 94% $β^-$ 1.07; 64% γ 0.308; 29% γ 0.296; 21% γ 0.112	−57.71	300 (γ) (r)	$\frac{5}{2}^-$

Table of the nuclides (*contd*)

Nuclide			Natural abundance or half-life	Decay modes and major radiations	Mass excess MeV	σ/b	Iπ
Z		A					
	Er	172	49 hr	3% β⁻ 0.48; 46% β⁻ 0.35; 48% β⁻ 0.28; 45% γ 0.610; 43% γ 0.407; 3% γ 0.068, 0.446	−56.49		0⁺
	Er	173	1.4 min	23% β⁻ 2.2; 64% β⁻ 1.3; 54% γ 0.895; 48% γ 0.199; 47% γ 0.193	−53.7		
69	Tm	153	1.6 s	α 5.10	−53.9		
	Tm	154	5 s	α 4.96	−54.5		
	Tm	154	3.0 s	α 5.03	−54.5		
	Tm	155	39 s	α 4.45	−56.5		
	Tm	156	80 s	α 4.23; β⁺; K; γ	−56.9		
	Tm	156	19 s	α 4.46	−56.9		
	Tm	157	3.6 min	β⁺; K; γ	−58.5		
	Tm	158	4.0 min	β⁺; K; 69% γ 0.192; 19% γ 0.335; 8% γ 1.150	−58.4		
	Tm	159	9.0 min	β⁺; K; 8% γ 0.038; 8% γ 0.085; 7% γ 0.271	−60.2		5/2
	Tm	160	9.2 min	85% K; β⁺; 35% γ 0.126; 13% γ 0.729; 9% γ 0.264, 1.369	−60.1		1⁻
	Tm	161	30 min	β⁺; K; 25% γ 0.046; 20% γ 1.648; 10% γ 0.084	−61.7		7/2
	Tm	162	21.8 min	93% K; β⁺ 3.8; 17% γ 0.102; 8% γ 0.799; 7% γ 0.227	−61.5		1⁻
	Tm	162m	24 s	90% IT; β⁺; K; 7% γ 0.067; 7% γ 0.812; 7% γ 0.900 (double)			
	Tm	163	1.8 hr	100% K; 20% γ 0.104; 20% γ 0.240 (double) 13% γ 1.434	−62.99		1/2⁺
	Tm	164	2.0 min	61% K; β⁺ 2.9; 7% γ 0.091; 2% γ 1.155; 1% γ 0.769	−61.98		1⁺
	Tm	164m	5.1 min	80% IT; β⁺; K; 17% γ 0.208; 11% γ 0.315; 9% γ 0.240			6
	Tm	165	30.06 hr	100% K; 35% γ 0.243; 25% γ 0.297 (double); 8% γ 0.807	−62.92		1/2⁺
	Tm	166	7.7 hr	98% K; β⁺ 1.9; 20% γ 2.052; 19% γ 0.779; 16% γ 0.184	−61.87		2⁺
	Tm	167	9.25 d	100% K; (98%m); 41% γ (m) 0.208; 2% γ 0.532	−62.54		1/2⁺
	Tm	168	93.1 d	98% K; β⁻; 50% γ 0.198; 46% γ 0.816; 22% γ 0.447	−61.31		3
	Tm	169	100%		−61.27	98 (γ)	1/2⁺
	Tm	170	128.6 d	76% β⁻ 0.97; 3% γ 0.084	−59.79	92 (γ) (s)	1⁻
	Tm	171	1.92 yr	98% β⁻ 0.097	−59.21	4.5 (γ) (s)	1/2⁺
	Tm	172	63.6 hr	29% β⁻ 1.87; 36% β⁻ 1.79; 7% γ 0.079; 6% γ 1.094; 6% γ 1.387	−57.38		2⁻
	Tm	173	8.2 hr	2% β⁻ 1.32; 77% β⁻ 0.92; 88% γ 0.399; 7% γ 0.460; 1% γ 0.063	−56.23		

Table of the nuclides (*contd*)

Nuclide		Natural abundance or half-life	Decay modes and major radiations	Mass excess MeV	σ/b	I^π
Z	A					
Tm	174	5.4 min	83% β^- 1.2; 100% γ 0.366; 90% γ 0.992; 87% γ 0.273	−53.9		
Tm	175	15 min	23% β^- 1.9; 36% β^- 0.9; 65% γ 0.515; 15% γ 0.941; 13% γ 0.364	−52.3		
Tm	176	1.9 min	β^-; 44% γ 0.190; 33% γ 1.069; 24% γ 0.382	−49.6		
70 Yb	154	0.39 s	α 5.32	−50.1		0^+
Yb	155	1.7 s	α 5.19	−50.5		
Yb	156	24 s	α 4.80	−53.1		0^+
Yb	157	34 s	α 4.50	−53.3		
Yb	158	1.1 min	β^+; K; γ	−55.5		0^+
Yb	160	4.8 min	β^+; K; γ	−57.6		0^+
Yb	161	4.2 min	β^+; K; γ	−57.4		
Yb	162	18.9 min	>98% K; β^+; 40% γ 0.163; 28% γ 0.119; 2% γ 0.045	−59.3		0^+
Yb	163	11.0 min	β^+; K; γ	−59.6		
Yb	164	76 min	100% K	−60.9		0^+
Yb	165	10 min	K; β^+ 1.58; 37% γ 0.080; 6% γ 0.069; 3% γ 1.090	−60.16		
Yb	166	56.7 hr	100% K; 15% γ 0.082	−61.58		0^+
Yb	167	17.5 min	100% K; 54% γ 0.113; 22% γ 0.106; 20% γ 0.176	−60.58		$\frac{5}{2}$
Yb	168	0.135%		−61.57	3500 (γ)	0^+
Yb	169	32.0 d	100% K; 44% γ 0.063; 36% γ 0.198; 22% γ 0.177	−60.36		$\frac{7}{2}^+$
Yb	169m	46 s	100% IT	−60.34		$\frac{1}{2}^-$
Yb	170	3.1%		−60.76	10 (γ)	0^+
Yb	171	14.4%		−59.30	53 (γ)	$\frac{1}{2}^-$
Yb	172	21.9%		−59.25	1.3 (γ)	0^+
Yb	173	16.2%		−57.55	17 (γ)	$\frac{5}{2}^-$
Yb	174	31.6%		−56.94	65 (γ)	0^+
Yb	175	4.19 d	87% β^- 0.47; 6% γ 0.396; 3% γ 0.283; 2% γ 0.114	−54.69		$\frac{7}{2}^-$
Yb	176	12.6%		−53.49	2.4 (γ)	0^+
Yb	176m	11.7 s	100% IT; 93% γ 0.293; 91% γ 0.390; 81% γ 0.190	−52.44		
Yb	177	1.9 hr	54% β^- 1.40; 20% γ 0.151; 6% γ 1.080; 3% γ 0.123	−50.99		$\frac{9}{2}^+$
Yb	177m	6.40 s	100% IT; 77% γ 0.104	−50.66		$\frac{1}{2}^-$
Yb	178	74 min	β^-; γ	−49.66		0^+
71 Lu	155	0.07 s	α 5.63	−42.6		
Lu	156	0.23 s	α 5.54	−43.8		
Lu	156	0.5 s	α 5.43	−43.8		
Lu	164	3.17 min	β^+; K; γ	−54.6		
Lu	165	11.8 min	β^+; K; 25% γ 0.121; 23% γ 0.132; 13% γ 0.174	−56.2		$\frac{1}{2}$
Lu	166	2.7 min	K; β^+; 77% γ 0.228; 41% γ 0.337; 32% γ 0.368	−56.1		
Lu	166m$_1$	1.4 min	β^+; K; 42% IT; 26% γ 0.228; 22% γ 0.102; 19% γ 0.285	−56.1		

Table of the nuclides (*contd*)

Z	A	Natural abundance or half-life	Decay modes and major radiations	Mass excess MeV	σ/b	I^π
Lu	166m$_2$	2.1 min	β^+; K; 23% γ 1.427; 16% γ 2.099; 15% γ 1.257	−56.1		
Lu	167	52 min	98% K; β^+ 2.1; 16% γ 0.030; 9% γ 0.239; 4% γ 0.213	−57.5		$\frac{7}{2}^+$
Lu	168	5.5 min	β^+; K; γ	−57.1		
Lu	168m	6.7 min	K; β^+; γ	−56.9		3^+
Lu	169	34.1 hr	99% K; β^+; (51%m); 23% γ 0.960; 22% γ 0.191; 9% γ 1.450	−57.88		$\frac{7}{2}^+$
Lu	169m	2.7 min	100% IT	−57.85		$\frac{1}{2}^-$
Lu	170	2.02 d	β^+; K; 9% γ 0.084; 8% γ 1.280; 6% γ 2.042	−57.32		0^+
Lu	170m	0.7 s	100% IT; 1% γ 0.045	−57.23		4^-
Lu	171	8.22 d	100% K; 51% γ 0.740; 12% γ 0.667; 6% γ 0.076	−57.82		$\frac{7}{2}^+$
Lu	171m	79 s	100% IT	−57.75		$\frac{1}{2}^-$
Lu	172	6.70 d	100% K; 63% γ 1.094; 28% γ 0.901; 19% γ 0.181	−56.73		4^-
Lu	172m	3.7 min	100% IT	−56.68		1^-
Lu	173	499 d	100% K; 13% γ 0.272; 8% γ 0.079; 3% γ 0.101	−56.87		$\frac{7}{2}^+$
Lu	174	3.3 yr	100% K; 6% γ 1.242; 4% γ 0.076	−55.56		
Lu	174m	142 d	99% IT; 1% K; 14% γ 0.045; 7% γ 0.067; 1% γ 0.273, 0.992	−55.39		
Lu	175	97.39%		−55.16	10 (γ); 16 (γm)	$\frac{7}{2}^+$
Lu	176	3.6×10^{10} yr 2.61%	99% β^- 0.59; 93% γ 0.307; 85% γ 0.202; 13% γ 0.088	−53.38	2000 (γ); 7 (γm)	7^-
Lu	176m	3.68 hr	40% β^- 1.31; 60% β^- 1.22; 9% γ 0.088	−53.25		1^-
Lu	177	6.71 d	78% β^- 0.50; 11% γ 0.208; 7% γ 0.113	−52.38		$\frac{7}{2}^+$
Lu	177m	160 d	78% β^- 0.15; (78%m$_1$); 22% IT; 65% γ (m) 0.208; 39% γ (m) 0.229; 30% γ (m) 0.379	−51.41		$\frac{23}{2}^-$
Lu	178	28.4 min	62% β^- 2.13; 28% β^- 2.04; 7% γ 0.093; 5% γ 1.341; 2% γ 1.310	−50.30		1^+
Lu	178m	22.9 min	83% β^- 1.3; (100% m$_1$); 97% γ (m) 0.426; 94% γ (m) 0.326; 81% γ (m) 0.213	−50.0		
Lu	179	4.6 hr	87% β^- 1.35; 12% γ 0.214	−49.11		
Lu	180	5.7 min	1% β^- 1.7; 91% β^- 1.5; 50% γ 0.408; 26% γ 1.200; 24% γ 1.106	−46.7		
72 Hf	157	0.12 s	α 5.68	−39.0		
Hf	158	3.0 s	α 5.27	−42.2		0^+
Hf	159	5.6 s	α 5.09	−42.8		
Hf	160	12 s	α 4.77	−45.8		0^+
Hf	161	17 s	α 4.60	−46.1		

Table of the nuclides (contd)

Nuclide		Natural abundance or half-life	Decay modes and major radiations	Mass excess MeV	σ/b	I^π
Z	A					
Hf	166	6.8 min	β^+; K; (100%m); 41% γ 0.079; 4% γ 0.342; 4% γ 0.408	−53.5		0^+
Hf	167	2.0 min	β^+; K; γ	−53.2		
Hf	168	26.0 min	98% K; β^+; γ	−55.1		0^+
Hf	169	3.26 min	86% K; 14% β^+ 1.9; 89% γ 0.493; 10% γ 0.370; 4% γ 0.124	−54.5		
Hf	170	15.9 hr	100% K; (5%m); 33% γ 0.165; 23% γ 0.621; 19% γ 0.120	−56.1		0^+
Hf	171	12.1 hr	β^+; K; γ	−55.3		$\frac{7}{2}^+$
Hf	172	1.87 yr	100% K; (100%m); 22% γ 0.024; 10% γ 0.126; 5% γ 0.067	−56.3		0^+
Hf	173	24.0 hr	100% K; 83% γ 0.124; 34% γ 0.297; 13% γ 0.140	−55.3		$\frac{1}{2}^-$
Hf	174	2×10^{15} yr 0.16%	α 2.50	−55.83	400 (γ)	0^+
Hf	175	70 d	100% K; 87% γ 0.343; 2% γ 0.089; 1% γ 0.433	−54.55		$\frac{5}{2}^-$
Hf	176	5.2%		−54.57	30 (γ)	0^+
Hf	177	18.6%		−52.88	390 (γ); 1.0 (γm_1)	$\frac{7}{2}^-$
Hf	177m₁	1.1 s	100% IT; 83% γ 0.208; 50% γ 0.229; 38% γ 0.379	−51.56		$\frac{23}{2}^+$
Hf	177m₂	51 min	100% IT; (100%m₁); 75% γ 0.277; 68% γ 0.295; 65% γ 0.327	−50.14		$\frac{37}{2}^-$
Hf	178	27.1%		−52.43	40 (γ); 50 (γm_1)	0^+
Hf	178m₁	4.3 s	100% IT; 97% γ 0.426; 94% γ 0.326; 81% γ 0.213	−51.29		8^-
Hf	178m₂	31 yr	100% IT; (100%m₁); 97% γ 0.426; 94% γ 0.326; 84% γ 0.574	−49.99		16^+
Hf	179	13.7%		−50.46	50 (γ); 0.4 (γm) (s)	$\frac{9}{2}^+$
Hf	179m₁	18.7 s	100% IT; 95% γ 0.214	−50.09		$\frac{1}{2}^-$
Hf	179m₂	25 d	100% IT; 65% γ 0.453; 38% γ 0.362; 27% γ 0.123	−49.36		$\frac{25}{2}^-$
Hf	180	35.2%		−49.78	14 (γ)	0^+
Hf	180m	5.5 hr	100% IT; 94% γ 0.332; 85% γ 0.443; 81% γ 0.215	−48.64		8^-
Hf	181	42.4 d	93% β^- 0.41; 81% γ 0.482; 40% γ 0.133; 13% γ 0.346	−47.40	30 (γ) (r)	$\frac{1}{2}^-$
Hf	182	9×10^6 yr	100% β^- 0.16; 80% γ 0.270; 7% γ 0.156; 3% γ 0.114	−46.0		0^+
Hf	182m	62 min	10% β^- 0.95; 43% β^- 0.48; (10%m₂); 46% IT; 46% γ 0.344; 38% γ 0.224; 24% γ 0.507	−44.8		
Hf	183	64 min	25% β^- 1.55; 68% β^- 1.15; 65% γ 0.784; 38% γ 0.073; 27% γ 0.459	−43.27		

Table of the nuclides (*contd*)

Nuclide Z	A	Natural abundance or half-life	Decay modes and major radiations	Mass excess MeV	σ/b	I^{π}
Hf	184	4.1 hr	46% β^- 1.1; 38% β^- 0.7; 48% γ 0.139; 38% γ 0.345; 15% γ 0.181	−41.5		0^+
73 Ta	166	32 s	β^+; K; γ	−46.1		
Ta	167	3 min	β^+; K	−48.0		
Ta	168	2.4 min	β^+; K; 37% γ 0.124; 28% γ 0.262; 10% γ 0.750	−48.4		
Ta	169	5 min	β^+; K; γ	−50.0		
Ta	170	6.8 min	β^+; K; 21% γ 0.101; 16% γ 0.221; 9% γ 0.987 (double)	−50.1		
Ta	171	23.3 min	β^+; K; γ	−51.6		
Ta	172	37 min	85% K; β^+; 54% γ 0.214; 17% γ 0.095; 15% γ 1.109	−51.4		
Ta	173	3.7 hr	K; β^+; 17% γ 0.172; 6% γ 0.070; 5% γ 0.090	−52.4		
Ta	174	1.1 hr	K; β^+; 57% γ 0.207; 16% γ 0.091; 5% γ 1.206	−52.0		3
Ta	175	10.5 hr	K; β^+; 14% γ 0.207; 12% γ 0.349; 11% γ 0.267 (double)	−52.4		$\frac{7}{2}^+$
Ta	176	8.1 hr	99% K; β^+ 2.1; 23% γ 1.159; 11% γ 0.088; 5% γ 0.202, 0.711	−51.5		
Ta	177	56.6 hr	100% K; 7% γ 0.113; 1% γ 0.208	−51.72		$\frac{7}{2}^+$
Ta	178	9.3 min	99% K; 1% β^+ 0.89; 7% γ 0.093; 1% γ 1.341; 1% γ 1.351	−50.5		1^+
Ta	178	2.4 hr	100% K; (100%m$_1$); 97% γ (m) 0.426; 94% γ (m) 0.326; 81% γ (m) 0.213			
Ta	179	665 d	100% K	−50.35		
Ta	180 (g)	> 10^{13} yr 0.0123%			700 (γ)	
Ta	180(m)	8.1 hr	87% K; 10% β^- 0.71; 3% β^- 0.61; 5% γ 0.093	−48.91		1
Ta	181	99.9877%		−48.43	21 (γ); 0.010 (γm)	$\frac{7}{2}^+$
Ta	182	115.0 d	5% β^- 0.59; 40% β^- 0.52; 29% β^- 0.26: 41% γ 0.068; 35% γ 1.121; 27% γ 1.221	−46.42	8200 (γ)	3^-
Ta	182m$_1$	0.283 s	100% IT	−46.40		5^+
Ta	182m$_2$	15.8 min	100% IT; 47% γ 0.172; 36% γ 0.147; 24% γ 0.185	−45.90		10^-
Ta	183	5.1 d	1% β^- 0.66; 91% β^- 0.62; (5%m); 27% γ 0.246; 12% γ 0.108; 12% γ 0.354	−45.28		$\frac{7}{2}^+$
Ta	184	8.7 hr	β^-; 74% γ 0.414; 45% γ 0.253; 33% γ 0.921	−42.82		
Ta	185	49 min	β^-; 26% γ 0.178; 22% γ 0.174; 4% γ 0.066, 0.244	−41.36		
Ta	186	10.5 min	2% β^- 2.9; 63% β^- 2.2; 58% γ 0.198; 49% γ 0.215; 43% γ 0.511	−38.6		

Table of the nuclides (*contd*)

Nuclide			Natural abundance or half-life	Decay modes and major radiations	Mass excess MeV	σ/b	I^π
Z		A					
74	W	162	<0.25 s	α 5.53	−34.1		0^+
	W	163	2.5 s	α 5.39	−35.3		
	W	164	6 s	α 5.15	−38.0		0^+
	W	165	5 s	α 4.91	−38.7		
	W	166	16 s	α 4.74	−41.5		0^+
	W	172	7 min	K; β^+; γ	−48.8		0^+
	W	173	16 min	β^+; K; γ	−48.5		
	W	174	29 min	100% K; γ	−50.1		0^+
	W	175	34 min	β^+; K; γ	−49.5		
	W	176	2.3 hr	100% K; γ	−50.6		0^+
	W	177	135 min	β^+; K; 58% γ 0.116 (double); 16% γ 0.186 (double); 13% γ 0.427	−49.7		
	W	178	21.5 d	100% K	−50.4		0^+
	W	179	38 min	100% K; 18% γ 0.031	−49.28		
	W	179m	6.4 min	100% IT; 9% γ 0.222	−49.06		
	W	180	0.13%		−49.62	10 (γ) (r)	0^+
	W	181	121 d	100% K	−48.24		$\frac{9}{2}^+$
	W	182	26.3%		−48.23	21 (γ)	0^+
	W	183	14.3%		−46.35	10.1 (γ)	$\frac{1}{2}^-$
	W	183m	5.4 s	100% IT; γ	−46.04		
	W	184	30.7%		−45.69	1.8 (γ); 0.002 (γm)	0^+
	W	185	75.1 d	100% β^- 0.43	−43.37		$\frac{3}{2}^-$
	W	185m	1.66 min	100% IT; 5% γ 0.066; 4% γ 0.132; 3% γ 0.174	−43.17		$\frac{11}{2}^+$
	W	186	28.6%		−42.50	38 (γ)	0^+
	W	187	23.9 hr	33% β^- 1.31; 53% β^- 0.63; 26% γ 0.686; 21% γ 0.480; 11% γ 0.072	−39.89	70 (γ)	$\frac{3}{2}^-$
	W	188	69 d	99% β^- 0.35	−38.66		0^+
	W	189	11.5 min	β^- 2.5; γ	−35.5		
	W	190	30 min	β^- 1.0; 39% γ 0.158; 11% γ 0.162	−34.2		0^+
75	Re	170	8 s	β^+; K; γ	−38.9		
	Re	172	30 s	β^+; K; γ	−41.5		
	Re	174	2.4 min	β^+; K; γ	−43.6		
	Re	175	4.6 min	β^+; K; γ	−45.2		
	Re	176	5.2 min	β^+; K; γ	−45.0		
	Re	177	14 min	β^+; K; γ	−46.1		
	Re	178	13.2 min	89% K; β^+; 45% γ 0.237; 23% γ 0.106; 9% γ 0.939	−45.8		
	Re	179	19.7 min	99% K; β^+; (27%m); 28% γ 0.430; 27% γ 0.290; 13% γ 1.680	−46.6		
	Re	180	2.4 min	92% K; 8% β^+ 1.8; 98% γ 0.902; 23% γ 0.104; 12% γ 0.825	−45.83		
	Re	181	20 hr	100% K; 57% γ 0.366; 12% γ 0.361; 6% γ 0.639	−46.4		$\frac{5}{2}^+$
	Re	182	64 hr	100% K; 27% γ 0.229; 21% γ 1.121; 17% γ 1.221	−45.4		

Table of the nuclides (*contd*)

Nuclide			Natural abundance or half-life	Decay modes and major radiations	Mass excess MeV	σ/b	I^π
	Z	A					
	Re	182	12.7 hr	100% K; 32% γ 1.121; 25% γ 1.221; 15% γ 1.189	−45.4		2^+
	Re	183	71 d	100% K; 24% γ 0.162; 8% γ 0.046; 3% γ 0.209, 0.292	−45.79		
	Re	184	38 d	100% K; 38% γ 0.903; 37% γ 0.792; 17% γ 0.111	−44.19		3^-
	Re	184m	169 d	75% IT; 25% K; 14% γ 0.105; 11% γ 0.253; 10% γ 0.217	−44.00		8^+
	Re	185	37.40%		−43.80	110 (γ); 0.3 (γm) (s)	$\frac{5}{2}^+$
	Re	186	90.6 hr	71% $β^-$ 1.08; 21% $β^-$ 0.94; 8% K; 9% γ 0.137; 1% γ 0.123	−41.91		1^-
	Re	186m	2×10^5 yr	100% IT; γ	−41.8		
	Re	187	4×10^{10} yr 62.60%	100% $β^-$ 0.003	−41.21	74 (γ); 1.0 (γm)	$\frac{5}{2}^+$
	Re	188	16.98 hr	70% $β^-$ 2.12; 15% γ 0.155; 1% γ 0.478; 1% γ 0.633	−39.01		1^-
	Re	188m	18.7 min	100% IT; 16% γ 0.064; 11% γ 0.106; 5% γ 0.092	−38.83		
	Re	189	24.3 hr	60% $β^-$ 1.0; (8%m); 6% γ 0.217; 4% γ 0.219; 3% γ 0.245	−37.97		
	Re	190	3.1 min	89% $β^-$ 1.8; 35% γ 0.558; 27% γ 0.829; 25% γ 0.569	−35.5		
	Re	190m	3.2 hr	$β^-$; 49% IT; 50% γ 0.187; 28% γ 0.558; 26% γ 0.569	−35.3		
	Re	192	16 s	$β^-$; γ	−32		
76	Os	169	3 s	α 5.56	−30.6		
	Os	170	7 s	α 5.40	−33.5		
	Os	171	8 s	α 5.24	−34.2		
	Os	172	19 s	$β^+$; K	−36.8		
	Os	173	16 s	$β^+$; K; γ	−37.4		
	Os	174	45 s	$β^+$; K; γ	−39.6		0^+
	Os	176	3.6 min	$β^+$; K; γ	−41.8		0^+
	Os	178	5.0 min	$β^+$; K; γ	−43.4		0^+
	Os	179	7 min	$β^+$; K; γ	−42.9		
	Os	180	22 min	100% K; 18% γ 0.020	−44.2		0^+
	Os	181	105 min	$β^+$; K; 46% γ 0.239; 21% γ 0.827; 13% γ 0.118			
	Os	181	2.7 min	$β^+$; K; 87% γ 0.145; 24% γ 0.118; 3% γ 1.119	−43.4		
	Os	182	22.0 hr	100% K; 52% γ 0.510; 33% γ 0.180; 7% γ 0.263	−44.6		0^+
	Os	183	13 hr	100% K; 86% γ 0.382; 20% γ 0.114; 8% γ 0.168	−43.5		$\frac{9}{2}^+$
	Os	183m	9.9 hr	89% K; 11% IT; 55% γ 1.102; 25% γ 1.108; 7% γ 1.035	−43.3		$\frac{1}{2}^-$
	Os	184	0.018%		−44.23	3000 (γ)	0^+
	Os	185	94 d	100% K; 81% γ 0.646; 7% γ 0.875; 5% γ 0.880	−42.79		$\frac{1}{2}^-$

Table of the nuclides (*contd*)

Nuclide		Natural abundance or half-life	Decay modes and major radiations	Mass excess MeV	σ/b	I^π
Z	A					
Os	186	2×10^{15} yr 1.6%	α 2.75	−42.99	80 (γ)	0^+
Os	187	1.6%		−41.21	330 (γ)	$\frac{1}{2}^-$
Os	188	13.3%		−41.13	<5 (γ)	0^+
Os	189	16.1%		−38.98	20 (γ); 3×10^{-4} (γm)	$\frac{3}{2}^-$
Os	189m	5.7 hr	100% IT	−38.95		$\frac{9}{2}^-$
Os	190	26.4%		−38.70	4 (γ); 9 (γm)	0^+
Os	190m	9.9 min	100% IT; 99% γ 0.502; 99% γ 0.616; 95% γ 0.361	−36.99		10^-
Os	191	15.4 d	100% β^- 0.14; (100%m); 26% γ(m) 0.129	−36.39		$\frac{9}{2}^-$
Os	191m	13.1 hr	100% IT	−36.31		$\frac{3}{2}^-$
Os	192	41.0%		−35.88	2.0 (γ)	0^+
Os	192m	6.1 s	100% IT; 70% γ 0.569; 66% γ 0.206; 59% γ 0.453	−33.86		
Os	193	30.6 hr	54% β^- 1.1; 4% γ 0.139; 4% γ 0.460; 3% γ 0.073	−33.39	1500 (γ) (r)	
Os	194	6.0 yr	67% β^- 0.10; 33% β^- 0.06; 3% γ 0.043	−32.42		0^+
Os	195	6.5 min	β^-	−29.7		
Os	196	35 min	β^-; 6% γ 0.408; 5% γ 0.126; 3% γ 0.315			0^+
77 Ir	172	2 s	α 5.81	−27.3		
Ir	173	3.0 s	α 5.67	−29.9		
Ir	174	4 s	α 5.48	−30.9		
Ir	175	5 s	α 5.39	−33.2		
Ir	176	8 s	α 5.12	−33.8		
Ir	177	21 s	α 5.01	−35.8		
Ir	178	12 s	β^+; K; γ	−36.3		
Ir	179	4 min	β^+; K	−37.9		
Ir	180	1.5 min	β^+; K; γ	−37.9		
Ir	181	5.0 min	β^+; K; γ	−39.3		
Ir	182	15 min	β^+; K; 45% γ 0.273; 35% γ 0.127; 9% γ 0.236, 0.912	−39.0		
Ir	183	55 min	β^+; K; γ	−40.1		
Ir	184	3.0 hr	K; β^+; 68% γ 0.264; 30% γ 0.120; 26% γ 0.390	−39.5		5
Ir	185	14 hr	β^+; K; γ	−40.3		$\frac{5}{2}$
Ir	186	16 hr	98% K; 2% β^+ 1.9; 62% γ 0.297; 41% γ 0.137; 34% γ 0.435	−39.16		5
Ir	186	1.7 hr	β^+; K; γ			
Ir	187	10.5 hr	100% K; 6% γ 0.913; 4% γ 0.401, 0.427, 0.611	−39.7		$\frac{3}{2}^+$
Ir	188	41.5 hr	100% K; 30% γ 0.155; 19% γ 2.215; 18% γ 0.633	−38.32		2^-
Ir	189	13.1 d	100% K; (8%m); 6% γ 0.245; 4% γ 0.069; 1% γ 0.059	−38.5		$\frac{3}{2}^+$
Ir	190	11.8 d	100% K; 52% γ 0.186; 39% γ 0.605; 35% γ 0.519	−36.7		

Table of the nuclides (*contd*)

Nuclide		Natural abundance or half-life	Decay modes and major radiations	Mass excess MeV	σ/b	I^{π}
Z	A					
Ir	190m$_1$	1.2 hr	100% IT	−36.7		
Ir	190m$_2$	3.2 hr	95% K; (95%m); 5% IT; 95% γ (m) 0.502; 94% γ (m) 0.616; 90% γ (m) 0.361	−36.5		
Ir	191	37.3%		−36.70	540 (γ); 400 (γm$_1$); 0.1 (γm$_2$)	$\frac{3}{2}^+$
Ir	191m	4.9 s	100% IT; 26% γ 0.129	−36.53		$\frac{11}{2}^-$
Ir	192	74.2 d	48% β⁻ 0.67; 41% β⁻ 0.54; 5% K; 83% γ 0.316; 48% γ 0.468; 30% γ 0.308	−34.83	1000 (t) (r)	4
Ir	192m$_1$	1.45 min	100% IT	−34.77		1
Ir	192m$_2$	241 yr	100% IT; γ	−34.67		9
Ir	193	62.7%		−34.52	110 (γ)	$\frac{3}{2}^+$
Ir	193m	10.6 d	100% IT	−34.44	0.05 (γm)	$\frac{11}{2}^-$
Ir	194	19.2 hr	86% β⁻ 2.25; 13% γ 0.328; 3% γ 0.294; 1% γ 0.645	−32.51		1⁻
Ir	194m	171 d	β⁻; 97% γ 0.483; 93% γ 0.329; 62% γ 0.601			
Ir	195	2.5 hr	13% β⁻ 1.1; 60% β⁻ 1.0; γ	−31.69		
Ir	195m	3.7 hr	33% β⁻ 1.0; 33% β⁻ 0.4; (48%m); 14% γ 0.320; 14% γ 0.433; 14% γ 0.685	−31.57		
Ir	196	52 s	80% β⁻ 3.2; 19% γ 0.356; 10% γ 0.779; 5% γ 0.447	−29.4		
Ir	196m	1.40 hr	80% β⁻ 1.1; 97% γ 0.394; 96% γ 0.521; 94% γ 0.356, 0.447	−29.0		
Ir	198	8 s	β⁻; γ	−25.5		
78 Pt	173	< 1 s	α 6.19	−21.8		
Pt	174	0.7 s	80% α (6.03); β⁺; K	−24.9		0⁺
Pt	175	2.5 s	75% α (5.95); β⁺; K	−25.6		
Pt	176	6.3 s	42% α 5.74; β⁺; K	−28.5		0⁺
Pt	177	7 s	6% α 5.53; 3% α 5.49; β⁺; K	−29.4		
Pt	178	21 s	7% α 5.46; β⁺; K	−31.6		0⁺
Pt	179	33 s	β⁺; K	−32.0		$\frac{1}{2}$
Pt	180	50 s	β⁺; K	−34.1		0⁺
Pt	181	51 s	β⁺; K	−34.1		
Pt	182	2.6 min	β⁺; K; γ	−36.0		0⁺
Pt	183	7 min	β⁺; K	−35.6		
Pt	184	17.3 min	β⁺; K; γ	−37.2		0⁺
Pt	185	71 min	β⁺; K; γ	−36.5		
Pt	185	33 min	β⁺; K; γ	−36.5		
Pt	186	2.0 hr	100% K; γ	−37.8		0⁺
Pt	187	2.35 hr	β⁺; K; γ	−36.8		$\frac{3}{2}$
Pt	188	10.2 d	100% K; 19% γ 0.188; 18% γ 0.195; 7% γ 0.382	−37.79		0⁺
Pt	189	10.9 hr	K; β⁺ 0.9; 6% γ 0.721; 5% γ 0.094; 5% γ 0.608	−36.6		$\frac{3}{2}^-$
Pt	190	6 × 10¹¹ yr 0.013%	α 3.18	−37.32	800 (γ)	0⁺

Table of the nuclides (*contd*)

Nuclide		Natural abundance or half-life	Decay modes and major radiations	Mass excess MeV	σ/b	Iπ
Z	A					
Pt	191	2.9 d	100% K; (1%m); 14% γ 0.539; 8% γ 0.409; 6% γ 0.360	−35.70		$\frac{3}{2}^-$
Pt	192	0.78%		−36.28	10 (γ); 2 (γm) (r)	0^+
Pt	193	50 yr	100% K	−34.46		
Pt	193 m	4.33 d	100% IT	−34.31		
Pt	194	32.9%		−34.77	1.1 (γ) (r); 0.09 (γm) (r)	0^+
Pt	195	33.8%		−32.80	27 (γ) (r)	$\frac{1}{2}^-$
Pt	195m	4.02 d	100% IT; 11% γ 0.099; 3% γ 0.130; 2% γ 0.031	−32.54		$\frac{13}{2}^+$
Pt	196	25.3%		−32.65	0.7 (γ) (r); 0.05 (γm) (r)	0^+
Pt	197	18.3 hr	11% β⁻ 0.72; 81% β⁻ 0.64; 17% γ 0.077; 3% γ 0.192	−30.43		$\frac{1}{2}^-$
Pt	197m	94 min	97% IT; 3% β⁻ 0.7; (3%m); 11% γ 0.346; 2% γ (m) 0.279	−30.03		$\frac{13}{2}^+$
Pt	198	7.2%		−29.92	3.7 (γ); 0.03 (γm)	0^+
Pt	199	30.8 min	63% β⁻ 1.7; 15% γ 0.543; 6% γ 0.494; 5% γ 0.317	−27.42	15 (γ) (r)	
Pt	199m	14 s	100% IT; 85% γ 0.392	−27.00		
Pt	200	12.6 hr	β⁻; γ	−26.6		0^+
Pt	201	2.5 min	β⁻ 2.7; γ	−23.7		
79 Au	175	0.1 s	α 6.44	−17.2		
Au	176	1.3 s	20% α 6.29; 80% α 6.26	−18.4		
Au	177	1.3 s	35% α 6.15; 65% α 6.11	−21.2		
Au	178	3 s	α 5.92	−22.4		
Au	179	7 s	α 5.85	−24.8		
Au	181	11 s	1% α 5.62; β⁺; K	−27.6		
Au	182	22 s	β⁺; K; 46% γ 0.155; 18% γ 0.265; 7% γ 0.855	−28.2		
Au	183	42 s	β⁺; K	−30.0		
Au	184	53 s	β⁺; K; γ	−30.2		
Au	185	4.3 min	β⁺; K; γ	−31.7		
Au	185	6.8 min	β⁺; K; γ	−31.7		
Au	186	11 min	β⁺; K; γ	−31.7		3
Au	186	< 2 min	β⁺; K	−31.7		
Au	187	8 min	β⁺; K; γ	−32.9		$\frac{1}{2}$
Au	188	8.8 min	β⁺; K; γ	−32.5		1
Au	189	28.7 min	β⁺; K; γ	−33.4		$\frac{1}{2}^+$
Au	189m	4.6 min	β⁺; K; γ	−33.2		$\frac{11}{2}^-$
Au	190	43 min	98% K; β⁺ 3.4; 71% γ 0.296; 25% γ 0.302; 10% γ 0.598	−32.88		1^-
Au	191	3.2 hr	100% K; γ	−33.9		$\frac{3}{2}^+$
Au	191m	0.9 s	100% IT; γ	−33.6		
Au	192	5.0 hr	99% K; β⁺ 2.5; γ	−32.77		1^-
Au	193	17.5 hr	K; β⁺; 11% γ 0.186; 7% γ 0.256; 4% γ 0.268	−33.4		$\frac{3}{2}^+$
Au	193m	3.9 s	100% IT; 65% γ 0.258	−33.1		$\frac{11}{2}^-$
Au	194	39.5 hr	97% K; β⁺ 1.5; 61% γ 0.328; 11% γ 0.294; 6% γ 1.469	−32.26		1^-

Table of the nuclides (*contd*)

Nuclide		Natural abundance or half-life	Decay modes and major radiations	Mass excess MeV	σ/b	I^π
Z	A					
Au	195	183 d	100% K; 11% γ 0.099; 1% γ 0.130	−32.57		$\frac{3}{2}^+$
Au	195m	30.6 s	100% IT; 67% γ 0.262; 1% γ 0.200	−32.25		$\frac{11}{2}^-$
Au	196	6.18 d	93% K; 7% β^- 0.25; 88% γ 0.356; 23% γ 0.333; 7% γ 0.426	−31.16		2^-
Au	196m$_1$	8.2 s	100% IT	−31.08		5^+
Au	196m$_2$	9.7 hr	100% IT; (100%m$_1$); 43% γ 0.148; 38% γ 0.188; 8% γ 0.168	−30.57		12^-
Au	197	100%		−31.15	98.8 (γ)	$\frac{3}{2}^+$
Au	197m	7.7 s	100% IT; 73% γ 0.279; 3% γ 0.131; 1% γ 0.202	−30.74		$\frac{11}{2}^-$
Au	198	2.70 d	99% β^- 0.96; 96% γ 0.412; 1% γ 0.676	−29.59	2.5×10^4 (γ)	2^-
Au	198m	2.3 d	100% IT; 77% γ 0.215; 70% γ 0.097; 51% γ 0.180	−28.78		
Au	199	3.14 d	6% β^- 0.45; 72% β^- 0.29; 42% γ 0.158; 9% γ 0.208	−29.10	30 (γ) (r)	$\frac{3}{2}^+$
Au	200	12.6 hr	77% β^- 2.2; 21% γ 0.368; 15% γ 1.226; 4% γ 1.263	−27.3		1
Au	200m	19 hr	β^-; 16% IT; 83% γ 0.368; 82% γ 0.498; 80% γ 0.579	−26		12
Au	201	26 min	82% β^- 1.3; 2% γ 0.543; 1% γ 0.167, 0.517, 0.613	−26.4		
Au	202	29 s	90% β^- 3.5; 10% γ 0.440; 3% γ 1.125; 2% γ 1.204, 1.307	−23.9		
Au	203	53 s	β^-; γ	−23.0		
Au	204	40 s	β^-; γ			
80 Hg	178	0.5 s	84% α (6.43); β^+; K	−15.9		0^+
Hg	179	1.09 s	53% α (6.27); β^+; K	−16.8		
Hg	180	2.9 s	α 6.12	−19.9		0^+
Hg	181	3.6 s	23% α 6.00; 3% α 5.92; β^+; K	−20.8		$\frac{1}{2}^-$
Hg	182	11 s	9% α 5.87; β^+; K; γ	−23.2		0^+
Hg	183	9 s	11% α 5.91; 1% α 5.83; 61% K; β^+	−23.7		$\frac{1}{2}^-$
Hg	184	30.6 s	1% α 5.54; β^+; K; γ	−26.0		0^+
Hg	185(g)	48 s	5% α 5.65; β^+; K; γ	−26.1		$\frac{1}{2}^-$
Hg	185(m)	17 s	α 5.38; IT?; γ			
Hg	186	1.4 min	96% K; β^+; γ	−28.4		0^+
Hg	187	1.6 min	β^+; K; γ	−28.1		
Hg	187	2.4 min	β^+; K; γ	−28.1		$\frac{3}{2}^-$
Hg	188	3.3 min	β^+; K; γ	−29.9		0^+
Hg	189	8.7 min	β^+; K; γ	−29.2		
Hg	189	7.5 min	β^+; K; γ	−29.2		$\frac{3}{2}^-$
Hg	190	20 min	100% K; γ	−31.0		0^+
Hg	191	4.9 min	β^+; K; γ	−30.5		
Hg	191m	51 min	β^+; K; (90%m); 55% γ 0.253; 18% γ 0.420; 17% γ 0.579	−30.3		

Table of the nuclides (*contd*)

Nuclide		Natural abundance or half-life	Decay modes and major radiations	Mass excess MeV	σ/b	I^π
Z	A					
Hg	192	4.9 hr	100% K; 40% γ 0.275; 6% γ 0.157; 4% γ 0.307	−32.0		0^+
Hg	193	4 hr	K; β^+; γ	−31.0		$\frac{3}{2}^-$
Hg	193m	11 hr	92% K; (92%m); 8% IT; 60% γ(m) 0.258; 25% γ 0.408; 14% γ 0.573	−30.9		$\frac{13}{2}^+$
Hg	194	260 yr	100% L	−32.21		0^+
Hg	195	10 hr	100% K; (3%m); 8% γ 0.779; 7% γ 0.061; 2% γ 0.180, 0.585, 0.600	−31.05		$\frac{1}{2}^-$
Hg	195m	41 hr	51% IT; 49% K; (49%m); 38% γ 0.262; 9% γ 0.560; 3% γ 0.388	−30.87		$\frac{13}{2}^+$
Hg	196	0.15%		−31.85	3000 (γ) (s); 120 (γm) (s)	0^+
Hg	197	64.1 hr	100% K; 19% γ 0.077; 1% γ 0.191	−30.74		$\frac{1}{2}^-$
Hg	197m	23.8 hr	94% IT; 6% K; (6%m); 34% γ 0.134; 5% γ (m) 0.279	−30.44		$\frac{13}{2}^+$
Hg	198	10.0%		−30.96	0.018 (γm)	0^+
Hg	199	16.8%		−29.56	2000 (γ)	$\frac{1}{2}^-$
Hg	199m	42.6 min	100% IT; 52% γ 0.158; 12% γ 0.374	−29.03		$\frac{13}{2}^+$
Hg	200	23.1%		−29.51	<60 (γ)	0^+
Hg	201	13.2%		−27.67	<60 (γ)	$\frac{3}{2}^-$
Hg	202	29.8%		−27.36	5.0 (γ)	0^+
Hg	203	46.8 d	100% β^- 0.21; 82% γ 0.279	−25.28		$\frac{5}{2}^-$
Hg	204	6.9%		−24.70	0.4 (γ) (r)	0^+
Hg	205	5.2 min	96% β^- 1.54; 2% γ 0.204	−22.30		$\frac{1}{2}^-$
Hg	206	8.2 min	63% β^- 1.31; 26% γ 0.305; 2% γ 0.650; 1% γ 0.344	−20.96		0^+
81 Tl	184	11 s	2% α 6.16; β^+; K; γ	−16.9		
Tl	185m	1.7 s	α 5.98; IT; γ	−18.7		
Tl	186	28 s	β^+; K; γ	−19.9		
Tl	186m	3 s	100% IT; 80% γ 0.374	−19.5		
Tl	187m	16 s	α 5.51; IT; γ	−21.6		
Tl	188	71 s	β^+; K; γ	−22.3		
Tl	189	1.4 min	β^+; K; γ	−24.0		
Tl	189	2.3 min	β^+; K; γ	−24.0		
Tl	190	2.6 min	β^+; K; γ	−24.2		
Tl	190	3.7 min	β^+; K; γ			
Tl	191	5.2 min	98% K; β^+; γ	−25.7		
Tl	192	10.8 min	β^+; K; γ	−25.6		
Tl	192	10.6 min	β^+; K; γ	−25.6		
Tl	193	21 min	>96% K; β^+; γ	−27.0		$\frac{1}{2}^+$
Tl	193m	2.1 min	100% IT; γ			
Tl	194	33 min	β^+; K; γ	−26.8		2^-
Tl	194m	32.8 min	β^+; K; 100% γ 0.428; 100% γ 0.636; 77% γ 0.749	−26.5		
Tl	195	1.16 hr	99% K; β^+; 4% γ 0.563; 3% γ 0.884; 3% γ 1.364	−27.9		$\frac{1}{2}^+$

Table of the nuclides (*contd*)

Nuclide			Natural abundance or half-life	Decay modes and major radiations	Mass excess MeV	σ/b	Iπ
	Tl	195m	3.6 s	100% IT; 91% γ 0.384	−27.4		$\frac{9}{2}-$
	Tl	196	1.84 hr	β^+; K; γ	−27.4		2
	Tl	196m	1.41 hr	β^+; K; 4% IT; γ	−27.0		
	Tl	197	2.84 hr	99% K; β^+; 12% γ 0.426; 8% γ 0.152; 4% γ 0.309 (double)	−28.3		$\frac{1}{2}+$
	Tl	197m	0.54 s	100% IT; 90% γ 0.385; 30% γ 0.222	−27.7		$\frac{9}{2}-$
	Tl	198	5.3 hr	99% K; β^+ 2.4; 78% γ 0.412; 10% γ 0.637; 10% γ 0.676	−27.5		$2-$
	Tl	198m	1.87 hr	β^+; K; 44% IT; 59% γ 0.412; 59% γ 0.637; 54% γ 0.587	−27.0		$7+$
	Tl	199	7.4 hr	100% K; 12% γ 0.208; 12% γ 0.455; 9% γ 0.247	−28.1		$\frac{1}{2}+$
	Tl	200	26.1 hr	100% K; 89% γ 0.368; 30% γ 1.206; 14% γ 0.579	−27.06		$2-$
	Tl	201	73 hr	100% K; 9% γ 0.167 (double); 2% γ 0.135	−27.19		$\frac{1}{2}+$
	Tl	202	12.23 d	100% K; 91% γ 0.440	−25.99		$2-$
	Tl	203	29.5%		−25.77	10 (γ)	$\frac{1}{2}+$
	Tl	204	3.77 yr	97% β^- 0.76; 3% K	−24.35	22 (γ) (r)	$2-$
	Tl	205	70.5%		−23.84	0.10 (γ) (r)	$\frac{1}{2}+$
(RaE^{II})	Tl	206	4.20 min	100% β^- 1.53	−22.27		$0-$
	Tl	206m	3.6 min	100% IT; γ	−19.63		
(AcC^{II})	Tl	207	4.77 min	100% β^- 1.42	−21.04		$\frac{1}{2}+$
	Tl	207m	1.3 s	100% IT; γ	−19.70		$\frac{11}{2}-$
(ThC^{II})	Tl	208	3.053 min	51% β^- 1.8; 22% β^- 1.5; 23% β^- 1.3; 100% γ 2.615; 86% γ 0.583; 22% γ 0.511	−16.77		
	Tl	209	2.2 min	100% β^- 1.8; 100% γ 1.566; 72% γ 0.467; 67% γ 0.117	−13.65		
(RaC^{II})	Tl	210	1.30 min	β^-; 99% γ 0.795; 79% γ 0.296; 21% γ 1.310	−9.25		
82	Pb	186	8 s	2% α 6.32; β^+; K	−14.3		$0+$
	Pb	187	17 s	1% α 6.19; 1% α 6.08; β^+; K	−14.9		
	Pb	188	25 s	3% α (5.98); β^+; K	−17.5		$0+$
	Pb	189	51 s	β^+; K	−17.9		
	Pb	190	1.2 min	β^+; K	−20.2		$0+$
	Pb	191	1.3 min	β^+; K	−20.2		
	Pb	192	2.3 min	β^+; K	−22.3		$0+$
	Pb	193	5.8 min	β^+; K; γ	−22.1		
	Pb	194	11 min	β^+; K; γ	−23.8		$0+$
	Pb	195	16.4 min	β^+; K; γ	−23.6		
	Pb	196	37 min	100% K; γ	−25.2		$0+$
	Pb	197m	42 min	β^+; K; (81%m); 19% IT; 137% γ 0.386 (double); 25% γ 0.222; 19% γ 0.773	−24.3		
	Pb	198	2.4 hr	100% K; 24% γ 0.173; 18% γ 0.290; 8% γ 0.865	−25.9		$0+$
	Pb	199	90 min	99% K; β^+; 62% γ 0.367; 13% γ 0.353; 11% γ 1.135	−25.3		$\frac{5}{2}-$

Table of the nuclides (*contd*)

Nuclide		Natural abundance or half-life	Decay modes and major radiations	Mass excess MeV	σ/b	I$^\pi$
Z	A					
Pb	199m	12.2 min	93% IT; β^+; K; 18% γ 0.424; 4% γ 0.369; 4% γ 0.382	−24.9		$\frac{13}{2}^+$
Pb	200	21.5 hr	100% K; 38% γ 0.148; 4% γ 0.236; 4% γ 0.257	−26.2		0^+
Pb	201	9.4 hr	100% K; 78% γ 0.331; 10% γ 0.361; 7% γ 0.946	−25.33		$\frac{5}{2}^-$
Pb	201m	61 s	100% IT; 55% γ 0.628	−24.70		$\frac{13}{2}^+$
Pb	202	3×10^5 yr	100% L	−25.94		0^+
Pb	202m	3.62 hr	91% IT; 9% K; 92% γ 0.961; 86% γ 0.422; 50% γ 0.787	−23.77		9^-
Pb	203	52.0 hr	100% K; 81% γ 0.279; 4% γ 0.401	−24.79		$\frac{5}{2}^-$
Pb	203m$_1$	6.1 s	100% IT; 70% γ 0.825; 8% γ 0.820	−23.97		$\frac{13}{2}^+$
Pb	203m$_2$	0.48 s	100% IT; γ	−21.84		$\frac{29}{2}^-$
Pb	204	1.42%		−25.12	0.66 (γ)	0^+
Pb	204m	66.9 min	100% IT; 99% γ 0.899; 96% γ 0.912; 89% γ 0.374	−22.93		9^-
Pb	205	1.4×10^7 yr	100% L	−23.78	3.8 (γ) (r)	$\frac{5}{2}^-$
Pb	206	24.1%		−23.80	0.03 (γ)	0^+
Pb	207	22.1%		−22.46	0.71 (γ)	$\frac{1}{2}^-$
Pb	207m	0.81 s	100% IT; 98% γ 0.570; 87% γ 1.064	−20.83		$\frac{13}{2}^+$
Pb	208	52.3%		−21.76	0.0005 (γ)	0^+
Pb	209	3.25 hr	100% β^- 0.64	−17.62		$\frac{9}{2}^+$
(RaD) Pb	210	22.3 yr	19% β^- 0.063; 81% β^- 0.016; 4% γ 0.046	−14.74	0.5 (γ)	0^+
(AcB) Pb	211	36.1 min	92% β^- 1.37; 3% γ 0.405; 3% γ 0.832; 2% γ 0.427	−10.492		
(ThB) Pb	212	10.64 hr	12% β^- 0.57; 83% β^- 0.33; 43% γ 0.239; 3% γ 0.300; 1% γ 0.115	−7.56		0^+
Pb	213	10.2 min	β^-	−3.1		
(RaB) Pb	214	26.8 min	6% β^- 1.02; 42% β^- 0.70; 48% β^- 0.65; 37% γ 0.352; 19% γ 0.295; 8% γ 0.242	−0.185		0^+
83 Bi	189	< 1.5 s	α 6.67	−9.9		
Bi	190	5 s	90% α (6.45); β^+; K	−10.9		
Bi	191	13 s	40% α (6.31); β^+; K	−13.1		
Bi	191m	20 s	α 6.86			
Bi	192	40 s	20% α (6.05); β^+; K	−13.7		
Bi	193	64 s	60% α (5.89); β^+; K	−15.6		
Bi	193m	3.5 s	24% α 6.48; 1% α 6.18; β^+; K			
Bi	194	1.7 min	β^+; K; 100% γ 0.965; 87% γ 0.575; 70% γ 0.280	−16.0		
Bi	195	2.8 min	β^+; K	−17.7		
Bi	195m	90 s	4% α (6.10); β^+; K			
Bi	196	4.5 min	β^+; K; 100% γ 1.049; 62% γ 0.688; 46% γ 0.372	−17.8		
Bi	197(m)	8 min	β^+; K			

Table of the nuclides (*contd*)

Nuclide Z	A	Natural abundance or half-life	Decay modes and major radiations	Mass excess MeV	σ/b	I^π
Bi	198	11.9 min	β^+; K; 100% γ 1.063; 80% γ 0.198; 79% γ 0.562	−19.3		
Bi	198m	8 s	100% IT; 39% γ 0.248	−19.1		
Bi	199(g)	27 min	100% K	−20.6		$\frac{9}{2}^-$
Bi	199(m)	24.7 min	α	−20.0		
Bi	200	36 min	K; β^+; 100% γ 1.026; 90% γ 0.462; 85% γ 0.420	−20.5		7
Bi	200m	0.4 s	100% IT; γ	−20.0		10
Bi	201	1.77 hr	β^+; K; γ	−21.4		$\frac{9}{2}^-$
Bi	201m	59 min	β^+; K; IT; γ	−20.6		
Bi	202	1.67 hr	99% K; β^+; 100% γ 0.961; 84% γ 0.422; 61% γ 0.657	−21.0		5
Bi	203	11.8 hr	100% K; (23% m_1); 29% γ 0.820; 16% γ(m) 0.825; 13% γ 0.897	−21.6		$\frac{9}{2}^-$
Bi	204	11.2 hr	100% K; (13% m); 100% γ 0.899; 74% γ 0.375; 59% γ 0.984	−20.8		6^+
Bi	205	15.31 d	100% K; 32% γ 1.764; 31% γ 0.703; 16% γ 0.988	−21.07		$\frac{9}{2}^-$
Bi	206	6.24 d	100% K; 99% γ 0.803; 67% γ 0.881; 41% γ 0.516	−20.03		6^+
Bi	207	38 yr	100% K; (83% m); 98% γ 0.570; 72% γ(m) 1.064; 7% γ 1.770	−20.06		$\frac{9}{2}^-$
Bi	208	3.7×10^5 yr	100% K; 100% γ 2.615	−18.88		
Bi	209	100%		−18.27	0.019 (γ) (r); 0.014 (γm) (r)	$\frac{9}{2}^-$
(RaE) Bi	210	5.01 d	100% β^- 1.16	−14.80	0.050 (γ) (s)	1^-
Bi	210m	3.0×10^6 yr	55% α 4.94; 40% α 4.90; 4% α 4.57; 50% γ 0.266; 28% γ 0.305; 4% γ 0.650	−14.53		9^-
(AcC) Bi	211	2.15 min	84% α 6.62; 16% α 6.28; 13% γ 0.351	−11.87		
(ThC) Bi	212	60.6 min	10% α 6.09; 26% α 6.05; 55% β^- 2.25; 6% γ 0.727; 1% γ 0.040, 0.785, 1.621	−8.14		1
Bi	$212m_1$	25 min	α 6.34; β^-	−7.88		
Bi	$212m_2$	9 min	β^-			
Bi	213	45.6 min	2% α 5.87; 66% β^- 1.42; 31% β^- 0.98; 21% γ 0.440	−5.24		
(RaC) Bi	214	19.7 min	18% β^- 3.27; 18% β^- 1.54; 18% β^- 1.51; 46% γ 0.609; 16% γ 1.765; 15% γ 1.120	−1.21		
Bi	215	7 min	β^-	1.7		
84 Po	194	0.6 s	α 6.85	−10.8		0^+
Po	195 (g)	4.5 s	α 6.61	−11.1		
Po	195(m)	2.0 s	α 6.70			
Po	196	5.5 s	α 6.52	−13.2		0^+
Po	197	58 s	90% α (6.28); β^+; K	−13.2		
Po	197m	26 s	α 6.38			

Table of the nuclides (*contd*)

Nuclide			Natural abundance or half-life	Decay modes and major radiations	Mass excess MeV	σ/b	Iπ
	Z	A					
	Po	198	1.8 min	70% α (6.18); β⁺; K	−15.1		0⁺
	Po	199	5.2 min	12% α (5.95); β⁺; K; 48% γ 1.034; 25% γ 1.021; 23% γ 0.362	−15.1		
	Po	199m	4.2 min	39% α (6.06); β⁺; K; 61% γ 1.002; 26% γ 0.500; 7% γ 0.274			
	Po	200	11.6 min	14% α (5.86); β⁺; K; γ	−16.7		0⁺
	Po	201	15.2 min	2% α (5.68); β⁺; K; 53% γ 0.890; 29% γ 0.905; 13% γ 0.848	−16.4		3/2
	Po	201m	9.0 min	3% α 5.79; 53% IT; β⁺; K; 43% γ 0.967; 19% γ 0.412; 9% γ 0.419	−16.0		
	Po	202	44 min	2% α (5.59); β⁺; K; 47% γ 0.687; 13% γ 0.316; 10% γ 0.166	−17.8		0⁺
	Po	203	33 min	β⁺; K; 57% γ 0.909; 20% γ 1.091; 19% γ 0.894	−17.4		5/2 −
	Po	203m	1.2 min	96% IT; β⁺; K; 51% γ 0.641; 4% γ 0.905; 2% γ 0.577	−16.7		
	Po	204	3.57 hr	1% α 5.38; 99% K; γ	−18.3		0⁺
	Po	205	1.80 hr	β⁺; K; 35% γ 0.872; 27% γ 1.001; 24% γ 0.850	−17.58		5/2 −
	Po	206	8.8 d	5% α (5.22); 95% K; 32% γ 1.032; 23% γ 0.286; 23% γ 0.511	−18.19		0⁺
	Po	207	5.7 hr	99% K; β⁺; 59% γ 0.992; 29% γ 0.743; 18% γ 0.912	−17.15		5/2 −
	Po	207m	2.8 s	100% IT; 99% γ 0.815; 43% γ 0.268; 30% γ 0.301	−15.77		19/2 −
	Po	208	2.898 yr	100% α 5.12	−17.48		0⁺
	Po	209	102 yr	100% α 4.88	−16.37		1/2 −
(RaF)	Po	210	138.38 d	100% α 5.30	−15.96	< 0.03 (γ) (r)	0⁺
(AcCᴵ)	Po	211	0.516 s	98% α 7.45; 1% α 6.89; 1% α 6.57; 1% γ 0.570, 0.898	−12.44		
	Po	211m	25.5 s	7% α 8.88; 2% α 8.00; 91% α 7.28; (91%)m; 80% γ (m) 0.570; 80% γ (m) 1.064	−10.98		
(ThCᴵ)	Po	212	2.96 × 10⁻⁷ s	α 8.78	−10.38		0⁺
	Po	212m	45 s	97% α 11.65; 1% α 9.10; 2% α 8.52; 3% γ 2.615; 2% γ 0.570	−7.48		
	Po	213	4 × 10⁻⁶ s	100% α 8.38	−6.66		9/2 +
(RaCᴵ)	Po	214	1.64 × 10⁻⁴ s	100% α 7.69	−4.48		0⁺
(AcA)	Po	215	0.00178 s	100% α 7.39	−0.541		
(ThA)	Po	216	0.145 s	100% α 6.78	1.77		0⁺
	Po	217	< 10 s	α 6.54	6.0		
(RaA)	Po	218	3.05 min	100% α 6.00	8.35		0⁺
85	At	196	0.3 s	α7.06	−4.1		
	At	197	0.4 s	α 6.96	−6.0		
	At	198	5 s	α 6.75	−6.7		

Table of the nuclides (*contd*)

Nuclide		Natural abundance or half-life	Decay modes and major radiations	Mass excess MeV	σ/b	I$^\pi$
Z	A					
At	198m	1.5 s	α 6.85			
At	199	7 s	α 6.64	−8.5		
At	200(g)	42 s	32% α 6.47; 21% α 6.42; β$^+$; K; γ	−8.7		
At	200(m)	4.3 s	α 6.54			
At	201	1.5 min	71% α (6.34); β$^+$; K; γ	−10.5		
At	202	3.0 min	9% α 6.23; 6% α 6.13; β$^+$; K; γ	−10.5		
At	203	7.3 min	31% α (6.09); β$^+$; K; γ	−12.0		
At	204	9.3 min	4% α (5.95); β$^+$; K; 94% γ 0.683; 91% γ 0.515; 71% γ 0.425	−12.0		
At	205	26 min	10% α 5.90; 87% K; β$^+$; 28% γ 0.719; 8% γ 0.669; 5% γ 0.629	−13.0		$\frac{9}{2}^-$
At	206	31 min	1% α 5.70; 82% K; 7% β$^+$ 3.3; 97% γ 0.700; 86% γ 0.477; 48% γ 0.396	−12.7		
At	207	1.80 hr	10% α 5.76; β$^+$; K; 49% γ 0.814; 22% γ 0.588; 14% γ 0.300	−13.3		$\frac{9}{2}^-$
At	208	1.63 hr	1% α 5.64; β$^+$; K; 98% γ 0.686; 90% γ 0.660; 46% γ 0.177	−12.6		
At	209	5.4 hr	4% α 5.65; 96% K; 94% γ 0.545; 86% γ 0.782; 66% γ 0.790	−12.89		$\frac{9}{2}^-$
At	210	8.3 hr	β$^+$; K; 99% γ 1.181; 79% γ 0.245; 47% γ 1.483	−11.98		5$^+$
At	211	7.21 hr	42% α 5.87; 58% K	−11.65		$\frac{9}{2}^-$
At	212	0.315 s	84% α 7.68; 15% α 7.62; 1% α 7.09	−8.63		
At	212m	0.122 s	33% α 7.90; 65% α 7.84; 2% α (dist)	−8.40		
At	213	1.1 × 10^{-7} s	100% α 9.08	−6.59		$\frac{9}{2}^-$
At	214	2 × 10^{-6} s	100% α 8.82	−3.39		
At	215	1.0 × 10^{-4} s	100% α 8.03	−1.26		
At	216	3.0 × 10^{-4} s	97% α 7.80; 2% α 7.70; 1% α 7.48	2.24		1
At	217	0.032 s	100% α 7.07	4.38		
At	218	1.7 s	4% α 6.76; 90% α 6.69; 6% α 6.65	8.10		
At	219	0.9 min	97% α (6.28); β$^-$	10.5		
86 Rn	200	1.0 s	α 6.91	−3.7		0$^+$
Rn	201(g)	7.0 s	α 6.72	−4.0		
Rn	201(m)	3.8 s	α 6.77			
Rn	202	9.9 s	α 6.64	−5.9		0$^+$
Rn	203	45 s	65% α (6.50); β$^+$; K	−6.0		
Rn	203m	28 s	α 6.55	−6.0		
Rn	204	75 s	72% α (6.42); β$^+$; K	−7.8		0$^+$
Rn	205	170 s	23% α (6.26); β$^+$; K; γ	−7.6		

Table of the nuclides (*contd*)

Nuclide Z	A	Natural abundance or half-life	Decay modes and major radiations	Mass excess MeV	σ/b	Iᵖ
Rn	206	5.7 min	64% α 6.26; β^+; K	−9.0		0^+
Rn	207	9.3 min	23% α 6.13; β^+; K; 63% γ 0.344; 21% γ 0.747; 16% γ 0.403, 0.674	−8.7		$\frac{5}{2}^-$
Rn	208	24.4 min	52% α 6.14; β^+; K	−9.6		0^+
Rn	209	29 min	17% α 6.04; 80% K; β^+; 47% γ 0.408; 21% γ 0.746; 14% γ 0.337	−8.99		$\frac{5}{2}^-$
Rn	210	2.4 hr	96% α 6.04; 4% K; 2% γ 0.458; 1% γ 0.073, 0.233	−9.61		0^+
Rn	211	14.6 hr	9% α 5.85; 16% α 5.78; 1% α 5.62; β^+; K; 46% γ 0.674; 33% γ 1.363; 29% γ 0.678	−8.76		$\frac{1}{2}^-$
Rn	212	23 min	100% α 6.26	−8.67		0^+
Rn	213	0.025 s	99% α 8.09; 1% α 7.55	−5.71		
Rn	214	2.7×10^{-7} s	α 9.04	−4.33		0^+
Rn	215	2.3×10^{-6} s	100% α 8.67	−1.18		
Rn	216	4×10^{-5} s	α 8.05	0.25		0^+
Rn	217	5×10^{-4} s	100% α 7.74	3.65		$\frac{9}{2}^+$
Rn	218	0.035 s	100% α 7.13	5.21		0^+
(An) Rn	219	3.96 s	81% α 6.82; 12% α 6.55; 7% α 6.42; 10% γ 0.271; 7% γ 0.402	8.83		
(Tn) Rn	220	55.6 s	100% α 6.29	10.60	< 0.2 (γ) (r)	0^+
Rn	221	25 min	20% α; β^-	14.4		
(Rn) Rn	222	3.824 d	100% α 5.49	16.37	0.7 (γ) (r)	0^+
Rn	223	43 min	β^-			
Rn	224	1.8 hr	β^-; γ	22.3		0^+
Rn	225	4.5 min	β^-			
Rn	226	6 min	β^-			0^+
87 Fr	203	0.7 s	α 7.13; β^+; K	1.2		
Fr	204	2.1 s	70% α 7.03; 30% α 6.97	0.9		
Fr	205	3.7 s	α 6.91	−1.0		
Fr	206	16.0 s	85% α (6.79); β^+; K	−1.2		
Fr	207	14.8 s	93% α (6.77); β^+; K	−2.7		
Fr	208	58.0 s	74% α (6.64); β^+; K	−2.8		
Fr	209	50.0 s	89% α 6.65; β^+; K	−3.8		$\frac{9}{2}^-$
Fr	210	3.2 min	α 6.57; β^+; K; γ	−3.6		
Fr	211	3.1 min	α 6.53; β^+; K; γ	−4.2		
Fr	212	19 min	9% α 6.41; 10% α 6.38; 18% α 6.26; 7% α (dist); β^+; K; γ	−3.7		
Fr	213	34.7 s	99% α 6.78; 1% K	−3.56		
Fr	214	0.0050 s	93% α 8.43; 5% α 8.36; 1% α 7.94; 1% α 7.61	−0.97		
Fr	214m	0.0034 s	46% α 8.55; 51% α 8.48; 3% α (dist)	−0.84		
Fr	215	1.2×10^{-7} s	100% α 9.36	0.31		$\frac{9}{2}^-$
Fr	216	7.0×10^{-7} s	100% α 9.01	2.98		
Fr	217	2×10^{-5} s	100% α 8.32	4.31		$\frac{9}{2}^-$
Fr	218	0.0007 s	93% α 7.87; 5% α 7.56; 1% α 7.53; 1% α (dist)	7.05		

Table of the nuclides (*contd*)

Nuclide			Natural abundance or half-life	Decay modes and major radiations	Mass excess MeV	σ/b	I$^\pi$
	Z	A					
	Fr	219	0.020 s	99% α 7.31; 1% α 6.97	8.62		
	Fr	220	27.4 s	61% α 6.69; 12% α 6.64; 10% α 6.58; 17% α (dist); γ	11.47		
	Fr	221	4.8 min	83% α 6.34; 1% α 6.24; 15% α 6.13; 1% α 5.98; 11% γ 0.218	13.27		
	Fr	222	14.4 min	β$^-$	16.34		
(AcK)	Fr	223	21.8 min	63% β$^-$ 1.09; 34% γ 0.050; 8% γ 0.080; 3% γ 0.235	18.38		
	Fr	224	2.7 min	β$^-$; γ	21.7		
	Fr	225	3.9 min	β$^-$	23.8		
	Fr	226	48 s	β$^-$; γ	27.5		
	Fr	227	2.4 min	β$^-$; γ	29.6		
	Fr	228	39 s	β$^-$			
	Fr	229	0.8 min	β$^-$			
88	Ra	206	0.4 s	α 7.27; β$^+$; K	4.0		0$^+$
	Ra	207	1.3 s	α 7.13; β$^+$; K	3.7		
	Ra	209	4.7 s	α 7.01	2.0		
	Ra	210	3.8 s	α 7.02	0.6		0$^+$
	Ra	211	15 s	α 6.91; β$^+$; K	0.8		
	Ra	212	13.0 s	α 6.90; β$^+$; K	−0.1		0$^+$
	Ra	213	2.7 min	36% α 6.73; 39% α 6.62; 5% α 6.52; β$^+$; K; 6% γ 0.110; 1% γ 0.215	0.29		
	Ra	213m	0.0021 s	1% α 8.47; 99% IT; 99% γ 0.546; 99% γ 1.063; 46% γ 0.161	2.06		
	Ra	214	2.46 s	α 7.14	0.09		0$^+$
	Ra	215	0.0016 s	96% α 8.70; 1% α 8.17; 3% α 7.88	2.53		
	Ra	216	1.8 × 10^{-7} s	α 9.35	3.29		0$^+$
	Ra	217	1.6 × 10^{-6} s	α 8.99	5.88		
	Ra	218	1.4 × 10^{-5} s	α 8.39	6.64		0$^+$
	Ra	219	0.010 s	35% α 7.98; 65% α 7.68	9.38		
	Ra	220	0.023 s	99% α 7.46; 1% α 6.99; 1% γ 0.465	10.26		0$^+$
	Ra	221	30 s	30% α 6.76; 20% α 6.67; 34% α 6.61; 16% α (dist); 15% γ 0.089; 13% γ 0.152; 2% γ 0.176	12.96		
	Ra	222	38 s	97% α 6.56; 3% α 6.24; 3% γ 0.324	14.31		0$^+$
(AcX)	Ra	223	11.435 d	9% α 5.75; 54% α 5.72; 24% α 5.61; 13% α (dist); 14% γ 0.269; 6% γ 0.154; 4% γ 0.324	17.235	134 (γ) (r)	$\frac{1}{2}^+$
(ThX)	Ra	224	3.67 d	95% α 5.69; 5% α 5.45; 4% γ 0.241	18.81	12 (γ) (r)	0$^+$
	Ra	225	14.8 d	33% β$^-$ 0.36; 67% β$^-$ 0.32; 29% γ 0.040	21.99		

Table of the nuclides (*contd*)

Nuclide			Natural abundance or half-life	Decay modes and major radiations	Mass excess MeV	σ/b	I^π
	Z	A					
(Ra)	Ra	226	1600 yr	94% α 4.78; 6% α 4.60; 3% γ 0.186	23.67	20 (γ) (r)	0^+
	Ra	227	42 min	33% β^- 1.31; 24% β^- 1.29; 17% γ 0.027; 5% γ 0.300; 5% γ 0.303	27.19		
(MsTh$_I$)	Ra	228	5.77 yr	60% β^- 0.039; 40% β^- 0.015	28.94	36 (γ) (s)	0^+
	Ra	229	4.0 min	β^-	32.7		
	Ra	230	93 min	β^-; γ	34.6		0^+
89	Ac	209	0.1 s	α 7.59	9.1		
	Ac	210	0.35 s	α 7.46	8.9		
	Ac	211	0.25 s	α 7.48; β^+; K	7.4		
	Ac	212	0.9 s	α 7.38	7.2		
	Ac	213	0.8 s	100% α 7.36	6.2		
	Ac	214	8.2 s	$\geqslant$45% α 7.21; $\geqslant$38% α 7.08; K	6.1		
	Ac	215	0.17 s	α 7.60	6.0		
	Ac	216	0.0003 s	90% α 9.07; 10% α 8.99	8.0		
	Ac	216m	0.0003 s	46% α 9.10; 50% α 9.02; 2% α 8.28; 2% α 8.19	8.0		
	Ac	217	1.1×10^{-7} s	100% α 9.65	8.70		
	Ac	218	2.7×10^{-7} s	100% α 9.20	10.84		
	Ac	219	7×10^{-6} s	100% α 8.66	11.56		
	Ac	220	0.026 s	24% α 7.85; 21% α 7.68; 23% α 7.61; 32% α (dist); γ	13.75		
	Ac	221	0.052 s	70% α 7.65; 20% α 7.44; 10% α 7.38	14.52		
	Ac	222	5.5 s	93% α 7.00; 6% α 6.96;	16.62		
	Ac	222m	66 s	α (7.00); IT; K			
	Ac	223	2.2 min	32% α 6.66; 45% α 6.65; 14% α 6.56; 8% α (dist); 1% K	17.83		
	Ac	224	2.9 hr	2% α 6.21; 3% α 6.14; 2% α 6.06; 3% α (dist); 90% K; 56% γ 0.217; 25% γ 0.133	20.22		
	Ac	225	10.0 d	51% α 5.83; 27% α 5.79 (double); 10% α 5.73; 12% α (dist); 2% γ 0.100; 1% γ 0.150	21.63		
	Ac	226	29 hr	24% β^- 1.11; 10% β^- 1.04; 49% β^- 0.88; 17% K; 27% γ 0.230; 18% γ 0.158; 6% γ 0.254	24.301		
(Ac)	Ac	227	21.77 yr	1% α 4.95 (double); 54% β^- 0.044; 35% β^- 0.035; 10% β^- 0.019	25.850	900 (γ) (s)	$\frac{3}{2}-$
(MsTh$_2$)	Ac	228	6.13 hr	11% β^- 2.14; 32% β^- 1.17; 27% γ 0.911; 16% γ 0.969; 12% γ 0.339	28.90		
	Ac	229	63 min	β^-; γ	30.7		
	Ac	230	122 s	β^-; γ	33.8		
	Ac	231	7.5 min	β^-; γ	35.9		
	Ac	232	35 s	β^-	39.2		

Table of the nuclides (*contd*)

Nuclide			Natural abundance or half-life	Decay modes and major radiations	Mass excess MeV	σ/b	Iπ
Z		A					
90	Th	215	1.2 s	40% α 7.52; 52% α 7.40; 8% α 7.33	10.9		
	Th	216	0.028 s	α 7.92	10.4		0+
	Th	217	2.5 × 10⁻⁴ s	α 9.25	12.14		
	Th	218	10⁻⁷ s	100% α 9.66	12.36		0+
	Th	219	1.0 × 10⁻⁶ s	α 9.34	14.47		
	Th	220	10⁻⁵ s	α 8.79	14.66		0+
	Th	221	0.0017 s	38% α 8.47; 56% α 8.15; 6% α 7.73	16.93		
	Th	222	0.0028 s	α 7.98	17.20		0+
	Th	223	0.66 s	40% α 7.32; 60% α 7.29	19.26		
	Th	224	1.0 s	79% α 7.17; 19% α 7.00; 1% α 6.76; 1% α 6.70; 9% γ 0.177; 1% γ 0.410	19.99		0+
	Th	225	8.0 min	8% α 6.80; 13% α 6.50; 39% α 6.48; 30% α (dist); 10% K; 27% γ 0.323	22.30		
	Th	226	30.9 min	75% α 6.34; 23% α 6.23; 2% α 6.10; 3% γ 0.111; 1% γ 0.242	23.19		0+
(RdAc)	Th	227	18.72 d	24% α 6.04; 23% α 5.98; 20% α 5.76; 33% α (dist); 11% γ 0.236; 7% γ 0.050; 6% γ 0.256	25.806	200 (γ) (r)	3/2+
(RdTh)	Th	228	1.913 yr	73% α 5.42; 27% α 5.34; 1% γ 0.084	26.76	100 (γ) (r)	0+
	Th	229	7340 yr	11% α 4.90; 56% α 4.85; 8% α 4.81; 25% α (dist); 5% γ 0.194; 4% γ 0.031; 4% γ 0.125 (double)	29.581	30 (f)	5/2+
(Io)	Th	230	8.0 × 10⁴ yr	76% α 4.69; 24% α 4.62	30.861	40 (γ)	0+
(UY)	Th	231	25.52 hr	37% β⁻ 0.31; 36% β⁻ 0.29; 15% γ 0.026; 7% γ 0.084; 1% γ 0.081, 0.090	33.812		5/2+
(Th)	Th	232	1.41 × 10¹⁰ yr 100%	77% α 4.02; 23% α 3.96	35.45	7.4 (γ)	0+
	Th	233	22.3 min	85% β⁻ 1.25; 3% γ 0.029; 3% γ 0.086; 1% γ 0.459	38.73	1400 (γ) (s); 15 (f)	1/2+
(UX₁)	Th	234	24.10 d	72% β⁻ 0.18; (100% m); 5% γ 0.093 (double); 4% γ 0.063 (double)	40.61	2.0 (γ)	0+
	Th	235	6.9 min	β⁻; γ	44.2		
	Th	236	38 min	46% β⁻ 1.1; 51% β⁻ 1.0; 5% γ 0.111; 1% γ 0.113, 0.132, 0.230	46.6		0+
91	Pa	216	0.20 s	α 7.92			
	Pa	222	0.006 s	30% α 8.54; 20% α 8.33; 50% α 8.18	21.96		
	Pa	223	0.007 s	45% α 8.20; 55% α 8.01	22.33		
	Pa	224	1.0 s	100% α 7.49	23.80		
	Pa	225	1.8 s	70% α 7.25; 30% α 7.20	24.32		

Table of the nuclides (*contd*)

Nuclide			Natural abundance or half-life	Decay modes and major radiations	Mass excess MeV	σ/b	Iπ
	Pa	226	1.8 min	38% α 6.86; 34% α 6.82; 2% α 6.73; 26% K	26.03		
	Pa	227	38.3 min	43% α 6.47; 23% α 6.42 (double); 8% α 6.40; 11% α (dist); 15% K; 5% γ 0.065	26.83		
	Pa	228	22 hr	2% α (6.12); 98% K; 16% γ 0.911; 13% γ 0.463; 13% γ 0.969	28.87		3^+
	Pa	229	1.4 d	100% K	29.89		
	Pa	230	18 d	91% K; 9% β^- 0.51; 28% γ 0.952; 8% γ 0.919; 6% γ 0.455	32.166	1500 (f) (r)	
(Pa)	Pa	231	3.28×10^4 yr	11% α 5.06; 25% α 5.01; 23% α 4.95; 41% α (dist); 10% γ 0.016, 0.027; 8% γ 0.013	33.423	200 (γ); 0.019 (f) (s)	$\frac{3}{2}^-$
	Pa	232	1.31 d	1% β^- 1.29; 71% β^- 0.32; 42% γ 0.969; 19% γ 0.894; 12% γ 0.150	35.93	800 (γ) (r); 700 (f) (r)	
	Pa	233	27.0 d	3% β^- 0.57; 36% β^- 0.23; 27% β^- 0.15; 37% γ 0.312; 7% γ 0.300; 4% γ 0.341	37.49	19 (γ); 20 (γm)	$\frac{3}{2}^-$
(UZ)	Pa	234	6.75 hr	β^-; 30% γ 0.880 (complex); 23% γ 0.926 (complex); 20% γ 0.131	40.35		4
(UX₂)	Pa	234m	1.175 min	99% β^- 2.29	40.4		
	Pa	235	24.2 min	97% β^- 1.4; (100%m); γ	42.3		
	Pa	236	9.1 min	10% β^- 3.1; 44% β^- 2.4; 30% γ 0.642; 8% γ 0.687; 5% γ 1.763	45.5		
	Pa	237	8.7 min	19% β^- 2.3; 52% β^- 1.4; 34% γ 0.854; 16% γ 0.865; 15% γ 0.529	47.6		
	Pa	238	2.3 min	β^-; γ	51.3		
92	U	226	0.5 s	α 7.43	27.19		0^+
	U	227	1.1 min	100% α 6.87	28.9		
	U	228	9.1 min	67% α 6.69; 28% α 6.60; ⩽5% K	29.22		0^+
	U	229	58 min	13% α 6.36; 4% α 6.33; 2% α 6.30; 1% α 6.22; 80% K	31.20		
	U	230	20.8 d	68% α 5.89; 32% α 5.82; 1% γ 0.072	31.61	20 (f) (r)	0^+
	U	231	4.2 d	100% K; 12% γ 0.026; 7% γ 0.084; 1% γ 0.220	33.78	300 (f) (r)	
	U	232	72 yr	68% α 5.32; 32% α 5.26	34.60	74 (γ); 76 (f)	0^+
	U	233	1.59×10^5 yr	85% α 4.82; 13% α 4.78; 2% α 4.73	36.915	46 (γ); 530 (f)	$\frac{5}{2}^+$
(U_II)	U	234	2.45×10^5 yr 0.0054%	72% α 4.77; 28% α 4.72	38.143	100 (γ)	0^+

Table of the nuclides *(contd)*

Nuclide			Natural abundance or half-life	Decay modes and major radiations	Mass excess MeV	σ/b	I^π
	Z	A					
(AcU)	U	235	7.04×10^8 yr 0.720%	5% α 4.60; 57% α 4.40; 18% α 4.37; 20% α (dist); 54% γ 0.186; 11% γ 0.144; 5% γ 0.163	40.916	98 (γ); 580 (f)	$\frac{7}{2}^-$
	U	235m	26 min (variable)	100% IT	40.917		$\frac{1}{2}^+$
	U	236	2.342×10^7 yr	74% α 4.49; 26% α 4.45	42.442	5 (γ)	0^+
	U	236f	1.2×10^{-7} s	88% IT; 12% SF	44.8		
	U	237	6.75 d	43% β^- 0.25; 53% β^- 0.24; 36% γ 0.060; 23% γ 0.208; 2% γ 0.026, 0.165	45.389	400 (γ)	$\frac{1}{2}^+$
(U₁)	U	238	4.47×10^9 yr 99.275%	77% α 4.20; 23% α 4.15	47.307	2.7 (γ); 2.7 (f)	0^+
	U	238f	2.0×10^{-7} s	IT; SF; γ	49.866		
	U	239	23.5 min	18% β^- 1.27; 73% β^- 1.20; 50% γ 0.075; 5% γ 0.044	50.572	22 (γ) (r); 15 (f) (r)	$\frac{5}{2}^+$
	U	240	14.1 hr	100% β^- 0.36; (100%m); 2% γ 0.044	52.71		0^+
93	Np	229	4.0 min	α 6.89; K	33.76		
	Np	230	4.6 min	α 6.66	35.23		
	Np	231	48.8 min	α 6.28; K; γ	35.63		
	Np	232	14.7 min	100% K; 52% γ 0.327; 33% γ 0.819; 24% γ 0.867	37.3		
	Np	233	36.2 min	100% K; 1% γ 0.312	38.0		
	Np	234	4.4 d	100% K; 19% γ 1.559; 12% γ 1.528; 10% γ 1.602	39.95	1000 (f) (r)	
	Np	235	396 d	100% K	41.04	160 (γ) (r)	$\frac{5}{2}^+$
	Np	236	1.2×10^5 yr	91% K; β^-; 27% γ 0.160; 7% γ 0.104		3000 (f)	
	Np	236	22.5 hr	50% K; 38% β^- 0.54; 12% β^- 0.50; 1% γ 0.642	43.43		I
	Np	237	2.14×10^6 yr	47% α 4.79; 25% α 4.77; 8% α 4.76; 20% α (dist); 14% γ 0.029; 13% γ 0.086	44.869	180 (γ); 0.02 (f)	$\frac{5}{2}^+$
	Np	237f	5×10^{-8} s	IT; SF	47.6		
	Np	238	2.117 d	35% β^- 1.25; 50% β^- 0.26; 28% γ 0.984; 21% γ 1.029; 9% γ 1.026	47.453	2100 (f) (s)	2^+
	Np	239	2.35 d	7% β^- 0.72; 45% β^- 0.43; 36% β^- 0.33; 28% γ 0.106; 14% γ 0.278; 11% γ 0.228	49.306	20 (γ) (r); 32 (γm) (r)	$\frac{5}{2}^+$
	Np	240	67 min	100% β^- 0.78; 29% γ 0.566; 23% γ 0.974; 22% γ 0.601	52.2		
	Np	240m	7.5 min	β^- 2.2; 22% γ 0.555; 13% γ 0.597; 1% γ 0.263, 0.303			I
	Np	241	16.0 min	β^- 1.4; γ	54.3		
94	Pu	232	34 min	α 6.60; >80% K	38.36		0^+
	Pu	233	20.9 min	100% K; γ	40.04		
	Pu	234	8.8 hr	4% α 6.20; 2% α 6.15; 94% K	40.34		0^+
	Pu	235	25.6 min	100% K; 2% γ 0.049	42.2		
	Pu	236	2.85 yr	69% α 5.77; 31% α 5.72	42.89	150 (f) (r)	0^+

Table of the nuclides (*contd*)

Nuclide Z	A	Natural abundance or half-life	Decay modes and major radiations	Mass excess MeV	σ/b	I^π
Pu	236f₁	3×10^{-11} s	SF			
Pu	236f₂	3×10^{-8} s	SF	46.4		
Pu	237	45.3 d	100% K; 3% γ 0.059	45.09	2100 (f) (r)	$\frac{7}{2}^-$
Pu	237m	0.18 s	100% IT; 2% γ 0.146	45.23		$\frac{1}{2}^+$
Pu	237f₁	1.1×10^{-7} s	SF	47.4		
Pu	237f₂	1.1×10^{-6} s	SF	47.7		
Pu	238	87.7 yr	72% α 5.50; 28% α 5.45	46.161	500 (γ) (r); 17 (f) (r)	0^+
Pu	238f₁	5×10^{-10} s	SF	48.6		
Pu	238f₂	7×10^{-9} s	SF	49.9		
Pu	239	2.413×10^4 yr	73% α 5.16; 15% α 5.14; 12% α 5.10; (100%m)	48.585	271 (γ); 742 (f)	$\frac{1}{2}^+$
Pu	239f₁	8×10^{-6} s	SF	50.8		
Pu	240	6570 yr	73% α 5.16; 27% α 5.12	50.123	290 (γ)	0^+
Pu	240f	3.8×10^{-9} s	SF	52.5		
Pu	241	14.36 yr	100% β^- 0.02	52.953	370 (γ); 1010 (f)	$\frac{5}{2}^+$
Pu	241f	2.3×10^{-5} s	SF	55.0		
Pu	242	3.76×10^5 yr	74% α 4.90; 26% α 4.86	54.715	19 (γ)	0^+
Pu	242f₁	4×10^{-9} s	SF			
Pu	242f₂	3×10^{-8} s	SF			
Pu	243	4.955 hr	60% β^- 0.58; 21% β^- 0.50; 23% γ 0.084; 1% γ 0.042	57.753	100 (γ) (r); 200 (f) (r)	$\frac{7}{2}^+$
Pu	243f	6×10^{-8} s	SF	59.6		
Pu	244	8.1×10^7 yr	80% α 4.59; 20% α 4.55	59.80	1.7 (γ)	0^+
Pu	244f	4×10^{-10} s	SF			
Pu	245	10.5 hr	β^-; 26% γ 0.327; 6% γ 0.560; 5% γ 0.308	63.16	150 (γ)	
Pu	246	10.85 d	β^-; 26% γ 0.044; 24% γ 0.224; 10% γ 0.180	65.3		0^+
95 Am	237	1.22 hr	100% K; 47% γ 0.280; 8% γ 0.438; 4% γ 0.474	46.6		$\frac{5}{2}$
Am	237f	5×10^{-9} s	SF	48.7		
Am	238	1.6 hr	100% K; 29% γ 0.963; 24% γ 0.919; 11% γ 0.561	48.42		1^+
Am	239	11.9 hr	100% K; 15% γ 0.278; 11% γ 0.228; 3% γ 0.210, 0.226	49.39		$\frac{5}{2}^-$
Am	239f	1.6×10^{-7} s	SF	51.9		
Am	240	50.8 hr	100% K; 73% γ 0.988; 25% γ 0.889; 2% γ 0.099	51.44		
Am	240f	9×10^{-4} s	SF	54.0		
Am	241	432.0 yr	85% α 5.49; 13% α 5.44; 2% α (dist); 36% γ 0.060	52.932	560 (γ); 62 (γm); 3.2 (f)	$\frac{5}{2}^-$
Am	241f	2×10^{-6} s	SF	55.1		
Am	242	16.01 hr	37% β^- 0.66; 46% β^- 0.62; 17% K	55.463	2100 (f) (r)	1^-
Am	242m	152 yr	100% IT	55.511	1650 (γ) (r); 7400 (f) (r)	5^-
Am	242f	0.0140 s	SF	57.8		
Am	243	7370 yr	88% α 5.28; 11% α 5.23; 1% α 5.18; 66% γ 0.075; 6% γ 0.044	57.170	6 (γ) (r); 80 (γm) (r)	$\frac{5}{2}^-$
Am	243f	5×10^{-6} s	SF	59.2		

Table of the nuclides (*contd*)

Nuclide		Natural abundance or half-life	Decay modes and major radiations	Mass excess MeV	σ/b	I^{π}
Z	A					
Am	244	10.1 hr	100% β^- 0.39; 66% γ 0.744; 28% γ 0.898; 19% γ 0.154	59.877	2200 (f) (r)	
Am	244m	26 min	80% β^- 1.50; 20% β^- 1.46	59.95	1600 (f) (r)	
Am	244f	0.0011 s	SF	61.5		
Am	245	2.05 hr	77% β^- 0.90; 6% γ 0.253	61.897		
Am	245f	6.4×10^{-7} s	SF			
Am	246	39 min	β^-; 52% γ 0.679; 35% γ 0.205; 25% γ 0.154			
Am	246	25.0 min	7% β^- 2.2; 37% β^- 1.2; 29% γ 1.079; 26% γ 0.799; 18% γ 1.062	64.9		
Am	246f	7×10^{-5} s	SF			
Am	247	24 min	76% β^- 1.4; 24% β^- 1.3; 23% γ 0.285; 6% γ 0.226	67.1		
96 Cm	238	2.3 hr	α 6.52; <90% K	49.40		0^+
Cm	239	2.9 hr	100% K; γ	51.1		
Cm	240	26.8 d	71% α 6.29; 29% α 6.25	51.71		0^+
Cm	240f	10^{-11} s	SF			
Cm	241	32.8 d	1% α 5.94; 99% K; 71% γ 0.472; 4% γ 0.132; 4% γ 0.431	53.70		$\frac{1}{2}^+$
Cm	241f	1.5×10^{-8} s	SF	55.7		
Cm	242	162.8 d	74% α 6.11; 26% α 6.07	54.802	20 (γ)	0^+
Cm	242f$_1$	4×10^{-11} s	SF			
Cm	242f$_2$	1.8×10^{-7} s	SF	57.6		
Cm	243	28.5 yr	2% α 6.07; 73% α 5.79; 11% α 5.74; 14% α (dist); 14% γ 0.278; 11% γ 0.228; 3% γ 0.210	57.177	1000 (t) (r); 130 (γ); 610 (f)	$\frac{5}{2}^+$
Cm	243f	4×10^{-8} s	SF	58.7		
Cm	244	18.10 yr	77% α 5.81; 23% α 5.76	58.450	14 (γ); 1 (f)	0^+
Cm	244f	$> 10^{-7}$ s	SF	61.5		
Cm	245	8540 yr	1% α 5.49; 1% α 5.45; 93% α 5.36; 5% α 5.30; 5% γ 0.133; 5% γ 0.172	61.001	350 (γ); 2000 (f)	$\frac{7}{2}^+$
Cm	245f	1.3×10^{-8} s	SF	62.7		
Cm	246	4710 yr	79% α 5.39; 21% α 5.34	62.616	1.3 (γ); 0.2 (f)	0^+
Cm	247	1.6×10^7 yr	14% α 5.27; 6% α 5.21; 71% α 4.87; 9% α (dist); 72% γ 0.402; 3% γ 0.278; 2% γ 0.288	65.53	60 (γ); 100 (f) (s)	$\frac{9}{2}^-$
Cm	248	3.40×10^5 yr	75% α 5.08; 17% α 5.03; 8% SF	67.39	4 (γ) (s); 0.3 (f)	0^+
Cm	249	65 min	96% β^- 0.90; 2% γ 0.634; 1% γ 0.560	70.75	2 (γ) (r)	$\frac{1}{2}^+$
Cm	250	1.1×10^4 yr	100% SF	72.99	80 (γ) (r)	0^+
Cm	251	16.8 min	73% β^- 1.42; 12% γ 0.543; 2% γ 0.530; 1% γ 0.390, 0.438	76.7		
97 Bk	242	7 min	100% K	57.8		
Bk	242f$_1$	6×10^{-7} s	SF			

Table of the nuclides (contd)

Nuclide		Natural abundance or half-life	Decay modes and major radiations	Mass excess MeV	σ/b	I^π
Z	A					
Bk	242f₂	10^{-8} s	SF			
Bk	243	4.5 hr	100% K; γ	58.69		
Bk	244	4.4 hr	100% K; γ	60.65		
Bk	244f	8×10^{-7} s	SF			
Bk	245	4.90 d	100% K; 31% γ 0.253; 2% γ 0.381	61.81		$\frac{3}{2}^-$
Bk	245f	2×10^{-9} s	SF			
Bk	246	1.80 d	100% K; 61% γ 0.799; 6% γ 1.081; 5% γ 0.834	64.0		
Bk	247	1.4×10^3 yr	17% α 5.71; 45% α 5.53; 13% α 5.69; 25% α (dist); 40% γ 0.084; 30% γ 0.265	65.48		
Bk	248	23.5 hr	65% β⁻ 0.7; 30% K; 5% γ 0.551	68.0		
Bk	248	>9 yr	?	68.0		
Bk	249	325 d	100% β⁻ 0.13	69.848	1000 (t) (r); <6 (f) (r)	$\frac{7}{2}^+$
Bk	250	3.22 hr	5% β⁻ 1.78; 83% β⁻ 0.75; 45% γ 0.989; 35% γ 1.032; 4% γ 1.029	72.95	1000 (f) (r)	2^-
Bk	251	56 min	β⁻; γ	75.3		
98 Cf	240	1.1 min	α 7.59	58.0		0^+
Cf	241	4 min	α 7.34	59.2		
Cf	242	3.5 min	80% α 7.39; 20% α 7.35	59.33		0^+
Cf	243	10 min	30% α (7.17); 70% K	60.9		
Cf	244	19 min	75% α 7.22; 25% α 7.18	61.47		0^+
Cf	245	44 min	α 7.14; 70% K	63.38		
Cf	246	36 hr	78% α 6.76; 22% α 6.72	64.096		0^+
Cf	247	3.15 hr	100% K; 1% γ 0.294	66.2		
Cf	248	334 d	82% α 6.26; 18% α 6.22	67.24		0^+
Cf	249	351 yr	2% α 6.19; 83% α 5.81; 5% α 5.76; 10% α (dist); 66% γ 0.388; 15% γ 0.333; 3% γ 0.253	69.722	480 (γ) (s); 1630 (f)	$\frac{9}{2}^-$
Cf	250	13.1 yr	83% α 6.03; 17% α 5.99	71.170	2000 (γ) (s); <350 (f) (r)	0^+
Cf	251	900 yr	12% α 6.01; 27% α 5.85; 35% α 5.68; 26% α (dist); 17% γ 0.177; 5% γ 0.227; 1% γ 0.285	74.13	2900 (γ) (s); 4500 (f) (s)	$\frac{1}{2}^+$
Cf	252	2.64 yr	82% α 6.12; 15% α 6.08; 3% SF	76.03	20 (γ); 32 (f) (s)	0^+
Cf	253	17.8 d	100% β⁻ 0.29	79.30	18 (γ) (r); 1300 (f) (r)	
Cf	254	60.5 d	100% SF	81.34	100 (t) (r)	0^+
Cf	256	12 min	100% SF			0^+
99 Es	243	21 s	α 7.89	64.8		
Es	244	37 s	4% α (7.57); β⁺; K	66.0		
Es	245	1.3 min	40% α (7.73); 60% K	66.4		
Es	246	8 min	10% α (7.36); β⁺; K	67.9		
Es	247	4.7 min	7% α (7.31); 93% K	68.55		
Es	248	28 min	100% K	70.2		

Table of the nuclides (*contd*)

Nuclide		Natural abundance or half-life	Decay modes and major radiations	Mass excess MeV	σ/b	Iπ
Z	A					
Es	249	1.70 hr	1% α 6.77; 99% K; 40% γ 0.379; 9% γ 0.813; 3% γ 0.375	71.12		
Es	250	8.6 hr	100% K; 74% γ 0.829; 22% γ 0.303; 20% γ 0.349	73.2		
Es	250	2.1 hr	100% K; 17% γ 0.989; 14% γ 1.032	73.2		
Es	251	33 hr	100% K; 2% γ 0.178	74.51		
Es	252	472 d	62% α 6.63; 11% α 6.56; 5% α (dist); 22% K; 16% γ 0.785; 12% γ 0.139; 2% γ 0.924	77.2		
Es	253	20.47 d	90% α 6.63; 7% α 6.59; 3% α (dist)	79.012	160 (γm); <60 (f) (r)	$\frac{7}{2}^+$
Es	254	276 d	93% α 6.43; 2% α 6.42; 3% α 6.36; 2% α (dist); 2% γ 0.063	81.99	2800 (f)	
Es	254m	39.3 hr	β⁻; 29% γ 0.649; 25% γ 0.694; 13% γ 0.689	82.07	1800 (f) (r)	2⁺
Es	255	38.3 d	7% α 6.30; 1% α 6.26; β⁻	84.1	65 (γ) (r)	
Es	256	7.6 hr	β⁻; γ	87.3		
Es	256	22 min	β⁻	87.3		
100 Fm	245	4 s	α 8.15	70.0		
Fm	246	1.2 s	92% α (8.24); 8% SF	70.13		0⁺
Fm	247	9 s	α 8.18	71.5		
Fm	247	35 s	α 7.93; α 7.87; <50% K	71.5		
Fm	248	36 s	80% α 7.87; 20% α 7.83	71.89		0⁺
Fm	249	3 min	α 7.53	73.5		
Fm	250	30 min	α 7.44	74.07		0⁺
Fm	250m	1.8 s	100% IT			
Fm	251	5.3 hr	2% α 6.83; 98% K; 2% γ 0.880; 1% γ 0.406; 1% γ 0.453	76.0		
Fm	252	25.4 hr	85% α 7.04; 15% α 7.00	76.82		0⁺
Fm	253	3.0 d	5% α 6.94; 3% α 6.67; 4% α (dist); 88% K; 3% γ 0.272	79.35		$\frac{1}{2}^+$
Fm	254	3.240 hr	85% α 7.19; 14% α 7.15; 1% α 7.05	80.90		0⁺
Fm	255	20.1 hr	93% α 7.02; 5% α 6.96; 1% α 7.08; 1% α 6.89	83.79	26 (γ) (r); 3300 (f) (s)	$\frac{7}{2}^+$
Fm	256	2.63 hr	8% α (6.92); 92% SF	85.48		0⁺
Fm	257	100.5 d	1% α 6.76; 4% α 6.70; 93% α 6.52; 2% α 6.44; 25% γ 0.115; 11% γ 0.241; 9% γ 0.179	88.59	5800 (t) (s); 3000 (f) (r)	
Fm	258	3.8 × 10⁻⁴ s	100% SF			0⁺
Fm	259	1.5 s	100% SF			
101 Md	248	7 s	5% α 8.36; 15% α 8.32; β⁺; K	77.0		
Md	249	24 s	>20% α (8.03); β⁺; K	77.3		
Md	250	52 s	2% α 7.82; 4% α 7.75; β⁺; K	78.6		
Md	252	2 min	β⁺; K	80.5		
Md	254	10 min	100% K	83.4		
Md	254	28 min	100% K	83.4		

Table of the nuclides (*contd*)

Nuclide			Natural abundance or half-life	Decay modes and major radiations	Mass excess MeV	σ/b	I^π
Z		A					
	Md	255	27 min	7% α 7.33; 93% K; 7% γ 0.430	84.8		
	Md	256	75 min	6% α 7.22; 2% α 7.15; 2% α (dist); 90% K	87.4		
	Md	257	5.0 hr	10% α (7.07); 90% K	89.0		
	Md	258	56 d	28% α 6.79; 72% α 6.72	91.8		
	Md	259	95 min	100% SF			
102	No	252	2.3 s	55% α 8.42; 18% α 8.37; 27% SF	82.87		0^+
	No	253	1.8 min	α 8.01	84.3		
	No	254	55 s	α 8.10	84.73		0^+
	No	254m	0.28 s	100% IT			
	No	255	3.3 min	28% α 8.12; 7% α 8.08; 7% α 7.93; 20% α (dist); 38% K	86.9		
	No	256	3.2 s	α 8.43	87.8		0^+
	No	257	26 s	19% α 8.32; 26% α 8.27; 55% α 8.22	90.22		
	No	259	58 min	11% α 7.59; 18% α 7.52; 30% α 7.49; 41% α (dist); 22% K	94.01		
103	Lr	255	22 s	40% α 8.43; 60% α 8.37	90.3		
	Lr	256	26 s	19% α 8.52; 38% α 8.43; 19% α 8.39; 24% α (dist)	91.8		
	Lr	257	0.65 s	85% α 8.86; 15% α 8.80	93.0		
	Lr	258	4.4 s	10% α 8.65; 35% α 8.61; 45% α 8.59; 10% α 8.54	94.8		
	Lr	259	5 s	100% α 8.46	96.0		
	Lr	260	3.0 min	100% α 8.04	98.1		
104		257	5 s	35% α 9.00; 30% α 8.95; 20% α 8.78; 15% α 8.70	96.0		
		259	3 s	α 8.86	98.5		
		261	65 s	α 8.29	101.3		
105		260	1.5 s	17% α 9.12; 25% α 9.07; 48% α 9.04; 10% SF	104		
		261	2 s	75% α (8.93); 25% SF	104		
		262	40 s	α 8.66	106		
106		263	0.9 s	α 9.25			

D.J.P.

3.6.2 The radioactive series and their precursors

The three naturally occurring radioactive series have great historical importance. Nowadays they, the 4n + 1 (Neptunium) series and all their precursors amongst the man-made nuclides are of considerable technological importance in the development of nuclear power. Precursors which feed the heads of the main chains are shown in Figs. A and B. Those which lead laterally (i.e by β-decay) into the main α-chains have been excluded. Lifetimes are given to two significant figures only. Where two lifetimes are given they are preceded by g (ground state) and i (isomeric state) if the order of the states is known.

Where α, β branching occurs, the intensity of the weaker branch is usually listed (see, for example ^{214}Bi). Where there is apparently no competitor to α-decay, e.g. ^{252}Cf, and a branch intensity is given, the competing decay process is spontaneous fission. All known branches have been included irrespective of their intensity. Prior to the discovery of fission only three α, β branches were known in the natural series. Now there are 11 cases in addition to those in the Neptunium series. Only the Thorium series has no new branches.

The classical names of the members of the natural radioactive series, though rarely used nowadays, are included because of their historical interest; they indicate the state of historical knowledge before 1939. The name, where applicable, is shown outside the bottom right-hand corner of the square representing the particular nuclide.

The source of the information is Lederer and Shirley 1978[1]. Only nuclides and isomers in this publication's categories A, B and C are included. Isomers which decay by α or β^- or β^+ or Electron Capture (but not by γ radiation alone) are included in the charts. The juxtaposition of lifetimes and the arrows indicating β^- and β^+ or EC (see key to the charts) in certain nuclides, e.g. ^{242}Am, is not intended to imply that the individual lifetimes are related to specific modes of decay.

Certain nuclides warrant specific mention. The isomer of ^{234}Pa (in the 4n + 2 series) is formed in 100% of transitions from its parent ^{234}Th. In 99.87% of cases the isomer decays by β^- emission with a half-life of 1.2 minutes. In the remainder of cases γ-emission to the ground state of ^{234}Pa takes place prior to β^- decay with a half-life of 6.8 hours. The isomeric state known to exist in ^{210}Bi is not excited in the natural radioactive decay chain and is not listed.

^{210}Tl (in the 4n + 2 series) is a weak delayed neutron emitter (in 0.007 + 0.007% − .004% of cases). It is the only example of a transition from one chain to another. The neutron emission is from an excited state of ^{210}Pb, leading to ^{209}Pb, a member of the Neptunium series (4n + 1), which β decays to stable ^{209}Bi.

The longest lived member and titular parent of the Neptunium series, ^{237}Np, is formed principally by α-decay of reactor produced ^{241}Am. The next member of the series, ^{233}U, is also abundantly produced in the sequence

$$^{232}\text{Th}(n,\gamma)\,^{233}\text{Th} \xrightarrow{\beta \text{ decay}} {}^{233}\text{Pa} \xrightarrow{\beta \text{ decay}} {}^{233}\text{U}$$

Reference

(1) Lederer and Shirley, *Table of Isotopes* (7th edn), John Wiley and Sons, Inc. 1978.

D.W.

3.6.2   Fig. A

A = 4n
Thorium Series

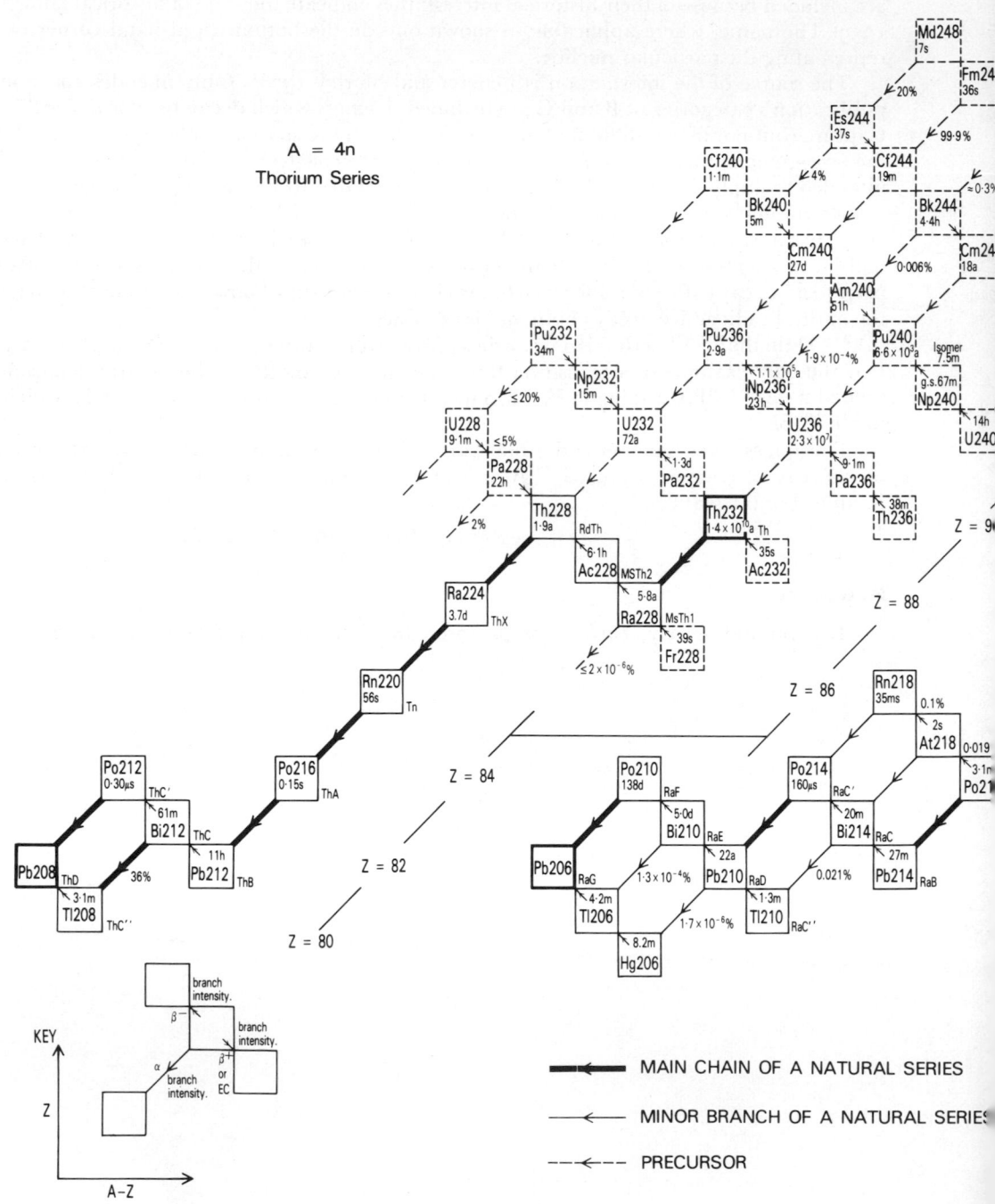

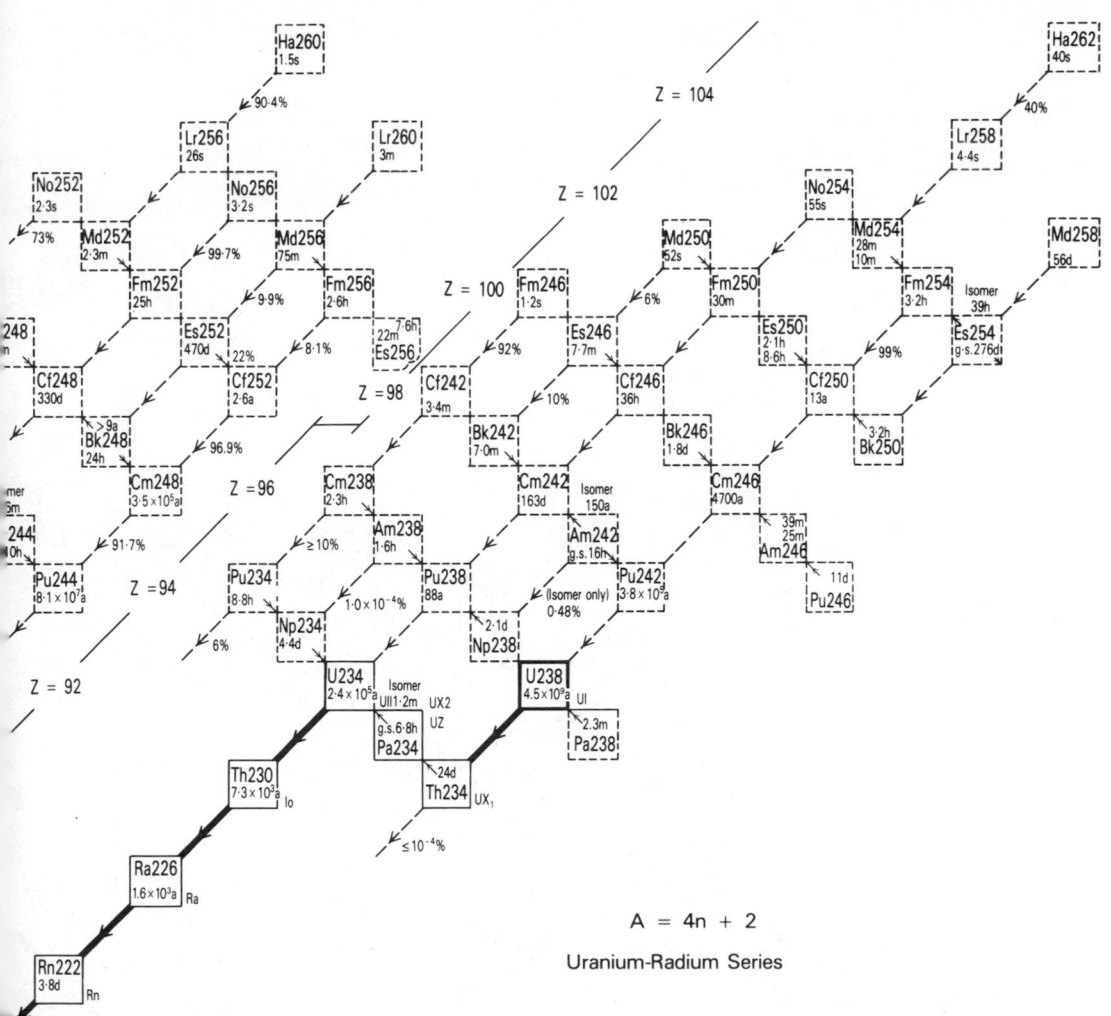

$A = 4n + 2$

Uranium-Radium Series

| STABLE OR LONG-LIVED NATURALLY OCCURRING NUCLIDE |
| MEMBER OF A NATURALLY OCCURRING SERIES |
| PRECURSOR |

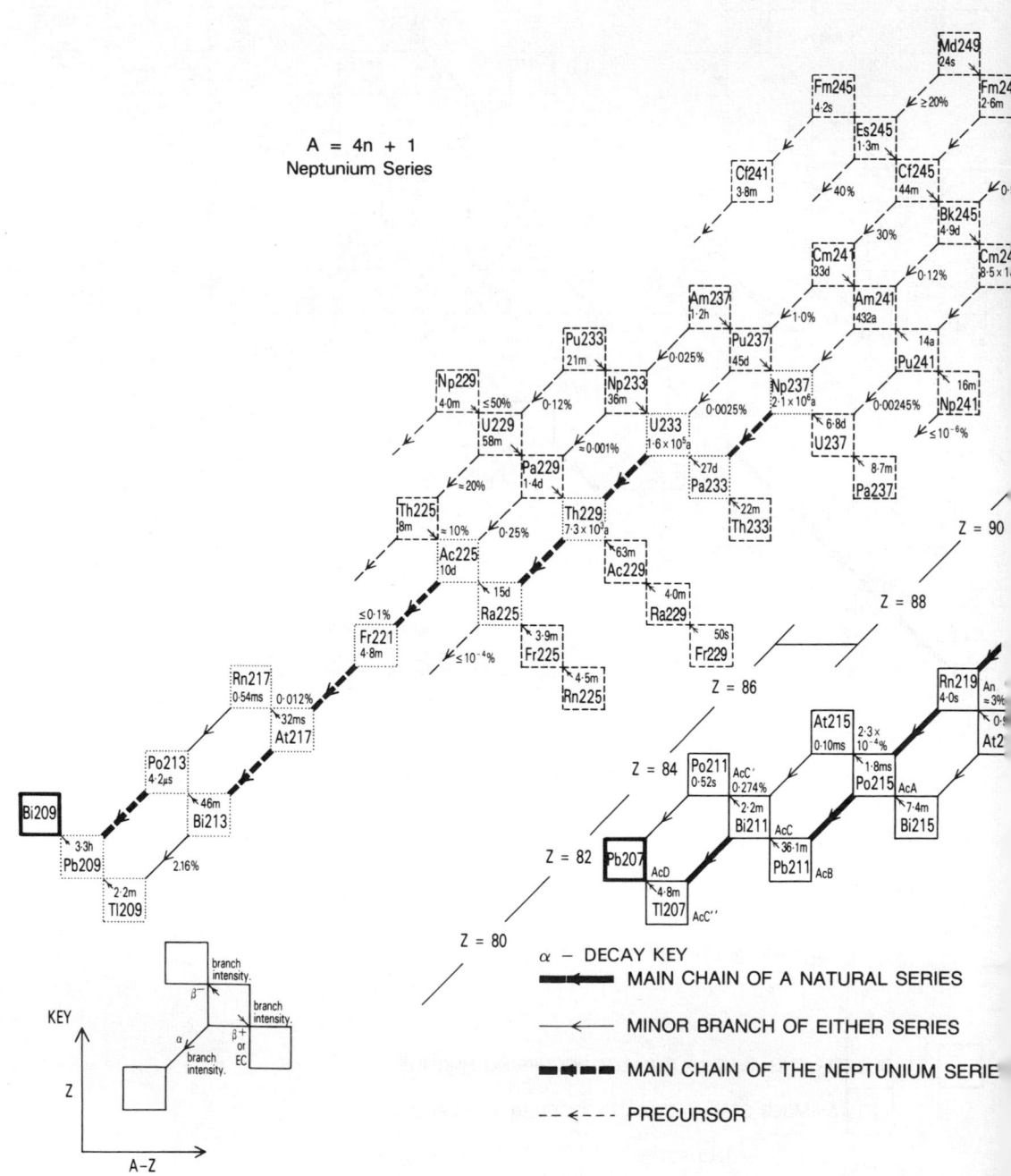

A = 4n + 1
Neptunium Series

α – DECAY KEY

    MAIN CHAIN OF A NATURAL SERIES

    MINOR BRANCH OF EITHER SERIES

    MAIN CHAIN OF THE NEPTUNIUM SERIE

- - ← - - PRECURSOR

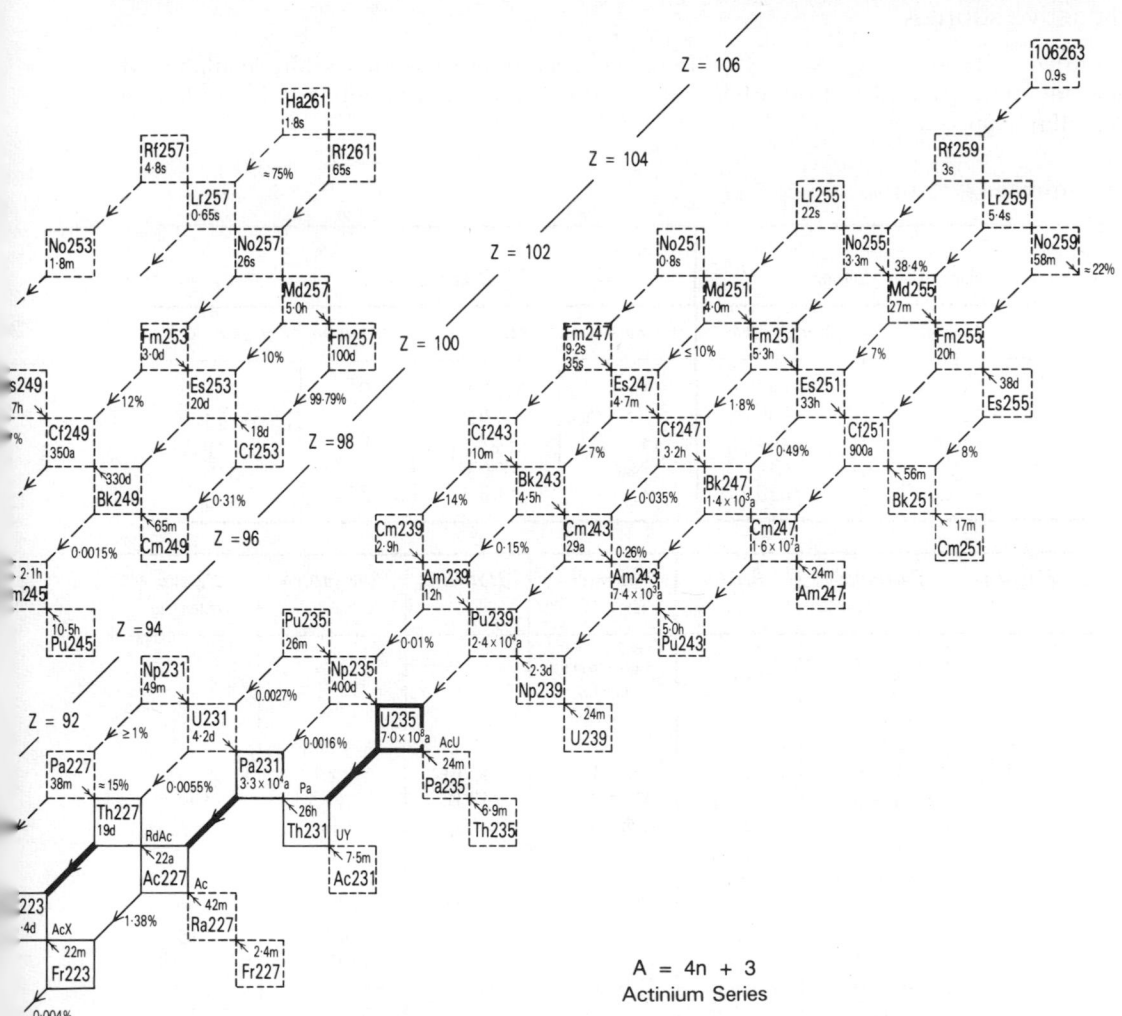

A = 4n + 3
Actinium Series

STABLE OR LONG-LIVED NATURALLY OCCURRING NUCLIDE

MEMBER OF A NATURALLY OCCURING RADIOACTIVE SERIES

MEMBER OF THE NEPTUNIUM SERIES

PRECURSOR

3.6.3 Radioactive sources

The sources listed in the table are a selection from the large number commercially available. It contains only the major radiations from each source and excludes 511 keV annihilation radiation generated by all β^+ sources.

Radioactive sources

Sources with monoenergetic electrons			Beta-ray sources			
Source	*Half-life*	*Energy/keV*	*Source*	*Half-life*	*Radiation*	*End point/keV*
^{137}Cs . . .	30.2 y	624	^{14}C	5730 y	β^-	156
		656	^{90}Sr/^{90}Y	29 y	β^-	2274
^{207}Bi . . .	38 y	482	^{3}H	12.3 y	β^-	19
		975	^{22}Na	2.60 y	β^+	546
		1048	^{58}Co	70.8 d	β^+	475

Source	Half-life	Energy/keV	Relative intensity	Source	Half-life	Energy/keV	Relative intensity
Gamma-ray source				*Low-energy photon source*			
^{22}Na . .	2.60 y	1274.6		^{55}Fe	2.69y	5.9–6.5	
^{54}Mn . .	312 d	834.8		^{109}Cd	453d	22–26	
^{56}Co . .	79 d	846.8	100	^{241}Am	433y	59.5	100
		977.5	1.4	*Alpha particle source*		26.3	~5
		1037.8	14.0				
		1175.1	2.3				
		1238.3	67.6	^{241}Am	433 y	5484	100
		1360.2	4.3			5442	15
		1771.5	15.7			5387	1.8
		2015.4	3.1	^{244}Cm	17.8 y	5806	100
		2034.9	7.9			5763	31
		2598.6	16.9	^{210}Pb	(22.3y)	5305	
		3010.2	1.0	^{212}Pb	(10.6h)	8784	
		3202.3	3.0			6090	
		3253.6	7.4			6050	
		3273.3	1.8				
		3451.6	0.9				
^{60}Co . .	5.27 y	1173.2	100				
		1332.5	100				
^{88}Y	107 d	898.0	92				
		1836.1	100				
^{137}Cs . .	30.2 y	661.6					
^{207}Bi . .	38 y	569.6	100	*Fission fragment source*			
		1063.6	78	^{252}Cf	2.65 y		
		1770.2	9				
^{208}Tl	(10.6h)	277.4	7				
		510.7	22				
		583.7	86				
		860.4	12				
		2614.5	100				

The alpha source ^{212}Pb is obtained by collecting recoil products, the so-called 'active deposit', from thoron (^{220}Em) gas above a radio thorium source; it gives rise in particular to 6090 and 6050 keV alphas from ^{212}Bi and 8784 keV alphas from ^{212}Po. The gamma source ^{208}Tl is obtained in the same way. The half life in brackets is that of the active deposit source itself, not the half life of the nuclide which is the source of the radiation.

Some extra information can be obtained from the table of nuclides (p. 341) and detailed information on γ-ray transitions from the published literature to 1977 is available in ref.(1). Table of γ-rays in order of increasing energy, half life and atomic number can be found in refs. (2) and (3), although in some cases the energies and half-lives have been superseded by more accurate values. Refs. (2) and (3) also catalogue the response of NaI and Ge(Li) detectors to radioactive sources.

References

(1) C. M. Lederer and V. S. Shirley (eds), *Table of Isotopes*, 1978.
(2) C. E. Crouthamel, *Applied Gamma-Ray Spectrometry*, 1970.
(3) R. L. Heath, *Scintillation Spectrometry*, IDO-15880 (1964).

N.C.

3.7 Nuclear fission and neutron interactions

3.7.1 Fission fundamentals

When nuclei of the heavy elements capture neutrons, fission takes place; that is to say, the nuclei divide usually into two parts, though in a small fraction ($< 1\%$) of cases a third light particle is emitted. Fission can also occur spontaneously but the half life for this is usually long (see Ewbank *et al.*, 1979). In the process of fission about 200 MeV of energy is released, largely in the form of kinetic energy of the two fission fragments. Fission is also accompanied by the emission of two or three neutrons and about ten gamma-rays, emitted from the fission fragments during their recoil. An approximate formula for the fission neutron spectrum is given in the section on Neutron Cross-sections (p. 428) which also contains information on the total number of neutrons per fission for various fissile nuclei in three different neutron spectra (table p. 430).

The mass partition between the two fragments is in general asymmetrical, particularly for fission induced by low energy neutrons. The special cases of the fission of U-233, U-235 and Pu-239 by thermal and fast neutrons are illustrated in the diagram and it will be observed that despite its characteristic shape the fission yield curve is by no means a smooth function of mass. For a given neutron energy the high mass peak is very similar for all the fissioning nuclei while the low mass peak varies in position. As the neutron energy increases the valley between the peaks fills in and the wings of the peaks rise, while at very high energies and also for very heavy elements the mass yield becomes symmetrical and single humped.

The products of fission are themselves generally neutron rich and hence radioactive and the Tables of Nuclides (p. 341) should be consulted for their properties. When some fission fragments have undergone a β^--decay they have sufficient excitation energy to emit a neutron. These neutrons, which are called delayed neutrons, are emitted with half lives of up to 55 s and are very important for the control of nuclear reactors. The β^- and gamma-ray decay energy of the fragments is emitted over a long period and the diagram shows for U-235 and Pu-239 the average β and gamma-ray energy emitted by fission products as a function of time after fission. In a reactor the actinides also make a significant contribution, particularly for decay times of $\sim 10^5$ s and greater than 10^8 s (see Tobias 1980).

The average total energy released by slow-neutron induced fission in the fissile materials U-233, U-235 and Pu-239 is given below.

	U-233	U-235	Pu-239
Instantaneously released energy			
Kinetic energy of fission fragments	168.2 MeV	169.1 MeV	175.8 MeV
Kinetic energy of prompt neutrons	4.9	4.8	5.9
Energy carried by prompt γ-rays	7.7	7.0	7.8
Energy from decaying fission products			
Energy of β^--particles	5.2	6.5	5.3
Energy of anti-neutrinos	6.9	8.8	7.1
Energy of delayed γ-rays	5.0	6.3	5.2
	197.9	202.5	207.1

In an operating reactor, all of this energy except that carried by the anti-neutrinos is converted into heat. There is, however, an additional source of energy arising from the binding energy released when those prompt neutrons which do not produce fission are finally captured. For thermal reactors this amounts to ~ 9.1 MeV for U-233, ~ 8.8 MeV for U-235 and ~ 11.5 MeV for Pu-239.

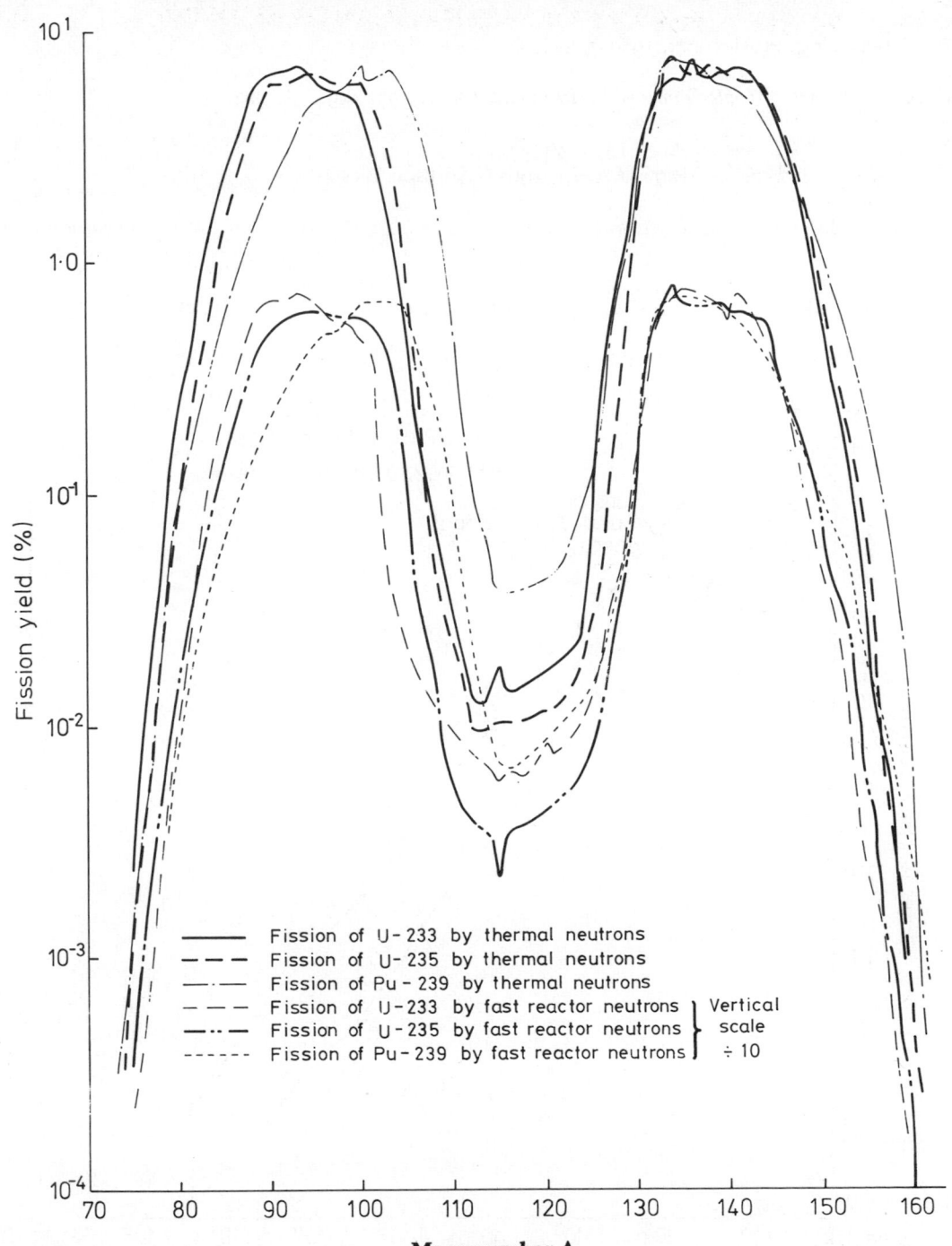

Fission yield (%)

Mass number A

Fission of U – 233 by thermal neutrons
Fission of U – 235 by thermal neutrons
Fission of Pu – 239 by thermal neutrons
Fission of U – 233 by fast reactor neutrons ⎱ Vertical
Fission of U – 235 by fast reactor neutrons ⎰ scale
Fission of Pu – 239 by fast reactor neutrons ⎰ ÷ 10

References

Energy from fission
R. Sher, BNL-NCS 51363, vol. II, p. 835, 1981.
M. F. James, *J.N.E.*, 1969, **23**, 517.

Decay heat
A. Tobias, *Prog. in Nucl. En.*, 1980. **5**, (1), 3.
Fission yields
E. A. C. Crouch, *Atomic Data and Nuclear Data Tables*, 1977, **19**, (5), 419.
Delayed neutrons
R. J. Tuttle, *Nucl. Sci. and Eng.*, 1975, **56**, 37.
G. R. Keepin, *Physics of Neutron Kinetics*, 1965, (Addison Wesley)
Nuclear fission
A. Michaudon (ed.), *Nuclear Fission and Neutron Induced Fission Cross-sections*, 1981, (Pergamon Press, Oxford).
Spontaneous fission
W. B. Ewbank, Y. A. Ellis and M. R. Schmorak, *Nuclear Data Sheets*, 1979, **26**, 1.

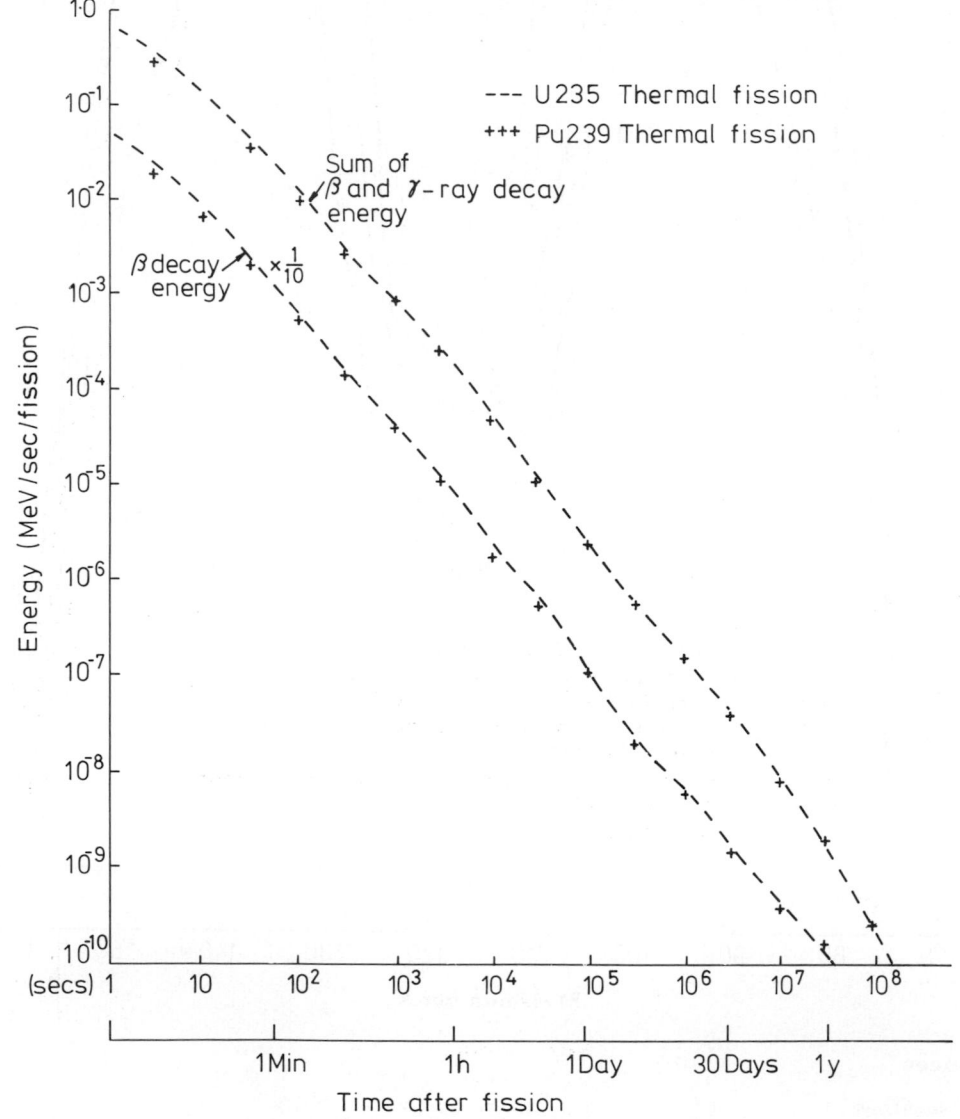

Beta and gamma-ray energy from fission product decay as a function of time after a fission.

M.G.S.

3.7.2 Neutron cross-sections

If $n(E)\,\mathrm{d}E$ is the number of neutrons per unit volume of a material having an energy between E and $E + \mathrm{d}E$, then the neutron flux, φ, is defined as $\varphi(E) = n(E)v$, where v is the neutron velocity. The number of neutrons giving a particular type of interaction per unit volume per unit time in a material containing N nuclei per unit volume is $\int N\phi(E)\sigma_x(E)\,\mathrm{d}E$, where $\sigma_x(R)$ is the nuclear cross-section for the particular type of interaction x. The product $N\sigma_x$ is called the macroscopic cross-section, Σ_x, and its reciprocal can be shown to be the mean free path.

After a neutron enters a nucleus the resulting compound nucleus may emit a charged particle (e.g. a proton or α particle), a γ-ray, a neutron or a pair of neutrons. In heavy materials fission may take place. These reactions are known respectively as (n, p) (n, α) (n, γ) (n, n) (n, 2n) (n, f) events. The (n, γ) reaction is referred to as radiative capture and the corresponding cross-section written σ_γ. The (n, n) interaction indicates scattering. If after the interaction the nucleus remains in the ground state the event is referred to as elastic scattering and the cross-section is written σ_n. If it is left in an excited state it is referred to as inelastic scattering and is denoted by (n, n′) and the cross-section is written $\sigma_{n'}$. The fission cross-section is written σ_f and the total cross-section for all events is written σ_T. The absorption cross-section, σ_A, is defined by $\sigma_A = \sigma_T - \sigma_n - \sigma_n$. A transport cross-section is defined as $\sigma_T - b\sigma_n$ where b is the average value of cos ψ and ψ is the angle of scatter of a neutron in a collision.

The energy range of interest is from 0.001 eV to 10 MeV and values of $\sigma(E)$ are shown graphically in BNL 325 (available from Office of Technical Services, Department of Commerce, Washington, D.C.) for most elements and for important individual isotopes. Point values are held in a number of nuclear data files prepared at several centres, e.g. the UK Nuclear Data Library (Ref. AEEW–M824 available from HM Stationery Office, London) gives a point by point energy representation (up to 4000 points) covering the range 0.0001 eV to 15 MeV and including all the important cross-sections of 88 nuclides. In Figures A, B, C, D, E and F a selection of cross-sections is shown. Certain features may be distinguished:

At low energies absorption cross-sections tend to be proportional to $1/v$ (i.e. to $E^{-1/2}$) and for ^{10}B and ^{6}Li (n, α) reactions this law is good to 5% in the region up to about 100 keV (see Figure A).

For light nuclides scattering is dominated by potential scattering which can be considered as the deflection of a neutron by the nuclear field without the formation of a compound nucleus, and the cross-section tends to be independent of energy. Departures, however, arise at low energies in crystalline materials (e.g. C and Be) when the neutron wavelength is comparable to the lattice spacing, and in molecular compounds (H_2, H_2O) when the collision energy is comparable to the vibrational quanta of the molecular oscillations (see Figure B).

At intermediate energies the cross-sections show sharp maxima at neutron energies at which the reacting particles have the same energy (including the binding energy) as one of the resonance levels in the compound nucleus. Light nuclei have larger energy level spacings than heavy nuclei (cf. Figures C and D). It is thus more probable to find resonances at low energies in heavy nuclei and such resonances can be seen in Figures A and B for U and Pu.

For heavy nuclei having an odd number of neutrons (e.g. ^{235}U) fission is possible at all energies and such materials are called fissile (see Figures A, D and E).

For heavy nuclei having an even number of neutrons (e.g. ^{238}U) fission is only significant above a threshold energy (see Figure F). Since neutron capture in such materials produces a fissile material they are called fertile.

All (n, n′) and (n, 2n) reactions have threshold energies. In some light nuclides (n, p) and (n, α) interactions have appreciable cross-sections at all energies but in the heavier nuclides these cross-sections always show an effective threshold at an MeV or so, even when the reaction is theoretically possible with low energy neutrons (see Figure F).

Because of the strong dependence of cross-section on energy it is usually necessary in practical calculations to make use of group cross-sections which are values averaged over a chosen energy region and weighted according to the energy spectrum in the system, i.e.

$$\overline{\sigma} = \frac{\displaystyle\int_{E_1}^{E_2} \sigma(E)\varphi(E)\,\mathrm{d}E}{\displaystyle\int_{E_1}^{E_2} \varphi(E)\,\mathrm{d}E}$$

For a general comparison of the interaction of neutrons in various materials it is often appropriate to consider three energy regions.

At high energies the cross-sections can be averaged over a fission neutron spectrum which is conventionally expressed as $\varphi_f = (E/b^2)\exp(-E/b)$, where $b = 2.0$ MeV. In the tables below the average values have been taken over the range 1 keV to 20 MeV although most of the contribution comes from a much narrower range around 2 MeV.

At intermediate energies the spectrum in a well-moderated system is nearly proportional to $1/E$ and the average cross-section is equal to:

$$\frac{\displaystyle\int_{E_1}^{E_2} (\sigma/E)\,\mathrm{d}E}{\ln(E_2/E_1)}$$

The numerator is referred to as the resonance integral (R.I.) since it gives the cumulative absorption in the various resonances as the neutron slows down from energy E_2 to E_1. For absorption processes it is customary to quote the resonance integrals rather than average values, and the tables show values where $E_2 = 0.55$ eV and $E_1 = 2$ MeV although the values are usually insensitive to the upper limit.

At lower energies, when the neutrons are in thermal equilibrium with their surroundings, $\varphi_{\mathrm{th}} = E\exp(-E/kT_n)/(kT_n)^2$, where T_n is the temperature of the surroundings. If the cross-section varies as $1/v$ then the average cross-section in a thermal flux can be shown to be equal to $\sigma_0(\pi T_0/4T_n)^{1/2}$, where σ_0 is the cross-section corresponding to an energy kT_0. It is customary to refer cross-sections to the value σ_0 at a standard neutron velocity of 2200 m s^{-1}, which corresponds to a neutron energy kT_0 of 0.02530 eV and a temperature of 293.59 K. Values of σ_0 have been given in the Table of Nuclides (p. 341) and values averaged over a thermal spectrum with $T_n = T_0$ are given in the following tables. The departure of these average values from the value of $\sigma_0(\pi/4)^{1/2}$ is indicative of the departure of the particular cross-section from the $1/v$ characteristic. The integrations were taken over the range 0.0001 eV to 1.0 eV although most of the contribution comes from a narrower range around 0.025 eV.

The tables on pages 429 and 430 give properties of nuclides which have been grouped according to their function in nuclear reactors. Cross-sections are given in barns (10^{-28} m^2). For fissionable materials the number of neutrons produced per fission and denoted by v are given as average values over the three spectral regions.

In systems which are not well moderated, such as fast nuclear reactors, the above division into three energy regions is not appropriate. There are virtually no thermal neutrons and no region at intermediate energies where the flux is proportional to E^{-1}. Detailed calculations require the neutron energy region to be split into a large number of groups with appropriate equations and corresponding cross-sections for each group and each nuclide present. For rough estimates and comparisons the average cross-sections given in the table on page 432 can be used for systems typical of a plutonium-uranium oxide fuelled, liquid metal cooled fast reactor with a fertile blanket.

Neutrons can be detected by both active and passive devices. Active detectors include fission chambers, BF$_3$ ion chambers, proton recoil counters and their response to neutrons is determined by the appropriate cross-section for the particular interaction employed in the device. The table of neutron cross-sections for reactor materials on pages 429 and 430 includes materials used in these detectors. The passive detectors contain materials which become radioactive when irradiated in a neutron flux and on removal from the flux the radioactivity can be measured. The activity $A(t)$

(particles s^{-1}) at a time t seconds after the removal of a thin foil from a neutron flux having an energy spectrum $\varphi(E)$ in which the foil has been irradiated for T seconds is given by

$$A(t) = \frac{V\rho N_A}{A}\,(1 - \exp(-\lambda T))\exp(-\lambda t)\int \sigma_{act}(E)\varphi(E)\,dE$$

where V, ρ and A are the volume, density and atomic weight of the foil material respectively, N_A is Avogadro's constant, λ (s^{-1}) is the decay constant of the radioactive species produced by the irradiation and $\sigma_{act}(E)$ is the activation cross-section of the material. Values of σ_{act} averaged over the thermal and fission energy regions are given in the table together with the activation resonance integral. The materials listed either have a predominant resonance or a threshold and may be used to detect neutrons over a narrow energy region. Additional decay properties of the isotopes listed may be found in the Table of Nuclides (p. 341).

The values in both tables are mostly taken from the UK Nuclear Data Library and the average values have been computed by A. L. Pope.

Activation cross-section for neutron foil detectors

Nuclide and reaction	Half-life of principal activity	Principal resonance energy (R) or mean response energy in fission spectrum $(T.D.)/eV$	Activation cross-section/b		
			Average over thermal spectrum	Resonance integral	Average over fission spectrum
^{164}Dy(n, γ)^{165}Dy . .	2.35 h	-1.89 (R)	2100†	300†	98 m†
^{176}Lu(n, γ)^{177}Lu . .	6.74 d	0.142 ,,	3120†	900†	
^{103}Rh(n, γ)^{104m}Rh .	4.4 min	1.26 ,,	9.8†	83†	15 m†
^{115}In(n, γ)^{116m}In .	54 min	1.46 ,,	142†	2600†	120 m†
^{197}Au(n, γ)^{198}Au .	2.7 d	4.9 ,,	87.9†	1560†	97 m
^{152}Sm(n, γ)^{153}Sm .	47 h	8.0 ,,	190†	3000†	
^{107}Ag(n, γ)^{108}Ag .	2.3 min	16.5 ,,	27†	74†	96 m†
^{186}W(n, γ)^{187}W .	23.7 h	18.8 ,,	37†	500†	41 m†
^{59}Co(n, γ)^{60}Co .	5.27 yr	130 ,,	33.2	71	3.1 m
^{55}Mn(n, γ)^{56}Mn .	2.58 h	337 ,,	11.7	14.1†	3.8 m
^{63}Cu(n, γ)^{64}Cu .	12.8 h	580 ,,	4.0	3.2	12 m
^{23}Na(n, γ)^{24}Na .	15.4 h	2.8 k ,,	0.47	0.31	0.26 m
^{115}In(n, n')^{115m}In .	4.5 h	2.8 M (T.D.)	0	70 m	173 m
^{31}P(n, p)^{31}S . . .	2.63 h	3.7 M ,,	0	0	34 m
^{32}S(n, p)^{32}P . . .	14.2 d	4.2 M ,,	0	0.087 m	65 m
^{58}Ni(n, p)^{58}Co . .	71.3 d	4.3 M ,,	0	9.7 m	109 m
^{27}Al(n, p)^{27}Mg . .	9.4 min	5.8 M ,,	0	0	3.8 m
^{56}Fe(n, p)^{56}Mn . .	2.58 h	7.4 M ,,	0	0	0.81 m
^{27}Al(n, α)^{24}Na . .	15 h	8.5 M ,,	0	0	0.62 m
^{63}Cu(n, 2n)^{62}Cu .	12.8 h	13.7 M ,,	0	0	0.05 m

(R) = Resonance Detector.   (T.D.) = Threshold Detector.
m indicates millibarns.
k and M represent keV and MeV respectively.
All values from UK Nuclear Data Library except those marked †.

Neutron cross-sections

Nuclide or element	Average over thermal spectrum				Slowing-down region				Average over fission neutron spectrum					
	σ_n/b	σ_γ/b	$\bar{\sigma}_f$/b	$\bar{\nu}$	σ_n/b	Capture R.I./b	Fission R.I./b	$\bar{\nu}$	$\bar{\sigma}_n$/b	σ_γ/b	σ_f/b	σ_n/b	$\bar{\nu}$	$\bar{\sigma}_A$/b
Fissile materials														
233U	11.6†	42.2†	472.6†	2.487†	10.3	144	740	2.487	4.44	0.043	1.89	1.08	2.673	
235U	17.0†	86.3†	502.2†	2.432†	10.7	139	264	2.424	4.73	0.081	1.24	1.46	2.658	
239Pu	8.6†	271.9†	693.2†	2.880†	11.5	171	280	2.882	4.41	0.049	1.78	1.79	3.169	
241Pu	12.0†	328.7†	936.1†	2.934†	10.8	165	555	2.934	4.76	0.040	1.66	1.33	3.24	
Fertile materials														
238U	7.94	2.42			19.4	269	0.17	2.62	5.06	0.087	0.300†	2.38	2.81	
240Pu	2.12	256			57.1	8011	2.41	3.055	5.26	0.0907	1.24	1.40	3.30	
232Th	12.1	6.53			10.6	88†	0.035	2.28†	4.75	0.12	0.074	2.4	2.28†	
234U			15		11.4	665†	1.78	2.46	5.14	0.028	1.15	1.46	2.62	
236U					11.3	320†	0.51	2.60	5.09	0.17	0.57	1.94	2.77	
238Pu	13.8	323			11.1	150†	72	2.88	4.7	0.079	2.19	0.63	3.11	
233Pa	11.0	36.8			11.1	918	1.3	2.62	5.0	0.12	0.96	2.17	2.78	
237Np		170†				720†	2.1				1.35			
Cladding and structural materials														
Al	1.4	0.20			2.3	0.18			2.9					4.8 m
Si	2.1	0.16†			2.5	0.010†			3.0					9.3 m
Cr	4.1	2.7			5.1	1.4			2.8					3.1 m
Mn	1.7	11.7			41	14.1†			2.8					5.8 m
Fe	11.3	2.26			10.3	1.28			2.5					7.9 m
Co		33.2				71								3.0 m
Ni	17.4	3.42			16	2.03			3.1					8.5 m
Cu	7.0	3.3			8.4	4.1			2.9					28 m
Zr	6.2	0.16			7.8	1.0			5.3					6.3 m
Nb	5.7	1.0			6.9	11			4.5					30 m
Mo	4.5	2.4			7.9	38			4.6					30 m
Ta	5.1	18			15	720			4.1					100 m
W	5.0	16			6.6	373†			4.6					51 m
Pb	8.6	0.15			10	0.20			5.6					2.9 m

Nuclide or element	Average over thermal spectrum				Slowing-down region				Average over fission neutron spectrum				
	$\bar{\sigma}_n$/b	$\bar{\sigma}_A$/b	$\bar{\sigma}_f$/b	$\bar{\nu}$	σ_n/b	Absorption R.I./b	Fission R.I./b	$\bar{\nu}$	$\bar{\sigma}_n$/b	$\bar{\sigma}_\gamma$/b	$\bar{\sigma}_f$/b	$\bar{\sigma}_{n'}$/b	$\bar{\nu}$
Components of coolants													
H in H₂O	45.3	0.294			17.2	0.1416			4.00	0.039 m		0	
D in D₂O	5.10	0.46 m			3.27	0.24 m			2.56	0.007 m		0	
He	0.73	6.1 m			1.12	3.0 m			3.66	0.001 m		0	
C	4.69	3.0 m			4.42	1.5 m			2.35	0.62 m		0.011	
N	11	1.69			7.7	0.90			1.9	0.13		5.8 m	
O	3.35	0.28 m			3.74	0.13 m			2.8	7.1 m		3.1 m	
Na	3.3	0.47			8.8	0.31			2.72	2.1 m		0.48	
K	2.2	1.8			2.2	120			2.6	0.1		0.10	
Absorbers and poisons													
⁶Li	0.72	833			0.99	405							
Li	1.04	63			1.2	30							
¹⁰B	2.13	3398			2.3	1640							
B	4.4	672			4.2	325							
Cd	10.1	2900			6.8	61							
¹³⁵Xe	3.8×10^5	2.7×10^6			560	5890							
Hf	8†	84†			16†	2000†							

m indicates that the cross-section is in mb (10^{-31} m^2).

From UK Nuclear Data Library except values marked †.

Values marked ‡ are recommended thermal cross-sections from the 1969 I.A.E.A. fit (G. C. Hanna et al., *Atomic Energy Review*, **7**, No. 4, p. 3).

Average neutron cross-sections in a fast reactor core and blanket

Material	Core				Blanket			
	σ_{tr}/b	σ_c/b	σ_f/b	ν	σ_{tr}/b	σ_c/b	σ_f/b	ν
^{10}B	4.8	2.3			6.0	3.6		
C	3.6	0.1 m			3.9	0.05 m		
O	3.5	0.6 m			3.6	0.3 m		
Na	3.7	1.4 m			3.9	1.9 m		
Cr	4.2	0.014			4.7	0.019		
Fe	3.8	8.6 m			4.5	10 m		
Ni	8.2	0.021			10.6	0.026		
Mo	7.2	0.11			7.6	0.16		
^{232}Th	9.2	0.36	8.9 m	2.30	10.7	0.58	4.5 m	2.28
^{233}U	8.8	0.22	2.6	2.52	10.5	0.32	3.2	2.51
^{235}U	9.7	0.49	1.8	2.45	11.6	0.74	2.3	2.43
^{238}U	9.4	0.25	0.041	2.73	10.3	0.34	0.021	2.71
^{239}Pu	9.5	0.40	1.7	2.93	11.2	0.73	2.0	2.90
^{240}Pu	9.4	0.42	0.37	3.06	10.8	0.67	0.25	3.02
^{241}Pu	10.1	0.40	2.3	2.98	12.0	0.63	3.0	2.95
^{242}Pu	9.7	0.29	0.29	3.02	11.8	0.51	0.18	2.99
Fission Product pair	12.7	0.39			14.0	0.65		

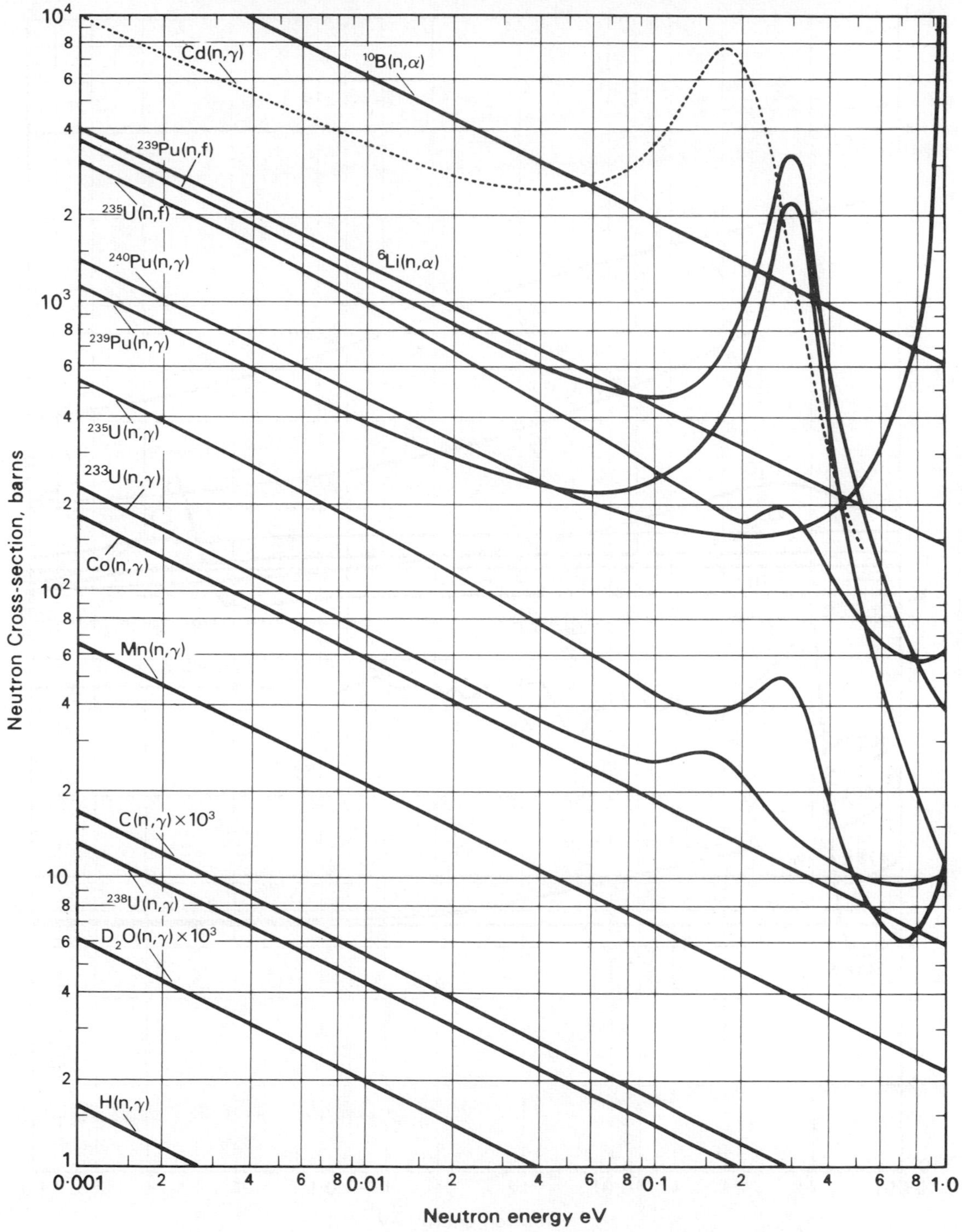

A   Absorption, capture and fission cross-sections in thermal region

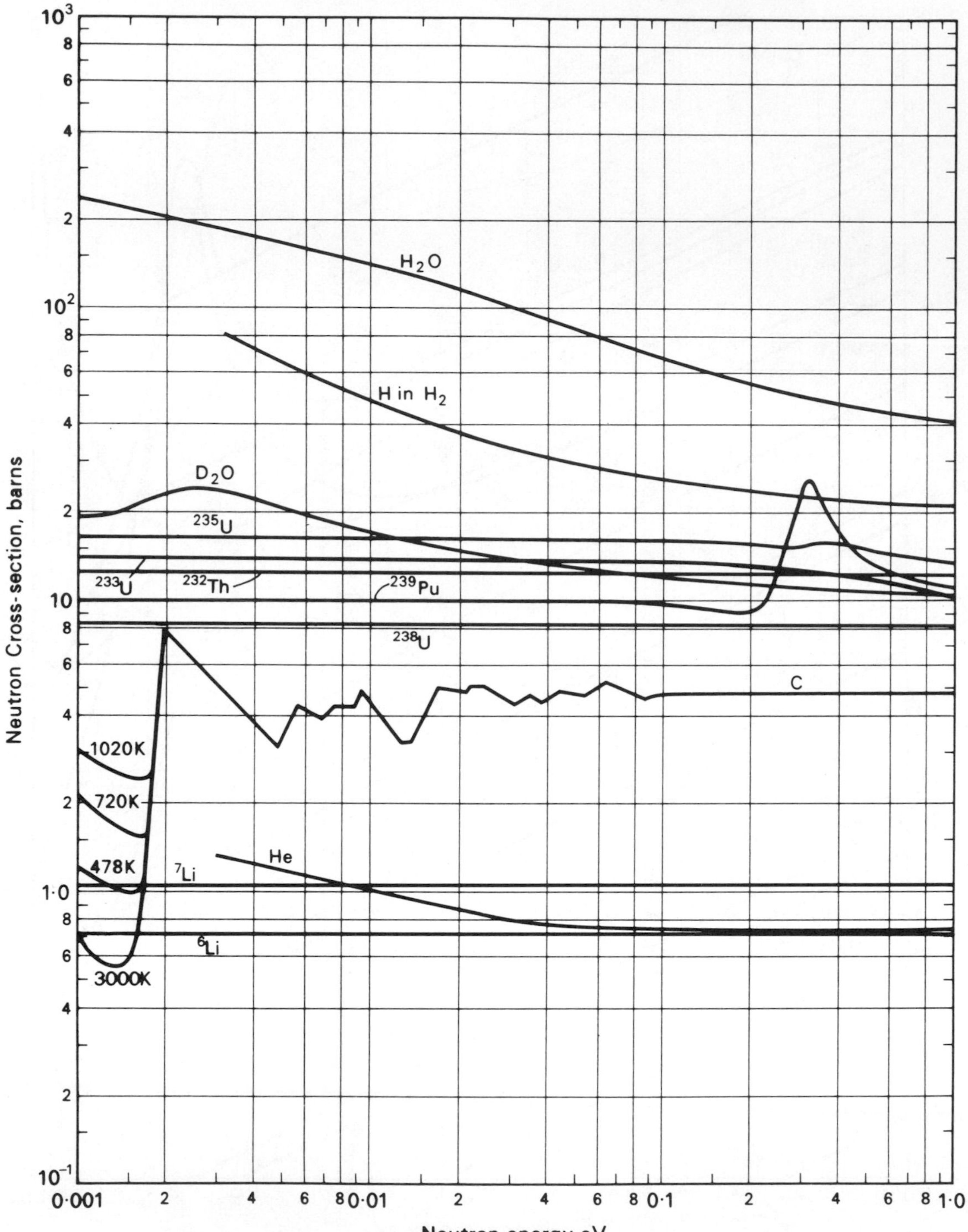

B   Elastic scattering cross-sections in thermal region

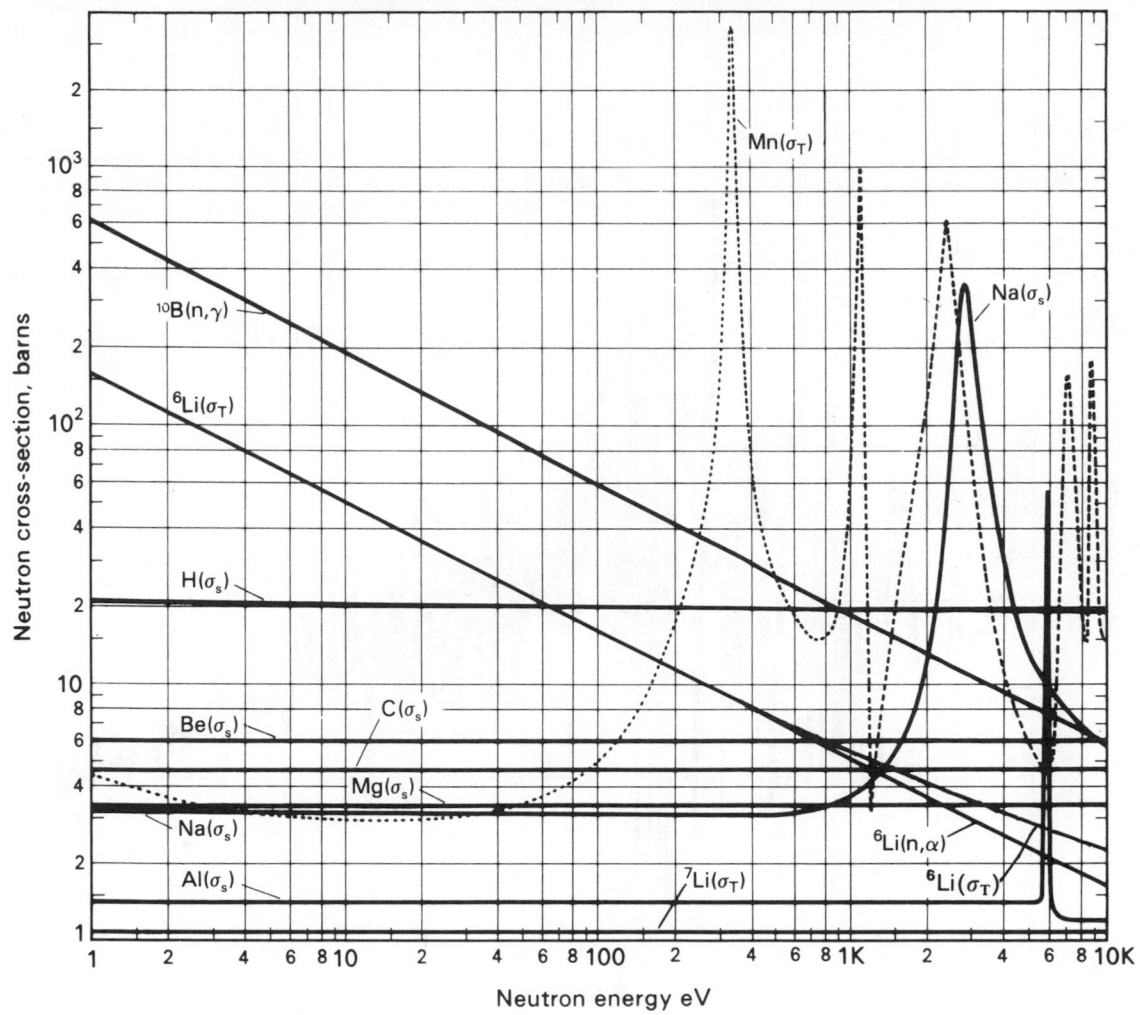

C   Cross-section of light nuclides in the slowing-down region

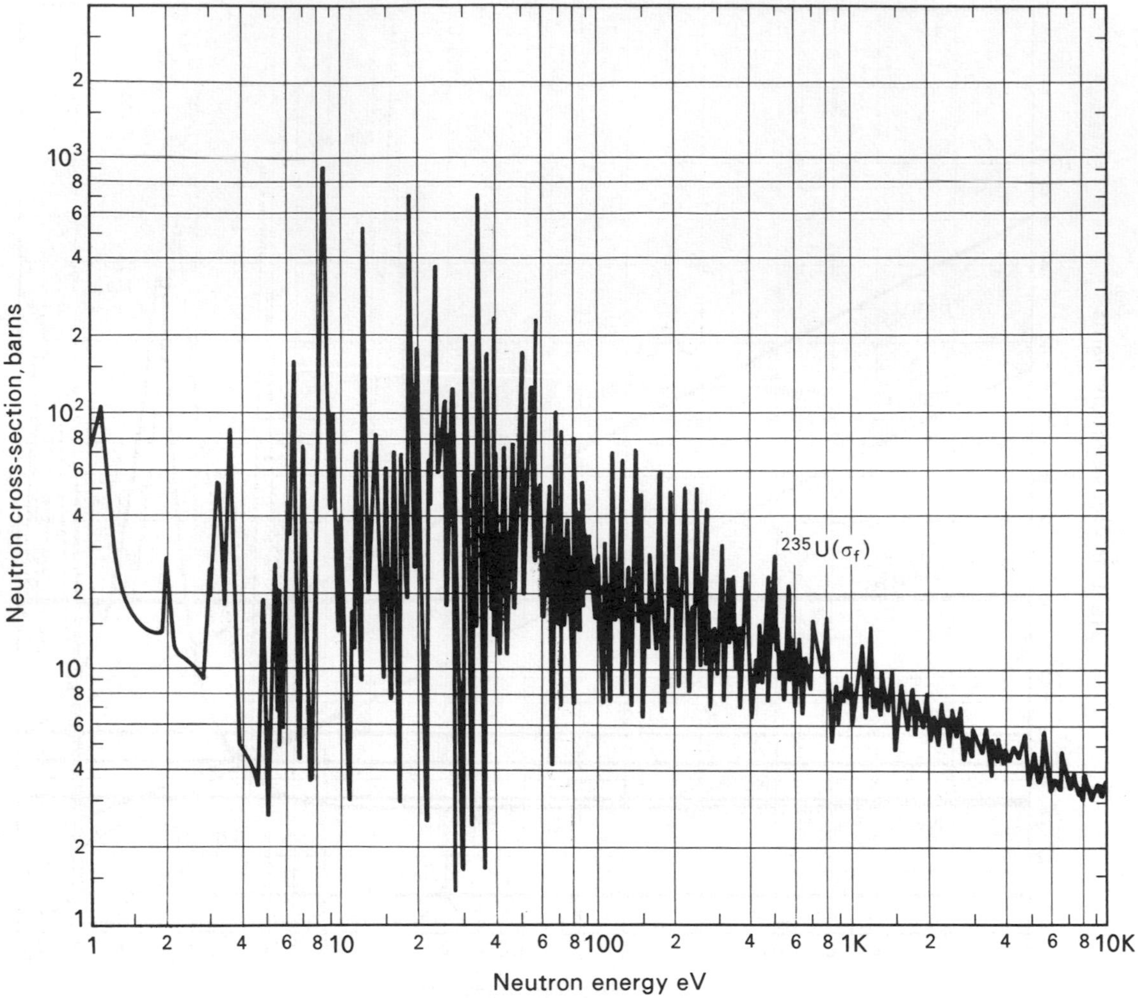

D   Cross-section of a heavy nuclide in the slowing down region.

(Above 20 eV the resonances are too numerous to be shown correctly on this scale.)

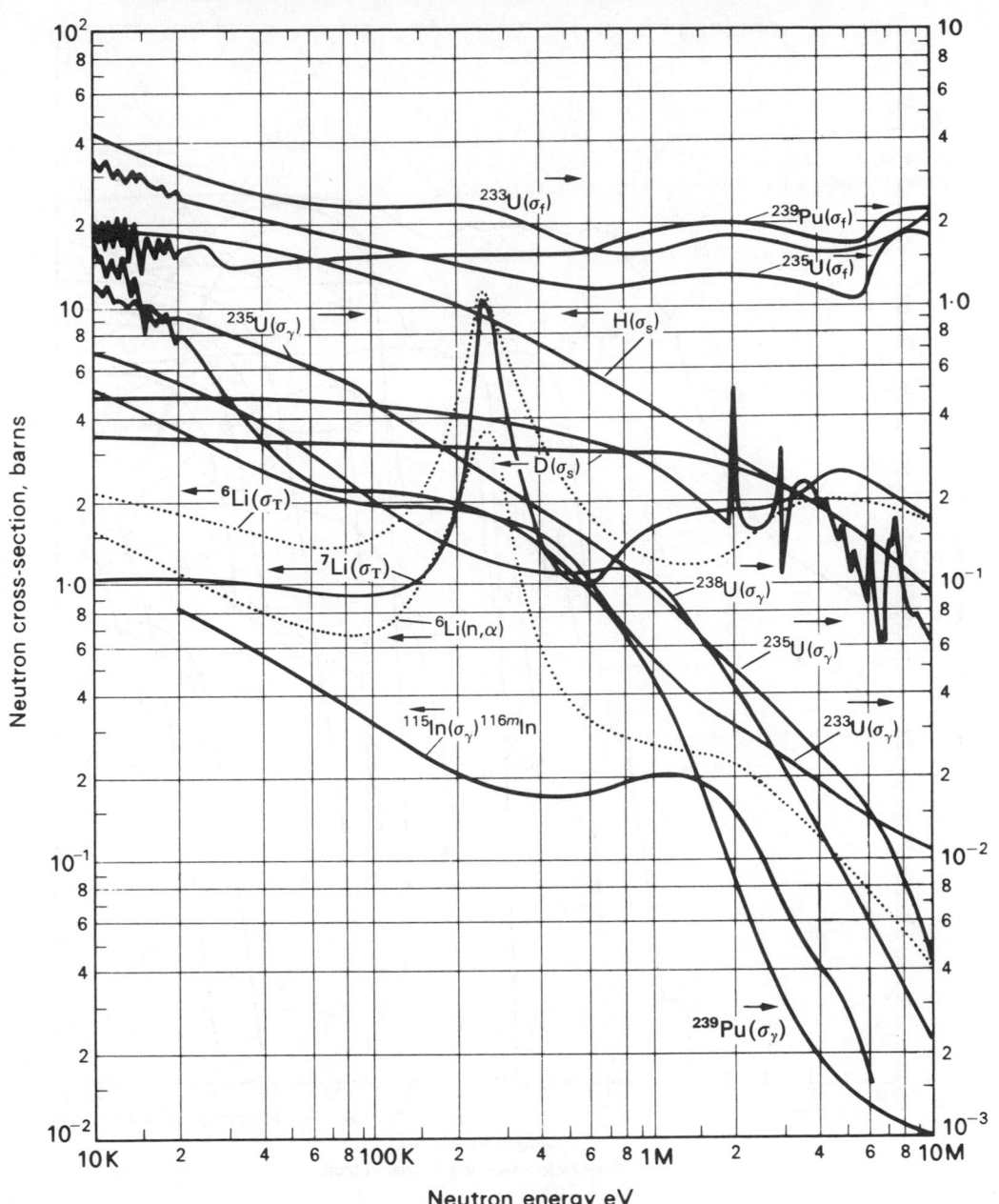

E   Non-threshold cross-sections in fast region
(Note different scales on left and right.)

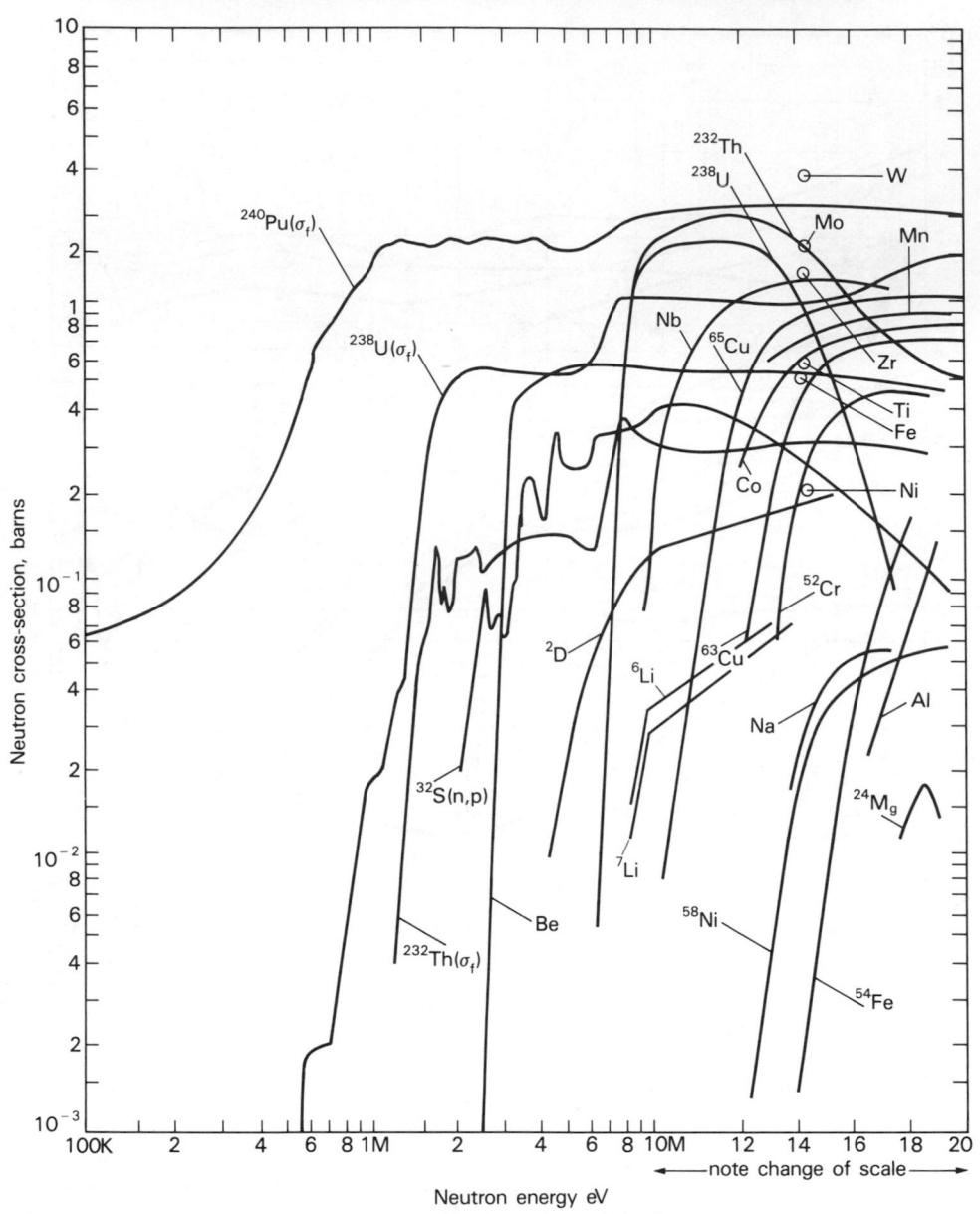

F Threshold reactions in fast region

F   Threshold reactions in the fast region
Curves refer to (n, 2n) reactions unless stated otherwise.
○ indicates spot values of (n, 2n) reactions at 14 MeV.

D.J.

3.7.3 Attenuation of fast neutrons: neutron moderation and diffusion

In materials containing atoms of low atomic mass, neutrons of all energies can lose a significant fraction of their energy in a single elastic collision and such materials are referred to as moderators. In heavy nuclei appreciable energy loss in a collision is only possible at high energies where inelastic scattering can occur. The neutron dose rate from a point source of fast neutrons falls off with distance r

approximately as $\exp{(-\Sigma_{rem}\,r)}/4\pi r^2$, where Σ_{rem} depends on the medium where Σ has been defined earlier. This macroscopic cross-section is called the removal cross-section and since all interactions tend to remove energy from the beam its value is not too different from the total macroscopic section $(N\sigma_{tot})$ of the material, but is slightly lower. This exponential fall off is only approximate and holds less well for media in which hydrogen is the principal fast neutron attenuator. In the following table the removal cross-section refers to a fission neutron source.

In the slowing-down region the average number of collisions, $\bar{n}$, to slow a neutron from energy E_1 to energy E_2 is equal to $\ln{(E_1/E_2)}/\xi$, where ξ is the average change per collision in the logarithm of the energy. At energies below that at which scattering becomes entirely elastic, ξ is independent of energy and is approximately equal to $2/(A+\frac{2}{3})$. The spatial distribution of neutrons of energy E_2 which have slowed down from a point source of energy E_1 is of the form $\exp{(-r^2/4\tau)^2}$ where τ is referred to as the Fermi Age and is the mean square distance a neutron migrates in slowing down from E_1 to E_2. It is given by:

$$\tau = \int_{E_2}^{E_1} \frac{D\;\mathrm{d}E}{E\xi\Sigma_s}$$

where D is the diffusion coefficient and equal to $(3\Sigma_{tot} - 3b\Sigma_s)^{-1}$ and b is the average value of $\cos\psi$ where ψ is the angle of scatter of a neutron in a collision. The table refers to the age of neutrons from a fission source slowing down to an energy of 1.46 eV. This value, which is just above the thermal region, is appropriate to the age determined from the measured spatial distribution of the resonance neutrons detected by indium foils.

The root mean square distance a neutron travels from the position where it is thermalized to the point where it is absorbed is the thermal diffusion length, L, and is equal to

$$(4/\pi)^{1/4}(D_{th}/\Sigma_a)^{1/2}$$

where D_{th} is the value of the diffusion coefficient averaged over the thermal neutron spectrum and Σ_a is assumed to have a $1/v$ dependence and is evaluated at an energy kT_n where T_n is the temperature of the medium.

Properties of moderators and shielding materials†

(At 20 °C unless stated otherwise)

Material	Density $\overline{10^3\,\mathrm{kg\,m^{-3}}}$	$\Sigma_{rem}/\mathrm{m^{-1}}$	ξ	$\tau/(10^3\,\mathrm{mm^2})$	L/mm	D_{th}/mm	$t_s/\mu s$	$t_{th}/\mu s$	$\bar{n}$
H_2O	1.00	9.0[a]	0.948	2.67[c]	27[a]	1.4[a]	6	205	20
D_2O (pure)	1.10	9.1[a]	0.570	11.7[c]	940	8.4	53	~10[5]	33
Diphenyl $(C_{12}H_{10})$ 85 °C .	0.99	7.1[b]	0.812	4.6[d]	48	2.6	13	354	23
Paraffin Wax $(C_{30}H_{62})$.	0.89	10.9[b]	0.913	1.8	21	1.1	7	160	21
Be	1.85	13.0[a]	0.209	7.32[e]	208[a]	5.0[a]	50	3460	90
BeO	3.00	14.3[b]	0.173	9.38[e]	290[a]	5.0[a]	102	7 000	109
Graphite	1.67	8.1[a]	0.158	29.8[c]	520[a]	8.5[a]	140	13 000	119
Concrete‡	2.3	8.8[b]	0.55	10.0	77	6.0	30	400	30
Al	2.70	7.9[a]	0.072	430	200	55	900	8 800	262
Fe	7.86	16.8[a]	0.035	33.0	12.7	3.4	360	19	540
Pb	11.35	11.6[a]	0.0096	600	121.8	9.2	2720	640	1960
Bi	9.75	9.8[a]	0.0095	800	320	11.2	3000	3 660	1990
U	18.9	1.7[a]	0.0084	§	13.7	7.0	2040	11	2250

† Data obtained from various sources: a, experimental value; b, derived from components; c, UK Nuclear Data File; d, taken from ANL 5800; e, ENDF/B data (this is a US data set—see BNL 50274). Values not marked are obtained from approximate formulae.

‡ Composition in 10^3 kg m^{-3}: H, 0.023; O, 1.22; C, 0.0023; Mg, 0.005; Al, 0.078; Si, 0.775; K, 0.03; Ca, 0.1; Fe, 0.03; Na, 0.037.

§ Because of its large absorption resonance integral and its small value of ξ, almost no neutrons slowing down in uranium reach thermal energies.

The slowing down time, t_s, is the time taken for neutrons to slow down from an energy E_s to the thermal energy E_0 and is independent of E_s when $E_s \gg E_0$. The mean thermal neutron lifetime, t_{th}, refers to thermal diffusion before the neutron is absorbed in the medium. In the table values of t_s and t_{th} have been obtained from the simple formulae $t_s = 2(\overline{1/\zeta\Sigma_s})/v_0$ and $t_{th} = 1/(\overline{\Sigma_a v})$ where the averages are taken over the slowing-down and thermal regions respectively.

D.J.

3.7.4 Nuclear fusion data

Fusion reactions

The first generation of fusion reactors producing electrical energy from nuclear fusion reactions between light ions will almost certainly exploit the D-T reaction with the ion plasma being confined magnetically, although a successful inertial confinement reactor cannot be ruled out. The D-T reaction is specified because of its high cross-section at low ion kinetic energies and its large energy release. However, even in a plasma containing equimolar quantities of deuterium and tritium there will occur fusion reactions between like as well as unlike ion species and between the energetic charged particle reaction products and the fuel ions. The important fusion reactions are listed below. Particle energies (in MeV) are quoted for the two-particle exothermic reactions, calculated relativistically for zero energy reactants; the total energy release is given for the other reactions.

Fusion reactions of CTR interest

1a. $D + D \xrightarrow{50\%} T (1.011) + p(3.002)$
1b. $D + D \xrightarrow{50\%} {}^3\text{He} (0.820) + n(2.449)$
2. $p + T \longrightarrow {}^3\text{He} + n - 0.764$
3. $D + T \longrightarrow {}^4\text{He} (3.560) + n(14.029)$
4. $T + T \longrightarrow {}^4\text{He} + 2n + 11.332$
5. $D + {}^3\text{He} \longrightarrow {}^4\text{He} (3.713) + p(14.640)$
6a. $T + {}^3\text{He} \xrightarrow{59\%} {}^4\text{He} + n + p + 12.096$
6b. $T + {}^3\text{He} \xrightarrow{41\%} {}^4\text{He} (4.800) + D(9.520)$
7. ${}^3\text{He} + {}^3\text{He} \longrightarrow {}^4\text{He} + 2p + 12.860$

Fusion reaction cross-sections

Plasma reactivity calculations require reaction cross-sections for energies well below those at which direct measurements are practicable, so it is necessary to extrapolate downwards using the theoretical formula

$$\sigma(E) = \frac{S(E)}{E} \exp(-R/\sqrt{E}) \text{ with } R = \pi \left(\frac{e^2}{\hbar c}\right) \sqrt{2mc^2} \, Z_1 Z_2,$$

where the cross-section is expressed in centre-of-mass units, $E = \frac{1}{2}mv^2$, $m = m_1 m_2/(m_1 + m_2)$ and v is the relative velocity of the interacting particles which have masses m_1 and m_2 and charges Z_1 and Z_2 respectively. The constants e, $\hbar$ and c have their usual meaning. $S(E) = A \exp(-\beta E)$ and the parameters A, β and R are given in the table below.

Low-energy cross-section parameterization

Reaction	A (barns–keV)	β (keV^{-1})	R (keV$^{1/2}$)
$D - D_p$	52.6	-5.8×10^{-3}	31.39
$D - D_n$	52.6	-5.8×10^{-3}	31.39
$D - T$	9821	-2.9×10^{-2}	34.37
$T - T$	175	9.6×10^{-3}	38.41
$D - {}^3He$	5666	-5.1×10^{-3}	68.74
$T - {}^3He$	2422	4.5×10^{-3}	76.82
${}^3He - {}^3He$	5500	-5.6×10^{-3}	153.7

Note that laboratory energies may be used if the substitution $E = (m/m_1)E_{\text{lab}}$ is made.

The theoretical formula quoted above applies only for energies well below the Coulomb barrier. For higher energies a more complex formula is required, such as that proposed by Peres (*J. Appl. Phys.*, 1979, **50**, 5569). For a recent review of the fusion cross-section data, see Jarmie (*Nucl. Sci. and Eng.*, 1981, **78**, 404). Cross-sections for the primary fusion reactions are plotted in Fig. 1, as a function of projectile energy (E_{lab}).

Fusion reaction rates

The reaction rate, r, between two species of ion with densities n_i and n_j in a plasma is given by $r = n_i n_j$ $(1 + \delta_{ij})^{-1} <\sigma v>$, where $<\sigma v>$ is the appropriate average of the fusion cross-section σ over the relative velocities v and δ_{ij} is The Krönecker delta function. The relative ion energy distribution in a plasma is customarily taken to be of Maxwellian form,

$$\mathcal{N}(E) = 2\pi^{-1/2}\theta^{-3/2}E^{1/2}\exp(-E/\theta),$$

where E is the relative energy and $\theta = kT$, so that

$$<\sigma v> = (8/\pi)^{1/2}\,m^{-1/2}\,\theta^{-3/2}\int_0^\infty E\sigma(E)\exp(-E/\theta)\mathrm{d}E.$$

For accurate calculations involving plasma reactivities, a numerical integration of the σv product over energy is required. Miley, Towner and Ivich (Coo-2218-17, 1974) have prepared a set of useful reactivity tabulations. Reactivity values for the primary fusion reactions are plotted in Fig. 2.

It is frequently convenient to utilize analytic expressions for $<\sigma v>$. Provided the cross-sections have the simple form $\sigma(E) = (A/E)\exp(-\beta E - R/\sqrt{E})$, the integration can be performed approximately by saddle-point evaluation to give

$$<\sigma v> = 0.8052 \times 10^{-22}\frac{AR^{1/3}\,m^{-1/2}\,\theta^{-2/3}}{(1 + \beta\theta)^{5/6}}\cdot\exp\left[-3\left(\frac{R^2}{4}\right)^{1/3}(1 + \beta\theta)^{1/3}\,\theta^{-1/3}\right]$$

in $m^3\,s^{-1}$, where A is in barns, θ in keV and m is the relative mass in a.m.u. Using the values of A, R and β from section 2, $<\sigma v>$ values for temperatures well below the Coulomb barrier may be calculated. It is customary to neglect the terms involving β. Fowler, Caughlan and Zimmerman (*Ann. Rev. Astron. and Astrophys.*, 1975, **13**, 69) present a comprehensive set of stellar thermonuclear reaction rate formulae applicable over a very wide range of temperatures. Particularly simple formulae for the four main fusion reactions are given by Hively (*Nuclear Technology/Fusion*, 1983, **3**, 199).

Thermonuclear neutron energy spectra

In a Maxwellian plasma containing two sets of particles at a common temperature θ, the mean energy of the centre-of-mass of colliding pairs is

$$<C> = \tfrac{1}{2}(m_1 + m_2)<V^2> = \tfrac{3}{2}\theta$$

and the mean relative kinetic energy is

$$<K> = \tfrac{1}{2}\left(\frac{m_1 m_2}{m_1 + m_2}\right)<v^2> = \tfrac{3}{2}\theta$$

where V is the centre of mass velocity and v is the relative velocity. The mean relative energy of reacting particles is obtained by weighting with the reactivity σv and Brysk (*Plasma Physics*, 1973, **15**, 611) has shown that there exists a convenient relationship

$$<K_r> \equiv <K\sigma v>/<\sigma v> = \theta^2(\mathrm{d}/\mathrm{d}\theta)\{\ln(<\sigma v>\theta^{3/2})\}.$$

Use of the simple Gamow cross-section weighted by a Maxwellian energy distribution

$$<\sigma v> = F\theta^{-2/3}\exp(-G\theta^{-1/3})\quad\text{gives}\quad <K_r> = \tfrac{1}{3}G\theta^{2/3} + \tfrac{5}{6}\theta.$$

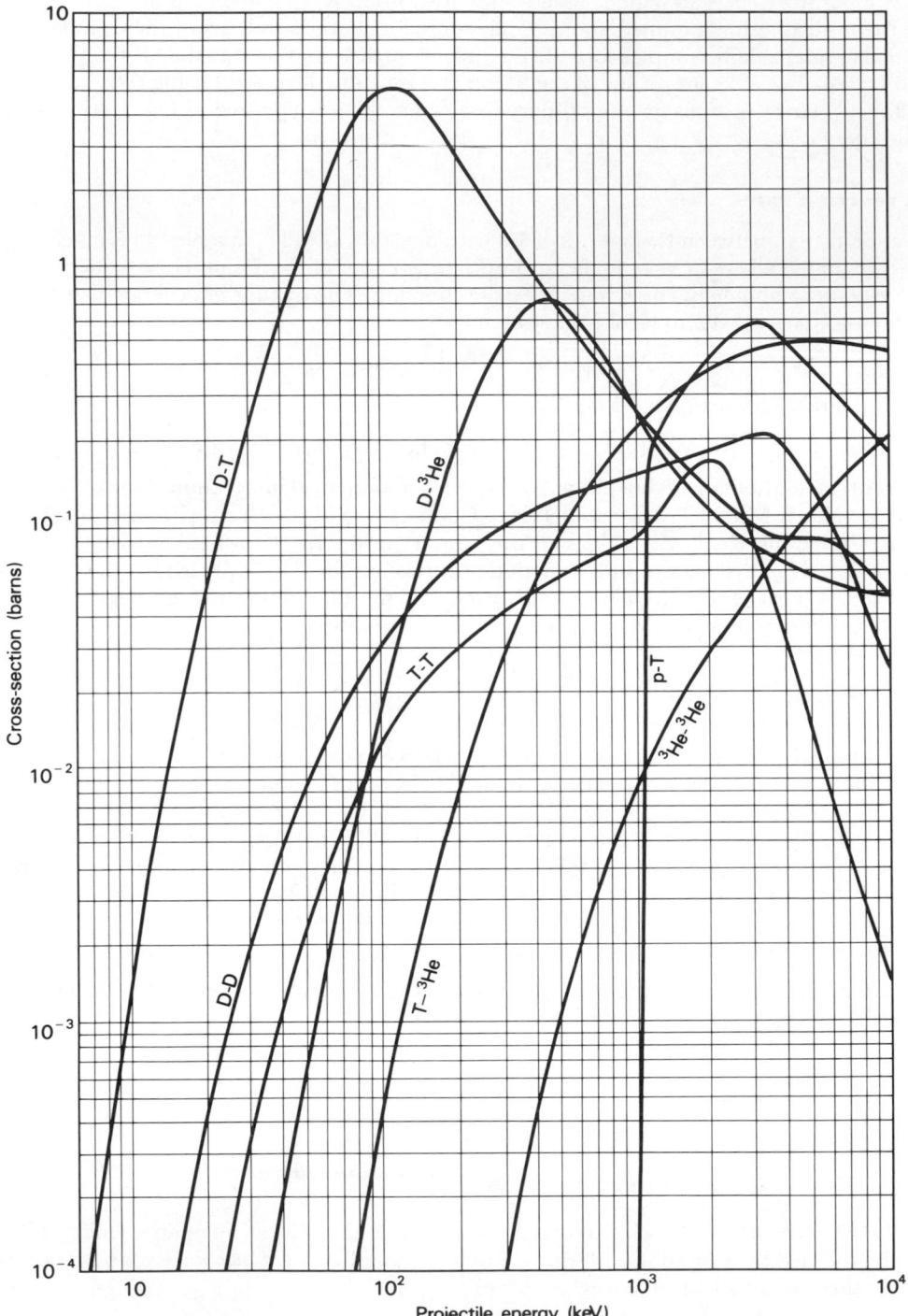

Fig. 1 Fusion reaction cross-sections.

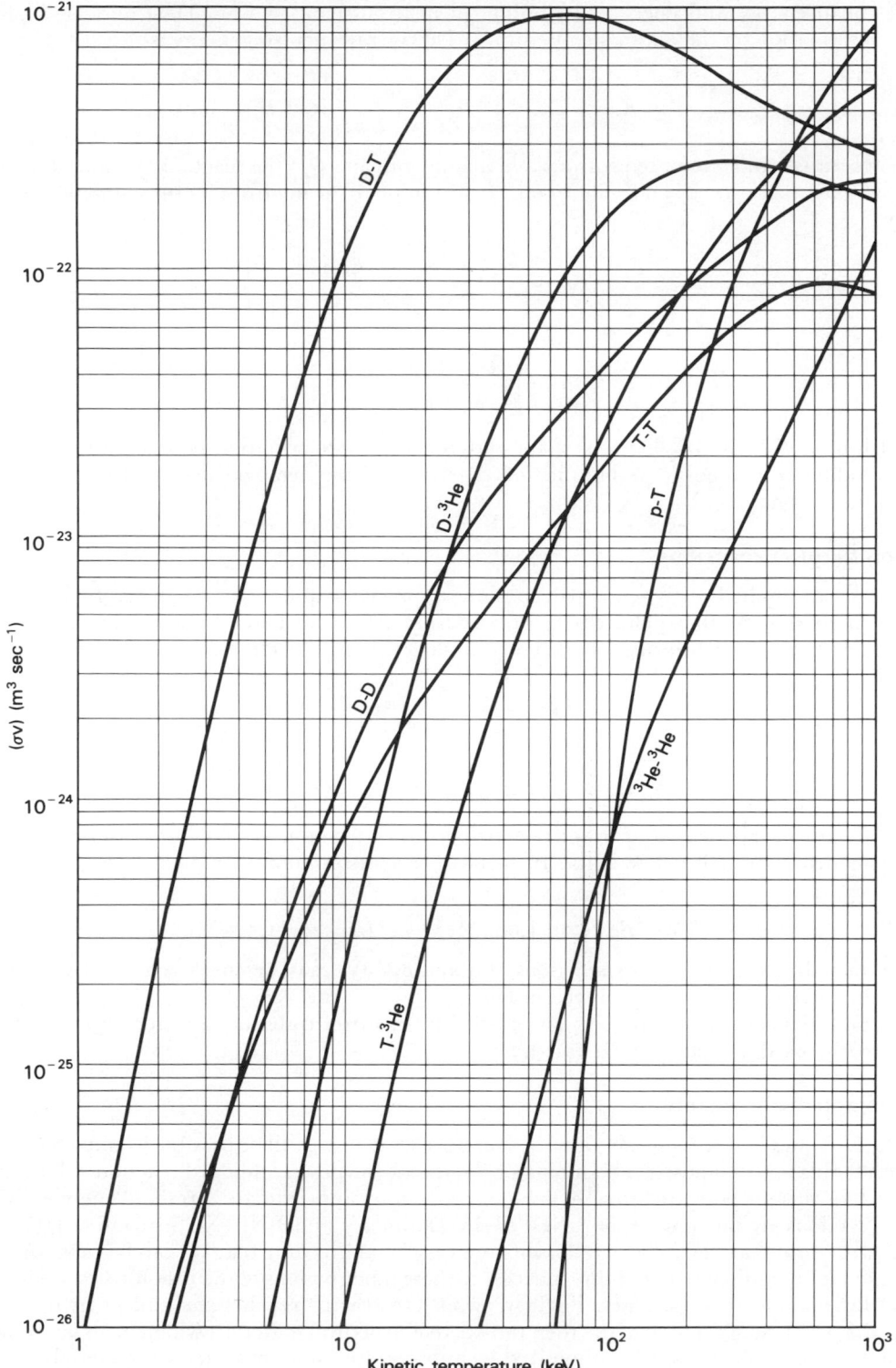

Fig. 2 Fusion reactivities.

The mean energy and energy distribution of neutrons emitted from D–D and D–T plasmas provide information on the plasma temperature. The average neutron energy is

$$< E_n > = \tfrac{1}{2} m_n < V^2 > + \frac{m_\alpha}{(m_n + m_\alpha)} (Q + K_r)$$

where subscripts n and α refer to neutrons and helium ions and Q is the reaction Q-value. The shift in $< E_n >$ with temperature is small. The energy distribution of the neutrons is approximately Gaussian in form

$$f(E_n) = \frac{1}{W \sqrt{\pi}} \exp[-\{E_n - < E_n >\}^2 / W^2]$$

where

$$W = \left[\frac{4 m_n < E_n > \theta}{m_n + m_\alpha} \right]^{1/2}.$$

The distribution *fwhm* is $2W\sqrt{\ln 2}$. The shape and spread of the distribution are measurable quantities, thus providing a diagnostic technique for determining the temperature of an approximately Maxwellian plasma.

Neutron blanket reactions

A reactor based on the D–T reaction will consume appreciable quantities of tritium. In a pure fusion reactor system, this tritium will have to be derived from nuclear reactions between the 14 MeV fusion neutrons and lithium in a breeding blanket surrounding the reaction chamber. The breeding reactions are

$$n + {}^6\mathrm{Li} \rightarrow {}^4He + T + 4.784\ \mathrm{MeV}$$

and

$$n + {}^7Li \rightarrow n' + {}^4He + T - 2.467\ \mathrm{MeV}.$$

Because of parasitic neutron capture and penetration losses, the number of tritium atoms produced per 14 MeV neutron may fall below unity. In this case a neutron multiplier such as beryllium will be needed:

$$n + {}^9Be \rightarrow {}^8Be + 2n - 1.665\ \mathrm{MeV} \rightarrow 2\ {}^4He + 2n - 1.573\ \mathrm{MeV}$$

Lead, with a threshold for (n,2n) reactions at about 8 MeV, may be employed instead. The cross-sections for these reactions are displayed in Fig. 3.

Reaction and scattering cross-sections needed for neutron transport and moderation considerations are discussed in sections 3.7.2 and 3.7.3.

Neutron activation reactions

Since the D–T and D–D plasmas are intense sources of neutrons it will be possible to employ activation foils both for plasma diagnostic and for reactor dosimetric purposes. Standard reactions which may be employed for this purpose are listed in Table 1, ordered by approximate threshold energies. Most of the cross sections are taken from the ENDF/B–IV Dosimetry File, BNL–NCS–50446 (1975).

The structural materials of which the vacuum vessel and breeding blanket can be constructed will become so highly radioactive that direct access to these inner regions of the fusion reactor will not be possible for maintenance purposes. Further, when the reactor reaches the end of its operational lifetime the problem of disposal of the then radioactive structural materials will have to be considered. These activation problems can be minimized by appropriate choice of structural material. The most important radionuclides produced in each of the chemical elements likely to be employed in a structural material are listed in Table 2.

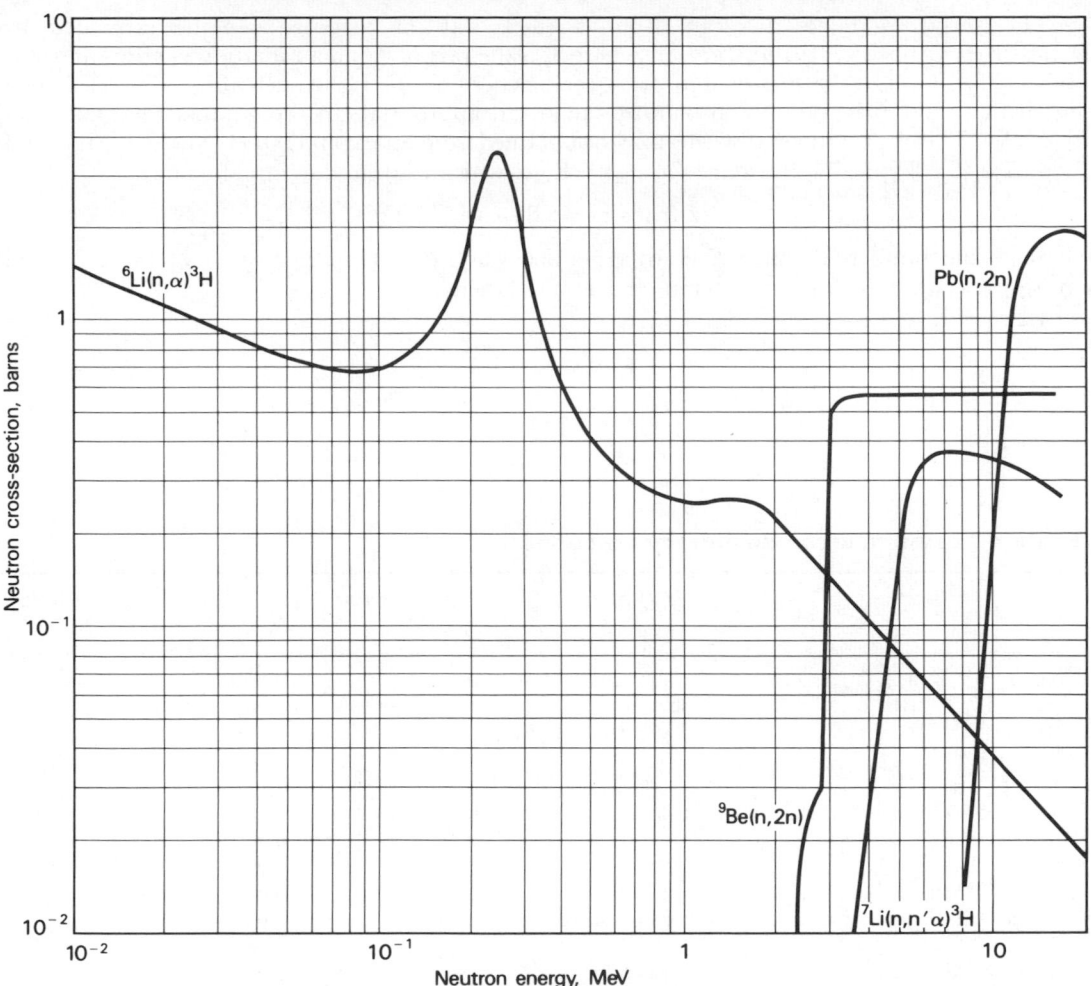

Fig. 3 Neutron cross-sections for tritium breeding.

The induced radioactivity R remaining after time T due to a particular reaction channel is given to first order by

$$R(Bq.kg^{-1}) = (1.32 \times 10^{-9}).\frac{F}{A}.\bar{\sigma}.\varphi.\frac{\exp(-0.69\,T/T_{1/2})}{T_{1/2}}$$

for an irradiation of duration brief compared with the decay time T and radionuclide half-life $T_{1/2}$. F is the fractional abundance of the target nuclide in the irradiated element, A is the atomic weight, $\bar{\sigma}$ is the energy-averaged neutron reaction cross-section in barns, φ is the neutron fluence (n.m^{-2}) and the times are in years. The above formula takes only first generation reaction products into account and ignores the burnup of target and daughter nuclides. For orientation purposes, $\bar{\sigma}$ can be taken to be the reaction cross-section averaged over the energy spectrum of neutrons emitted from a D–T plasma at 20 keV ion temperature as provided in Table 3 for the significant particle-emitting reactions. More detailed cross-section information can be found in the data tape associated with the report by Gardner and Howerton, UCRL–50400 Vol. 18, Lawrence Radiation Laboratory, California (1978).

The surface gamma dose-rate from a thick slab of material can be estimated from $D(\text{rads/h}) = 5.76 \times 10^{-8} (\mu_a/\mu_m)(B/2)S_v$, where S_v is the rate of gamma radiation energy emission (in $\text{MeV kg}^{-1}\text{s}^{-1}$), μ_a is the energy absorption coefficient of air, μ_m is the mass absorption coefficient of the element and B is the photon buildup factor, all appropriate to the gamma radiation energy $E_\gamma(\text{MeV})$. The absorption coefficients may be obtained from Storm and Israel, Nuclear Data Tables, 1970, **A7**, 565. Typically, $B \approx 2$ and $(\mu_a \, \mu_m) \approx E\gamma/2$, which leads to the simplified form

$$D(\text{rads/h}) = 2.88 \times 10^{-8} f R E_\gamma^2$$

where f is the number of quanta E_γ emitted per decay and R is the specific activity. If several gamma emissions occur, then D must be summed over all of them.

Table 1 Fusion neutron dosimetry reactions

Reaction	Half-life	Cross-section (barns)		Threshold (MeV)
		2.5 MeV	14.0 MeV	
$^{93}\text{Nb}(n,n')^{93m}\text{Nb}$	16.4 y	0.127	0.036	0.1
$^{103}\text{Rh}(n,n')^{103m}\text{Rh}$	56.1 min	1.0	0.25	0.1
$^{115}\text{In}(n,n')^{115m}\text{In}$	4.486 h	0.28	0.068	0.4
$^{238}\text{U}(n,f)\text{F.P.}$	—	0.55	1.14	1.0
$^{47}\text{Ti}(n,p)^{47}\text{Sc}$	3.35 d	0.030	0.115	1.0
$^{58}\text{Ni}(n,p)^{58}\text{Co}$	70.8 d	0.12	0.38	1.0
$^{232}\text{Th}(n,f)\text{F.P.}$	—	0.125	0.36	1.2
$^{32}\text{S}(n,p)^{32}\text{P}$	14.29 d	0.080	0.220	1.2
$^{54}\text{Fe}(n,p)^{54}\text{Mn}$	312.5 d	0.05	0.36	1.5
$^{31}\text{P}(n,p)^{31}\text{Si}$	157.3 min	0.02	0.09	1.5
$^{64}\text{Zn}(n,p)^{64}\text{Cu}$	12.7 h	0.05	0.16	1.8
$^{27}\text{Al}(n,p)^{27}\text{Mg}$	9.462 min	—	0.080	2.5
$^{46}\text{Ti}(n,p)^{46}\text{Sc}$	83.83 d	—	0.26	3.0
$^{90}\text{Zr}(n,p)^{90}\text{Y}$	64.0 h	—	0.09	3.5
$^{60}\text{Ni}(n,p)^{60}\text{Co}$	5.27 yr	—	0.13	4.3
$^{56}\text{Fe}(n,p)^{56}\text{Mn}$	2.578 h	—	0.110	4.5
$^{48}\text{Ti}(n,p)^{48}\text{Sc}$	43.7 h	—	0.063	4.5
$^{59}\text{Co}(n,\alpha)^{56}\text{Mn}$	2.578 h	—	0.030	5.0
$^{27}\text{Al}(n,\alpha)^{24}\text{Na}$	15.00 h	—	0.12	5.2
$^{24}\text{Mg}(n,p)^{24}\text{Na}$	15.00 h	—	0.18	5.5
$^{63}\text{Cu}(n,\alpha)^{60}\text{Co}$	5.27 yr	—	0.036	5.5
$^{232}\text{Th}(n,2n)^{231}\text{Th}$	25.52 h	—	1.3	6.5
$^{127}\text{I}(n,2n)^{126}\text{I}$	13.02 d	—	1.6	9.3
$^{65}\text{Cu}(n,2n)^{64}\text{Cu}$	12.7 h	—	0.93	10.1
$^{55}\text{Mn}(n,2n)^{54}\text{Mn}$	312.5 d	—	0.80	10.4
$^{59}\text{Co}(n,2n)^{58}\text{Co}$	70.8 d	—	0.50	10.5
$^{63}\text{Cu}(n,2n)^{62}\text{Cu}$	9.74 min	—	0.53	10.9
$^{90}\text{Zr}(n,2n)^{89}\text{Zr}$	78.4 h	—	0.65	12.1
$^{64}\text{Zn}(n,2n)^{63}\text{Zn}$	38.1 min	—	0.16	12.2
$^{58}\text{Ni}(n,2n)^{57}\text{Ni}$	36.08 h	—	0.026	12.5
$^{47}\text{Ti}(n,np)^{46}\text{Sc}$	83.83 d	—	0.012	13.5
$^{50}\text{Cr}(n,2n)^{49}\text{Cr}$	41.93 min	—	0.01	13.6
$^{54}\text{Fe}(n,2n)^{53}\text{Fe}$	8.51 min	—	0.005	13.9

Table 2 Important contributing radionuclides, by half-life†

Element	$T_{1/2} <$ day	1 day–30 days	30 days–5 years	5–100 years	1000 years $< T_{1/2}$
C	—	—	—	—	^{14}C, ^{10}Be
Mg . . .	^{24}Na	—	^{22}Na	—	—
Al . . .	^{27}Mg, ^{24}Na	—	—	—	^{26}Al
Si	^{28}Al, ^{27}Mg, . ^{31}Si, ^{24}Na	—	^{22}Na	—	^{26}Al
Ti . . .	—	^{48}Sc, ^{47}Sc	^{46}Sc, ^{45}Ca	^{42}Ar	—
V	^{52}V, ^{51}Ti	^{48}Sc	^{48}V, ^{49}V, ^{46}Sc	—	—
Cr . . .	^{52}V	^{51}Cr	^{54}Mn, ^{49}V	—	—
Mn . . .	^{56}Mn	—	^{54}Mn	—	^{53}Mn
Fe . . .	^{56}Mn	—	^{54}Mn, ^{55}Fe	—	^{53}Mn
Co . . .	^{58m}Co	—	^{59}Fe, ^{58}Co	^{60}Co	^{60}Fe
Ni . . .	^{60m}Co	—	^{58}Co, ^{57}Co, ^{55}Fe	^{60}Co, ^{63}Ni	^{59}Ni
Cu . . .	^{62}Cu, ^{64}Cu	—	—	^{60}Co, ^{63}Ni	^{60}Fe
Zn . . .	^{63}Zn, ^{64}Cu	—	^{65}Zn	^{60}Co, ^{63}Ni	^{60}Fe
Zr . . .	—	^{90}Y, ^{89}Zr	^{95}Nb, ^{95}Zr, ^{88}Y	^{90}Sr	^{94}Nb, ^{93}Zr
Nb . . .	^{94m}Nb	^{92m}Nb	—	^{93m}Nb	^{94}Nb, ^{93}Zr
Mo . . .	^{99m}Tc	^{99}Mo, ^{92m}Nb	^{95}Nb	^{93m}Nb	^{93}Mo, ^{94}Nb, ^{99}Tc, ^{98}Tc
W	^{187}W	—	^{185}W, ^{181}W	—	^{182}Hf

† Note the lack of important nuclides with 100 years $< T_{1/2} <$ 1000 years.

Table 3 Reaction cross-sections (in barns), averaged for 14 MeV fusion neutrons

Target element	Reaction type				
	(n,2n)	(n,α)	(n,p)	(n,n′p)	(n,n′α)
C	0.0	0.026	0.002	0.0	0.262
Mg	0.034	0.087	0.149	0.004	0.0
Al	0.035	0.123	0.077	0.340	0.0
Si	0.010	0.109	0.295	0.002	0.0
Ti	0.284	0.034	0.072	0.053	0.001
V	0.630	0.017	0.036	0.063	0.0
Cr	0.298	0.038	0.082	0.109	0.001
Mn	0.774	0.033	0.047	0.152	0.001
Fe	0.427	0.045	0.123	0.072	0.001
Co	0.642	0.029	0.081	0.078	0.001
Ni	0.152	0.095	0.292	0.402	0.016
Cu	0.647	0.030	0.087	0.149	0.008
Zn	0.374	0.018	0.101	0.210	0.001
Zr	0.880	0.009	0.031	0.038	0.001
Nb	0.487	0.009	0.040	0.011	0.003
Mo	0.961	0.023	0.028	0.040	0.001
W	2.140	0.001	0.001	0.001	0.001
Pb	1.979	0.002	0.002	0.001	0.001

O.N.J.

3.8 Nuclei and particles

3.8.1 Size of atomic nuclei

The majority of atomic nuclei are approximately spherical in shape, and it is sufficient for many purposes to use spherically symmetrical distributions to describe them. The mean square radius of the matter distribution is given approximately by $r_0 A^{1/3}$, where $r_0 \simeq 1.2$ fm and A is the mass number. For medium and heavy nuclei the density distribution ρ as a function of distance r from the nuclear centre is customarily approximated by the form

$$\rho = \rho_0 \left(1 + \exp\left(\frac{r-c}{a} \right) \right)^{-1}$$

where c, the radius at which the density drops to $\frac{1}{2}\rho_0$, is $\simeq 1.1 A^{1/3}$ fm, the nuclear surface thickness parameter $a \simeq 0.6$ fm is little dependent on nuclear size, and $\rho_0 \simeq 0.17$ nucleon fm^{-3}.

See A. Bohr and B. Mottelson, *Nuclear Structure*, 1969, Vol. 1 (Benjamin, New York).

B.R.

3.8.2 Subatomic particles

Classification

Subatomic particles are classified according to whether they do or do not respond to the strong nuclear force. Those that do are named 'hadrons', of which the protons and neutron are particular examples, while those that do not respond to the strong force are called 'leptons', and the positive and negative electrons are examples.

The leptons appear to be pointlike even when probed to the highest resolution currently available ($\simeq 10^{-18}$ m). The hadrons are known to be extended objects ($\simeq 10^{-15}$ m diameter) which are built from pointlike particles known as 'quarks'. Five varieties of quark have been identified which are distinguished by their electrical charges and further intrinsic properties named 'strangeness', 'charm', 'bottom' (or 'beauty') which are additive quantum numbers. Thus a hadron containing two charmed quarks will have net 2 units of charm and is referred to as a charmed hadron; the more strange quarks there are contained in a hadron, the more will be its net strangeness. The total electrical charge of a hadron is the sum of the electrical charges of its constituent quarks.

The quarks respond to the strong nuclear force but otherwise behave very much like leptons. This has led to conjectures that leptons and quarks may be fundamental and related to one another, though the precise nature of this relationship is still unclear. Recent evidence for a sixth variety of quark (variously called the 'top' quark or sometimes 'truth') completes a parallelism with the six leptons.

For each variety of particle listed nature also has an antiparticle with the same mass and spin but opposite charge, strangeness and other additive quantum numbers. Their symbols are the same as those for their particle equivalents but for a bar above them, e.g. p is a proton and $\bar{p}$ is the antiproton. The anti-electron is denoted $\bar{e}^-$, or most commonly e^+, and named the 'positron'.

The data on antiparticles are much poorer than for particles. General theorems require that antiparticles have the same mass and lifetimes as their particle equivalents: the poor data are all consistent with this and are not listed here.

In addition, there are particles classified by the family name 'gauge bosons'. These are carriers of the fundamental forces and have spin. The photon is the carrier of the electromagnetic force; the W and Z bosons are the carriers of the weak forces.

Fundamental particles with spin $\frac{1}{2}$

Leptons

Name and symbol		Charge	Mass/MeV	Mean life/s†	Principal decay modes
Electron	e	$-e$	0.511 0034 ± 0.000 0014	stable ($>6 \times 10^{28}$)	stable
Electron-neutrino	ν_e	0	$<4.6 \times 10^{-5}$	stable	stable
Muon	μ	$-e$	105.659 32 ± 0.000 29	$2.197 09 \times 10^{-6}$ ± 0.000 05	$e\nu\bar{\nu}$
Muon-neutrino	ν_μ	0	<0.52	stable	stable
Tau	τ	$-e$	1784.2 ± 3.2	3.4×10^{-13} ± 0.5	Hadron + neutrals $\pi^- \pi^0 \nu$, $\mu\nu\nu$, $e\nu\nu$
Tau-neutrino	ν_τ	0	<74	stable	stable

† The observed mean life of a particle in flight is longer than its mean life at rest by a factor

$$(1 - \beta^2)^{-\frac{1}{2}} = 1 + \frac{\text{Kinetic energy}}{\text{Rest energy}}$$

where β is the ratio of its velocity in the observer's frame to the velocity of light.

Quarks

Name and symbol		Charge	Z-component (I_Z) of isospin	Baryon number	Other non-zero quantum numbers	Mass/GeV
Down	d	$-\frac{1}{3}e$	$-\frac{1}{2}$	$\frac{1}{3}$	0	0.35
Up	u	$\frac{2}{3}e$	$+\frac{1}{2}$	$\frac{1}{3}$	0	0.35
Strange	s	$-\frac{1}{3}e$	0	$\frac{1}{3}$	Strangeness -1	0.5
Charm	c	$\frac{2}{3}e$	0	$\frac{1}{3}$	Charm $+1$	1.5
Bottom	b	$-\frac{1}{3}e$	0	$\frac{1}{3}$	Bottom $+1$	4.7
Top	t	$\frac{2}{3}e$	0	$\frac{1}{3}$	Top $+1$	30–50

† No individual quarks have been isolated so the concept of mass is not well defined. The listing is merely a qualitative guide extracted from the masses of the lightest hadrons built from the respective quarks. For more details see F. E. Close, *An Introduction to Quarks and Partons*, 1979, (Academic Press).

Stable and metastable hadrons

These data refer only to particles immune to decay via the strong interaction; they are derived by the Particle Data Group (Rev. Mod. Phys, April 1984) which contains a considerable expansion of the data of this table together with data on unstable mesons and baryons. At present these data are updated biennially by the Particle Data Group.

The quark content is listed by symbols, e.g. the proton built of two up quarks and one down quark is denoted uud. Several metastable heavy particles have been found built from a charmed quark (c)

and charmed antiquark ($\bar{c}$) or from a bottom quark (b) and a bottom antiquark ($\bar{b}$). The resulting bound states of these quarks appear to be effectively non-relativistic systems with spin 0 or 1 and orbital angular momentum similar to positronium. By analogy they are known as 'charmonium' and 'bottomonium' respectively. The spectrum is listed in Section 4. The charmonium and bottomonium states all have zero baryon number, strangeness, charm and bottom quantum numbers.

The charm (C), strangeness (S) and baryon number (B) of the hadrons which appear in the subsequent tables are as follows:

Hadron type	Mesons ($B=0$)	S	C	Baryons ($B=1$)	S	C
non-strange . .	π, η	0	0	p, n	0	0
strange	K^+	+1	0	Λ, Σ	−1	0
	K^-	−1	0	Ξ	−2	0
	$K_S^0, K_L^0 \begin{cases} \sim 50\% \\ \sim 50\% \end{cases}$	+1	0	Ω^-	−3	0
		−1	0			
charmed	D^+, D^0	0	1	Λ_c^+	0	+1
	$D^-, \bar{D}^0$	0	−1			
	F^+	+1	+1			
	F^-	−1	−1			

Mesons consist of a quark and an antiquark; baryons ($B=1$) consist of three quarks (each with $B=1/3$). The superscripts denote charges in units of the proton charge.

Nonstrange hadrons

Name and symbol	Quark content	Spin	Mass/MeV	Mean life/s	Principal modes of decay
Pion π^+, (π^-) . .	$u\bar{d}$ ($d\bar{u}$)	0	139.5673 ± 0.0007	2.6030×10^{-8} ± 0.0023	$\mu^{\pm} \nu$
π^0	$u\bar{u}$ and $d\bar{d}$	0	134.9630 ± 0.0038	0.83×10^{-16} ± 0.06	$\gamma\gamma$
Eta η^0	$u\bar{u}, d\bar{d}$ and $s\bar{s}$	0	548.8 ± 0.6	7.93×10^{-19} ± 1.1	$\gamma\gamma, \pi^0\pi^0\pi^0$ $\pi^+\pi^-\pi^0$
Proton p	uud	$\frac{1}{2}$	938.2796† ± 0.0027	stable $(> 2.5 \times 10^{38})$	stable
Neutron n . . .	ddu	$\frac{1}{2}$	939.5731† ± 0.0027	898 ± 16	$pe^-\bar{\nu}$

† The difference ($m_p - m_n$) between the proton and neutron masses is known very accurately to be $-1.293\,323 \pm 0.000\,016$ MeV.

Strange hadrons

Name and symbol	Quark content	Spin	Mass/MeV	Mean life/s	Principal modes of decay
K-mesons K^+ (K^-)	$u\bar{s}$, ($s\bar{u}$)	0	493.667 ± 0.015	$1.237\ 1 \times 10^{-8}$ $\pm 0.002\ 6$	$\mu^{\pm}\nu$, $\pi^{\pm}\pi^0$
K^0_S	$\sim 50\% s\bar{d}$	0	497.67 ± 0.13	$0.892\ 3 \times 10^{-10}$ $\pm 0.002\ 2$	$\pi^+\pi^-$, $\pi^0\pi^0$
K^0_L	$\sim 50\% d\bar{s}$			5.183×10^{-8} ± 0.040	$\pi^0\pi^0\pi^0$, $\pi^+\pi^-\pi^0$ $\pi^{\pm}ev$, $\pi^{\pm}\mu\nu$
Hyperons Λ	uds	$\frac{1}{2}$	1115.60 ± 0.05	2.632×10^{-10} ± 0.020	$p\pi^-$, $n\pi^0$
Σ^+	uus	$\frac{1}{2}$	1189.36 ± 0.06	0.800×10^{-10} ± 0.004	$p\pi^0$, $n\pi^+$
Σ^-	dds	$\frac{1}{2}$	1197.34 ± 0.05	1.482×10^{-10} ± 0.011	$n\pi^-$
Σ^0	uds	$\frac{1}{2}$	1192.46 ± 0.08	5.8×10^{-20} ± 1.3	$\Lambda\gamma$
Ξ^-	dss	$\frac{1}{2}$	1321.32 ± 0.13	1.641×10^{-10} ± 0.016	$\Lambda\pi^-$
Ξ^0	uss	$\frac{1}{2}$	1314.9 ± 0.6	2.90×10^{-10} ± 0.10	$\Lambda\pi^0$
Ω^-	sss	$\frac{3}{2}$	1672.45 ± 0.32	0.819×10^{-10} ± 0.027	ΛK^-, $\Xi^0\pi^-$ $\Xi^-\pi^0$

The difference $(M_{kL} - M_{kS})$ between the K^0_L and K^0_S masses is known very accurately to be

$$M_{kL} - M_{kS} = (3.521 \pm 0.014) \times 10^{-12}\ \text{MeV}$$

Charmed hadrons

Name and symbol	Quark content	Spin	Mass/MeV	Mean life/s	Principal modes of decay
D^+ (D^-)	$c\bar{d}$ ($d\bar{c}$)	0	1869.4 ± 0.6	$(9.2\,^{+1.7}_{-1.2}) \times 10^{-13}$	$\bar{K}^0\pi^{\pm}\pi^0$ $\bar{K}^0\pi^{\pm}\pi^+\pi^-$
D^0 ($\bar{D}^0$)	$c\bar{u}$ ($u\bar{c}$)	0	1864.7 ± 0.6	$(4.4\,^{+0.8}_{-0.6}) \times 10^{-13}$	$K^-\pi^+\pi^0$ $K^-\pi^+\pi^0$
F^+ (F^-)	$c\bar{s}$ ($\bar{c}s$)	0	2021 ± 15	$(1.9\,^{+1.3}_{-0.7}) \times 10^{-13}$	$\eta\pi$, $\eta\pi\pi\pi$
Λ^+_c	cud	$\frac{1}{2}$	2282.2 ± 3.1	$(1.1\,^{+0.9}_{-0.4}) \times 10^{-13}$	$pK^-\pi^+$

Heavy quark spectroscopy

The massive charmed and bottom quarks form non-relativistic bound states with their corresponding antiquarks. The resulting spectroscopy is similar to that of positronium and is known as 'charmonium' (charmed quark and charmed antiquark) or 'bottomonium'. The low-lying energy levels are metastable. These energy levels yield important information on the nature of the interquark forces. The quark and antiquark couple their spins to a total of spin 0 or 1 and are in a state of relative orbital angular momentum L. The spectrum is listed in standard $^{2S+1}L_J$ notation.

Charmonium spectroscopy

State	$c\bar{c}$ configuration	Spin	Mass/MeV	Width/MeV	Principal modes of decay†
$\eta_c(2980)$	$^1S_0$	0	2981 ± 6	20	unknown
$J/\psi(3100)$	$^3S_1$	1	3096.9 ± 0.1	0.063 ± 0.009	$e^+e^-, \mu^+\mu^-$ γ^+ hadrons
$\chi(3415)$	$^3P_0$	0	3415.0 ± 1.0		$2(\mu^+\pi^-)$
P_c or $\chi(3510)$	$^3P_1$	1	3510.0 ± 0.6		$\gamma J/\psi(3100)$
$\chi(3555)$	$^3P_2$	2	3555.8 ± 0.6		$\gamma J/\psi(3100)$
$\eta_c(3590)$	$^1S_0$	0	3594.0 ± 5.0		unknown
$\psi(3685)$	$^3S_1$	1	3686.0 ± 0.1	0.215 ± 0.040	$J/\psi\pi^+\pi^-$, $J/\psi\pi^0\pi^0$ $\gamma\chi(3415)$ $\gamma\chi(3510)$ $\gamma\chi(3555)$
$\psi(3770)$	$^3D_1$	1	3770 ± 3	25 ± 3	$D\bar{D}$

Heavier charmonium states decay rapidly to charmed hadrons.

† $\gtrsim 5\%$.

Bottomonium spectroscopy

State	$b\bar{b}$ configuration	Spin	Mass/MeV†	Width/MeV	Principal modes of decay
$\Upsilon(9460)$	1^3S_1	1	9456	0.042 ± 0.015	$e^+e^-, \mu^+\mu^-$ hadrons
$\chi_b(9873)$	1^3P_0	0	9873		$\gamma\Upsilon(9460)$
$\chi_b(9894)$	1^3P_1	1	9894	unknown	$\gamma\Upsilon(9460)$
$\chi_b(9914)$	1^3P_2	2	9914		$\gamma\Upsilon(9460)$
$\Upsilon(10020)$	2^3S_1	1	10016	0.030 ± 0.010	$\Upsilon(9460)$
$\chi_b'(10231)$	2^3P_0	0	10231		
$\chi_b'(10249)$	2^3P_1	1	10249	unknown	unknown
$\chi_b'(10264)$	2^3P_2	2	10264		
$\Upsilon(10350)$	3^3S_1	1	10347	unknown	unknown
$\Upsilon(10570)$	4^3S_1	1	10569	14 ± 5	unknown

† All masses have an overall ± 10 MeV error.

Gauge bosons with spin 1

Name and symbol	Charge	Force transmitted	Mass/GeV	Mean life (width Γ)	Principal modes of decay
Photon γ	0	Electromagnetic	0	stable	stable
Gluon g	0	Interquark colour and strong forces	0	stable	stable
Weak bosons $W^\pm$	$\pm e$	Charged weak (radioactivity)	80.8 ± 2.7	$\Gamma 7$ GeV	$e \pm \nu, \mu \pm \nu, \tau \pm \nu$, hadrons
Z^0	0	Neutral weak	92.9 ± 1.6	$\Gamma 8.5$ GeV	$e^+e^-, \mu^+\mu^-, \tau^+\tau$, hadrons

F.E.C.

Mathematical Functions

4.1 Functions

The main purpose of this book is to tabulate values of physical and chemical constants. Previous editions included, in addition, short tables of logarithms, trigonometrical and other elementary mathematical functions as an aid to the calculation of experimental results. The widespread use of pocket calculators and of computers has now made these unnecessary. However, although the equations of mathematical physics can now readily be solved directly in numerical terms, there remain occasions when it is helpful to express the solutions using familiar tabulated functions; we therefore give references to tables where values of these functions can be found.

Exponential and trigonometric functions appear as the solutions of linear differential equations with constant coefficients. Many other functions appear in the solution of differential equations with stated boundary conditions, particularly in the determination of fields (electromagnetic, hydrodynamic, or gravitational, for example). **Bessel functions** occur in the solution of potential problems in two dimensions, or in three dimensions with cylindrical symmetry, and in a great variety of other problems. In three-dimensional potential problems involving spheres or spheroids—the Earth's gravitational and magnetic fields, for example—it is convenient to work in terms of **surface harmonics** and **Legendre functions**, as on pp. 159 and 160. **Elliptic functions** and related integrals occur in the solution of certain non-linear differential equations. The references below include tables of these functions and of elementary functions to numbers of decimal places beyond those usually available on pocket calculators, and also tables for group theory which find extensive application in many areas of physics and chemistry.

Certain constants and their logarithms

Constant	Value	Logarithm	Constant	Value	Logarithm
π	3.141 59	0.497 15	$\sqrt{3}$	1.732 05	0.238 56
π^2	9.869 60	0.994 30	$\sqrt{10}$	3.162 28	0.500 00
$1/\pi$	0.318 31	$\bar{1}$.502 85	e	2.718 28	0.434 29
$1/\pi^2$	0.101 32	$\bar{1}$.005 70	$\log_e 10$	2.302 59	0.362 22
$\sqrt{\pi}$	1.772 45	0.248 57	1 rad	57.295 78°	1.758 12
$\sqrt{2}$	1.414 21	0.150 51	1°	0.017 45 rad	$\bar{2}$.241 88

References

(1) Jahnke and Emde (revised by Lösch), *Tables of Higher Functions*, 1960 (McGraw-Hill). Includes, among others, gamma function, exponential, sine and cosine integrals, elliptic functions, orthogonal polynomials, Legendre functions, Bessel functions, Mathieu functions and confluent hypergeometric functions.

(2) National Bureau of Standards (Federal Works Agency, Works Projects Administration of the City of New York): *Tables of the Exponential Function*, 1939 (12–18 decimal places).
Tables of Sines and Cosines, 1940 (radian arguments, 8 places, increasing to 15 for small arguments).
Tables of Sine, Cosine and Exponential Integrals, 1940 (9 places).

(3) M. Abramowitz and I. A. Stegun (eds), *Handbook of Mathematical Functions with Formulas, Graphs and Mathematical Tables*, NBS Applied Mathematics Series No. 55, 1965. Includes extensive tables of functions to many decimal places, e.g. exponentials to 15 places, sines and cosines to 23 places.

(4) P. W. Atkins, M. S. Child and C. S. G. Phillips: *Tables for Group Theory*, 1970 (Oxford University Press).

A.E.B.

4.2 Statistical methods for the treatment of experimental data

The result of an experiment to measure a physical quantity is never exact but is subject to errors. It is convenient, though sometimes an oversimplification, to divide these into two classes: (a) *random errors*, which vary unpredictably from one repetition of the measurement to another, and (b) *systematic errors* due to consistent but unknown inaccuracies in the equipment used or in the calculations. The effect of random errors can be reduced by repeating the experiment. Statistical methods for treating the data are well known, although their logical bases are still to some extent a subject of controversy. Some elementary formulae are given below. More detailed treatments can be found in the references. Systematic errors are more difficult to handle: each case has to be treated on its merits. Sometimes it is straightforward, as when a constant (such as the velocity of light) used in the calculations has a known uncertainty. Often a detailed analysis of the possible sources of error in the apparatus is needed and may involve subsidiary experiments.

Suppose that the object of the experiment is to measure the value of some quantity, x. If the experiment is carried out n times, there will be n results, $x_1 \ldots x_i \ldots x_n$, varying in value because of random errors. Merely to list these results is not very informative: the purpose of statistical treatment is to reduce them to a form whose significance can be readily appreciated. A first step which is often useful when n is large is to plot a *histogram* of the results, dividing the range of values taken by the x_i into a number of equal intervals and plotting the number of readings falling in each interval. In a typical case the histogram will have a single peak, which we may assume to be somewhere near the true value of x, and will have a spread about this peak which is an indication of the precision of the measurements.

As best estimate of the true value of x it is usual to take the *mean* of the x_i, given by

$$\bar{x} = n^{-1} \sum_{i=1}^{n} x_i$$

The *standard deviation*, s, is taken as a quantitative measure of the spread of the readings. Its square is known as the *variance* and is given by

$$v = s^2 = (n-1)^{-1} \sum_{i=1}^{n} (x_i - \bar{x})^2$$

Another important quantity is the *standard deviation of the mean*, sometimes called the *standard error* and equal to $n^{-1/2}s$. The mean, $\bar{x}$, and either s or $n^{-1/2}s$ are quantities which, with an estimate of the systematic errors, are often all that is needed to summarize the results of a physical measurement. To go further, and fully to appreciate their significance, needs some underlying theory. The unqualified statement that the result of a series of experiments is $\bar{x} \pm a$ is meaningless because the reader does not know if a is to be taken as s, $n^{-1/2}s$ or some other quantity. It is essential that what is stated should be made clear.

In biology and the social sciences it is often the object of an investigation to make an estimate of some property of a large but finite *population* of individuals, too numerous for each to be measured, by means of measurements on a limited *sample*. There is a unique true value to the property being estimated, which could be ascertained given sufficient time and effort. A range of statistical techniques is available to estimate the properties of the parent population from sample measurements with calculable uncertainties and confidence limits.

In the physical sciences a finite parent population is relatively uncommon. Instead it is convenient to postulate that our series of measurements forms a finite sample from an infinite population. We can speak of this population having a *probability distribution function* $F(x)$ with the following properties:

(i)
$$\int_{-\infty}^{\infty} F(x)\, \mathrm{d}x = 1$$

(ii) The true value of the quantity being measured is the mean of the population,

$$\mu = \int_{-\infty}^{\infty} x \, F(x) \, \mathrm{d}x$$

(iii) A measure of the 'true' uncertainty of the measurement (due to random errors) is given by the standard deviation, σ, of the population; its square, the variance, is given by

$$\sigma^2 = \int_{-\infty}^{\infty} (x-\mu)^2 \, F(x) \, \mathrm{d}x$$

The histogram, mean and standard deviation derived from our set of measurements can be regarded as sample approximations to the probability distribution, mean and standard deviation of the parent population.

It is sometimes necessary to combine the results of a number of independent determinations of a quantity into a single estimate. The best precision (minimum variance) is obtained if the individual results are 'weighted' inversely to their variances. Suppose that there are m separate results X_j, each the mean of n_j readings with variance s_j^2. Then the best estimate of the quantity is

$$\bar{X} = \sum_{j=1}^{m} X_j n_j s_j^{-2} \Big/ \sum_{j=1}^{m} n_j s_j^{-2}$$

and the expected variance of this will be

$$\sum_{j=1}^{m} n_j s_j^{-2}$$

In most physical measurements, where the random errors can be thought of as made up of a number of small independent contributions, the probability distribution has the form of the *normal error function* or *Gaussian*,

$$F(x) \, \mathrm{d}x = (2\pi)^{-1/2} \sigma^{-1} \exp\left\{-(x-\mu)^2/2\sigma^2\right\} \mathrm{d}x$$

Much of the further statistical treatment of the data is based on theory which assumes that the distribution function is Gaussian. If the histogram has a form which differs widely from the Gaussian it is a warning to proceed with caution. However, the *Central Limit Theorem* states that the sample means from a non-Gaussian population have a distribution which approximates to the Gaussian, and the larger the number of observations the better the approximation. Consequently, many tests are valid even with non-Gaussian populations.

The Gaussian and its integral, commonly known as the error function, are tabulated in a number of the references below.

In an important class of experiments the object is to test the *significance* of an observation or some combination of observations. As a very simple example we may take the following: given a sample observation x, test the hypothesis that it comes from a parent population with Gaussian distribution having mean μ and standard deviation σ. We form the test function $c = |x-\mu|/\sigma$. By integrating the Gaussian function we can calculate the probability $p(C)$ that a sample observation taken at random will have a value of c less than C, and construct the following table:

C	1.65	1.96	2.58	3.29
$p(C)$	90%	95%	99%	99.9%

If in our experiment $c > 2.58$ we can say that the difference between the observation and the assumed distribution is significant at the 1% level of probability, meaning that there is less than a 1% chance that the result is due to random causes.

In practice we do not usually know the standard deviation σ of the parent population but have to work with the standard deviation s of the observations. This causes an important change in the method as the following illustration will show: given a set of n observations with mean $\bar{x}$ and standard deviation s, test the hypothesis that the true value is μ (i.e. that they are a sample from a parent population with Gaussian distribution having mean μ). We now form the test function

$$t = |\bar{x} - \mu|/(s/n^{1/2})$$

This has a distribution which can be calculated for a Gaussian population and is known as the '*Student' t distribution*. It involves a parameter known as the number of *degrees of freedom* which is, loosely, the number of independent observations, $n-1$ in our case (not n because for a given $\bar{x}$, once $n-1$ values are known, the nth is determinate). Values of t^2 for given significance levels P and numbers of degrees of freedom are tabulated in the literature and in the following short table where values of t^2 are given against parameter values $N_1 = 1$, $N_2 = n - 1$ for three significance levels $P = 0.05$, 0.01 and 0.001 (corresponding to values of our earlier p of 95%, 99% and 99.9%).

As an illustration of the use of this we may return to our original experiment, to measure x, and improve on our earlier statement that $\bar{x}$ is an estimate of the true value of x by adding *confidence limits*. We can say that the value of x is

$$\bar{x} \pm tsn^{-1/2}$$

at the confidence level $1 - P$, where t^2 is the entry in the following table at the appropriate values of P and the other parameters. We are then asserting that the probability is $1 - P$ that the true value of x lies between the limits specified. A statement of the final result of our measurement would then include the following items:

(i) The estimated value $\bar{x}$.

(ii) The confidence limits due to random errors (calculated as above from the data and desirably including a statement that the Gaussian nature of the distribution has been checked, perhaps by plotting a histogram).

(iii) The confidence limits due to systematic errors (the way in which these are calculated will depend on the circumstances of the experiment).

In some cases it may be necessary or desirable to state results in a less complete form. However they are presented, exactly what is being stated should be made clear.

In a similar way, the estimate of the standard deviation σ of the parent population can be improved by using the test function

$$\chi^2 = (n-1)s^2/\sigma^2$$

The distribution function of χ^2, the *chi-squared distribution*, is tabulated in the literature and can be used to set confidence limits for σ. It can also be used to check whether a series of observations agrees with an assumed distribution function. In the following table χ^2 for $k-1$ degrees of freedom is given by the entries with $N_1 = k - 1$, $N_2 = \infty$.

Another test function, used in checking whether two independent estimates s_1 and s_2 (with degrees of freedom $n_1 - 1$ and $n_2 - 1$ respectively) of the standard deviation of a measured quantity are significantly different, is

$$F = s_1^2/s_2^2$$

The *F distribution* is also tabulated. Values of F can be obtained from the following table with $N_1 = n_1 - 1$, $N_2 = n_2 - 1$.

References

(1) J. T. Richardson, *The Reduction and Presentation of Experimental Results*, BS 2846: 1957.

(2) National Bureau of Standards (Federal Works Agency, Works Projects Administration of the City of New York), *Tables of Probability Functions*, 1942. Tables of the Gaussian function and its integral, the error function, up to 15 decimal places.

(3) R. A. Fisher, *Statistical Methods for Research Workers*, 1958 (Oliver and Boyd).

(4) C. Mack, *Essentials of Statistics for Scientists and Technologists*, 1967 (Plenum Press).

(5) O. L. Davies (ed.), *Statistical Methods in Research and Production*, 1957 (Oliver and Boyd).

(6) R. Hammond and P. S. McCullagh, *Quantitative Techniques in Geography*, 2nd edn, 1978 (Clarendon Press, Oxford).

(7) H. H. Ku (ed.), *Precision Measurement and Calibration—Statistical Concepts and Procedures*, National Bureau of Standards Special Publication 300, Vol. 1, 1969.

(8) O. L. Davies (ed.), *The Design and Analysis of Industrial Experiments*, 1960 (Oliver and Boyd).

A.E.B.

Table for significance tests

Nine commonly needed tests are listed in column 1. Use of the transformations of the observations given in column 2 enables a single table to be used. To make a test calculate the function of the

1 *To test whether :*	2 *Function of observations to be calculated*	3 *Cell of table*	
		N_1	N_2
1. A single, randomly chosen,† observation differs significantly from a given mean (μ) (standard deviation (σ) known).	$\left(\dfrac{x-\mu}{\sigma}\right)^2$	1	∞
2. The mean of n observations differs significantly from a given value (μ) (standard deviation (σ) known).	$n\left(\dfrac{\bar{x}-\mu}{\sigma}\right)^2$	1	∞
3. The mean of n observations differs significantly from a given value (μ) (standard deviation estimated (s) from sample).	$n\left(\dfrac{\bar{x}-\mu}{s}\right)^2$ $s^2 = \{\text{Sum } (x-\bar{x})^2\}/(n-1)$	1	$n-1$
4. Two means, of n_1 and n_2 observations respectively, differ significantly (standard deviation estimated (s) from samples).	$\dfrac{n_1 n_2}{n_1+n_2}\left(\dfrac{\bar{x}_1-\bar{x}_2}{s}\right)^2$ $s^2 = \dfrac{1}{n_1+n_2-2}\left\{\begin{matrix}\text{Sum}\\ \text{over } (x-\bar{x}_1)^2\\ n_1\end{matrix} + \begin{matrix}\text{Sum}\\ \text{over } (x-\bar{x}_2)^2\\ n_2\end{matrix}\right\}$	1	n_1+n_2-2
5. The numbers (o) of observations falling into k classes differ significantly from expected numbers (e) (total number expected made to agree with observation).	Sum over $\{(o-e)^2/e\}/(k-1)$ k No e should be less than 5	$k-1$	∞
6. The numbers (o) of observations falling into k classes differ significantly from expected numbers (e) (expected numbers calculated from observations via a function of l parameters).	Sum over $\{(o-e)^2/e\}/(k-l-1)$ k No e should be less than 5	$k-l-1$	∞
7. One variance (v_1) estimated from n_1 observations is significantly larger than another (v_2) estimated from n_2 observations.	v_1/v_2	n_1-1	n_2-1
8. One variance estimated from n_1 observations is significantly different from another estimated from n_2 observations.	v_1/v_2 where v_1 is the larger of the two estimates, v_2 the smaller.‡	(n_1-1)	(n_2-1)
9. An observed proportion, r out of n, differs significantly from a given proportion p. (a) $p<r/n$ (b) $p>r/n$	$(1-p)r/p(n-r+1)$ $p(n-r)/(1-p)(r+1)$	$2(n-r+1)$ $2(r+1)$	$2r$ $2(n-r)$

† E.g. this test would not be legitimate for testing the largest of a set.

‡ For this test the values of P are to be doubled, the three lines of the table are then for levels of significance $P = 0.10, 0.02, 0.002$.

observations given in column 2 and compare its value with those given in that cell of the table identified in column 3. Greater values should be adjudged significant; smaller values are not conclusive, a larger experiment might show significance. P of the table is the *level of significance* to be quoted.

The three entries in each cell of the table are for three levels of significance ($P = 0.05$, 0.01, 0.001), P being the risk of a wrong decision *when no difference exists*; the risk of a wrong decision in other cases obviously cannot be stated generally because the differences may be of any magnitude. The smaller the value of P used the larger will a real difference have to be before it makes itself apparent by these tests. The choice of P must therefore be made by balancing this risk against the magnitude of the difference which will just escape detection, i.e. the table value.

Except for test 9, the table is calculated on the assumption that the error or random sampling variation referred to above results in observations being normally distributed. Small departures from normality will not usually affect the decisions because their effect on P is small.

This table abridged, by kind permission of the authors and publishers, from Table V of *Statistical Tables for Biological, Agricultural and Medical Research* by R. A. Fisher and F. Yates (Oliver and Boyd).

The three rows for each value of N_2 correspond to values of $P = 0.05$, 0.01 and 0.001 respectively; the values for $P = 0.01$ are printed in bold type.

Table for significance tests

N_2	N_1									
	1	2	3	4	5	6	8	12	24	∞
1	161.4	199.5	215.7	224.6	230.2	234.0	238.9	243.9	249.0	254.3
	4052	**4999**	**5403**	**5625**	**5764**	**5859**	**5981**	**6106**	**6234**	**6366**
	Values for $P = 0.001$ too large for entry									
2	18.5	19.0	19.2	19.2	19.3	19.3	19.4	19.4	19.5	19.5
	98.5	**99.0**	**99.2**	**99.2**	**99.3**	**99.3**	**99.4**	**99.4**	**99.5**	**99.5**
	998.5	999.0	999.2	999.2	999.3	999.3	999.4	999.4	999.5	999.5
3	10.1	9.6	9.3	9.1	9.0	8.9	8.8	8.7	8.6	8.5
	34.1	**30.8**	**29.5**	**28.7**	**28.2**	**27.9**	**27.5**	**27.1**	**26.6**	**26.1**
	167.5	148.5	141.1	137.1	134.6	132.8	130.6	128.3	125.9	123.5
4	7.7	6.9	6.6	6.4	6.3	6.2	6.0	5.9	5.8	5.6
	21.2	**18.0**	**16.7**	**16.0**	**15.5**	**15.2**	**14.8**	**14.4**	**13.9**	**13.5**
	74.1	61.2	56.2	53.4	51.7	50.5	49.0	47.4	45.8	44.1
5	6.6	5.8	5.4	5.2	5.1	5.0	4.8	4.7	4.5	4.4
	16.3	**13.3**	**12.1**	**11.4**	**11.0**	**10.7**	**10.3**	**9.9**	**9.5**	**9.0**
	47.0	36.6	33.2	31.1	29.7	28.8	27.6	26.4	25.1	23.8
6	6.0	5.1	4.8	4.5	4.4	4.3	4.1	4.0	3.8	3.7
	13.7	**10.9**	**9.8**	**9.1**	**8.7**	**8.5**	**8.1**	**7.7**	**7.3**	**6.9**
	35.5	27.0	23.7	21.9	20.8	20.0	19.0	18.0	16.9	15.7
7	5.6	4.7	4.3	4.1	4.0	3.9	3.7	3.6	3.4	3.2
	12.2	**9.5**	**8.5**	**7.8**	**7.5**	**7.2**	**6.8**	**6.5**	**6.1**	**5.6**
	29.2	21.7	18.8	17.2	16.2	15.5	14.6	13.7	12.7	11.7
8	5.3	4.5	4.1	3.8	3.7	3.6	3.4	3.3	3.1	2.9
	11.3	**8.6**	**7.6**	**7.0**	**6.6**	**6.4**	**6.0**	**5.7**	**5.3**	**4.9**
	25.4	18.5	15.8	14.4	13.5	12.9	12.0	11.2	10.3	9.3

Table for significance tests (*contd*)

N_2	N_1									
	1	2	3	4	5	6	8	12	24	∞
9	5.1	4.3	3.9	3.6	3.5	3.4	3.2	3.1	2.9	2.7
	10.6	8.0	7.0	6.4	6.1	5.8	5.5	5.1	4.7	4.3
	22.9	16.4	13.9	12.6	11.7	11.1	10.4	9.6	8.7	7.8
10	5.0	4.1	3.7	3.5	3.3	3.2	3.1	2.9	2.7	2.5
	10.0	7.6	6.6	6.0	5.6	5.4	5.1	4.7	4.3	3.9
	21.0	14.9	12.6	11.3	10.5	9.9	9.2	8.4	7.6	6.8
11	4.7	4.0	3.6	3.4	3.2	3.1	2.9	2.8	2.6	2.4
	9.6	7.2	6.2	5.7	5.3	5.1	4.7	4.4	4.0	3.6
	19.7	13.8	11.6	10.3	9.6	9.0	8.4	7.6	6.8	6.0
12	4.8	3.9	3.5	3.3	3.1	3.0	2.8	2.7	2.5	2.3
	9.3	6.9	6.0	5.4	5.1	4.8	4.5	4.2	3.8	3.4
	18.6	13.0	10.8	9.6	8.9	8.4	7.7	7.0	6.2	5.4
13	4.7	3.8	3.4	3.2	3.0	2.9	2.8	2.6	2.4	2.2
	9.1	6.7	5.7	5.2	4.9	4.6	4.3	4.0	3.6	3.2
	17.8	12.3	10.2	9.1	8.4	7.9	7.2	6.5	5.8	5.0
14	4.6	3.7	3.3	3.1	3.0	2.8	2.7	2.5	2.3	2.1
	8.9	6.5	5.6	5.0	4.7	4.5	4.1	3.8	3.4	3.0
	17.1	11.8	9.7	8.6	7.9	7.4	6.8	6.1	5.4	4.6
15	4.5	3.7	3.3	3.1	2.9	2.8	2.6	2.5	2.3	2.1
	8.7	6.4	5.4	4.9	4.6	4.3	4.0	3.7	3.3	2.9
	16.6	11.3	9.3	8.3	7.6	7.1	6.5	5.8	5.1	4.3
16	4.5	3.6	3.2	3.0	2.9	2.7	2.6	2.4	2.2	2.0
	8.5	6.2	5.3	4.8	4.4	4.2	3.9	3.6	3.2	2.8
	16.1	11.0	9.0	7.9	7.3	6.8	6.2	5.5	4.8	4.1
17	4.5	3.6	3.2	3.0	2.8	2.7	2.5	2.4	2.2	2.0
	8.4	6.1	5.2	4.7	4.3	4.1	3.8	3.5	3.1	2.7
	15.7	10.7	8.7	7.7	7.0	6.6	6.0	5.3	4.6	3.8
18	4.4	3.6	3.2	2.9	2.8	2.7	2.5	2.3	2.1	1.9
	8.3	6.0	5.1	4.6	4.2	4.0	3.7	3.4	3.0	2.6
	15.4	10.4	8.5	7.5	6.8	6.4	5.8	5.1	4.4	3.7
19	4.4	3.5	3.1	2.9	2.7	2.6	2.5	2.3	2.1	1.9
	8.2	5.9	5.0	4.5	4.2	3.9	3.6	3.3	2.9	2.5
	15.1	10.2	8.3	7.3	6.6	6.2	5.6	5.0	4.3	3.5
20	4.4	3.5	3.1	2.9	2.7	2.6	2.4	2.3	2.1	1.8
	8.1	5.8	4.9	4.4	4.1	3.9	3.6	3.2	2.9	2.4
	14.8	10.0	8.1	7.1	6.5	6.0	5.4	4.8	4.1	3.4
22	4.3	3.4	3.0	2.8	2.7	2.5	2.4	2.2	2.0	1.8
	7.9	5.7	4.8	4.3	4.0	3.8	3.5	3.1	2.7	2.3
	16.4	9.6	7.8	6.8	6.2	5.8	5.2	4.6	3.9	3.2

N_2	N_1									
	1	2	3	4	5	6	8	12	24	∞
24	4.3	3.4	3.0	2.8	2.6	2.5	2.4	2.2	2.0	1.7
	7.8	**5.6**	**4.7**	**4.2**	**3.9**	**3.7**	**3.4**	**3.0**	**2.7**	**2.2**
	14.0	9.3	7.6	6.6	6.0	5.6	5.0	4.4	3.7	3.0
26	4.2	3.4	3.0	2.7	2.6	2.5	2.3	2.1	1.9	1.7
	7.7	**5.5**	**4.6**	**4.1**	**3.8**	**3.6**	**3.3**	**3.0**	**2.6**	**2.1**
	13.7	9.1	7.4	6.4	5.8	5.4	4.8	4.2	3.6	2.8
28	4.2	3.3	2.9	2.7	2.6	2.4	2.3	2.1	1.9	1.7
	7.6	**5.5**	**4.6**	**4.1**	**3.8**	**3.5**	**3.2**	**2.9**	**2.5**	**2.1**
	13.5	8.9	7.2	6.3	5.7	5.2	4.7	4.1	3.5	2.7
30	4.2	3.3	2.9	2.7	2.5	2.4	2.3	2.1	1.9	1.6
	7.6	**5.4**	**4.5**	**4.0**	**3.7**	**3.5**	**3.2**	**2.8**	**2.5**	**2.0**
	13.3	8.8	7.0	6.1	5.5	5.1	4.6	4.0	3.4	2.6
40	4.1	3.2	2.8	2.6	2.4	2.3	2.2	2.0	1.8	1.5
	7.3	**5.2**	**4.3**	**3.8**	**3.5**	**3.3**	**3.0**	**2.7**	**2.3**	**1.8**
	12.6	8.3	6.6	5.7	5.1	4.7	4.2	3.6	3.0	2.2
60	4.0	3.2	2.8	2.5	2.4	2.3	2.1	1.9	1.7	1.4
	7.1	**5.0**	**4.1**	**3.6**	**3.3**	**3.1**	**2.8**	**2.5**	**2.1**	**1.6**
	12.0	7.8	6.2	5.3	4.8	4.4	3.9	3.3	2.7	1.9
120	3.9	3.1	2.7	2.4	2.3	2.2	2.0	1.8	1.6	1.3
	6.9	**4.8**	**3.9**	**3.5**	**3.2**	**3.0**	**2.7**	**2.3**	**1.9**	**1.4**
	11.4	7.3	5.8	4.9	4.4	4.0	3.6	3.0	2.4	1.6
∞	3.8	3.0	2.6	2.4	2.2	2.1	1.9	1.8	1.5	1.0
	6.6	**4.6**	**3.8**	**3.3**	**3.0**	**2.8**	**2.5**	**2.2**	**1.8**	**1.0**
	10.8	6.9	5.4	4.6	4.1	3.7	3.3	2.7	2.1	1.0

E.D.v.R.

INDEX